HANDBOOK OF FLUORESCENT

DYES AND PROBES

HANDBOOK OF FLUORESCENT DYES AND PROBES

R. W. Sabnis, Ph.D.

WILEY

Published by John Wiley & Sons, Inc., Hoboken, New Jersey
Published simultaneously in Canada

For general information on our other products and services or for technical support, please contact our Customer Care Department within the United States at (800) 762-2974, outside the United States at (317) 572-3993 or fax (317) 572-4002.

Wiley also publishes its books in a variety of electronic formats. Some content that appears in print may not be available in electronic formats. For more information about Wiley products, visit our web site at www.wiley.com.

Library of Congress Cataloging-in-Publication Data:

Sabnis, R. W. (Ram Wasudeo)
 Handbook of fluorescent dyes and probes / R.W. Sabnis.
    pages cm
 Includes index.
 ISBN 978-1-118-02869-8 (cloth)
1. Optical brighteners–Handbooks, manuals, etc. 2. Dyes and dyeing–Handbooks, manuals, etc. 3. Stains and staining–Handbooks, manuals, etc. 4. Dyes and dyeing–Chemistry. 5. Fluorescent probes. I. Title.
 TP910.S23 2015
 667′.2–dc23
                                                                                                2014043255

Cover image: iStock © specular
         iStock ©bananahuman
         iStock ©nini
         iStock ©malyugin
         iStock ©burneingphoto

Printed in the United States of America

10 9 8 7 6 5 4 3 2 1

1   2015

Dedicated To
My Wife
Mrs. Madhuri R. Sabnis

Contents

Preface

Fluorescence has been a fascination for individuals for a long time. This book is intended as a reference guide on fluorescent dyes used in medicine, life science, imaging science, cell biology, labeling technology, clinical science, biological chemistry, dye chemistry, biological staining, color chemistry, environmental science, forensic science, organic chemistry, histochemistry, cytochemistry, medicinal chemistry, adhesives, agriculture, coatings, devices, electronics, petroleum, photography, plastics, polymers, security, textile and toys. There are hundreds of fluorescent dyes reported but this book mainly focuses on those dyes, which are widely used in various industrial and academic research.

There is no book available in the market directly on fluorescent dyes and probes, which provides information (such as CAS Registry Numbers, synthesis, various properties, safety/toxicity data and a wide variety of applications) in one source, even though use of fluorescent dyes is wide spread, growing rapidly and has exploded in the past decade. There was a need to publish a book that provided an immediate incentive for compiling the notes to update the scientific community with the wealth of information on fluorescent dyes and probes. To remedy this situation, we have undertaken an ambitious and monumental task of assembling in one publication all the critical data relevant in the field of fluorescent dyes. The dyestuff literature, particularly on fluorescent dyes, is largely in patents. The book provides systematic and up-to-date library of information on 150+ fluorescent dyes and probes as a reference handbook. The book is compiled as a resource guide for chemist and non-chemist in industry as well as in university.

Apart from supplying specific data, the comprehensive, interdisciplinary and comparative nature of the book will provide the user with an easy overview of the state of the art, pinpointing the gaps in the fluorescent dyes knowledge and providing a basis for further research. In addition, it will enable the researcher to use the book in most facile and user-friendly manner.

Fluorescent dyes and probes are arranged alphabetically by the most commonly used name. Again, the choice of primary name is somewhat arbitrary, but an effort has been made to strike a balance between names that are easily recognizable and names that are chemically informative. The detail information of each fluorescent dye or probe is covered in the following order: CAS registry number, chemical structure, CA index name, other names, Merck index number (Merck Index 15th Edition, 2013), chemical/dye class, molecular formula, molecular weight, physical form, solubility, melting point, boiling point, pK_a, absorption (λ max), emission (λ max), molar extinction coefficient, quantum yield, synthesis, imaging/labeling applications, biological/medical applications, industrial applications, safety/toxicity and references. Where there are discrepancies between different values, the author used his judgment on selecting the most likely value.

Numerous recent references have been provided on various synthetic methods, imaging/labeling applications, biological/medical applications, industrial applications and safety/toxicity data. Space and format limitations prevent giving all the references for each dye. This is the first ever book which provides safety/toxicity data with reference to acute toxicity, aquatic toxicity, carcinogenicity, cytotoxicity, chronic toxicity, ecotoxicity, genotoxicity, hematotoxicity, hepatotoxicity, immunotoxicity, microbial toxicity, mutagenicity, nephrotoxicity, neurotoxicity, nucleic acid damage, oral toxicity, phototoxicity, phytotoxicity, skin toxicity, reproductive toxicity, and so on.

Several appendixes have been provided at the end of the book for scientists to conveniently and easily find a dye as per their need. These appendixes include CAS Registry Numbers, Acridines, Anthracenes, Boron co-ordination compounds/dyes, Coumarins, Cyanines/Styryls, Heterocycles, Pyrenes and Xanthenes.

Omissions as well as errors of fact and interpretation are inevitable in dealing with so vast a subject as fluorescent dyes. I shall be glad to have my attention drawn to errors and to incorporate suggestions for improvement when a revision becomes possible.

I express my profound respect and appreciation to my Guru/Mentor/Advisor, Prof. D. W. Rangnekar, who brought me to this wonderful world of Color Science in the Department of Dyestuffs Technology, Institute of Chemical Technology (ICT), where I laid the foundation stone for my research career in Dye Chemistry.

It is a pleasure to make grateful acknowledgement to Dr. Alan Fanta, Dr. Doina Ene, and Dr. Jeffrey Talkington for extremely useful discussions, encouragement and inspiration.

Words are inadequate to express my sincere appreciation to my wife Madhuri and daughter Anika. It would not have been possible to write this book without their encouragement and patience. It is a great pleasure to express my gratitude and appreciation to John Wiley & Sons Inc., for giving me an opportunity to write this book.

R. W. Sabnis

E-mail: ramsabnis@yahoo.com

About the Author

Ram W. Sabnis is a Senior Patent Agent in USA. His interests include dyes, pigments, organic chemistry, heterocycles, polymers, synthesis, formulations, coatings and patents. Presently, he focuses on drafting and prosecuting US and international patents. He is a registered patent agent with US Patent and Trademark Office (USPTO). Prior to entering the legal (patents) field, he was a research chemist for Ascadia, General Electric, Brewer Science, U.S. Textiles and Thermo Fisher (Molecular Probes) in USA. He also worked as a Patent Agent at Squire Patton Boggs L.L.P. and Senior Manager at Pfizer Inc. (Wyeth).

Dr. Sabnis was born and raised in Mumbai, India. He received his Ph.D. in Organic Chemistry (Dye Chemistry) from Institute of Chemical Technology (ICT) (formerly UDCT), Mumbai, India. He is a Chartered Colourists, Fellow of Society of Dyers and Colourists (CCol FSDC).

Dr. Sabnis is one of the world's foremost experts in dyes, inventing world's first colored bubbles (non-staining) and color changing dye system with many applications. He has immensely contributed to color science and technology for the past 25 years, particularly, dyes for biomedicine, personal care products, health/beauty products, electronics (displays, OLEDs), inks, paints, plastics, textiles and toys/bubbles.

He has over 200 publications which include books, book chapters, encyclopedia chapters, patents, reviews, papers, and symposia presentations. He is also an inventor of several US and international patents (issued/published). Dr. Sabnis is the recipient of Pfish Award, Perkin Innovation Award, Grand Innovation Award, Competitive Spirit Award and Best Doctoral Thesis Award. Dr. Sabnis is awarded the Gold Medal for "Outstanding service to the coloration industry" by the Society of Dyers and Colourists, Bradford, U.K.

He has written two books on color chemistry, namely, "*Handbook of Biological Dyes and Stains*" and "*Handbook of Acid-Base Indicators.*" He will continue to focus his activities on fascinating dye chemistry as well as demanding intellectual property in the years to come.

ACRIDINE HOMODIMER

CAS Registry Number   57576-49-5

Chemical Structure

CA Index Name   1,4-Butanediamine, N,N'-bis[3-[(6-chloro-2-methoxy-9-acridinyl)amino]propyl]-

Other Names   Acridine homodimer; NSC 219743

Merck Index Number   Not listed

Chemical/Dye Class   Acridine

Molecular Formula $C_{38}H_{42}Cl_2N_6O_2$

Molecular Weight   685.69

Physical Form   Orange-brown powder or yellow solid

Solubility   Soluble in water, N,N-dimethylformamide, dimethyl sulfoxide, methanol

Melting Point   169–170 °C[2]

Boiling Point (Calcd.)   885.4 ± 65.0 °C   Pressure: 760 Torr

pKa (Calcd.)   10.63±0.19 Most Basic Temperature: 25 °C

Absorption (λmax)   431 nm (H_2O/DNA); 418 nm (MeOH)

Emission (λmax)   498 nm (H_2O/DNA); 500 nm (MeOH)

Molar Extinction Coefficient   12,000 cm^{-1} M^{-1} (MeOH)

Synthesis   Synthetic methods[1–3]

Imaging/Labeling Applications   Nucleic acids;[1–9] chromosomes[10]

Biological/Medical Applications   Detecting nucleic acids;[1–9] diagnosis and selective tissue necrosis;[11] treating cancer,[11] malformed proteins causing neurodegenerative disease,[13] prion disease[12]

Industrial Applications   Not reported

Safety/Toxicity   Neurotoxicity[13]

REFERENCES

1. Sabnis, R. W. *Handbook of Biological Dyes and Stains*; John Wiley & Sons Inc.: Hoboken, **2010**; pp 3–4.

2. Canellakis, E. S.; Shaw, Y. H.; Hanners, W. E.; Schwartz, R. A. Diacridines: Bifunctional intercalators. I. Chemistry, physical chemistry and growth inhibitory properties. *Biochim. Biophys. Acta* **1976**, *418*, 277–289.

3. Barbet, J.; Roques, B. P.; Le Pecq, J. B. Compounds from polyintercalating DNA. Synthesis of acridine dimers. *Compt. Rend. Seances Acad. Sci., Ser. D* **1975**, *281*, 851–853.

4. Park, H. O.; Kim, H. B.; Chi, S. M. Detection method of DNA amplification using probes labeled with intercalating dye. PCT Int. Appl. WO 2006004267, 2006.

5. Markovits, J.; Garbay-Jaureguiberry, C.; Roques, B. P.; Le Pecq, J. B. Acridine dimers: influence of the intercalating ring and of the linking-chain nature on the equilibrium and kinetic DNA-binding parameters. *Eur. J. Biochem.* **1989**, *180*, 359–366.

6. Bottiroli, G.; Giordano, P.; Prosperi, E. Fluorescent probes in nucleic acid research. *Acta Histochem., Suppl.* **1982**, *26*, 189–194.

7. Bottiroli, G.; Giordano, P.; Doglia, S.; Cionini, P. G. Employment of bis-intercalating dyes for the "*in situ*" study of DNA composition. *Basic Appl. Histochem.* **1979**, *23*, 59–63.

8. Le Bret, M.; Le Pecq, J. B.; Barbet, J.; Roques, B. P. A reexamination of the problem of resonance energy transfer between DNA intercalated chromophores using bisintercalating compounds. *Nucleic Acids Res.* **1977**, *4*, 1361–1379.

9. Le Pecq, J. B.; Le Bret, M.; Barbet, J.; Roques, B. DNA polyintercalating drugs. DNA binding of diacridine derivatives. *Proc. Natl. Acad. Sci. U.S.A.* **1975**, *72*, 2915–2919.

10. Van de Sande, J. H.; Lin, C. C.; Deugau, K. V. Clearly differentiated and stable chromosome bands produced by a spermine bis-acridine, a bifunctional intercalating analog of quinacrine. *Exp. Cell Res.* **1979**, *120*, 439–444.

11. Mills, R. L. Pharmaceuticals and apparatus based on Moessbauer isotopic resonant absorption of γ emission (MIRAGE) providing diagnosis and selective tissue necrosis. Can. Pat. Appl. CA 2005039, 1991.

12. May, B. C. H.; Fafarman, A. T.; Hong, S. B.; Rogers, M.; Deady, L., W.; Prusiner, S. B.; Cohen, F. E. Potent inhibition of scrapie prion replication in cultured cells by bis-acridines. *Proc. Natl. Acad. Sci. U.S.A.* **2003**, *100*, 3416–3421.

13. Prusiner, S. B.; Korth, C.; May, B. C. H. Cyclic *bis*-compounds clearing malformed proteins. U.S. Pat. Appl. Publ. US 2004229898, 2004.

ACRIDINE ORANGE (AO)

CAS Registry Number   65-61-2

Chemical Structure

CA Index Name   3,6-Acridinediamine, N^3,N^3,N^6, N^6-tetramethyl-, hydrochloride (1:1)

Other Names   3,6-Acridinediamine, N,N,N',N'-tetramethyl-, monohydrochloride; Acridine Orange R; Acridine, 3,6-bis(dimethylamino)-, hydrochloride; Acridine, 3,6-bis(dimethylamino)-, monohydrochloride; 3,6-Bis(dimethylamino)acridine hydrochloride; Acridine Orange; AO; Acridine Orange N; Acridine Orange NO; Acridine Orange NS; Basic Orange 14; Basic Orange 3RN; C.I. 46005; C.I. Basic Orange 14; Rhoduline Orange NO; Sumitomo Acridine Orange NO; Sumitomo AcridineOrange RK conc

Merck Index Number   Not listed

Chemical/Dye Class   Acridine

Molecular Formula $C_{17}H_{20}ClN_3$

Molecular Weight   301.82

Physical Form   Orange solid

Solubility   Soluble in water, dimethyl sulfoxide, ethanol, methanol

Absorption (λ_{max}) 500 nm (H_2O/DNA); 460 nm (H_2O/RNA); 489 nm (MeOH)

Emission (λ_{max}) 526 nm (H_2O/DNA); 650 nm (H_2O/RNA); 520 nm (MeOH)

Molar Extinction Coefficient   53,000 cm^{-1} M^{-1} (H_2O/DNA); 64,000 cm^{-1} M^{-1} (MeOH)

Synthesis   Synthetic methods[1–8]

Imaging/Labeling Applications   Bacteria;[9–15] blood smears;[16–18] casein;[19] cells/tissues;[20–23] chromosomes;[24,25] endospores;[26] lignin;[27,107] liposomes;[28] lysosomes;[29–35] micronucleus;[36–38] microorganisms;[39–49] mucin;[50] nuclei;[51] nucleic acids;[52–74] parasites;[75–79] sperms;[80,81] tumors;[82,83] yeast[84–88]

Biological/Medical Applications   Analyzing/counting/measuring microorganisms;[39–49] analyzing/detecting/identifying nucleic acids;[52–74] counting/detecting cells/tissues;[20–23] detecting parasites;[75–79] measuring phagosome-lysosome fusion;[31,33] for photodynamic therapy;[82,83] monitoring atmospheric/indoor bioaerosols;[89,90] apoptosis assay;[91–97] cytotoxicity assay;[98,99] genotoxicity assay;[99] as temperature sensor;[100,101] dental materials for crowns and bridges[102]

Industrial Applications   Adhesives;[103] aluminophosphate crystalline materials;[104] detecting clay particles;[105] display device;[106] evaluating fiber surface characteristics;[107] glass matrixes;[108] imaging material;[109] inks;[110,111] lasers;[112] recording materials;[113–115] photoresists;[116,117] textiles;[118] thin films;[119,120] tracers for hydrology;[121] wiring boards[122]

Safety/Toxicity   Carcinogenicity;[123–125] cytotoxicity;[126,127] DNA damage;[128] embryotoxicity;[129] genotoxicity;[130–134] mutagenicity;[135–138] photodynamic toxicity;[139] phototoxicity[140–142]

REFERENCES

1. Sabnis, R. W. *Handbook of Biological Dyes and Stains*; John Wiley & Sons Inc.: Hoboken, **2010**; pp 5–7.

2. Albert, A. *The Acridines: Their Preparation, Physical, Chemical, and Biological Properties and Uses*; St. Martin's Press: New York, **1966**; p 113.

3. Acheson, R. M. *Acridines*; Interscience Publishers, Inc.: New York, **1956**; pp 26–35.

4. Glushko, V. N.; Parbuzina, I. L.; Petrova, G. S. Acridine orange hydrochloride. *Khim. Promyshl., Ser.: Reakt. Osobo Chistye Veshch.* **1980**, 3–4.

5. Glushko, V. N.; Parbuzina, I. L.; Petrova, G. S. 3,6-Bis(dimethylamino)acridine hydrochloride (Acridine Orange). U.S.S.R. SU 694525, 1979.

6. Albert, A. Acridine syntheses and reactions. III. Synthesis of aminoacridines from formic acid and amines. *J. Chem. Soc.* **1947**, 244–250.

7. Karr, A. E. Acridine oranges. *Text. Colorist* **1940**, *62*, 604–607, 634, 676–679, 763–767, 836–837, 852.

8. Biehringer, J. Ueber die farbstoffe der pyroningruppe. *J. Prakt. Chem.* **1897**, *54*, 217–258

9. Araki, H.; Okasawa, Y.; Kaneno, M. Acid-fast bacteria acridine orange fluorescent staining

method. Jpn. Kokai Tokkyo Koho JP 2004069341, 2004.

10. Rychlik, I.; Cardova, L.; Sevcik, M.; Barrow, P. A. Flow cytometry characterization of *Salmonella typhimurium* mutants defective in proton translocating proteins and stationary-phase growth phenotype. *J. Microbiol. Methods* **2000**, *42*, 255–263.

11. Yano, R.; Nogami, T. A method and an apparatus for detecting live bacteria using fluorescence-labeled bacteriophage. Jpn. Kokai Tokkyo Koho JP 11318499, 1999.

12. Back, J. P.; Kroll, R. G. The differential fluorescence of bacteria stained with acridine orange and the effects of heat. *J. Appl. Bacteriol.* **1991**, *71*, 51–58.

13. Maki, J. S.; LaCroix, S. J.; Hopkins, B. S.; Staley, J. T. Recovery and diversity of heterotrophic bacteria from chlorinated drinking waters. *Appl. Environ. Microbiol.* **1986**, *51*, 1047–1055.

14. Bergstrom, I.; Heinanen, A.; Salonen, K. Comparison of acridine orange, acriflavine, and bisbenzimide stains for enumeration of bacteria in clear and humic waters. *Appl. Environ. Microbiol.* **1986**, *51*, 664–667.

15. Meseguer, M.; de Rafael, L.; Baquero, M.; Martinez, F. M; Lopez-Brea, M. Acridine orange stain in the early detection of bacteria in blood cultures. *Eur. J. Clin. Microbiol.* **1984**, *3*, 113–115.

16. Kvetnaya, A. S.; Zhelezova, L. I. Prognosis of clinical course of acute intestinal infection in child based on evaluation of functional state of polymorphonuclear leukocytes. Russ. RU 2275634, 2006.

17. Sciotto, C. G.; Lauer, B. A.; White, W. L.; Istre, G. R. Detection of *Borrelia* in acridine orange-stained blood smears by fluorescence microscopy. *Arch. Pathol. Lab. Med.* **1983**, *107*, 384–386.

18. Mirrett, S.; Lauer, B. A.; Miller, G. A.; Reller, L. B. Comparison of acridine orange, methylene blue, and gram stains for blood cultures. *J. Clin. Microbiol.* **1982**, *15*, 562–566.

19. Das, M.; Roy, B. R.; Chattoraj, D. K. Binding of dyes to casein. *Indian J. Technol.* **1986**, *24*, 95–100.

20. Skyggebjerg, O.; Glensbjerg, M. A method and a system for counting cells from a plurality of species. PCT Int. Appl. WO 2002101087, 2002.

21. Foglieni, C.; Meoni, C.; Davalli, A. M. Fluorescent dyes for cell viability: an application on prefixed conditions. *Histochem. Cell Biol.* **2001**, *115*, 223–229.

22. Bartzatt, R. Acridine orange staining I. *In vitro* derived cells lines. *J. Histotechnol.* **1987**, *10*, 91–93.

23. Schmitz-Moormann, P. Tissue staining by basic dyes. I. Influence of the pH of the staining solution and of the dye-affinity on the adsorption of the dye. *Histochemie* **1968**, *16*, 23–35.

24. Lin, C. C.; Jorgenson, K. F.; Van de Sande, J. H. Specific fluorescent bands on chromosomes produced by acridine orange after prestaining with base specific non-fluorescent DNA ligands. *Chromosoma* **1980**, *79*, 271–286.

25. Forabosco, A.; Couturier, J.; Dutrillaux, B. Effects of pH on the staining of human chromosomes with acridine orange. *Exp. Cell Res.* **1974**, *88*, 418–421.

26. Schichnes, D.; Nemson, J. A.; Ruzin, S. E. Fluorescent staining method for bacterial endospores. *Microscope* **2006**, *54*, 91–93.

27. Perdih, F.; Perdih, A. Lignin selective dyes: quantum-mechanical study of their characteristics. *Cellulose* **2011**, *18*, 11392–1150.

28. Van Rooijen, N.; Van Nieuwmegen, R. Fluorochrome staining of multilamellar liposomes. *Stain Technol.* **1978**, *53*, 307–310.

29. Krolenko, S. A.; Adamyan, S. Ya.; Belyaeva, T. N.; Mozhenok, T. P. Acridine orange bioaccumulation in acid organelles of normal and vacuolated frog skeletal muscle fibres. *Cell Biol. Int.* **2006**, *30*, 933–939.

30. Traganos, F.; Darzynkiewicz, Z. Lysosomal proton pump activity: Supravital cell staining with acridine orange differentiates leukocyte subpopulations. *Methods Cell Biol.* **1994**, *41*, 185–194.

31. Steinberg, T. H.; Swanson, J. A. Measurement of phagosome-lysosome fusion and phagosomal pH. *Methods Enzymol.* **1994**, *236*, 147–160.

32. Rashid, F.; Horobin, R. W.; Williams, M. A. Predicting the behavior and selectivity of fluorescent probes for lysosomes and related structures by means of structure-activity models. *Histochem. J.* **1991**, *23*, 450–459.

33. Kielian, M. Assay of phagosome-lysosome fusion. *Methods Enzymol.* **1986**, *132*, 257–267.

34. Moriyama, Y.; Takano, T.; Ohkuma, S. Acridine orange as a fluorescent probe for lysosomal proton pump. *J. Biochem.* **1982**, *92*, 1333–1336.

35. Wilson, C. L.; Jumper, G. A.; Mason, D. L. Acridine orange as a lysosome marker in fungal spores. *Phytopathology* **1978**, *68*, 1564–1567.

36. Polard, T.; Jean, S.; Merlina, G.; Laplanche, C.; Pinelli, E.; Gauthier, L. Giemsa versus acridine orange staining in the fish micronucleus assay and validation for use in water quality monitoring. *Ecotoxicol. Environ. Saf.* **2011**, *74*, 144–149.

37. Nersesyan, A.; Kundi, M.; Atefie, K.; Schulte-Hermann, R.; Knasmueller, S. Effect of staining procedures on the results of micronucleus assays with exfoliated oral mucosa cells. *Cancer Epidemiol., Biomarkers Prev.* **2006**, *15*, 1835–1840.

38. Hayashi, M.; Sofuni, T.; Ishidate, M., Jr., An application of acridine orange fluorescent staining to the micronucleus test. *Mutat. Res. Lett.* **1983**, *120*, 241–247.

39. Li, C. S.; Chia, W. C.; Chen, P. S. Fluorochrome and flow cytometry to monitor microorganisms in treated hospital wastewater. *J. Environ. Sci. Health, Part A* **2007**, *42*, 195–203.

40. Horikiri, S. Microorganism cell detection method using multiple fluorescent indicators. Jpn. Kokai Tokkyo Koho JP 2006238779, 2006.

41. Noda, N.; Mizutani, T. Microorganism-measuring method using multiple staining. Jpn. Kokai Tokkyo Koho JP 2006340684, 2006.

42. Chen, P.; Li, C. Real-time quantitative PCR with gene probe, fluorochrome and flow cytometry for microorganism analysis. *J. Environ. Monit.* **2005**, *7*, 257–262.

43. Besson, F. I.; Hermet, J. P.; Ribault, S. Reaction medium and process for universal detection of microorganisms. Fr. Demande FR 2847589, 2004.

44. Sunamura, T.; Maruyama, A.; Kurane, R. Method for detecting and counting microorganism. Jpn. Kokai Tokkyo Koho JP 2002291499, 2002.

45. Giorgio, A.; Rambaldi, M.; Maccario, P.; Ambrosone, L.; Moles, D. A. Detection of microorganisms in clinical specimens using slides prestained with acridine orange (AOS). *Microbiologica* **1989**, *12*, 97–100.

46. Tsuji, T.; Karasawa, M. Method for counting of microorganisms. Jpn. Kokai Tokkyo Koho JP 60210997, 1985.

47. Lauer, B. A.; Reller, L. B.; Mirrett, S. Comparison of acridine orange and gram stains for detection of microorganisms in cerebrospinal fluid and other clinical specimens. *J. Clin. Microbiol.* **1981**, *14*, 201–205.

48. McCarthy, L. R.; Senne, J. E. Evaluation of acridine orange stain for detection of microorganisms in blood cultures. *J. Clin. Microbiol.* **1980**, *11*, 281–285.

49. Scholefield, J. Staining microorganisms. Ger. Offen. DE 2728077, 1978.

50. Hicks, J. D.; Matthaei, E. A selected fluorescence stain for mucin. *J. Pathol. Bacteriol.* **1958**, *75*, 473–476.

51. Horobin, R. W.; Stockert, J. C.; Rashid-Doubell, F. Fluorescent cationic probes for nuclei of living cells: Why are they selective? A quantitative structure-activity relations analysis. *Histochem. Cell Biol.* **2006**, *126*, 165–175.

52. Kjaerulff, S.; Glensbjerg, M. Method for analysis of cellular DNA content. PCT Int. Appl. WO 2011098085, 2011.

53. Miyamoto, S.; Kato, T.; Tomono, J. Nucleic acid identification method. Jpn. Kokai Tokkyo Koho JP 2010233530, 2010.

54. Lai, S.; Chang, X.; Tian, L.; Wang, S.; Bai, Y.; Zhai, Y. Fluorometric determination of DNA using nano-SiO$_2$ particles as an effective dispersant and stabilizer for acridine orange. *Microchim. Acta* **2007**, *156*, 225–230.

55. Bi, S.; Qiao, C.; Song, D.; Tian, Y.; Gao, D.; Sun, Y.; Zhang, H. Study of interactions of flavonoids with DNA using acridine orange as a fluorescence probe. *Sens. Actuators, B: Chem.* **2006**, *B119*, 199–208.

56. Martens-Habbena, W.; Sass, H. Sensitive determination of microbial growth by nucleic acid staining in aqueous suspension. *Appl. Environ. Microbiol.* **2006**, *72*, 87–95.

57. Park, H. O.; Kim, H. B.; Chi, S. M. Detection method of DNA amplification using probes labeled with intercalating dye. PCT Int. Appl. WO 2006004267, 2006.

58. Ovadekova, R.; Jantova, S.; Labuda, J. Detection of the effective DNA protection by quinazolines using a DNA-based electrochemical biosensor. *Anal. Lett.* **2005**, *38*, 2625–2638.

59. El-Naggar, A. K. Concurrent flow cytometric analysis of DNA and RNA. *Methods Mol. Biol.* **2004**, *263*, 371–384.

60. Lauretti, F.; Lucas de Melo, F.; Benati, F. J.; de Mello, V. E.; Santos, N.; Linhares, R. E. C.; Nozawa, C. Use of acridine orange staining for the detection of rotavirus RNA in polyacrylamide gels. *J. Virol. Methods* **2003**, *114*, 29–35.

61. Tomita, N.; Mori, Y. Method for efficiently detecting double-stranded nucleic acid. PCT Int. Appl. WO 2002103053, 2002.

62. Gonzalez, K.; McVey, S.; Cunnick, J.; Udovichenko, I. P.; Takemoto, D. J. Acridine orange differential staining of total DNA and RNA in normal and galactosemic lens epithelial cells in culture using flow cytometry. *Curr. Eye Res.* **1995**, *14*, 269–273.

63. Evenson, D.; Darzynkiewicz, Z.; Jost, L.; Janca, F.; Ballachey, B. Changes in accessibility of DNA to various fluorochromes during spermatogenesis. *Cytometry* **1986**, *7*, 45–53.

64. Villanueva, A.; Stockert, J. C.; Armas-Portela, R. A simple method for the fluorescence analysis of nucleic acid-dye complexes in cytological preparations. *Histochemistry* **1984**, *81*, 103–104.

65. Wallen, C. A.; Higashikubo, R.; Dethlefsen, L. A. Comparison of two flow cytometric assays for cellular RNA - acridine orange and propidium iodide. *Cytometry* **1982**, *3*, 155–160.

66. Curtis, S. K.; Cowden, R. R. Four fluorochromes for the demonstration and microfluorometric estimation of RNA. *Histochemistry* **1981**, *72*, 39–48.

67. Dutt, M. K. Acridine orange - its use in the specific staining of DNA in mammalian tissue sections. *Microsc. Acta* **1981**, *84*, 37–42.

68. Carmichael, G. G.; McMaster, G. K. The analysis of nucleic acids in gels using glyoxal and acridine orange. *Methods Enzymol.* **1980**, *65*, 380–391.

69. Taylor, I. W.; Milthorpe, B. K. An evaluation of DNA fluorochromes, staining techniques, and analysis for flow cytometry. I. Unperturbed cell populations. *J. Histochem. Cytochem.* **1980**, *28*, 1224–1232.

70. Coulson, P. B.; Bishop, A. O.; Lenarduzzi, R. Quantitation of cellular deoxyribonucleic acid by flow microfluorometry. *J. Histochem. Cytochem.* **1977**, *25*, 1147–1153.

71. Tomita, G. Molecular complexes of acridine orange and nucleosides. *Biophysik* **1967**, *4*, 118–128.

72. Yamabe, S. A spectrophotometric study on binding of acridine orange with DNA. *Mol. Pharmacol.* **1967**, *3*, 556–560.

73. Kasten, F. H. Cytochemical studies with acridine orange and the influence of dye contaminants in the staining of nucleic acids. *Int. Rev. Cytol.* **1967**, *21*, 141–202.

74. Lecatsas, G. Detection of plant nucleic acids with acridine orange. *S. Afr. J. Sci.* **1967**, *63*, 61.

75. Xu, L.; Chaudhuri, A. *Plasmodium yoelii*: a differential fluorescent technique using acridine orange to identify infected erythrocytes and reticulocytes in Duffy knockout mouse. *Exp. Parasitol.* **2005**, *110*, 80–87.

76. Saito-Ito, A.; Akai, Y.; He, S.; Kimura, M.; Kawabata, M. A rapid, simple and sensitive flow cytometric system for detection of *Plasmodium falciparum. Parasitol. Int.* **2001**, *50*, 249–257.

77. Freedman, D. O.; Berry, R. S. Rapid diagnosis of *Bancroftian filariasis* by acridine orange staining of centrifuged parasites. *Am. J. Trop. Med. Hyg.* **1992**, *47*, 787–793.

78. Rickman, L. S.; Long, G. W.; Oberst, R.; Cabanban, A.; Sangalang, R.; Smith, J. I.; Chulay, J. D.; Hoffman, S. L. Rapid diagnosis of malaria by acridine orange staining of centrifuged parasites. *Lancet* **1989**, *1*, 68–71.

79. Wongsrichanalai, C.; Webster, H. K.; Brown, A. E. Rapid diagnosis of malaria by acridine orange staining of centrifuged parasites. *Lancet* **1989**, *1*, 967.

80. Yaniz, Y. L.; Palacin, I.; Vicente-Fiel, S.; Gosalvez, J.; Lopez-Fernandez, C.; Santolaria, P. Comparison of membrane-permeant fluorescent probes for sperm viability assessment in the ram. *Reprod. Dom. Anim.* **2013**, *48*, 598–603.

81. Chohan, K. R.; Griffin, J. T.; Lafromboise, M.; De Jonge, C. J.; Carrell, D. T. Comparison of chromatin assays for DNA fragmentation evaluation in human sperm. *J. Androl.* **2006**, *27*, 53–59.

82. Matsubara, T.; Kusuzaki, K.; Matsumine, A.; Shintani, K.; Satonaka, H.; Uchida, A. Acridine orange used for photodynamic therapy accumulates in malignant musculoskeletal tumors depending on pH gradient. *Anticancer Res.* **2006**, *26*, 187–193.

83. Ueda, H.; Murata, H.; Takeshita, H.; Minami, G.; Hashiguchi, S.; Kubo, T. Unfiltered xenon light is useful for photodynamic therapy with acridine orange. *Anticancer Res.* **2005**, *25*, 3979–3983.

84. Muro, K.; Izumi, K. Method for evaluating yeast activity. Jpn. Kokai Tokkyo Koho JP 2004357618, 2004.

85. Bogen, H. J.; Elste, U. Physiology of the effect of acridine orange (AO). Studies of yeast cells. *Planta* **1955**, *45*, 325–375.

86. Bogen, H. J. Staining, damaging, and destroying of yeast cells by acridine orange. *Arch. Mikrobiol.* **1953**, *18*, 170–197.

87. Kolbel, H. Quantitative investigation of acridine orange uptake of living and dead yeast cells and correlation with electrical conditions of the cell. *Z. Naturforsch.* **1947**, *26*, 382–392.

88. Strugger, S. Investigations on the vital fluorochroming (fluorostaining) of yeast cells. *Flora* **1943**, *37*, 73–94.

89. Chi, M.; Li, C. Fluorochrome in monitoring atmospheric bioaerosols and correlations with meteorological factors and air pollutants. *Aerosol Sci. Technol.* **2007**, *41*, 672–678.

90. Li, C.; Huang, T. Fluorochrome in monitoring indoor bioaerosols. *Aerosol Sci. Technol.* **2006**, *40*, 237–241.

91. Verduzco, D.; Amatruda, J. F. Analysis of cell proliferation, senescence, and cell death in zebrafish embryos. *Methods Cell Biol.* **2011**, *101*, 19–38.

92. Alvarez, M.; Villanueva, A.; Acedo, P.; Canete, M.; Stockert, J. C. Cell death causes relocalization of photosensitizing fluorescent probes. *Acta Histochem.* **2011**, *113*, 363–368.

93. Kajstura, M.; Halicka, H. D.; Pryjma, J.; Darzynkiewicz, Z. Discontinuous fragmentation of nuclear DNA during apoptosis revealed by discrete "sub-G1" peaks on DNA content histograms. *Cytometry, Part A* **2007**, *71A*, 125–131.

94. Hiruma, H.; Katakura, T.; Takenami, T.; Igawa, S.; Kanoh, M.; Fujimura, T.; Kawakami, T. Vesicle disruption, plasma membrane bleb formation, and acute cell death caused by illumination with blue light in acridine orange-loaded malignant melanoma cells. *J. Photochem. Photobiol., B: Biol.* **2007**, *86*, 1–8.

95. Baskic, D.; Popovic, S.; Ristic, P.; Arsenijevic, N. N. Analysis of cycloheximide-induced apoptosis in human leukocytes: Fluorescence microscopy using annexin V/propidium iodide versus acridine orange/ethidium bromide. *Cell Biol. Int.* **2006**, *30*, 924–932.

96. Giuliano, M.; Bellavia, G.; Lauricella, M.; D'Anneo, A.; Vassallo, B.; Vento, R.; Tesoriere, G. Staurosporine-induced apoptosis in Chang liver cells is associated with down-regulation of Bcl-2 and Bcl-XL. *Int. J. Mol. Med.* **2004**, *13*, 565–571.

97. Loweth, A. C.; Morgan, N. G. Methods for the study of NO-induced apoptosis in cultured cells. *Methods Mol. Biol.* **1998**, *100*, 311–320.

98. Odawara, K. Visible light cytotoxicity expression ability assay method, and its use. Jpn. Kokai Tokkyo Koho JP 2007143465, 2007.

99. Chreno, O. Method for testing cytotoxicity and genotoxicity of chemical substances. Slovakia SK 278857, 1998.

100. Bousseksou, A.; Salmon, L.; Molnar, G.; Cobo, S. Materials with thermochromic spin transition doped with one or more fluorescent agents for use as temperature sensor. Fr. Demande FR 2952371, 2011.

101. Bousseksou, A.; Salmon, L.; Molnar, G.; Cobo, S. Heat-sensitive spin-transition materials doped with one or more fluorescent agents for use as temperature sensor. PCT Int. Appl. WO 2011058277, 2011.

102. Klemm, E.; Hoerhold, H. H.; Doms, I. Light-curing dental materials for crowns and bridges. Ger. (East) DD 215699, 1984.

103. Hoerhold, H. H.; Klemm, E.; Flammersheim, H. J.; Maertin, R.; Wolf, H. Adhesives. Ger. Offen. DE 3431440, 1985.

104. Gandara, F.; Lopez-Arbeloa, F.; Ruiz-Hitzky, E.; Camblor, M. A. "Bottle-around-a-ship" confinement of high loadings of acridine orange in new aluminophosphate crystalline materials. *J. Mater. Chem.* **2006**, *16*, 1765–1771.

105. Jain, R. K.; D'Hoore, J. Detection and determination of clay particles in natural waters. *J. Indian Soc. Soil Sci.* **1982**, *30*, 415–417.

106. Liu, T. Ion-color controlling electrophoresis display device. Faming Zhuanli Shenqing CN 1461967, 2003.

107. Drnovsek, T.; Perdih, A.; Perdih, M. Fiber surface characteristics evaluated by principal component analysis. *J. Wood Sci.* **2005**, *51*, 507–513.

108. Gaponenko, S. V.; Germanenko, I. N.; Stupak, A. P.; Eyal, M.; Brusilovsky, D.; Reisfeld, R.; Graham, S.; Klingshirn, C. Fluorescence of acridine orange in inorganic glass matrixes. *Appl. Phys. B: Lasers Opt.* **1994**, *B58*, 283–288.

109. Namiki, T.; Shinozaki, F.; Ikeda, T. Light-sensitive imaging material. Ger. Offen. DE 2831101, 1979.

110. Ueda, T.; Yasutomi, H. Aqueous ink-jet inks with mildew growth prevention. Jpn. Kokai Tokkyo Koho JP 10007960, 1998.

111. Ishii, K.; Takahashi, H.; Watanabe, K. Water-thinned fluorescent inks for ballpoint pens. Jpn. Kokai Tokkyo Koho JP 06049405, 1994.

112. Volkin, H. C. Direct solar pumped laser. U.S. Patent 4281294, 1981.

113. Ikoma, K.; Miura, K.; Kawade, I.; Oguchi, Y.; Myagawa, M. Optical recording material containing light-emitting dye, and recording method. Jpn. Kokai Tokkyo Koho JP 63062792, 1988.

114. Baumann, R.; Meisel, A.; Singer, W.; Fritzsche, K.; Bauriegel, L. Laser-sensitive information recording material. Ger. (East) DD 223834, 1985.

115. Ikegami, K.; Okuyama, H. Holographic recording material and process for producing holograms. Eur. Pat. Appl. EP 0084452, 1983.

116. Endo, M.; Sasako, M.; Tani, Y.; Ogawa, K. Resist compositions containing acridine or its derivatives. Jpn. Kokai Tokkyo Koho JP 01010236, 1989.

117. Farid, S. Y.; Haley, N. F.; Moody, R. E.; Specht, D. P. Negative working photoresists responsive to shorter visible wavelengths and novel coated articles. U.S. Patent 4743529, 1988.

118. Gamblin, R. L. Surfactant enhanced dyeing. U.S. Patent 5593459, 1997.

119. Machida, S.; Wakamatsu, T.; Masuo, S.; Jinnai, H.; Itaya, A. Morphology and photophysical properties of polymer thin films dispersed with dye nanoparticle. *Thin Solid Films* **2008**, *516*, 2615–2619.

120. Li, F.; Pfeiffer, M.; Werner, A.; Harada, K.; Leo, K.; Hayashi, N.; Seki, K.; Liu, X.; Dang, X. D. Acridine orange base as a dopant for n doping of C60 thin films. *J. Appl. Phys.* **2006**, *100*, 023716/1–023716/9.

121. Viriot, M. L.; Andre, J. C. Fluorescent dyes: a search for new tracers for hydrology. *Analusis* **1989**, *17*, 97–111.

122. Mori, K.; Ito, K. Method for pattern formation of metal deposition layers, and manufacture of wiring boards. Jpn. Kokai Tokkyo Koho JP 2007177322, 2007.

123. Schehrer, L.; Regan, J. D.; Westendorf, J. UDS induction by an array of standard carcinogens in human and rodent hepatocytes: effect of cryopreservation. *Toxicology* **2000**, *147*, 177–191.

124. Kowalski, L. A.; Laitinen, A. M.; Martazavi-Asl, B.; Wee, R. K. H.; Erb, H. E.; Assi, K. P.; Madden, Z. *In vitro* determination of carcinogenicity of sixty-four compounds using a bovine papillomavirus DNA-carrying C3H/10T1/2 cell line. *Environ. Mol. Mutagen.* **2000**, *35*, 300–311.

125. Heil, J.; Reifferscheid, G. Detection of mammalian carcinogens with an immunological DNA synthesis-inhibition test. *Carcinogenesis* **1992**, *13*, 2389–2394.

126. Bradburne, C. E.; Delehanty, J. B.; Gemmill, K. B.; Mei, B. C.; Mattoussi, H.; Susumu, K.; Blanco-Canosa, J. B.; Dawson, P. E.; Medintz, I. L. Cytotoxicity of quantum dots used for *in vitro* cellular labeling: Role of QD surface ligand, delivery modality, cell type, and direct comparison to organic fluorophores. *Bioconjugate Chem.* **2013**, *24*, 1570–1583.

127. Chang, Y. S.; Wu, C. L.; Tseng, S. H.; Kuo, P. Y.; Tseng, S. Y. Cytotoxicity of triamcinolone acetonide on human retinal pigment epithelial cells. *Invest. Ophthalmol. Vis. Sci.* **2007**, *48*, 2792–2798.

128. McCarroll, N. E.; Piper, C. E.; Keech, B. H. An *E. coli* microsuspension assay for the detection of DNA damage induced by direct-acting agents and promutagens. *Environ. Mutagen.* **1981**, *3*, 429–444.

129. Kohler, M.; Kundig, A.; Reist, H. W.; Michel, C. Modification of *in vitro* mouse embryogenesis by x-rays and fluorochromes. *Radiat. Environ. Biophys.* **1994**, *33*, 341–351.

130. Mascini, M. Determination of the genotoxicity of aqueous samples by using a DNA-based biosensor and dedicated apparatus. Ital. Appl. IT 2004RM0559, 2005.

131. Gonzalez, N. V.; Soloneski, S.; Larramendy, M. L. Dicamba-induced genotoxicity in Chinese hamster ovary (CHO) cells is prevented by vitamin E. *J. Hazard. Mater.* **2009**, *163*, 337–343.

132. Knight, A. W.; Billinton, N.; Cahill, P. A.; Scott, A.; Harvey, J. S.; Roberts, K. J.; Tweats, D. J.; Keenan, P. O.; Walmsley, R. M. An analysis of results from 305 compounds tested with the yeast RAD54-GFP genotoxicity assay (GreenScreen GC) - including relative predictivity of regulatory tests and rodent carcinogenesis and performance with autofluorescent and colored compounds. *Mutagenesis* **2007**, *22*, 409–416.

133. He, L.; Jurs, P. C.; Custer, L. L.; Durham, S. K.; Pearl, G. M. Predicting the genotoxicity of polycyclic aromatic compounds from molecular structure with different classifiers. *Chem. Res. Toxicol.* **2003**, *16*, 1567–1580.

134. Fernandez, M.; Gauthier, L.; Jaylet, A. Use of newt larvae for *in vivo* genotoxicity testing of water: Results on 19 compounds evaluated by the micronucleus test. *Mutagenesis* **1989**, *4*, 17–26.

135. Lee, I. E.; Nguyen, V. C.; Hayase, F.; Kato, H. Desmutagenicity of melanoidins against various kinds of mutagens and activated mutagens. *Biosci., Biotechnol., Biochem.* **1994**, *58*, 18–23.

136. Klopman, G.; Frierson, M. R.; Rosenkranz, H. S. The structural basis of the mutagenicity of chemicals in *Salmonella typhimurium*: The Gene-Tox data base. *Mutat. Res.* **1990**, *228*, 1–50.

137. Xamena, N.; Creus, A.; Marcos, R. Mutagenic activity of some intercalating compounds in the

Drosophila zeste somatic eye mutation test. *Mutat. Res.* **1984**, *138*, 169–173.

138. Rogers, A. M.; Back, K. C. Comparative mutagenicity of 4 DNA-intercalating agents in L5178Y mouse lymphoma cells. *Mutat. Res.* **1982**, *102*, 447–455.

139. Herkovits, J.; Perez-Coll, C. S.; Stockert, J. C.; Blazquez, A. The screening of photodynamic toxicity of dyes by means of a bioassay using amphibian embryos. *Res. J. Chem. Environ.* **2007**, *11*, 86–91.

140. Harris, A. G.; Sinitsina, I.; Messmer, K. Intravital fluorescence microscopy and phototocicity: effects on leukocytes. *Eur. J. Med. Res.* **2002**, *7*, 117–124.

141. Saetzler, R. K.; Jallo, J.; Lehr, H. A.; Philips, C.M.; Vasthare, U.; Arfors, K. E.; Tuma, R. F. Intravital fluorescence microscopy: impact of light-induced phototoxicity on adhesion of fluorescently labeled leukocytes. *J. Histochem. Cytochem.* **1997**, *45*, 505–573.

142. Lukasiak-Bachurzewska, B.; Dulczewska-Klopotowska, M. Studies on the phototoxic properties of some coal derivatives. *Przeglad Dermatol.* **1981**, *68*, 33–37.

ACRIDINE ORANGE 10-DODECYL BROMIDE (DODECYL-ACRIDINE ORANGE (DAO))

CAS Registry Number   41387-42-2

Chemical Structure

CA Index Name   Acridinium, 3,6-bis(dimethylamino)-10-dodecyl-, bromide (1:1)

Other Names   Acridinium, 3,6-bis(dimethylamino)-10-dodecyl-, bromide; 10-Dodecylacridine Orange Bromide; 3,6-Bis(dimethylamino)-10-dodecylacridinium bromide; AO 10 Dodecylbromide; Acridine orange 10-dodecyl bromide; Dodecyl-Acridine Orange; BDA; D 455; DADAB; DAO

Merck Index Number   Not listed

Chemical/Dye Class   Acridine

Molecular Formula $C_{29}H_{44}BrN_3$

Molecular Weight   514.59

Physical Form   Orange solid

Solubility   Soluble in dimethyl sulfoxide, ethanol, methanol

Melting Point   >250 °C

Absorption (λ_{max})   495 nm (MeOH)

Emission (λ_{max})   520 nm (MeOH)

Molar Extinction Coefficient   87,000 cm^{-1} M^{-1} (MeOH)

Synthesis   Synthetic methods[1–4]

Imaging/Labeling Applications   Chloride ions;[5] keratin fibers/hairs;[6,7] Langmuir-Blodgett (LB) monolayers;[23] micelles;[2–4,8,9] mitochondrial membranes;[10] proteins[11–14]

Biological/Medical Applications   Analyzing chloride ions;[5] characterizing drug binding sites on glycoproteins;[14] opthalmic devices (intraocular lenses (IOL))[15]

Industrial Applications   Determining cationic surfactants;[16] electroluminescent devices;[17–19] Langmuir-Blodgett films;[20,23] photographic imaging system;[21] photoresists;[22] semiconductor electrodes;[23–25] silica-surfactant composite films[26]

Safety/Toxicity   No data available

REFERENCES

1. Haugland, R. P. *Handbook of Fluorescent Probes and Research Chemicals*; Molecular Probes Inc.: Eugene, **1996**; pp 314–316.

2. Usui, Y.; Saga, K. The photoreduction and photosensitized reduction of dyes bound to a surfactant micellar surface. *Bull. Chem. Soc. Jpn.* **1982**, *55*, 3302–3307.

3. Yamagishi, A.; Masui, T.; Watanabe, F. Selective activation of reactant molecules by reversed micelles. *J. Phys. Chem.* **1981**, *85*, 281–285.

4. Kubota, Y.; Kodama, M.; Miura, M. Second CMC [critical micelle concentration] of an aqueous solution of sodium dodecyl sulfate. IV. Fluorescence depolarization. *Bull. Chem. Soc. Jpn.* **1973**, *46*, 100–103.

5. Kawabata, Y.; Toge, Y. Fluorometric analysis of chloride ion and chemical sensor therefor. U.S. Patent 5691205, 1997.

6. Schaefer, K. Microscopic investigation of the diffusion of dyes in keratin fibers. *Melliand Textilber.*

1993, *74*, E382–E385, 1138, 1141–1142, 1145–1148, 1151–1152.

7. Schaefer, K.; Koch, U. Investigation of the diffusion of dyes in keratin fibers by fluorescence microscopy (AiF 8415). *DWI Rep.* **1992**, *109*, 699–716.

8. Mitsuzuka, M.; Kikuchi, K.; Kokubun, H.; Usui, Y. The primary processes of the photosensitized reduction of methylene blue in an aqueous sodium dodecylsulfate micellar solution. *J. Photochem.* **1985**, *29*, 363–373.

9. Usui, Y.; Gotou, A. Migration of singlet excitation energy from acridine orange-10-dodecyl bromide to methylene blue in micelles. *Photochem. Photobiol.* **1979**, *29*, 165–167.

10. Hedley, D.; Chow, S. Flow cytometric measurement of lipid peroxidation in vital cells using parinaric acid. *Cytometry* **1992**, *13*, 686–692.

11. Fitos, I.; Visy, J.; Zsila, F.; Bikadi, Z.; Mady, G.; Simonyi, M. Specific ligand binding on genetic variants of human α 1-acid glycoprotein studied

by circular dichroism spectroscopy. *Biochem. Pharmacol.* **2004**, *67*, 679–688.

12. Morii, H.; Ichimura, K.; Uedaira, H. Asymmetric inclusion by de novo designed proteins: fluorescence probes studies on amphiphilic α-helix bundles. *Proteins: Struct., Funct., Genet.* **1991**, *11*, 133–141.

13. Morii, H.; Ichimura, K.; Uedaira, H. Amphiphilic tailor-made proteins as novel chiral hosts. *Chem. Lett.* **1990**, 1987–1990.

14. Maruyama, T.; Otagiri, M.; Takadate, A. Characterization of drug binding sites on α 1-acid glycoprotein. *Chem. Pharm. Bull.* **1990**, *38*, 1688–1691.

15. Mentak, K. Ultra violet, violet, and blue light filtering polymers for ophthalmic applications. U.S. Pat. Appl. Publ. US 20060252844, 2006.

16. Masadome, T. Flow injection fluorometric determination of cationic surfactants using 3,6-bis(dimethylamino)-10-dodecylacridinium bromide. *Anal. Lett.* **1998**, *31*, 1071–1079.

17. Eguchi, T.; Kawada, H.; Nishimura, Y. Electroluminescent device. Jpn. Kokai Tokkyo Koho JP 61037883, 1986.

18. Eguchi, T.; Kawada, H.; Nishimura, Y. Electroluminescent device. Jpn. Kokai Tokkyo Koho JP 61037882, 1986.

19. Eguchi, T.; Kawada, H.; Nishimura, Y. Electroluminescent device. Jpn. Kokai Tokkyo Koho JP 61037888, 1986.

20. Baltuska, A.; Gadonas, R.; Pugzlys, A. Laser induced photobleaching and anisotropy of polymethine dyes in Langmuir-Blodgett films. *Ser. Nonlinear Opt.* **1996**, *3*, 595–600.

21. Farid, S. Y.; Haley, N. F.; Moody, R. E.; Specht, D. P. Dye-sensitized photographic imaging system. U.S. Patent 4743531, 1988.

22. Farid, S. Y.; Haley, N. F.; Moody, R. E.; Specht, D. P. Negative working photoresists responsive to shorter wavelength visible light and novel coated articles. U.S. Patent 4743529, 1988.

23. Nishiyama, Y.; Azuma, T.; Obata, N.; Kasatani, K.; Sato, H. Photoelectric conversion and fluorescence quenching in mixed-dye monolayers: inhomogeneous distribution of dyes. *J. Photochem. Photobiol., A: Chem.* **1991**, *59*, 341–355.

24. Sato, H.; Kawasaki, M.; Kasatani, K.; Higuchi, Y.; Azuma, T.; Nishiyama, Y. Light-harvesting effect in photoelectric conversion with dye multilayers on a semiconductor electrode. *J. Phys. Chem.* **1988**, *92*, 754–759.

25. Higuchi, Y.; Kasatani, K.; Kawasaki, M.; Sato, H. Photoelectric conversion with the dye multilayer on a semiconductor electrode. *Chem. Lett.* **1986**, 1651–1654.

26. Hayakawa, K.; Fujiyama, N.; Satake, I. Fluorescent solubilizates in the silica-surfactant composite films. *Stud. Surf. Sci. Catal.* **2001**, *132*, 813–816.

ACRIDINE ORANGE 10-NONYL BROMIDE (NONYL-ACRIDINE ORANGE (NAO))

CAS Registry Number   75168-11-5

Chemical Structure

CA Index Name   Acridinium, 3,6-bis(dimethylamino)-10-nonyl-, bromide (1:1)

Other Names   Acridinium, 3,6-bis(dimethylamino)-10-nonyl-, bromide; Acridine Orange 10-nonyl bromide; 3,6-bis-(dimethylamino)-10-nonylacridinium bromide; 10-nonyl acridine orange; A 1372; Nonyl-Acridine Orange; NAO

Merck Index Number   Not listed

Chemical/Dye Class   Acridine

Molecular Formula $C_{26}H_{38}BrN_3$

Molecular Weight   472.50

Physical Form   Orange solid or brown solid[2]

Solubility   Soluble in dichloromethane, N,N-dimethylformamide, dimethyl sulfoxide, ethanol, methanol; slightly soluble in water; insoluble in hexanes, toluene

Melting Point $\geq 250\,°C$

Absorption (λ_{max})   495 nm (MeOH)

Emission (λ_{max})   519 nm (MeOH)

Molar Extinction Coefficient $84,000\,cm^{-1}\,M^{-1}$ (MeOH)

Synthesis   Synthetic methods[1–3]

Imaging/Labeling Applications   Mitochondria;[1–23,29,36,37] cardiolipin;[22–29] blood cells;[30] cells;[31] liposomes;[32] lipid membranes;[33] nerve terminals;[34] peptides[35]

Biological/Medical Applications   Analyzing mitochondria;[1–23,29,36,37] assessing/measuring cardiolipin (content);[22–29] assessing/recording membrane potential;[17–22] apoptosis assay;[36,37] hematotoxicity assay;[30] detecting prostate cancer;[14] diagnosing/treating arthritic disorders,[15] type 2 diabetes;[16] testing cardiotoxicity of compounds;[31] opthalmic devices (intraocular lenses (IOL))[38]

Industrial Applications   Not reported

Safety/Toxicity   Mitochondrial toxicity[39]

REFERENCES

1. Sabnis, R. W. *Handbook of Biological Dyes and Stains*; John Wiley & Sons Inc.: Hoboken, **2010**; pp 339–341.

2. Rodriguez, M. E.; Azizuddin, K.; Zhang, P.; Chiu, S.; Lam, M.; Kenney, M. E.; Burda, C.; Oleinick, N. L. Targeting of mitochondria by 10-N-alkyl acridine orange analogues: Role of alkyl chain length in determining cellular uptake and localization. *Mitochondrion* **2008**, *8*, 237–246.

3. Septinus, M.; Seiffert, W.; Zimmermann, H. W. Hydrophobic acridine dyes for fluorescence staining of mitochondria in living cells. 1. Thermodynamic and spectroscopic properties of 10-n-alkylacridine orange chlorides. *Histochemistry* **1983**, *79*, 443–456.

4. Zhang, S.; Zhu, S.; Yang, L.; Zheng, Y.; Gao, M.; Wang, S.; Zeng, J.; Yan, X. High-throughput multiparameter analysis of individual mitochondria. *Anal. Chem.* **2012**, *84*, 6421–6428.

5. Zhao, W.; Waisum, O.; Fung, Y.; Cheung, M. P. Analysis of mitochondria by capillary electrophoresis: cardiolipin levels decrease in response to carbonyl cyanide 4-(trifluoromethoxy) phenylhydrazone. *Eur. J. Lipid Sci. Technol.* **2010**, *112*, 1058–1066.

6. Hattori, F.; Fukuda, K. Method of selecting myocardial cells using intracellular mitochondria as indication. PCT Int. Appl. WO 2006022377, 2006.

7. Ahmadzadeh, H.; Thompson, L. V.; Arriaga, E. A. On-column labeling for capillary electrophoretic analysis of individual mitochondria directly sampled from tissue cross sections. *Anal. Bioanal. Chem.* **2006**, *384*, 169–174.

8. Ahmadzadeh, H.; Johnson, R. D.; Thompson, L.; Arriaga, E. A. Direct sampling from muscle cross sections for electrophoretic analysis of individual mitochondria. *Anal. Chem.* **2004**, *76*, 315–321.

9. Benel, L.; Ronot, X.; Mounolou, J. C.; Gaudemer, F.; Adolphe, M. Compared flow cytometric analysis of mitochondria using 10-n-nonyl acridine orange and rhodamine 123. *Basic Appl. Histochem.* **1989**, *33*, 71–80.

10. Maftah, A.; Petit, J. M.; Ratinaud, M. H.; Julien, R. 10-N-nonyl-acridine orange: A fluorescent probe which stains mitochondria independently of their energetic state. *Biochem. Biophys. Res. Commun.* **1989**, *164*, 185–190.

11. Ratinaud, M. H.; Leprat, P.; Julien, R. *In situ* flow cytometric analysis of nonyl acridine orange-stained mitochondria from splenocytes. *Cytometry* **1988**, *9*, 206–212.

12. Septinus, M.; Berthold, T.; Naujok, A.; Zimmermann, H. W. Hydrophobic acridine dyes for fluorescent staining of mitochondria in living cells. 3. Specific accumulation of the fluorescent dye NAO on the mitochondrial membranes in HeLa cells by hydrophobic interaction. Depression of respiratory activity, changes in the ultrastructure of mitochondria due to NAO. Increase of fluorescence in vital stained mitochondria *in situ* by irradiation. *Histochemistry* **1985**, *82*, 51–66.

13. Erbrich, U.; Septinus, M.; Naujok, A.; Zimmermann, H. W. Hydrophobic acridine dyes for fluorescence staining of mitochondria in living cells. 2. Comparison of staining of living and fixed HeLa-cells with NAO and DPPAO. *Histochemistry* **1984**, *80*, 385–388.

14. Dickman, D. Methods of detecting prostate cancer. PCT Int. Appl. WO 2006054296, 2006.

15. Murphy, A. N.; Dykens, J. A.; Ghosh, S. S.; Davis, R. E.; Granston, A. E., Jr.; Terkeltaub, R. Methods and compositions for diagnosing and treating arthritic disorders and regulating bone mass. PCT Int. Appl. WO 2001020018, 2001.

16. Anderson, C. M.; Davis, R. E. Indicators of altered mitochondrial function in predictive methods for determining risk of type 2 diabetes mellitus. U.S. Patent 6140067, 2000.

17. Quarato, G.; Piccoli, C.; Scrima, R.; Capitanio, N. Functional imaging of membrane potential at the single mitochondrion level: Possible application for diagnosis of human diseases. *Mitochondrion* **2011**, *11*, 764–773.

18. Widlansky, M. E.; Wang, J.; Shenouda, S. M.; Hagen, T. M.; Smith, A. R.; Kizhakekuttu, T. J.; Kluge, M. A.; Weihrauch, D.; Gutterman, D. D.; Vita, J. A. Altered mitochondrial membrane potential, mass, and morphology in the mononuclear cells of humans with type 2 diabetes. *Transl. Res.* **2010**, *156*, 15–25.

19. Dykens, J. A.; Fleck, B.; Ghosh, S.; Lewis, M.; Velicelebi, G.; Ward, M. W. High-throughput assessment of mitochondrial membrane potential *in situ* using fluorescence resonance energy transfer. *Mitochondrion* **2002**, *1*, 461–473.

20. Keij, J. F.; Bell-Prince, C.; Steinkamp, J. A. Staining of mitochondrial membranes with 10-nonylacridine orange, MitoFluor Green, and MitoTracker Green is affected by mitochondrial membrane potential altering drugs. *Cytometry* **2000**, *39*, 203–210.

21. Fujii, H.; Cody, S. H.; Seydel, U.; Papadimitriou, J. M.; Wood, D. J.; Zheng, M. H. Recording of mitochondrial transmembrane potential and volume in cultured rat osteoclasts by confocal laser scanning microscopy. *Histochem. J.* **1997**, *29*, 571–581.

22. Jacobson, J.; Duchen, M. R.; Heales, S. J. R. Intracellular distribution of the fluorescent dye nonyl acridine orange responds to the mitochondrial membrane potential: implications for assays of cardiolipin and mitochondrial mass. *J. Neurochem.* **2002**, *82*, 224–233.

23. Zhao, W.; Chen, Q.; Wu, R.; Wu, H.; Fung, Y.; Waisum, O. Capillary electrophoresis with LIF detection for assessment of mitochondrial number based on the cardiolipin content. *Electrophoresis* **2011**, *32*, 3025–3033.

24. Mileykovskaya, E.; Dowhan, W. Cardiolipin membrane domains in prokaryotes and eukaryotes. *Biochim. Biophys. Acta, Biomembr.* **2009**, *1788*, 2084–2091.

25. Gohil, V. M.; Gvozdenovic-Jeremic, J.; Schlame, M.; Greenberg, M. L. Binding of 10-N-nonylacridine orange to cardiolipin-deficient yeast cells: Implications for assay of cardiolipin. *Anal. Biochem.* **2005**, *343*, 350–352.

26. Mileykovskaya, E.; Dowhan, W.; Birke, R. L.; Zheng, D.; Lutterodt, L.; Haines, T. H. Cardiolipin binds nonyl acridine orange by aggregating the dye at exposed hydrophobic domains on bilayer surfaces. *FEBS Lett.* **2001**, *507*, 187–190.

27. Kaewsuya, P.; Danielson, N. D.; Ekhterae, D. Fluorescent determination of cardiolipin using 10-N-nonyl acridine orange. *Anal. Bioanal. Chem.* **2007**, *387*, 2775–2782.

28. Garcia Fernandez, M. I.; Ceccarelli, D.; Muscatello, U. Use of the fluorescent dye 10-N-nonyl acridine orange in quantitative and location assays of cardiolipin: A study on different experimental models. *Anal. Biochem.* **2004**, *328*, 174–180.

29. Fuller, K. M.; Duffy, C. F.; Arriaga, E. A. Determination of the cardiolipin content of individual mitochondria by capillary electrophoresis with laser-induced fluorescence detection. *Electrophoresis* **2002**, *23*, 1571–1576.

30. Dertinger, S. D.; Bemis, J. C.; Bryce, S. M. Method for measuring *in vivo* hematotoxicity with an

emphasis on radiation exposure assessment. U.S. Pat. Appl. Publ. US 2008311586, 2008.

31. Crane, P. D.; Orlandi, C. Method for detecting the cardiotoxicity of compounds. U.S. Patent 5604112, 1997.

32. Agafonov, A. V.; Gritsenko, E. N.; Shlyapnikova, E. A.; Kharakoz, D. P.; Belosludtseva, N. V.; Lezhnev, E. I.; Saris, N. E. L.; Mironova, G. D. Ca2+−Induced phase separation in the membrane of palmitate-containing liposomes and its possible relation to membrane permeabilization. *J. Membr. Biol.* **2007**, *215*, 57–68.

33. Lobasso, S.; Saponetti, M. S.; Polidoro, F.; Lopalco, P.; Urbanija, J.; Kralj-Iglic, V.; Corcelli, A. Archaebacterial lipid membranes as models to study the interaction of 10-N-nonyl acridine orange with phospholipids. *Chem. Phys. Lipids* **2009**, *157*, 12–20.

34. Herrera, A. A.; Banner, L. R. The use and effects of vital fluorescent dyes: Observation of motor nerve terminals and satellite cells in living frog muscles. *J Neurocytol.* **1990**, *19*, 67–83.

35. Oreopoulos, J.; Epand, R. F.; Epand, R. M.; Yip, C. M. Peptide-induced domain formation in supported lipid bilayers: direct evidence by combined atomic force and polarized total internal reflection fluorescence microscopy. *Biophys. J.* **2010**, *98*, 815–823.

36. Ferlini, C.; Scambia, G. Assay for apoptosis using the mitochondrial probes, Rhodamine 123 and 10-N-nonyl acridine orange. *Nat. Protoc.* **2007**, *2*, 3111–3114.

37. King, M. A.; Eddaoudi, A.; Davies, D. C. A comparison of three flow cytometry method for evaluating mitochondrial damage during staurosporine-induced apoptosis in Jurkat cells. *Cytometry* **2007**, *71A*, 668–674.

38. Mentak, K. Ultra violet, violet, and blue light filtering polymers for ophthalmic applications. U.S. Pat. Appl. Publ. US 2006252844, 2006.

39. Zhang, H.; Chen, Q.; Xiang, M.; Ma, C.; Huang, Q.; Yang, S. In silico prediction of mitochondrial toxicity by using GA-CG-SVM approach. *Toxicol. in Vitro* **2009**, *23*, 134–140.

ALEXA FLUOR 350 CARBOXYLIC ACID SUCCINIMIDYL ESTER (AMCA-S)

CAS Registry Number   200554-19-4
Chemical Structure

CA Index Name   2*H*-1-Benzopyran-3-acetic acid, 7-amino-4-methyl-2-oxo-6-sulfo-, 3-(2,5-dioxo-1-pyrrolidinyl) ester

Other Names   2*H*-1-Benzopyran-6-sulfonic acid, 7-amino-3-[2-[(2,5-dioxo-1-pyrrolidinyl)oxy]-2-oxoethyl]-4-methyl-2-oxo-; 7-amino-3-((((succinimidyl)oxy)carbonyl)methyl)-4-methylcoumarin-6-sulfonic acid; AMCA-S; Alexa 350 carboxylic acid succinimidyl ester; Alexa Fluor 350 carboxy acid succinimidyl ester; Alexa Fluor 350 succinimidyl ester; Alexa Fluor 350NHS; Sulfosuccinimidyl-7-amino-4-methylcoumarin-3-acetic acid

Merck Index Number   Not listed

Chemical/Dye Class   Coumarin

Molecular Formula $C_{16}H_{14}N_2O_9S$

Molecular Weight   410.35

Physical Form   Pale yellow precipitate[2] or yellow/orange solid

Solubility   Soluble in water, dimethyl sulfoxide, methanol

pKa (Calcd.)   -0.46±0.20 Most Basic Temperature: 25 °C

Absorption (λ_{max})   346 nm (Buffer pH 7); 353 nm (MeOH)

Emission (λ_{max})   445 nm (Buffer pH 7); 438 nm (MeOH)

Molar Extinction Coefficient   19,000 cm^{-1} M^{-1} (Buffer pH 7); 20,000 cm^{-1} M^{-1} (MeOH)

Synthesis   Synthetic methods[1,2]

Imaging/Labeling Applications   Amines;[3,7,24] neomycin;[4] nucleic acids;[5–7] peptides;[8–11] proteins;[1,2,10,12–19] silane coupling agent;[20–24] silica particles/beads[20–24]

Biological/Medical Applications   Analyzing peptides;[8–11] detecting mutations;[25] detecting/identifying nucleic acids;[5–7] detecting/isolating/purifying proteins;[1,2,10,12–19] investigating location of phosphatidylinositol-4,5-bisphosphate (PI(4,5)P$_2$) in cell membranes;[4] colloidal diagnostic devices;[20] multiplex genetic analysis;[25] as nucleic acid hybridization probe;[7] useful for synthesizing nucleic acid,[21,22] peptides[20,22]

Industrial Applications   Poly(vinyl alcohol)-coated microfluidic devices;[3] self-assembled monolayers;[26] silica particles/beads[20–24]

Safety/Toxicity   No data available

REFERENCES

1. Leung, W.; Trobridge, P. A.; Haugland, R. P.; Haugland, R. P.; Mao, F. 7-amino-4-methyl-6-sulfo-coumarin-3-acetic acid: a novel blue fluorescent dye for protein labeling. *Bioorg. Med. Chem. Lett.* **1999**, 9, 2229–2232.

2. Wang, H.; Leung, W.; Mao, F. Sulfonated derivatives of 7-aminocoumarin. U.S. Patent 5696157, 1997.

3. Belder, D.; Deege, A.; Koehler, F.; Ludwig, M. Poly(vinyl alcohol)-coated microfluidic devices for high-performance microchip electrophoresis. *Electrophoresis* **2002**, 23, 3567–3573.

4. Arbuzova, A.; Martushova, K.; Hangyas-Mihalyne, G.; Morris, A. J.; Ozaki, S.; Prestwich, G. D.; McLaughlin, S. Fluorescently labeled neomycin as a probe of phosphatidylinositol-4,5-bisphosphate in

membranes. *Biochim. Biophys. Acta, Biomembr.* **2000**, *1464*, 35–48.

5. Sergeev, N. V.; Brevnov, M. G.; Furtado, M. R. Identification of nucleic acids. PCT Int. Appl. WO 2011143478, 2011.

6. Caputo, D.; de Cesare, G.; Nascetti, A.; Negri, R. Spectral tuned amorphous silicon p-i-n for DNA detection. *J. Non-Cryst. Solids* **2006**, *352*, 2004–2006.

7. Cox, W. G.; Singer, V. L. Fluorescent DNA hybridization probe preparation using amine modification and reactive dye coupling. *BioTechniques* **2004**, *36*, 114–122.

8. Bahadduri, P. M.; Ray, A.; Khandelwal, A.; Swaan, P. W. Design of high-affinity peptide conjugates with optimized fluorescence quantum yield as markers for small peptide transporter PEPT1 (SLC15A1). *Bioorg. Med. Chem. Lett.* **2008**, *18*, 2555–2557.

9. Wilson, J. J.; Brodbelt, J. S. MS/MS simplification by 355 nm ultraviolet photodissociation of chromophore-derivatized peptides in a quadrupole ion trap. *Anal. Chem.* **2007**, *79*, 7883–7892.

10. Pashkova, A.; Chen, H.; Rejtar, T.; Zang, X.; Giese, R.; Andreev, V.; Moskovets, E.; Karger, B. L. Coumarin tags for analysis of peptides by MALDI-TOF MS and MS/MS. 2. Alexa Fluor 350 tag for increased peptide and protein identification by LC-MALDI-TOF/TOF MS. *Anal. Chem.* **2005**, *77*, 2085–2096.

11. Pashkova, A.; Moskovets, E.; Karger, B. L. Coumarin tags for improved analysis of peptides by MALDI-TOF MS and MS/MS. 1. Enhancement in MALDI MS signal intensities. *Anal. Chem.* **2004**, *76*, 4550–4557.

12. Rothschild, K. J.; Gite, S.; Olejnik, J. Methods for the detection, analysis and isolation of nascent proteins. U.S. Pat. Appl. Publ. US 20110250609, 2011.

13. Zauner, G.; Lonardi, E.; Bubacco, L.; Aartsma, T. J.; Canters, G. W.; Tepper, A. W. J. W. Tryptophan-to-dye fluorescence energy transfer applied to oxygen sensing by using type-3 copper proteins. *Chem.- Eur. J.* **2007**, *13*, 7085–7090.

14. Gee, K. R.; Hart, C. R.; Haugland, R.; Patton, W. F.; Whitney, S. Site-specific labeling of affinity tags in fusion proteins. U.S. Pat. Appl. Publ. US 20060141554, 2006.

15. Nguyen, T.; Joshi, N. S.; Francis, M. B. An affinity-based method for the purification of fluorescently-labeled biomolecules. *Bioconjugate Chem.* **2006**, *17*, 869–872.

16. Gee, K.; Hart, C.; Haugland, R.; Patton, W.; Whitney, S. Site-specific labeling of affinity tags in fusion proteins. PCT Int. Appl. WO 2005038460, 2005.

17. Kremser, L.; Petsch, M.; Blaas, D.; Kenndler, E. Labeling of capsid proteins and genomic RNA of human rhinovirus with two different fluorescent dyes for selective detection by capillary electrophoresis. *Anal. Chem.* **2004**, *76*, 7360–7365.

18. Diwu, Z.; Gee, K.; Hart, C.; Haugland, R.; Leung, W.; Patton, W.; Rukavishnikov, A. Site-specific labeling of affinity tags in fusion proteins. PCT Int. Appl. WO 2004025259, 2004.

19. Panchuk-Voloshina, N.; Haugland, R. P.; Bishop-Stewart, J.; Bhalgat, M. K.; Millard, P. J.; Mao, F.; Leung, W.; Haugland, R. P. Alexa dyes, a series of new fluorescent dyes that yield exceptionally bright, photostable conjugates. *J. Histochem. Cytochem.* **1999**, *47*, 1179–1188.

20. Lawrie, G.; Grondahl, L.; Battersby, B.; Keen, I.; Lorentzen, M.; Surawski, P.; Trau, M. Tailoring surface properties to build colloidal diagnostic devices: Controlling interparticle associations. *Langmuir* **2006**, *22*, 497–505.

21. Lawrie, G. A.; Battersby, B. J.; Trau, M. Synthesis of optically complex core-shell colloidal suspensions: Pathways to multiplexed biological screening. *Adv. Funct. Mater.* **2003**, *13*, 887–896.

22. Trau, M.; Johnston, A. Synthesis and use of organosilica particles. PCT Int. Appl. WO 2003002633, 2003.

23. Matthews, D. C.; Grondahl, L.; Battersby, B. J.; Trau, M. Multi-fluorescent silica colloids for encoding large combinatorial libraries. *Aust. J. Chem.* **2001**, *54*, 649–656.

24. Saleh, S. M.; Mueller, R.; Mader, H. S.; Duerkop, A.; Wolfbeis, O. S. Novel multicolor fluorescently labeled silica nanoparticles for interface fluorescence resonance energy transfer to and from labeled avidin. *Anal. Bioanal. Chem.* **2010**, *398*, 1615–1623.

25. Richardson, J. A.; Gerowska, M.; Shelbourne, M.; French, D.; Brown, T. Six-colour HyBeacon probes for multiplex genetic analysis. *ChemBioChem* **2010**, *11*, 2530–2533.

26. Scrimgeour, J.; Kodali, V. K.; Kovari, D. T.; Curtis, J. E. Photobleaching-activated micropatterning on self-assembled monolayers. *J. Phys.: Condens. Matter* **2010**, *22*, 194103/1–194103/6.

ALEXA FLUOR 430 CARBOXYLIC ACID SUCCINIMIDYL ESTER

CAS Registry Number   467233-94-9

Chemical Structure

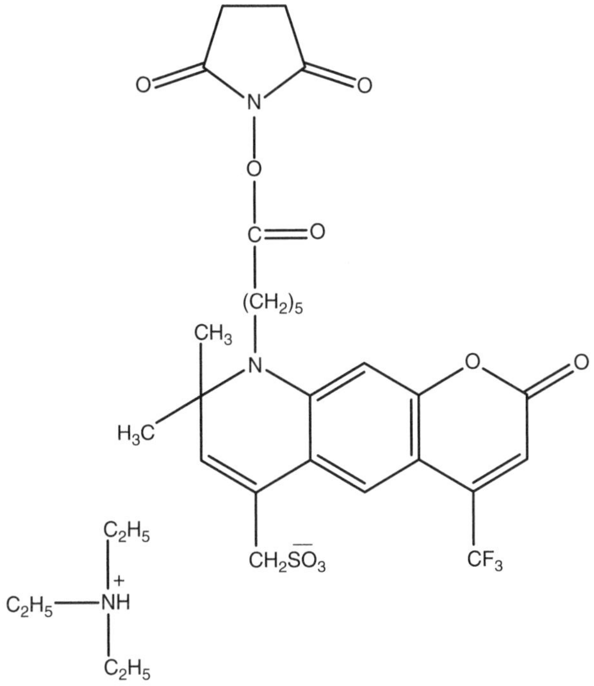

CA Index Name *2H*-Pyrano[3,2-*g*]quinoline-9(8*H*)-hexanoic acid, 8,8-dimethyl-2-oxo-6-(sulfomethyl)-4-(trifluoromethyl)-, 2,5-dioxo-1-pyrrolidinyl ester, compd. with *N,N*-diethylethanamine (1:1)

Other Names *2H*-Pyrano[3,2-*g*]quinoline-6-methane-sulfonic acid, 9-[6-[(2,5-dioxo-1-pyrrolidinyl)oxy]-6-oxohexyl]-8,9-dihydro-8,8-dimethyl-2-oxo-4-(trifluoro-methyl)-, compd. with *N,N*-diethylethanamine (1:1);

Alexa 430; Alexa 430 succinimidyl ester; Alexa Fluor 430; Alexa Fluor 430 NHS; Alexa Fluor 430 NHS ester; Alexa Fluor 430 carboxylic acid SE; Alexa Fluor 430 carboxylic acid succinimidyl ester; Alexa Fluor 430 SE; Alexa Fluor 430 succinimidyl ester

Merck Index Number   Not listed

Chemical/Dye Class   Heterocycle

Molecular Formula $C_{32}H_{42}F_3N_3O_9S$

Molecular Weight   701.75

Physical Form   Yellow solid

Solubility   Soluble in water, dimethyl sulfoxide, methanol

Absorption (λ_{max})   430 nm (Buffer pH 7); 425 nm (MeOH)

Emission (λ_{max})   545 nm (Buffer pH 7); 530 nm (MeOH)

Molar Extinction Coefficient   15,000 cm^{-1} M^{-1} (Buffer pH 7); 16,200 cm^{-1} M^{-1} (MeOH)

Synthesis   Synthetic method[1]

Imaging/Labeling Applications   Amines/amino groups;[16,24] antibodies;[2,3] bacteria;[4] cells;[5,6] chromosomes;[7,8] glucose;[9] histone deacetylases;[10] microorganisms;[2,11] nucleic acids/nucleotides;[12–16] proteins/peptides;[17–25] quantum dots;[26] silane coupling agents;[27–29] viruses[30,31]

Biological/Medical Applications   Analyzing sugar flux;[9] assaying retinol-binding protein-transthyretin interaction;[21] detecting chromosomal aberration;[7,8] detecting/quantifying biotinylated proteins;[23] identifying microorganisms;[2,11] monitoring reorganization of matrigel[22]

Industrial Applications   Explosives detection polymers;[32] nanowalled polymeric microtubes;[33] photoresists;[34] silica particles/beads;[27–29] optical waveguides[35,36]

Safety/Toxicity   No data available

REFERENCES

1. Haugland, R. P. *Handbook of Fluorescent Probes and Research Products*; Molecular Probes Inc.: Eugene, **2002**; pp 20–35.

2. Tsilivakos, V.; Gritzapis, A. Method of intracellular infectious agent detection in sperm cells. PCT Int. Appl. WO 2013144662, 2013.

3. Arcangeli, A.; Becchetti, A.; Pillozzi, S.; Masselli, M.; De Lorenzo, E. Method and kit for the prevention and/or the monitoring of chemoresistance of leukemia forms. PCT Int. Appl. WO 2011058509, 2011.

4. Kim, S. U.; Kim, S. J.; Lee, S. H.; Lee, D. H.; Ryu, H. S. Kit and method for detecting food-borne bacteria. Repub. Korean Kongkae Taeho Kongbo KR 2013065337, 2013.

5. Klimanskaya, I.; Gay, R. J. Methods for detection of rare subpopulations of cells and highly purified

compositions of cells. PCT Int. Appl. WO 2012012803, 2012.

6. Bestvater, F.; Spiess, E.; Stobrawa, G.; Hacker, M.; Feurer, T.; Porwol, T.; Berchner-Pfannschmidt, U.; Wotzlaw, C.; Acker, H. Two-photon fluorescence absorption and emission spectra of dyes relevant for cell imaging. *J. Microsc.* **2002**, *208*, 108–115.

7. Hauke, S. Method for detecting a chromosomal aberration. PCT Int. Appl. WO 2012150022, 2012.

8. Poulsen, T. S.; Poulsen, S. M.; Petersen, K. H. Methods for detecting chromosome aberrations. PCT Int. Appl. WO 2005111235, 2005.

9. Chaudhuri, B.; Hoermann, F.; Frommer, W. B. Dynamic imaging of glucose flux impedance using FRET sensors in wild-type *Arabidopsis* plants. *J. Exp. Bot.* **2011**, *62*, 2411–2417.

10. Heidebrecht, R. W., Jr.,; Kral, A. M.; Miller, T. A. Fluorescent compounds that bind to histone deacetylase. U.S. Pat. Appl. Publ. US 20090156825, 2009.

11. Poetter, K. F.; Vandegraaff, N. Compositions and methods of detecting respiratory pathogens using nucleic acid probes and subsets of beads. PCT Int. Appl. WO 2013049891, 2013.

12. Di Pasquale, F.; Mueller, D. Methods for sequencing, amplification and detection of nucleic acids comprising internally labelled primer. PCT Int. Appl. WO 2012152698, 2012.

13. Zhang, B.; Wang, W.; Qu, D. Method and kit for labeling nucleic acid in living cells. Faming Zhuanli Shenqing CN 101921835, 2010.

14. Al Attar, H. A.; Monkman, A. P. FRET and competing processes between conjugated polymer and dye substituted DNA strands: A comparative study of probe selection in DNA detection. *Biomacromolecules* **2009**, *10*, 1077–1083.

15. Pimentel, A. C.; Prazeres, D. M. F.; Chu, V.; Conde, J. P. Fluorescence detection of DNA using an amorphous silicon p-i-n photodiode. *J. Appl. Phys.* **2008**, *104*, 054913/1–054913/10.

16. Marras, S. A. E.; Kramer, F. R.; Tyagi, S. Efficiencies of fluorescence resonance energy transfer and contact-mediated quenching in oligonucleotide probes. *Nucleic Acids Res.* **2002**, *30*, e122/1–e122/8.

17. Hoeoek, M.; Ganesh, V. K.; Ross, C. L.; Liang, X. Engineered collagen-binding MSCRAMM with enhanced affinities for collagen. PCT Int. Appl. WO 2013159021, 2013.

18. Regino, C. A. S.; McBride, W. J.; Chang, C.; Goldenberg, D. M. Dye conjugated peptides for fluorescent imaging. U.S. Pat. Appl. Publ. US 20130039861, 2013.

19. Ningsih, Z.; Hossain, M. A.; Wade, J. D.; Clayton, A. H. A.; Gee, M. L. Slow insertion kinetics during interaction of a model antimicrobial peptide with unilamellar phospholipid vesicles. *Langmuir* **2012**, *28*, 2217–2224.

20. Rapson, A. C.; Hossain, M. A.; Wade, J. D.; Nice, E. C.; Smith, T. A.; Clayton, A. H. A.; Gee, M. L. Structural dynamics of a lytic peptide interacting with a supported lipid bilayer. *Biophys. J.* **2011**, *100*, 1353–1361.

21. Mata, N. L.; Phan, K.; Han, Y. Assay of retinol-binding protein-transthyretin interaction and techniques to identify competing ligands. *Methods Mol. Biol.* **2010**, *652*, 209–227.

22. Lockwood, N. A.; Mohr, J. C.; Ji, L.; Murphy, C. J.; Palecek, S. P.; de Pablo, J. J.; Abbott, N. L. Thermotropic liquid crystals as substrates for imaging the reorganization of matrigel by human embryonic stem cells. *Adv. Funct. Mater.* **2006**, *16*, 618–624.

23. Lewis, B.; Rathman, S.; McMahon, R. J. Detection and quantification of biotinylated proteins using the storm 840 optical scanner. *J. Nutr. Biochem.* **2003**, *14*, 196–202.

24. Alroy, I.; Moskowitz, H.; Reiss, Y.; Shoham, B. A. Methods and compositions related to tagging of membrane surface proteins. PCT Int. Appl. WO 2002099077, 2002.

25. Panchuk-Voloshina, N.; Haugland, R. P.; Bishop-Stewart, J.; Bhalgat, M. K.; Millard, P. J.; Mao, F.; Leung, W.; Haugland, R. P. Alexa dyes, a series of new fluorescent dyes that yield exceptionally bright, photostable conjugates. *J. Histochem. Cytochem.* **1999**, *47*, 1179–1188.

26. Giraud, G.; Schulze, H.; Bachmann, T. T.; Campbell, C. J.; Mount, A. R.; Ghazal, P.; Khondoker, M. R.; Ross, A. J.; Ember, S. W. J.; Ciani, I.; Tlili, C.; Walton, A. J.; Terry, J. G.; Crain, J. Fluorescence lifetime imaging of quantum dot labeled DNA microarrays. *Int. J. Mol. Sci.* **2009**, *10*, 1930–1941.

27. Kim, D. H.; Lee, G. G.; Lee, S. J.; Park, T. J. Preparation of gold-doped fluorescent silica nanoparticle composite. Repub. Korean Kongkae Taeho Kongbo KR 2012089928, 2012.

28. Lawrie, G. A.; Battersby, B. J.; Trau, M. Synthesis of optically complex core-shell colloidal suspensions: Pathways to multiplexed biological screening. *Adv. Funct. Mater.* **2003**, *13*, 887–896.

29. Matthews, D. C.; Grondahl, L.; Battersby, B. J.; Trau, M. Multi-fluorescent silica colloids for encoding large combinatorial libraries. *Aust. J. Chem.* **2001**, *54*, 649–656.

30. In, G. H.; Kim, W. G.; Kim, S. U.; Kim, S. J.; Lee, S. H. PCR device for detecting new influenza a virus, and method using the same for detecting influenza A virus. Repub. Korean Kongkae Taeho Kongbo KR 2013091025, 2013.

31. Kim, S. U.; Kim, S. J.; Lee, D. H.; Kim, D. J.; Ryu, H. S. Kit and method for influenza A virus detection. Repub. Korean Kongkae Taeho Kongbo KR 2013081948, 2013.

32. Lei, Y.; Wang, Y. Explosives detection polymer comprising functionalized polyamine polymers and methods of using the same. PCT Int. Appl. WO 2013165625, 2013.

33. Todaro, M. T.; Blasi, L.; Giordano, C.; Rizzo, A.; Cingolani, R.; Gigli, G.; Passaseo, A.; de Vittorio, M. Nanowalled polymer microtubes fabricated by using strained semiconductor templates. *Nanotechnology* **2010**, *21*, 245305/1–245305/5.

34. Garza, C. M.; Cho, S. Metrology of bilayer photoresist processes. U.S. Pat. Appl. Publ. US 20090220895, 2009.

35. Llobera, A.; Cadarso, V. J.; Carregal-Romero, E.; Brugger, J.; Dominguez, C.; Fernandez-Sanchez, C. Fluorophore-doped xerogel antiresonant reflecting optical waveguides. *Opt. Express* **2011**, *19*, 5026–5039.

36. Carregal-Romero, E.; Llobera, A.; Cadarso, V. J.; Darder, M.; Aranda, P.; Dominguez, C.; Ruiz-Hitzky, E.; Fernandez-Sanchez, C. One-step patterning of hybrid xerogel materials for the fabrication of disposable solid-state light emitters. *ACS Appl. Mater. Interfaces* **2012**, *4*, 5029–5037.

ALEXA FLUOR 488 CARBOXYLIC ACID SUCCINIMIDYL ESTER

CAS Registry Number   222164-96-7
Chemical Structure

CA Index Name   Xanthylium, 3,6-diamino-9-[2-carboxy-4(*or* 5)-[[(2,5-dioxo-1-pyrrolidinyl)oxy]carbonyl]phenyl]-4,5-disulfo-, inner salt, lithium salt (1:2)

Other Names   Xanthylium, 3,6-diamino-9-[2-carboxy-4 (*or* 5)-[[(2,5-dioxo-1-pyrrolidinyl)oxy]carbonyl]phenyl]-4,5-disulfo-, inner salt, dilithium salt; Alexa 488; Alexa 488 succinimidyl ester; Alexa Fluor 488; Alexa Fluor 488 NHS; Alexa Fluor 488 NHS ester; Alexa Fluor 488 carboxylic acid SE; Alexa Fluor 488 carboxylic acid succinimidyl ester; Alexa Fluor 488 SE; Alexa Fluor 488 succinimidyl ester

Merck Index Number   Not listed

Chemical/Dye Class   Xanthene

Molecular Formula $C_{25}H_{15}Li_2N_3O_{13}S_2$

Molecular Weight   643.41

Physical Form   Orange solid

Solubility   Soluble in water, dimethyl sulfoxide

Absorption (λmax)   494 nm (Buffer pH 7)

Emission (λmax)   517 nm (Buffer pH 7)

Molar Extinction Coefficient   73,000 cm^{-1} M^{-1} (Buffer pH 7)

Quantum Yield   0.92 (Buffer pH 7.2)

Synthesis   Synthetic method[1]

Imaging/Labeling Applications   Amines/amino groups;[2–5,21,29,42] antibodies;[6–12] cells/tissues;[12–15] chromosomes;[16,17] cyclodextrins;[18] histone deacetylases;[19] insulin-saccharide conjugates;[20] liposomes/lysozymes;[34] nucleic acids/nucleotides;[21–29] polymersomes;[30] proteins/peptides;[3–5,9,11,31–44] receptor-like kinases (RLKs) microbeads;[45] silane coupling agents;[46] cowpea mosaic virus (CPMV);[47] dengue virus[48]

Biological/Medical Applications   Analyzing/profiling amyloid-β peptides;[3,4] assessing/evaluating protein-lipid interactions;[35] characterizing submicron protein aggregates/conjugates;[32,33] detecting/quantifying proteins;[41] detecting chromosomal aberration;[16,17] labeling/detecting nucleic acids;[21–29] measuring protein dynamics;[37,38] patterning wide range of proteins;[36] tracing adenosine A2A receptors *in vitro*;[39] visualizing vasculature/blood flow in living mouse and chick embryos[47] polymersomes in biomedical or cosmetic applications[30]

Industrial Applications   Photoresists;[49] silica particles/beads[46]

Safety/Toxicity   No data available

REFERENCES

1. Mao, F.; Leung, W.; Haugland, R. P. Sulfonated xanthene derivatives. PCT Int. Appl. WO 9915517, 1999.

2. Sueyoshi, K.; Hashiba, K.; Kawai, T.; Kitagawa, F.; Otsuka, K. Hydrophobic labeling of amino acids: Transient trapping-capillary/microchip electrophoresis. *Electrophoresis* **2011**, *32*, 1233–1240.

3. Verpillot, R.; Essellmann, H.; Mohamadi, M. R.; Klafki, H.; Poirier, F.; Lehnert, S.; Otto, M.; Wiltfang, J.; Jean Louis, V.; Taverna, M. Analysis of amyloid-β peptides in cerebrospinal fluid samples by capillary electrophoresis coupled with LIF detection. *Anal. Chem.* **2011**, *83*, 1696–1703.

4. Mohamadi, M. R.; Svobodova, Z.; Verpillot, R.; Esselmann, H.; Wiltfang, J.; Otto, M.; Taverna, M.; Bilkova, Z.; Viovy, J. Microchip electrophoresis profiling of Aβ peptides in the cerebrospinal fluid of patients with Alzheimer's disease. *Anal. Chem.* **2010**, *82*, 7611–7617.

5. Cheng, K.; Blumen, S. R.; MacPherson, M. B.; Steinbacher, J. L.; Mossman, B. T.; Landry, C. C. Enhanced uptake of porous silica microparticles by bifunctional surface modification with a targeting antibody and a biocompatible polymer. *ACS Appl. Mater. Interfaces* **2010**, *2*, 2489–2495.

6. Karver, M. R.; Weissleder, R.; Hilderbrand, S. A. Bioorthogonal reaction pairs enable simultaneous, selective, multi-target imaging. *Angew. Chem., Int. Ed.* **2012**, *51*, 920–922.

7. Arcangeli, A.; Becchetti, A.; Pillozzi, S.; Masselli, M.; De Lorenzo, E. Method and kit for the prevention and/or the monitoring of chemoresistance of leukemia forms. PCT Int. Appl. WO 2011058509, 2011.

8. Kokko, T.; Liljenback, T.; Peltola, M. T.; Kokko, L.; Soukka, T. Homogeneous dual-parameter assay for prostate-specific antigen based on fluorescence resonance energy transfer. *Anal. Chem.* **2008**, *80*, 9763–9768.

9. Uhlen, M.; Svahn, H. A.; Lundberg, E. Solid phase labeling method. U.S. Pat. Appl. Publ. US 20080233660, 2008.

10. Lundberg, E.; Sundberg, M.; Graeslund, T.; Uhlen, M.; Svahn, H. A. A novel method for reproducible fluorescent labeling of small amounts of antibodies on solid phase. *J. Immunol. Methods* **2007**, *322*, 40–49.

11. Albizu, L.; Teppaz, G.; Seyer, R.; Bazin, H.; Ansanay, H.; Manning, M.; Mouillac, B.; Durroux, T. Toward efficient drug screening by homogeneous assays based on the development of new fluorescent vasopressin and oxytocin receptor ligands. *J. Med. Chem.* **2007**, *50*, 4976–4985.

12. Fuller, C. J.; Straight, A. F. Imaging nanometre-scale structure in cells using *in situ* aberration correction. *J. Microsc.* **2012**, *248*, 90–101.

13. Klimanskaya, I.; Gay, R. J. Methods for detection of rare subpopulations of cells and highly purified compositions of cells. PCT Int. Appl. WO 2012012803, 2012.

14. Bestvater, F.; Spiess, E.; Stobrawa, G.; Hacker, M.; Feurer, T.; Porwol, T.; Berchner-Pfannschmidt, U.; Wotzlaw, C.; Acker, H. Two-photon fluorescence absorption and emission spectra of dyes relevant for cell imaging. *J. Microsc.* **2002**, *208*, 108–115.

15. Tsurui, H.; Nishimura, H.; Hattori, S. Hirose, S. Okumura, K.; Shirai, T. Seven-color fluorescence imaging of tissue samples based on fourier spectroscopy and singular value decomposition. *J. Histochem. Cytochem.* **2000**, *48*, 653–662.

16. Hauke, S. Method for detecting a chromosomal aberration. PCT Int. Appl. WO 2012150022, 2012.

17. Poulsen, T. S.; Poulsen, S. M.; Petersen, K. H. Methods for detecting chromosome aberrations. PCT Int. Appl. WO 2005111235, 2005.

18. Granadero, D.; Bordello, J.; Perez-Alvite, M. J.; Novo, M.; Al-Soufi, W. Host-guest complexation studied by fluorescence correlation spectroscopy: adamantane-cyclodextrin inclusion. *Int. J. Mol. Sci.* **2010**, *11*, 173–188.

19. Heidebrecht, R. W., Jr.,; Kral, A. M.; Miller, T. A. Fluorescent compounds that bind to histone deacetylase. U.S. Pat. Appl. Publ. US 20090156825, 2009.

20. Murikipudi, S.; Lancaster, T. M.; Zion, T. C. Uses of macrophage mannose receptor to screen compounds and uses of these compounds. PCT Int. Appl. WO 2012050822, 2012.

21. Ball, R. W. Process for conjugation of NHS esters with oligonucleotides. U.S. Pat. Appl. Publ. US 20130158243, 2013.

22. Korlach, J.; Bibillo, A.; Wegener, J.; Peluso, P.; Pham, T. T.; Park, I.; Clark, S.; Otto, G. A.; Turner, S. W. Long, processive enzymatic DNA synthesis using 100% dye-labeled terminal phosphate-linked nucleotides. *Nucleosides, Nucleotides Nucleic Acids* **2008**, *27*, 1072–1082.

23. Agnew, B.; Gee, K.; Kumal, K.; Ford, M. Labeling and detection of nucleic acids. PCT Int. Appl. WO 2008101024, 2008.

24. Agnew, B.; Ford, M. J.; Gee, K. R.; Kumar, K. Labeling and detection of nucleic acids. U.S. Pat. Appl. Publ. US 20080050731, 2008.

25. Ichikawa, K.; Nakamura, K.; Kurata, S. Real-time PCR method using the novel type of oligonucleotide probes. Jpn. Kokai Tokkyo Koho JP 2008182974, 2008.

26. Chiuman, W.; Li, Y. Efficient signaling platforms built from a small catalytic DNA and doubly labeled fluorogenic substrates. *Nucleic Acids Res.* **2007**, *35*, 401–405.

27. Cox, W. G.; Singer, V. L. Fluorescent DNA hybridization probe preparation using amine modification and reactive dye coupling. *BioTechniques* **2004**, *36*, 114–122.

28. Gordon, K. M.; Duckett, L.; Daul, B.; Petrie, H. T. A simple method for detecting up to five immunofluorescent parameters together with DNA staining for cell cycle or viability on a benchtop flow cytometer. *J. Immunol. Methods* **2003**, *275*, 113–121.

29. Marras, S. A. E.; Kramer, F. R.; Tyagi, S. Efficiencies of fluorescence resonance energy transfer and contact-mediated quenching in oligonucleotide probes. *Nucleic Acids Res.* **2002**, *30*, e122/1–e122/8.

30. Marguet, M.; Edembe, L.; Lecommandoux, S. Polymersomes in polymersomes: Multiple loading and permeability control. *Angew. Chem., Int. Ed.* **2012**, *51*, 1173–1176.

31. Fabian, A.; Horvath, G.; Vamosi, G.; Vereb, G.; Szoellosi, J. Triple FRET measurements in flow cytometry. *Cytometry, Part A* **2013**, *83A*, 375–385.

32. Skinner, J. P.; Chi, L.; Ozeata, P. F.; Ramsay, C. S.; O'Hara, R. L.; Calfin, B. B.; Tetin, S. Y. Introduction of the mass spread function for characterization of protein conjugates. *Anal. Chem.* **2012**, *84*, 1172–1177.

33. Filipe, V.; Poole, R.; Kutscher, M.; Forier, K.; Braeckmans, K.; Jiskoot, W. Fluorescence single particle tracking for the characterization of submicron protein aggregates in biological fluids and complex formulations. *Pharm. Res.* **2011**, *28*, 1112–1120.

34. Melo, A. M.; Prieto, M.; Coutinho, A. The effect of variable liposome brightness on quantifying lipid-protein interactions using fluorescence correlation spectroscopy. *Biochim. Biophys. Acta, Biomembr.* **2011**, *1808*, 2559–2568.

35. Kobashigawa, Y.; Harada, K.; Yoshida, N.; Ogura, K.; Inagaki, F. Phosphoinositide-incorporated lipid-protein nanodiscs: A tool for studying protein-lipid interactions. *Anal. Biochem.* **2011**, *410*, 77–83.

36. Wylie, R. G.; Ahsan, S.; Aizawa, Y.; Maxwell, K. L.; Morshead, C. M.; Shoichet, M. S. Spatially controlled simultaneous patterning of multiple growth factors in three-dimensional hydrogels. *Nat. Mater.* **2011**, *10*, 799–806.

37. Komatsu, T.; Johnsson, K.; Okuno, H.; Bito, H.; Inoue, T.; Nagano, T.; Urano, Y. Real-time measurements of protein dynamics using fluorescence activation-coupled protein labeling method. *J. Am. Chem. Soc.* **2011**, *133*, 6745–6751.

38. Nettels, D.; Hoffmann, A.; Schuler, B. Unfolded protein and peptide dynamics investigated with single-molecule FRET and correlation spectroscopy from picoseconds to seconds. *J. Phys. Chem. B* **2008**, *112*, 6137–6146.

39. Brand, F.; Klutz, A. M.; Jacobson, K. A.; Fredholm, B. B.; Schulte, G. Adenosine A2A receptor dynamics studied with the novel fluorescent agonist Alexa488-APEC. *Eur. J. Pharmacol.* **2008**, *590*, 36–42.

40. Mutch, S. A.; Fujimoto, B. S.; Kuyper, C. L.; Kuo, J. S.; Bajjalieh, S. M.; Chiu, D. T. Deconvolving single-molecule intensity distributions for quantitative microscopy measurements. *Biophys. J.* **2007**, *92*, 2926–2943.

41. Klenerman, D.; Li, H. Protein detection. PCT Int. Appl. WO 2005119265, 2005.

42. Alroy, I.; Moskowitz, H.; Reiss, Y.; Shoham, B. A. Methods and compositions related to tagging of membrane surface proteins. PCT Int. Appl. WO 2002099077, 2002.

43. Hodgson, L.; Qiu, W.; Dong, C.; Henderson, A. J. Use of green fluorescent protein-conjugated β-aactin as a novel molecular marker for *in vitro* tumor cell chemotaxis assay. *Biotechnol. Prog.* **2000**, *16*, 1106–1114.

44. Panchuk-Voloshina, N.; Haugland, R. P.; Bishop-Stewart, J.; Bhalgat, M. K.; Millard, P. J.; Mao, F.; Leung, W.; Haugland, R. P. Alexa dyes, a series of new fluorescent dyes that yield exceptionally bright, photostable conjugates. *J. Histochem. Cytochem.* **1999**, *47*, 1179–1188.

45. Shinohara, H.; Matsubayashi, Y. Functional immobilization of plant receptor-like kinase onto microbeads towards receptor array construction and receptor-based ligand fishing. *Plant J.* **2007**, *52*, 175–184.

46. Lawrie, G.; Grondahl, L.; Battersby, B.; Keen, I.; Lorentzen, M.; Surawski, P.; Trau, M. Tailoring surface properties to build colloidal diagnostic devices: Controlling interparticle associations. *Langmuir* **2006**, *22*, 497–505.

47. Lewis, J. D.; Destito, G.; Zijlstra, A.; Gonzalez, M. J.; Quigley, J. P.; Manchester, M.; Stuhlmann, H. Viral nanoparticles as tools for intravital vascular imaging. *Nat. Med.* **2006**, *12*, 354–360.

48. Zhang, S. L.; Tan, H.; Hanson, B. J.; Ooi, E. E. A simple method for Alexa Fluor dye labelling of dengue virus. *J. Virol. Methods* **2010**, *167*, 172–177.

49. Garza, C. M.; Cho, S. Metrology of bilayer photoresist processes. U.S. Pat. Appl. Publ. US 20090220895, 2009.

ALEXA FLUOR 514 CARBOXYLIC ACID SUCCINIMIDYL ESTER

CAS Registry Number   918946-23-3

Chemical Structure

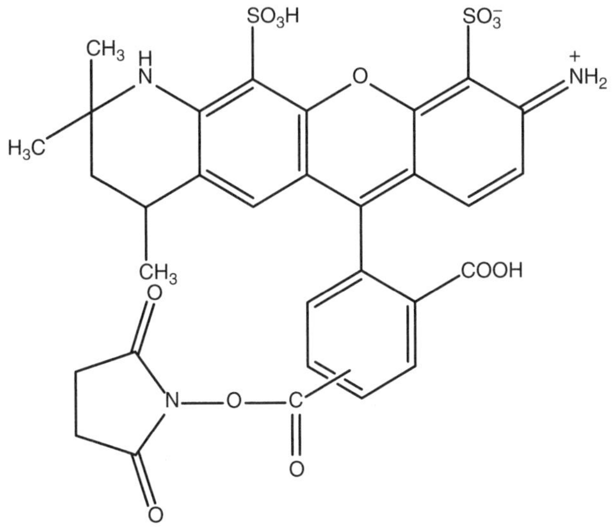

CA Index Name   [1]Benzopyrano[3,2-*g*]quinolin-11-ium,   9-amino-6-[2-carboxy-4(*or*   5)-[[(2,5-dioxo-1-pyrrolidinyl)-oxy]carbonyl]-phenyl]-1,2,3,4-tetrahydro-2,2,4-trimethyl-10,12-disulfo-, inner salt

Other Names   Alexa 514; Alexa 514 succinimidyl ester; Alexa Fluor 514; Alexa Fluor 514 NHS; Alexa Fluor 514 NHS ester; Alexa Fluor 514 carboxylic acid SE; Alexa Fluor 514 carboxylic acid succinimidyl ester; Alexa Fluor 514 SE; Alexa Fluor 514 succinimidyl ester

Merck Index Number   Not listed

Chemical/Dye Class   Xanthene

Molecular Formula $C_{31}H_{27}N_3O_{13}S_2$

Molecular Weight   713.69

Physical Form   Red solid

Solubility   Soluble in water, dimethyl sulfoxide

Absorption (λ_{max})   517 nm (Buffer pH 7)

Emission (λ_{max})   542 nm (Buffer pH 7)

Molar Extinction Coefficient   80,000 cm^{-1} M^{-1} (Buffer pH 7)

Synthesis   Synthetic method[1]

Imaging/Labeling Applications   Amines/amino groups;[2] antibodies;[3,4] cells;[5-7] chromosomes;[8,9] histone deacetylases;[10] microorganisms;[11] nucleic acids;[4,12] proteins[4]

Biological/Medical Applications   Analyzing/detecting analytes;[4] detecting chromosomal aberration;[8,9] identifying/characterizing microorganisms;[11] imaging cells;[5-7] labeling amines/amino groups[2]

Industrial Applications   Photoresists[13]

Safety/Toxicity   No data available

REFERENCES

1. Haugland, R. P. *The Handbook: A Guide to Fluorescent Probes and Labeling Technologies*; Molecular Probes Inc.: Eugene, **2005**; pp 26–46.

2. Wayment, J. R.; Harris, J. M. Controlling binding site densities on glass surfaces. *Anal. Chem.* **2006**, *78*, 7841–7849.

3. Arcangeli, A.; Becchetti, A.; Pillozzi, S.; Masselli, M.; De Lorenzo, E. Method and kit for the prevention and/or the monitoring of chemoresistance of leukemia forms. PCT Int. Appl. WO 2011058509, 2011.

4. Green, D. P. L.; Rawle, C. B. Analysis system and method. PCT Int. Appl. WO 2009082242, 2009.

5. Klimanskaya, I.; Gay, R. J. Methods for detection of rare subpopulations of cells and highly purified compositions of cells. PCT Int. Appl. WO 2012012803, 2012.

6. Tadross, M. R.; Park, S. A.; Veeramani, B.; Yue, D. T. Robust approaches to quantitative ratiometric FRET imaging of CFP/YFP fluorophores under confocal microscopy. *J. Microsc.* **2009**, *233*, 192–204.

7. Bestvater, F.; Spiess, E.; Stobrawa, G.; Hacker, M.; Feurer, T.; Porwol, T.; Berchner-Pfannschmidt, U.; Wotzlaw, C.; Acker, H. Two-photon fluorescence absorption and emission spectra of dyes relevant for cell imaging. *J. Microsc.* **2002**, *208*, 108–115.

8. Hauke, S. Method for detecting a chromosomal aberration. PCT Int. Appl. WO 2012150022, 2012.

9. Poulsen, T. S.; Poulsen, S. M.; Petersen, K. H. Methods for detecting chromosome aberrations. PCT Int. Appl. WO 2005111235, 2005.

10. Heidebrecht, R. W., Jr.,; Kral, A. M.; Miller, T. A. Fluorescent compounds that bind to histone deacetylase. U.S. Pat. Appl. Publ. US 20090156825, 2009.

11. Borisy, G. Methods and compositions for identifying cells by combinatorial fluoroscence imaging. PCT Int. Appl. WO 2009102844, 2009.

12. Choi, H. M.; Chang, J. Y.; Trinhle, A.; Padilla, J. E.; Pierce, N. A. Programmable *in situ* amplification for multiplexed imaging of mRNA expression. *Nat. Biotechnol.* **2010**, *28*, 1208–1212.

13. Garza, C. M.; Cho, S. Metrology of bilayer photoresist processes. U.S. Pat. Appl. Publ. US 20090220895, 2009.

ALEXA FLUOR 532 CARBOXYLIC ACID SUCCINIMIDYL ESTER

CAS Registry Number 477876-64-5

Chemical Structure

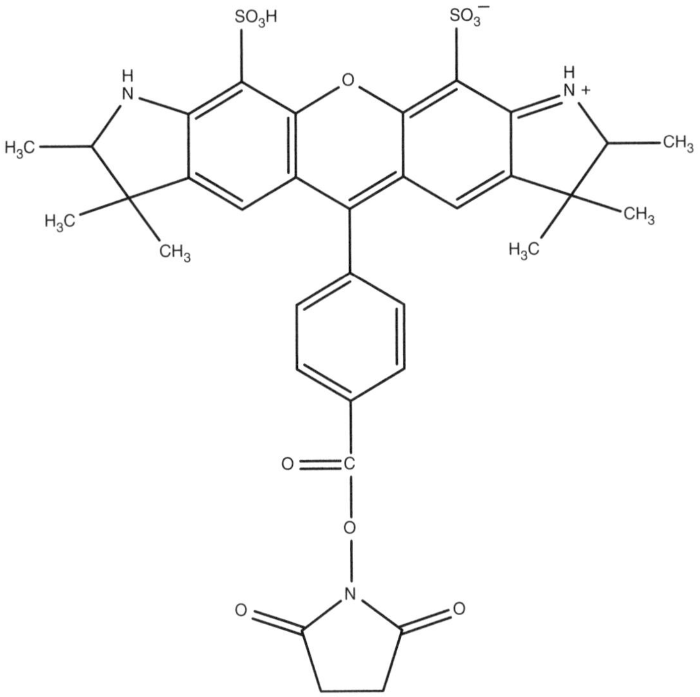

CA Index Name Pyrano[3,2-*f*:5,6-*f'*]diindol-11-ium, 5-[4-[[(2,5-dioxo-1-pyrrolidinyl)oxy]carbonyl]phenyl]-1,2,3,7,8,9-hexahydro-2,3,3,7,7,8-hexamethyl-10,12-disulfo-, inner salt

Other Names Alexa 532; Alexa 532 succinimidyl ester; Alexa Fluor 532; Alexa Fluor 532 NHS; Alexa Fluor 532 NHS ester; Alexa Fluor 532 carboxylic acid SE; Alexa Fluor 532 carboxylic acid succinimidyl ester; Alexa Fluor 532 SE; Alexa Fluor 532 succinimidyl ester

Merck Index Number Not listed

Chemical/Dye Class Xanthene

Molecular Formula $C_{34}H_{33}N_3O_{11}S_2$

Molecular Weight 723.77

Physical Form Red solid

Solubility Soluble in water, dimethyl sulfoxide, methanol

Absorption (λmax) 530 nm (Buffer pH 7); 532 nm (MeOH)\

Emission (λmax) 555 nm (Buffer pH 7); 558 nm (MeOH)

Molar Extinction Coefficient 81,000 cm^{-1} M^{-1} (Buffer pH 7); 89,000 cm^{-1} M^{-1} (MeOH)

Quantum Yield 0.61 (Buffer pH 7.2)

Synthesis Synthetic method[1]

Imaging/Labeling Applications Amines/amino groups;[22,23,27,29,30,32,37] antibodies;[2–4,33,34] benzodiazepines;[4,5] cells/tissues;[6–8,24] chromosomes;[9,10] flagellar filaments;[11] histone deacetylases;[12] microorganisms;[13] muscimol;[14,15] nucleic acids/nucleotides;[3,16–23] proteins/peptides[3,24–38]

Biological/Medical Applications Analyzing/detecting analytes;[3] counting cells;[24] detecting chromosomal aberration;[9,10] identifying/characterizing microorganisms;[13] labeling/detecting nucleic acids;[3,16–23] measuring albumin diffusion rates and concentration profiles;[28] probing the charge-transfer dynamics in DNA;[17] quantifying proteins[24]

Industrial Applications Photoresists[39]

Safety/Toxicity No data available

REFERENCES

1. Haugland, R. P. *Handbook of Fluorescent Probes and Research Products*; Molecular Probes Inc.: Eugene, **2002**; pp 20–35.

2. Arcangeli, A.; Becchetti, A.; Pillozzi, S.; Masselli, M.; De Lorenzo, E. Method and kit for the prevention and/or the monitoring of chemoresistance of leukemia forms. PCT Int. Appl. WO 2011058509, 2011.

3. Green, D. P. L.; Rawle, C. B. Analysis system and method. PCT Int. Appl. WO 2009082242, 2009.

4. Hegener, O.; Jordan, R.; Haeberlein, H. Dye-labeled benzodiazepines: Development of small ligands for receptor binding studies using fluorescence correlation spectroscopy. *J. Med. Chem.* **2004**, *47*, 3600–3605.

5. Hegener, O.; Jordan, R.; Haberlein, H. Benzodiazepine binding studies on living cells: application of small ligands for fluorescence correlation spectroscopy. *Biol. Chem.* **2002**, *383*, 1801–1807.

6. Klimanskaya, I.; Gay, R. J. Methods for detection of rare subpopulations of cells and highly purified compositions of cells. PCT Int. Appl. WO 2012012803, 2012.

7. Bestvater, F.; Spiess, E.; Stobrawa, G.; Hacker, M.; Feurer, T.; Porwol, T.; Berchner-Pfannschmidt, U.; Wotzlaw, C.; Acker, H. Two-photon fluorescence absorption and emission spectra of dyes relevant for cell imaging. *J. Microsc.* **2002**, *208*, 108–115.

8. Tsurui, H.; Nishimura, H.; Hattori, S.; Hirose, S.; Okumura, K.; Shirai, T. Seven-color fluorescence imaging of tissue samples based on fourier spectroscopy and singular value decomposition. *J. Histochem. Cytochem.* **2000**, *48*, 653–662.

9. Hauke, S. Method for detecting a chromosomal aberration. PCT Int. Appl. WO 2012150022, 2012.

10. Poulsen, T. S.; Poulsen, S. M.; Petersen, K. H. Methods for detecting chromosome aberrations. PCT Int. Appl. WO 2005111235, 2005.

11. Turner, L.; Ryu, W. S.; Berg, H. C. Real-time imaging of fluorescent flagellar filaments. *J. Bacteriol.* **2000**, *182*, 2793–2801.

12. Heidebrecht, R. W., Jr.,; Kral, A. M.; Miller, T. A. Fluorescent compounds that bind to histone deacetylase. U.S. Pat. Appl. Publ. US 20090156825, 2009.

13. Borisy, G. Methods and compositions for identifying cells by combinatorial fluoroscence imaging. PCT Int. Appl. WO 2009102844, 2009.

14. Meissner, O.; Haeberlein, H. Lateral mobility and specific binding to GABA (A) receptors on hippocampal neurons monitored by fluorescence correlation spectroscopy. *Biochemistry* **2003**, *42*, 1667–1672.

15. Halbsguth, C.; Meissner, O.; Haeberlein, H. Positive cooperation of protoberberine type 2 alkaloids from *Corydalis cava* on the GABA (A) binding site. *Planta Med.* **2003**, *69*, 305–309.

16. Stupi, B. P.; Li, H.; Wu, W.; Hersh, M. N.; Hertzog, D.; Morris, S. E.; Metzker, M. L. 5-Methoxy,3′-hydroxy unblocked, fast photocleavable terminating nucleotides and methods for nucleic acid sequencing. PCT Int. Appl. WO 2013040257, 2013.

17. Kawai, K.; Matsutani, E.; Maruyama, A.; Majima, T. Probing the charge-transfer dynamics in DNA at the single-molecule level. *J. Am. Chem. Soc.* **2011**, *133*, 15568–15577.

18. Ichikawa, K.; Nakamura, K.; Kurata, S. Real-time PCR method using the novel type of oligonucleotide probes. Jpn. Kokai Tokkyo Koho JP 2008182974, 2008.

19. Agnew, B.; Gee, K.; Kumal, K.; Ford, M. Labeling and detection of nucleic acids. PCT Int. Appl. WO 2008101024, 2008.

20. Agnew, B.; Ford, M. J.; Gee, K. R.; Kumar, K. Labeling and detection of nucleic acids. U.S. Pat. Appl. Publ. US 20080050731, 2008.

21. Li, J.; Lee, J. Y.; Yeung, E. S. Quantitative screening of single copies of human papilloma viral DNA without amplification. *Anal. Chem.* **2006**, *78*, 6490–6496.

22. Cox, W. G.; Singer, V. L. Fluorescent DNA hybridization probe preparation using amine modification and reactive dye coupling. *BioTechniques* **2004**, *36*, 114–122.

23. Marras, S. A. E.; Kramer, F. R.; Tyagi, S. Efficiencies of fluorescence resonance energy transfer and contact-mediated quenching in oligonucleotide probes. *Nucleic Acids Res.* **2002**, *30*, e122/1–e122/8.

24. Pihlasalo, S.; Engbert, A.; Martikkala, E.; Ylander, P.; Hanninen, P.; Harma, H. Nonspecific particle-based method with two-photon excitation detection for sensitive protein quantification and cell counting. *Anal. Chem.* **2013**, *85*, 2689–2696.

25. Blazer, L. L.; Roman, D. L.; Chung, A.; Larsen, M. J.; Greedy, B. M.; Husbands, S. M.; Neubig, R. R. Reversible, allosteric small-molecule inhibitors

of regulator of G protein signaling proteins. *Mol. Pharmacol.* **2010**, *78*, 524–533.

26. Ma, C.; Yeung, E. S. Single molecule imaging of protein molecules in nanopores. *Anal. Chem.* **2010**, *82*, 478–482.

27. Weller Roska, R. L.; Lama, T. G. S.; Hennes, J. P.; Carlson, R. E. Small molecule-based binding environments: Combinatorial construction of microarrays for multiplexed affinity screening. *J. Am. Chem. Soc.* **2009**, *131*, 16660–16662.

28. Stevens, A. P.; Hlady, V.; Dull, R. O. Fluorescence correlation spectroscopy can probe albumin dynamics inside lung endothelial glycocalyx. *Am. J. Physiol.* **2007**, *293*, L328–L335.

29. Groll, J.; Haubensak, W.; Ameringer, T.; Moeller, M. Biofunctional patterning of ultrathin star-PEG coatings. *Polym. Prepr.* **2005**, *46*, 1215–1216.

30. Groll, J.; Haubensak, W.; Ameringer, T.; Moeller, M. Ultrathin coatings from isocyanate terminated star PEG prepolymers: Patterning of proteins on the layers. *Langmuir* **2005**, *21*, 3076–3083.

31. Naqvi, T.; Rouhani, R.; Fung, P.; Eglen, R.; Singh, R. IP3 protein binding assay. PCT Int. Appl. WO 2004038369, 2004.

32. Alroy, I.; Moskowitz, H.; Reiss, Y.; Shoham, B. A. Methods and compositions related to tagging of membrane surface proteins. PCT Int. Appl. WO 2002099077, 2002.

33. Hahn, K. M.; Toutchkine, A.; Muthyala, R.; Kraynov, V.; Bark, S. J.; Burton, D. R.; Chamberlain, C. Labeled peptides, proteins and antibodies and processes and intermediates useful for their preparation. U.S. Pat. Appl. Publ. US 20020055133, 2002.

34. Hahn, K. M.; Toutchkine, A.; Muthyala, R.; Kraynov, V.; Bark, S. J.; Burton, D. R.; Chamberlain, C. Labeled peptides, proteins and antibodies and processes and intermediates useful for their preparation. PCT Int. Appl. WO 2002008245, 2002.

35. Oksvold, M. P.; Skarpen, E.; Widerberg, J.; Huitfeldt, H. S. Fluorescent histochemical techniques for analysis of intracellular signaling. *J. Histochem. Cytochem.* **2002**, *50*, 289–303.

36. Von Eggeling, F.; Gawriljuk, A.; Fiedler, W.; Ernst, G.; Claussen, U.; Klose, J.; Romer, I. Fluorescent dual colour 3D-protein gel electrophoresis for rapid detection of differences in protein pattern with standard image analysis software. *Int. J. Mol. Med.* **2001**, *8*, 373–377.

37. Bark, S. J.; Schmid, S.; Hahn, K. M. A highly efficient method for site-specific modification of unprotected peptides after chemical synthesis. *J. Am. Chem. Soc.* **2000**, *122*, 3567–3573.

38. Panchuk-Voloshina, N.; Haugland, R. P.; Bishop-Stewart, J.; Bhalgat, M. K.; Millard, P. J.; Mao, F.; Leung, W.; Haugland, R. P. Alexa dyes, a series of new fluorescent dyes that yield exceptionally bright, photostable conjugates. *J. Histochem. Cytochem.* **1999**, *47*, 1179–1188.

39. Garza, C. M.; Cho, S. Metrology of bilayer photoresist processes. U.S. Pat. Appl. Publ. US 20090220895, 2009.

ALEXA FLUOR 555 CARBOXYLIC ACID SUCCINIMIDYL ESTER

CAS Registry Number   886046-94-2

Chemical Structure   Not reported

CA Index Name   Alexa Fluor 555 carboxylic acid succinimidyl ester

Other Names   Alexa 555; Alexa 555 succinimidyl ester; Alexa Fluor 555; Alexa Fluor 555 NHS; Alexa Fluor 555 NHS ester; Alexa Fluor 555 carboxylic acid SE; Alexa Fluor 555 carboxylic acid succinimidyl ester; Alexa Fluor 555 SE; Alexa Fluor 555 succinimidyl ester

Merck Index Number   Not listed

Chemical/Dye Class   Not reported

Molecular Formula   Not reported

Molecular Weight   ~1250

Physical Form   Purple solid

Solubility   Soluble in water, dimethyl sulfoxide, methanol

Absorption (λmax)   555 nm (MeOH)

Emission (λmax)   572 nm (MeOH)

Molar Extinction Coefficient $155,000 \, cm^{-1} \, M^{-1}$ (MeOH)

Quantum Yield   0.10 (Buffer pH 7.2)

Synthesis   Synthetic method[1]

Imaging/Labeling Applications   Antibodies;[2–6] bicyclononynes;[7] cells;[8,9] chromosomes;[10,11] dendrimers;[12] endosomes;[13] histone deacetylases;[14] nucleic acids;[4,15] proteins;[4,5,7,16–21] cowpea mosaic virus (CPMV)[22]

Biological/Medical Applications   Analyzing/detecting analytes;[4] detecting chromosomal aberration;[10,11] encapsulating proteins inside surface-tethered liposomes;[18] investigating late endosome-lysosome retrograde fusion events;[13] visualizing living melanoma cells,[7] vasculature/blood flow in living mouse and chick embryos[22]

Industrial Applications   Not reported

Safety/Toxicity   No data available

REFERENCES

1. Haugland, R. P. *Handbook of Fluorescent Probes and Research Products*; Molecular Probes Inc.: Eugene, **2002**; pp 20–35.

2. Arcangeli, A.; Becchetti, A.; Pillozzi, S.; Masselli, M.; De Lorenzo, E. Method and kit for the prevention and/or the monitoring of chemoresistance of leukemia forms. PCT Int. Appl. WO 2011058509, 2011.

3. Devaraj, N. K.; Upadhyay, R.; Haun, J. B.; Hilderbrand, S. A.; Weissleder, R. Fast and sensitive pretargeted labeling of cancer cells through a tetrazine/trans-cyclooctene cycloaddition. *Angew. Chem., Int. Ed.* **2009**, *48*, 7013–7016.

4. Green, D. P. L.; Rawle, C. B. Analysis system and method. PCT Int. Appl. WO 2009082242, 2009.

5. Uhlen, M.; Svahn, H. A.; Lundberg, E. Solid phase labeling method. U.S. Pat. Appl. Publ. US 20080233660, 2008.

6. Lundberg, E.; Sundberg, M.; Graeslund, T.; Uhlen, M.; Svahn, H. A. A novel method for reproducible fluorescent labeling of small amounts of antibodies on solid phase. *J. Immunol. Methods* **2007**, *322*, 40–49.

7. Dommerholt, J.; Schmidt, S.; Temming, R.; Hendriks, L. J. A.; Rutjes, F. P. J. T.; van Hest, J. C. M.; Lefeber, D. J.; Friedl, P.; van Delft, F. L. Readily accessible bicyclononynes for bioorthogonal labeling and three-dimensional imaging of living cells. *Angew. Chem., Int. Ed.* **2010**, *49*, 9422–9425.

8. Klimanskaya, I.; Gay, R. J. Methods for detection of rare subpopulations of cells and highly purified compositions of cells. PCT Int. Appl. WO 2012012803, 2012.

9. Bestvater, F.; Spiess, E.; Stobrawa, G.; Hacker, M.; Feurer, T.; Porwol, T.; Berchner-Pfannschmidt, U.; Wotzlaw, C.; Acker, H. Two-photon fluorescence absorption and emission spectra of dyes relevant for cell imaging. *J. Microsc.* **2002**, *208*, 108–115.

10. Hauke, S. Method for detecting a chromosomal aberration. PCT Int. Appl. WO 2012150022, 2012.

11. Poulsen, T. S.; Poulsen, S. M.; Petersen, K. H. Methods for detecting chromosome aberrations. PCT Int. Appl. WO 2005111235, 2005.

12. Opitz, A. W.; Czymmek, K. J.; Wickstrom, E.; Wagner, N. J. Uptake, efflux, and mass transfer coefficient of fluorescent PAMAM dendrimers into pancreatic cancer cells. *Biochim. Biophys. Acta, Biomembr.* **2013**, *1828*, 294–301.

13. Kaufmann, A. M.; Goldman, S. D. B.; Krise, J. P. A fluorescence resonance energy transfer-based approach for investigating late endosome-lysosome

retrograde fusion events. *Anal. Biochem.* **2009**, *386*, 91–97.

14. Heidebrecht, R. W., Jr.,; Kral, A. M.; Miller, T. A. Fluorescent compounds that bind to histone deacetylase. U.S. Pat. Appl. Publ. US 20090156825, 2009.

15. Cox, W. G.; Beaudet, M. P.; Angew, J. Y.; Ruth, J. L. Possible source of dye-related signal correlation bias in two-color DNA microarray assays. *Anal. Biochem.* **2004**, *331*, 243–254.

16. Johnston, S.; Diehnelt, C.; Belcher, P.; Arntzen, C.; Sutherland, R. Peptide ligands. PCT Int. Appl. WO 2012134416, 2012.

17. Loparo, J. J.; Kulczyk, A. W.; Richardson, C. C.; van Oijen, A. M. Simultaneous single-molecule measurements of phage T7 replisome composition and function reveal the mechanism of polymerase exchange. *Proc. Natl. Acad. Sci. U.S.A.* **2011**, *108*, 3584–3589.

18. Liu, B.; Mazouchi, A.; Gradinaru, C. C. Trapping single molecules in liposomes: Surface interactions and freeze-thaw effects. *J. Phys. Chem. B* **2010**, *114*, 15191–15198.

19. Blois, T. M.; Hong, H.; Kim, T. H.; Bowie, J. U. Protein unfolding with a steric trap. *J. Am. Chem. Soc.* **2009**, *131*, 13914–13915.

20. Clayton, A. H.; Walker, F.; Orchard, S. G.; Henderson, C.; Fuch, D.; Rothacker, J.; Nice, E. C.; Burgess, A. W. Ligand-induced dimer-tetramer transition during the activation of the cell surface epidermal growth factor receptor-A multidimentional microscopy analysis. *J. Biol. Chem.* **2005**, *280*, 30392–30399.

21. Berlier, J. E.; Rothe, A.; Buller, G.; Bradford, J.; Gray, D. R.; Filanoski, B. J.; Telford, W. G.; Yue, S.; Liu, J.; Cheung, C.; Chang, W.; Hirsch, J. D.; Beechem, J. M.; Haugland, R. P.; Haugland, R. P. Quantitative comparison of long-wavelength Alexa Fluor dyes to Cy dyes: Fluorescence of the dyes and their bioconjugates. *J. Histochem. Cytochem.* **2003**, *51*, 1699–1712.

22. Lewis, J. D.; Destito, G.; Zijlstra, A.; Gonzalez, M. J.; Quigley, J. P.; Manchester, M.; Stuhlmann, H. Viral nanoparticles as tools for intravital vascular imaging. *Nat. Med.* **2006**, *12*, 354–360.

ALEXA FLUOR 594 CARBOXYLIC ACID SUCCINIMIDYL ESTER

CAS Registry Number   295348-87-7

Chemical Structure

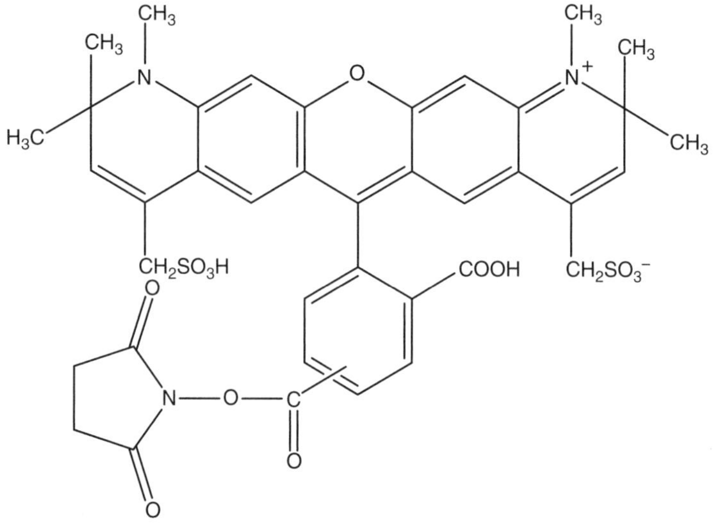

CA Index Name   Pyrano[3,2-*g*:5,6-*g*']diquinolin-13-ium, 6-[2-carboxy-4(*or* 5)-[[(2,5-dioxo-1-pyrrolidinyl)-oxy]carbonyl]phenyl]-1,2,10,11-tetrahydro-1,2,2,10,10,11-hexamethyl-4,8-bis(sulfomethyl)-, inner salt

Other Names   Alexa 594; Alexa 594 succinimidyl ester; Alexa Fluor 594; Alexa Fluor 594 NHS; Alexa Fluor 594 NHS ester; Alexa Fluor 594 carboxylic acid SE; Alexa Fluor 594 carboxylic acid succinimidyl ester; Alexa Fluor 594 SE; Alexa Fluor 594 succinimidyl ester

Merck Index Number   Not listed

Chemical/Dye Class   Xanthene

Molecular Formula $C_{39}H_{37}N_3O_{13}S_2$

Molecular Weight   819.85

Physical Form   Blue solid

Solubility   Soluble in water, dimethyl sulfoxide

Absorption (λ_{max})   590 nm (Buffer pH 7)

Emission (λ_{max})   617 nm (Buffer pH 7)

Molar Extinction Coefficient   92,000 cm^{-1} M^{-1} (Buffer pH 7)

Quantum Yield   0.66 (Buffer pH 7.2)

Synthesis   Synthetic methods[1-3]

Imaging/Labeling Applications   Amines/amino groups;[21,25] antibodies;[4,5] biotin;[6] cells/tissues;[7-10] chromosomes;[11,12] cyclodextrin/dextran;[13-15] dendrimers;[16] glucose;[17] histone deacetylases;[18] hydrogen peroxide;[19] nucleic acids/nucleotides;[1,5,20,21] proteins/peptides;[5,22-26] quantum dots;[27,28] dengue virus[29]

Biological/Medical Applications   Analyzing/detecting analytes;[5] characterizing submicron protein aggregates in human serum/plasma;[23] detecting chromosomal aberration;[11,12] detecting/quantifying biotin derivatives;[6] measuring blood glucose;[17] hydrogen peroxide sensor[19]

Industrial Applications   Photoresists[30]

Safety/Toxicity   No data available

REFERENCES

1. Turcatti, G.; Romieu, A.; Fedurco, M.; Tairi, A. A new class of cleavable fluorescent nucleotides: synthesis and optimization as reversible terminators for DNA sequencing by synthesis. *Nucleic Acids Res.* **2008**, *36*, e25/1–e25/13.

2. Haugland, R. P. *Handbook of Fluorescent Probes and Research Products*; Molecular Probes Inc.: Eugene, **2002**; pp 20–35.

3. Mao, F.; Leung, W.; Haugland, R. P. Sulfonated xanthene derivatives. PCT Int. Appl. WO 9915517, 1999.

4. Arcangeli, A.; Becchetti, A.; Pillozzi, S.; Masselli, M.; De Lorenzo, E. Method and kit for the prevention and/or the monitoring of chemoresistance of leukemia forms. PCT Int. Appl. WO 2011058509, 2011.

5. Green, D. P. L.; Rawle, C. B. Analysis system and method. PCT Int. Appl. WO 2009082242, 2009.

6. Wilbur, D. S.; Pathare, P. M.; Hamlin, D. K.; Frownfelter, M. B.; Kegley, B. B.; Leung, W.; Gee, K. R. Evaluation of biotin-dye conjugates for use in an HPLC assay to assess relative binding of biotin derivatives with avidin and streptavidin. *Bioconjugate Chem.* **2000**, *11*, 584–598.

7. Klimanskaya, I.; Gay, R. J. Methods for detection of rare subpopulations of cells and highly purified compositions of cells. PCT Int. Appl. WO 2012012803, 2012.

8. Hollenberg, M. D.; Renaux, B.; Hyun, E.; Houle, S.; Vergnolle, N.; Saifeddine, M.; Ramachandran, R. Derivatized 2-furoyl-LIGRLO-amide, a versatile and selective probe for proteinase-activated receptor 2: binding and visualization. *J. Pharmacol. Exp. Ther.* **2008**, *326*, 453–462.

9. Bestvater, F.; Spiess, E.; Stobrawa, G.; Hacker, M.; Feurer, T.; Porwol, T.; Berchner-Pfannschmidt, U.; Wotzlaw, C.; Acker, H. Two-photon fluorescence absorption and emission spectra of dyes relevant for cell imaging. *J. Microsc.* **2002**, *208*, 108–115.

10. Tsurui, H.; Nishimura, H.; Hattori, S. Hirose, S. Okumura, K.; Shirai, T. Seven-color fluorescence imaging of tissue samples based on fourier spectroscopy and singular value decomposition. *J. Histochem. Cytochem.* **2000**, *48*, 653–662.

11. Hauke, S. Method for detecting a chromosomal aberration. PCT Int. Appl. WO 2012150022, 2012.

12. Poulsen, T. S.; Poulsen, S. M.; Petersen, K. H. Methods for detecting chromosome aberrations. PCT Int. Appl. WO 2005111235, 2005.

13. Shah, S.; Solanki, A.; Sasmal, P. K.; Lee, K. Single vehicular delivery of siRNA and small molecules to control stem cell differentiation. *J. Am. Chem. Soc.* **2013**, *135*, 15682–15685.

14. Kim, C.; Shah, B. P.; Subramaniam, P.; Lee, K. Synergistic induction of apoptosis in brain cancer cells by targeted codelivery of siRNA and anticancer drugs. *Mol. Pharm.* **2011**, *8*, 1955–1961.

15. del Mercato, L. L.; Abbasi, A. Z.; Parak, W. J. Synthesis and characterization of ratiometric ion-sensitive polyelectrolyte capsules. *Small* **2011**, *7*, 351–363.

16. Boswell, C. A.; Eck, P. K.; Regino, C. A. S.; Bernardo, M.; Wong, K. J.; Milenic, D. E.; Choyke, P. L.; Brechbiel, M. W. Synthesis, characterization, and biological evaluation of integrin α v β 3-targeted PAMAM dendrimers. *Mol. Pharm.* **2008**, *5*, 527–539.

17. Ibey, B. L.; Yadavalli, V. K.; Thomas, H. R.; Rounds, R. M.; Pishko, M. V.; Cote, G. L. Implantable fluorescence-based glucose sensor development. *Proc. SPIE-Int. Soc. Opt. Eng.* **2005**, *5702*, 1–6.

18. Heidebrecht, R. W., Jr.,; Kral, A. M.; Miller, T. A. Fluorescent compounds that bind to histone deacetylase. U.S. Pat. Appl. Publ. US 20090156825, 2009.

19. Weinstain, R.; Savariar, E. N.; Felsen, C. N.; Tsien, R. Y. *In vivo* targeting of hydrogen peroxide by activatable cell-penetrating peptides. *J. Am. Chem. Soc.* **2014**, *136*, 874–877.

20. Murata, A.; Sato, S.; Kawazoe, Y.; Uesugi, M. Small-molecule fluorescent probes for specific RNA targets. *Chem. Commun.* **2011**, *47*, 4712–4714.

21. Cox, W. G.; Singer, V. L. Fluorescent DNA hybridization probe preparation using amine modification and reactive dye coupling. *BioTechniques* **2004**, *36*, 114–122.

22. Ferez, L.; Thami, T.; Akpalo, E.; Flaud, V.; Tauk, L.; Janot, J.; Dejardin, P. Interface of covalently bonded phospholipids with a phosphorylcholine head: Characterization, protein nonadsorption, and further functionalization. *Langmuir* **2011**, *27*, 11536–11544.

23. Filipe, V.; Poole, R.; Kutscher, M.; Forier, K.; Braeckmans, K.; Jiskoot, W. Fluorescence single particle tracking for the characterization of submicron protein aggregates in biological fluids and complex formulations. *Pharm. Res.* **2011**, *28*, 1112–1120.

24. Janot, J.; Boissiere, M.; Thami, T.; Tronel-Peyroz, E.; Helassa, N.; Noinville, S.; Quiquampoix, H.; Staunton, S.; Dejardin, P. Adsorption of Alexa-labeled Bt toxin on mica, glass, and hydrophobized glass: Study by normal scanning confocal fluorescence. *Biomacromolecules* **2010**, *11*, 1661–1666.

25. Alroy, I.; Moskowitz, H.; Reiss, Y.; Shoham, B. A. Methods and compositions related to tagging of membrane surface proteins. PCT Int. Appl. WO 2002099077, 2002.

26. Panchuk-Voloshina, N.; Haugland, R. P.; Bishop-Stewart, J.; Bhalgat, M. K.; Millard, P. J.; Mao, F.; Leung, W.; Haugland, R. P. Alexa dyes, a series of new fluorescent dyes that yield exceptionally bright, photostable conjugates. *J. Histochem. Cytochem.* **1999**, *47*, 1179–1188.

27. Han, H.; Hilderbrand, S. A.; Devaraj, N. K.; Weissleder, R.; Bawendi, M. G. Compositions and methods for bioconjugation to quantum dots. PCT Int. Appl. WO 2011112970, 2011.

28. Han, H.; Devaraj, N. K.; Lee, J.; Hilderbrand, S. A.; Weissleder, R.; Bawendi, M. G. Development of a bioorthogonal and highly efficient conjugation method for quantum dots using tetrazine-norbornene cycloaddition. *J. Am. Chem. Soc.* **2010**, *132*, 7838–7839.

29. Zhang, S. L.; Tan, H.; Hanson, B. J.; Ooi, E. E. A simple method for Alexa Fluor dye labelling of dengue virus. *J. Virol. Methods* **2010**, *167*, 172–177.

30. Garza, C. M.; Cho, S. Metrology of bilayer photoresist processes. U.S. Pat. Appl. Publ. US 20090220895, 2009.

ALEXA FLUOR 633 CARBOXYLIC ACID SUCCINIMIDYL ESTER

CAS Registry Number   1132773-86-4

Chemical Structure   Not reported

CA Index Name   Alexa Fluor 633 carboxylic acid succinimidyl ester

Other Names   Alexa 633; Alexa 633 succinimidyl ester; Alexa Fluor 633; Alexa Fluor 633 NHS; Alexa Fluor 633 NHS ester; Alexa Fluor 633 carboxylic acid SE; Alexa Fluor 633 carboxylic acid succinimidyl ester; Alexa Fluor 633 SE; Alexa Fluor 633 succinimidyl ester

Merck Index Number   Not listed

Chemical/Dye Class   Not reported

Molecular Formula   Not reported

Molecular Weight   ~1200

Physical Form   Blue solid

Solubility   Soluble in water, dimethyl sulfoxide, methanol

Absorption (λ_{max})   621 nm (MeOH)

Emission (λ_{max})   639 nm (MeOH)

Molar Extinction Coefficient   159,000 cm^{-1} M^{-1} (MeOH)

Synthesis   Synthetic method[1]

Imaging/Labeling Applications   Amines/amino groups;[11,18] antibodies;[2,3,18] cells;[4,5] chromosomes;[6,7] histone deacetylases;[8] liposomes;[9,10] nanoparticles;[11] nucleic acids/nucleotides;[3,12–17] polymersomes;[18] proteins[3,9,18–20]

Biological/Medical Applications   Analyzing/detecting analytes;[3] detecting chromosomal aberration;[6,7] investigating degradation of liposomal subcompartments in PEGylated capsosomes;[9] labeling/detecting nucleic acids;[12–17] patterning wide range of proteins;[19] studying molecule-membrane interactions or molecule-molecule interactions[10]

Industrial Applications   Not reported

Safety/Toxicity   No data available

REFERENCES

1. Haugland, R. P. *Handbook of Fluorescent Probes and Research Products*; Molecular Probes Inc.: Eugene, **2002**; pp 20–35.

2. Arcangeli, A.; Becchetti, A.; Pillozzi, S.; Masselli, M.; De Lorenzo, E. Method and kit for the prevention and/or the monitoring of chemoresistance of leukemia forms. PCT Int. Appl. WO 2011058509, 2011.

3. Green, D. P. L.; Rawle, C. B. Analysis system and method. PCT Int. Appl. WO 2009082242, 2009.

4. Klimanskaya, I.; Gay, R. J. Methods for detection of rare subpopulations of cells and highly purified compositions of cells. PCT Int. Appl. WO 2012012803, 2012.

5. Bestvater, F.; Spiess, E.; Stobrawa, G.; Hacker, M.; Feurer, T.; Porwol, T.; Berchner-Pfannschmidt, U.; Wotzlaw, C.; Acker, H. Two-photon fluorescence absorption and emission spectra of dyes relevant for cell imaging. *J. Microsc.* **2002**, *208*, 108–115.

6. Hauke, S. Method for detecting a chromosomal aberration. PCT Int. Appl. WO 2012150022, 2012.

7. Poulsen, T. S.; Poulsen, S. M.; Petersen, K. H. Methods for detecting chromosome aberrations. PCT Int. Appl. WO 2005111235, 2005.

8. Heidebrecht, R. W., Jr.,; Kral, A. M.; Miller, T. A. Fluorescent compounds that bind to histone deacetylase. U.S. Pat. Appl. Publ. US 20090156825, 2009.

9. Chandrawati, R.; Chong, S.; Zelikin, A. N.; Hosta-Rigau, L.; Staedler, B.; Caruso, F. Degradation of liposomal subcompartments in PEGylated capsosomes. *Soft Matter* **2011**, *7*, 9638–9646.

10. Ehrlich, N.; Christensen, A. L.; Stamou, D. Fluorescence anisotropy based single liposome assay to measure molecule-membrane interactions. *Anal. Chem.* **2011**, *83*, 8169-8176.

11. Sun, H.; Almdal, K.; Andresen, T. L. Expanding the dynamic measurement range for polymeric nanoparticle pH sensors. *Chem. Commun.* **2011**, *47*, 5268–5270.

12. Nikiforov, T.; Beechem, J. Methods and apparatus for single molecule sequencing using energy transfer detection. PCT Int. Appl. WO 2010111674, 2010.

13. Agnew, B.; Gee, K.; Kumal, K.; Ford, M. Labeling and detection of nucleic acids. PCT Int. Appl. WO 2008101024, 2008.

14. Agnew, B.; Ford, M. J.; Gee, K. R.; Kumar, K. Labeling and detection of nucleic acids. U.S. Pat. Appl. Publ. US 20080050731, 2008.

15. Korlach, J.; Bibillo, A.; Wegener, J.; Peluso, P.; Pham, T. T.; Park, I.; Clark, S.; Otto, G. A.; Turner, S. W. Long, processive enzymatic DNA synthesis using 100% dye-labeled terminal phosphate-linked nucleotides. *Nucleosides, Nucleotides Nucleic Acids* **2008**, *27*, 1072–1082.

16. Giller, G.; Tasara, T.; Angerer, B.; Muhlegger, K.; Amacker, M.; Winter, H. Incorporation of reported molecule-labeled nucleotides by DNA polymerases. I. Chemical synthesis of various reporter group-labeled 2'-deoxyribonucleoside-5'-triphosphates. *Nucleic Acids Res.* **2003**, *31*, 2630–2635.

17. Gordon, K. M.; Duckett, L.; Daul, B.; Petrie, H. T. A simple method for detecting up to five immunofluorescent parameters together with DNA staining for cell cycle or viability on a benchtop flow cytometer. *J. Immunol. Methods* **2003**, *275*, 113–121.

18. Egli, S.; Nussbaumer, M. G.; Balasubramanian, V.; Chami, M.; Bruns, N.; Palivan, C.; Meier, W. Biocompatible functionalization of polymersome surfaces: A new approach to surface immobilization and cell targeting using polymersomes. *J. Am. Chem. Soc.* **2011**, *133*, 4476–4483.

19. Wylie, R. G.; Ahsan, S.; Aizawa, Y.; Maxwell, K. L.; Morshead, C. M.; Shoichet, M. S. Spatially controlled simultaneous patterning of multiple growth factors in three-dimensional hydrogels. *Nat. Mater.* **2011**, *10*, 799–806.

20. Berlier, J. E.; Rothe, A.; Buller, G.; Bradford, J.; Gray, D. R.; Filanoski, B. J.; Telford, W. G.; Yue, S.; Liu, J.; Cheung, C.; Chang, W.; Hirsch, J. D.; Beechem, J. M.; Haugland, R. P.; Haugland, R. P. Quantitative comparison of long-wavelength Alexa Fluor dyes to Cy dyes: Fluorescence of the dyes and their bioconjugates. *J. Histochem. Cytochem.* **2003**, *51*, 1699–1712.

ALEXA FLUOR 660 CARBOXYLIC ACID SUCCINIMIDYL ESTER

CAS Registry Number   422309-89-5

Chemical Structure   Not reported

CA Index Name   Alexa Fluor 660 carboxylic acid succinimidyl ester

Other Names   Alexa 660; Alexa 660 succinimidyl ester; Alexa Fluor 660; Alexa Fluor 660 NHS; Alexa Fluor 660 NHS ester; Alexa Fluor 660 carboxylic acid SE; Alexa Fluor 660 carboxylic acid succinimidyl ester; Alexa Fluor 660 SE; Alexa Fluor 660 succinimidyl ester

Merck Index Number   Not listed

Chemical/Dye Class   Not reported

Molecular Formula   Not reported

Molecular Weight   ~1100

Physical Form   Blue solid

Solubility   Soluble in water, dimethyl sulfoxide, methanol

Absorption (λ_{max})   668 nm (MeOH); 660 nm (Buffer pH 7)

Emission (λ_{max})   698 nm (MeOH); 689 nm (Buffer pH 7)

Molar Extinction Coefficient   132,000 cm^{-1} M^{-1} (MeOH)

Quantum Yield   0.37 (Buffer pH 7.2)

Synthesis   Synthetic method[1]

Imaging/Labeling Applications   Amines/amino groups;[15,25,27] antibodies;[2–6,30] bacteria;[7] blood vessels;[8] cells;[9–11] chromosomes;[12,13] Cu(II);[14] dendrimers;[15,16] epidermal growth factor receptor (EGFR);[30] folate;[17,18] histone deacetylases;[19] kinases;[20] nucleic acids/nucleotides;[21–27] proteins;[5,28–32] tumors;[30,33,34] Zn(II) cytochrome c[35]

Biological/Medical Applications   Analyzing/detecting nucleic acids;[21–27] assessing epithelial neoplasia;[30] detecting chromosomal aberration;[12,13] detecting/confirming lymph nodes;[36,37] fluorescent labels in kinases (FLiK) assay;[20] glucose sensor for diabetic monitoring;[16] identifying stent thrombosis;[8] measuring/quantifying folate;[17,18] understanding lymph node metastasis;[34] as temperature sensor[38,39]

Industrial Applications   Not reported

Safety/Toxicity   No data available

REFERENCES

1. Haugland, R. P. *Handbook of Fluorescent Probes and Research Products*; Molecular Probes Inc.: Eugene, **2002**; pp 20–35.

2. Baklaushev, V. P.; Yusubalieva, G. M.; Tsitrin, E. B.; Gurina, O. I.; Grinenko, N. P.; Victorov, I. V.; Chekhonin, V. P. Visualization of Connexin 43-positive cells of glioma and the periglioma zone by means of intravenously injected monoclonal antibodies. *Drug Deliv.* **2011**, *18*, 331–337.

3. Arcangeli, A.; Becchetti, A.; Pillozzi, S.; Masselli, M.; De Lorenzo, E. Method and kit for the prevention and/or the monitoring of chemoresistance of leukemia forms. PCT Int. Appl. WO 2011058509, 2011.

4. Mullins, J. M. Fluorochromes: properties and characteristics. *Methods Mol. Biol.* **2010**, *588*, 123–134.

5. Nichkova, M.; Dosev, D.; Gee, S. J.; Hammock, B. D.; Kennedy, I. M. Multiplexed immunoassays for proteins using magnetic luminescent nanoparticles for internal calibration. *Anal. Biochem.* **2007**, *369*, 34–40.

6. Boushaba, R.; Kaminski, C. F.; Slater, N. K. H. Dual fluorescence confocal imaging of the accessibility and binding of F(ab')2 to an EBA resin with various immobilized antigen densities. *Process Biochem.* **2007**, *42*, 812–819.

7. Kim, S. U.; Kim, S. J.; Lee, S. H.; Lee, D. H.; Ryu, H. S. Kit and method for detecting food-borne bacteria. Repub. Korean Kongkae Taeho Kongbo KR 2013065337, 2013.

8. Jaffer, F.; Rajopadhye, M. Methods and compositions for identifying subjects at risk of developing stent thrombosis. U.S. Pat. Appl. Publ. US 20100268070, 2010.

9. Klimanskaya, I.; Gay, R. J. Methods for detection of rare subpopulations of cells and highly purified compositions of cells. PCT Int. Appl. WO 2012012803, 2012.

10. Peterson, J. D.; Rajopadhye, M. Viable near-infrared fluorochrome labeled cells and methods of making and using the same. U.S. Pat. Appl. Publ. US 20100172841, 2010.

11. Bestvater, F.; Spiess, E.; Stobrawa, G.; Hacker, M.; Feurer, T.; Porwol, T.; Berchner-Pfannschmidt, U.; Wotzlaw, C.; Acker, H. Two-photon fluorescence absorption and emission spectra of dyes relevant for cell imaging. *J. Microsc.* **2002**, *208*, 108–115.

12. Hauke, S. Method for detecting a chromosomal aberration. PCT Int. Appl. WO 2012150022, 2012.

13. Poulsen, T. S.; Poulsen, S. M.; Petersen, K. H. Methods for detecting chromosome aberrations. PCT Int. Appl. WO 2005111235, 2005.

14. Thompson, R. B.; Zeng, H.; Fierke, C. A.; Fones, G.; Moffett, J. Real-time *in-situ* determination of free Cu(II) at picomolar levels in sea water using a fluorescence lifetime-based fiber optic biosensor. *Proc. SPIE-Int. Soc. Opt. Eng.* **2002**, *4625*, 137–143.

15. Kobayashi, H.; Koyama, Y.; Barrett, T.; Hama, Y.; Regino, C. A. S.; Shin, I. S.; Jang, B.; Le, N.; Paik, C. H.; Choyke, P. L.; Urano, Y. Multimodal nanoprobes for radionuclide and five-color near-infrared optical lymphatic imaging. *ACS Nano* **2007**, *1*, 258–264.

16. Ibey, B. L.; Beier, H. T.; Rounds, R. M.; Pishko, M. V.; Cote, G. L. Dendrimer based fluorescent glucose sensor for diabetic monitoring. *Proc. SPIE-Int. Soc. Opt. Eng.* **2006**, *6094*, 609401/1609401/8.

17. Martin, H.; Comeskey, D. Folate measurement in mammalian tissues by fluorescence polarization. *Pteridines* **2011**, *22*, 105–110.

18. Martin, H.; Comeskey, D.; Simpson, R. M.; Laing, W. A.; McGhie, T. K. Quantification of folate in fruits and vegetables: A fluorescence-based homogeneous assay. *Anal. Biochem.* **2010**, *402*, 137–145.

19. Heidebrecht, R. W., Jr.,; Kral, A. M.; Miller, T. A. Fluorescent compounds that bind to histone deacetylase. U.S. Pat. Appl. Publ. US 20090156825, 2009.

20. Schneider, R.; Gohla, A.; Simard, J. R.; Yadav, D. B.; Fang, Z.; van Otterlo, W. A. L.; Rauh, D. Overcoming compound fluorescence in the FLiK screening assay with red-shifted fluorophores. *J. Am. Chem. Soc.* **2013**, *135*, 8400–8408.

21. Ohara, T.; Imai, K.; Saito, T.; Takahashi, S. Nucleic acid analysis device, nucleic acid analytical apparatus, and nucleic acid analysis method. Jpn. Kokai Tokkyo Koho JP 2012055250, 2012.

22. Zhang, B.; Wang, W.; Qu, D. Method and kit for labeling nucleic acid in living cells. Faming Zhuanli Shenqing CN 101921835, 2010.

23. Nikiforov, T.; Beechem, J. Methods and apparatus for single molecule sequencing using energy transfer detection. PCT Int. Appl. WO 2010111674, 2010.

24. Okagbare, P. I.; Soper, S. A. High throughput single molecule detection for monitoring biochemical reactions. *Analyst* **2009**, *134*, 97–106.

25. Cox, W. G.; Singer, V. L. Fluorescent DNA hybridization probe preparation using amine modification and reactive dye coupling. *BioTechniques* **2004**, *36*, 114–122.

26. Gordon, K. M.; Duckett, L.; Daul, B.; Petrie, H. T. A simple method for detecting up to five immunofluorescent parameters together with DNA staining for cell cycle or viability on a benchtop flow cytometer. *J. Immunol. Methods* **2003**, *275*, 113–121.

27. Marras, S. A. E.; Kramer, F. R.; Tyagi, S. Efficiencies of fluorescence resonance energy transfer and contact-mediated quenching in oligonucleotide probes. *Nucleic Acids Res.* **2002**, *30*, e122/1–e122/8.

28. Fujiyoshi, S.; Furuya, Y.; Iseki, M.; Watanabe, M.; Matsushita, M. Vibrational microspectroscopy of single proteins. *J. Phys. Chem. Lett.* **2010**, *1*, 2541–2545.

29. Endoh, T.; Sisido, M.; Ohtsuki, T. Spatial regulation of specific gene expression through photoactivation of RNAi. *J. Controlled Release* **2009**, *137*, 241–245.

30. Hsu, E. R.; Anslyn, E. V.; Dharmawardhane, S.; Alizadeh-Naderi, R.; Aaron, J. S.; Sokolov, K. V.; El-Naggar, A. K.; Gillenwater, A. M.; Richards-Kortum, R. R. A far-red fluorescent contrast agent to image epidermal growth factor receptor expression. *Photochem. Photobiol.* **2004**, *79*, 272–279.

31. Berlier, J. E.; Rothe, A.; Buller, G.; Bradford, J.; Gray, D. R.; Filanoski, B. J.; Telford, W. G.; Yue, S.; Liu, J.; Cheung, C.; Chang, W.; Hirsch, J. D.; Beechem, J. M.; Haugland, R. P.; Haugland, R. P. Quantitative comparison of long-wavelength Alexa Fluor dyes to Cy dyes: Fluorescence of the dyes and their bioconjugates. *J. Histochem. Cytochem.* **2003**, *51*, 1699–1712.

32. Swift, K. M.; Anderson, S.; Matayoshi, E. D. Dual-laser fluorescence correlation spectroscopy as a biophysical probe of binding interactions: evaluation of new red fluorescent dyes. *Proc. SPIE-Int. Soc. Opt. Eng.* **2001**, *4252*, 47–58.

33. Gerber, H.; Marrinucci, D.; Pirie-Shepherd, S.; Tucker, E. Methods for detecting 5 T4-positive circulating tumor cells and methods of diagnosis of 5 T4-positive cancer in a mammalian subject. PCT Int. Appl. WO 2013111054, 2013.

34. Kobayashi, H.; Ogawa, M.; Kosaka, N.; Choyke, P. L.; Urano, Y. Multicolor imaging of lymphatic function with two nanomaterials: quantum dot-labeled cancer cells and dendrimer-based optical agents. *Nanomedicine* **2009**, *4*, 411–419.

35. Lee, A. J.; Ensign, A. A.; Krauss, T. D.; Bren, K. L. Zinc porphyrin as a donor for FRET in Zn(II)cytochrome c. *J. Am. Chem. Soc.* **2010**, *132*, 1752–1753.

36. Lim, Y. T.; Noh, Y. U. Fluorescent polymer nanogel for lymph node detection and lymph node confirming method using it. Repub. Korean Kongkae Taeho Kongbo KR 2013085294, 2013.

37. Sung, M.; Lim, Y. T.; Lee, Il H. Optical-imaging probe for detecting sentinel lymph nodes which contains a composite of poly-gamma-glutamic acid and an optical-imaging dye. PCT Int. Appl. WO 2012067458, 2012.

38. Bousseksou, A.; Salmon, L.; Molnar, G.; Cobo, S. Materials with thermochromic spin transition doped with one or more fluorescent agents for use as temperature sensor. Fr. Demande FR 2952371, 2011.

39. Bousseksou, A.; Salmon, L.; Molnar, G.; Cobo, S. Heat-sensitive spin-transition materials doped with one or more fluorescent agents for use as temperature sensor. PCT Int. Appl. WO 2011058277, 2011.

ALEXA FLUOR 680 CARBOXYLIC ACID SUCCINIMIDYL ESTER

CAS Registry Number   948558-33-6

Chemical Structure   Not reported

CA Index Name   Alexa Fluor 680 carboxylic acid succinimidyl ester

Other Names   Alexa 680; Alexa 680 succinimidyl ester; Alexa Fluor 680; Alexa Fluor 680 NHS; Alexa Fluor 680 NHS ester; Alexa Fluor 680 carboxylic acid SE; Alexa Fluor 680 carboxylic acid succinimidyl ester; Alexa Fluor 680 SE; Alexa Fluor 680 succinimidyl ester; Sulfo-NHS-AF680

Merck Index Number   Not listed

Chemical/Dye Class   Not reported

Molecular Formula   Not reported

Molecular Weight   ~1150

Physical Form   Blue solid

Solubility   Soluble in water, dimethyl sulfoxide, methanol

Absorption (λ_{max})   684 nm (MeOH)

Emission (λ_{max})   707 nm (MeOH)

Molar Extinction Coefficient   183,000 cm^{-1} M^{-1} (MeOH)

Quantum Yield   0.36 (Buffer pH 7.2)

Synthesis   Synthetic method[1]

Imaging/Labeling Applications   Amines/amino groups;[16,21,28,29] antibodies;[2–6] blood vessels;[7] cells;[8–13,23,26] chromosomes;[14,15] dendrimers;[16] histone deacetylase;[17] nucleic acids/nucleotides;[4,18–22] proteins/peptides;[4,23–34] tetanus toxin (TTc)[34]

Biological/Medical Applications   Analyzing/detecting analytes;[4] counting cells;[23,26] detecting chromosomal aberration;[14,15] identifying stent thrombosis;[7] imaging of fast retrograde axonal transport in living animals;[34] monitoring mTORC1 activation;[25] quantifying proteins[23,27,30]

Industrial Applications   Not reported

Safety/Toxicity   No data available

REFERENCES

1. Haugland, R. P. *Handbook of Fluorescent Probes and Research Products*; Molecular Probes Inc.: Eugene, **2002**; pp 20–35.

2. Arcangeli, A.; Becchetti, A.; Pillozzi, S.; Masselli, M.; De Lorenzo, E. Method and kit for the prevention and/or the monitoring of chemoresistance of leukemia forms. PCT Int. Appl. WO 2011058509, 2011.

3. Ogawa, M.; Regino, C. A.; Choyke, P. L.; Kobayashi, H. *In vivo* target-specific activatable near-infrared optical labeling of humanized monoclonal antibodies. *Mol. Cancer Ther.* **2009**, *8*, 232–239.

4. Green, D. P. L.; Rawle, C. B. Analysis system and method. PCT Int. Appl. WO 2009082242, 2009.

5. Chang, S. K.; Rizvi, I.; Solban, N.; Hasan, T. *In vivo* optical molecular imaging of vascular endothelial growth factor for monitoring cancer treatment. *Clin. Cancer Res.* **2008**, *14*, 4146–4153.

6. Kokko, T.; Liljenback, T.; Peltola, M. T.; Kokko, L.; Soukka, T. Homogeneous dual-parameter assay for prostate-specific antigen based on fluorescence resonance energy transfer. *Anal. Chem.* **2008**, *80*, 9763–9768.

7. Jaffer, F.; Rajopadhye, M. Methods and compositions for identifying subjects at risk of developing stent thrombosis. U.S. Pat. Appl. Publ. US 20100268070, 2010.

8. Klimanskaya, I.; Gay, R. J. Methods for detection of rare subpopulations of cells and highly purified compositions of cells. PCT Int. Appl. WO 2012012803, 2012.

9. Peterson, J. D.; Rajopadhye, M. Viable near-infrared fluorochrome labeled cells and methods of making and using the same. U.S. Pat. Appl. Publ. US 20100172841, 2010.

10. Pittet, M. J.; Swirski, F. K.; Reynolds, F.; Josephson, L.; Weissleder, R. Labeling of immune cells for *in vivo* imaging using magnetofluorescent nanoparticles. *Nat. Protoc.* **2006**, *1*, 73–79.

11. Schellenberger, E. A.; Reynolds, F.; Weissleder, R.; Josephson, L. Surface-functionalized nanoparticle library yields probes for apoptotic cells. *ChemBioChem* **2004**, *5*, 275–279.

12. Muczynski, K. A.; Ekle, D. M.; Coder, D. M.; Anderson, S. K. Normal human kidney HLA-DR-expressing renal microvascular endothelial cells: characterization, isolation, and regulation of MHC class II expression. *J. Am. Soc. Nephrol.* **2003**, *14*, 1336–1348.

13. Bestvater, F.; Spiess, E.; Stobrawa, G.; Hacker, M.; Feurer, T.; Porwol, T.; Berchner-Pfannschmidt, U.;

Wotzlaw, C.; Acker, H. Two-photon fluorescence absorption and emission spectra of dyes relevant for cell imaging. *J. Microsc.* **2002**, *208*, 108–115.

14. Hauke, S. Method for detecting a chromosomal aberration. PCT Int. Appl. WO 2012150022, 2012.

15. Poulsen, T. S.; Poulsen, S. M.; Petersen, K. H. Methods for detecting chromosome aberrations. PCT Int. Appl. WO 2005111235, 2005.

16. Kobayashi, H.; Koyama, Y.; Barrett, T.; Hama, Y.; Regino, C. A. S.; Shin, I. S.; Jang, B.; Le, N.; Paik, C. H.; Choyke, P. L.; Urano, Y. Multimodal nanoprobes for radionuclide and five-color near-infrared optical lymphatic imaging. *ACS Nano* **2007**, *1*, 258–264.

17. Heidebrecht, R. W., Jr.,; Kral, A. M.; Miller, T. A. Fluorescent compounds that bind to histone deacetylase. U.S. Pat. Appl. Publ. US 20090156825, 2009.

18. Nikiforov, T.; Beechem, J. Methods and apparatus for single molecule sequencing using energy transfer detection. PCT Int. Appl. WO 2010111674, 2010.

19. Meade, S. O.; Chen, M. Y.; Sailor, M. J.; Miskelly, G. M. Multiplexed DNA detection using spectrally encoded porous SiO_2 photonic crystal particles. *Anal. Chem.* **2009**, *81*, 2618–2625.

20. Korlach, J.; Bibillo, A.; Wegener, J.; Peluso, P.; Pham, T. T.; Park, I.; Clark, S.; Otto, G. A.; Turner, S. W. Long, processive enzymatic DNA synthesis using 100% dye-labeled terminal phosphate-linked nucleotides. *Nucleosides, Nucleotides Nucleic Acids* **2008**, *27*, 1072–1082.

21. Cox, W. G.; Singer, V. L. Fluorescent DNA hybridization probe preparation using amine modification and reactive dye coupling. *BioTechniques* **2004**, *36*, 114–122.

22. Gordon, K. M.; Duckett, L.; Daul, B.; Petrie, H. T. A simple method for detecting up to five immunofluorescent parameters together with DNA staining for cell cycle or viability on a benchtop flow cytometer. *J. Immunol. Methods* **2003**, *275*, 113–121.

23. Pihlasalo, S.; Puumala, P.; Hanninen, P.; Harma, H. Sensitive method for determination of protein and cell concentrations based on competitive adsorption to nanoparticles and time-resolved luminescence resonance energy transfer between labeled proteins. *Anal. Chem.* **2012**, *84*, 4950–4956.

24. Karhunen, U.; Rosenberg, J.; Lamminmaki, U.; Soukka, T. Homogeneous detection of avidin based on switchable lanthanide luminescence. *Anal. Chem.* **2011**, *83*, 9011–9016.

25. Hoffman, G. R.; Moerke, N. J.; Hsia, M.; Shamu, C. E.; Blenis, J. A high-throughput, cell-based screening method for siRNA and small molecule inhibitors of mTORC1 signaling using the in cell western technique. *Assay Drug Dev. Technol.* **2010**, *8*, 186–199.

26. Pihlasalo, S.; Pellonperae, L.; Martikkala, E.; Haenninen, P.; Haermae, H. Sensitive fluorometric nanoparticle assays for cell counting and viability. *Anal. Chem.* **2010**, *82*, 9282–9288.

27. Valanne, A.; Suojanen, J.; Peltonen, J.; Soukka, T.; Hanninen, P.; Harma, H. Multiple sized europium(III) chelate-dyed polystyrene particles as donors in FRET - an application for sensitive protein quantification utilizing competitive adsorption. *Analyst* **2009**, *134*, 980–986.

28. Jin, X.; Newton, J. R.; Montgomery-Smith, S.; Smith, G. P. A generalized kinetic model for amine modification of proteins with application to phage display. *BioTechniques* **2009**, *46*, 175–182.

29. Pihlasalo, S.; Hara, M.; Haenninen, P.; Slotte, J. P.; Peltonen, J.; Haermae, H. Liposome-based homogeneous luminescence resonance energy transfer. *Anal. Biochem.* **2009**, *384*, 231–237.

30. Harma, H.; Dahne, L.; Pihlasalo, S.; Suojanen, J.; Peltonen, J.; Hanninen, P. Sensitive quantitative protein concentration method using luminescent resonance energy transfer on a layer-by-layer europium(III) chelate particle sensor. *Anal. Chem.* **2008**, *80*, 9781–9786.

31. Rantanen, T.; Paekkilae, H.; Jaemsen, L.; Kuningas, K.; Ukonaho, T.; Loevgren, T.; Soukka, T. Tandem dye acceptor used to enhance upconversion fluorescence resonance energy transfer in homogeneous assays. *Anal. Chem.* **2007**, *79*, 6312–6318.

32. Ma, L.; Yu, P.; Veerendra, B.; Rold, T. L.; Retzloff, L.; Prasanphanich, A.; Sieckman, G.; Hoffman, T. J.; Volkert, W. A.; Smith, C. J. *In vitro* and *in vivo* evaluation of Alexa Fluor 680-bombesin[7,14]NH2 peptide conjugate, a high-affinity fluorescent probe with high selectivity for the gastrin-releasing peptide receptor. *Mol. Imaging* **2007**, *6*, 171–180.

33. Berlier, J. E.; Rothe, A.; Buller, G.; Bradford, J.; Gray, D. R.; Filanoski, B. J.; Telford, W. G.; Yue, S.; Liu, J.; Cheung, C.; Chang, W.; Hirsch, J. D.; Beechem, J. M.; Haugland, R. P.; Haugland, R. P. Quantitative comparison of long-wavelength Alexa Fluor dyes to Cy dyes: Fluorescence of the dyes and their bioconjugates. *J. Histochem. Cytochem.* **2003**, *51*, 1699–1712.

34. Schellingerhout, D.; Le Roux, L. G.; Bredow, S.; Gelovani, J. G. Fluorescence imaging of fast retrograde axonal transport in living animals. *Mol. Imaging* **2009**, *8*, 319–329.

ALEXA FLUOR 700 CARBOXYLIC ACID SUCCINIMIDYL ESTER

CAS Registry Number   1246956-22-8

Chemical Structure   Not reported

CA Index Name   Alexa Fluor 700 carboxylic acid succinimidyl ester

Other Names   Alexa 700; Alexa 700 succinimidyl ester; Alexa Fluor 700; Alexa Fluor 700 NHS; Alexa Fluor 700 NHS ester; Alexa Fluor 700 carboxylic acid SE; Alexa Fluor 700 carboxylic acid succinimidyl ester; Alexa Fluor 700 SE; Alexa Fluor 700 succinimidyl ester

Merck Index Number   Not listed

Chemical/Dye Class   Not reported

Molecular Formula   Not reported

Molecular Weight   ~1400

Physical Form   Blue solid

Solubility   Soluble in water, dimethyl sulfoxide, methanol

Absorption (λ_{max})   702 nm (MeOH)

Emission (λ_{max})   723 nm (MeOH)

Molar Extinction Coefficient   205,000 $cm^{-1} M^{-1}$ (MeOH)

Quantum Yield   0.25 (Buffer pH 7.2)

Synthesis   Synthetic method[1]

Imaging/Labeling Applications   Amines/amino groups;[11] antibodies;[2,3] blood vessels;[4] cells;[5–8] chromosomes;[9,10] dendrimers;[11] histone deacetylases;[12] Hodgkin and Reed Sternberg (HRS) cells;[13] nucleic acids;[3,14,15] proteins[3,16,17]

Biological/Medical Applications   Analyzing/detecting analytes;[3] detecting chromosomal aberration;[9,10] diagnosing classical Hodgkin lymphoma (CHL) in lymph nodes;[13] fluorescent cell barcoding;[7] identifying stent thrombosis[4]

Industrial Applications   Not reported

Safety/Toxicity   No data available

REFERENCES

1. Haugland, R. P. *Handbook of Fluorescent Probes and Research Products*; Molecular Probes Inc.: Eugene, **2002**; pp 20–35.

2. Arcangeli, A.; Becchetti, A.; Pillozzi, S.; Masselli, M.; De Lorenzo, E. Method and kit for the prevention and/or the monitoring of chemoresistance of leukemia forms. PCT Int. Appl. WO 2011058509, 2011.

3. Green, D. P. L.; Rawle, C. B. Analysis system and method. PCT Int. Appl. WO 2009082242, 2009.

4. Jaffer, F.; Rajopadhye, M. Methods and compositions for identifying subjects at risk of developing stent thrombosis. U.S. Pat. Appl. Publ. US 20100268070, 2010.

5. Klimanskaya, I.; Gay, R. J. Methods for detection of rare subpopulations of cells and highly purified compositions of cells. PCT Int. Appl. WO 2012012803, 2012.

6. Peterson, J. D.; Rajopadhye, M. Viable near-infrared fluorochrome labeled cells and methods of making and using the same. U.S. Pat. Appl. Publ. US 20100172841, 2010.

7. Krutzik, P. O.; Nolan, G. P. Fluorescent cell barcoding in flow cytometry allows high-throughput drug screening and signaling profiling. *Nat. Methods* **2006**, *3*, 361–368.

8. Bestvater, F.; Spiess, E.; Stobrawa, G.; Hacker, M.; Feurer, T.; Porwol, T.; Berchner-Pfannschmidt, U.; Wotzlaw, C.; Acker, H. Two-photon fluorescence absorption and emission spectra of dyes relevant for cell imaging. *J. Microsc.* **2002**, *208*, 108–115.

9. Hauke, S. Method for detecting a chromosomal aberration. PCT Int. Appl. WO 2012150022, 2012.

10. Poulsen, T. S.; Poulsen, S. M.; Petersen, K. H. Methods for detecting chromosome aberrations. PCT Int. Appl. WO 2005111235, 2005.

11. Kobayashi, H.; Koyama, Y.; Barrett, T.; Hama, Y.; Regino, C. A. S.; Shin, I. S.; Jang, B.; Le, N.; Paik, C. H.; Choyke, P. L.; Urano, Y. Multimodal nanoprobes for radionuclide and five-color near-infrared optical lymphatic imaging. *ACS Nano* **2007**, *1*, 258–264.

12. Heidebrecht, R. W., Jr.,; Kral, A. M.; Miller, T. A. Fluorescent compounds that bind to histone deacetylase. U.S. Pat. Appl. Publ. US 20090156825, 2009.

13. Fromm, J. R.; Thomas, A.; Wood, B. L. Flow cytometry can diagnose classical Hodgkin lymphoma in lymph nodes with high sensitivity and specificity. *Am. J. Clin. Pathol.* **2009**, *131*, 322–332.

14. Nikiforov, T.; Beechem, J. Methods and apparatus for single molecule sequencing using energy

transfer detection. PCT Int. Appl. WO 2010111674, 2010.

15. Choi, H. M.; Chang, J. Y.; Trinhle, A.; Padilla, J. E.; Pierce, N. A. Programmable *in situ* amplification for multiplexed imaging of mRNA expression. *Nat. Biotechnol.* **2010**, *28*, 1208–1212.

16. Rantanen, T.; Paekkilae, H.; Jaemsen, L.; Kuningas, K.; Ukonaho, T.; Loevgren, T.; Soukka, T. Tandem dye acceptor used to enhance upconversion fluorescence resonance energy transfer in homogeneous assays. *Anal. Chem.* **2007**, *79*, 6312–6318.

17. Berlier, J. E.; Rothe, A.; Buller, G.; Bradford, J.; Gray, D. R.; Filanoski, B. J.; Telford, W. G.; Yue, S.; Liu, J.; Cheung, C.; Chang, W.; Hirsch, J. D.; Beechem, J. M.; Haugland, R. P.; Haugland, R. P. Quantitative comparison of long-wavelength Alexa Fluor dyes to Cy dyes: Fluorescence of the dyes and their bioconjugates. *J. Histochem. Cytochem.* **2003**, *51*, 1699–1712.

ALEXA FLUOR 750 CARBOXYLIC ACID SUCCINIMIDYL ESTER

CAS Registry Number   697795-06-5

Chemical Structure   Not reported

CA Index Name   Alexa Fluor 750

Other Names   Alexa 750; Alexa 750 NHS; Alexa 750 succinimidyl ester; Alexa Fluor 750; Alexa Fluor 750 NHS; Alexa Fluor 750 NHS ester; Alexa Fluor 750 carboxylic acid SE; Alexa Fluor 750 carboxylic acid succinimidyl ester; Alexa Fluor 750 SE; Alexa Fluor 750 succinimidyl ester

Merck Index Number   Not listed

Chemical/Dye Class   Not reported

Molecular Formula   Not reported

Molecular Weight   ~1300

Physical Form   Blue solid

Solubility   Soluble in water, dimethyl sulfoxide, methanol

Absorption (λ_{max})   753 nm (MeOH)

Emission (λ_{max})   782 nm (MeOH)

Molar Extinction Coefficient $290,000\,cm^{-1}\,M^{-1}$ (MeOH)

Quantum Yield   0.12 (Buffer pH 7.2)

Synthesis   Synthetic method[1]

Imaging/Labeling Applications   Amines/amino groups;[2,3,11,21] amyloid-β (Aβ) peptides;[4] atherosclerotic plaque;[5] antibodies;[6–12] blood vessels;[13] caries in tooth enamel;[14] cells;[15–18] chromosomes;[19,20] dendrimers;[21] glucose;[22–24] histone deacetylases;[25] liposomes;[26] nanoparticles;[27–29] nucleic acids;[10,30–34] phospholipid micelles;[43] proteins/peptides;[10,23,24,26,35–41] quantum dots;[42] tumors[26,43,44]

Biological/Medical Applications   Analyzing/detecting analytes;[10] detecting chromosomal aberration;[19,20] fluorescent cell barcoding;[17] glucose sensor;[22–24] homing/trafficking human hematopoietic stem cells (HSC);[28] identifying carious lesion;[14] imaging/identifying stent thrombosis;[13] quantifying proteins;[39] understanding lymph node metastasis;[44] use in agriculture and plant cultivation[45]

Industrial Applications   Not reported

Safety/Toxicity   No data available

REFERENCES

1. Haugland, R. P. *Handbook of Fluorescent Probes and Research Products*; Molecular Probes Inc.: Eugene, **2002**; pp 20–35.

2. Karver, M. R.; Weissleder, R.; Hilderbrand, S. A. Bioorthogonal reaction pairs enable simultaneous, selective, multi-target imaging. *Angew. Chem., Int. Ed.* **2012**, *51*, 920–922.

3. Karver, M. R.; Weissleder, R.; Hilderbrand, S. A. Synthesis and evaluation of a series of 1,2,4,5-tetrazines for bioorthogonal conjugation. *Bioconjugate Chem.* **2011**, *22*, 2263–2270.

4. Skoch, J.; Dunn, A.; Hyman, B. T.; Bacskai, B. J. Development of an optical approach for noninvasive imaging of Alzheimer's disease pathology. *J. Biomed. Opt.* **2005**, *10*, 011007/1–011007/7.

5. Zhu, B.; Jaffer, F. A.; Ntziachristos, V.; Weissleder, R. Development of a near infrared fluorescence catheter: Operating characteristics and feasibility for atherosclerotic plaque detection. *J. Phys. D: Appl. Phys.* **2005**, *38*, 2701–2707.

6. Alata, W.; Paris-Robidas, S.; Emond, V.; Bourasset, F.; Calon, F. Brain uptake of a fluorescent vector targeting the transferrin receptor: A novel application of *in situ* brain perfusion. *Mol. Pharm.* **2014**, *11*, 243–253.

7. Paris-Robidas, S.; Emond, V.; Tremblay, C.; Soulet, D.; Calon, F. *In vivo* labeling of brain capillary endothelial cells after intravenous injection of monoclonal antibodies targeting the transferrin receptor. *Mol. Pharmacol.* **2011**, *80*, 32–39.

8. Arcangeli, A.; Becchetti, A.; Pillozzi, S.; Masselli, M.; De Lorenzo, E. Method and kit for the prevention and/or the monitoring of chemoresistance of leukemia forms. PCT Int. Appl. WO 2011058509, 2011.

9. Paudyal, P.; Paudyal, B.; Iida, Y.; Oriuchi, N.; Hanaoka, H.; Tominaga, H.; Ishikita, T.; Yoshioka, H.; Higuchi, T.; Endo, K. Dual functional molecular imaging probe targeting CD20 with PET and optical imaging. *Oncol. Rep.* **2009**, *22*, 115–119.

10. Green, D. P. L.; Rawle, C. B. Analysis system and method. PCT Int. Appl. WO 2009082242, 2009.

11. Bhattacharyya, S.; Wang, S.; Reinecke, D.; Kiser, W., Jr.,; Kruger, R. A.; DeGrado, T. R. Synthesis and evaluation of near-Infrared (NIR) dye-herceptin conjugates as photoacoustic computed tomography

(PCT) probes for HER2 expression in breast cancer. *Bioconjugate Chem.* **2008**, *19*, 1186–1193.

12. Hassan, M.; Riley, J.; Chernomordik, V.; Smith, P.; Pursley, R.; Lee, S. B.; Capala, J.; Gandjbakhche, A. H. Fluorescence lifetime imaging system for *in vivo* studies. *Mol. Imaging* **2007**, *6*, 229–236.

13. Jaffer, F.; Rajopadhye, M. Methods and compositions for identifying subjects at risk of developing stent thrombosis. U.S. Pat. Appl. Publ. US 20100268070, 2010.

14. Nagai, S. Methods and kits with fluorescent probes for caries detection in tooth enamel. U.S. Patent 8647119, 2014.

15. Klimanskaya, I.; Gay, R. J. Methods for detection of rare subpopulations of cells and highly purified compositions of cells. PCT Int. Appl. WO 2012012803, 2012.

16. Peterson, J. D.; Rajopadhye, M. Viable near-infrared fluorochrome labeled cells and methods of making and using the same. U.S. Pat. Appl. Publ. US 20100172841, 2010.

17. Krutzik, P. O.; Nolan, G. P. Fluorescent cell barcoding in flow cytometry allows high-throughput drug screening and signaling profiling. *Nat. Methods* **2006**, *3*, 361–368.

18. Bestvater, F.; Spiess, E.; Stobrawa, G.; Hacker, M.; Feurer, T.; Porwol, T.; Berchner-Pfannschmidt, U.; Wotzlaw, C.; Acker, H. Two-photon fluorescence absorption and emission spectra of dyes relevant for cell imaging. *J. Microsc.* **2002**, *208*, 108–115.

19. Hauke, S. Method for detecting a chromosomal aberration. PCT Int. Appl. WO 2012150022, 2012.

20. Poulsen, T. S.; Poulsen, S. M.; Petersen, K. H. Methods for detecting chromosome aberrations. PCT Int. Appl. WO 2005111235, 2005.

21. Kobayashi, H.; Koyama, Y.; Barrett, T.; Hama, Y.; Regino, C. A. S.; Shin, I. S.; Jang, B.; Le, N.; Paik, C. H.; Choyke, P. L.; Urano, Y. Multimodal nanoprobes for radionuclide and five-color near-infrared optical lymphatic imaging. *ACS Nano* **2007**, *1*, 258–264.

22. Shah, R.; Kristensen, J. S.; Wolfe, K. T.; Aasmul, S.; Bansal, A. Orthogonally redundant sensor systems and methods. U.S. Pat. Appl. Publ. US 20130060105, 2013.

23. Dweik, M. Glucose binding protein as a novel optical glucose nanobiosensor. *Sens. Transducers J.* **2009**, *110*, 1–8.

24. Dweik, M.; Milanick, M.; Grant, S. Development of a glucose binding protein biosensor. *Proc. SPIE-Int. Soc. Opt. Eng.* **2007**, *6759*, 67590I/1-67590I/8.

25. Heidebrecht, R. W., Jr.,; Kral, A. M.; Miller, T. A. Fluorescent compounds that bind to histone deacetylase. U.S. Pat. Appl. Publ. US 20090156825, 2009.

26. Lowery, A.; Onishko, H.; Hallahan, D. E.; Han, Z. Tumor-targeted delivery of liposome-encapsulated doxorubicin by use of a peptide that selectively binds to irradiated tumors. *J. Controlled Release* **2011**, *150*, 117–124.

27. Herz, E.; Ow, H.; Bonner, D.; Burns, A.; Wiesner, U. Dye structure-optical property correlations in near-infrared fluorescent core-shell silica nanoparticles. *J. Mater. Chem.* **2009**, *19*, 6341–6347.

28. Maxwell, D. J.; Bonde, J.; Hess, D. A.; Hohm, S. A.; Lahey, R.; Zhou, P.; Creer, M. H.; Piwnica-Worms, D.; Nolta, J. A. Fluorophore-conjugated iron oxide nanoparticle labeling and analysis of engrafting human hematopoietic stem cells. *Stem Cells* **2008**, *26*, 517–524.

29. McCarthy, J. R.; Jaffer, F. A.; Weissleder, R. A macrophage-targeted theranostic nanoparticle for biomedical applications. *Small* **2006**, *2*, 983–987.

30. Zhang, X.; Song, Y.; Shah, A. Y.; Lekova, V.; Raj, A.; Huang, L.; Behlke, M. A.; Tsourkas, A. Quantitative assessment of ratiometric bimolecular beacons as a tool for imaging single engineered RNA transcripts and measuring gene expression in living cells. *Nucleic Acids Res.* **2013**, *41*, e152.

31. Suzuki, S.; Kirimura, H.; Seike, M.; Iwanaga, S.; Hori, N.; Uraoka, Y.; Bin, Z. Method for electrochemically detecting analyte. Eur. Pat. Appl. EP 2515112, 2012.

32. Stein, I. H.; Steinhauer, C.; Tinnefeld, P. Single-molecule four-color FRET visualizes energy-transfer paths on DNA origami. *J. Am. Chem. Soc.* **2011**, *133*, 4193–4195.

33. Nikiforov, T.; Beechem, J. Methods and apparatus for single molecule sequencing using energy transfer detection. PCT Int. Appl. WO 2010111674, 2010.

34. Zhang, B.; Wang, W.; Qu, D. Method and kit for labeling nucleic acid in living cells. Faming Zhuanli Shenqing CN 101921835, 2010.

35. Kanesaki, K.; Yamauchi, F.; Ogawa, K. Compounds having improved biological stability, and imaging agents containing the compounds. Jpn. Kokai Tokkyo Koho JP 2012017281, 2012.

36. Ma, L.; Zhang, M.; Yu, P. Imaging site-specific peptide-targeting in tumor tissues using spectral-domain optical coherence tomography. *Proc. SPIE* **2011**, *7890*, 78900V/1–78900V/8.

37. Erbse, A. H.; Berlinberg, A. J.; Cheung, C.; Leung, W.; Falke, J. J. OS-FRET: A new one-sample method for improved FRET measurements. *Biochemistry* **2011**, *50*, 451–457.

38. Chernomordik, V.; Hassan, M.; Lee, S. B.; Zielinski, R.; Gandjbakhche, A.; Capala, J. Quantitative analysis of HER2 receptor expression *in vivo* by near-infrared optical imaging. *Mol. Imaging* **2010**, *9*, 192–200.

39. Chen, A. K.; Cheng, Z.; Behlke, M. A.; Tsourkas, A. Assessing the sensitivity of commercially available fluorophores to the intracellular environment. *Anal. Chem.* **2008**, *80*, 7437–7444.

40. Zhang, J.; Matveeva, E.; Gryczynski, I.; Leonenko, Z.; Lakowicz, J. R. Metal-enhanced fluoroimmunoassay on a silver film by vapor deposition. *J. Phys. Chem. B* **2005**, *109*, 7969–7975.

41. Berlier, J. E.; Rothe, A.; Buller, G.; Bradford, J.; Gray, D. R.; Filanoski, B. J.; Telford, W. G.; Yue, S.; Liu, J.; Cheung, C.; Chang, W.; Hirsch, J. D.; Beechem, J. M.; Haugland, R. P.; Haugland, R. P. Quantitative comparison of long-wavelength Alexa Fluor dyes to Cy dyes: Fluorescence of the dyes and their bioconjugates. *J. Histochem. Cytochem.* **2003**, *51*, 1699–1712.

42. Muro, E.; Vermeulen, P.; Ioannou, A.; Skourides, P.; Dubertret, B.; Fragola, A.; Loriette, V. Single-shot optical sectioning using two-color probes in HiLo fluorescence microscopy. *Biophys. J.* **2011**, *100*, 2810–2819.

43. Papagiannaros, A.; Kale, A.; Levchenko, T. S.; Mongayt, D.; Hartner, W. C.; Torchilin, V. P. Near infrared planar tumor imaging and quantification using nanosized Alexa 750-labeled phospholipid micelles. *Int. J. Nanomed.* **2009**, *4*, 123–131.

44. Kobayashi, H.; Ogawa, M.; Kosaka, N.; Choyke, P. L.; Urano, Y. Multicolor imaging of lymphatic function with two nanomaterials: quantum dot-labeled cancer cells and dendrimer-based optical agents. *Nanomedicine* **2009**, *4*, 411–419.

45. Kuznetsov, Y. P.; Yasinskii, A. M. Luminescent greenhouse material for use in agriculture and plant cultivation. Russ. RU 2248386, 2005.

ALEXA FLUOR 790 CARBOXYLIC ACID SUCCINIMIDYL ESTER

CAS Registry Number 950891-33-5

Chemical Structure Not reported

CA Index Name Alexa Fluor 790 carboxylic acid succinimidyl ester

Other Names Alexa 790; Alexa 790 succinimidyl ester; Alexa Fluor 790; Alexa Fluor 790 NHS; Alexa Fluor 790 NHS ester; Alexa Fluor 790 carboxylic acid SE; Alexa Fluor 790 carboxylic acid succinimidyl ester; Alexa Fluor 790 SE; Alexa Fluor 790 succinimidyl ester

Merck Index Number Not listed

Chemical/Dye Class Not reported

Molecular Formula Not reported

Molecular Weight ~1750

Physical Form Blue solid

Solubility Soluble in water, dimethyl sulfoxide, methanol

Absorption (λ_{max}) 784 nm (MeOH)

Emission (λ_{max}) 814 nm (MeOH)

Molar Extinction Coefficient 260,000 cm^{-1} M^{-1} (MeOH)

Synthesis Synthetic method[1]

Imaging/Labeling Applications Antibodies;[2] blood vessels;[3] cells;[4–6] chromosomes;[7,8] histone deacetylases;[9] proteins;[10] tetanus toxin (TTc)[10]

Biological/Medical Applications Detecting chromosomal aberration;[7,8] detecting/imaging cells;[4–6] identifying stent thrombosis;[3] imaging retrograde axonal transport in living animals;[10] treating ocular diseases[11]

Industrial Applications Not reported

Safety/Toxicity No data available

REFERENCES

1. *The Molecular Probes Handbook: A Guide to Fluorescent Probes and Labeling Technologies*; Life Technologies Corporation: Eugene, **2010**; pp 35–56.

2. Arcangeli, A.; Becchetti, A.; Pillozzi, S.; Masselli, M.; De Lorenzo, E. Method and kit for the prevention and/or the monitoring of chemoresistance of leukemia forms. PCT Int. Appl. WO 2011058509, 2011.

3. Jaffer, F.; Rajopadhye, M. Methods and compositions for identifying subjects at risk of developing stent thrombosis. U.S. Pat. Appl. Publ. US 20100268070, 2010.

4. Klimanskaya, I.; Gay, R. J. Methods for detection of rare subpopulations of cells and highly purified compositions of cells. PCT Int. Appl. WO 2012012803, 2012.

5. Peterson, J. D.; Rajopadhye, M. Viable near-infrared fluorochrome labeled cells and methods of making and using the same. U.S. Pat. Appl. Publ. US 20100172841, 2010.

6. Bestvater, F.; Spiess, E.; Stobrawa, G.; Hacker, M.; Feurer, T.; Porwol, T.; Berchner-Pfannschmidt, U.; Wotzlaw, C.; Acker, H. Two-photon fluorescence absorption and emission spectra of dyes relevant for cell imaging. *J. Microsc.* **2002**, *208*, 108–115.

7. Hauke, S. Method for detecting a chromosomal aberration. PCT Int. Appl. WO 2012150022, 2012.

8. Poulsen, T. S.; Poulsen, S. M.; Petersen, K. H. Methods for detecting chromosome aberrations. PCT Int. Appl. WO 2005111235, 2005.

9. Heidebrecht, R. W., Jr.,; Kral, A. M.; Miller, T. A. Fluorescent compounds that bind to histone deacetylase. U.S. Pat. Appl. Publ. US 20090156825, 2009.

10. Schellingerhout, D.; Le Roux, L. G.; Bredow, S.; Gelovani, J. G. Fluorescence imaging of fast retrograde axonal transport in living animals. *Mol. Imaging* **2009**, *8*, 319–329.

11. Schulze, B.; Michaelis, U.; Agostini, H.; Hua, J.; Guenzi, E.; Gottfried, M.; Hansen, L. Use of a cationic colloidal preparation for the diagnosis and treatment of ocular diseases. PCT Int. Appl. WO 2008006535, 2008.

9-AMINO-6-CHLORO-2-METHOXYACRIDINE (ACMA)

CAS Registry Number   3548-09-2

Chemical Structure

CA Index Name   9-Acridinamine, 6-chloro-2-methoxy-

Other Names   Acridine, 9-amino-6-chloro-2-methoxy-; 2-Methoxy-6-chloro-9-aminoacridine; 3-Chloro-7-methoxy-9-aminoacridine; 6-Chloro-9-amino-2-methoxyacridine; 9-Amino-3-chloro-7-methoxyacridine; 9-Amino-6-chloro-2-methoxyacridine; G 185; NSC 15300

Merck Index Number   Not listed

Chemical/Dye Class   Acridine

Molecular Formula $C_{14}H_{11}ClN_2O$

Molecular Weight   258.70

Physical Form   Yellow crystals;[15] Yellow cubic crystals;[17] Orange-yellow crystals;[16] Yellow solid

Solubility   Insoluble in water; soluble in *N,N*-dimethylformamide, dimethyl sulfoxide, methanol

Melting Point   341 °C;[3] 281 °C;[16] 274 °C;[4,10,17] 273 °C;[15] 272 °C;[9] 265–267 °C;[13] 250 °C[8]

Boiling Point (Calcd.)   475.1±35.0 °C, pressure: 760 Torr

pK$_a$ 8.6, temperature: 22 °C

pK$_a$ (Calcd.)   8.69 ± 0.10, most basic, temperature: 25 °C

Absorption (λ_{max})   412 nm (MeOH acidified with a trace of HCl)

Emission (λ_{max})   471 nm (MeOH acidified with a trace of HCl)

Molar Extinction Coefficient   8200 cm^{-1} M^{-1} (MeOH acidified with a trace of HCl)

Synthesis   Synthetic methods[1–17]

Imaging/Labeling Applications   Cancer cells;[18] chromosome;[19] membranes;[20–28] microorganisms;[29] nucleic acids/nucleic acid bases[30–49]

Biological/Medical Applications   Analyzing membrane architecture;[28] antimalarial activity;[13,50,51] anti-prion activity;[3] antiviral activity;[54] bactericidal activity;[52,53] detecting cancer cells,[18] measuring microorganisms;[29] studying proton-pumping activity of various membrane-bound ATPases,[20,21,26,27] structure and function of membranes;[28] treating malformed forms of proteins causing neurodegenerative diseases;[55] as temperature sensor[56,57]

Industrial Applications   Not reported

Safety/Toxicity   Genotoxicity;[58–60] mutagenicity[58,61–64]

REFERENCES

1. Sabnis, R. W. *Handbook of Biological Dyes and Stains*; John Wiley & Sons Inc.: Hoboken, **2010**; pp 22–23.

2. Sabnis, R. W. *Handbook of Acid–base Indicators*; CRC Press: Boca Raton, **2008**; pp 22–23.

3. Thi, H. T. N.; Lee, C.; Teruya, K.; Ong, W.; Doh-ura, K.; Go, M. Antiprion activity of functionalized 9-aminoacridines related to quinacrine. *Bioorg. Med. Chem.* **2008**, *16*, 6737–6746.

4. Bonse, S.; Santelli-Rouvier, C.; Barbe, J.; Krauth-Siegel, R. L. Inhibition of *Trypanosoma cruzi* trypanothione reductase by acridines: Kinetic studies and structure-activity relationships. *J. Med. Chem.* **1999**, *42*, 5448–5454.

5. Mansour, M.; Thaller, S.; Parlar, H.; Korte, F. Photoinduced reaction of Atebrine. *Z. Naturforsch., B* **1984**, *39B*, 1626–1628.

6. Shibnev, V. A.; Finogenova, M. P.; Gazumyan, A. K.; Poletaev, A. I.; Mar'yash, L. I. 2-Methoxy-6,9-dichloroacridine in peptide synthesis as a fluorescent label. *Bioorg. Khim.* **1984**, *10*, 610–617.

7. Albert, A. Acridine syntheses and reactions. VI. A new dehalogenation of 9-chloroacridine and its derivatives. Further acridine ionization constants and ultraviolet spectra. *J. Chem. Soc.* **1965**, 4653–4657.

8. Nechaeva, O. N.; Pushkareva, Z. V. Heterocyclic N-oxides. VI. Polarographic reduction of some N-oxides of phenazine and acridine series. *Zh. Obshch. Khim.* **1958**, *28*, 2693–2701.

9. Kitani, K. The syntheses of 9-substituted acridines. III. Reaction of 9-amino-and 9-(alkylamino)acridines with amines. *Nippon Kagaku Kaishi* **1954**, *75*, 477–480.

10. Gerchuk, M. P.; Livshits, D. A.; Taits, S. Z. Exchange reactions in the series of urea derivatives. *Zh. Obshch. Khim.* **1950**, *20*, 924–930.

11. Barber, H. J. Amino-substituted acridines. U.S. Patent 2450367, 1948.

12. Barber, H. J.; Wilkinson, J. H.; Edwards, W. G. H. The reaction of 9-alkoxyacridines with bases and their salts. *J. Soc. Chem. Ind.* **1947**, *66*, 411–415.

13. Guha, P. C.; Mukherjee, S. P. Synthesis of new antimalarial drugs related to atebrin. II. *J. Indian Inst. Sci.* **1946**, *28A*, 70–74.

14. Barber, H. J. 9-Aminoacridines. Brit. GB 581695, 1946.

15. Shionogi Drug Manufg. Co. 2-Alkoxy-6-halo-9-aminoacridines. JP 162716, 1944.

16. Albert, A.; Goldacre, R.; Heymann, E. Amino acridines: Some partition and surface phenomena. *J. Chem. Soc.* **1943**, 651–654.

17. Gerchuk, M. P.; Arbuzova, P. G.; Kel'manskaya, I. A. Synthesis of new chemotherapeutic pyroplasmocidic compounds. II. Synthesis of acridyl-substituted ureas. *Zh. Obshch. Khim.* **1941**, *11*, 948–953.

18. Schwarz, G.; Wittekind, D. Selected aminoacridines as fluorescent probes in cytochemistry in general and in the detection of cancer cells in particular. *Anal. Quant. Cytol.* **1982**, *4*, 44–54.

19. Tsou, K. C.; Giles, B.; Kohn, G. Chemical basis of chromosome banding patterns. *Stain Technol.* **1975**, *50*, 293–295.

20. McCarty R. E. The decay of the ATPase activity of light plus thiol-activated thylakoid membranes in the dark. *J. Bioenerg. Biomembr.* **2006**, *38*, 67–74.

21. Rottenberg, H.; Moreno-Sanchez, R. The proton pumping activity of H+−ATPases: an improved fluorescence assay. *Biochim. Biophys. Acta, Bioenerg.* **1993**, *1183*, 161–170.

22. Casadio, R. Measurements of transmembrane pH differences of low extents in bacterial chromatophores. Study with the fluorescent probe 9-amino-6-chloro-2-methoxyacridine. *Eur. Biophys. J.* **1991**, *19*, 189–201.

23. Kopacz, S. J.; Mueller, D. M.; Lee, C. P. Photoaffinity labeling of submitochondrial membranes with the 3-azido analog of 9-amino-3-chloro-7-methoxyacridine. *Biochim. Biophys. Acta, Bioenerg.* **1985**, *807*, 177–188.

24. Torres-Pereira, J. M. G.; Sang, H. W. W. F.; Theuvenet, A. P. R.; Kraayenhof, R. Electric surface charge dynamics of chloroplast thylakoid membranes. Temperature dependence of electrokinetic potential and aminoacridine interaction. *Biochim. Biophys. Acta, Bioenerg.* **1984**, *767*, 295–303.

25. Huang, C. S.; Kopacz, S. J.; Lee, C. P. Mechanistic differences in the energy-linked fluorescence decreases of 9-aminoacridine dyes associated with bovine heart submitochondrial membranes. *Biochim. Biophys. Acta, Bioenerg.* **1983**, *722*, 107–115.

26. Dufour, J. P.; Goffeau, A.; Tsong, T. Y. Active proton uptake in lipid vesicles reconstituted with the purified yeast plasma membrane ATPase. Fluorescence quenching of 9-amino-6-chloro-2-methoxyacridine. *J. Biol. Chem.* **1982**, *257*, 9365–9371.

27. Blasco, F.; Gidrol, X. The proton-translocating ATPase of *Candida tropicalis* plasma membrane. *Biochimie* **1982**, *64*, 531–536.

28. Kraayenhof, R. Analysis of membrane architecture: fluorimetric approach. *Methods Enzymol.* **1980**, *69*, 510–520.

29. Noda, N.; Mizutani, T. Microorganism-measuring method using multiple staining. Jpn. Kokai Tokkyo Koho JP 2006340684, 2006.

30. Busto, N.; Garcia, B.; Leal, J. M.; Secco, F.; Venturini, M. The mode of binding ACMA-DNA relies on the base-pair nature. *Org. Biomol. Chem.* **2012**, *10*, 2594–2602.

31. Park, H. O.; Kim, H. B.; Chi, S. M. Detection method of DNA amplification using probes labeled with intercalating dye. PCT Int. Appl. WO 2006004267, 2006.

32. Shi, Y.; Machida, K.; Kuzuya, A.; Komiyama, M. Design of phosphoramidite monomer for optimal incorporation of functional intercalator to main chain of oligonucleotide. *Bioconjugate Chem.* **2005**, *16*, 306–311.

33. MacFarlane, D. E. Method for inhibiting immunostimulatory DNA associated responses. U.S. Pat. Appl. Publ. US 20030232856, 2003.

34. Hess, S.; Davis, W. B.; Voityuk, A. A.; Rosch, N.; Michel-Beyerle, M. E.; Ernsting, N. P.; Kovalenko, S. A.; Lustres, J. L. P. Excited-state photophysics of an acridine derivative selectively intercalated in duplex DNA. *ChemPhysChem* **2002**, *3*, 452–455.

35. McNally, A. J.; Wu, R. S.; Li, Z. Immunoassay based on DNA replication using labeled primer. U.S. Pat. Appl. Publ. US 20020072053, 2002.

36. Chin, A. M. A library of modified primers for nucleic acid sequencing, and method of use thereof. PCT Int. Appl. WO 2000028087, 2000.

37. Fukui, K.; Tanaka, K. The acridine ring selectively intercalated into a DNA helix at various types

of abasic sites: double strand formation and photophysical properties. *Nucleic Acids Res.* **1996**, *24*, 3962–3967.

38. Mergny, J. L.; Boutorine, A. S.; Garestier, T.; Belloc, F.; Rougee, M.; Bulychev, N. V.; Koshkin, A. A.; Bourson, J.; Lebedev, A. V.; Valeur, B. Fluorescence energy transfer as a probe for nucleic acid structures and sequences. *Nucleic Acids Res.* **1994**, *22*, 920–928.

39. Asakawa, M.; Endo, K.; Kobayashi, K.; Toi, H.; Aoyama, Y. Enantioselectivity in the interaction of calf thymus DNA with acridine derivatives having an amino ester or amino alcohol substituent. *Bull. Chem. Soc. Jpn.* **1992**, *65*, 2050–2055.

40. Constant, J. F.; Fkyerat, A.; Demeunynck, M.; Laval, J.; O'Connor, T. R.; Lhomme, J. Design of molecules which specifically cleave abasic sites in DNA. *Anti-Cancer Drug Des.* **1990**, *5*, 59–62.

41. Constant, J. F.; Laugaa, P.; Roques, B. P.; Lhomme, J. Heterodimeric molecules including nucleic acid bases and 9-aminoacridine. Spectroscopic studies, conformations, and interactions with DNA. *Biochemistry* **1988**, *27*, 3997–4003.

42. Verspieren, P.; Cornelissen, A. W. C. A.; Thuong, N. T.; Helene, C.; Toulme, J. J. An acridine-linked oligodeoxynucleotide targeted to the common 5' end of trypanosome mRNAs kills cultured parasites. *Gene* **1987**, *61*, 307–315.

43. Cazenave, C.; Loreau, N.; Thuong, N. T.; Toulme, J. J.; Helene, C. Enzymic amplification of translation inhibition of rabbit β-globin mRNA mediated by anti-messenger oligodeoxynucleotides covalently linked to intercalating agents. *Nucleic Acids Res.* **1987**, *15*, 4717–4736.

44. Helene, C.; Montenay-Garestier, T.; Saison-Behmoaras, T.; Toulme, J. J.; Boidot-Forget, M.; Cazenave, C.; Asseline, U.; Lancelot, G.; Maurizot, J. C. Oligodeoxynucleotides covalently linked to intercalating agents: a new family of gene regulatory substances. *Biochem. Soc. Trans.* **1986**, *14*, 201–202.

45. Toulme, J. J.; Krisch, H. M.; Loreau, N.; Thuong, N. T.; Helene, C. Specific inhibition of mRNA translation by complementary oligonucleotides covalently linked to intercalating agents. *Proc. Natl. Acad. Sci. U.S.A.* **1986**, *83*, 1227–1231.

46. Helene, C.; Montenay-Garestier, T.; Saison, T.; Takasugi, M.; Toulme, J. J.; Asseline, U.; Lancelot, G.; Maurizot, J. C.; Toulme, F.; Thuong, N. T. Oligodeoxynucleotides covalently linked to intercalating agents: a new class of gene regulatory substances. *Biochimie* **1985**, *67*, 777–783.

47. Asseline, U.; Toulme, F.; Nguyen, T. T.; Delarue, M.; Montenay-Garestier, T.; Helene, C. Oligodeoxynucleotides covalently linked to intercalating dyes as base sequence-specific ligands. Influence of dye attachment site. *EMBO J.* **1984**, *3*, 795–800.

48. Gaugain, B.; Markovits, J.; Le Pecq, J. B.; Roques, B. P. Hydrogen bonding in deoxyribonucleic acid base recognition. 1. Proton nuclear magnetic resonance studies of dinucleotide-acridine alkylamide complexes. *Biochemistry* **1981**, *20*, 3035–3042.

49. Zeleznick, L. D.; Crim, J. A.; Gray, G. D. Immunosuppression by compounds which complex with deoxyribonucleic acid. *Biochem. Pharmacol.* **1969**, *18*, 1823–1827.

50. Winter, R. W.; Kelly, J. X.; Smilkstein, M. J.; Dodean, R.; Bagby, G. C.; Rathbun, R. K.; Levin, J. I.; Hinrichs, D.; Riscoe, M. K. Evaluation and lead optimization of antimalarial acridones. *Exp. Parasitol.* **2006**, *114*, 47–56.

51. Girault, S.; Delarue, S.; Grellier, P.; Berecibar, A.; Maes, L.; Quirijnen, L.; Lemiere, P.; Debreu-Fontaine, M. A.; Sergheraert, C. Antimalarial *in-vivo* activity of *bis*(9-amino-6-chloro-2-methoxy-acridines). *J. Pharm. Pharmacol.* **2001**, *53*, 935–938.

52. Wainwright, M.; Phoenix, D. A.; Marland, J.; Wareing, D. R. A.; Bolton, F. J. A comparison of the bactericidal and photobactericidal activities of aminoacridines and *bis*(aminoacridines). *Lett. Appl. Microbiol.* **1998**, *26*, 404–406.

53. Wainwright, M.; Phoenix, D. A.; Marland, J.; Wareing, D. R. A.; Bolton, F. J. *In vitro* photobactericidal activity of aminoacridines. *J. Antimicrob. Chemother.* **1997**, *40*, 587–589.

54. Greenhalgh, N.; Hull, R.; Hurst, E. W. The antiviral activity of acridines in eastern equine encephalomyelitis, Rift Valley fever, and psittacosis in mice, and lymphogranuloma venereum in chick embryos. *Br. J. Pharmacol. Chemother.* **1956**, *11*, 220–224.

55. Prusiner, S. B.; Korth, C.; May, B. C. H. Cyclic *bis*-compounds clearing malformed proteins. U.S. Pat. Appl. Publ. US 20040229898, 2004.

56. Bousseksou, A.; Salmon, L.; Molnar, G.; Cobo, S. Materials with thermochromic spin transition doped with one or more fluorescent agents for use as temperature sensor. Fr. Demande FR 2952371, 2011.

57. Bousseksou, A.; Salmon, L.; Molnar, G.; Cobo, S. Heat-sensitive spin-transition materials doped with one or more fluorescent agents for use as temperature sensor. PCT Int. Appl. WO 2011058277, 2011.

58. Busto, N.; Garcia, B.; Leal, J. M.; Gaspar, J. F.; Martins, C.; Boggioni, A.; Secco, F. ACMA (9-amino-6-chloro-2-methoxy acridine) forms three complexes in the presence of DNA. *Phys. Chem. Chem. Phys.* **2011**, *13*, 19534–19545.

59. He, L.; Jurs, P. C.; Custer, L. L.; Durham, S. K.; Pearl, G. M. Predicting the genotoxicity of polycyclic aromatic compounds from molecular structure with different classifiers. *Chem. Res. Toxicol.* **2003**, *16*, 1567–1580.

60. Mattioni, B. E.; Kauffman, G. W.; Jurs, P. C.; Custer, L. L.; Durham, S. K.; Pearl, G. M. Predicting the genotoxicity of secondary and aromatic amines using data subsetting to generate a model ensemble. *J. Chem. Inf. Comput. Sci.* **2003**, *43*, 949–963.

61. Henry, D. R.; Lavine, B. K.; Jurs, P. C. Electronic factors and acridine frameshift mutagenicity - a pattern recognition study. *Mutat. Res.* **1987**, *179*, 115–121.

62. Ferguson, L. R.; Denny, W. A.; MacPhee, D. G. Three consistent patterns of response to substituted acridines in a variety of bacterial tester strains used for mutagenicity testing. *Mutat. Res., Genet. Toxicol. Test.* **1985**, *157*, 29–37.

63. Brown, B. R.; Firth, W. J., III,; Yielding, L. W. Acridine structure correlated with mutagenic activity in *Salmonella. Mutat. Res.* **1980**, *72*, 373–388.

64. D'Amato, F. Mutagenic activity of acridines. XXXIII-LI. *Caryologia* **1952**, *4*, 388–413.

4-AMINO-5-METHYLAMINO-2′,7′-DIFLUOROFLUORESCEIN (DAF-FM)

CAS Registry Number 254109-20-1
Chemical Structure

CA Index Name Spiro[isobenzofuran-1(3H),9′-[9H]xanthen]-3-one, 4-amino-2′,7′-difluoro-3′,6′-dihydroxy-5-(methylamino)-

Other Names 4-Amino-5-methylamino-2′,7′-difluorofluorescein; DAF-FM

Merck Index Number Not listed
Chemical/Dye Class Xanthene
Molecular Formula $C_{21}H_{14}F_2N_2O_5$
Molecular Weight 412.35
Physical Form Brown solid; White to beige powder
Solubility Soluble in dimethyl sulfoxide, methanol
Melting Point 265 °C[2]
Boiling Point (Calcd.) 693.7 ± 55.0 °C, pressure: 760 Torr
pK$_a$ (Calcd.) 8.19 ± 0.20, most acidic, temperature: 25 °C; 4.23 ± 0.20, most basic, temperature: 25 °C
Absorption (λ_{max}) 487 nm (Buffer pH 8.0)
Emission (λ_{max}) Fluorescence is very weak; 515 nm (Buffer pH 7.4) (of benzotriazole)
Molar Extinction Coefficient 84,000 cm^{-1} M^{-1} (Buffer pH 8.0)
Quantum Yield 0.005 (Buffer pH 7.4)
Synthesis Synthetic methods[1,2]
Imaging/Labeling Applications Bacteria;[3] microorganisms;[3] nitric oxide ions[1,2,4–18]
Biological/Medical Applications Detecting bacteria/microorganisms;[3] nitric oxide indicator[1,2,4–18]
Industrial Applications Not reported
Safety/Toxicity No data available

REFERENCES

1. Sabnis, R. W. *Handbook of Biological Dyes and Stains*; John Wiley & Sons Inc.: Hoboken, **2010**; pp 123–124.

2. Kojima, H.; Urano, Y.; Kikuchi, K.; Higuchi, T.; Hirata, Y.; Nagano, T. Fluorescent indicators for imaging nitric oxide production. *Angew. Chem., Int. Ed.* **1999**, *38*, 3209–3212.

3. Yoshimi, K.; Ogawa, N. Method and apparatus for detecting microorganism by fluorometry. Jpn. Kokai Tokkyo Koho JP 2003144193, 2003.

4. Cortese-Krott, M. M.; Rodriguez-Mateos, A.; Kuhnle, G. G. C.; Brown, G.; Feelisch, M.; Kelm, M. A multilevel analytical approach for detection and visualization of intracellular NO production and nitrosation events using diaminofluoresceins. *Free Radical Biol. Med.* **2012**, *53*, 2146–2158.

5. Gan, N.; Hondou, T.; Miyata, H. Spontaneous increases in the fluorescence of 4,5-diaminofluorescein and its analogs: their impact on the fluorometry of nitric oxide production in endothelial cells. *Biol. Pharm. Bull.* **2012**, *35*, 1454–1459.

6. Chen, Z.; Li, Q.; Sun, Q.; Chen, H.; Wang, X.; Li, N.; Yin, M.; Xie, Y.; Li, H.; Tang, B. Simultaneous determination of reactive oxygen and nitrogen species in mitochondrial compartments of apoptotic HepG2 cells and PC12 cells based on microchip electrophoresis-laser-induced fluorescence. *Anal. Chem.* **2012**, *84*, 4687–4694.

7. Mur, L. A. J.; Mandon, J.; Cristescu, S. M.; Harren, F. J. M.; Prats, E. Methods of nitric oxide detection in plants: A commentary. *Plant Sci.* **2011**, *181*, 509–519.

8. Schultz, G. S.; Gibson, D. J. Materials and methods for measuring nitric oxide levels in a biological fluid. PCT Int. Appl. WO 2010141719, 2010.

9. Halpin, S. T.; Spence, D. M. Direct plate-reader measurement of nitric oxide released from hypoxic

erythrocytes flowing through a microfluidic device. *Anal. Chem.* **2010**, *82*, 7492–7497.

10. Toda, T.; Kashiwagi, M.; Kosuga, J. Method and apparatus for measuring of nitrogen monoxide by fluorescent indicators. Jpn. Kokai Tokkyo Koho JP 2010078426, 2010.

11. Vitecek, J.; Reinohl, V.; Jones, R. L. Measuring NO production by plant tissues and suspension cultured cells. *Mol. Plant* **2008**, *1*, 270–284.

12. Zguris, J.; Pishko, M. V. Nitric oxide sensitive fluorescent poly(ethylene glycol) hydrogel microstructures. *Sens. Actuators, B* **2006**, *B115*, 503–509.

13. Kim, W. S.; Ye, X.; Rubakhin, S. S.; Sweedler, J. V. Measuring nitric oxide in single neurons by capillary electrophoresis with laser-induced fluorescence: Use of ascorbate oxidase in diaminofluorescein measurements. *Anal. Chem.* **2006**, *78*, 1859–1865.

14. Balcerczyk, A.; Soszynski, M.; Bartosz, G. On the specificity of 4-amino-5-methylamino-2′,7′-difluorofluorescein as a probe for nitric oxide. *Free Radical Biol. Med.* **2005**, *39*, 327–335.

15. Lacza, Z.; Snipes, J. A.; Zhang, J.; Horvath, E. M.; Figueroa, J. P.; Szabo, C.; Busija, D. W. Mitochondrial nitric oxide synthase is not eNOS, nNOS or iNOS. *Free Radical Biol. Med.* **2003**, *35*, 1217–1228.

16. Li, N.; Sul, J. Y.; Haydon, P. G. A calcium-induced calcium influx factor, nitric oxide, modulates the refilling of calcium stores in astrocytes. *J. Neurosci.* **2003**, *23*, 10302–10310.

17. Itoh, Y.; Ma, F. H.; Hoshi, H.; Oka, M.; Noda, K.; Ukai, Y.; Kojima, H.; Nagano, T.; Toda, N. Determination and bioimaging method for nitric oxide in biological specimens by diaminofluorescein fluorometry. *Anal. Biochem.* **2000**, *287*, 203–209.

18. Kojima, H.; Nagano, T. Fluorescent indicators for nitric oxide. *Adv. Mater.* **2000**, *12*, 763–765.

4-AMINO-5-METHYLAMINO-2′,7′-DIFLUOROFLUORESCEIN DIACETATE (DAF-FM DA)

CAS Registry Number   254109-22-3

Chemical Structure

CA Index Name   Spiro[isobenzofuran-1(3H),9′-[9H]-xanthen]-3-one,   3′,6′-bis(acetyloxy)-4-amino-2′,7′-difluoro-5-(methylamino)-

Other Names   4-Amino-5-methylamino-2′,7′-difluorofluorescein diacetate; DAF-FM DA

Merck Index Number   Not listed

Chemical/Dye Class   Xanthene

Molecular Formula $C_{25}H_{18}F_2N_2O_7$

Molecular Weight   496.42

Physical Form   Orange/yellow solid

Solubility   Soluble in dimethyl sulfoxide

Melting Point   135 °C[1]

Boiling Point (Calcd.)   695.7 ± 55.0 °C, pressure: 760 Torr

pK$_a$ (Calcd.)   4.20 ± 0.20, most basic, temperature: 25 °C

Absorption (λ$_{max}$)   365 nm (MeOH)

Emission (λ$_{max}$)   None

Synthesis   Synthetic method[1]

Imaging/Labeling Applications   Bacteria;[2] microorganisms;[2] nitric oxide ions[1,3–20]

Biological/Medical Applications   Detecting bacteria/microorganisms;[2] nitric oxide indicator[1,3–20]

Industrial Applications   Not reported

Safety/Toxicity   No data available

REFERENCES

1. Kojima, H.; Urano, Y.; Kikuchi, K.; Higuchi, T.; Hirata, Y.; Nagano, T. Fluorescent indicators for imaging nitric oxide production. *Angew. Chem., Int. Ed.* **1999**, *38*, 3209–3212.

2. Yoshimi, K.; Ogawa, N. Method and apparatus for detecting microorganism by fluorometry. Jpn. Kokai Tokkyo Koho JP 2003144193, 2003.

3. Wang, A.; Xian, J.; Miao, Y.; Pan, X.; Li, B.; Guo, H.; Zhang, S. Detection method of nitrogen oxide content of shrimp blood cells by flow cytometry. Faming Zhuanli Shenqing CN 103063634, 2013.

4. Duong, H. T. T.; Kamarudin, Z. M.; Erlich, R. B.; Li, Y.; Jones, M. W.; Kavallaris, M.; Boyer, C.; Davis, T. P. Intracellular nitric oxide delivery from stable NO-polymeric nanoparticle carriers. *Chem. Commun.* **2013**, *49*, 4190–4192.

5. Mainz, E. R.; Gunasekara, D. B.; Caruso, G.; Jensen, D. T.; Hulvey, M. K.; Fracassi da Silva, J. A.; Metto, E. C.; Culbertson, A. H.; Culbertson, C. T.; Lunte, S. M. Monitoring intracellular nitric oxide production using microchip electrophoresis and laser-induced fluorescence detection. *Anal. Methods* **2012**, *4*, 414–420.

6. Paul, D. M.; Vilas, S. P.; Kumar, J. M. A flow-cytometry assisted segregation of responding and non-responding population of endothelial cells for enhanced detection of intracellular nitric oxide production. *Nitric Oxide* **2011**, *25*, 31–40.

7. Takahama, U.; Hirota, S.; Kawagishi, S. Effects of pH on nitrite-induced formation of reactive nitrogen oxide species and their scavenging by phenolic antioxidants in human oral cavity. *Free Radical Res.* **2009**, *43*, 250–261.

8. Ederli, L.; Reale, L.; Madeo, L.; Ferranti, F.; Gehring, C.; Fornaciari, M.; Romano, B.; Pasqualini, S. NO release by nitric oxide donors *in vitro* and *in planta*. *Plant Physiol. Biochem.* **2009**, *47*, 42–48.

9. Namkoong, S.; Chung, B. H.; Ha, K. S.; Lee, H.; Kwon, Y. G.; Kim, Y. M. Microscopic technique for the detection of nitric oxide-dependent angiogenesis

in an animal model. *Methods Enzymol.* **2008**, *441*, 393–402.

10. Pye, D.; Palomero, J.; Kabayo, T.; Jackson, M. J. Real-time measurement of nitric oxide in single mature mouse skeletal muscle fibres during contractions. *J. Physiol.* **2007**, *581*, 309–318.

11. Gong, X.; Fu, Y.; Jiang, D.; Li, G.; Yi, X.; Peng, Y. L-Arginine is essential for conidiation in the filamentous fungus *Coniothyrium minitans*. *Fungal Genet. Biol.* **2007**, *44*, 1368–1379.

12. Lepiller, S.; Laurens, V.; Bouchot, A.; Herbomel, P.; Solary, E.; Chluba, J. Imaging of nitric oxide in a living vertebrate using a diaminofluorescein probe. *Free Radical Biol. Med.* **2007**, *43*, 619–627.

13. Carroll, J. S.; Ku, C.; Karunarathne, W.; Spence, D. M. Red blood cell stimulation of platelet nitric oxide production indicated by quantitative monitoring of the communication between cells in the bloodstream. *Anal. Chem.* **2007**, *79*, 5133–5138.

14. Ku, C.; Karunarathne, W.; Kenyon, S.; Root, P.; Spence, D. Fluorescence determination of nitric oxide production in stimulated and activated platelets. *Anal. Chem.* **2007**, *79*, 2421–2426.

15. Oblak, T. D.; Root, P.; Spence, D. M. Fluorescence monitoring of ATP-stimulated, endothelium-derived nitric oxide production in channels of a poly(dimethylsiloxane)-based microfluidic device. *Anal. Chem.* **2006**, *78*, 3193–3197.

16. Sheng, J.; Wang, D.; Braun, A. P. DAF-FM (4-amino-5-methylamino-2′,7′-difluorofluorescein) diacetate detects impairment of agonist-stimulated nitric oxide synthesis by elevated glucose in human vascular endothelial cells: Reversal by vitamin C and L-sepiapterin. *J. Pharmacol. Exp. Ther.* **2005**, *315*, 931–940.

17. Zeidler, D.; Zajringer, U.; Gerber, I.; Dubrey, I.; Hartung, T.; Bors, W.; Hutzler, P.; Durner, J. Innate immunity in *Arabidopsis thaliana*: lipopolysaccharides activate nitric oxide synthase (NOS) and induce defense genes. *Proc. Natl. Acad. Sci. U.S.A.* **2004**, *101*, 15811–15816.

18. Li, N.; Sul, J. Y.; Haydon, P. G. A calcium-induced calcium influx factor, nitric oxide, modulates the refilling of calcium stores in astrocytes. *J. Neurosci.* **2003**, *23*, 10302–10310.

19. Itoh, Y.; Ma, F. H.; Hoshi, H.; Oka, M.; Noda, K.; Ukai, Y.; Kojima, H.; Nagano, T.; Toda, N. Determination and bioimaging method for nitric oxide in biological specimens by diaminofluorescein fluorometry. *Anal. Biochem.* **2000**, *287*, 203–209.

20. Kojima, H.; Nagano, T. Fluorescent indicators for nitric oxide. *Adv. Mater.* **2000**, *12*, 763–765.

7–AMINO-4-METHYLCOUMARIN (AMC) (COUMARIN 120) (COUMARIN 440)

CAS Registry Number 26093-31-2

Chemical Structure

CA Index Name 2*H*-1-Benzopyran-2-one, 7-amino-4-methyl-

Other Names Coumarin, 7-amino-4-methyl-; (4-Methyl-2-oxo-2*H*-chromen-7-yl)amine; 1: PN: JP2013135660 PAGE: 3 claimed sequence; 4-Methyl-7-aminocoumarin; 7-AMC; 7-Amino-4-methyl-2*H*-chromen-2-one; 7-Amino-4-methylcoumarin; C 120; Coumarin 120; Coumarin 440; NSC 45796; AMC

Merck Index Number Not listed

Chemical/Dye Class Coumarin

Molecular Formula $C_{10}H_9NO_2$

Molecular Weight 175.18

Physical Form Yellow needles;[49] Yellow crystals;[1,26] Yellow solid; Brown crystals;[14] Brown solid;[12] Colorless solid;[19] Cream colored glistening needles[51]

Solubility Soluble in acetone, chloroform, *N,N*-dimethylformamide, dimethyl sulfoxide, ethanol, methanol

Melting Point 226–227 °C;[30,31,50] 226 °C;[48,51] 224–226 °C;[15,21] 223–226 °C;[4,17] 223–225 °C;[2] 223 °C;[44] 222–224 °C (decompose);[47] 222–224 °C;[6] 222–223 °C;[16] 221–225 °C;[26] 221–224 °C;[35] 221–223 °C;[19,22,28] 220–224 °C;[13,42,49] 220–222 °C;[14,20,38] 219–222 °C;[29] 219–220 °C[43]

Boiling Point (Calcd.) 378.3 ± 37.0 °C, pressure: 760 Torr

pK$_a$ (Calcd.) 1.89 ± 0.20, most basic, temperature: 25 °C

Absorption (λ_{max}) 351 nm (MeOH); 342 nm (Buffer pH 7)

Emission (λ_{max}) 430 nm (MeOH); 441 nm (Buffer pH 7)

Molar Extinction Coefficient 18,000 cm^{-1} M^{-1} (MeOH); 16,000 cm^{-1} M^{-1} (Buffer pH 7)

Synthesis Synthetic methods[1–51]

Imaging/Labeling Applications Amyloid β peptide;[52] bacteria;[53] carbohydrates/sugars;[54–63] cells;[64] cellulose nanocrystals;[65] clay mineral;[66] metal ions (copper ions,[67,68] mercury ions,[69,70] zinc ions;[71]) nitrite;[72] nucleic acids/nucleotides;[73,74] proteins;[75–78] silica nanoparticles[79,80]

Biological/Medical Applications Identifying bacteria;[53] measuring urinary kallikrein levels;[81] monitoring platinum(iv) reduction and platinum(ii) reactions in cancer cells;[82] as a substrate for measuring adipinylamidases activity,[83] aromatic amino acid decarboxylases activity,[84] histone deacetylases activity,[85–93] hydrolases/esterases activity,[94–97] β-lactamases activity,[98] microbial nitroreductases activity,[99] penicillin G acylases activity,[100,101] proteases/proteinases/peptidases activity,[26,102–135] glutamyltransferases activity,[136] tryptophanases activity;[137] as an effective antitubercular agent;[138] use as temperature sensor[139,140]

Industrial Applications For authentification;[141] dye lasers;[45,142–163,172] color display devices;[164–166] inks;[167] monitoring underlying corrosion on coated aluminum;[168] photoinitiating systems;[169] recording materials;[170] semiconductor devices;[171] dye-sensitized solar cells (DSCs);[172,173] mesoporous silica fibers;[174] SiO$_2$ sonogel hybrid materials;[175] textiles;[176] waveguides;[177,178] whitening agents[179,180]

Safety/Toxicity Mutagenicity[181]

REFERENCES

1. Behl, G.; Sikka, M.; Chhikara, A.; Chopra, M. PEG-coumarin based biocompatible self-assembled fluorescent nanoaggregates synthesized via click reactions and studies of aggregation behavior. *J. Colloid Interface Sci.* **2014**, *416*, 151–160.

2. Rafiee, E.; Fakhri, A.; Joshaghani, M. Coumarins: Facile and expeditious synthesis via Keggin-type heteropoly compounds under solvent-free condition. *J. Heterocycl. Chem.* **2013**, *50*, 1121–1128.

3. Yu, J.; Wang, Y.; Zhang, P.; Wu, J. Direct amination of phenols under metal-free conditions. *Synlett* **2013**, *24*, 1448–1454.

4. Karami, B.; Jamshidi, M.; Khodabakshi, S. Modified Paal-Knorr synthesis of novel and known

pyrroles using tungstate sulfuric acid as a recyclable catalyst. *Lett. Org. Chem.* **2013**, *10*, 12–16.

5. Sharma, R. K.; Monga, Y.; Puri, A. Zirconium(IV)-modified silica@magnetic nanocomposites: Fabrication, characterization and application as efficient, selective and reusable nanocatalysts for Friedel-Crafts, Knoevenagel and Pechmann condensation reactions. *Catal. Commun.* **2013**, *35*, 110–114.

6. Rahmatpour, A.; Mohammadian, S. An environmentally friendly, chemoselective, and efficient protocol for the preparation of coumarin derivatives by Pechman condensation reaction using new and reusable heterogeneous Lewis acid catalyst polystyrene-supported $GaCl_3$. *C. R. Chim.* **2013**, *16*, 271–278.

7. Khaligh, N. G. Ultrasound-assisted one-pot synthesis of substituted coumarins catalyzed by poly(4-vinylpyridinium) hydrogen sulfate as an efficient and reusable solid acid catalyst. *Ultrason. Sonochem.* **2013**, *20*, 1062–1068.

8. Khaligh, N. G.; Shirini, F. Introduction of poly(4-vinylpyridinium) perchlorate as a new, efficient, and versatile solid acid catalyst for one-pot synthesis of substituted coumarins under ultrasonic irradiation. *Ultrason. Sonochem.* **2013**, *20*, 26–31.

9. Song, S. J.; Liu, W. H.; Ma, J. J.; Qi, J. M. [HSO_3-pmim][CH_3SO_3] catalyzed one-pot protocol for the synthesis of coumarins under solvent-free conditions. *Asian J. Chem.* **2012**, *24*, 4433–4435.

10. Khaligh, N. G. Synthesis of coumarins via Pechmann reaction catalyzed by 3-methyl-1-sulfonic acid imidazolium hydrogen sulfate as an efficient, halogen-free and reusable acidic ionic liquid. *Catal. Sci. Technol.* **2012**, *2*, 1633–1636.

11. Vadola, P. A.; Sames, D. Catalytic coupling of arene C-H bonds and alkynes for the synthesis of coumarins: Substrate scope and application to the development of neuroimaging agents. *J. Org. Chem.* **2012**, *77*, 7804–7814.

12. van Kalkeren, H. A.; Bruins, J. J.; Rutjes, F. P. J. T.; van Delft, F. L. Organophosphorus-catalyzed Staudinger reduction. *Adv. Synth. Catal.* **2012**, *354*, 1417–1421.

13. Soleimani, E.; Khodaei, M. M.; Batooie, N.; Samadi, S. Tetrakis(acetonitrile)copper(I) hexafluorophosphate catalyzed coumarin synthesis via Pechmann condensation under solvent-free condition. *J. Heterocycl. Chem.* **2012**, *49*, 409–412.

14. Kathuria, A.; Priya, N.; Chand, K.; Singh, P.; Gupta, A.; Jalal, S.; Gupta, S.; Raj, H. G.; Sharma, S. K. Substrate specificity of acetoxy derivatives of coumarins and quinolones towards calreticulin mediated transacetylation: Investigations on antiplatelet function. *Bioorg. Med. Chem.* **2012**, *20*, 1624–1638.

15. Al-Rifai, A. A.; Ayoub, M. T.; Shakya, A. K.; Abu Safieh, K. A.; Mubarak, M. S. Synthesis, characterization, and antimicrobial activity of some new coumarin derivatives. *Med. Chem. Res.* **2012**, *21*, 468–476.

16. Borah, K. J.; Borah, R. Poly(4-vinyl-pyridine)-supported sulfuric acid, an efficient solid acid catalyst for the synthesis of coumarin derivatives under solvent-free conditions. *Monatsh. Chem.* **2011**, *142*, 1253–1257.

17. Karami, B.; Kiani, M. $ZrOCl_2.8H_2O/SiO_2$: An efficient and recyclable catalyst for the preparation of coumarin derivatives by Pechmann condensation reaction. *Catal. Commun.* **2011**, *14*, 62–67.

18. Karimi, B.; Behzadnia, H. Periodic mesoporous silica chloride (PMSCl) as an efficient and recyclable catalyst for the Pechmann reaction. *Catal. Commun.* **2011**, *12*, 1432–1436.

19. Montazeri, N.; Khaksar, S.; Nazari, A.; Alavi, S. S.; Vahdat, S. M.; Tajbakhsh, M. Pentafluorophenylammonium triflate (PFPAT): An efficient, metal-free and reusable catalyst for the von Pechmann reaction. *J. Fluorine Chem.* **2011**, *132*, 450–452.

20. Niralwad, K. S.; Shingate, B. B.; Shingare, M. S. Microwave-induced one-pot synthesis of coumarins using potassium dihydrogen phosphate as a catalyst under solvent-free condition. *J. Korean Chem. Soc.* **2011**, *55*, 486–489.

21. Mustafa, M. S.; El-Abadelah, M. M.; Zihlif, M. A.; Naffa, R. G.; Mubarak, M. S. Synthesis, and antitumor activity of some N1-(7-coumarinyl)amidrazones and related congeners. *Molecules* **2011**, *16*, 4305–4317.

22. Gao, S.; Li, C.; Wang, Y.; Ma, J.; Wang, C.; Zhang, J. $NbCl_5$-catalyzed, solvent-free, one-pot synthesis of coumarins. *Synth. Commun.* **2011**, *41*, 1486–1491.

23. Sinhamahapatra, A.; Sutradhar, N.; Pahari, S.; Bajaj, H. C.; Panda, A. B. Mesoporous zirconium phosphate: An efficient catalyst for the synthesis of coumarin derivatives through Pechmann condensation reaction. *Appl. Catal., A: Gen.* **2011**, *394*, 93–100.

24. Zhang, Y.; Zou, B.; Wang, K. Microwave-assisted synthesis of 4-methyl-coumarin derivatives. *Huaxue Shiji* **2010**, *32*, 787–789, 831.

25. Zhang, S. Method for preparing laser dye 7-amino-4-methylcoumarin. Faming Zhuanli Shenqing Gongkai Shuomingshu CN 101712666, 2010.

26. Reszka, P.; Schulz, R.; Methling, K.; Lalk, M.; Bednarski, P. J. Synthesis, enzymatic evaluation, and docking studies of fluorogenic caspase 8 tetrapeptide substrates. *ChemMedChem* **2010**, *5*, 103–117.

27. Pawar, O. B.; Chavan, F. R.; Shinde, N. D. EPZ 10 as a reusable heterogeneous catalyst for the synthesis of coumarins. *Org. Chem.: Indian J.* **2009**, *5*, 449–451.

28. Mandhane, P. G.; Joshi, R. S.; Ghawalkar, A. R.; Jadhav, G. R.; Gill, C. H. Ammonium metavanadate: a mild and efficient catalyst for the synthesis of coumarins. *Bull. Korean Chem. Soc.* **2009**, *30*, 2969–2972.

29. Goswami, P. Dually activated organo- and nano-cocatalyzed synthesis of coumarin derivatives. *Synth. Commun.* **2009**, *39*, 2271–2278.

30. Ronad, P. M.; Hunashal, R. D.; Darbhamalla, S.; Maddi, V. S. Synthesis and evaluation of anti-inflammatory and analgesic activities of a novel series of substituted-N-(4-methyl-2-oxo-2H-chromen-7-yl)benzamides. *Arzneim. Forsch.* **2008**, *58*, 641–646.

31. Ronad, P.; Dharbamalla, S.; Hunshal, R.; Maddi, V. Synthesis of novel substituted 7-(benzylideneamino)-4-methyl-2H-chromen-2-one derivatives as anti-inflammatory and analgesic agents. *Arch. Pharm.* **2008**, *341*, 696–700.

32. Yang, J.; Ji, C.; Zhao, Y. NaHSO$_4$•H$_2$O catalyzed synthesis of coumarin derivatives under solvent-free conditions. *Youji Huaxue* **2008**, *28*, 1740–1743.

33. Karimi, B.; Zareyee, D. Design of a highly efficient and water-tolerant sulfonic acid nanoreactor based on tunable ordered porous silica for the von Pechmann reaction. *Org. Lett.* **2008**, *10*, 3989–3992.

34. Prajapati, D.; Gohain, M. A novel process for the production of 4-methylcoumarin derivatives. Indian Pat. Appl. IN 2003DE01481, 2008.

35. Prajapati, D.; Gohain, M. Iodine a simple, effective and inexpensive catalyst for the synthesis of substituted coumarins. *Catal. Lett.* **2007**, *119*, 59–63.

36. Tyagi, B.; Mishra, M. K.; Jasra, R. V. Synthesis of 7-substituted 4-methyl coumarins by Pechmann reaction using nano-crystalline sulfated-zirconia. *J. Mol. Catal. A: Chem.* **2007**, *276*, 47–56.

37. Prajapati, D.; Boruah, R. C.; Gohain, M. An improved process for the preparation of substituted 4-methylcoumarins. Indian Pat. Appl. IN 2005DE00536, 2007.

38. Sun, P.; Hu, Z. Gallium triiodide-catalyzed organic reaction. A convenient procedure for the synthesis of coumarins. *Synth. Commun.* **2005**, *35*, 1875–1880.

39. De, S. K.; Gibbs, R. A. An efficient and practical procedure for the synthesis of 4-substituted coumarins. *Synthesis* **2005**, 1231–1233.

40. Feng, G.; Yang, M.; Li, T.; Jin, T. A solvent-free procedure for synthesis of coumarins catalyzed by zirconium sulfate tetrahydrate-silica gel under microwave irradiation. *Chem.: Indian J.* **2003**, *1*, 104–108.

41. Williams, A. C.; Camp, N. Product class 4: benzopyranones and benzopyranthiones. *Sci. Synth.* **2003**, *14*, 347–638.

42. Frere, S.; Thiery, V.; Besson, T. Microwave acceleration of the Pechmann reaction on graphite/montmorillonite K10: application to the preparation of 4-substituted 7-aminocoumarins. *Tetrahedron Lett.* **2001**, *42*, 2791–2794.

43. Pozdnev, V. F. An improved method of synthesis of 7-amino-4-methylcoumarin. *Khim. Geterotsikl. Soedin.* **1990**, 312–314.

44. Kirpichenok, M. A.; Mel'nikova, L. M.; Denisov, L. K.; Grandberg, I. I. Photochemical reactions of 7-aminocoumarins. 4. 4-Methyl-7-diethylaminocoumarin reactions with photolytically labile compounds. *Khim. Geterotsikl. Soedin.* **1989**, 460–469.

45. Li, S.; Gao, Y. Series of coumarin laser dyes. *Kexue Tongbao* **1988**, *33*, 1617–1623.

46. Pozdnev, V. F. 7-Amino-4-methylcoumarin. U.S.S.R. SU 1325050, 1987.

47. Strakova, I.; Tetere, Z.; Rijkure, I.; Zicane, D.; Ravina, I.; Gudriniece, E. Enzyme substrates. IV. Synthesis of 7-amino-4-methylcoumarin. *Latv. PSR Zinatnu Akad. Vestis, Kim. Ser.* **1985**, 621–624.

48. Krejcoves, J.; Drobnik, J.; Jokl, J.; Kalal, J. The preparation and characterization of some novel fluorescence labels derived from 7-substituted 2H-1-benzopyran-2-ones. *Collect. Czech. Chem. Commun.* **1979**, *44*, 2211–2220.

49. Atkins, R. L.; Bliss, D. E. Substituted coumarins and azacoumarins. Synthesis and fluorescent properties. *J. Org. Chem.* **1978**, *43*, 1975–1980.

50. Pretka, J. E. 4-Substituted-7-carbalkoxyaminocoumarins. U.S. Patent 3008969, 1961.

51. Rao, N. V. S.; Sundaramurthy, V. Search for physiologically active compounds. II. Synthesis of 7-amino and 7-halo-4-methylcoumarins. *Proc. Indian Acad. Sci., Sect. A* **1956**, *43A*, 149–151.

52. Taniguchi, A.; Skwarczynski, M.; Sohma, Y.; Okada, T.; Ikeda, K.; Prakash, H.; Mukai, H.; Hayashi, Y.; Kimura, T.; Hirota, S.; Matsuzaki, K.; Kiso, Y. Controlled production of amyloid β peptide from a phototriggered, water-soluble precursor "click peptide". *ChemBioChem* **2008**, *9*, 3055–3065.

53. Godsey, J. H.; Matteo, M. R.; Shen, D.; Tolman, G.; Gohlke, J. R. Rapid identification of Enterobacteriaceae with microbial enzyme activity profiles. *J. Clin. Microbiol.* **1981**, *13*, 483–490.

54. Abe, A.; Shimaoka, H. Monosaccharide analysis sample production method. Jpn. Kokai Tokkyo Koho JP 2013076649, 2013.

55. Abe, K.; Shimaoka, H.; Abe, A.; Aihara, O. Method for determining sugar chains using substrate having sugar chains immobilized thereon. Jpn. Kokai Tokkyo Koho JP 2012211817, 2012.

56. Yodoshi, M.; Tani, A.; Ohta, Y.; Suzuki, S. Optimized conditions for high-performance liquid chromatography analysis of oligosaccharides using 7-amino-4-methylcoumarin as a reductive amination reagent. *J. Chromatogr., A* **2008**, *1203*, 137–145.

57. Wilson, J. J.; Brodbelt, J. S. Ultraviolet photodissociation at 355 nm of fluorescently labeled oligosaccharides. *Anal. Chem.* **2008**, *80*, 5186–5196.

58. Miyamura, T.; Burke, T. J.; Bolger, R. E.; Kato, I. Method for measuring interaction between sugar and target using fluorescent-labeled sugars. Eur. Pat. Appl. EP 867722, 1998.

59. Lakowicz, J. R.; Maliwal, B. Optical sensing of glucose using phase-modulation fluorimetry. *Anal. Chim. Acta* **1992**, *271*, 155–164.

60. Khorlin, A. Y.; Shiyan, S. D.; Nasonov, V. V.; Markin, V. A. Fluorescent derivatives of carbohydrates in structural studies of glycoconjugates. N-(4-Methylcoumarin-7-yl) glycamines in studying asparagine-linked carbohydrate chains of glycoproteins. *Bioorg. Khim.* **1987**, *13*, 1266–1274.

61. Khorlin, A. Y.; Shiyan, S. D.; Markin, V. A.; Nasonov, V. V.; Mirzayanova, M. N. Fluorescent derivatives of carbohydrates in struct-ural studies of glycoconjugates. N-(4-methyl-7-coumarinyl)glycamines: synthesis, characterization, and use in carbohydrate analysis. *Bioorg. Khim.* **1986**, *12*, 1203–1212.

62. Prakash, C.; Vijay, I. K. A new fluorescent tag for labeling of saccharides. *Anal. Biochem.* **1983**, *128*, 41–46.

63. Yalpani, M.; Hall, L. D. Synthesis of fluorescent probe - carbohydrate conjugates. *Can. J. Chem.* **1981**, *59*, 2934–2939.

64. Pastor, M. V. D. Direct immunofluorescent labeling of cells. *Methods Mol. Biol.* **2010**, *588*, 135–142.

65. Huang, J.; Li, C.; Gray, D. G. Cellulose nanocrystals incorporating fluorescent methylcoumarin groups. *ACS Sustainable Chem. Eng.* **2013**, *1*, 1160–1164.

66. Windsor, S. A.; Harrison, N. J.; Tinker, M. H. Electro-fluorescence studies of the binding of fluorescent dyes to sepiolite. *Clay Miner.* **1996**, *31*, 81–94.

67. Wang, M.; Huang, S.; Meng, X.; Zhu, M.; Guo, Q. Coumarin-coupled receptor as a membrane-permeable, Cu2+-selective fluorescent chemosensor for imaging copper(II) in HEPG-2 cell. *Chem. Lett.* **2008**, *37*, 462–463.

68. Wang, M. X.; Meng, X. M.; Zhu, M. Z.; Guo, Q. X. A novel selective fluorescent chemosensor for Cu(II). *Chin. Chem. Lett.* **2007**, *18*, 1403–1406.

69. Voutsadaki, S.; Tsikalas, G. K.; Klontzas, E.; Froudakis, G. E.; Pergantis, S. A.; Demadis, K. D.; Katerinopoulos, H. E. A cyclam-type "turn on" fluorescent sensor selective for mercury ions in aqueous media. *RSC Adv.* **2012**, *2*, 12679–12682.

70. Shiraishi, Y.; Sumiya, S.; Hirai, T. A coumarin-thiourea conjugate as a fluorescent probe for Hg(II) in aqueous media with a broad pH range 2-12. *Org. Biomol. Chem.* **2010**, *8*, 1310–1314.

71. Dakanali, M.; Roussakis, E.; Kay, A. R.; Katerinopoulos, H. E. Synthesis and photophysical properties of a fluorescent TREN-type ligand incorporating the coumarin chromophore and its zinc complex. *Tetrahedron Lett.* **2005**, *46*, 4193–4196.

72. Diallo, S.; Bastard, P.; Prognon, P.; Dauphin, C.; Hamon, M. A new spectrofluorometric microdetermination of nitrite in water after derivatization with 4-methyl-7-aminocoumarin. *Talanta* **1996**, *43*, 359–364.

73. Lerga, T. M.; O'Sullivan, C. K. Rapid determination of total hardness in water using fluorescent

molecular aptamer beacon. *Anal. Chim. Acta* **2008**, *610*, 105–111.

74. Chen, Y.; Lu, Z. Dye sensitized luminescent europium nanoparticles and its time-resolved fluorometric assay for DNA. *Anal. Chim. Acta* **2007**, *587*, 180–186.

75. Davies, K. J. A.; Pickering, A. M. Labeling of proteins with the fluorophore, 7-amino-4-methylcoumarin (AMC), generated novel proteolytic substrates. U.S. Pat. Appl. Publ. US 20130059321, 2013.

76. Pickering, A. M.; Davies, K. J. A. A simple fluorescence labeling method for studies of protein oxidation, protein modification, and proteolysis. *Free Radical Biol. Med.* **2012**, *52*, 239–246.

77. Shobini, J.; Mishra, A. K.; Sandhya, K.; Chandra, N. Interaction of coumarin derivatives with human serum albumin: investigation by fluorescence spectroscopic technique and modeling studies. *Spectrochim. Acta, Part A: Mol. Biomol. Spectrosc.* **2001**, *57A*, 1133–1147.

78. Forster, Y.; Haas, E. Preparation and characterization of three fluorescent labels for proteins, suitable for structural studies. *Anal. Biochem.* **1993**, *209*, 9–14.

79. Rosen, J. E.; Jones, L.; Gu, F. X. Light-induced aggregation of nanoparticles functionalized with 7-amino-4-methylcoumarin. *Nano LIFE* **2012**, *2*, 1241007/1–1241007/8.

80. Liu, Q.; De Shong, P.; Zachariah, M. R. One-step synthesis of dye-incorporated porous silica particles. *J. Nanopart. Res.* **2012**, *14*, 923/1–923/8.

81. Iwanaga, S.; Kato, H.; Sakakibara, S. Determination of kallikrein in urine. Jpn. Kokai Tokkyo Koho JP 55036759, 1980.

82. New, E. J.; Duan, R.; Zhang, J. Z.; Hambley, T. W. Investigations using fluorescent ligands to monitor platinum(iv) reduction and platinum(ii) reactions in cancer cells. *Dalton Trans.* **2009**, 3092–3101.

83. Aretz, W.; Hedtmann, U. Preparation of coumarin derivative as substrate for determining α-adipinylamidase activity. Ger. Offen. DE 3834597, 1990.

84. Tang, L.; Frank, G. Using fluorimetric assays for detecting aromatic amino acid decarboxylase activity *in vitro*. *J. Biochem., Mol. Biol. Biophys.* **2001**, *5*, 45–54.

85. Dose, A.; Jost, J. O.; Spiess, A. C.; Henklein, P.; Beyermann, M.; Schwarzer, D. Facile synthesis of colorimetric histone deacetylase substrates. *Chem. Commun.* **2012**, *48*, 9525–9527.

86. Singh, R. K.; Mandal, T.; Balsubramanian, N.; Viaene, T.; Leedahl, T.; Sule, N.; Cook, G.; Srivastava, D. K. Histone deacetylase activators: N-acetylthioureas serve as highly potent and isozyme selective activators for human histone deacetylase-8 on a fluorescent substrate. *Bioorg. Med. Chem. Lett.* **2011**, *21*, 5920–5923.

87. Riester, D.; Hildmann, C.; Gruenewald, S.; Beckers, T.; Schwienhorst, A. Factors affecting the substrate specificity of histone deacetylases. *Biochem. Biophys. Res. Commun.* **2007**, *357*, 439–445.

88. Heltweg, B.; Dequiedt, F.; Marshall, B. L.; Brauch, C.; Yoshida, M.; Nishino, N.; Verdin, E.; Jung, M. Subtype selective substrates for histone deacetylases. *J. Med. Chem.* **2004**, *47*, 5235–5243.

89. Wegener, D.; Hildmann, C.; Riester, D.; Schwienhorst, A. Improved fluorogenic histone deacetylase assay for high-throughput-screening applications. *Anal. Biochem.* **2003**, *321*, 202–208.

90. Yoshida, M.; Nishino, N. Non-isotopic histone deacetylase activity assay using fluorescent or chromogenic ε-acetyl lysine peptide derivative substrate. Jpn. Kokai Tokkyo Koho JP 2003221398, 2003.

91. Heltweg, B.; Dequiedt, F.; Verdin, E.; Jung, M. Nonisotopic substrate for assaying both human zinc and NAD+-dependent histone deacetylases. *Anal. Biochem.* **2003**, *319*, 42–48.

92. Wegener, D.; Wirsching, F.; Riester, D.; Schwienhorst, A. A fluorogenic histone deacetylase assay well suited for high-throughputa screening. *Chem. Biol.* **2003**, *10*, 61–68.

93. Tamai, K.; Miyazaki, T.; Wada, E.; Tatsuzawa, A. Non-isotopic histone deacetylase activity assay using ε-acetyl lysine peptide aminocoumarin derivative substrate. Jpn. Kokai Tokkyo Koho JP 2001149081, 2001.

94. Kage, K. L.; Richardson, P. L.; Traphagen, L.; Severin, J.; Pereda-Lopez, A.; Lubben, T.; Davis-Taber, R.; Vos, M. H.; Bartley, D.; Walter, K.; Harlan, J.; Solomon, L.; Warrior, U.; Holzman, T. F.; Faltynek, C.; Surowy, C. S.; Scott, V. E. A high throughput fluorescent assay for measuring the activity of fatty acid amide hydrolase. *J. Neurosci. Methods* **2007**, *161*, 47–54.

95. Cummins, I.; Landrum, M.; Steel, P. G.; Edwards, R. Structure activity studies with xenobiotic substrates using carboxylesterases isolated from *Arabidopsis thaliana*. *Phytochemistry* **2007**, *68*, 811–818.

96. Weder, J. K. P.; Kaiser, K. Fluorogenic substrates for hydrolase detection following electrophoresis. *J. Chromatogr., A* **1995**, *698*, 181–201.

97. Goodfellow, M.; Thomas, E. G.; James, A. L. Characterization of rhodococci using peptide hydrolase substrates based on 7-amino-4-methylcoumarin. *FEMS Microbiol. Lett.* **1987**, *44*, 349–355.

98. Quante, M. J.; Hoke, R. A.; Mize, P. D.; Woodward, D. L.; Millner, E. O. Preparation of fluorogenic and chromogenic β-lactamase substrates. Eur. Pat. Appl. EP 553741, 1993.

99. James, A. L.; Perry, J. D.; Jay, C.; Monget, D.; Rasburn, J. W.; Gould, F. K. Fluorogenic substrates for the detection of microbial nitroreductases. *Lett. Appl. Microbiol.* **2001**, *33*, 403–408.

100. Ninkovic, M.; Riester, D.; Wirsching, F.; Dietrich, R.; Schwienhorst, A. Fluorogenic assay for penicillin G acylase activity. *Anal. Biochem.* **2001**, *292*, 228–233.

101. Scheper, T.; Weiss, M.; Schuegerl, K. Two new fluorogenic substrates for the detection of penicillin-G-acylase activity. *Anal. Chim. Acta* **1986**, *182*, 203–206.

102. Jang, S. H.; Lee, C. U.; Choi, M. G. Synthetic fluorogenic substrate for measuring activity of dipeptidyl peptidase IV. Repub. Korean Kongkae Taeho Kongbo KR 2012092360, 2012.

103. Orcutt, S. J.; Wu, J.; Eddins, M. J.; Leach, C. A.; Strickler, J. E. Bioluminescence assay platform for selective and sensitive detection of Ub/Ubl proteases. *Biochim. Biophys. Acta, Mol. Cell Res.* **2012**, *1823*, 2079–2086.

104. Mustafa, M. S.; El-Abadelah, M. M.; Mubarak, M. S.; Chibueze, I.; Shao, D.; Agu, R. U. Synthesis and fluorogenic properties of some 1-(coumarin-7-yl)-4,5-dihydro-1,2,4-triazin-6(1H)-ones. *Int. J. Chem.* **2011**, *3*, 89–103.

105. Aoki, N.; Omura, T.; Takahashi, Y. Intestine protease and its use. Jpn. Kokai Tokkyo Koho JP 2010024143, 2010.

106. Agu, R. U.; Obimah, D. U.; Lyzenga, W. J.; Jorissen, M.; Massoud, E.; Verbeke, N. Specific aminopeptidases of excised human nasal epithelium and primary culture: a comparison of functional characteristics and gene transcripts expression. *J. Pharm. Pharmacol.* **2009**, *61*, 599–606.

107. Meier, H. J.; Hofer, H. W. Assays for cellular proteinases using sorptive removal of endogenous inhibitors. Ger. Offen. DE 102008036329, 2009.

108. Hassiepen, U.; Eidhoff, U.; Meder, G.; Bulber, J.; Hein, A.; Bodendorf, U.; Lorthiois, E.; Martoglio, B. A sensitive fluorescence intensity assay for deubiquitinating proteases using ubiquitin-rhodamine110-glycine as substrate. *Anal. Biochem.* **2007**, *371*, 201–207.

109. Hennig, A.; Roth, D.; Enderle, T.; Nau, W. M. Nanosecond time-resolved fluorescence protease assays. *ChemBioChem* **2006**, *7*, 733–737.

110. Tran, T. V.; Ellis, K. A.; Kam, C.; Hudig, D.; Powers, J. C. Dipeptidyl peptidase I: importance of progranzyme activation sequences, other dipeptide sequences, and the N-terminal amino group of synthetic substrates for enzyme activity. *Arch. Biochem. Biophys.* **2002**, *403*, 160–170.

111. Sheppeck, J. E., II,; Kar, H.; Gosink, L.; Wheatley, J. B.; Gjerstad, E.; Loftus, S. M.; Zubiria, A. R.; Janc, J. W. Synthesis of a statistically exhaustive fluorescent peptide substrate library for profiling protease specificity. *Bioorg. Med. Chem. Lett.* **2000**, *10*, 2639–2642.

112. Mack, A.; Furmann, C.; Hacker, G. Detection of caspase-activation in intact lymphoid cells using standard caspase substrates and inhibitors. *J. Immunol. Methods* **2000**, *241*, 19–31.

113. Lee, D.; Adams, J. L.; Brandt, M.; DeWolf, W. E., Jr.,; Keller, P. M.; Levy, M. A. A substrate combinatorial array for caspases. *Bioorg. Med. Chem. Lett.* **1999**, *9*, 1667–1672.

114. Yokoi, S.; Shigyo, T.; Tamaki, T. A fluorometric assay for proteinase A in beer and its application for the investigation of enzymic effects on foam stability. *J. Inst. Brew.* **1996**, *102*, 33–37.

115. Midwinter, R. G.; Pritchard, G. G. Aminopeptidase N from *Streptococcus salivarius* subsp. thermophilus NCDO 573: Purification and properties. *J. Appl. Bacteriol.* **1994**, *77*, 288–295.

116. Sato, E.; Matsuhisa, A.; Sakashita, M.; Kanaoka, Y. Organic fluorescent reagents. Part XIII. New water-soluble fluorogenic amine. 7-Amino-coumarin-4-methanesulfonic acid (ACMS) and related substrates for proteinases. *Chem. Pharm. Bull.* **1988**, *36*, 3496–3502.

117. Alekseenko, L. P.; Pozdnev, V. F.; Orekhovich, V. N. Cystylaminopeptidase from human erythrocytes. *Dokl. Akad. Nauk SSSR* **1987**, *293*, 728–731.

118. Nagatsu, T.; Harada, M.; Sakakibara, S. Measurment of dipeptidyl aminopetidase II activity. Jpn. Kokai Tokkyo Koho JP 61056100, 1986.

119. Kokotos, G.; Tzougraki, C. Fluorogenic substrates for chymotrypsin with new fluorescent markers. *Int. J. Pept. Protein Res.* **1986**, *28*, 186–191.

120. Kanaoka, Y.; Takahashi, T.; Nakayama, H.; Tanizawa, K. New fluorogenic substrates for subtilisin. *Chem. Pharm. Bull.* **1985**, *33*, 1721–1724.

121. Sinha, P.; Gossrau, R. Isoelectric focusing (IEF) and band detection with fluorogenic protease substrates. *Histochemistry* **1984**, *81*, 167–169.

122. Sinha, P.; Gossrau, R.; Smith, R. E.; Lojda, Z. Fluorescence detection of proteases with AFC, AMC and MNA peptides using isoelectric focusing. *Adv. Exp. Med. Biol.* **1984**, *167*, 219–226.

123. Kanaoka, Y.; Takahashi, T.; Nakayama, H.; Ueno, T.; Sekine, T. Organic fluorescence reagents. 8. Synthesis of a new fluorogenic substrate for cystine aminopeptidase. *Chem. Pharm. Bull.* **1982**, *30*, 1485–1487.

124. Ashe, B. M.; Zimmerman, M. Fluorogenic substrates for human leukocyte and porcine pancreatic elastase. *J. Appl. Biochem.* **1980**, *2*, 445–447.

125. Yoshimoto, T.; Ogita, K.; Walter, R.; Koida, M.; Tsuru, D. Post-proline cleaving enzyme. Synthesis of a new fluorogenic substrate and distribution of the endopeptidase in rat tissues and body fluids of man. *Biochim. Biophys. Acta, Enzymol.* **1979**, *569*, 184–192.

126. Kojima, K.; Kinoshita, H.; Kato, T.; Nagatsu, T.; Takada, K.; Sakakibara, S. A new and highly sensitive fluorescence assay for collagenase-like peptidase activity. *Anal. Biochem.* **1979**, *100*, 43–50.

127. Sakakibara, S. Peptide derivatives of 4-methyl-coumarin. Jpn. Kokai Tokkyo Koho JP 54003074, 1979.

128. Sakakibara, S. Peptide derivatives of 4-methyl-coumarin. Jpn. Kokai Tokkyo Koho JP 54003073, 1979.

129. Sakakibara, S. Peptide derivatives of 4-methyl-coumarin. Jpn. Kokai Tokkyo Koho JP 54003072, 1979.

130. Fujiwara, K.; Tsuru, D. New chromogenic and fluorogenic substrates for pyrrolidonyl peptidase. *J. Biochem.* **1978**, *83*, 1145–1149.

131. Nagatsu, T.; Sakakibara, S. Dipeptide derivatives and their use in measuring enzyme activity. Ger. Offen. DE 2808111, 1978.

132. Kato, T.; Nagatsu, T.; Kimura, T.; Sakakibara, S. Fluorescence assay of X-prolyl dipeptidyl-aminopeptidase activity with a new fluorogenic substrate. *Biochem. Med.* **1978**, *19*, 351–359.

133. Kanaoka, Y.; Takahashi, T.; Nakayama, H. Organic fluorescence reagents. III. A new fluorogenic substrate for aminopeptidase. *Chem. Pharm. Bull.* **1977**, *25*, 362–363.

134. Kanaoka, Y.; Takahashi, T.; Nakayama, H.; Takada, K.; Kimura, T.; Sakakibara, S. Organic fluorescence reagent. IV. Synthesis of a key fluorogenic amide, L-arginine-4-methylcoumaryl-7-amide (L-Arg-MCA) and its derivatives. Fluorescence assays for trypsin and papain. *Chem. Pharm. Bull.* **1977**, *25*, 3126-3128.

135. Zimmerman, M.; Yurewicz, E.; Patel, G. A new fluorogenic substrate for chymotrypsin. *Anal. Biochem.* **1976**, *70*, 258–262.

136. Prusak, E.; Siewinski, M.; Szewczuk, A. A new fluorimetric method for the determination of γ-glutamyltransferase activity in blood serum. *Clin. Chim. Acta* **1980**, *107*, 21-26.

137. Linn, C. P.; Mize, P. D.; Hoke, R. A.; Quante, J. M.; Pitner, J. B. Synthesis of serine-AMC-carbamate: a fluorogenic tryptophanase substrate. *Anal. Biochem.* **1992**, *200*, 400–404.

138. Tandon, R.; Ponnan, P.; Aggarwal, N.; Pathak, R.; Baghel, A. S.; Gupta, G.; Arya, A.; Nath, M.; Parmar, V. S.; Raj, H. G.; Prasad, A. K.; Bose, M. Characterization of 7-amino-4-methylcoumarin as an effective antitubercular agent: structure-activity relationships. *J. Antimicrob. Chemother.* **2011**, *66*, 2543–2555.

139. Bousseksou, A.; Salmon, L.; Molnar, G.; Cobo, S. Materials with thermochromic spin transition doped with one or more fluorescent agents for use as temperature sensor. Fr. Demande FR 2952371, 2011.

140. Bousseksou, A.; Salmon, L.; Molnar, G.; Cobo, S. Heat-sensitive spin-transition materials doped with one or more fluorescent agents for use as temperature sensor. PCT Int. Appl. WO 2011058277, 2011.

141. Isler, U.; Hoehener, K.; Meier, W.; Poux, S. Procedure for the treatment of yarn with encapsulated marking substance for authentification and device for proof of authenticity. Ger. Offen. DE 102005047786, 2007.

142. Yang, Y.; Lin, G.; Xu, H.; Cui, Y.; Wang, Z.; Qian, G. Energy transfer mechanisms among various laser dyes co-doped into gel glasses. *Dyes Pigm.* **2013**, *96*, 242–248.

143. Fan, R. W.; Jiang, Y. G.; Xia, Y. Q.; Chen, D. Y. High pulse energy output from tunable solid-state

dye laser based on MPMMA doped with PM5671. *Laser Phys.* **2011**, *21*, 652–655.

144. Yang, Y.; Zou, J.; Rong, H.; Qian, G. D.; Wang, Z. Y.; Wang, M. Q. Influence of various coumarin dyes on the laser performance of laser dyes co-doped into ORMOSILs. *Appl. Phys. B: Lasers Opt.* **2007**, *86*, 309–313.

145. Ghazy, R.; Zim, S. A.; Shaheen, M.; El-Mekawey, F. Experimental investigations on energy-transfer characteristics and performance of some laser dye mixtures. *Opt. Laser Technol.* **2002**, *34*, 99–105.

146. Nemkovich, N. A.; Reis, H.; Baumann, W. Ground and excited state dipole moments of coumarin laser dyes: Investigation by electrooptical absorption and emission methods. *J. Luminesc.* **1997**, *71*, 255–263.

147. Kunjappu, J. T. Photophysical properties of five laser dyes (C120, C1, C102, C1F and C153) in homogeneous, surfactant and membrane media. *J. Photochem. Photobiol., A: Chem.* **1993**, *71*, 269–273.

148. Kunjappu, J. T.; Rao, K. N. Photodegradation of laser dyes: studies on 7-amino-4-methylcoumarin (C120). *Indian J. Chem., Sect. A* **1987**, *26A*, 453–457.

149. Mukherjee, T.; Rao, K. N.; Mittal, J. P. Photodecomposition of coumarin laser dyes in different solvent systems. *Indian J. Chem., Sect. A* **1986**, *25A*, 993–1000.

150. Jasny, J. Novel method for wavelength tuning of distributed feedback dye lasers. *Opt. Commun.* **1985**, *53*, 238–242.

151. Eschrich, T. C.; Morgan, T. J. Dye laser radiation in the 370-760-nm region pumped by a xenon monofluoride excimer laser. *Appl. Opt.* **1985**, *24*, 937–938.

152. Uchino, O.; Mizunami, T.; Maeda, M.; Miyazoe, Y. Efficient dye lasers pumped by a xenon chloride (XeCl) excimer laser. *Appl. Phys.* **1979**, *19*, 35–37.

153. Cox, A. J.; Scott, G. W. Short-cavity picosecond dye laser design. *Appl. Opt.* **1979**, *18*, 532–535.

154. Huppert, D.; Rentzepis, P. M. A high-efficiency tunable picosecond dye laser. *J. Appl. Phys.* **1978**, *49*, 543–548.

155. Cox, A. J.; Scott, G. W.; Talley, L. D. Tunable blue picosecond pulses from a dye laser. *Appl. Phys. Lett.* **1977**, *31*, 389–391.

156. Drexhage, K. H.; Erikson, G. R.; Hawks, G. H.; Reynolds, G. A. Water-soluble coumarin dyes for flashlamp-pumped dye lasers. *Opt. Commun.* **1975**, *15*, 399–403.

157. Reynolds, G. A.; Drexhage, K. H. New coumarin dyes with rigidized structure for flashlamp-pumped dye lasers. *Opt. Commun.* **1975**, *13*, 222–225.

158. Lin, C.; Shank, C. V. Subnanosecond tunable dye laser pulse generation by controlled resonator transients. *Appl. Phys. Lett.* **1975**, *26*, 389–391.

159. Jain, R. K.; Dienes, A. Polychromatic molecular nitrogen laser-pumped dye lasers. *Spectrosc. Lett.* **1974**, *7*, 491–501.

160. Dunning, F. B.; Stebbings, R. F. Efficient generation of tunable near uv radiation using a molecular nitrogen pumped dye laser. *Opt. Commun.* **1974**, *11*, 112–114.

161. Bell, M. I.; Tyte, R. N. Pulsed dye laser system for Raman and luminescence spectroscopy. *Appl. Opt.* **1974**, *13*, 1610–1614.

162. Yarborough, J. M. A cw [continuous wave] dye laser emission spanning the visible spectrum. *Appl. Phys. Lett.* **1974**, *24*, 629–630.

163. Tuccio, S. A.; Drexhage, K. H.; Reynolds, G. A. CW [continuous wave] laser emission from coumarin dyes in the blue and green. *Opt. Commun.* **1973**, *7*, 248–252.

164. Kijima, Y. Blue-light emitting electric field luminescent device containing coumarin dye. Jpn. Kokai Tokkyo Koho JP 08298342, 1996.

165. Pope, E. J. A. Optically active sol-gel microspheres for flat-panel color displays. *Proc. SPIE-Int. Soc. Opt. Eng.* **1994**, *2288*, 536–545.

166. Pope, E. J. A. Solid-state luminescent color displays. *Mater. Res. Soc. Symp. Proc.* **1994**, *345*, 331-336.

167. Mitera, J.; Stejskal, M.; Kozak, P.; Mostecky, J. Fluorescent inks for ink-jet printers. Czech. CS 239658, 1986.

168. Liu, G.; Wheat, H. G. Use of a fluorescent indicator in monitoring underlying corrosion on coated aluminum 2024-T4. *J. Electrochem. Soc.* **2009**, *156*, C160–C166.

169. Allonas, X.; Fouassier, J. P.; Kaji, M.; Miyasaka, M.; Hidaka, T. Two and three component photoinitiating systems based on coumarin derivatives. *Polymer* **2001**, *42*, 7627–7634.

170. Kitagawa, S.; Shinkai, M.; Kobe, E.; Inoue, T. Optical recording material containing coumarin dye. Jpn. Kokai Tokkyo Koho JP 2001096918, 2001.

171. Ushiyama, F. Semiconductor device and production thereof. Jpn. Kokai Tokkyo Koho JP 2000294560, 2000.

172. Liu, X.; Cole, J. M.; Waddell, P. G.; Lin, T.; Radia, J.; Zeidler, A. Molecular origins of optoelectronic

properties in coumarin dyes: Toward designer solar cell and laser applications. *J. Phys. Chem. A* **2012**, *116*, 727–737.

173. Busenbender, I.; Aguero, G. Dye-sensitized solar cell with long-term stability. Ger. Offen. DE 19640065, 1998.

174. Telbiz, G.; Shvets, O.; Boron, S.; Vozny, V.; Brodyn, M.; Stucky, G. Laser dye doped mesoporous silica fibers: host-guest interaction and fluorescence properties. *Stud. Surf. Sci. Catal.* **2001**, *135*, 3564–3570.

175. Morales-Saavedra, O. G.; Rivera, E.; Flores-Flores, J. O.; Castaneda, R.; Banuelos, J. G.; Saniger, J. M. Preparation and optical characterization of catalyst free SiO$_2$ sonogel hybrid materials. *J. Sol-Gel Sci. Technol.* **2007**, *41*, 277–289.

176. Naqvi, K. R.; Clark, M. Synthesis of an amino coumarin-based fluorescent reactive dye and its application to wool fibres. *Color. Technol.* **2011**, *127*, 62–68.

177. Sugimoto, S.; Kawaguchi, T.; Yamashita, K. Integrated light sources based on self-formed polymer waveguide doped with active medium. *Proc. SPIE* **2012**, *8435*, 84351O/1–84351O/8.

178. Tomizawa, T.; Itoh, K.; Ishii, K.; Mito, K.; Sasaki, K. Optically active dye doped thin film waveguides. *MCLC S&T, Sec.t B: Nonlinear Opt.* **1995**, *14*, 321–326.

179. Naik, N. M.; Desai, K. R. 3-Phenyl-4-methyl-7-[4-(arylurea)-6-amino-s-triazin-2-ylamino] coumarin synthesis as optical whitening agent. *J. Inst. Chem.* **1988**, *60*, 179–180.

180. Umemoto, H.; Kitao, T.; Konishi, K. Fluorescent whitening agents for synthetic fibers. XIII. Fluorescence of some coumarins as whitening agents. *Kogyo Kagaku Zasshi* **1970**, *73*, 1146–1151.

181. Wuebbles, B. J. Y.; Felton, J. S. Evaluation of laser dye mutagenicity using the Ames/*Salmonella* microsome test. *Environ. Mutagen.* **1985**, *7*, 511–522.

8-AMINOPYRENE-1,3,6-TRISULFONIC ACID TRISODIUM SALT (APTS)

CAS Registry Number　196504-57-1

Chemical Structure

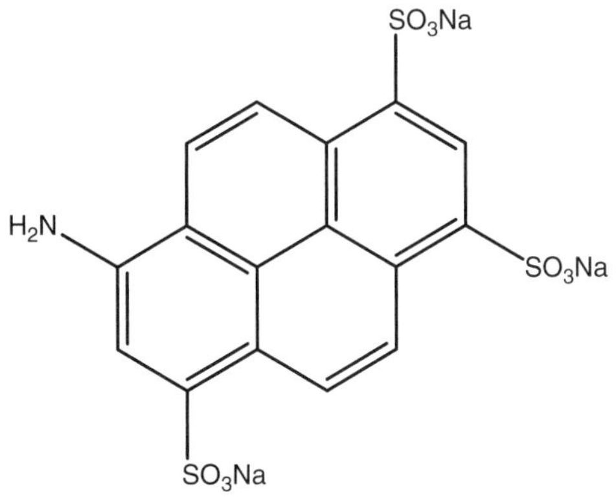

CA Index Name　1,3,6-Pyrenetrisulfonic acid, 8-amino-, sodium salt (1:3)

Other Names　8-Aminopyrene-1,3,6-trisulfonic acid, trisodium salt; APTS; 1,3,6-Pyrenetrisulfonic acid, 8-amino-, trisodium salt; Trisodium 8-aminopyrene-1,3,6-trisulfonate

Merck Index Number　Not listed

Chemical/Dye Class　Pyrene

Molecular Formula　$C_{16}H_8NNa_3O_9S_3$

Molecular Weight　523.39

Physical Form　Yellow solid;[2] Brown powder (crude); Orange powder (pure);[3] Orange solid

Solubility　Soluble in water, N,N-dimethylformamide, dimethyl sulfoxide, methanol

Melting Point　$\geq 250\,^{\circ}C$

Absorption (λ_{max})　424 nm (Buffer pH 7.0); 424 nm, 383 nm (H_2O)

Emission (λ_{max})　505 nm (Buffer pH 7.0)

Molar Extinction Coefficient　19,000 $cm^{-1}\,M^{-1}$ (Buffer pH 7.0); 20,600 $cm^{-1}\,M^{-1}$ (H_2O)

Synthesis　Synthetic methods[1–3]

Imaging/Labeling Applications　Cyclen 1;[4] glycoproteins;[5] glucose/monosaccharides;[3,6] oligosaccharides;[7–10] polysaccharides[11,12]

Biological/Medical Applications　Analysizing monosaccharides;[3,6] detecting oligosaccharides;[7–10] separating polysaccharides;[11,12] ultrathin protein films[13]

Industrial Applications　Not reported

Safety/Toxicity　No data available

REFERENCES

1. Tietze, E.; Bayer, O. Sulfonic acids of pyrene and their derivatives. *Justus Liebigs Ann. Chem.* **1939**, *540*, 189–210.

2. Bhatt, R.; Conrad, M. J.; Bencheikh, A.; Xiong, Y. Novel green and orange fluorescent labels and their uses. PCT Int. Appl. WO 2004027388, 2004.

3. Sharrett, Z.; Gamsey, S.; Hirayama, L.; Vilozny, B.; Suri, J. T.; Wessling, R. A.; Singaram, B. Exploring the use of APTS as a fluorescent reporter dye for continuous glucose sensing. *Org. Biomol. Chem.* **2009**, *7*, 1461–1470.

4. Winschel, C. A.; Kalidindi, A.; Zgani, I.; Magruder, J. L.; Sidorov, V. Receptor for anionic pyrene derivatives provides the basis for new biomembrane assays. *J. Am. Chem. Soc.* **2005**, *127*, 14704–14713.

5. Szabo, Z.; Guttman, A.; Karger, B. L. Rapid release of N-linked glycans from glycoproteins by pressure-cycling technology. *Anal. Chem.* **2010**, *82*, 2588–2593.

6. Dang, F.; Chen, Y.; Guo, Q. Analysis of saccharides as 8-aminopyrene-1,3,6-trisulfonate derivatives by capillary electrophoresis with laser-induced fluorescence detection. *Fenxi Huaxue* **2000**, *28*, 80–83.

7. Partyka, J.; Foret, F. Cationic labeling of oligosaccharides for electrophoretic preconcentration and separation with contactless conductivity detection. *J. Chromatogr., A* **2012**, *1267*, 116–120.

8. Dang, F.; Chen, Y.; Guo, Q.; Xu, G. Ultrasensitive fluorometric detection of oligosaccharides as their APTS derivatives by capillary electrophoresis. *Gaodeng Xuexiao Huaxue Xuebao* **2000**, *21*, 206–209.

9. Suzuki, H.; Mueller, O.; Guttman, A.; Karger, B. L. Analysis of 1-aminopyrene-3,6,8-trisulfonate-derivatized oligosaccharides by capillary electrophoresis with matrix-assisted laser desorption/ionization time-of-flight mass spectrometry. *Anal. Chem.* **1997**, *69*, 4554–4559.

10. Guttman, A. Capillary gel electrophoresis of 8-aminopyrene-3,6,8-trisulfonate-labeled oligosaccharides. *Tech. Glycobiol.* **1997**, *377–389.*

11. Wiedmer, S. K.; Cassely, A.; Hong, M.; Novotny, M. V.; Riekkola, M. Electrophoretic studies of polygalacturonate oligomers and their interactions with metal ions. *Electrophoresis* **2000**, *21*, 3212–3219.

12. Roberts, M. A.; Zhong, H. J.; Prodolliet, J.; Goodall, D. M. Separation of high-molecular-mass carrageenan polysaccharides by capillary electrophoresis with laser-induced fluorescence detection. *J. Chromatogr., A* **1998**, *817*, 353–366.

13. Peng, X.; Yu, Q.; Ye, Z.; Ichinose, I. Flexible ultrathin free-standing fluorescent films of CdSexS1-x/ZnS nanocrystalline and protein. *J. Mater. Chem.* **2011**, *21*, 4424–4431.

7-ANILINOCOUMARIN-4-ACETIC ACID (ACAA)

CAS Registry Number 82412-15-5
Chemical Structure

CA Index Name 2H-1-Benzopyran-4-acetic acid, 2-oxo-7-(phenylamino)-

Other Names 2H-1-Benzopyran-4-acetic acid, 7-anilino-2-oxo-; 7-Anilinocoumarin-4-acetic acid; 7-Phenylaminocoumarin-4-acetic acid; ACAA

Merck Index Number Not listed

Chemical/Dye Class Coumarin

Molecular Formula $C_{17}H_{13}NO_4$

Molecular Weight 295.29

Physical Form Pale yellow needles[2]

Solubility Soluble in water, cyclohexane, dimethyl sulfoxide, ethanol

Melting Point 186–189 °C (decompose);[2] 183.5–184.5 °C;[1] >160 °C (decompose)[3]

Boiling Point (Calcd.) 545.5 ± 50.0 °C, pressure: 760 Torr

pK$_a$ (Calcd.) 4.22 ± 0.10, most acidic, temperature: 25 °C; −1.59 ± 0.20, most basic, temperature: 25 °C

Absorption (λ_{max}) 262 nm, 375 nm (H_2O); 269 nm, 376 nm (EtOH); 272 nm, 378 nm (Cyclohexane)

Emission (λ_{max}) 459 nm (H_2O); 458 nm (EtOH); 404 nm (Cyclohexane); 475 nm (Phosphate Buffer pH 7.4 in the presence of human serum albumin (HSA))

Molar Extinction Coefficient 4,150 cm^{-1} M^{-1}, 4,350 cm^{-1} M^{-1} (H_2O); 4,220 cm^{-1} M^{-1} (EtOH); 4,160 cm^{-1} M^{-1}, 4,440 cm^{-1} M^{-1} (Cyclohexane)

Quantum Yield <0.01 (H_2O); 0.01 (EtOH); 0.32 (Cyclohexane); 0.06 (Phosphate Buffer pH 7.4 in the presence of human serum albumin (HSA))

Synthesis Synthetic methods[1–3]

Imaging/Labeling Applications Human serum albumin (HSA);[2,4–6] iron chelators[1]

Biological/Medical Applications Evaluating iron chelators;[1] as a probe for human serum albumin (HAS) binding sites;[2] studying binding sites of bile acids on human serum albumin,[6] drug-albumin interactions;[2] ultrafiltration assays of ligand binding to human serum albumin[5]

Industrial Applications Not reported

Safety/Toxicity No data available

REFERENCES

1. Ma, Y.; Luo, W.; Quinn, P. J.; Liu, Z.; Hider, R. C. Design, synthesis, physicochemical properties, and evaluation of novel iron chelators with fluorescent sensors. *J. Med. Chem.* **2004**, *47*, 6349–6362.

2. Goya, S.; Takadate, A.; Fujino, H.; Otagiri, M.; Uekama, K. New fluorescence probes for drug-albumin interaction studies. *Chem. Pharm. Bull.* **1982**, *30*, 1363–1369.

3. Woods, L. L.; Sapp, J. Ethyl esters of coumarin-4-acetic acids. *J. Chem. Eng. Data* **1963**, *8*, 235–236.

4. Liu, J.; Zhai, H.; Zhang, J.; Tian, J.; Hu, Z. QSPR study on the binding constants of coumarins and human serum albumin. *Guangpu Shiyanshi* **2006**, *23*, 602–605.

5. Gu, C.; Nikolic, D.; Lai, J.; Xu, X.; Van Breemen, R. B. Assays of ligand-human serum albumin binding using pulsed ultrafiltration and liquid chromatography-mass spectrometry. *Comb. Chem. High Throughput Screen.* **1999**, *2*, 353–359.

6. Takikawa, H.; Sugiyama, Y.; Hanano, M.; Kurita, M.; Yoshida, H.; Sugimoto, T. A novel binding site for bile acids on human serum albumin. *Biochim. Biophys. Acta* **1987**, *926*, 145–153.

9-ANTHROYLNITRILE

CAS Registry Number   85985-44-0

Chemical Structure

CA Index Name   9-Anthraceneacetonitrile, α-oxo-

Other Names   9-Anthracenecarbonyl cyanide; 9-Anthroylnitrile; 9-AN; α-Oxo-anthracene-9-acetonitrile; ANN

Merck Index Number   Not listed

Chemical/Dye Class   Anthracene

Molecular Formula $C_{16}H_9NO$

Molecular Weight   231.25

Physical Form   Orange-yellow needles;[1] Orange powder[2]

Solubility   Soluble in acetonitrile, chloroform, *N,N*-dimethylformamide, methanol

Melting Point   143–144 °C[1]

Boiling Point (Calcd.)   426.8±14.0 °C, pressure: 760 Torr

Absorption (λ_{max})   361 nm, 380 nm, 344 nm (MeOH)

Emission (λ_{max})   470 nm (MeOH)

Molar Extinction Coefficient   7500 cm⁻¹ M⁻¹ (MeOH)

Molar Extinction Coefficient $7500\ cm^{-1}\ M^{-1}$ (MeOH)

Synthesis   Synthetic methods[1,2]

Imaging/Labeling Applications   Alcohols;[1,2,7,17,19] aloesin;[3] analytes;[4] bile acid *N*-acetylglucosaminides;[5,6] carnitine enantiomers;[7] corticosteroids;[8] cortisol;[9] cortisone;[9] farnesyl pyrophosphate (FPP);[10] hydroxy steroids;[1,2] archaeal ether core lipids;[11] *N*-terminal peptide segment of myosin subfragment-1 (S-1);[12] serine-181;[13,14] thyroid hormones;[15] triamcinolone (TMC)[16]

Biological/Medical Applications   Analyzing primary/secondary/tertiary alcohols;[17] detecting aloesin,[3] analytes,[4] cortisol/cortisone in biological fluids,[9] farnesyl pyrophosphate (FPP),[10] archaeal ether core lipids,[11] thyroid hormones in pharmaceutical preparations,[15] small quantities of triamcinolone (TMC);[16] identifying serine 181;[13,14] reagents for hydroxy steroids;[1,2] separating/characterizing bile acid *N*-acetylglucosaminides;[5,6] separating carnitine enantiomers in pharmaceuticals;[7] reagents for alcohols/hydroxyl groups[1,2,7,17,19]

Industrial Applications   Analyzing chemical compositions for carbonized distillate of coal-tar refined soft pitch;[18] detecting absolute amount of resin-bound hydroxyl groups[19]

Safety/Toxicity   No data available

REFERENCES

1. Goto, J.; Goto, N.; Shamsa, F.; Saito, M.; Komatsu, S.; Suzaki, K.; Nambara, T. New sensitive derivatization of hydroxy steroids for high-performance liquid chromatography with fluorescence detection. *Anal. Chim. Acta* **1983**, *147*, 397–400.

2. Shimadzu Seisakusho Ltd. Anthroylnitriles. Jpn. Kokai Tokkyo Koho JP 58057356, 1983.

3. Kim, K. H.; Lee, J. G.; Park, J. H.; Shin, Y. G.; Lee, S. K.; Cho, T. H.; Oh, S. T. Determination of aloesin in plasma by high-performance liquid chromatography as fluorescent 9-anthroyl derivative. *Arch. Pharm. Res.* **1998**, *21*, 651–656.

4. Kohn, B. A.; Radlo, J. L. Generic signalling mechanism for detection of analytes. PCT Int. Appl. WO 9932886, 1999.

5. Niwa, T.; Fujita, K.; Goto, J.; Nambara, T. Separation and characterization of ursodeoxycholate 7-N-acetylglucosaminides in human urine by high-performance liquid chromatography with fluorescence detection. *J. Liq. Chromatogr.* **1993**, *16*, 2531–2544.

6. Niwa, T.; Fujita, K.; Goto, J.; Nambara, T. Separation of bile acid *N*-acetylglucosaminides by high-performance liquid chromatography with precolumn fluorescence labeling. *Anal. Sci.* **1992**, *8*, 659–662.

7. Takahashi, M.; Terashima, K.; Nishijima, M.; Kamata, K. Separation of carnitine enantiomers as the 9-anthroylnitrile derivatives and high-performance liquid chromatographic analysis on an ovomucoid-conjugated column. *J. Pharm. Biomed. Anal.* **1996**, *14*, 1579–1584.

8. Neufeld, E.; Chayen, R.; Stern, N. Fluorescence derivatization of urinary corticosteroids for high-performance liquid chromatographic analysis. *J. Chromatogr., B* **1998**, *718*, 273–277.

9. Glowka, F. K.; Kosicka, K.; Karazniewicz-Lada, M. HPLC method for determination of fluorescence derivatives of cortisol, cortisone and their tetrahydro- and allo-tetrahydro-metabolites in biological fluids. *J. Chromatogr., B* **2010**, *878*, 283–289.

10. Saisho, Y.; Morimoto, A.; Umeda, T. Determination of farnesyl pyrophosphate in dog and human plasma by high-performance liquid chromatography with fluorescence detection. *Anal. Biochem.* **1997**, *252*, 89–95.

11. Bai, Q. Y.; Zelles, L. A method for determination of archaeal ether-linked glycerolipids by high performance liquid chromatography with fluorescence detection as their 9-anthroyl derivatives. *Chemosphere* **1997**, *35*, 263–274.

12. Hiratsuka, T. Conformational changes in the 23-kilodalton amino-terminal peptide segment of myosin ATPase associated with ATP hydrolysis. *J. Biol. Chem.* **1990**, *265*, 18786–18790.

13. Hiratsuka, T.; Katoh, T. Chemical identification of serine 181 at the ATP-binding site of myosin as a residue esterified selectively by the fluorescent reagent 9-anthroylnitrile. *J. Biol. Chem.* **2003**, *278*, 31891–31894.

14. Szarka, K.; Bodis, E.; Visegrady, B.; Nyitrai, M.; Kilar, F.; Somogyi, B. 9-Anthroylnitrile binding to Serine-181 in myosin subfragment 1 as revealed by FRET spectroscopy and molecular modeling. *Biochemistry* **2001**, *40*, 14806–14811.

15. Takahashi, M.; Nagashima, M.; Shigeoka, S.; Kamimura, H.; Kamata, K. Determination of thyroid hormones in pharmaceutical preparations, after derivatization with 9-anthroylnitrile, by high-performance liquid chromatography with fluorescence detection. *J. Chromatogr., A* **2002**, *958*, 299–303.

16. Glowka, F. K.; Karazniewicz, M.; Lipnicka, E. RP-HPLC method with fluorescence detection for determination of small quantities of triamcinolone in plasma in presence of endogenous steroids after derivatization with 9-anthroyl nitrile; pharmacokinetic studies. *J. Chromatogr., B* **2006**, *839*, 54–61.

17. Nelson, T. J. Fluorescent high-performance liquid chromatography assay for lipophilic alcohols. *Anal. Biochem.* **2011**, *419*, 40–45.

18. Gao, L.; Lai, S.; Zhang, X.; Zhao, X.; Lu, Y. Analysis of chemical compositions for carbonized distillate of coal-tar refined soft pitch with GC-MS-RI. *Liaoning Keji Daxue Xuebao* **2008**, *31*, 455–459.

19. Yan, B.; Liu, L.; Astor, C. A.; Tang, Q. Determination of the absolute amount of resin-bound hydroxyl or carboxyl groups for the optimization of solid-phase combinatorial and parallel organic synthesis. *Anal. Chem.* **1999**, *71*, 4564–4571.

BOBO 1

CAS Registry Number   169454-13-1

Chemical Structure

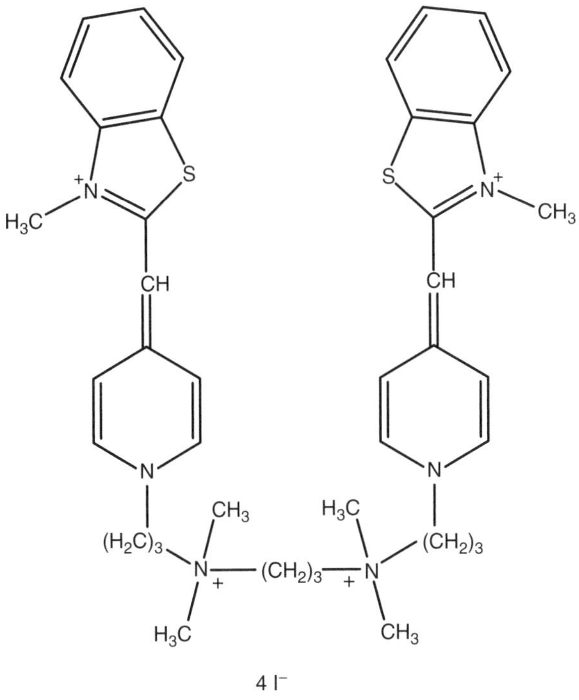

4 I⁻

CA Index Name   Benzothiazolium, 2,2′-[1,3-propanediylbis-[(dimethyliminio)-3,1-propanediyl-1(4*H*)-pyridinyl-4-ylidenemethylidyne]]bis[3-methyl-, iodide (1:4)

Other Names   Benzothiazolium, 2,2′-[1,3-propanediyl-bis-[(dimethyliminio)-3,1-propanediyl-1(4*H*)-pyridinyl-4-ylidenemethylidyne]]bis[3-methyl-, tetraiodide; BOBO 1; BOBO 1 iodide

Merck Index Number   Not listed

Chemical/Dye Class   Cyanine

Molecular Formula $C_{41}H_{54}I_4N_6S_2$

Molecular Weight   1202.66

Physical Form   Yellow-brown powder[2]

Solubility   Soluble in dimethyl sulfoxide

Absorption (λ_{max})   462 nm (H_2O/DNA)

Emission (λ_{max})   481 nm (H_2O/DNA)

Molar Extinction Coefficient   113,600 $cm^{-1} M^{-1}$ (H_2O/DNA)

Quantum Yield   0.22 (H_2O/DNA)

Synthesis   Synthetic methods[1,2]

Imaging/Labeling Applications   Nucleic acids;[1–12] cells;[2,11,12] bacteria;[12] chromosome spreads;[2] nuclei;[13–16] micronuclei;[17] megakaryocyte;[18] microorganisms;[19–21] sperms;[11,22] hairs[23]

Biological/Medical Applications   Detecting nucleic acids,[1–12] cells;[2,11,12] classifying/counting megakaryocytes;[18] counting erythroblasts,[15,16] sperms;[11,22] characterizing DNA/lipid complexes (lipoplexes);[9,10] detecting/measuring microorganisms;[19–21] as temperature sensor[24,25]

Industrial Applications   Not reported

Safety/Toxicity   No data available

REFERENCES

1. Sabnis, R. W. *Handbook of Biological Dyes and Stains*; John Wiley & Sons Inc.: Hoboken, **2010**; pp 51–52.

2. Yue, S. T.; Haugland, R. P. Dimers of unsymmetrical cyanine dyes containing pyridinium moieties. U.S. Patent 5410030, 1995.

3. Sergeev, N. V.; Brevnov, M. G.; Furtado, M. R. Identification of nucleic acids. PCT Int. Appl. WO 2011143478, 2011.

4. Exner, M.; Rogers, A. Methods for detecting nucleic acids using multiple signals. U.S. Pat. Appl. Publ. US 2007172836, 2007.

5. Wittwer, C. T.; Dujols, V. E.; Reed, G.; Zhou, L. Amplicon melting analysis with saturation dyes. PCT Int. Appl. WO 2004038038, 2004.

6. Erikson, G. H.; Daksis, J. I.; Kandic, I.; Picard, P. Nucleic acid multiplex formation. PCT Int. Appl. WO 2002103051, 2002.

7. Erikson, G. H.; Daksis, J. I. Pre-incubation method to improve signal/noise ratio of nucleic acid assays. U.S. Pat. Appl. Publ. US 2004180345, 2004.

8. Erikson, G. H. Method for modifying transcription and/or translation in an organism for therapeutic, prophylactic and/or analytic uses. U.S. Pat. Appl. Publ. US 2003181412, 2003.

Handbook of Fluorescent Dyes and Probes, First Edition. R. W. Sabnis.
© 2015 John Wiley & Sons, Inc. Published 2015 by John Wiley & Sons, Inc.

9. Madeira, C.; Fedorov, A.; Aires-Barros, M. R.; Prieto, M.; Loura, L. M. S. Photophysical behavior of a dimeric cyanine dye (BOBO-1) within cationic liposomes. *Photochem. Photobiol.* **2005**, *81*, 1450–1459.

10. Madeira, C.; Loura, L. M. S.; Aires-barros, M. R.; Fedorov, A.; Prieto, M. Characterization of DNA/lipid complexes by fluorescence resonance energy transfer. *Biophys. J.* **2003**, *85*, 3106–3119.

11. Anderson, A. L.; Knutson, C. R.; Mueth, D.; Plewa, J.; Tanner, E. Methods for staining cells for identification and sorting. U.S. Pat. Appl. Publ. US 2006172315, 2006.

12. Millard, P. J.; Roth, B. L.; Yue, S. T.; Haugland, R. P. Fluorescent viability assay using cyclic-substituted unsymmetrical cyanine dyes. U.S. Patent 5534416, 1996.

13. Gauer, C.; Mann, W.; Alunni-Fabbroni, M. Method for carrying out an enzymic reactions. Ger. DE 102006056694, 2010.

14. Gauer, C.; Mann, W.; Alunni-Fabbroni, M. Method for carrying out an enzymic reaction. PCT Int. Appl. WO 2008064730, 2008.

15. Heuven, B.; Wong, F.; Tsuji, T.; Sakata, T.; Hamaguchi, I. Method for classifying and counting erythroblasts by flow cytometry. Jpn. Kokai Tokkyo Koho JP 11326323, 1999.

16. Heuven, B.; Wong, F.; Tsuji, T.; Sakata, T.; Hamaguchi, I. Process for discriminating and counting erythroblasts. U.S. Pat. Appl. Publ. US 20020006631, 2002.

17. Dertinger, S. D.; Cairns, S. E.; Avlasevich, S. L.; Torous, D. K. Method for enumerating mammalian cell micronuclei with an emphasis on differentially staining micronuclei and the chromatin of dead and dying cells. PCT Int. Appl. WO 2006007479, 2006.

18. Minakami, T.; Mori, Y.; Tsuji, T.; Ikeuchi, Y. Megakaryocyte classification/counting method by double fluorescent staining and flow cytometry. Jpn.Kokai Tokkyo Koho JP 2006275985, 2006.

19. Eckert, R. H.; Kaplan, C.; He, J.; Yarbrough, D. K.; Anderson, M.; Sim, J. Methods and devices for the selective detection of microorganisms. U.S. Pat. Appl. Publ. US 20120003661, 2012.

20. Noda, N.; Mizutani, T. Microorganism-measuring method using multiple staining. Jpn. Kokai Tokkyo Koho JP 2006340684, 2006.

21. Vannier, E. Methods for detection of pathogens in red blood cells. PCT Int. Appl. WO 2006031544, 2006.

22. Matsumoto, T.; Okada, H.; Hamaguchi, Y. Method and reagent for counting sperm by flow cytometry. Jpn. Kokai Tokkyo Koho JP 2001242168, 2001.

23. Lagrange, A. Hair dye compositions containing a polycationic direct dye. Fr. Demande FR 2848840, 2004.

24. Bousseksou, A.; Salmon, L.; Molnar, G.; Cobo, S. Materials with thermochromic spin transition doped with one or more fluorescent agents for use as temperature sensor. Fr. Demande FR 2952371, 2011.

25. Bousseksou, A.; Salmon, L.; Molnar, G.; Cobo, S. Heat-sensitive spin-transition materials doped with one or more fluorescent agents for use as temperature sensor. PCT Int. Appl. WO 2011058277, 2011.

BOBO 3

CAS Registry Number   169454-17-5

Chemical Structure

Emission (λ_{max})   602 nm (H_2O/DNA)

Molar Extinction Coefficient   147,800 cm^{-1} M^{-1} (H_2O/DNA)

Quantum Yield   0.39 (H_2O/DNA)

CA Index Name   Benzothiazolium, 2,2′-[1,3-propanediylbis-[(dimethyliminio)-3,   1-propanediyl-1(4H)-pyridinyl-4-ylidene-1-propen-1-yl-3-ylidene]]bis[3-methyl-, iodide (1:4)

Other Names   Benzothiazolium, 2,2′-[1,3-propanediyl-bis-[(dimethyliminio)-3,1-propanediyl-1(4H)-pyridinyl-4-ylidene-1-propen-1-yl-3-ylidene]]bis[3-methyl-, tetraiodide; BOBO 3; BOBO 3 iodide

Merck Index Number   Not listed

Chemical/Dye Class   Cyanine

Molecular Formula $C_{45}H_{58}I_4N_6S_2$

Molecular Weight   1254.73

Physical Form   Yellow-brown powder; Purple powder[2]

Solubility   Soluble in dimethyl sulfoxide

Absorption (λ_{max})   570 nm (H_2O/DNA)

Synthesis   Synthetic methods[1,2]

Imaging/Labeling Applications   Nucleic acids;[1–13] cells;[1,2,12,13] bacteria;[13,14] chromosome spreads;[2] nuclei;[15–18] megakaryocytes;[19] microorganisms;[20,21] sperms;[12,22] hairs[23]

Biological/Medical Applications   Detecting nucleic acids,[1–13] cells,[1,2,12,13] microorganisms;[20,21] IDH1/IDH2 mutations;[24] JAK2 gene mutation;[25] PIK3CA mutation;[26] B-type Raf Kinase (BRAF) mutation;[27] Mucin 1;[28] thrombin;[29] classifying/counting megakaryocytes;[19] counting erythroblasts,[17,18] sperms;[12,22] nucleic acid sequencing;[30] as temperature sensor;[31,32] biophotonic wire assemblies[33]

Industrial Applications   Semiconductor nanocrystal[34]

Safety/Toxicity   No data available

REFERENCES

1. Sabnis, R. W. *Handbook of Biological Dyes and Stains; John Wiley & Sons Inc.*: Hoboken, **2010**; pp 53–54.

2. Yue, S. T.; Haugland, R. P. Dimers of unsymmetrical cyanine dyes containing pyridinium moieties. U.S. Patent 5410030, 1995.

3. Sergeev, N. V.; Brevnov, M. G.; Furtado, M. R. Identification of nucleic acids. PCT Int. Appl. WO 2011143478, 2011.

4. Exner, M.; Rogers, A. Methods for detecting nucleic acids using multiple signals. U.S. Pat. Appl. Publ. US 2007172836, 2007.

5. Wittwer, C. T.; Dujols, V. E.; Reed, G.; Zhou, L. Amplicon melting analysis with saturation dyes. PCT Int. Appl. WO 2004038038, 2004.

6. Erikson, G. H.; Daksis, J. I.; Kandic, I.; Picard, P. Nucleic acid multiplex formation. PCT Int. Appl. WO 2002103051, 2002.

7. Kricka, L. J. Stains, labels and detection strategies for nucleic acids assays. *Ann. Clin. Biochem.* **2002**, *39*, 114–129.

8. Erikson, G. H.; Daksis, J. I. Pre-incubation method to improve signal/noise ratio of nucleic acid assays. U.S. Pat. Appl. Publ. US 2004180345, 2004.

9. Erikson, G. H. Method for modifying transcription and/or translation in an organism for therapeutic, prophylactic and/or analytic uses. U.S. Pat. Appl. Publ. US 2003181412, 2003.

10. Ruedas-Rama, M. J.; Alvarez-Pez, J. M.; Paredes, J. M.; Talavera, E. M.; Orte, A. Binding of BOBO-3 intercalative dye to DNA homo-oligonucleotides with different base compositions. *J. Phys. Chem. B* **2010**, *114*, 6713-6721.

11. Jiang, S.; Zhang, Y. Upconversion nanoparticle-based FRET system for study of siRNA in live cells. *Langmuir* **2010**, *26*, 6689-6694.

12. Anderson, A. L.; Knutson, C. R.; Mueth, D.; Plewa, J.; Tanner, E. Methods for staining cells for identification and sorting. U.S. Pat. Appl. Publ. US 2006172315, 2006.

13. Millard, P. J.; Roth, B. L.; Yue, S. T.; Haugland, R. P. Fluorescent viability assay using cyclic-substituted unsymmetrical cyanine dyes. U.S. Patent 5534416, 1996.

14. Kitaguchi, A.; Yamaguchi, N.; Nasu, M. Simultaneous enumeration of viable *Enterobacteriaceae* and *Pseudomonas* spp. within three hours by multicolor fluorescence *in situ* hybridization with vital staining. *J. Microbiol. Methods* **2006**, *65*, 623–627.

15. Gauer, C.; Mann, W.; Alunni-Fabbroni, M. Method for carrying out an enzymic reactions. Ger. DE 102006056694, 2010.

16. Gauer, C.; Mann, W.; Alunni-Fabbroni, M. Method for carrying out an enzymic reaction. PCT Int. Appl. WO 2008064730, 2008.

17. Heuven, B.; Wong, F.; Tsuji, T.; Sakata, T.; Hamaguchi, I. Method for classifying and counting erythroblasts by flow cytometry. Jpn. Kokai Tokkyo Koho JP 11326323, 1999.

18. Heuven, B.; Wong, F.; Tsuji, T.; Sakata, T.; Hamaguchi, I. Process for discriminating and counting erythroblasts. U.S. Pat. Appl. Publ. US 20020006631, 2002.

19. Minakami, T.; Mori, Y.; Tsuji, T.; Ikeuchi, Y. Megakaryocyte classification/counting method by double fluorescent staining and flow cytometry. Jpn. Kokai Tokkyo Koho JP 2006275985, 2006.

20. Eckert, R. H.; Kaplan, C.; He, J.; Yarbrough, D. K.; Anderson, M.; Sim, J. Methods and devices for the selective detection of microorganisms. U.S. Pat. Appl. Publ. US 20120003661, 2012.

21. Vannier, E. Methods for detection of pathogens in red blood cells. PCT Int. Appl. WO 2006031544, 2006.

22. Matsumoto, T.; Okada, H.; Hamaguchi, Y. Method and reagent for counting sperm by flow cytometry. Jpn. Kokai Tokkyo Koho JP 2001242168, 2001.

23. Lagrange, A. Hair dye compositions containing a polycationic direct dye. Fr. Demande FR 2848840, 2004.

24. Park, H. G.; Choi, J. J.; Kim, H. S.; Jung, S. U. Method and kit for detecting IDH1 and IDH2 mutations using peptide nucleic acid (PNA)-based real-time polymerase chain reaction (PCR) clamping. Repub. Korean Kongkae Taeho Kongbo KR 2012127679, 2012.

25. Park, H. G.; Choi, J. J.; Kim, H. S.; Jung, S. U. JAK2 gene mutation detection kit based on PNA mediated real-time PCR clamping. Repub. Korean Kongkae Taeho Kongbo KR 2012119571, 2012.

26. Park, H. K.; Choi, J. J.; Kim, H. S. PIK3CA mutation detection method and kit using real-time PCR clamping of PNA. PCT Int. Appl. WO 2012020965, 2012.

27. Park, H. K.; Choi, J. J.; Cho, M. H. Detection of mutation of BRAF by real-time PCR using PNA clamping probe for diagnosis of cancers. PCT Int. Appl. WO 2011093606, 2011.

28. Shin, S.; Nam, H. Y.; Lee, E. J.; Jung, W.; Hah, S. S. Molecular beacon-based quantitiation of epithelial tumor marker mucin 1. *Bioorg. Med. Chem. Lett.* **2012**, *22*, 6081–6084.

29. Chi, C.; Lao, Y.; Li, Y.; Chen, L. A quantum dot-aptamer beacon using a DNA intercalating dye as the FRET reporter: Application to label-free thrombin detection. *Biosens. Bioelectron.* **2011**, *26*, 3346–3352.

30. Williams, J. G. K.; Anderson, J. P. Field-switch sequencing. PCT Int. Appl. WO 2005111240, 2005.

31. Bousseksou, A.; Salmon, L.; Molnar, G.; Cobo, S. Materials with thermochromic spin transition doped with one or more fluorescent agents for use as temperature sensor. Fr. Demande FR 2952371, 2011.

32. Bousseksou, A.; Salmon, L.; Molnar, G.; Cobo, S. Heat-sensitive spin-transition materials doped with one or more fluorescent agents for use as temperature sensor. PCT Int. Appl. WO 2011058277, 2011.

33. Boeneman, K.; Prasuhn, D. E.; Blanco-Canosa, J. B.; Dawson, P. E.; Melinger, J. S.; Ancona, M.; Stewart, M. H.; Susumu, K.; Huston, A.; Medintz, I. L. Self-assembled quantum dot-sensitized multivalent DNA photonic wires. *J. Am. Chem. Soc.* **2010**, *132*, 18177–18190.

34. Ruedas-Rama, M. J.; Orte, A.; Hall, E. A. H.; Alvarez-Pez, J. M.; Talavera, E. M. Effect of surface modification on semiconductor nanocrystal fluorescence lifetime. *ChemPhysChem* **2011**, *12*, 919–929.

BODIPY FL C$_5$-CERAMIDE (C$_5$-DMB-CERAMIDE)

CAS Registry Number   133867-53-5

Chemical Structure

CA Index Name   Boron, [5-[(3,5-dimethyl-2*H*-pyrrol-2-ylidene-κ*N*)methyl]-*N*-[(1*S*,2*R*,3*E*)-2-hydroxy-1-(hydroxymethyl)-3-heptadecen-1-yl]-1*H*-pyrrole-2-pentanamidato-κ*N*^{1}]difluoro-, (*T*-4)-

Other Names   Boron, [5-[(3,5-dimethyl-2*H*-pyrrol-2-ylidene)-methyl]-*N*-[2-hydroxy-1-(hydroxymethyl)-3-heptadecenyl]-1*H*-pyrrole-2-pentanamidato-*N*^{1},*N*^{5}] difluoro-, [*T*-4-[*R*-[*R**,*S**-(*E*)]]]-; Boron, [5-[(3,5-dimethyl-2*H*-pyrrol-2-ylidene-κ*N*)methyl]-*N*-[(1*S*,2*R*,3*E*)-2-hydroxy-1-(hydroxymethyl)-3-heptadecenyl]-1*H*-pyrrole-2-pentanamidato-κ*N*^{1}]difluoro-,   (*T*-4)-;   1*H*-Pyrrole-2-pentanamide,   5-[(3,5-dimethyl-2*H*-pyrrol-2-ylidene)methyl]-*N*-[2-hydroxy-1-(hydroxymethyl)-3-heptadecenyl]-, boron complex, [*R*-[*R**,*S**-(*E*)]]-; BODIPY FL C$_5$-ceramide; BODIPY FL C$_5$-cer; C$_5$-DMB-ceramide; C$_5$-DMB-cer

Merck Index Number   Not listed

Chemical/Dye   Class   Boron   co-ordination compound/dye

Molecular Formula   C$_{34}$H$_{54}$BF$_2$N$_3$O$_3$

Molecular Weight   601.63

Physical Form   Orange solid

Solubility   Soluble   in   acetonitrile,   chloroform, *N*,*N*-dimethylformamide, dimethyl sulfoxide, methanol

Absorption (λ$_{max}$)   505 nm (MeOH)

Emission (λ$_{max}$)   511 nm (MeOH)

Molar   Extinction   Coefficient   91,000 cm^{-1} M^{-1} (MeOH)

Synthesis   Synthetic methods[1,2]

Imaging/Labeling Applications   Golgi apparatus;[1-9,28] antigen   presenting   cells   (APCs);[10]   bacteria;[11] cytoplasmic   membranes;[28]   exosome   subpopulations;[12] glycoproteins;[13]   lipids;[14-17]   lipid   bilayers;[18] lipoproteins;[19] live cells;[20] Madin-Darby canine kidney (MDCK)   cells;[21]   myelin;[22]   nucleic   acids;[27]   Pc 4;[28] sphingolipid[23-26]

Biological/Medical   Applications   Analyzing lipoproteins;[19] characterizing exosome subpopulations;[12] detecting trogocytosis;[10] identifying nucleic acids;[27] inhibiting   glycoprotein   traffic;[13]   investigating sphingolipid transport and metabolism;[23-26] measuring rate of myelination;[22] modulating peptidolytic activity of cathepsin D;[29] monitoring live cells;[20] as a substrate for measuring hydrolytic enzyme activity,[30] inositol phosphorylceramide synthase activity[31]

Industrial Applications   Not reported

Safety/Toxicity   No data available

REFERENCES

1. Sabnis, R. W. *Handbook of Biological Dyes and Stains*; John Wiley & Sons Inc.: Hoboken, **2010**; pp 55–56.

2. Pagano, R. E.; Martin, O. C.; Kang, H. C.; Haugland, R. P. A novel fluorescent ceramide analog for studying membrane traffic in animal cells: Accumulation at the Golgi apparatus results in altered spectral properties of the sphingolipid precursor. *J. Cell Biol.* **1991**, *113*, 1267–1279.

3. Teiten, M. H.; Bezdetnaya, L.; Morlière, P.; Santus, R.; Guillemin, F. Endoplasmic reticulum and Golgi apparatus are the preferential sites of Foscan

localisation in cultured tumour cells. *Br. J. Cancer* **2003**, *88*, 146–152.

4. Moreno, R. D.; Schatten, G.; Ramalho-Santos, J. Golgi apparatus dynamics during mouse oocyte *in vitro* maturation: Effect of the membrane trafficking inhibitor brefeldin A. *Biol. Reprod.* **2002**, *66*, 1259–1266.

5. Roth, M. G. Inheriting the Golgi. *Cell* **1999**, *99*, 559–562.

6. Ladinsky, M. S.; Kremer, J. R.; Furcinitti, P. S.; McIntosh, J. R.; Howell, K. E. HVEM tomography of the trans-Golgi network: structural insights and identification of a lace-like vesicle coat. *J. Cell Biol.* **1994**, *127*, 29–38.

7. Takizawa, P. A.; Yucel, J. K.; Veit, B.; Faulkner, D. J.; Deerinck, T.; Soto, G.; Ellisman, M.; Malhotra, V. Complete vesiculation of Golgi membranes and inhibition of protein transport by a novel sea sponge metabolite, ilimaquinone. *Cell* **1993**, *73*, 1079–1090.

8. Ralston, E. Changes in architecture of the Golgi complex and other subcellular organelles during myogenesis. *J. Cell Biol.* **1993**, *120*, 399–409.

9. Hidalgo, J.; Garcia-Navarro, R.; Garcia-Navarro, F.; Perez-Vilar, J. Velasco, A. Presence of Golgi remnant membranes in the cytoplasm of brefeldin A-treated cells. *Eur. J. Cell Biol.* **1992**, *58*, 214–227.

10. Daubeuf, S.; Bordier, C.; Hudrisier, D.; Joly, E. Suitability of various membrane lipophilic probes for the detection of trogocytosis by flow cytometry. *Cytometry, Part A* **2009**, *75A*, 380–389.

11. Boleti, H.; Ojcius, D. M.; Dautry-Varsat, A. Fluorescent labelling of intracellular bacteria in living host cells. *J. Microbiol. Methods* **2000**, *40*, 265–274.

12. Laulagnier, K.; Vincent-Schneider, H.; Hamdi, S.; Subra, C.; Lankar, D.; Record, M. Characterization of exosome subpopulations from RBL-2H3 cells using fluorescent lipids. *Blood Cells, Mol. Dis.* **2005**, *35*, 116–121.

13. Rosenwald, A. G.; Pagano, R. E. Inhibition of glycoprotein traffic through the secretory pathway by ceramide. *J. Biol. Chem.* **1993**, *268*, 4577–4579.

14. Kuerschner, L.; Ejsing, C. S.; Ekroos, K.; Shevchenko, A.; Anderson, K. I.; Thiele, C. Polyene-lipids: a new tool to image lipids. *Nat. Methods* **2005**, *2*, 39–45.

15. Allan, D. Lipid metabolic changes caused by short-chain ceramides and the connection with apoptosis. *Biochem. J.* **2000**, *345*, 603–610.

16. Bai, J.; Pagano, R. E. Measurement of spontaneous transfer and transbilayer movement of BODIPY-labeled lipids in lipid vesicles. *Biochemistry* **1997**, *36*, 8840–8848.

17. Redman, C. A.; Kusel, J. R. Distribution and biophysical properties of fluorescent lipids on the surface of adult *Schistosoma mansoni*. *Parasitology* **1996**, *113*, 137–143.

18. Johnson, M. E.; Berk, D. A.; Blankschtein, D.; Golan, D. E.; Jain, R. K.; Langer, R. S. Lateral diffusion of small compounds in human stratum corneum and model lipid bilayer systems. *Biophys. J.* **1996**, *71*, 2656–2668.

19. Ping, G.; Zhu, B.; Jabasini, M.; Xu, F.; Oka, H.; Sugihara, H.; Baba, Y. Analysis of lipoproteins by microchip electrophoresis with high speed and high reproducibility. *Anal. Chem.* **2005**, *77*, 7282–7287.

20. Battaglia, G. Method of monitoring live cells. U.S. Pat. Appl. Publ. US 20090286274, 2009.

21. Iida-Tanaka, N.; Namekata, I.; Tamura, M.; Kawamata, Y.; Kawanishi, T.; Tanaka, H. Membrane-labeled MDCK cells and confocal microscopy for the analyses of cellular volume and morphology. *Biol. Pharm. Bull.* **2008**, *31*, 731–734.

22. Bilderback, T. R.; Chan, J. R.; Harvey, J. J.; Glaser, M. Measurement of the rate of myelination using a fluorescent analog of ceramide. *J. Neurosci. Res.* **1997**, *49*, 497–507.

23. Marks, D. L.; Bittman, R.; Pagano, R. E. Use of Bodipy-labeled sphingolipid and cholesterol analogs to examine membrane microdomains in cells. *Histochem. Cell Biol.* **2008**, *130*, 819–832.

24. Marks, D. L.; Singh, R. D.; Choudhury, A.; Wheatley, C. L.; Pagano, R. E. Use of fluorescent sphingolipid analogs to study lipid transport along the endocytic pathway. *Methods* **2005**, *36*, 186–195.

25. Pagano, R. E.; Watanabe, R.; Wheatley, C.; Dominguez, M. Applications of BODIPY-sphingolipid analogs to study lipid traffic and metabolism in cells. *Methods Enzymol.* **2000**, *312*, 523–534.

26. Pagano, R. E.; Chen, C. S. Use of BODIPY-labeled sphingolipids to study membrane traffic along the endocytic pathway. *Ann. N.Y. Acad. Sci.* **1998**, *845*, 152–160.

27. Sergeev, N. V.; Brevnov, M. G.; Furtado, M. R. Identification of nucleic acids. PCT Int. Appl. WO 2011143478, 2011.

28. Trivedi, N. S.; Wang, H.; Nieminen, A.; Oleinick, N. L.; Izatt, J. A. Quantitative analysis of Pc 4 localization in mouse lymphoma (LY-R) cells via double-label confocal fluorescence microscopy. *Photochem. Photobiol.* **2000**, *71*, 634–639.

29. Zebrakovska, I.; Masa, M.; Srp, J.; Horn, M.; Vavrova, K.; Mares, M. Complex modulation of peptidolytic activity of cathepsin D by sphingolipids. *Biochim. Biophys. Acta, Mol. Cell Biol. Lipids* **2011**, *1811*, 1097–1104.

30. Karuso, P. H.; Choi, H. Method for monitoring hydrolytic activity. PCT Int. Appl. WO 2007051257, 2007.

31. Elhammer, A. Novel assay for inositol phosphorylceramide synthase activity. U.S. Pat. Appl. Publ. US 20070269844, 2007.

BODIPY FL C$_5$-GANGLIOSIDE GM1

CAS Registry Number   908143-55-5

Chemical Structure

CA Index Name   Boron, [5-[(3,5-dimethyl-2H-pyrrol-2-ylidene-κN)methyl]-N-[(1S,2R,3E)-1-[[[O-β-D-galacto-pyranosyl-(1→4)-2-O-(acetylamino)-2-deoxy-β-D-galac-topyranosyl-(1→4)-O-[N-acetyl-α-neuraminosyl-(2→3)]-O-β-D-galactopyranosyl-(1→4)-β-D-glucopyranosyl]oxy]methyl]-2-hydroxy-3-heptadecenyl]-1H-pyrrole-2-pentanamidato-κN^1]difluoro-, (T-4)-

Other Names   BODIPY FL C$_5$-ganglioside GM$_1$; BODIPY FL C$_5$-GM$_1$; BO-GM1

Merck Index Number   Not listed

Chemical/Dye Class   Boron co-ordination compound/dye

Molecular Formula   C$_{71}$H$_{114}$BF$_2$N$_5$O$_{31}$

Molecular Weight   1582.50

Physical Form   Orange solid

Solubility   Soluble in dimethyl sulfoxide, ethanol, methanol

Absorption (λ_{max})   505 nm (MeOH)

Emission (λ_{max})   512 nm (MeOH)

Molar Extinction Coefficient   80,000 cm^{-1} M^{-1} (MeOH)

Synthesis   Synthetic method[1]

Imaging/Labeling Applications   Amyloid beta-protein;[2] antigen presenting cells (APCs);[3] cholera toxin (CT);[4,5] bilayer lipid membranes;[1] laminin-1;[8] lipid rafts[5–9]

Biological/Medical Applications   Detecting lipid rafts in live red blood cells (RBCs) membranes,[6] protein toxin,[4,5] trogocytosis;[3] investigating binding of laminin-1 to BO-GM1 for neurite outgrowth[8]

Industrial Applications   Not reported

Safety/Toxicity   No data available

REFERENCES

1. Samsonov, A. V.; Mihalyov, I.; Cohen, F. C. Characterization of cholesterol-sphingomyelin domains and their dynamics in bilayer membranes. *Biophys. J.* **2001**, *81*, 1486–1500.

2. Kakio, A.; Nishimoto, S.; Yanagisawa, K.; Kozutsumi, Y.; Matsuzaki, K. Cholesterol-dependent formation of GM1 ganglioside-bound amyloid β-protein, an endogenous seed for Alzheimer amyloid. *J. Biol. Chem.* **2001**, *276*, 24985–24990.

3. Daubeuf, S.; Bordier, C.; Hudrisier, D.; Joly, E. Suitability of various membrane lipophilic probes for the detection of trogocytosis by flow cytometry. *Cytometry, Part A* **2009**, *75A*, 380–389.

4. Ma, G.; Cheng, Q. Manipulating FRET with polymeric vesicles: Development of a "mix-and-detect" type fluorescence sensor for bacterial toxin. *Langmuir* **2006**, *22*, 6743–6745.

5. Shaw, J. E.; Epand, R. F.; Epand, R. M.; Li, Z.; Bittman, R. Correlated fluorescence-atomic force microscopy of membrane domains: Structure of fluorescence probes determines lipid localization. *Biophys. J.* **2006**, *90*, 2170–2178.

6. Mikhalyov, I.; Samsonov, A. Lipid raft detecting in membranes of live erythrocytes. *Biochim. Biophys. Acta, Biomembr.* **2011**, *1808*, 1930–1939.

7. Giurisato, E.; McIntosh, D. P.; Tassi, M.; Gamberucci, A. Benedetti, A. T cell receptor can be recruited to a subset of plasma membrane rafts, independently of cell signaling and attendantly to raft clustering. *J. Biol. Chem.* **2003**, *278*, 6771–6778.

8. Ichikawa, N.; Iwabuchi, K.; Kurihara, H.; Ishii, K.; Kobayashi, T.; Sasaki, T.; Hattori, N.; Mizuno, Y.; Hozumi, K.; Yamada, Y.; Arikawa-Hirasawa, E. Binding of laminin-1 to monosialoganglioside GM1 in lipid rafts is crucial for neurite outgrowth. *J. Cell Sci.* **2009**, *122*, 289–299.

9. Janes, P. W.; Ley, S. L.; Magee, A. I. Aggregation of lipid rafts accompanies signaling via the T cell antigen receptor. *J. Cell Biol.* **1999**, *147*, 447–461.

BODIPY FL C$_5$-LACTOSYLCERAMIDE

CAS Registry Number 251955-11-0

Chemical Structure

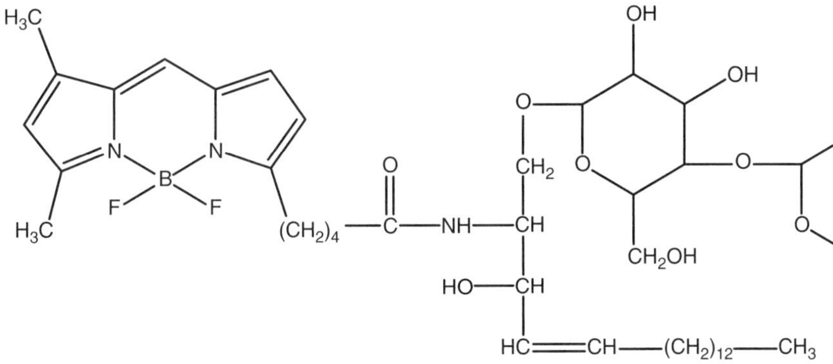

Emission (λ_{max}) 511 nm (MeOH)

Molar Extinction Coefficient 80,000 cm^{-1} M^{-1} (MeOH)

CA Index Name Boron, [5-[(3,5-dimethyl-2*H*-pyrrol-2-ylidene-κ*N*)methyl]-*N*-[(1*S*,2*R*,3*E*)-1-[[(4-*O*-β-D-galactopyranosyl-β-D-glucopyranosyl)oxy]methyl]-2-hydroxy-3-heptadecenyl]-1*H*-pyrrole-2-pentanamidato-κ*N^1*]difluoro-, (*T*-4)-

Other Names BODIPY FL C$_5$-lactosylceramide; BODIPY FL C$_5$-LacCer; BODIPY LacCer; *N*-(4,4-Difluoro-5,7-dimethyl-4-bora-3a,4a-diaza-*s*-indacene-3-pentanoyl)sphingosyl 1-β-D-lactoside

Merck Index Number Not listed

Chemical/Dye Class Boron co-ordination compound/dye

Molecular Formula C$_{46}$H$_{74}$BF$_2$N$_3$O$_{13}$

Molecular Weight 925.91

Physical Form Orange solid

Solubility Soluble in dimethyl sulfoxide, ethanol, methanol

Absorption (λ_{max}) 505 nm (MeOH)

Synthesis Synthetic methods[1–3]

Imaging/Labeling Applications Golgi apparatus;[1,3,5,11,12] antigen presenting cells (APCs);[4] caveolae;[5,6] cellular organelles;[8] endosomes;[11] glycosphingolipids (GSLs);[1,2,6,7,9] lipids;[10] lysosomes;[11] oligodendrocytes;[15] plasma membrane micrometric domains[16]

Biological/Medical Applications Analyzing glycosphingolipid transport;[1,2] detecting Niemann-Pick C (NPC) variants,[11] trogocytosis;[4] diagnosing diseases/disorders associated with an alteration in sialyltransferase activity such as neurological disorders, inflammatory disorders, and cancers;[9] differentiating oligodendrocytes;[15] investigating mechanism of endocytosis,[6] sphingolipid transport/metabolism;[12–14] monitoring glycosphingolipid metabolism;[7] as a substrate for measuring sialyltransferase activity;[1,9] visualizing cellular organelles[8]

Industrial Applications Not reported

Safety/Toxicity No data available

REFERENCES

1. Daikoku, S.; Ono, Y.; Ohtake, A.; Hasegawa, Y.; Fukusaki, E.; Suzuki, K.; Ito, Y.; Goto, S.; Kanie, O. Fluorescence-monitored zero dead-volume nanoLC-microESI-QIT-TOF MS for analysis of fluorescently tagged glycosphingolipids. *Analyst* **2011**, *136*, 1046–1050.

2. Liu, Y.; Bittman, R. Synthesis of fluorescent lactosylceramide stereoisomers. *Chem. Phys. Lipids* **2006**, *142*, 58–69.

3. Martin, O. C.; Pagano, R. E. Internalization and sorting of a fluorescent analog of glucosylceramide to the Golgi apparatus of human skin fibroblasts: utilization of endocytic and nonendocytic transport mechanisms. *J. Cell Biol.* **1994**, *125*, 769–781.

4. Daubeuf, S.; Bordier, C.; Hudrisier, D.; Joly, E. Suitability of various membrane lipophilic probes for the detection of trogocytosis by flow cytometry. *Cytometry, Part A* **2009**, *75A*, 380–389.

5. Puri, V.; Watanabe, R.; Singh, R. D.; Dominguez, M.; Brown, J. C.; Wheatley, C. L.; Marks, D. L.; Pagano, R. E. Clathrin-dependent and -independent intennalization of plasma membrane sphingolipids initiates two Golgi targeting pathways. *J. Cell Biol.* **2001**, *154*, 535–547.

6. Singh, R. D.; Liu, Y.; Wheatley, C. L.; Holicky, E. L.; Makino, A.; Marks, D. L.; Kobayashi, T.; Subramaniam, G.; Bittman, R.; Pagano, R. E. Caveolar endocytosis and microdomain association of a glycosphingolipid analog is dependent on its sphingosine stereochemistry. *J. Biol. Chem.* **2006**, *281*, 30660–30668.

7. Essaka, D. C.; Prendergast, J.; Keithley, R. B.; Palcic, M. M.; Hindsgaul, O.; Schnaar, R. L.; Dovichi, N. J. Metabolic cytometry: Capillary electrophoresis with two-color fluorescence detection for the simultaneous study of two glycosphingolipid metabolic pathways in single primary neurons. *Anal. Chem.* **2012**, *84*, 2799–2804.

8. Suzuki, K.; Tobe, A.; Adachi, S.; Daikoku, S.; Hasegawa, Y.; Shioiri, Y.; Kobayashi, M.; Kanie, O. N-Hexyl-4-aminobutyl glycosides for investigating structures and biological functions of carbohydrates. *Org. Biomol. Chem.* **2009**, *7*, 4726–4733.

9. Sommer-Knudsen, J.; Sun, C. Q.; Hubl, U. A method for the determination of sialyltransferase activity. PCT Int. Appl. WO 2006049519, 2006.

10. D'Auria, L.; Van Der Smissen, P.; Bruyneel, F.; Courtoy, P. J.; Tyteca, D. Segregation of fluorescent membrane lipids into distinct micrometric domains: evidence for phase compartmentation of natural lipids? *PLoS One* **2011**, *6*, e17021.

11. Sun, X.; Marks, D. L.; Park, W. D.; Wheatley, C. L.; Puri, V.; O'Brien, J. F.; Kraft, D. L.; Lundquist, P. A.; Patterson, M. C.; Pagano, R. E.; Snow, K. Niemann-Pick C variant detection by altered sphingolipid trafficking and correlation with mutations within a specific domain of NPC1. *Am. J. Hum. Genet.* **2001**, *68*, 1361–1372.

12. Chen, C. S.; Patterson, M. C.; Wheatley, C. L.; O'Brien, J. F.; Pagano, R. E. Broad screening test for sphingolipid-storage diseases. *Lancet* **1999**, *354*, 901–905.

13. Puri, V.; Watanabe, R.; Dominguez, M.; Sun, X.; Wheatley, C. L.; Marks, D. L.; Pagano, R. E. Cholesterol modulates membrane traffic along the endocytic pathway in sphingolipid-storage diseases. *Nat. Cell Biol.* **1999**, *1*, 386–388.

14. Pagano, R. E.; Chen, C. S. Use of BODIPY-labeled sphingolipids to study membrane traffic along the endocytic pathway. *Ann. N.Y. Acad. Sci.* **1998**, *845*, 152–160.

15. Watanabe, R.; Asakura, K.; Rodriguez, M.; Pagano, R. E. Internalization and sorting of plasma membrane sphingolipid analogues in differentiating oligodendrocytes. *J. Neurochem.* **1999**, *73*, 1375–1383.

16. Tyteca, D.; D'Auria, L.; Van Der Smissen, P.; Medts, T.; Carpentier, S.; Monbaliu, J. C.; de Diesbach, P.; Courtoy, P. J. Three unrelated sphingomyelin analogs spontaneously cluster into plasma membrane micrometric domains. *Biochim. Biophys. Acta, Biomembr.* **2010**, *1798*, 909–927.

BODIPY FL C$_5$-SPHINGOMYELIN (C$_5$-DMB-SPHINGOMYELIN)

CAS Registry Number 191853-29-9

Chemical Structure

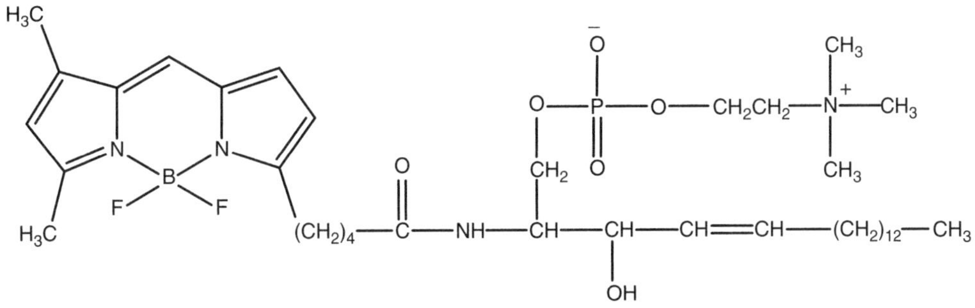

Physical Form Orange solid

Solubility Soluble in chloroform, methanol

Absorption (λ_{max}) 505 nm (MeOH)

Emission (λ_{max}) 512 nm (MeOH)

Molar Extinction Coefficient 77,000 cm^{-1} M^{-1} (MeOH)

CA Index Name Boron, [(7S)-13-[2-[(3,5-dimethyl-1H-pyrrol-2-yl-κN)-methylene]-2H-pyrrol-5-yl-κN]-4-hydroxy-7-[(1R,2E)-1-hydroxy-2-hexadecenyl]-N,N,N-trimethyl-9-oxo-3,5-dioxa-8-aza-4-phosphatridecan-1-aminium 4-oxidato(2-)]-difluoro-, (T-4)-

Other Names BODIPY FL C$_5$-sphingomyelin; BODIPY FL C$_5$-SM; C$_5$-DMB-sphingomyelin; C$_5$-DMB-SM; N-(4,4-difluoro-5,7-dimethyl-4-bora-3a,4a-diaza-s-indacene-3-pentanoyl)sphingosyl phospho choline

Merck Index Number Not listed

Chemical/Dye Class Boron co-ordination compound/dye

Molecular Formula C$_{39}$H$_{66}$BF$_2$N$_4$O$_6$P

Molecular Weight 766.75

Synthesis Synthetic methods[1-3]

Imaging/Labeling Applications Golgi apparatus;[4,17] antigen presenting cells (APCs);[5] endosomes;[6,11] immobilized enzymes;[7] lipids;[8-13] lipid bilayer membranes;[14] low-density lipoproteins;[6] lysosomes;[11] plasma membrane micrometric domains[15,16]

Biological/Medical Applications Detecting trogocytosis;[5] distinguishing unique populations of early endosomes from one another;[6,11] examing lipid bilayer membranes;[14] investigating sphingolipid transport and metabolism;[1,9-12,17,18] measuring spontaneous transfer and transbilayer movement of lipids in lipid vesicles;[13] studying enzymic reactions,[7] membrane lipid traffic along the endocytic pathway[9-12]

Industrial Applications Not reported

Safety/Toxicity No data available

REFERENCES

1. Koval, M.; Pagano, R. E. Lipid recycling between plasma membrane and intracellular compartments: Transport and metabolism of fluorescent sphingomyelin analogues and cultured fibroblasts. *J. Cell Biol.* **1989**, *108*, 2169–2181.

2. Lipsky, N. G.; Pagano, R. E. Fluorescent sphingomyelin labels the plasma membrane of cultured fibroblasts. *Ann. N.Y. Acad. Sci.* **1984**, *435*, 306–308.

3. Kishimoto, Y. A facile synthesis of ceramides. *Chem. Phys. Lipids* **1975**, *15*, 33–36.

4. Martin, O. C.; Pagano, R. E. Internalization and sorting of a fluorescent analog of glucosylceramide to the Golgi apparatus of human skin fibroblasts: utilization of endocytic and nonendocytic transport mechanisms. *J. Cell Biol.* **1994**, *125*, 769–781.

5. Daubeuf, S.; Bordier, C.; Hudrisier, D.; Joly, E. Suitability of various membrane lipophilic probes for the detection of trogocytosis by flow cytometry. *Cytometry, Part A* **2009**, *75A*, 380–389.

6. Chen, C. S.; Martin, O. C.; Pagano, R. E. Changes in the spectral properties of a plasma membrane lipid analog during the first seconds of endocytosis in living cells. *Biophys. J.* **1997**, *72*, 37–50.

7. Nurminen, T. A.; Holopainen, J. M.; Zhao, H.; Kinnunen, P. K. J. Observation of topical catalysis by sphingomyelinase coupled to microspheres. *J. Am. Chem. Soc.* **2002**, *124*, 12129–12134.

8. Kuerschner, L.; Ejsing, C. S.; Ekroos, K.; Shevchenko, A.; Anderson, K. I.; Thiele, C. Polyene-lipids: a new tool to image lipids. *Nat. Methods* **2005**, *2*, 39–45.

9. Marks, D. L.; Singh, R. D.; Choudhury, A.; Wheatley, C. L.; Pagano, R. E. Use of fluorescent sphingolipid analogs to study lipid transport along the endocytic pathway. *Methods* **2005**, *36*, 186–195.

10. Pagano, R. E.; Watanabe, R.; Wheatley, C.; Dominguez, M. Applications of BODIPY-sphingolipid analogs to study lipid traffic and metabolism in cells. *Methods Enzymol.* **2000**, *312*, 523–534.

11. Pagano, R. E.; Watanabe, R.; Wheatley, C.; Chen, C.-S. Use of N-5-(5,7-dimethyl boron dipyrromethene difluoride)-sphingomyelin to study membrane traffic along the endocytic pathway. *Chem. Phys. Lipids* **1999**, *102*, 55–63.

12. Pagano, R. E.; Chen, C. S. Use of BODIPY-labeled sphingolipids to study membrane traffic along the endocytic pathway. *Ann. N.Y. Acad. Sci.* **1998**, *845*, 152–160.

13. Bai, J.; Pagano, R. E. Measurement of spontaneous transfer and transbilayer movement of BODIPY-labeled lipids in lipid vesicles. *Biochemistry* **1997**, *36*, 8840–8848.

14. Baumgart, T.; Hunt, G.; Farkas, E. R.; Webb, W. W.; Feigenson, G. W. Fluorescence probe partitioning between Lo/Ld phases in lipid membranes. *Biochim. Biophys. Acta, Biomembr.* **2007**, *1768*, 2182–2194.

15. Tyteca, D.; D'Auria, L.; Van Der Smissen, P.; Medts, T.; Carpentier, S.; Monbaliu, J. C.; de Diesbach, P.; Courtoy, P. J. Three unrelated sphingomyelin analogs spontaneously cluster into plasma membrane micrometric domains. *Biochim. Biophys. Acta, Biomembr.* **2010**, *1798*, 909–927.

16. Marks, D. L.; Bittman, R.; Pagano, R. E. Use of Bodipy-labeled sphingolipid and cholesterol analogs to examine membrane microdomains in cells. *Histochem. Cell Biol.* **2008**, *130*, 819–832.

17. Chen, C. S.; Patterson, M. C.; Wheatley, C. L.; O'Brien, J. F.; Pagano, R. E. Broad screening test for sphingolipid-storage diseases. *Lancet* **1999**, *354*, 901–905.

18. Puri, V.; Watanabe, R.; Dominguez, M.; Sun, X.; Wheatley, C. L.; Marks, D. L.; Pagano, R. E. Cholesterol modulates membrane traffic along the endocytic pathway in sphingolipid-storage diseases. *Nat. Cell Biol.* **1999**, *1*, 386–388.

BODIPY FL C$_5$-SUCCINIMIDYL ESTER

CAS Registry Number 133867-52-4

Chemical Structure

CA Index Name Boron, [1-[[5-[5-[(3,5-dimethyl-2*H*-pyrrol-2-ylidene-κ*N*)methyl]-1*H*-pyrrol-2-yl-κ*N*]-1-oxopentyl]oxy]-2,5-pyrrolidinedionato]difluoro-, (*T*-4)-

Other Names Boron, [1-[[5-[5-[(3,5-dimethyl-2*H*-pyrrol-2-ylidene)methyl]-1*H*-pyrrol-2-yl]-1-oxopentyl]oxy]-2,5-pyrrolidinedionato]difluoro-, (*T*-4)-; 2,5-Pyrrolidinedione, 1-[[5-[5-[(3,5-dimethyl-2*H*-pyrrol-2-ylidene)methyl]-1*H*-pyrrol-2-yl]-1-oxopentyl]oxy]-, boron complex; BODIPY FL-C$_5$-SE; BODIPY FL C$_5$-succinimidyl ester; BODIPY FL C$_5$-N-hydroxysuccinimidyl (NHS) ester; BODIPY FL C$_5$- NHS ester; BODIPY-NHS ester

Merck Index Number Not listed

Chemical/Dye Class Boron co-ordination compound/dye

Molecular Formula C$_{20}$H$_{22}$BF$_2$N$_3$O$_4$

Molecular Weight 417.22

Physical Form Orange solid

Solubility Soluble in acetonitrile, dimethyl sulfoxide, methanol

Absorption (λ_{max}) 504 nm (MeOH)

Emission (λ_{max}) 511 nm (MeOH)

Molar Extinction Coefficient 87,000 cm^{-1} M^{-1} (MeOH)

Synthesis Synthetic method[1]

Imaging/Labeling Applications Glycosphingolipids (GSLs);[2,3] nucleoside triphosphates;[4–6] oligonucleotides;[7] pladienolide[8]

Biological/Medical Applications Dyed polymer microparticles;[9,10] purifying oligonucleotides;[7] regulating meiosis;[11] as a substrate for measuring lysophospholipase D (lysoPLD) activity,[12–14] phospholipases A (PLA) activity[15]

Industrial Applications For synthesizing boron dipyrromethene difluoride (BODIPY) ceramide analogs;[16] for synthesizing lactosylceramide stereoisomers[3]

Safety/Toxicity No data available

REFERENCES

1. Haugland, R. P. *Handbook of Fluorescent Probes and Research Chemicals*; Molecular Probes Inc.: Eugene, **1996**; pp 13–17.

2. Sarver, S. A.; Keithley, R. B.; Essaka, D. C.; Tanaka, H.; Yoshimura, Y.; Palcic, M. M.; Hindsgaul, O.; Dovichi, N. J. Preparation and electrophoretic separation of Bodipy-Fl-labeled glycosphingolipids. *J. Chromatogr., A* **2012**, *1229*, 268–273.

3. Liu, Y.; Bittman, R. Synthesis of fluorescent lactosylceramide stereoisomers. *Chem. Phys. Lipids* **2006**, *142*, 58–69.

4. Benner, S. A.; Hutter, D.; Leal, N. A.; Chen, F. Reagents for reversibly terminating primer extension. U.S. Pat. Appl. Publ. US 20110275124, 2011.

5. Hutter, D.; Kim, M.; Karalkar, N.; Leal, N. A.; Chen, F.; Guggenheim, E.; Visalakshi, V.; Olejnik, J.; Gordon, S.; Benner, S. A. Labeled nucleoside triphosphates with reversibly terminating aminoalkoxyl groups. *Nucleosides, Nucleotides Nucleic Acids* **2010**, *29*, 879–895.

6. Benner, S. A.; Hutter, D.; Leal, N. A.; Karalkar, N. B. Reagents for reversibly terminating primer extension. PCT Int. Appl. WO 2010110775, 2010.

7. Hanning, A.; Hellstroem, J. Method for the purification of synthetic oligonucleotides containing one or several labels. PCT Int. Appl. WO 2006132588, 2006.

8. Kotake, Y.; Sagane, K.; Owa, T.; Mizui, Y.; Shimizu, H.; Kiyosue, Y. Pladienolide target molecule, compound capable of binding to the target molecule, and screening method for the compound. PCT Int. Appl. WO 2008016187, 2008.

9. Banerjee, S.; Georgescu, C.; Seul, M. Method for controlling solute loading of polymer microparticles. U.S. Pat. Appl. Publ. US 20040142102, 2004.

10. Banerjee, S.; Georgescu, C.; Daniels, E. S.; Dimonie, V. L.; Seul, M. Production of dyed polymer microparticles. U.S. Pat. Appl. Publ. US 20040139565, 2004.

11. Blume, T.; Esperling, P.; Kuhnke, J.; Hegele-Hartung, C.; Lessl, M. Unsaturated cholestane derivatives and their use for the preparation of meiosis regulating medicaments. PCT Int. Appl. WO 2000047604, 2000.

12. Ferguson, C.; Prestwich, G.; Madan, D. Compounds and methods of use thereof for assaying lysophospholipase D activity. U.S. Pat. Appl. Publ. US 20100260682, 2010.

13. Ferguson, C. G.; Bigman, C. S.; Richardson, R. D.; Van Meeteren, L. A.; Moolenaar, W. H.; Prestwich, G. D. Fluorogenic phospholipid substrate to detect lysophospholipase D/autotaxin activity. *Org. Lett.* **2006**, *8*, 2023–2026.

14. Ferguson, C.; Prestwich, G. Compounds and methods of use thereof for assaying lysophospholipase D activity. PCT Int. Appl. WO 2004053459, 2004.

15. Rose, T. M.; Prestwich, G. D. Fluorogenic phospholipids as head group-selective reporters of phospholipase A activity. *ACS Chem. Biol.* **2006**, *1*, 83–92.

16. Pagano, R. E.; Martin, O. C.; Kang, H. C.; Haugland, R. P. A novel fluorescent ceramide analog for studying membrane traffic in animal cells: Accumulation at the Golgi apparatus results in altered spectral properties of the sphingolipid precursor. *J. Cell Biol.* **1991**, *113*, 1267–1279.

BODIPY TR CERAMIDE

CAS Registry Number   571186-05-5

Chemical Structure

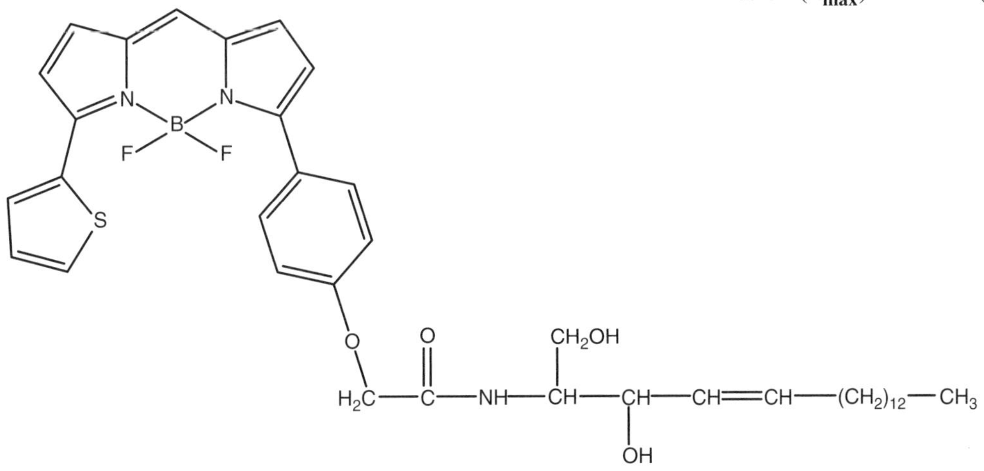

CA Index Name   Boron, difluoro[N-[(1S,2R,3E)-2-hydroxy-1-(hydroxymethyl)-3-heptadecen-1-yl]-2-[4-[5-[[5-(2-thienyl)-2H-pyrrol-2-ylidene-κN]methyl]-1H-pyrrol-2-yl-κN]phenoxy]acetamidato]-, (T-4)-

Other Names   Boron, difluoro[N-[(1S,2R,3E)-2-hydroxy-1-(hydroxymethyl)-3-heptadecenyl]-2-[4-[5-[[5-(2-thienyl)-2H-pyrrol-2-ylidene-κN]methyl]-1H-pyrrol-2-yl-κN]phenoxy]acetamidato]-, (T-4)-; BODIPY TR ceramide; Bodipy TR C5 ceramide

Merck Index Number   Not listed

Chemical/Dye Class   Boron co-ordination compound/dye

Molecular Formula $C_{39}H_{50}BF_2N_3O_4S$

Molecular Weight   705.71

Physical Form   Purple solid

Solubility   Soluble in chloroform, dimethyl sulfoxide, methanol

Absorption (λ_{max})   589 nm (MeOH)

Emission (λ_{max})   616 nm (MeOH)

Molar Extinction Coefficient   65,000 cm^{-1} M^{-1} (MeOH)

Synthesis   Synthetic method[1]

Imaging/Labeling Applications   Golgi apparatus;[1–6] cells;[7,9,10] liposomes;[8] live malaria blood stage parasites;[9,10] PC-3 prostate cancer cells;[11] sphingomyelin[2]

Biological/Medical Applications   Demonstrating sphingomyelin intracellular distribution;[2] mapping whole cell architecture;[10] revealing modular features of exomembrane system of malaria parasite;[10] studying malaria blood stages biology[9]

Industrial Applications   Not reported

Safety/Toxicity   No data available

REFERENCES

1. Haugland, R. P. *Handbook of Fluorescent Probes and Research Chemicals*; Molecular Probes Inc.: Eugene, **1996**; pp 279–282.

2. Bakrac, B.; Kladnik, A.; Macek, P.; McHaffie, G.; Werner, A.; Lakey, J. H.; Anderluh, G. A toxin-based probe reveals cytoplasmic exposure of golgi sphingomyelin. *J. Biol. Chem.* **2010**, *285*, 22186–22195.

3. Puthenveedu, M. A.; Bachert, C.; Puri, S.; Lanni, F.; Linstedt, A. D. GM130 and GRASP65-dependent lateral cisternal fusion allows uniform Golgi-enzyme distribution. *Nat. Cell. Biol.* **2006**, *8*, 238–248.

4. Verma, A.; Davis, G. E.; Ihler, G. M. Infection of human endothelial cells with *Bartonella bacilliformis* in dependent on Rho and results in activation of Rho. *Infect. Immun.* **2000**, *68*, 5960–5969.

5. Field, H.; Sherwin, T.; Smith, A. C.; Gull, K.; Field, M. C. Cell-cycle and developmental regulation of TbRAB31 localisation, a GTP-locked Rab protein from *Trypanosoma brucei. Mol. Biochem. Parasitol.* **2000**, *106*, 21–35.

6. Dohrman, D. P.; Diamond, I.; Gordon, A. S. Ethanol causes translocation of cAMP-dependent protein kinase catalytic subunit to the nucleus. *Proc. Natl. Acad. Sci. U.S.A.* **1996**, *93*, 10217–10221.

7. Bestvater, F.; Spiess, E.; Stobrawa, G.; Hacker, M.; Feurer, T.; Porwol, T.; Berchner-Pfannschmidt, U.; Wotzlaw, C.; Acker, H. Two-photon fluorescence absorption and emission spectra of dyes relevant for cell imaging. *J. Microsc.* **2002**, *208*, 108–115.

8. Reppy, M. A. Enhancing the emission of polydiacetylene sensing materials through fluorophore addition and energy transfer. *J. Fluoresc.* **2008**, *18*, 461–471.

9. Gruering, C.; Spielmann, T. Imaging of live malaria blood stage parasites. *Methods Enzymol.* **2012**, *506*, 81–92.

10. Hanssen, E.; Carlton, P.; Deed, S.; Klonis, N.; Sedat, J.; DeRisi, J.; Tilley, L. Whole cell imaging reveals novel modular features of the exomembrane system of the malaria parasite, *Plasmodium falciparum. Int. J. Parasitol.* **2010**, *40*, 123–134.

11. Bhushan, K. R.; Liu, F.; Misra, P.; Frangioni, J. V. Microwave-assisted synthesis of near-infrared fluorescent sphingosine derivatives. *Chem. Commun.* **2008**, 4419–4421.

BO-PRO 1

CAS Registry Number 157199-57-0

Chemical Structure

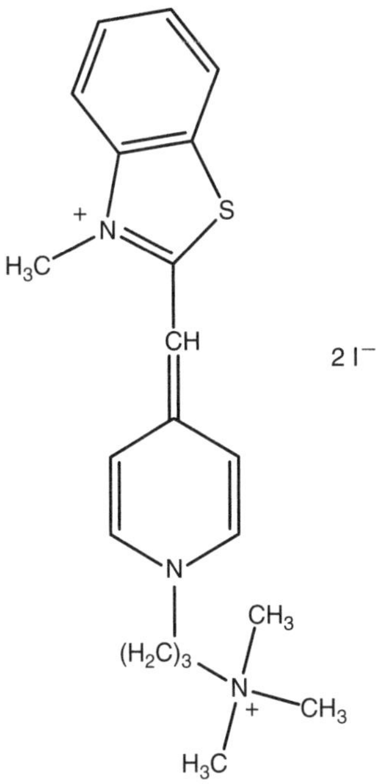

CA Index Name Benzothiazolium, 3-methyl-2-[[1-[3-(trimethylammonio)propyl]-4(1H)-pyridinylidene]methyl]-,iodide (1:2)

Other Names Benzothiazolium, 3-methyl-2-[[1-[3-(trimethylammonio)propyl]-4(1H)-pyridinylidene]methyl]-, diiodide; BO-PRO 1; BO-PRO 1 iodide

Merck Index Number Not listed

Chemical/Dye Class Cyanine

Molecular Formula $C_{20}H_{27}I_2N_3S$

Molecular Weight 595.32

Physical Form Solid

Solubility Soluble in dimethyl sulfoxide

Absorption (λ_{max}) 462 nm (H$_2$O/DNA)

Emission (λ_{max}) 481 nm (H$_2$O/DNA)

Molar Extinction Coefficient 58,100 cm^{-1} M^{-1} (H$_2$O/DNA)

Quantum Yield 0.16 (H$_2$O/DNA)

Synthesis Synthetic methods[1,2]

Imaging/Labeling Applications Nucleic acids;[1–10] cells;[1,2,10] bacteria;[2] nuclei;[11–13] cytoplasm;[13] megakaryocyte;[14] microorganisms;[15] sperms[10,16]

Biological/Medical Applications Detecting nucleic acids,[1–10] cells,[1,2,10] microorganisms,[15] classifying/counting megakaryocyte;[14] counting erythroblasts,[11,12] sperms;[10,16] nucleic acids amplification;[9] nucleic acid sequencing;[17] as temperature sensor[18,19]

Industrial Applications Not reported

Safety/Toxicity No data available

REFERENCES

1. Yue, S. T.; Johnson, I. D.; Huang, Z.; Haugland, R. P. Unsymmetrical cyanine dyes with a cationic side chain. U.S. Patent 5321130, 1994.

2. Millard, P. J.; Roth, B. L.; Yue, S. T.; Haugland, R. P. Fluorescent viability assay using cyclic-substituted unsymmetrical cyanine dyes. U.S. Patent 5534416, 1996.

3. Sergeev, N. V.; Brevnov, M. G.; Furtado, M. R. Identification of nucleic acids. PCT Int. Appl. WO 2011143478, 2011.

4. Exner, M.; Rogers, A. Methods for detecting nucleic acids using multiple signals. U.S. Pat. Appl. Publ. US 2007172836, 2007.

5. Wittwer, C. T.; Dujols, V. E.; Reed, G.; Zhou, L. Amplicon melting analysis with saturation dyes. PCT Int. Appl. WO 2004038038, 2004.

6. Erikson, G. H.; Daksis, J. I. Pre-incubation method to improve signal/noise ratio of nucleic acid assays. U.S. Pat. Appl. Publ. US 2004180345, 2004.

7. Erikson, G. H. Method for modifying transcription and/or translation in an organism for therapeutic, prophylactic and/or analytic uses. U.S. Pat. Appl. Publ. US 2003181412, 2003.

8. Erikson, G. H.; Daksis, J. I.; Kandic, I.; Picard, P. Nucleic acid multiplex formation. PCT Int. Appl. WO 2002103051, 2002.

9. Sutherland, J. W.; Patterson, D. R. Homogeneous method for assay of double-stranded nucleic acids using fluorescent dyes and kit useful therein. Eur. Pat. Appl. EP 684316, 1995.

10. Anderson, A. L.; Knutson, C. R.; Mueth, D.; Plewa, J.; Tanner, E. Methods for staining cells for

identification and sorting. U.S. Pat. Appl. Publ. US 2006172315, 2006.

11. Heuven, B.; Wong, F.; Tsuji, T.; Sakata, T.; Hamaguchi, I. Method for classifying and counting erythroblasts by flow cytometry. Jpn. Kokai Tokkyo Koho JP 11326323, 1999.

12. Heuven, B.; Wong, F.; Tsuji, T.; Sakata, T.; Hamaguchi, I. Process for discriminating and counting erythroblasts. U.S. Pat. Appl. Publ. US 20020006631, 2002.

13. Bink, K.; Walch, A.; Feuchtinger, A.; Eisenmann, H.; Hutzler, P.; Hofler, H.; Werner, M. TO-PRO-3 is an optimal fluorescent dye for nuclear counterstaining in dual-colour FISH on paraffin sections. *Histochem. Cell Biol.* **2001**, *115*, 293–299.

14. Minakami, T.; Mori, Y.; Tsuji, T.; Ikeuchi, Y. Megakaryocyte classification/counting method by double fluorescent staining and flow cytometry. Jpn.Kokai Tokkyo Koho JP 2006275985, 2006.

15. Eckert, R. H.; Kaplan, C.; He, J.; Yarbrough, D. K.; Anderson, M.; Sim, J. Methods and devices for the selective detection of microorganisms. U.S. Pat. Appl. Publ. US 20120003661, 2012.

16. Matsumoto, T.; Okada, H.; Hamaguchi, Y. Method and reagent for counting sperm by flow cytometry. Jpn. Kokai Tokkyo Koho JP 2001242168, 2001.

17. Hoser, M. J. Nucleic acid sequencing methods, kits and reagents. PCT Int. Appl. WO 2004074503, 2004.

18. Bousseksou, A.; Salmon, L.; Molnar, G.; Cobo, S. Materials with thermochromic spin transition doped with one or more fluorescent agents for use as temperature sensor. Fr. Demande FR 2952371, 2011.

19. Bousseksou, A.; Salmon, L.; Molnar, G.; Cobo, S. Heat-sensitive spin-transition materials doped with one or more fluorescent agents for use as temperature sensor. PCT Int. Appl. WO 2011058277, 2011.

BO-PRO 3

CAS Registry Number   173357-16-9

Chemical Structure

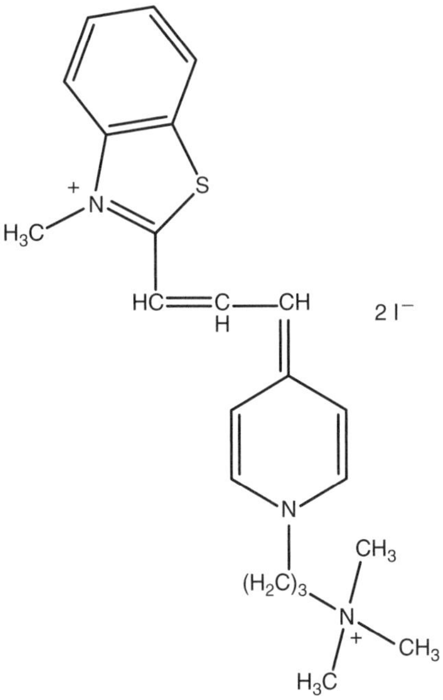

CA Index Name   Benzothiazolium, 3-methyl-2-[3-[1-[3-(trimethylammonio)propyl]-4(1H)-pyridinylidene]-1-propen-1-yl]-, iodide (1:2)

Other Names   Benzothiazolium, 3-methyl-2-[3-[1-[3-(trimethylammonio)propyl]-4(1H)-pyridinylidene]-1-propenyl]-, diiodide; BO-PRO 3; BO-PRO 3 iodide

Merck Index Number   Not listed

Chemical/Dye Class   Cyanine

Molecular Formula $C_{22}H_{29}I_2N_3S$

Molecular Weight   621.36

Physical Form   Solid

Solubility   Soluble in dimethyl sulfoxide

Absorption (λ_{max})   575 nm (H_2O/DNA)

Emission (λ_{max})   599 nm (H_2O/DNA)

Molar Extinction Coefficient   80,900 cm^{-1} M^{-1} (H_2O/DNA)

Quantum Yield   0.62 (H_2O/DNA)

Synthesis   Synthetic method[1]

Imaging/Labeling Applications   Nucleic acids;[1–10] cells;[1,10] bacteria;[1] microorganisms;[11] sperms[10]

Biological/Medical Applications   Detecting nucleic acids,[1–10] cells,[1,10] microorganisms;[11] nucleic acids amplification;[9] as temperature sensor[12,13]

Industrial Applications   Photonic fabric display[14]

Safety/Toxicity   No data available

REFERENCES

1. Millard, P. J.; Roth, B. L.; Yue, S. T.; Haugland, R. P. Fluorescent viability assay using cyclic-substituted unsymmetrical cyanine dyes. U.S. Patent 5534416, 1996.

2. Exner, M.; Rogers, A. Methods for detecting nucleic acids using multiple signals. U.S. Pat. Appl. Publ. US 2007172836, 2007.

3. Wittwer, C. T.; Dujols, V. E.; Reed, G.; Zhou, L. Amplicon melting analysis with saturation dyes. PCT Int. Appl. WO 2004038038, 2004.

4. Erikson, G. H.; Daksis, J. I. Pre-incubation method to improve signal/noise ratio of nucleic acid assays. U.S. Pat. Appl. Publ. US 2004180345, 2004.

5. Marin, V.; Hansen, H. F.; Koch, T.; Armitage, B. A. Effect of LNA modifications on small molecule binding to nucleic acids. *J. Biomol. Struct. Dyn.* **2003**, *21*, 841–850.

6. Erikson, G. H. Method for modifying transcription and/or translation in an organism for therapeutic, prophylactic and/or analytic uses. U.S. Pat. Appl. Publ. US 2003181412, 2003.

7. Kang, J. S.; Piszczek, G.; Lakowicz, J. R. Enhanced emission induced by FRET from a long-lifetime, low quantum yield donor to a long-wavelength, high quantum yield acceptor. *J. Fluoresc.* **2002**, *12*, 97–103.

8. Erikson, G. H.; Daksis, J. I.; Kandic, I.; Picard, P. Nucleic acid multiplex formation. PCT Int. Appl. WO 2002103051, 2002.

9. Sutherland, J. W.; Patterson, D. R. Homogeneous method for assay of double-stranded nucleic acids using fluorescent dyes and kit useful therein. Eur. Pat. Appl. EP 684316, 1995.

10. Anderson, A. L.; Knutson, C. R.; Mueth, D.; Plewa, J.; Tanner, E. Methods for staining cells for identification and sorting. U.S. Pat. Appl. Publ. US 2006172315, 2006.

11. Eckert, R. H.; Kaplan, C.; He, J.; Yarbrough, D. K.; Anderson, M.; Sim, J. Methods and devices for the selective detection of microorganisms. U.S. Pat. Appl. Publ. US 20120003661, 2012.

12. Bousseksou, A.; Salmon, L.; Molnar, G.; Cobo, S. Materials with thermochromic spin transition doped with one or more fluorescent agents for use as temperature sensor. Fr. Demande FR 2952371, 2011.

13. Bousseksou, A.; Salmon, L.; Molnar, G.; Cobo, S. Heat-sensitive spin-transition materials doped with one or more fluorescent agents for use as temperature sensor. PCT Int. Appl. WO 2011058277, 2011.

14. Tao, X.; Cheng, X.; Yu, J.; Liu, L.; Wong, W.; Tam, W. Photonic fabric display with controlled pattern, color, luminescence intensity, scattering intensity and light self-amplification. U.S. Pat. Appl. Publ. US 20070281155, 2007.

4-BROMOMETHYL-6,7-DIMETHOXYCOUMARIN

CAS Registry Number 88404-25-5

Chemical Structure

CA Index Name 2*H*-1-Benzopyran-2-one, 4-(bromomethyl)-6,7-dimethoxy-

Other Names 4-Bromomethyl-6,7-dimethoxycoumarin; 4-(Bromomethyl)-6,7-dimethoxycoumarin; 4-(Bromomethyl)-6,7-dimethoxy-2-oxo-2*H*-benzopyran; BrCU; 4-Bdmc; BDMC; BrDMC; Brmdmc; Br-Mdmc; Brdmc-4,6,7

Merck Index Number Not listed

Chemical/Dye Class Coumarin

Molecular Formula $C_{12}H_{11}BrO_4$

Molecular Weight 299.12

Physical Form Yellow crystals or powder

Solubility Soluble in acetonitrile, dichloromethane, *N*,*N*-dimethylformamide, methanol

Melting Point 208 °C[1]

Boiling Point (Calcd.) 417.1 ± 45.0 °C, pressure: 760 Torr

Absorption (λ_{max}) 348 nm (MeOH)

Emission (λ_{max}) 425 nm (MeOH)

Molar Extinction Coefficient 10,631 $cm^{-1} M^{-1}$ (MeOH)

Quantum Yield 0.24 (MeOH)

Synthesis Synthetic method[1]

Imaging/Labeling Applications Bacteria;[2] bile acid;[3] calcium-magnesium-ATPase;[4] myosin ATPase;[5] carboxylic acids;[1,6,7,9–11] cyclosporine;[8] fatty acids;[1,9–11], lauric acid metabolites;[10,11] metail ions;[12–17] proteins;[18–20] sulfhydryl groups[5,19,20]

Biological/Medical Applications Analyzing bile acid;[3] anticancer agents;[21,22] identifying bacteria;[2] identifying/measuring lauric acid metabolites;[10,11] measuring cyclosporine;[8] studying/monitoring early stages of protein folding;[18] treating gastrointestinal diseases;[23] derivatization reagent for carboxylic acids;[1,6,7,9–11] as a substrate for measuring lauric acid hydroxylase activity[10,11]

Industrial Applications Electrostatic imaging process;[24] resists;[25] printed circuit boards;[26] semiconductor devices[26]

Safety/Toxicity No data available

REFERENCES

1. Farinotti, R.; Siard, P.; Bourson, J.; Kirkiacharian, S.; Valeur, B.; Mahuzier, G. 4-Bromomethyl-6,7-dimethoxycoumarin as a fluorescent label for carboxylic acids in chromatographic detection. *J. Chromatogr.* **1983**, *269*, 81–90.

2. Simon, V. A.; Hale, Y.; Taylor, A. Identification of *Mycobacterium tuberculosis* and related organisms by HPLC with fluorescence detection. *LC-GC* **1993**, *11*, 444, 446, 448.

3. Budai, K.; Javitt, N. B. Bile acid analysis in biological fluids: a novel approach. *J. Lipid Res.* **1997**, *38*, 1906–1912.

4. Stefanova, H. I.; East, J. M.; Gore, M. G.; Lee, A. G. Labeling the calcium-magnesium-ATPase of sarcoplasmic reticulum with 4-(bromomethyl)-6,7-dimethoxycoumarin: detection of conformational changes. *Biochemistry* **1992**, *31*, 6023–6031.

5. Hiratsuka, T. Selective fluorescent labeling of the 50-, 26-, and 20-kilodalton heavy chain segments of myosin ATPase. *J. Biochem.* **1987**, *101*, 1457–1462.

6. Peng, M.; Wang, M.; Cui, Y.; Chen, C.; Zhi, Z. Method for determining surface terminal carboxyl group content of magnetic nanoparticle with spectrophotometry. Faming Zhuanli Shenqing Gongkai Shuomingshu CN 101706420, 2010.

7. Coenen, A. J. J. M.; Kerkhoff, M. J. G.; Heringa, R. M.; Van der Wal, S. Comparison of several methods for the determination of trace amounts of polar aliphatic monocarboxylic acids by high-performance liquid chromatography. *J. Chromatogr.* **1992**, *593*, 243–252.

8. French, M. T.; Miller, J. N.; Seare, N. J.; Lachno, D. R.; Yacoub, M. H. New fluorescent derivatives of cyclosporin for use in immunoassays. *J. Pharm. Biomed. Anal.* **1992**, *10*, 23–30.

9. Alvarez, J. G.; Fang, X. G.; Grob, R. L.; Touchstone, J. C. Determination of free fatty acids by diphasic-two dimensional TLC-fluorescence spectrodensitometry. *J. Liq. Chromatogr.* **1990**, *13*, 2727–2735.

10. Amet, Y.; Berthou, F.; Menez, J. F. Simultaneous radiometric and fluorimetric detection of lauric acid metabolites using high-performance liquid chromatography following esterification with 4-bromomethyl-6,7-dimethoxycoumarin in human and rat liver microsomes. *J. Chromatogr., B: Biomed. Appl.* **1996**, *681*, 233–239.

11. Dirven, H. A. A. M.; De Bruijn, A. A. G. M.; Sessink, P. J. M.; Jongeneelen, F. J. Determination of the cytochrome P-450 IV marker ω-hydroxylauric acid, by high-performance liquid chromatography and fluorometric detection. *J. Chromatogr., Biomed. Appl.* **1991**, *564*, 266–271.

12. Jang, Y. J.; Moon, B.; Park, M. S.; Kang, B.; Kwon, J. Y.; Hong, J. S. J.; Yoon, Y. J.; Lee, K. D.; Yoon, J. New cavitand derivatives bearing four coumarin groups as fluorescent chemosensors for Cu2+ and recognition of dicarboxylates utilizing Cu2+ complex. *Tetrahedron Lett.* **2006**, *47*, 2707–2710.

13. Ryu, D. H.; Noh, J. H.; Chang, S. Selective ratiometric signaling of Hg2+ ions by a fluorescein-coumarin chemodosimeter. *Bull. Korean Chem. Soc.* **2010**, *31*, 246–249.

14. Kim, H.; Park, J.; Choi, M.; Ahn, S.; Chang, S. Selective chromogenic and fluorogenic signalling of Hg2+ ions using a fluorescein-coumarin conjugate. *Dyes Pigm.* **2010**, *84*, 54–58.

15. Lim, N. C.; Schuster, J. V.; Porto, M. C.; Tanudra, M. A.; Yao, L.; Freake, H. C.; Brueckner, C. Coumarin-based chemosensors for zinc(II): Toward the determination of the design algorithm for CHEF-type and ratiometric probes. *Inorg. Chem.* **2005**, *44*, 2018–2030.

16. Lim, N. C.; Brueckner, C. DPA-substituted coumarins as chemosensors for zinc(II): modulation of the chemosensory characteristics by variation of the position of the chelate on the coumarin. *Chem. Commun.* **2004**, 1094–1095.

17. Lim, N. C.; Yao, L.; Freake, H. C.; Bruckner, C. Synthesis of a fluorescent chemosensor suitable for the imaging of zinc(II) in live cells. *Bioorg. Med. Chem. Lett.* **2003**, *13*, 2251–2254.

18. Chen, H.; Hsu, J. C.; Viet, M. H.; Li, M. S.; Hu, C.; Liu, C.; Luh, F. Y.; Chen, S. S.; Chang, E. S.; Wang, A. H. Studying submicrosecond protein folding kinetics using a photolabile caging strategy and time-resolved photoacoustic calorimetry. *Proteins: Struct., Funct., Bioinf.* **2010**, *78*, 2973–2983.

19. Baranowska-Kortylewicz, J.; Kassis, A. I. Labeling of sulfhydryl groups in intact mammalian cells with coumarins. *Bioconjugate Chem.* **1993**, *4*, 305–307.

20. Baranowska-Kortylewicz, J.; Kassis, A. I. Labeling of immunoglobulins with bifunctional, sulfhydryl-selective, and photoreactive coumarins. *Bioconjugate Chem.* **1993**, *4*, 300–304.

21. Nam, N.; Ngoc, N. A.; Zun, A. B. Synthesis of (6,7-dimethoxy-2-oxo-2H-chromen-4-yl)methyl 3-arylacrylates as water soluble antitumor agents. *Lett. Drug Des. Discov.* **2011**, *8*, 312–316.

22. Goto, J. Anticancer agents containing coumarins. Jpn. Kokai Tokkyo Koho JP 05000945, 1993.

23. Jung, S. N. Coumarin derivatives for preventing and treating gastrointestinal diseases. Repub. Korean Kongkae Taeho Kongbo KR 2010059036, 2010.

24. Malhotra, S. L. Electrostatic imaging process. U.S. Patent 5663029, 1997.

25. Uchino, M.; Utaka, S.; Ueno, T.; Sasaki, M.; Hashimoto, M. Resist and its patterning with high resolution. Jpn. Kokai Tokkyo Koho JP 07159992, 1995.

26. Otaka, T.; Saito, T. Photosensitive resin composition containing polyimide precursor and base generating compound. Jpn. Kokai Tokkyo Koho JP 2009265294, 2009.

5-(BROMOMETHYL)FLUORESCEIN

CAS Registry Number 148942-72-7

Chemical Structure

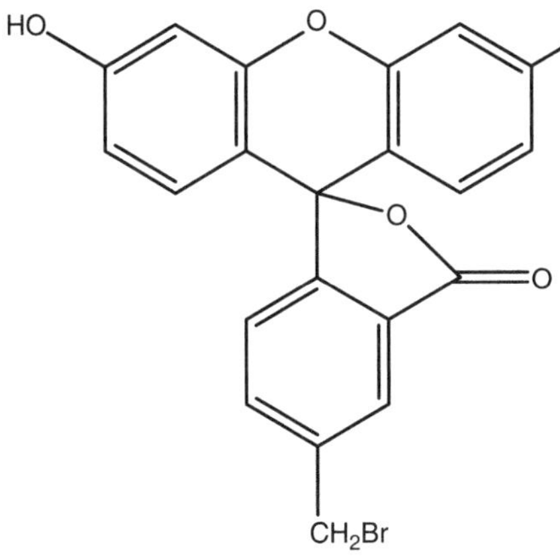

CA Index Name Spiro[isobenzofuran-1(3H),9′-[9H] xanthen]-3-one, 5-(bromomethyl)-3′,6′-dihydroxy-

Other Names 5-(Bromomethyl)fluorescein; 5-BMF; BMF; BrF

Merck Index Number Not listed

Chemical/Dye Class Xanthene

Molecular Formula $C_{21}H_{13}BrO_5$

Molecular Weight 425.23

Physical Form Orange solid

Solubility Soluble in water, N,N-dimethylformamide, dimethyl sulfoxide, methanol

Boiling Point (Calcd.) 681.8 ± 55.0 °C, pressure: 760 Torr

pK$_a$ (Calcd.) 9.33 ± 0.20, most acidic, temperature: 25 °C

Absorption (λ_{max}) 492 nm (Buffer pH 9.0)

Emission (λ_{max}) 515 nm (Buffer pH 9.0)

Molar Extinction Coefficient 81,000 cm^{-1} M^{-1} (Buffer pH 9.0)

Synthesis Synthetic method[1]

Imaging/Labeling Applications Acidic fibroblast growth factor;[2] calcium-ATPase/calcium-magnesium-ATPase of sarcoplasmic reticulum at Glu-439;[3,4] cefuroxime;[5] γ-cholesteryloxybutyric acid (CBA);[5,6] fatty acids;[5,7–9] jasmonic acid (JA);[10] oleic acid;[5] palmitic acid;[5,8,9] prostaglandin E$_2$ (PGE$_2$);[5] retinoic acid;[11,12] nucleic acids;[13–15] proteins;[16–20] thiols[11,12,14,21,22]

Biological/Medical Applications Analyzing fatty acids;[5,7–9] analyzing/detecting/quantitating proteins;[16–20] analyzing/measuring retinoic acid concentration in biological samples;[11,12] determining/separating/quantifying jasmonic acid (JA);[10] inhibiting/monitoring glutathione S-transferase omega 1 (GSTO1) activity;[23] labeling of thiols;[11,12,14,21,22] labeling/purifying nucleic acids;[13–15] derivatization reagent for carboxylic acids[3–12]

Industrial Applications Not reported

Safety/Toxicity No data available

REFERENCES

1. Haugland, R. P. *Handbook of Fluorescent Probes and Research Chemicals*; Molecular Probes Inc.: Eugene, **1996**; pp 76–79.

2. Feito, M. J.; Jimenez, M.; Fernandez-Cabrera, C.; Rivas, G.; Gimenez-Gallego, G.; Lozano, R. M. Strategy for fluorescent labeling of human acidic fibroblast growth factor without impairment of mitogenic activity: A bona fide tracer. *Anal. Biochem.* **2011**, *411*, 1–9.

3. Baker, K. J.; East, J. M.; Lee, A. G. Localization of the hinge region of the Ca(2+)-ATPase of sarcoplasmic reticulum using resonance energy transfer. *Biochim. Biophys. Acta* **1994**, *1192*, 53–60.

4. Stefanova, H. I.; Mata, A. M.; Gore, M. G.; East, J. M.; Lee, A. G. Labeling the calcium-magnesium-ATPase of sarcoplasmic reticulum at Glu-439 with 5-(bromomethyl)fluorescein. *Biochemistry* **1993**, *32*, 6095–6103.

5. Mukherjee, P. S.; Karnes, H. T. Reaction of 5-bromomethylfluorescein (5-BMF) with cefuroxime and other carboxyl-containing analytes to form derivatives suitable for laser-induced fluorescence detection. *Analyst* **1996**, *121*, 1573–1579.

6. Mukherjee, P. S.; Karnes, H. T. Analysis of γ-(cholesteryloxy)butyric acid in biologic samples by derivatization with 5-(bromomethyl)fluorescein followed by high-performance liquid chromatography with laser-Induced fluorescence detection. *Anal. Chem.* **1996**, *68*, 327–332.

7. Zuriguel, V.; Causse, E.; Bounery, J. D.; Nouadje, G.; Simeon, N.; Nertz, M.; Salvayre, R.; Couderc, F. Short chain fatty acids analysis by capillary

electrophoresis and indirect UV detection or laser-induced fluorescence. *J. Chromatogr., A* **1997**, *781*, 233–238.

8. Mukherjee, P. S.; DeSilva, K. H.; Karnes, H. T. An argon-ion laser fluorometer optimal for concentration detection of 5-Bromomethyl fluorescein derivatized carboxylic acids. *Mikrochim. Acta* **1996**, *124*, 99–109.

9. Mukherjee, P. S.; DeSilva, K. H.; Karnes, H. T. 5-Bromomethyl fluorescein (5-BMF) for derivatization of carboxyl containing analytes for use with laser-induced fluorescence detection. *Pharm. Res.* **1995**, *12*, 930–936.

10. Zhang, Z.; Liu, X.; Li, D.; Lu, Y. Determination of jasmonic acid in bark extracts from *Hevea brasiliensis* by capillary electrophoresis with laser-induced fluorescence detection. *Anal. Bioanal. Chem.* **2005**, *382*, 1616–1619.

11. Donato, L. J.; Noy, N. Fluorescence-based technique for analyzing retinoic acid. *Methods Mol. Biol.* **2010**, *652*, 177–187.

12. Donato, L. J.; Noy, N. A fluorescence-based method for analyzing retinoic acid in biological samples. *Anal. Biochem.* **2006**, *357*, 249–256.

13. Laayoun, A.; Menou, L.; Ginot, F. Method for labeling and purifying nucleic acids of interest present in a biological sample to be treated in a single reaction vessel. PCT Int. Appl. WO 2006018572, 2006.

14. Short, J. M. Modified nucleotides and methods useful for nucleic acid sequencing. PCT Int. Appl. WO 9949082, 1999.

15. Monnot, V.; Tora, C.; Lopez, S.; Menou, L.; Laayoun, A. Labeling during cleavage (LDC), a new labeling approach for RNA. *Nucleosides, Nucleotides Nucleic Acids* **2001**, *20*, 1177–1179.

16. Lee, J.; Son, J.; Ha, H.; Chang, Y. Fluorescent labeling of membrane proteins on the surface of living cells by a self-catalytic glutathione S-transferase omega 1 tag. *Mol. BioSys.* **2011**, *7*, 1270–1276.

17. Vogel, K.; Newman, R.; Riddle, S. Sensor proteins and assay methods. U.S. Pat. Appl. Publ. US 20080213811, 2008.

18. Singh, S.; Salimi-Moosavi, H.; Tahir, S. H.; Wallweber, G. J.; Kirakossian, H.; Matray, T. J.; Hernandez, V. S. Methods and compositions for analyzing proteins. PCT Int. Appl. WO 2002095356, 2002.

19. Singh, S.; Zivin, R. A. Analyzing phosphorylated proteins. PCT Int. Appl. WO 2002094998, 2002.

20. Koo, J.; Boschetti, E. Online detection of a desired solute in an effluent stream using fluorescence spectroscopy. PCT Int. Appl. WO 9640398, 1996.

21. Lochman, P.; Adam, T.; Friedecky, D.; Hlidkova, E.; Skopkova, Z. High-throughput capillary electrophoretic method for determination of total aminothiols in plasma and urine. *Electrophoresis* **2003**, *24*, 1200–1207.

22. Hart, J. J.; Welch, R. M.; Norvell, W. A.; Kochian, L. V. Measurement of thiol-containing amino acids and phytochelatin (PC2) via capillary electrophoresis with laser-induced fluorescence detection. *Electrophoresis* **2002**, *23*, 81–87.

23. Son, J.; Lee, J.; Lee, J.; Schuller, A.; Chang, Y. Isozyme-specific fluorescent inhibitor of glutathione S-transferase omega 1. *ACS Chem. Biol.* **2010**, *5*, 449–453.

4-BROMOMETHYL-7-METHOXYCOUMARIN

CAS Registry Number   35231-44-8

Chemical Structure

CA Index Name   2H-1-Benzopyran-2-one, 4-(bromomethyl)-7-methoxy-

Other Names   4-(Bromomethyl)-7-methoxy-2H-chromen-2-one; 4-(Bromomethyl)-7-methoxy-2H-chromene-2-one; 4-Bromomethyl-7-methoxy-2-oxo-2H-benzopyran; 4-Bromomethyl-7-methoxycoumarin; 7-Methoxy-4-(bromomethyl)coumarin; BMC; Br-Mmc; BrMMC

Merck Index Number   Not listed

Chemical/Dye Class   Coumarin

Molecular Formula $C_{11}H_9BrO_3$

Molecular Weight   269.09

Physical Form   Pale yellow crystals/needles[8]

Solubility   Soluble in acetone, acetonitrile, N,N-dimethylformamide, dimethyl sulfoxide, methanol

Melting Point   214–215 °C;[4] 209–210 °C (decompose);[10] 208–209 °C;[8] 206–208 °C (decompose);[5] 205–207 °C;[6] 204 °C[11]

Boiling Point (Calcd.)   385.3 ± 42.0 °C, pressure: 760 Torr

Absorption (λ_{max})   330 nm (MeOH)

Emission (λ_{max})   None

Molar Extinction Coefficient   13,000 $cm^{-1} M^{-1}$ (MeOH)

Synthesis   Synthetic methods[1–11]

Imaging/Labeling Applications   Amino acids;[12,13] bile acids;[14–17] biotin and its analogs;[18] carboxylic acids;[19–37] cyclosporins;[38] dalapon;[39] ethosuximide;[40,41] fatty acids;[42–52] fusilade;[53,54] immunoglobulins;[55] luminol derivatives;[56] mercury ions;[57,58] nitric oxide;[59,60] penicillins;[61,62] phosphoric acid;[63] picomole;[64] prostaglandins;[65–70] pyrimidine nucleobases;[8,71–82] steviol;[83] surfactants;[84–86] thromboxanes[87,88]

Biological/Medical Applications   Analyzing/detecting carboxylic acids;[19–37] analyzing/labeling fatty acids;[42–52] detecting nitric oxide;[59,60] detecting/separating bile acids;[14–17] labeling pyrimidine nucleobases;[8,71–82] anticancer agents;[4,89,90] biomedical trace analysis;[37,49] herbicides;[34,39,53,54,64] as a substrate for measuring histone acetyltransferases (HATs) activity,[91] peptidases activity,[92,93] sirtuin activity[94]

Industrial Applications   Detection of nitrated explosives;[95] measuring ultraviolet ray power;[56] photoacid generators;[96] polyurethane foams;[97] photoresists;[98] silica nanoparticles[99]

Safety/Toxicity   Mutagenicity[100]

REFERENCES

1. Revankar, H. M.; Kulkarni, M. V.; Joshi, S. D.; More, U. A. Synthesis, biological evaluation and docking studies of 4-aryloxymethyl coumarins derived from substructures and degradation products of vancomycin. *Eur. J. Med. Chem.* **2013**, *70*, 750–757.

2. Jeyachandran, M.; Ramesh, P.; Sriram, D.; Senthilkumar, P.; Yogeeswari, P. Synthesis and *in vitro* antitubercular activity of 4-aryl/alkylsulfonylmethylcoumarins as inhibitors of *Mycobacterium tuberculosis. Bioorg. Med. Chem. Lett.* **2012**, *22*, 4807–4809.

3. Atta, S.; Jana, A.; Ananthakirshnan, R.; Dhuleep, P. S. N. Fluorescent caged compounds of 2,4-dichlorophenoxyacetic acid (2,4-D):

Photorelease technology for controlled release of 2,4-D. *J. Agric. Food Chem.* **2010**, *58*, 11844–11851.

4. Belluti, F.; Fontana, G.; Dal Bo, L.; Carenini, N.; Giommarelli, C.; Zunino, F. Design, synthesis and anticancer activities of stilbene-coumarin hybrid compounds: identification of novel proapoptotic agents. *Bioorg. Med. Chem.* **2010**, *18*, 3543–3550.

5. Mubarak, M. S.; Peters, D. G. Electrochemical reduction of 4-(bromomethyl)-2H-chromen-2-ones at carbon cathodes in dimethylformamide. *J. Electrochem. Soc.* **2008**, *155*, F184–F189.

6. Friebe, W.; Schaefer, W.; Scheuer, W.; Tibes, U. Use of coumarins and carbostyrils as PLA₂ inhibitors, new coumarins and carbostyrils, processes for the

production thereof and pharmaceutical agent. U.S. Patent 5716983, 1998.

7. Kulkarni, M. V.; Patil, V. D. Studies on coumarins. I. *Arch. Pharm.* **1981**, *314*, 708–711.

8. Secrist, J. A., III,; Barrio, J. R.; Leonard, N. J. Attachment of a fluorescent label to 4-thiouracil and 4-thiouridine. *Biochem. Biophys. Res. Commun.* **1971**, *45*, 1262–1270.

9. Sehgal, J. M.; Seshadri, T. R. Benzopyrone series. XXXVII. 4-(Hydroxymethyl)coumarins and their derivatives. *J. Scient. Ind. Res.* **1953**, 12B, 346–349.

10. Seshadri, T. R.; Varadarajan, S. Halogenation of hydroxy coumarins. I. Bromination of umbelliferones. *J. Scient. Ind. Res.* **1952**, *11B*, 39–49.

11. Dey, B. B.; Radhabai, K. Action of chlorine and bromine on 3- and 4-coumarin-acetic acids. Halo-3- and -4-coumarinacetic acids and 4-halomethylcoumarins. *J. Indian Chem. Soc.* **1934**, *11*, 635–650.

12. Kele, P.; Sui, G.; Huo, Q.; Leblanc, R. M. Highly enantioselective synthesis of a fluorescent amino acid. *Tetrahedron: Asymmetry* **2000**, *11*, 4959–4963.

13. Andreae, F. Preparation of α-amino acids marked with a group which is detectable by fluorescence or absorption spectroscopy. Austrian AT 401773, 1996.

14. Maitra, U.; Nath, S. Bile acid derived PET-based cation sensors: molecular structure dependence of their sensitivity. *Chem. Asian J.* **2009**, *4*, 989–997.

15. Guldutuna, S.; You, T.; Kurts, W.; Leuschner, U. High performance liquid chromatographic determination of free and conjugated bile acids in serum, liver biopsies, bile, gastric juice and feces by fluorescence labeling. *Clin. Chim. Acta* **1993**, *214*, 195–207.

16. Wang, G.; Stacey, N. H.; Earl, J. Determination of individual bile acids in serum by high performance liquid chromatography. *Biomed. Chromatogr.* **1990**, *4*, 136–140.

17. Okuyama, S.; Uemura, D.; Hirata, Y. The improved method of high performance liquid chromatographic separation of individual bile acids: free and glycine-conjugated bile acids. *Chem. Lett.* **1979**, 461–462.

18. Desbene, P. L.; Coustal, S.; Frappier, F. Separation of biotin and its analogs by high-performance liquid chromatography: convenient labeling for ultraviolet or fluorimetric detection. *Anal. Biochem.* **1983**, *128*, 359–362.

19. Pobozy, E.; Krol, E.; Wojcik, L.; Wachowicz, M.; Trojanowicz, M. HPLC determination of perfluorinated carboxylic acids with fluorescence detection. *Microchim. Acta* **2011**, *172*, 409–417.

20. Zacharis, C. K.; Raikos, N.; Giouvalakis, N.; Tsoukali-Papadopoulou, H.; Theodoridis, G. A. A new method for the HPLC determination of gamma-hydroxybutyric acid (GHB) following derivatization with a coumarin analogue and fluorescence detection. *Talanta* **2008**, *75*, 356–361.

21. Knauer, A.; Wintersteiger, R.; Markl, R.; Sametz, W.; Juan, H. Derivatization of carboxylic acids with fluorescent reagents. *J. Planar Chromatogr.-Mod. TLC* **1999**, *12*, 211–214.

22. He, X.; Li, Y.; Li, L.; Wang, Y.; Zhou, T. Analysis of ethacrynic acid in urine by HPLC-fluorescence detector. *Chin. Chem. Lett.* **1991**, *2*, 953–956.

23. Van der Horst, F. A. L.; Reijn, J. M.; Post, M. H.; Bult, A.; Holthuis, J. J. M.; Brinkman, U. A. T. Mechanistic study on the derivatization of aliphatic carboxylic acids in aqueous non-ionic micellar systems. *J. Chromatogr.* **1990**, *507*, 351–366.

24. Van der Horst, F. A. L.; Post, M. H.; Holthuis, J. J. M. Influence of aqueous micellar systems on the derivatization of undecylenic acid with 4-(bromomethyl)-7-methoxycoumarin. *J. Chromatogr.* **1988**, *456*, 201–218.

25. Ertel, K. D.; Carstensen, J. T. Quantitative determination of octanoic acid by high-performance liquid chromatography following derivatization with 4-bromomethyl-7-methoxycoumarin. *J. Chromatogr.* **1987**, *411*, 297–304.

26. Chakir, S.; Leroy, P.; Nicolas, A.; Ziegler, J. M.; Labory, P. High-performance liquid chromatographic analysis of glucuronic acid conjugates after derivatization with 4-bromomethyl-7-methoxycoumarin. *J. Chromatogr.* **1987**, *395*, 553–561.

27. Grayeski, M. L.; DeVasto, J. K. Coumarin derivatizing agents for carboxylic acid detection using peroxyoxalate chemiluminescence with liquid chromatography. *Anal. Chem.* **1987**, *59*, 1203–1206.

28. McGuffin, V. L.; Zare, R. N. Laser fluorescence detection in microcolumn liquid chromatography: application to derivatized carboxylic acids. *Appl. Spectrosc.* **1985**, *39*, 847–853.

29. Bousquet, E.; Romeo, G.; Giannola, L. I. High-performance liquid chromatography with fluorometric detection in biological tissues of the 4-bromomethyl-7-methoxycoumarin ester

derivative of 5-pyrrolidone-2-carboxylic acid. *J. Chromatogr., Biomed. Appl.* **1985**, *344*, 325–331.

30. Elbert, W.; Breitenbach, S.; Neftel, A.; Hahn, J. 4-Methyl-7-methoxycoumarin as a fluorescent label for high-performance liquid chromatographic analysis of dicarboxylic acids. *J. Chromatogr.* **1985**, *328*, 111–120.

31. Roseboom, H.; Herbold, H. A.; Berkhoff, C. J. Determination of phenoxy carboxylic acid pesticides by gas and liquid chromatography. *J. Chromatogr.* **1982**, *249*, 323–331.

32. Tsuchiya, H.; Hayashi, T.; Naruse, H.; Takagi, N. High-performance liquid chromatography of carboxylic acids using 4-bromomethyl-7-acetoxycoumarin as fluorescence reagent. *J. Chromatogr.* **1982**, *234*, 121–130.

33. Gonnet, C.; Marichy, M.; Philippe, N. Comparison of different chromatographic techniques for analysis of short chain mono- and dicarboxylic acids. *Analusis* **1979**, *7*, 370–375.

34. Duenges, W.; Mueller, K. E.; Mueller, M. Conditions for fluorescence labeling of acids with 4-bromomethyl-7-methoxycoumarin. *Methodol. Surv. Biochem.* **1978**, *7*, 257–268.

35. Grushka, E.; Lam, S.; Chassin, J. Fluorescence labeling of dicarboxylic acids for high performance liquid chromatographic separation. *Anal. Chem.* **1978**, *50*, 1398–1399.

36. Duenges, W.; Meyer, A.; Mueller, K. E.; Mueller, M.; Pietschmann, R.; Plachetta, C.; Sehr, R.; Tuss, H. Fluorescence labeling of organic acidic compounds with 4-bromomethyl-7-methoxycoumarin (Br-Mmc). *Fresenius' Z. Anal. Chem.* **1977**, *288*, 361–368.

37. Dunges, W. The HPLC and TLC determination of acidic compounds after their fluorescence labeling with 4-bromomethyl-7-methoxycoumarin: a new tool for biomedical trace analysis. *UV Spectrom. Group Bull.* **1977**, *5*, 38–45.

38. French, M. T.; Miller, J. N.; Seare, N. J.; Lachno, D. R.; Yacoub, M. H. New fluorescent derivatives of cyclosporin for use in immunoassays. *J. Pharm. Biomed. Anal.* **1992**, *10*, 23–30.

39. Reichert, J.; Gernikeites, T.; Winkler, M. Application of HPLC with fluorescence detection to environmental analysis exemplified with the herbicide dalapon. *Vom Wasser* **1994**, *82*, 37–48.

40. Chen, S.; Wu, H.; Wu, J.; Kou, H.; Wu, S. Determination of ethosuximide in plasma by derivatization and high performance liquid chromatography. *J. Liq. Chromatogr. Relat. Technol.* **1997**, *20*, 1579–1589.

41. Wu, J.; Chen, S.; Kou, H.; Wu, S.; Wu, H. Derivatization and high performance liquid chromatographic determination of ethosuximide. *Chin. Pharm. J.* **1994**, *46*, 413–421.

42. Chang, C.; Wu, H.; Wu, S.; Chen, S.; Kou, H. Trace analysis of very long chain free fatty acids in plasma by fluorogenic derivatization and liquid chromatography. *J. Liq. Chromatogr. Relat. Technol.* **1998**, *21*, 669–679.

43. Abushufa, R.; Reed, P.; Weinkove, C. Fatty acids in erythrocytes measured by isocratic HPLC. *Clin. Chem.* **1994**, *40*, 1707–1712.

44. Noda, H.; Hata, H. Method for analysis of fatty acids by HPLC after derivatization. Jpn. Kokai Tokkyo Koho JP 05034327, 1993.

45. Jansen, E. H. J. M.; De Fluiter, P. Determination of lauric acid metabolites in peroxisome proliferation after derivatization and HPLC analysis with fluorimetric detection. *J. Liq. Chromatogr.* **1992**, *15*, 2247–2260.

46. Wolf, J. H.; Korf, J. Improved automated precolumn derivatization reaction of fatty acids with bromomethylmethoxycoumarin as label. *J. Chromatogr.* **1990**, *502*, 423–430.

47. Wolf, J. H.; Korf, J. Automated solid-phase catalyzed pre-column derivatization of fatty acids for reversed-phase high-performance liquid chromatographic analysis with fluorescence detection. *J. Chromatogr.* **1988**, *436*, 437–445.

48. Hordijk, K. A.; Cappenberg, T. E. Quantitative high-pressure liquid chromatography-fluorescence determination of some important lower fatty acids in lake sediments. *Appl. Environ. Microbiol.* **1983**, *46*, 361–369.

49. Voelter, W.; Huber, R.; Zech, K. Fluorescence labeling in trace analysis of biological samples. Simultaneous determination of free fatty acids and related carboxylic compounds. *J. Chromatogr.* **1981**, *217*, 491–507.

50. Zelenski, S. G.; Huber, J. W., III,. Application of 4-methyl-7-methoxycoumarin derivatives to the high pressure liquid chromatographic analysis of fatty acids. *Chromatographia* **1978**, *11*, 645–646.

51. Lam, S.; Grushka, E. Labeling of fatty acids with 4-bromomethyl-7-methoxycoumarin via crown ether catalyst for fluorimetric detection in high-performance liquid chromatography. *J. Chromatogr.* **1978**, *158*, 207–214.

52. Duenges, W. 4-Bromomethyl-7-methoxycoumarin as a new fluorescence label for fatty acids. *Anal. Chem.* **1977**, *49*, 442–445.

53. Brennan, J. D.; Brown, R. S.; Cohen, H.; Egamino, J.; Semchyschyn, D.; Williams, R. E.; Krull, U. J. Towards a homogeneous fluorescence assay for a herbicide: characterization of the interactions of fusilade, bromomethylmethoxy coumarin derivatized fusilade, and anti-fusilade IgG antibody. *Anal. Chim. Acta* **1996**, *336*, 157–166.

54. Cohen, H.; Boutin-Muma, B. High-performance liquid chromatographic determination of Fusilade using a fluorescence reagent, 4-bromomethyl-7-methoxycoumarin. *J. Liq. Chromatogr.* **1991**, *14*, 313–326.

55. Baranowska-Kortylewicz, J.; Kassis, A. I. Labeling of immunoglobulins with bifunctional, sulfhydryl-selective, and photoreactive coumarins. *Bioconjugate Chem.* **1993**, *4*, 300–304.

56. Nakazono, M.; Hino, T.; Zaitsu, K. Photosensitive luminol derivatives and measurement of ultraviolet ray power. *J. Photochem. Photobiol., A: Chem.* **2007**, *186*, 99–105.

57. Bazzicalupi, C.; Caltagirone, C.; Cao, Z.; Chen, Q.; Di Natale, C.; Garau, A.; Lippolis, V.; Lvova, L.; Liu, H.; Lundstroem, I. Multimodal use of new coumarin-based fluorescent chemosensors: towards highly selective optical sensors for Hg^{2+} probing. *Chem. Eur. J.* **2013**, *19*, 14639–14653.

58. Saha, S.; Mahato, P.; Baidya, M.; Ghosh, S. K.; Das, A. An interrupted PET coupled TBET process for the design of a specific receptor for Hg^{2+} and its intracellular detection in MCF7 cells. *Chem. Commun.* **2012**, *48*, 9293–9295.

59. Soh, N.; Imato, T.; Kawamura, K.; Maeda, M.; Katayama, Y. Ratiometric direct detection of nitric oxide based on a novel signal-switching mechanism. *Chem. Commun.* **2002**, 2650–2651.

60. Plater, M. J.; Greig, I.; Helfrich, M. H.; Ralston, S. H. The synthesis and evaluation of o-phenylenediamine derivatives as fluorescent probes for nitric oxide detection. *J. Chem. Soc., Perkin Trans. 1* **2001**, 2553–2559.

61. Ushimizu, T.; Sato, T.; Saito, T.; Itoh, T. A new HPLC analysis of residual penicillins in edible animal tissues by pre-column fluorescence derivatization. *Anim. Sci. J.* **2001**, *72*, J570–J578.

62. Berger, K.; Petz, M. Fluorescence HPLC determination of penicillins with 4-bromomethyl -7-methoxycoumarin as a precolumn labeling agent. *Dtsch. Lebens.-Rundsch.* **1991**, *87*, 137–141.

63. McKenna, C. E.; Kashemirov, B. A.; Blazewska, K. M. Product class 16: phosphoric acid and derivatives. *Sci. Synth.* **2009**, *42*, 779–921.

64. Duenges, W. Fluorescence labeling of picomole amounts of acidic herbicides for toxicological analysis. *Chromatographia* **1976**, *9*, 624–626.

65. Imbs, A. B.; Vologodskaya, A. V.; Nevshupova, N. V.; Khotimchenko, S. V.; Titlyanov, E. A. Response of prostaglandin content in the red alga *Gracilaria verrucosa* to season and solar irradiance. *Phytochemistry* **2001**, *58*, 1067–1072.

66. Wintersteiger, R.; Knauer, A.; Friesenbichler, K.; Juan, H.; Sametz, W. Determination of prostaglandins by LC and fluorescence detection. *Pharm. Pharmacol. Lett.* **2000**, *10*, 41–44.

67. Bruno, P.; Caselli, M.; Traini, A. HPTLC and OPLC separation and detection of prostaglandin esters using 4-bromomethyl-7-methoxycoumarin (BrMMC). *J. Planar Chromatogr.-Mod. TLC* **1988**, *1*, 299–303.

68. Wintersteiger, R.; Juan, H. Prostaglandin determination with fluorescent reagents. *Prostaglandins, Leukotrienes Med.* **1984**, *14*, 25–40.

69. Alekseev, S. M.; Pomoinitskii, V. D.; Sarycheva, I. K.; Evstigneeva, R. P. Use of 4-bromomethyl-7-methoxycoumarin for the quantitative fluorometric analysis of prostaglandins. *Khim.-Farm. Zh.* **1981**, *15*, 115–118.

70. Turk, J.; Weiss, S. J.; Davis, J. E.; Needleman, P. Fluorescent derivatives of prostaglandins and thromboxanes for liquid chromatography. *Prostaglandins* **1978**, *16*, 291–309.

71. Liu, K.; Zhong, D.; Zou, H.; Chen, X. Determination of tegafur, 5-fluorouracil, gimeracil and oxonic acid in human plasma using liquid chromatography-tandem mass spectrometry. *J. Pharm. Biomed. Anal.* **2010**, *52*, 550–556.

72. Wang, K.; Nano, M.; Mulligan, T.; Bush, E. D.; Edom, R. W. Derivatization of 5-fluorouracil with 4-bromomethyl-7-methoxycoumarin for determination by liquid chromatography-mass spectrometry. *J. Am. Soc. Mass Spectrom.* **1998**, *9*, 970–976.

73. Stratford, M. R. L.; Dennis, M. F. Measurement of incorporation of bromodeoxyuridine into DNA by high perfomrance liquid chromatography using a novel fluorescent labeling technique. *Int. J. Radiat. Oncol., Biol., Phys.* **1992**, *22*, 485–487.

74. Yoshida, S.; Urakami, K.; Kito, M.; Takeshima, S.; Hirose, S. Precolumn derivatization for

the determination of fluoropyrimidines by liquid chromatography with chemiluminescence detection. *Anal. Chim. Acta* **1990**, *239*, 181–187.

75. Yoshida, S.; Urakami, K.; Kito, M.; Takeshima, S.; Hirose, S. High-performance liquid chromatography with chemiluminescence detection of serum levels of pre-column derivatized fluoropyrimidine compounds. *J. Chromatogr., Biomed. Appl.* **1990**, *530*, 57–64.

76. Kindberg, C. G.; Slavik, M.; Riley, C. M.; Stobaugh, J. F. High-performance liquid chromatography of 5-fluorouracil after derivatization with 4-bromomethyl-7-methoxycoumarin. Characterization of the derivative and the use of column switching for the improvement of resolution and the enhancement of sensitivity. *J. Pharm. Biomed. Anal.* **1989**, *7*, 459–469.

77. Yoshida, S.; Hirose, S.; Iwamoto, M. Use of 4-bromomethyl-7-methoxycoumarin for derivatization of pyrimidine compounds in serum analyzed by high-performance liquid chromatography with fluorimetric detection. *J. Chromatogr., Biomed. Appl.* **1986**, *383*, 61–68.

78. Iwamoto, M.; Yoshida, S.; Hirose, S. Fluorescence labeling of pyrimidine nucleobases and their related compounds for high-performance liquid chromatography. *Nucleic Acids Symp. Ser.* **1984**, *15*, 21–24.

79. Iwamoto, M.; Yoshida, S.; Hirose, S. High performance liquid chromatographic determination of pyrimidine nucleobases and 5-fluorouracil labeled by fluorescent reagent. *Yakugaku Zasshi* **1984**, *104*, 1251–1256.

80. Iwamoto, M.; Yoshida, S.; Chow, T.; Hirose, S. Labeling of 5-fluorouracil and pyrimidine nucleosides with 4-bromomethyl-7-methoxycoumarin for fluorimetric detection in high performance liquid chromatography. *Yakugaku Zasshi* **1983**, *103*, 967–973.

81. Yang, C.; Soell, D. Covalent attachment of fluorescent groups to transfer ribonucleic acid. Reactions with 4-bromomethyl-7-methoxy-2-oxo-2H-benzopyran. *Biochemistry* **1974**, *13*, 3615–3621.

82. Yang, C. H.; Soell, D. Covalent attachment of a fluorescent group to 4-thiouridine in transfer RNA. *J. Biochem.* **1973**, *73*, 1243–1247.

83. Ceunen, S.; Geuns, J. M. C. Spatio-temporal variation of the diterpene steviol in *Stevia rebaudiana* grown under different photoperiods. *Phytochemistry* **2013**, *89*, 32–38.

84. Trojanowicz, M.; Wojcik, L.; Musijowski, J.; Koc, M.; Pobozy, E.; Krol, E. New analytical methods developed for determination of perfluorinated surfactants in waters and wastes. *Croat. Chem. Acta* **2011**, *84*, 439–446.

85. Kondo, M.; Takano, S. Analysis of carboxy betaine amphoteric surfactants. Jpn. Kokai Tokkyo Koho JP 62006166, 1987.

86. Kondoh, Y.; Takano, S. Determination of carboxybetaine amphoteric surfactants in household and cosmetic products by high performance liquid chromatography with prelabeling. *Anal. Sci.* **1986**, *2*, 467–471.

87. Ben Gueddour, R.; Matt, M.; Muller, S.; Donner, M.; El Haloui, N.; Nicolas, A. Fluorescent derivatives of thromboxane B2: synthesis, spectroscopic and immunologic properties. *Talanta* **1994**, *41*, 485–493.

88. Ben Gueddour, R.; Matt, M.; Nicolas, A.; Donner, M.; Stoltz, J. F. Determination of thromboxane B2 by normal-phase high-performance liquid chromatography after pre-column derivatization with fluorescent reagents. *Anal. Lett.* **1993**, *26*, 429–443.

89. Al-Soud, Y. A.; Al-Sa'doni, H. H.; Amajaour, H. A. S.; Salih, K. S. M.; Mubarak, M. S.; Al-Masoudi, N. A.; Jaber, I. H. Synthesis, characterization and anti-HIV and antitumor activities of new coumarin derivatives. *Z. Naturforsch., B: Chem. Sci.* **2008**, *63*, 83–89.

90. Goto, J. Anticancer agents containing coumarins. Jpn. Kokai Tokkyo Koho JP 05000945, 1993.

91. Chen, Z.; Zheng, Y. G. Synthesis of a coumarin-histone conjugate for HAT fluorescent assay. *Heterocycl. Commun.* **2007**, *13*, 343–346.

92. Knight, C. G. Stereospecific synthesis of L-2-amino-3-(7-methoxy-4-coumaryl)propionic acid, an alternative to tryptophan in quenched fluorescent substrates for peptidases. *Lett. Pept. Sci.* **1998**, *5*, 1–4.

93. Knight, C. G. A quenched fluorescent substrate for thimet peptidase containing a new fluorescent amino acid, DL-2-amino-3-(7-methoxy-4-coumaryl)propionic acid. *Biochem. J.* **1991**, *274*, 45–48.

94. Abromeit, H.; Kannan, S.; Sippl, W.; Scriba, G. K. E. A new nonpeptide substrate of human sirtuin in a capillary electrophoresis-based assay. Investigation of the binding mode by docking experiments. *Electrophoresis* **2012**, *33*, 1652–1659.

95. Meaney, M. S.; McGuffin, V. L. Investigation of common fluorophores for the detection of nitrated explosives by fluorescence quenching. *Anal. Chim. Acta* **2008**, *610*, 57–67.

96. Tsushima, N.; Uesugi, T. Photoacid generators. Jpn. Kokai Tokkyo Koho JP 2006126694, 2006.

97. Velencoso, M. M.; Gonzalez, A. S. B.; Garcia-Martinez, J. C.; Ramos, M. J.; De Lucas, A.; Rodriguez, J. F. Click-ligation of coumarin to polyether polyols for polyurethane foams. *Polym. Int.* **2013**, *62*, 783–790.

98. Uchino, M.; Utaka, S.; Ueno, T.; Sasaki, M.; Hashimoto, M. Resist and its patterning with high resolution. Jpn. Kokai Tokkyo Koho JP 07159992, 1995.

99. Rampazzo, E.; Brasola, E.; Marcuz, S.; Mancin, F.; Tecilla, P.; Tonellato, U. Surface modification of silica nanoparticles: a new strategy for the realization of self-organized fluorescence chemosensors. *J. Mater. Chem.* **2005**, *15*, 2687–2696.

100. Azuma, S.; Kishino, S.; Katayama, S.; Akahori, Y.; Matsushita, H. Mutagenicity of 12 HPLC labeling reagents for analyzing carboxyl compounds. *Kankyo Kagaku* **1997**, *7*, 249–255.

CALCEIN BLUE AM

CAS Registry Number   168482-84-6

Chemical Structure

CA Index Name   Glycine, N-[2-[(acetyloxy)methoxy]-2-oxoethyl]-N-[(7-hydroxy-4-methyl-2-oxo-2H-1-benzopyran-8-yl)methyl]-, (acetyloxy)methyl ester

Other Names   Calcein Blue AM, Calcein Blue acetoxymethyl ester; CellTrace Calcein Blue AM

Merck Index Number   Not listed

Chemical/Dye Class   Coumarin

Molecular Formula $C_{21}H_{23}NO_{11}$

Molecular Weight   465.41

Physical Form   Colorless Solid

Solubility   Soluble in acetonitrile, chloroform, N,N-dimethylformamide, dimethyl sulfoxide, ethyl acetate, methanol

Boiling Point (Calcd.)   617.0 $\pm$ 55.0 °C, pressure: 760 Torr

pK$_a$ (Calcd.)   7.76 $\pm$ 0.20, most acidic, temperature: 25 °C; 2.51 $\pm$ 0.50, most basic, temperature: 25 °C

Absorption (λ_{max})   322 nm (DMSO)

Emission (λ_{max})   437 nm (DMSO)

Molar Extinction Coefficient   13,000 cm^{-1} M^{-1} (DMSO)

Synthesis   Synthetic methods[1,2]

Imaging/Labeling Applications   Bacteria;[1,3] cells[1,2]

Biological/Medical Applications   Cell viability assay;[1,2] cytotoxicity assay;[1,4] monitoring bacterial transport;[1,3] as a substrate for measuring esterases activity[1,2]

Industrial Applications   Not reported

Safety/Toxicity   No data available

REFERENCES

1. Sabnis, R. W. *Handbook of Biological Dyes and Stains*; John Wiley & Sons Inc.: Hoboken, **2010**; p 80.

2. Millard, P. J.; Roth, B. L.; Yue, S. T.; Haugland, R. P. Fluorescent viability assay using cyclic-substituted unsymmetrical cyanine dyes. U.S. Patent 5534416, 1996.

3. Fuller, M. E.; Streger, S. H.; Rothmel, R. K.; Mailloux, B. J.; Hall, J. A.; Onstott, T. C.; Fredrickson, J. K.; Balkwill, D. L.; DeFlaun, M. F. Development of a vital fluorescent staining method for monitoring bacterial transport in subsurface environments. *Appl. Environ. Microbiol.* **2000**, *66*, 4486–4496.

4. Liminga, G.; Nygren, P.; Dhar, S.; Nilsson, K.; Larsson, R. Cytotoxic effect of calcein acetoxymethyl ester on human tumor cell lines: drug delivery by intracellular trapping. *Anti-Cancer Drugs* **1995**, *6*, 578–585.

Handbook of Fluorescent Dyes and Probes, First Edition. R. W. Sabnis.
© 2015 John Wiley & Sons, Inc. Published 2015 by John Wiley & Sons, Inc.

5-CARBOXY-2′,7′-DICHLOROFLUORESCEIN

CAS Registry Number   142975-81-3

Chemical Structure

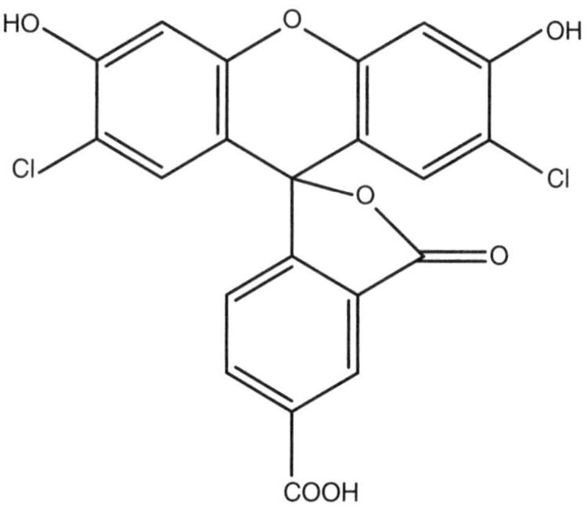

CA Index Name   Spiro[isobenzofuran-1(3H),9′-[9H]xanthene]-5-carboxylic acid, 2′,7′-dichloro-3′,6′-dihydroxy-3-oxo-

Other Names   5-Carboxy-2′,7′-dichlorofluorescein; Carboxy-DCF; 5-CDCF

Merck Index Number   Not listed

Chemical/Dye Class   Xanthene

Molecular Formula $C_{21}H_{10}Cl_2O_7$

Molecular Weight   445.21

Physical Form   Orange solid[4]

Solubility   Insoluble in water, soluble in N,N-dimethylformamide, dimethyl sulfoxide, methanol

Melting Point   >250°C

Boiling Point (Calcd.)   732.3 ± 60.0°C, pressure: 760 Torr

pK$_a$ 4.8, temperature: 22°C

pK$_a$ (Calcd.)   3.72 ± 0.20, most acidic, temperature: 25°C

Absorption (λ_{max})   495 nm (Buffer pH 4.0); 504 nm (Buffer pH 8.0)

Emission (λ_{max})   529 nm (Buffers pH 4.0, pH 8.0)

Molar Extinction Coefficient   38,000 cm^{-1} M^{-1} (Buffer pH 4.0); 107,000 cm^{-1} M^{-1} (Buffer pH 8.0)

Synthesis   Synthetic methods[1–5]

Imaging/Labeling Applications   Bacteria;[6,7] dideoxynucleotides;[8] lysosomes;[2] mitochondria;[9] nucleic acids;[10] protein-2 (MRP2, ABCC2);[11–13] protein-5 (MRP5, ABCC5)[13]

Biological/Medical Applications   Assessing reactive oxygen species (ROS) generation in isolated-perfused rat heart tissue;[14] detecting nucleic acids;[10] detecting/quantifying bacteria;[6,7] evaluating multimeric complexes;[15] monitoring blood gas;[16] monitoring/measuring analytes concentration;[17] multidrug resistance-associated protein-2 (MRP2, ABCC2) substrates;[11–13] multidrug resistance-associated protein-5 (MRP5, ABCC5) substrates;[13] screening mitochondrial permeability transition;[9] studying hepatic disposition;[18,19] a substrate for measuring aryl sulfatase activity,[2] acid phosphatase activity,[2] β-galactosidase activity,[2] lipase activity[2]

Industrial Applications   Fiber optic sensor for sol–gel material;[20] planar laser-induced fluorescence (PLIF) measurements in aqueous flows[21]

Safety/Toxicity   Metabolic toxicity[22]

REFERENCES

1. Feng, S.; Sun, X.; Chen, J.; Zheng, H.; Wang, Y. Method for preparing 2′,7′-dichloro-5(6)-carboxyfluorescein isomer. Faming Zhuanli Shenqing CN 103058980, 2013.

2. Coleman, D. J.; Naleway, J. J. Fluorogenic enzyme substrates for visualizing acidic organelles. U.S. Pat. Appl. Publ. US 20100233744, 2010.

3. Ge, F.; Chen, L. Synthesis and fluorescent properties of 2′,7′-dichloro-5(6)-carboxyfluorescein. *Huaxue Tongbao* **2009**, *72*, 78–81.

4. Woodroofe, C. C.; Masalha, R.; Barnes, K. R.; Frederickson, C. J.; Lippard, S. J. Membrane-permeable and -impermeable sensors of the Zinpyr family and their application to imaging of hippocampal zinc *in vivo*. *Chem. Biol.* **2004**, *11*, 1659–1666.

5. Lippard, S. J.; Woodroofe, C. C. Sensors, and methods of making and using the same. U.S. Pat. Appl. Publ. US 20040224420, 2004.

6. Ribeiro Pinto de Oliveira Azevedo, N. F.; Macieira Cerqueira, L. I.; Torres Faria, N. R.; Lopes da Costa Vieira, M. J. Methods and peptide nucleic acid probes for specific detection of *Helicobacter pylori* and its response to clarithromycin. PCT Int. Appl. WO 2011030319, 2011.

7. Ribeiro Pinto de Oliveira Azevedo, N. F.; Lopes da Costa Vieira, M. J.; Almeida, C. M. F.; Keevil, C. W. Peptide nucleic acid probe, kit and method for detection and/or quantification of *Salmonella* spp. and applications thereof. PCT Int. Appl. WO 2011121544, 2011.

8. Lee, L. G.; Connell, C. R.; Woo, S. L.; Cheng, R. D.; McArdle, B. F.; Fuller, C. W.; Halloran, N. D.; Wilson, R. K. DNA sequencing with dye-labeled terminators and T7 DNA polymerase: effect of dyes and dNTPs on incorporation of dye-terminators and probability of termination fragments. *Nucleic Acids Res.* **1992**, *20*, 2471–2483.

9. Blattner, J. R.; He, L.; Lemasters, J. J. Screening assays for the mitochondrial permeability transition using a fluorescence multiwell plate reader. *Anal. Biochem.* **2001**, *295*, 220–226.

10. Sergeev, N. V.; Brevnov, M. G.; Furtado, M. R. Identification of nucleic acids. PCT Int. Appl. WO 2011143478, 2011.

11. Munic, V.; Hlevnjak, M.; Haber, V. E. Characterization of rhodamine-123, calcein and 5(6)-carboxy-2′,7′-dichlorofluorescein (CDCF) export via MRP2 (ABCC2) in MES-SA and A549 cells. *Eur. J. Pharm. Sci.* **2011**, *43*, 359–369.

12. Pelis, R. M.; Shahidullah, M.; Ghosh, S.; Coca-Prados, M.; Wright, S. H.; Delamere, N. A. Localization of multidrug resistance-associated protein 2 in the nonpigmented ciliary epithelium of the eye. *J. Pharmacol. Exp. Ther.* **2009**, *329*, 479–485.

13. Pratt, S.; Chen, V.; Perry, W. I.; Starling, J. J.; Dantzig, A. H. Kinetic validation of the use of carboxydichlorofluorescein as a drug surrogate for MRP5-mediated transport. *Eur. J. Pharm. Sci.* **2006**, *27*, 524–532.

14. Kehrer, J. P.; Paraidathathu, T. The use of fluorescent probes to assess oxidative processes in isolated-perfused rat heart tissue. *Free Radical Res. Commun.* **1992**, *16*, 217–225.

15. Yang, X.; Fitz, L. J.; Lee, J. M.; Brooks, J.; Wolf, S. F. Assays and methods using homogeneous proximity-based detection methods for evaluating multimeric complexes. PCT Int. Appl. WO 2009061911, 2009.

16. Bentsen, J. G.; Wood, K. B. Sensor with improved drift stability. U.S. Patent 5403746, 1995.

17. Sharma, A. Optical probe and method for monitoring analyte concentration. PCT Int. Appl. WO 9212424, 1992.

18. Chandra, P.; Johnson, B. M.; Zhang, P.; Pollack, G. M.; Brouwer, K. L. R. Modulation of hepatic canalicular or basolateral transport proteins alters hepatobiliary disposition of a model organic anion in the isolated perfused rat liver. *Drug Metab. Dispos.* **2005**, *33*, 1238–1243.

19. Zamek-Gliszczynski, M. J.; Xiong, H.; Patel, N. J.; Turncliff, R. Z.; Pollack, G. M.; Brouwer, K. L. R. Pharmacokinetics of 5 (and 6)-carboxy-2′,7′-dichlorofluorescein and its diacetate promoiety in the liver. *J. Pharmacol. Exp. Ther.* **2003**, *304*, 801–809.

20. Manyam, U. H.; Shahriari, M. R.; Morris, M. J. A complete spectrum fiber optic pH sensor based on a novel fluorescent indicator doped porous sol–gel material. *Proc. SPIE- Int. Soc. Opt. Eng.* **1999**, *3540*, 10–18.

21. Karasso, P. S.; Mungal, M. G. PLIF measurements in aqueous flows using the Nd:YAG laser. *Exp. Fluids* **1997**, *23*, 382–387.

22. Sinclair, J. A.; Henderson, C.; Tettey, J. N. A.; Grant, M. H. The influence of the choice of digestion enzyme used to prepare rat hepatocytes on xenobiotic uptake and efflux. *Toxicol. In Vitro* **2013**, *27*, 451–457.

6-CARBOXY-2',7'-DICHLOROFLUORESCEIN

CAS Registry Number   144316-86-9

Chemical Structure

CA Index Name   Spiro[isobenzofuran-1(3H),9'-[9H]xanthene]-6-carboxylic acid, 2',7'-dichloro-3',6'-dihydroxy-3-oxo-

Other Names   6-Carboxy-2',7'-dichlorofluorescein; Carboxy-DCF

Merck Index Number   Not listed

Chemical/Dye Class   Xanthene

Molecular Formula $C_{21}H_{10}Cl_2O_7$

Molecular Weight   445.21

Physical Form   Orange solid;[6] Yellow solid[3]

Solubility   Insoluble in water; soluble in N,N-dimethylformamide, dimethyl sulfoxide, methanol

Melting Point   >250 °C

Boiling Point (Calcd.)   742.4 ± 60.0 °C pressure: 760 Torr

pK$_a$ 4.8 Temperature: 22 °C

pK$_a$ (Calcd.)   3.51 ± 0.20 most acidic, temperature: 25 °C

Absorption (λ_{max})   495 nm (Buffer pH 4.0); 504 nm (Buffer pH 8.0)

Emission (λ_{max})   529 nm (Buffers pH 4.0, pH 8.0)

Molar Extinction Coefficient   38,000 cm^{-1} M^{-1} (Buffer pH 4.0); 107,000 cm^{-1} M^{-1} (Buffer pH 8.0)

Quantum Yield   0.67 (Buffer pH 7.4)

Synthesis   Synthetic methods[1–7]

Imaging/Labeling Applications   Bacteria;[8,9] lysosomes;[4] mitochondria;[10] proteins;[3] protein-2 (MRP2, ABCC2);[11–13] protein-5 (MRP5, ABCC5)[13]

Biological/Medical Applications   Assessing reactive oxygen species (ROS) generation in isolated-perfused rat heart tissue;[14] detecting/quantifying bacteria;[8,9] monitoring blood gas;[15] monitoring/measuring analytes concentration;[16] multidrug resistance-associated protein-2 (MRP2, ABCC2) substrates;[11–13] multidrug resistance-associated protein-5 (MRP5, ABCC5) substrates;[13] screening mitochondrial permeability transition;[10] studying hepatic disposition;[17,18] a substrate for measuring aryl sulfatase activity,[4] acid phosphatase activity,[4] β-galactosidase activity,[4] lipase activity[4]

Industrial Applications   Fiber optic sensor for sol–gel material;[19] planar laser-induced fluorescence (PLIF) measurements in aqueous flows[20]

Safety/Toxicity   Metabolic toxicity[21]

REFERENCES

1. Feng, S.; Sun, X.; Chen, J.; Zheng, H.; Wang, Y. Method for preparing 2',7'-dichloro-5(6)-carboxyfluorescein isomer. Faming Zhuanli Shenqing CN 103058980, 2013.

2. Shi, P. Process for preparation 6-carboxyfluorescein and derivatives. Faming Zhuanli Shenqing CN 102942553. 2013.

3. Hirabayashi, K.; Hanaoka, K.; Shimonishi, M.; Terai, T.; Komatsu, T.; Ueno, T.; Nagano, T. Selective two-step labeling of proteins with an off/on fluorescent probe. Chem.-Eur. J. 2011, 17, 14763–14771.

4. Coleman, D. J.; Naleway, J. J. Fluorogenic enzyme substrates for visualizing acidic organelles. U.S. Pat. Appl. Publ. US 20100233744, 2010.

5. Ge, F.; Chen, L. Synthesis and fluorescent properties of 2',7'-dichloro-5(6)-carboxyfluorescein. Huaxue Tongbao 2009, 72, 78–81.

6. Woodroofe, C. C.; Masalha, R.; Barnes, K. R.; Frederickson, C. J.; Lippard, S. J. Membrane-permeable and -impermeable sensors of the Zinpyr family and their application to imaging of hippocampal zinc in vivo. Chem. Biol. 2004, 11, 1659–1666.

7. Lippard, S. J.; Woodroofe, C. C. Sensors, and methods of making and using the same. U.S. Pat. Appl. Publ. US 20040224420, 2004.

8. Ribeiro Pinto de Oliveira Azevedo, N. F.; Macieira Cerqueira, L. I.; Torres Faria, N. R.; Lopes da Costa Vieira, M. J. Methods and peptide nucleic acid probes

for specific detection of *Helicobacter pylori* and its response to clarithromycin. PCT Int. Appl. WO 2011030319, 2011.

9. Ribeiro Pinto de Oliveira Azevedo, N. F.; Lopes da Costa Vieira, M. J.; Almeida, C. M. F.; Keevil, C. W. Peptide nucleic acid probe, kit and method for detection and/or quantification of *Salmonella* spp. and applications thereof. PCT Int. Appl. WO 2011121544, 2011.

10. Blattner, J. R.; He, L.; Lemasters, J. J. Screening assays for the mitochondrial permeability transition using a fluorescence multiwell plate reader. *Anal. Biochem.* **2001**, *295*, 220–226.

11. Munic, V.; Hlevnjak, M.; Haber, V. E. Characterization of rhodamine-123, calcein and 5(6)-carboxy-2′,7′-dichlorofluorescein (CDCF) export via MRP2 (ABCC2) in MES-SA and A549 cells. *Eur. J. Pharm. Sci.* **2011**, *43*, 359–369.

12. Pelis, R. M.; Shahidullah, M.; Ghosh, S.; Coca-Prados, M.; Wright, S. H.; Delamere, N. A. Localization of multidrug resistance-associated protein 2 in the nonpigmented ciliary epithelium of the eye. *J. Pharmacol. Exp. Ther.* **2009**, *329*, 479–485.

13. Pratt, S.; Chen, V.; Perry, W. I.; Starling, J. J.; Dantzig, A. H. Kinetic validation of the use of carboxydichlorofluorescein as a drug surrogate for MRP5-mediated transport. *Eur. J. Pharm. Sci.* **2006**, *27*, 524–532.

14. Kehrer, J. P.; Paraidathathu, T. The use of fluorescent probes to assess oxidative processes in isolated-perfused rat heart tissue. *Free Radical Res. Commun.* **1992**, *16*, 217–225.

15. Bentsen, J. G.; Wood, K. B. Sensor with improved drift stability. U.S. Patent 5403746, 1995.

16. Sharma, A. Optical probe and method for monitoring analyte concentration. PCT Int. Appl. WO 9212424, 1992.

17. Chandra, P.; Johnson, B. M.; Zhang, P.; Pollack, G. M.; Brouwer, K. L. R. Modulation of hepatic canalicular or basolateral transport proteins alters hepatobiliary disposition of a model organic anion in the isolated perfused rat liver. *Drug Metab. Dispos.* **2005**, *33*, 1238–1243.

18. Zamek-Gliszczynski, M. J.; Xiong, H.; Patel, N. J.; Turncliff, R. Z.; Pollack, G. M.; Brouwer, K. L. R. Pharmacokinetics of 5 (and 6)-carboxy-2′,7′-dichlorofluorescein and its diacetate promoiety in the liver. *J. Pharmacol. Exp. Ther.* **2003**, *304*, 801–809.

19. Manyam, U. H.; Shahriari, M. R.; Morris, M. J. A complete spectrum fiber optic pH sensor based on a novel fluorescent indicator doped porous sol–gel material. *Proc. SPIE- Int. Soc. Opt. Eng.* **1999**, *3540*, 10–18.

20. Karasso, P. S.; Mungal, M. G. PLIF measurements in aqueous flows using the Nd:YAG laser. *Exp. Fluids* **1997**, *23*, 382–387.

21. Sinclair, J. A.; Henderson, C.; Tettey, J. N. A.; Grant, M. H. The influence of the choice of digestion enzyme used to prepare rat hepatocytes on xenobiotic uptake and efflux. *Toxicol. In Vitro* **2013**, *27*, 451–457.

5-CARBOXY-2′,7′-DICHLORO-FLUORESCEIN DIACETATE

CAS Registry Number   144489-09-8

Chemical Structure

CA Index Name   Spiro[isobenzofuran-1(3H),9′-[9H]xanthene]-5-carboxylic acid, 3′,6′-bis(acetyloxy)-2′,7′-dichloro-3-oxo-

Other Names   5-Carboxy-2′,7′-dichlorofluorescein diacetate; Carboxy-DCFDA; 5-CDCF DA

Merck Index Number   Not listed

Chemical/Dye Class   Xanthene

Molecular Formula $C_{25}H_{14}Cl_2O_9$

Molecular Weight   529.28

Physical Form   Light brown solid;[1] Off-white solid

Solubility   Insoluble in water; soluble acetonitrile, chloroform, N,N-dimethylformamide, dimethyl sulfoxide, methanol

Melting Point   178–180 °C[1]

Boiling Point (Calcd.)   743.3 ± 60.0 °C, pressure: 760 Torr

pK$_a$ (Calcd.)   3.70 ± 0.20, most acidic, temperature: 25 °C

Absorption (λ_{max})   <300 nm (Buffer pH 4.0); 292 nm (MeOH)

Emission (λ_{max})   None

Molar Extinction Coefficient   6,000 (MeOH) cm^{-1} M^{-1}

Synthesis   Synthetic methods[1,2]

Imaging/Labeling Applications   Caco-2 cell;[11] lysosomes;[3] microparticles;[4] protein-2 (MRP2);[10] vacuoles;[5] yeast cells[6]

Biological/Medical Applications   Assessing reactive oxygen species (ROS) generation;[7,8] counting yeast cells;[6] identifying vacuolar system;[5] monitoring blood gas;[9] multidrug resistance-associated protein-2 (MRP2) substrates;[10,11] studying hepatic disposition;[12] a substrate for measuring esterase activity[3]

Industrial Applications   Not reported

Safety/Toxicity   Metabolic toxicity[13]

REFERENCES

1. Woodroofe, C. C.; Masalha, R.; Barnes, K. R.; Frederickson, C. J.; Lippard, S. J. Membrane-permeable and -impermeable sensors of the Zinpyr family and their application to imaging of hippocampal zinc *in vivo*. *Chem. Biol.* **2004**, *11*, 1659–1666.

2. Lippard, S. J.; Woodroofe, C. C. Sensors, and methods of making and using the same. U.S. Pat. Appl. Publ. US 20040224420, 2004.

3. Coleman, D. J.; Naleway, J. J. Fluorogenic enzyme substrates for visualizing acidic organelles. U.S. Pat. Appl. Publ. US 20100233744, 2010.

4. Haugland, R. P.; Haugland, R. P.; Brinkley, J. M.; Kang, H. C.; Kuhn, M.; Wells, K. S.; Zhang, Y. Z. Dipyrrometheneboron difluoride labeled fluorescent microparticles. U.S. Patent 5723218, 1998.

5. Inselman, A. L.; Gathman, A. C.; Lilly, W. W. Two fluorescent markers identify the vacuolar system of *Schizophyllum commune. Curr. Microbiol.* **1999**, *38*, 295–299.

6. Sugata, K.; Ueda, R.; Doi, T.; Onishi, T.; Matsumoto, K. Fluorescein-derivative method and apparatus for counting living microbial cells. U.S. Patent 5389544, 1995.

7. Pogue, A. I.; Jones, B. M.; Bhattacharjee, S.; Percy, M. E.; Zhao, Y.; Lukiw, W. J. Metal-sulfate induced generation of ROS in human brain cells: detection using an isomeric mixture of 5-and 6-carboxy-2′,7′-dichlorofluorescein diacetate (Carboxy-DCFDA) as a cell permeant tracer. *Int. J. Mol. Sci.* **2012**, *13*, 9615–9626.

8. Kehrer, J. P.; Paraidathathu, T. The use of fluorescent probes to assess oxidative processes in isolated-perfused rat heart tissue. *Free Radical Res. Commun.* **1992**, *16*, 217–225.

9. Bentsen, J. G.; Wood, K. B. Sensor with improved drift stability. U.S. Patent 5403746, 1995.

10. Nakanishi, T.; Shibue, Y.; Fukuyama, Y.; Yoshida, K.; Fukuda, H.; Shirasaka, Y.; Tamai, I. Quantitative time-lapse imaging-based analysis of drug-drug interaction mediated by hepatobiliary transporter, multidrug resistance-associated protein 2, in sandwich-cultured rat hepatocytes. *Drug Metab. Dispos.* **2011**, *39*, 984–991.

11. Siissalo, S.; Hannukainen, J.; Kolehmainen, J.; Hirvonen, J.; Kaukonen, A. M. A Caco-2 cell based screening method for compounds interacting with MRP2 efflux protein. *Eur. J. Pharm. Biopharm.* **2009**, *71*, 332–338.

12. Zamek-Gliszczynski, M. J.; Xiong, H.; Patel, N. J.; Turncliff, R. Z.; Pollack, G. M.; Brouwer, K. L. R. Pharmacokinetics of 5 (and 6)-carboxy-2′,7′-dichlorofluorescein and its diacetate promoiety in the liver. *J. Pharmacol. Exp. Ther.* **2003**, *304*, 801–809.

13. Sinclair, J. A.; Henderson, C.; Tettey, J. N. A.; Grant, M. H. The influence of the choice of digestion enzyme used to prepare rat hepatocytes on xenobiotic uptake and efflux. *Toxicol. In Vitro* **2013**, *27*, 451–457.

6-CARBOXY-2′,7′-DICHLORO-FLUORESCEIN DIACETATE

CAS Registry Number   144489-10-1

Chemical Structure

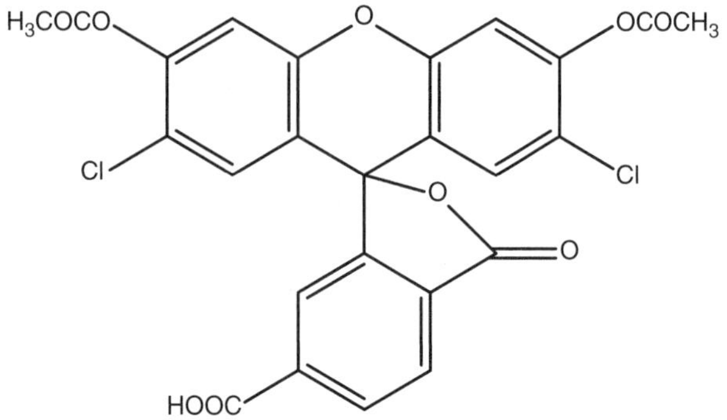

CA Index Name   Spiro[isobenzofuran-1(3H),9′-[9H]xanthene]-6-carboxylic acid, 3′,6′-bis(acetyloxy)-2′,7′-dichloro-3-oxo-

Other Names   3′,6′-Diacetyl-2′,7′-dichloro-6-carboxy-fluorescein;   6-Carboxy-2′,7′-dichlorofluorescein diacetate; Carboxy-DCFDA; 6-CDCF DA

Merck Index Number   Not listed

Chemical/Dye Class   Xanthene

Molecular Formula $C_{25}H_{14}Cl_2O_9$

Molecular Weight   529.28

Physical Form   Off-white solid

Solubility   Insoluble in water; soluble in dimethyl sulfoxide

Melting Point   >300 °C (decompose)[2]

Boiling Point (Calcd.)   751.5 ± 60.0 °C, pressure: 760 Torr

pK$_a$ (Calcd.)   3.49 ± 0.20, most acidic, temperature: 25 °C

Absorption (λ_{max})   <300 nm (Buffer pH 4.0); 292 nm (MeOH)

Emission (λ_{max})   None

Molar Extinction Coefficient   6,000   (MeOH) cm^{-1} M^{-1}

Synthesis   Synthetic methods[1–3]

Imaging/Labeling Applications   Caco-2 cell;[13] lysosomes;[4] proteins;[1] protein-2 (MRP2);[12] vacuoles;[5] yeast cells[6,7]

Biological/Medical Applications   Assessing reactive oxygen species (ROS) generation;[8,9] counting/detecting yeast cells;[6,7] diagnosing Cowden syndrome;[10] identifying vacuolar system;[5] monitoring blood gas;[11] multidrug resistance-associated protein-2 (MRP2) substrates;[12,13] studying hepatic disposition;[14] a substrate for measuring esterase activity[4]

Industrial Applications   Not reported

Safety/Toxicity   Metabolic toxicity[15]

REFERENCES

1. Hirabayashi, K.; Hanaoka, K.; Shimonishi, M.; Terai, T.; Komatsu, T.; Ueno, T.; Nagano, T. Selective two-step labeling of proteins with an off/on fluorescent probe. *Chem.-Eur. J.* **2011**, *17*, 14763–14771.

2. Woodroofe, C. C.; Masalha, R.; Barnes, K. R.; Frederickson, C. J.; Lippard, S. J. Membrane-permeable and -impermeable sensors of the Zinpyr family and their application to imaging of hippocampal zinc *in vivo*. *Chem. Biol.* **2004**, *11*, 1659–1666.

3. Lippard, S. J.; Woodroofe, C. C. Sensors, and methods of making and using the same. U.S. Pat. Appl. Publ. US 20040224420, 2004.

4. Coleman, D. J.; Naleway, J. J. Fluorogenic enzyme substrates for visualizing acidic organelles. U.S. Pat. Appl. Publ. US 20100233744, 2010.

5. Inselman, A. L.; Gathman, A. C.; Lilly, W. W. Two fluorescent markers identify the vacuolar system of *Schizophyllum commune*. *Curr. Microbiol.* **1999**, *38*, 295–299.

6. Abe, F.; Horikoshi, H. The microorganism detection by the flow cytometry method. Jpn. Kokai Tokkyo Koho JP 2000069995, 2000.

7. Sugata, K.; Ueda, R.; Doi, T.; Onishi, T.; Matsumoto, K. Fluorescein-derivative method and apparatus for counting living microbial cells. U.S. Patent 5389544, 1995.

8. Pogue, A. I.; Jones, B. M.; Bhattacharjee, S.; Percy, M. E.; Zhao, Y.; Lukiw, W. J. Metal-sulfate induced generation of ROS in human brain cells: detection using an isomeric mixture of 5-and 6-carboxy-2′,7′-dichlorofluorescein diacetate (Carboxy-DCFDA) as a cell permeant tracer. *Int. J. Mol. Sci.* **2012**, *13*, 9615–9626.

9. Kehrer, J. P.; Paraidathathu, T. The use of fluorescent probes to assess oxidative processes in isolated-perfused rat heart tissue. *Free Radical Res. Commun.* **1992**, *16*, 217–225.

10. Eng, C. Diagnosis of Cowden syndrome based on detection of mutated succinate dehydrogenase isoenzymes B and D. U.S. Pat. Appl. Publ. US 20100092961, 2010.

11. Bentsen, J. G.; Wood, K. B. Sensor with improved drift stability. U.S. Patent 5403746, 1995.

12. Nakanishi, T.; Shibue, Y.; Fukuyama, Y.; Yoshida, K.; Fukuda, H.; Shirasaka, Y.; Tamai, I. Quantitative time-lapse imaging-based analysis of drug-drug interaction mediated by hepatobiliary transporter, multidrug resistance-associated protein 2, in sandwich-cultured rat hepatocytes. *Drug Metab. Dispos.* **2011**, *39*, 984–991.

13. Siissalo, S.; Hannukainen, J.; Kolehmainen, J.; Hirvonen, J.; Kaukonen, A. M. A Caco-2 cell based screening method for compounds interacting with MRP2 efflux protein. *Eur. J. Pharm. Biopharm.* **2009**, *71*, 332–338.

14. Zamek-Gliszczynski, M. J.; Xiong, H.; Patel, N. J.; Turncliff, R. Z.; Pollack, G. M.; Brouwer, K. L. R. Pharmacokinetics of 5 (and 6)-carboxy-2′,7′-dichlorofluorescein and its diacetate promoiety in the liver. *J. Pharmacol. Exp. Ther.* **2003**, *304*, 801–809.

15. Sinclair, J. A.; Henderson, C.; Tettey, J. N. A.; Grant, M. H. The influence of the choice of digestion enzyme used to prepare rat hepatocytes on xenobiotic uptake and efflux. *Toxicol. In Vitro* **2013**, *27*, 451–457.

5-CARBOXYFLUORESCEIN

CAS Registry Number   76823-03-5

Chemical Structure

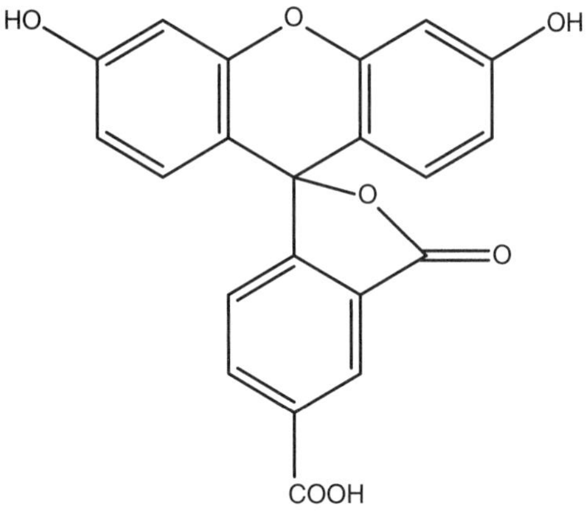

CA Index Name   Spiro[isobenzofuran-1(3H),9'-[9H] xanthene]-5-carboxylic acid, 3',6'-dihydroxy-3-oxo-

Other Names   5-Carboxyfluorescein; 5-FAM; FAM; FAM (dye); FAM 494/520

Merck Index Number   Not listed

Chemical/Dye Class   Xanthene

Molecular Formula $C_{21}H_{12}O_7$

Molecular Weight   376.32

Physical Form   Orange solid

Solubility   Soluble in N,N-dimethylformamide, dimethyl sulfoxide, methanol

Melting Point   368–372 °C[4]

Boiling Point (Calcd.)   725.2 ± 60.0 °C, pressure: 760 Torr

pK$_a$ (Calcd.)   3.80 ± 0.20, most acidic, temperature: 25 °C

Absorption (λ_{max})   492 nm (Buffer pH 9.0)

Emission (λ_{max})   518 nm (Buffer pH 9.0)

Molar Extinction Coefficient   79,000 cm^{-1} M^{-1} (Buffer pH 9.0)

Synthesis   Synthetic methods[1–5]

Imaging/Labeling Applications   β-Amyloid peptides;[6,7] bacteria;[8–10] cell;[11] dendrimers;[12] liposomes;[13–17] metal ions (lead ions,[18] mercury ions,[19–21] silver ions;[22]) microorganisms;[23,24] nucleic acids/nucleotides;[25–44] proteins/peptides;[45–55] quantum dots;[56] viruses[57,58]

Biological/Medical Applications   Authenticating canned products;[59] detecting/measuring citrate;[60,61] as a substrate for measuring phospholipase activity;[62] use in ophthalmology[63]

Industrial Applications   Colored or luminescent products;[64] as geothermal tracers;[65] microfabricated electrophoresis devices;[66,67] porous polymer monoliths;[68] separation devices;[69] thin-film polymer light emitting diodes[70]

Safety/Toxicity   Genotoxicity;[71] membrane toxicity;[72] ophthalmic toxicity;[73] pulmonary toxicity[74]

REFERENCES

1. Sabnis, R. W. *Handbook of Acid–base Indicators*; CRC Press: Boca Raton, **2008**; pp 64–65.

2. Chaudhari, A. S.; Parab, Y. S.; Patil, V.; Sekar, N.; Shukla, S. R. Intrinsic catalytic activity of Bronsted acid ionic liquids for the synthesis of triphenylmethane and phthalein under microwave irradiation. *RSC Adv.* **2012**, *2*, 12112–12117.

3. Oeberg, C. T.; Carlsson, S.; Fillion, E.; Leffler, H.; Nilsson, U. J. Efficient and expedient two-step pyranose-retaining fluorescein conjugation of complex reducing oligosaccharides: Galectin oligosaccharide specificity studies in a fluorescence polarization assay. *Bioconjugate Chem.* **2003**, *14*, 1289–1297.

4. Rossi, F. M.; Kao, J. P. Y. Practical method for the multigram separation of the 5- and 6- isomers of carboxyfluorescein. *Bioconjugate Chem.* **1997**, *8*, 495–497.

5. Drechsler, G.; Smagin, S. Preparation of trimelliteins and 5'-carboxyfluorescein. *J. Prakt. Chem.* **1965**, *28*, 315–324.

6. Edwin, N. J.; Hammer, R. P.; McCarley, R. L.; Russo, P. S. Reversibility of β-amyloid self-assembly: Effects of pH and added salts assessed by fluorescence photobleaching recovery. *Biomacromolecules* **2010**, *11*, 341–347.

7. Edwin, N. J.; Bantchev, G. B.; Russo, P. S.; Hammer, R. P.; McCarley, R. L. Elucidating the kinetics of β-amyloid fibril formation. *ACS Symp. Ser.* **2005**, *916*, 106–118.

8. Wu, Q.; Zhang, Y. Detection of *Salmonella* by real-time PCR. Faming Zhuanli Shenqing CN 103233070, 2013.

9. Matveeva, T. V.; Bogomaz, D. I.; Lutova, L. A. Real-time PCR-based diagnostic technique for detecting agrobacteria in plants and biomaterials. Russ. RU 2458142, 2012.

10. Hourfar, K. M.; Schmidt, M.; Seifried, E. Procedure for detection of bacteria in blood-derived samples with real-time PCR. Ger. Offen. DE 102006026963, 2007.

11. Jing, C.; Cornish, V. W. A fluorogenic TMP-tag for high signal-to-background intracellular live cell imaging. *ACS Chem. Biol.* **2013**, *8*, 1704–1712.

12. Rainwater, J. C.; Anslyn, E. V. Amino-terminated PAMAM dendrimers electrostatically uptake numerous anionic indicators. *Chem. Commun.* **2010**, *46*, 2904–2906.

13. Locascio-Brown, L.; Plant, A. L.; Horvath, V.; Durst, R. A. Liposome flow injection immunoassay: implications for sensitivity, dynamic range, and antibody regeneration. *Anal. Chem.* **1990**, *62*, 2587–2593.

14. Barber, R. F.; Shek, P. N. Tear-induced release of liposome-entrapped agents. *Int. J. Pharm.* **1990**, *60*, 219–227.

15. Domingo, J. C.; Rosell, F.; Mora, M.; De Madariaga, M. A. Importance of the purification grade of 5(6)-carboxyfluorescein on the stability and permeability properties of N-acyl-phosphatidylethanolamine liposomes. *Biochem. Soc. Trans.* **1989**, *17*, 997–999.

16. Chen, R. F.; Knutson, J. R. Mechanism of fluorescence concentration quenching of carboxyfluorescein in liposomes: energy transfer to nonfluorescent dimers. *Anal. Biochem.* **1988**, *172*, 61–77.

17. Barber, R. F.; Shek, P. N. Liposomes and tear fluid. I. Release of vesicle-entrapped carboxyfluorescein. *Biochim. Biophys. Acta, Lipids Lipid Metab.* **1986**, *879*, 157–163.

18. Zhan, S.; Wu, Y.; Liu, L.; Xing, H.; He, L.; Zhan, X.; Luo, Y.; Zhou, P. A simple fluorescent assay for lead(II) detection based on lead(ii)-stabilized G-quadruplex formation. *RSC Adv.* **2013**, *3*, 16962–16966.

19. Li, H.; Zhai, J.; Sun, X. Nano-C60 as a novel, effective fluorescent sensing platform for mercury(II) ion detection at critical sensitivity and selectivity. *Nanoscale* **2011**, *3*, 2155–2157.

20. Lin, Y.; Chang, H. Detection of mercury and phenylmercury ions using DNA-based fluorescent probe. *Analyst* **2011**, *136*, 3323–3328.

21. Wang, Z.; Heon Lee, J.; Lu, Y. Highly sensitive "turn-on" fluorescent sensor for Hg^{2+} in aqueous solution based on structure-switching DNA. *Chem. Commun.* **2008**, 6005–6007.

22. Chen, X.; Chen, Y.; Zhou, X.; Hu, J. Detection of Ag^+ ions and cysteine based on chelation actions between Ag+ ions and guanine bases. *Talanta* **2013**, *107*, 277–283.

23. Beimfohr, C.; Thelen, K.; Snaidr, J. A fluorescence in situ hybridization method for detection of microorganisms using FRET probes to increase signal:noise ratio. Ger. Offen. DE 102010012421, 2010.

24. Filipenko, M. L.; Khripko, Yu. I.; Afonyushkin, V. N.; Boyarskikh, U. A.; Dudareva, E. V.; Stepnov, V. A. Quantitative PCR-based method of detecting microorganisms of genus *Salmonella* in livestock and poultry. Russ. RU 2360004, 2009.

25. Riahi, R.; Dean, Z.; Wu, T.; Teitell, M. A.; Chiou, P.; Zhang, D. D.; Wong, P. K. Detection of mRNA in living cells by double-stranded locked nucleic acid probes. *Analyst* **2013**, *138*, 4777–4785.

26. Demina, A. V.; Ternovoi, V. A.; Netesov, S. V. Set of oligodeoxyribonucleotide primers and fluorescent-marked probes for enterovirus RNA identification by RT-PCR with hybridisation fluorescent detection. Russ. RU 2459830, 2012.

27. Renciuk, D.; Zhou, J.; Beaurepaire, L.; Guedin, A.; Bourdoncle, A.; Mergny, J. A FRET-based screening assay for nucleic acid ligands. *Methods* **2012**, *57*, 122–128.

28. Yang, X.; Zhu, Y.; Liu, P.; He, L.; Li, Q.; Wang, Q.; Wang, K.; Huang, J.; Liu, J. G-quadruplex fluorescence quenching ability: a simple and efficient strategy to design a single-labeled DNA probe. *Anal. Methods* **2012**, *4*, 895–897.

29. Zhang, L.; Guo, S.; Dong, S.; Wang, E. Pd nanowires as new biosensing materials for magnified fluorescent detection of nucleic acid. *Anal. Chem.* **2012**, *84*, 3568–3573.

30. Graser, E.; Hildebrand, T. Use of guanine quenching of fluorescence in the detection of nucleic acid hybrids in real time. Ger. Offen. DE 102010052524, 2012.

31. Li, H.; Luo, Y.; Sun, X. Fluorescence resonance energy transfer dye-labeled probe for fluorescence-enhanced DNA detection: An effective strategy to greatly improve discrimination ability toward single-base mismatch. *Biosens. Bioelectron.* **2011**, *27*, 167–171.

32. Doi, T.; Okada, Y.; Takano, T.; Yamamoto, T.; Kiyono, S.; Nakagawa, S. Method for detecting nucleic acid molecule using gold magnetic particles. Jpn. Kokai Tokkyo Koho JP 2010178724, 2010.

33. Nakayama, S.; Yan, L.; Sintim, H. O. Junction probes - sequence specific detection of nucleic acids via template enhanced hybridization processes. *J. Am. Chem. Soc.* **2008**, *130*, 12560–12561.

34. Kostina, E. V.; Ryabinin, V. A.; Maksakova, G. A.; Sinyakov, A. N. TaqMan probes based on oligonucleotide-hairpin minor groove binder conjugates. *Russ. J. Bioorg. Chem.* **2007**, *33*, 614–616.

35. Kawai, R.; Kimoto, M.; Mitsui, T.; Yokoyama, S.; Hirao, I. Site-specific fluorescent labeling of RNA by a base-pair expanded transcription system. *Nucleic Acids Symp. Ser.* **2004**, 35–36.

36. Marti, A. A.; Li, X.; Jockusch, S.; Stevens, N.; Li, Z.; Raveendra, B.; Kalachikov, S.; Morozova, I.; Russo, J. J.; Akins, D. L.; Ju, J.; Turro, N. J. Design and characterization of two-dye and three-dye binary fluorescent probes for mRNA detection. *Tetrahedron* **2007**, *63*, 3591–3600.

37. Weizmann, Y.; Cheglakov, Z.; Pavlov, V.; Willner, I. Autonomous fueled mechanical replication of nucleic acid templates for the amplified optical detection of DNA. *Angew. Chem., Int. Ed.* **2006**, *45*, 2238–2242.

38. Faulds, K.; Smith, W. E.; Graham, D. Evaluation of surface-enhanced resonance raman scattering for quantitative DNA analysis. *Anal. Chem.* **2004**, *76*, 412–417.

39. Koehler, T.; Rost, A. Nucleic acids for use as internal standards in quantitative assays and their preparation and use. Ger. Offen. DE 10209071, 2003.

40. Nishimura, M.; Ueda, N.; Naito, S. FRET detection of human liver function marker protein mRNAs using PCR primers and fluorescent dye pair labeled probes. Jpn. Kokai Tokkyo Koho JP 2003225091, 2003.

41. Ota, N.; Sato, T.; Taira, K.; Ohkawa, J. Molecular tryst peeping: Detection of interactions between nonlabeled nucleic acids by fluorescence resonance energy transfer. *Biochem. Biophys. Res. Commun.* **2001**, *289*, 1067–1074.

42. Torimura, M.; Kurata, S.; Yamada, K.; Yokomaku, T.; Kamagata, Y.; Kanagawa, T.; Kurane, R. Fluorescence-quenching phenomenon by photoinduced electron transfer between a fluorescent dye and a nucleotide base. *Anal. Sci.* **2001**, *17*, 155–160.

43. Menchen, S.; Benson, S.; Upadhya, K.; Theisen, P.; Hauser, J.; Kenney, P. Fluorescent labeling of DNA for genetic analysis. *Proc. SPIE-Int. Soc. Opt. Eng.* **2000**, *3926*, 88–94.

44. Oefner, P. J.; Huber, C. G.; Umlauft, F.; Berti, G.; Stimpfl, E.; Bonn, G. K. High-resolution liquid chromatography of fluorescent dye-labeled nucleic acids. *Anal. Biochem.* **1994**, *223*, 39–46.

45. Brun, M. A.; Tan, K.; Griss, R.; Kielkowska, A.; Reymond, L.; Johnsson, K. A semisynthetic fluorescent sensor protein for glutamate. *J. Am. Chem. Soc.* **2012**, *134*, 7676–7678.

46. Nokihara, K.; Sogon, T.; Hirata, A. Peptide intracellular uptake quantity evaluation method, and cell identification method using it. Jpn. Kokai Tokkyo Koho JP 2012050390, 2012.

47. Helms, B. A.; Reulen, S. W. A.; Nijhuis, S.; de Graaf-Heuvelmans, P. T. H. M.; Merkx, M.; Meijer, E. W. High-affinity peptide-based collagen targeting using synthetic phage mimics: From phage display to dendrimer display. *J. Am. Chem. Soc.* **2009**, *131*, 11683–11685.

48. Brun, M. A.; Tan, K.; Nakata, E.; Hinner, M. J.; Johnsson, K. Semisynthetic fluorescent sensor proteins based on self-labeling protein tags. *J. Am. Chem. Soc.* **2009**, *131*, 5873–5884.

49. Hosaka, T. Preparation of N-terminal labeled proteins using modified initiator tRNA. Jpn. Kokai Tokkyo Koho JP 2008193911, 2008.

50. Ye, G.; Nam, N.; Kumar, A.; Saleh, A.; Shenoy, D. B.; Amiji, M. M.; Lin, X.; Sun, G.; Parang, K. Synthesis and evaluation of tripodal peptide analogues for cellular delivery of phosphopeptides. *J. Med. Chem.* **2007**, *50*, 3604–3617.

51. Onoue, S.; Liu, B.; Nemoto, Y.; Hirose, M.; Yajima, T. Chemical synthesis and application of C-terminally 5-carboxyfluorescein-labeled thymopentin as a fluorescent probe for thymopoietin receptor. *Anal. Sci.* **2006**, *22*, 1531–1535.

52. Shen, D.; Liang, K.; Ye, Y.; Tetteh, E.; Achilefu, S. Preliminary study on the inhibition of nuclear internalization of Tat peptides by conjugation with a receptor-specific peptide and fluorescent dyes. *Proc. SPIE-Int. Soc. Opt. Eng.* **2006**, *6097*, 609704/1–609704/9.

53. Wu, X.; Pawliszyn, J. Whole-column fluorescence-imaged capillary electrophoresis. *Electrophoresis* **2004**, *25*, 3820–3824.

54. Fischer, R.; Mader, O.; Jung, G.; Brock, R. Extending the applicability of carboxyfluorescein in solid-phase synthesis. *Bioconjugate Chem.* **2003**, *14*, 653–660.

55. Rockey, J. H.; Li, W.; Eccleston, J. F. Binding of fluorescein and carboxyfluorescein by human serum proteins: significance of kinetic and equilibrium

parameters of association in ocular fluorometric studies. *Exp. Eye Res.* **1983**, *37*, 455–466.

56. Chen, H.; Zou, P.; Connarn, J.; Paholak, H.; Sun, D. Intracellular dissociation of a polymer coating from nanoparticles. *Nano Res.* **2012**, *5*, 815–825.

57. Li, X. Method and test kit for detecting 5 hepatitis C virus (HCV) subtypes by RT-PCR. Faming Zhuanli Shenqing CN 102676693, 2012.

58. Shikov, A. N.; Ternovoi, V. A.; Sementsova, A. O.; Agafonov, A. P.; Sergeev, A. N. Set of oligonucleotide primers and fluorescent-marked probes for species-specific instant identification of marburg virus by polymerase chain reaction. Russ. RU 2458143, 2012.

59. Infante, C.; Catanese, G.; Ponce, M.; Manchado, M. Novel method for the authentication of frigate tunas (*Auxis thazard* and *Auxis rochei*) in commercial canned products. *J. Agric. Food Chem.* **2004**, *52*, 7435–7443.

60. Schmuck, C.; Schwegmann, M. A naked-eye sensing ensemble for the selective detection of citrate-but not tartrate or malate-in water based on a tris-cationic receptor. *Org. Biomol. Chem.* **2006**, *4*, 836–838.

61. Metzger, A.; Anslyn, E. V. A chemosensor for citrate in beverages. *Angew. Chem., Int. Ed.* **1998**, *37*, 649–652.

62. Graham, R. J. Fluorescent phospholipase assays and compositions. PCT Int. Appl. WO 2005005977, 2005.

63. Laties, A. M.; Stone, R. A. Ophthalmic use of carboxyfluorescein. U.S. Patent 4350676, 1982.

64. Miyoshi, K. Colorant compositions with good adhesion and resistance to dry conditions, and colored or luminescent products containing them. Jpn. Kokai Tokkyo Koho JP 2006265306, 2006.

65. Wong, Y. L.; Rose, P. E. The testing of fluorescein derivatives as candidate geothermal tracers. *GRC Trans.* **2000**, *24*, 637–640.

66. Tian, H.; Emrich, C. A.; Scherer, J. R.; Mathies, R. A.; Andersen, P. S.; Larsen, L. A.; Christiansen, M. High-throughput single-strand conformation polymorphism analysis on a microfabricated capillary array electrophoresis device. *Electrophoresis* **2005**, *26*, 1834–1842.

67. Mogensen, K. B.; Petersen, N. J.; Hubner, J.; Kutter, J. P. Monolithic integration of optical waveguides for absorbance detection in microfabricated electrophoresis devices. *Electrophoresis* **2001**, *22*, 3930–3938.

68. Rohr, T.; Yu, C.; Davey, M. H.; Svec, F.; Frechet, J. M. J. Porous polymer monoliths: simple and efficient mixers prepared by direct polymerization in the channels of microfluidic chips. *Electrophoresis* **2001**, *22*, 3959–3967.

69. Cezar de Andrade Costa, R.; Mogensen, K. B.; Kutter, J. P. Microfabricated porous glass channels for electrokinetic separation devices. *Lab Chip* **2005**, *5*, 1310–1314.

70. Edel, J. B.; Beard, N. P.; Hofmann, O.; DeMello, J. C.; Bradley, D. D. C.; DeMello, A. J. Thin-film polymer light emitting diodes as integrated excitation sources for microscale capillary electrophoresis. *Lab Chip* **2004**, *4*, 136–140.

71. He, L.; Jurs, P. C.; Custer, L. L.; Durham, S. K.; Pearl, G. M. Predicting the genotoxicity of polycyclic aromatic compounds from molecular structure with different classifiers. *Chem. Res. Toxicol.* **2003**, *16*, 1567–1580.

72. Hamid K. A.; Katsumi H.; Sakane T.; Yamamoto A. The effects of common solubilizing agents on the intestinal membrane barrier functions and membrane toxicity in rats. *Int. J. Pharm.* **2009**, *379*, 100–108.

73. Vehanen, K.; Hornof, M.; Urtti, A.; Uusitalo, H. Peribulbar poloxamer for ocular drug delivery. *Acta Ophthalmol.* **2008**, *86*, 91–96.

74. He, L.; Gao, Y.; Lin, Y.; Katsumi, H.; Fujita, T.; Yamamoto, A. Improvement of pulmonary absorption of insulin and other water-soluble compounds by polyamines in rats. *J. Controlled Release* **2007**, *122*, 94–101.

6-CARBOXYFLUORESCEIN

CAS Registry Number 3301-79-9

Chemical Structure

CA Index Name Spiro[isobenzofuran-1(3H),9'-[9H] xanthene]-6-carboxylic acid, 3',6'-dihydroxy-3-oxo-

Other Names Spiro[phthalan-1,9'-xanthene]-6-carboxylic acid, 3',6'-dihydroxy-3-oxo-; Terephthalic acid, (3,6,9-trihydroxyxanthen-9-yl)-, γ-lactone; 6-Carboxyfluorescein; 6-FAM; Fluorescein, 6-carboxy-

Merck Index Number Not listed

Chemical/Dye Class Xanthene

Molecular Formula $C_{21}H_{12}O_7$

Molecular Weight 376.32

Physical Form Orange solid

Solubility Soluble in acetone, chloroform, dichloromethane, N,N-dimethylformamide, dimethyl sulfoxide, methanol

Melting Point >360 °C;[5] 352–356 °C[4]

Boiling Point (Calcd.) 736.4 ± 60.0 °C, pressure: 760 Torr

pK$_a$ (Calcd.) 3.58 ± 0.20, most acidic, temperature: 25 °C

Absorption (λ_{max}) 492 nm (Buffer pH 9.0)

Emission (λ_{max}) 515 nm (Buffer pH 9.0)

Molar Extinction Coefficient 81,000 cm^{-1} M^{-1} (Buffer pH 9.0)

Synthesis Synthetic methods[1–5]

Imaging/Labeling Applications Ammonia gas;[6,7] antibodies;[8,9] axons;[10] bacteria;[11,12] cornea;[13] cytoplasm;[14] insulin;[15] islet cells;[16] liposomes;[17–33,74] myosin subfragment 1;[34] nucleic acids/nucleotides;[35–44] proteins/peptides;[45–48] protoplasts;[49] silver nanoparticles;[50] symplast;[51–54] viruses[55–69]

Biological/Medical Applications Diagnosing bladder cancer;[70] evaluating corneal endothelial barrier function;[13] as a substrate for measuring protein kinase activity,[71] methyltransferase activity,[72] phospholipase activity,[73,74] rhodanese activity,[75] sulfotransferase activity;[76] use in ophthalmology[77]

Industrial Applications Not reported

Safety/Toxicity Cytotoxicity;[78] genotoxicity;[79,80] immunotoxicity;[80] membrane toxicity;[81–83] neurotoxicity;[84] nucleic acid damage;[85] ophthalmic toxicity;[86–90] pulmonary toxicity[91,92]

REFERENCES

1. Sabnis, R. W. *Handbook of Acid–base Indicators*; CRC Press: Boca Raton, **2008**; pp 66–67.

2. Shi, P. Process for preparation 6-carboxyfluorescein and derivatives. Faming Zhuanli Shenqing CN 102942553, 2013.

3. Oeberg, C. T.; Carlsson, S.; Fillion, E.; Leffler, H.; Nilsson, U. J. Efficient and expedient two-step pyranose-retaining fluorescein conjugation of complex reducing oligosaccharides: Galectin oligosaccharide specificity studies in a fluorescence polarization assay. *Bioconjugate Chem.* **2003**, *14*, 1289–1297.

4. Rossi, F. M.; Kao, J. P. Y. Practical method for the multigram separation of the 5- and 6- isomers of carboxyfluorescein. *Bioconjugate Chem.* **1997**, *8*, 495–497.

5. Drechsler, G.; Smagin, S. Preparation of trimelliteins and 5'-carboxyfluorescein. *J. Prakt. Chem.* **1965**, *28*, 315–324.

6. Kar, S.; Arnold, M. A. Air-gap fiber-optic ammonia gas sensor. *Talanta* **1993**, *40*, 757–760.

7. Rhines, T. D.; Arnold, M. A. Fiber-optic biosensor for urea based on sensing of ammonia gas. *Anal. Chim. Acta* **1989**, *227*, 387–396.

8. Schenk, J. A.; Sellrie, F.; Boettger, V.; Menning, A.; Stoecklein, W. F. M.; Micheel, B. Generation and application of a fluorescein-specific single chain antibody. *Biochimie* **2007**, *89*, 1304–1311.

9. Micheel, B.; Scharte, G.; Jantscheff, P. Production of monoclonal antibodies to FITC. Ger. (East) DD 260082, 1988.

10. Brink, P. R.; Ramanan, S. V. A model for the diffusion of fluorescent probes in the septate giant axon of earthworm. Axoplasmic diffusion and junctional membrane permeability. *Biochem. J.* **1985**, *48*, 299–309.

11. Ruppel, S.; Ruehlmann, J.; Merbach, W. Quantification and localization of bacteria in plant tissues using quantitative real-time PCR and online emission fingerprinting. *Plant Soil* **2006**, *286*, 21–35.

12. Huijsdens, X. W.; Linskens, R. K.; Mak, M.; Meuwissen, S. G. M.; Vandenbroucke-Grauls, C. M. J. E.; Savelkoul, P. H. M. Quantification of bacteria adherent to gastrointestinal mucosa by real-time PCR. *J. Clin. Microbiol.* **2002**, *40*, 4423–4427.

13. Araie, M. Carboxyfluorescein. A dye for evaluating the corneal endothelial barrier function *in vivo. Exp. Eye Res.* **1986**, *42*, 141–150.

14. Fushimi, K.; Verkman, A. S. Low viscosity in the aqueous domain of cell cytoplasm measured by picosecond polarization microfluorometry. *J. Cell Biol.* **1991**, *112*, 719–725.

15. Touitou, E.; Alhaique, F.; Fisher, P.; Memoli, A.; Riccieri, F. M.; Santucci, E. Prevention of molecular self-association by sodium salicylate: effect on insulin and 6-carboxyfluorescein. *J. Pharm. Sci.* **1987**, *76*, 791–793.

16. Meda, P. Tracer microinjections into islet cells. *Methods Diabetes Res.* **1984**, *1(A)*, 193–204.

17. Domingo, J. C.; Rosell, F.; Mora, M.; De Madariaga, M. A. Importance of the purification grade of 5(6)-carboxyfluorescein on the stability and permeability properties of N-acyl-phosphatidylethanolamine liposomes. *Biochem. Soc. Trans.* **1989**, *17*, 997–999.

18. Tsuneyoshi, T.; Ishimori, Y. Sensitized liposomes for analysis by complement lysis immunoassay. Jpn. Kokai Tokkyo Koho JP 63151858, 1988.

19. Iga, K.; Hamaguchi, S.; Ogawa, T. Manufacture of liposomes containing pharmaceuticals. Jpn. Kokai Tokkyo Koho JP 63112512, 1988.

20. Chen, R. F.; Knutson, J. R. Mechanism of fluorescence concentration quenching of carboxyfluorescein in liposomes: energy transfer to nonfluorescent dimers. *Anal. Biochem.* **1988**, *172*, 61–77.

21. Woolfrey, S. G.; Taylor, G.; Kellaway, I. W.; Smith, A. Pulmonary absorption of liposome-encapsulated 6-carboxyfluorescein. *J. Controlled Release* **1988**, *5*, 203–209.

22. Ruiz, J.; Goni, F. M.; Alonso, A. Surfactant-induced release of liposomal contents. A survey of methods

and results. *Biochim. Biophys. Acta, Biomembr.* **1988**, *937*, 127–134.

23. Fransen, G. J.; Salemink, P. J. M.; Crommelin, D. J. A. Critical parameters in freezing of liposomes. *Int. J. Pharm.* **1986**, *33*, 27–35.

24. Weinstein, J. N.; Blumenthal, R.; Klausner, R. D. Carboxyfluorescein leakage assay for lipoprotein-liposome interaction. *Methods Enzymol.* **1986**, *128*, 657–668.

25. Lee, C. H.; Choi, M. U. A permeability measurement of small unilamellar vesicles by 6-carboxyfluorescein. *Bull. Korean Chem. Soc.* **1984**, *5*, 154–158.

26. Truneh, A.; Machy, P.; Barbet, J.; Mishal, Z.; Lemonnier, F. A.; Leserman, L. D. Endocytosis of HLA and H-2 molecules on transformed murine cells measured by fluorescence dequenching of liposome-encapsulated carboxyfluorescein. *EMBO J.* **1983**, *2*, 2285–2291.

27. Truneh, A.; Mishal, Z.; Barbet, J.; Machy, P.; Leserman, L. D. Endocytosis of liposomes bound to cell surface proteins measured by flow cytofluorometry. *Biochem. J.* **1983**, *214*, 189–194.

28. Straubinger, R. M.; Hong, K.; Friend, D. S.; Papahadjopoulos, D. Endocytosis of liposomes and intracellular fate of encapsulated molecules: encounter with a low pH compartment after internalization in coated vesicles. *Cell* **1983**, *32*, 1069–1079.

29. Norrie, D. H.; Pietrowski, R. A.; Stephen, J. Screening the efficiency of intracytoplasmic delivery of materials to HeLa cells by liposomes. *Anal. Biochem.* **1982**, *127*, 276–281.

30. Senior, J.; Gregoriadis, G. Stability of small unilamellar liposomes in serum and clearance from the circulation: the effect of the phospholipid and cholesterol components. *Life Sci.* **1982**, *30*, 2123–2136.

31. Yasuda, T.; Naito, Y.; Tsumita, T.; Tadakuma, T. A simple method to measure antiglycolipid antibody by using complement-mediated immune lysis of fluorescent dye-trapped liposomes. *J. Immunol. Methods* **1981**, *44*, 153–158.

32. Parker, R. J.; Sieber, S. M.; Weinstein, J. N. Effect of liposome encapsulation of a fluorescent dye on its uptake by the lymphatics of the rat. *Pharmacology* **1981**, *23*, 128–136.

33. Leung, J. G. M. Liposomes. Effect of temperature on their mode of action on single frog skeletal muscle

fibers. *Biochim. Biophys. Acta, Biomembr.* **1980**, *597*, 427–432.

34. Mornet, D.; Ue, K. Incorporation of 6-carboxyfluorescein into myosin subfragment 1. *Biochemistry* **1985**, *24*, 840–846.

35. Chiryasova, E. A. Oligonucleotide primers and probes for quantitative evaluation of terminal nucleotides of G-rich strand of human telomeric RNA using PCR and duplex-specific analysis. Russ. RU 2508407, 2014.

36. Matsuno, T. Method for detecting target nucleic acid using molecular beacon-type probe. Jpn. Kokai Tokkyo Koho JP 2013215167, 2013.

37. Demkin, V. V.; Kirillova, N. V. Oligonucleotides and PCR-based method for detecting DNA of bacteria from the family of *Chlamydiaceae*. Russ. RU 2486255, 2013.

38. Chateigner-Boutin, A.; Small, I. A rapid high-throughput method for the detection and quantification of RNA editing based on high-resolution melting of amplicons. *Nucleic Acids Res.* **2007**, *35*, e114/1-e114/8.

39. Zhu, L.; Soper, S. A. Multiplexed fluorescence detection for DNA sequencing. *Rev. Fluoresc.* **2006**, *3*, 525–574.

40. Guo, D.; Milewicz, D. M. Methodology for using a universal primer to label amplified DNA segments for molecular analysis. *Biotechnol. Lett.* **2003**, *25*, 2079–2083.

41. Fountain, K. J.; Gilar, M.; Budman, Y.; Gebler, J. C. Purification of dye-labeled oligonucleotides by ion-pair reversed-phase high-performance liquid chromatography. *J. Chromatogr., B* **2003**, *783*, 61–72.

42. Gardner, A. F.; Jack, W. E. Acyclic and dideoxy terminator preferences denote divergent sugar recognition by archaeon and Taq DNA polymerases. *Nucleic Acids Res.* **2002**, *30*, 605–613.

43. Townsend, M. S.; Henning, J. A.; Moore, D. L. AFLP analysis of DNA from dried hop cones. *Crop Sci.* **2000**, *40*, 1383–1386.

44. Lauter, F.; Grohmann, L.; Staesche, R. Test kit and procedure for the quantitative detection of DNA from genetically manipulated organisms in food using fluorescence-coupled PCR. Ger. Offen. DE 19906169, 2000.

45. Fischer, R.; Mader, O.; Jung, G.; Brock, R. Extending the applicability of carboxyfluorescein in solid-phase synthesis. *Bioconjugate Chem.* **2003**, *14*, 653–660.

46. Galluzzi, J. R.; Ordovas, J. M. Genotyping method for point mutation detection in the intestinal fatty

acid binding protein, using fluorescent probes. *Clin. Chem.* **1999**, *45*, 1092–1094.

47. Rockey, J. H.; Li, W.; Eccleston, J. F. Binding of fluorescein and carboxyfluorescein by human serum proteins: significance of kinetic and equilibrium parameters of association in ocular fluorometric studies. *Exp. Eye Res.* **1983**, *37*, 455–466.

48. Rockey, J. H.; Li, W. Binding of fluorescein and carboxyfluorescein by normal and glycosylated human serum proteins. *Ophthalmic Res.* **1982**, *14*, 416–427.

49. Glimelius, K.; Djupsjoebacka, M.; Fellner-Feldegg, H. Selection and enrichment of plant protoplast heterokaryons of Brassicaceae by flow sorting. *Plant Sci.* **1986**, *45*, 133–141.

50. Yang, C.; Ge, F.; Li, J.; Cai, Z.; Qin, F. Silver nanoparticles with enhanced fluorescence effects on fluorescein derivative. *Adv. Mater. Res.* **2013**, *602–604*, 187–191.

51. Erwee, M. G.; Goodwin, P. B. Symplast domains in extrastelar tissues of *Egeria densa* Planch. *Planta* **1985**, *163*, 9–19.

52. Erwee, M. G.; Goodwin, P. B. Characterization of the *Egeria densa* leaf symplast: response to plasmolysis, deplasmolysis and to aromatic amino acids. *Protoplasma* **1984**, *122*, 162–168.

53. Erwee, M. G.; Goodwin P. B., Characterization of the *Egeria densa* Planch. leaf symplast. Inhibition of the intercellular movement of fluorescent probes by group II ions. *Planta* **1983**, *158*, 320–328.

54. Goodwin, P. B. Molecular size limit for movement in the symplast of the *Elodea* leaf. *Planta* **1983**, *157*, 124–130.

55. Marova, A. A.; Nikonova, A. A.; Zverev, V. V.; Egorova, O. V.; Kalinkina, M. A.; Oksanich, A. S.; Faizuloev, E. B. Method for detection of intestinal viruses in clinical samples by real-time multiplex PCR and oligonucleotide sequences for its application. Russ. RU 2506317, 2014.

56. Berillo, S. A.; Ternovoi, V. A.; Sergeeva, E. I.; Shikov, A. N.; Demina, O. K.; Agafonov, A. P.; Sergeev, A. N. Kit of oligonucleotide primers and fluorescent-marked probes for subtype-specific identification of dengue virus RNA based on multiplex PCR in real time. Russ. RU 2483115, 2013.

57. Panferova, A. V.; Tsybanov, S. Z.; Guzalova, A. G.; Lunitsin, A. V.; Kolbasov, D. V. Oligonucleotide primers, probes and a method for detecting bluetongue virus serotypes 1, 4 and 16 by real time RT-PCR. Russ. RU 2481404, 2013.

58. Farkas, T.; Szekely, E.; Belak, S.; Kiss, I. Real-time PCR-based pathotyping of Newcastle disease virus by use of TaqMan minor groove binder probes. *J. Clin. Microbiol.* **2009**, *47*, 2114–2123.

59. Zeng, S. Q.; Halkosalo, A.; Salminen, M.; Szakal, E. D.; Puustinen, L.; Vesikari, T. One-step quantitative RT-PCR for the detection of rotavirus in acute gastroenteritis. *J. Virol. Methods* **2008**, *153*, 238–240.

60. Balinsky, C. A.; Delhon, G.; Smoliga, G.; Prarat, M.; French, R. A.; Geary, S. J.; Rock, D. L.; Rodriguez, L. L. Rapid preclinical detection of Sheeppox virus by a real-time PCR assay. *J. Clin. Microbiol.* **2008**, *46*, 438–442.

61. Fedele, C. G.; Negredo, A.; Molero, F.; Sanchez-Seco, M. P.; Tenorio, A. Use of internally controlled real-time genome amplification for detection of variola virus and other orthopoxviruses infecting humans. *J. Clin. Microbiol.* **2006**, *44*, 4464–4470.

62. Jothikumar, N.; Cromeans, T. L.; Sobsey, M. D.; Robertson, B. H. Development and evaluation of a broadly reactive TaqMan assay for rapid detection of hepatitis A virus. *Appl. Environ. Microbiol.* **2005**, *71*, 3359–3363.

63. Ward, C. L.; Dempsey, M. H.; Ring, C. J. A.; Kempson, R. E.; Zhang, L.; Gor, D.; Snowden, B. W.; Tisdale, M. Design and performance testing of quantitative real time PCR assays for influenza A and B viral load measurement. *J. Clin. Virol.* **2004**, *29*, 179–188.

64. Rasmussen, T. B.; Uttenthal, A.; de Stricker, K.; Belak, S.; Storgaard, T. Development of a novel quantitative real-time RT-PCR assay for the simultaneous detection of all serotypes of Foot-and-mouth disease virus. *Arch. Virol.* **2003**, *148*, 2005–2021.

65. Risatti, G. R.; Callahan, J. D.; Nelson, W. M.; Borca, M. V. Rapid detection of classical swine fever virus by a portable real-time reverse transcriptase PCR assay. *J. Clin. Microbiol.* **2003**, *41*, 500–505.

66. Smith, I. L.; Northill, J. A.; Harrower, B. J.; Smith, G. A. Detection of Australian bat lyssavirus using a fluorogenic probe. *J. Clin. Virol.* **2002**, *25*, 285–291.

67. Cook, R. F.; Cook, S. J.; Li, F.; Montelaro, R. C.; Issel, C. J. Development of a multiplex real-time reverse transcriptase-polymerase chain reaction for equine infectious anemia virus (EIAV). *J. Virol. Methods* **2002**, *105*, 171–179.

68. Moen, E. M.; Sleboda, J.; Grinde, B. Real-time PCR methods for independent quantitation of TTV and TLMV. *J. Virol. Methods* **2002**, *104*, 59–67.

69. Smith, I. L.; Halpin, K.; Warrilow, D.; Smith, G. A. Development of a fluorogenic RT-PCR assay (TaqMan) for the detection of Hendra virus. *J. Virol. Methods* **2001**, *98*, 33–40.

70. Funae, Y. Method and kit for diagnosing bladder cancer. Jpn. Kokai Tokkyo Koho JP 2002238599, 2002.

71. Katayama, Y.; Koga, H.; Toita, R.; Niidome, T.; Mori, T. A novel fluorometric assay method for the detection of protein kinase activities. Jpn. Kokai Tokkyo Koho JP 2012254938, 2012.

72. Nishimura, M.; Yaguchi, H.; Naito, S.; Hiraoka, I. Real-time RT-PCR quantitative assay for methyltransferase. Jpn. Kokai Tokkyo Koho JP 2002058481, 2002.

73. Graham, R. J. Fluorescent phospholipase assays and compositions. PCT Int. Appl. WO 2005005977, 2005.

74. Chen, R. F. Enzyme assay by fluorescence quenching release: a novel fluorometric method. *Anal. Lett.* **1977**, *10*, 787–795.

75. Nishimura, M.; Yaguchi, H.; Naito, S.; Hiraoka, I. Real-time RT-PCR quantitative assay for rhodanese. Jpn. Kokai Tokkyo Koho JP 2002058482, 2002.

76. Nishimura, M.; Yaguchi, H.; Naito, S.; Hiraoka, I. Real-time RT-PCR quantitative assay for sulfotransferase. Jpn. Kokai Tokkyo Koho JP 2002085067, 2002.

77. Laties, A. M.; Stone, R. A. Ophthalmic use of carboxyfluorescein. U.S. Patent 4350676, 1982.

78. Pokorny, A.; Almeida, P. F. F. Permeabilization of raft-containing lipid vesicles by delta lysin: a mechanism for cell sensitivity to cytotoxic peptides. *Biochemistry* **2005**, *44*, 9538–9544.

79. He, L.; Jurs, P. C.; Custer, L. L.; Durham, S. K.; Pearl, G. M. Predicting the genotoxicity of polycyclic aromatic compounds from molecular structure with different classifiers. *Chem. Res. Toxicol.* **2003**, *16*, 1567–1580.

80. Strauss, G. H. Non-random cell killing in cryopreservation: implications for performance of the battery of leukocyte tests (BLT), I. Toxic and immunotoxic effects. Mutat. Res. **1991**, *252*, 1–15.

81. Hamid K. A.; Katsumi H.; Sakane T.; Yamamoto A. The effects of common solubilizing agents on the intestinal membrane barrier functions and membrane toxicity in rats. *Int. J. Pharm.* **2009**, *379*, 100–108.

82. Catania, J. M.; Pershing, A. M.; Gandolfi, A. J. Precision-cut tissue chips as an *in vitro* toxicology system. *Toxicology In Vitro* **2007**, *21*, 956–961.

83. Sugiyama, N.; Araki, M.; Ishida, M.; Nagashima, Y.; Shiomi, K. Further isolation and characterization of grammistins from the skin secretion of the soapfish *Grammistes sexlineatus*. *Toxicon* **2005**, *45*, 595–601.

84. Anders, J. J.; Woolery, S. Microbeam laser-injured neurons increase in vitro astrocytic gap junctional communication as measured by fluorescence recovery after laser photobleaching. *Lasers Surg. Med.* **1992**, *12*, 51–62.

85. Zhou, J.; Lu, Q.; Tong, Y.; Wei, W.; Liu, S. Detection of DNA damage by using hairpin molecular beacon probes and graphene oxide. *Talanta* **2012**, *99*, 625–630.

86. Nakagawa, S.; Usui, T.; Yokoo, S.; Omichi, S.; Kimakura, M.; Mori, Y.; Miyata, K.; Aihara, M.; Amano, S.; Araie, M. Toxicity evaluation of antiglaucoma drugs using stratified human cultivated corneal epithelial sheets. *Invest. Ophthalmol. Vis. Sci.* **2012**, *53*, 5154–5160

87. Kakkar, S.; Kaur, I. P. Spanlastics–a novel nanovesicular carrier system for ocular delivery. *Int. J. Pharm.* **2011**, *413*, 202–210.

88. Vehanen, K.; Hornof, M.; Urtti, A.; Uusitalo, H. Peribulbar poloxamer for ocular drug delivery. *Acta Ophthalmol.* **2008**, *86*, 91–96.

89. Chen, W.; Zhao, K.; Li, X.; Yoshitomi, T. Keratoconjunctivitis sicca modifies epithelial stem cell proliferation kinetics in conjunctiva. *Cornea* **2007**, *26*, 1101–1106.

90. Ogura, Y.; Guran, T.; Shahidi, M.; Mori, M. T.; Zeimer, R. C. Feasibility of targeted drug delivery to selective areas of the retina. *Invest. Ophthalmol. Vis. Sci.* **1991**, *32*, 2351–2356.

91. van der Deen, M.; de Vries, E. G. E.; Visserman, H.; Zandbergen, W.; Postma, D. S.; Timens, W.; Timmer-Bosscha, H. Cigarette smoke extract affects functional activity of MRP1 in bronchial epithelial cells. *J. Biochem. Mol. Toxicol.* **2007**, *21*, 243–251

92. He, L.; Gao, Y.; Lin, Y.; Katsumi, H.; Fujita, T.; Yamamoto, A. Improvement of pulmonary absorption of insulin and other water-soluble compounds by polyamines in rats. *J. Controlled Release* **2007**, *122*, 94–101.

5-CARBOXYFLUORESCEIN DIACETATE

CAS Registry Number   79955-27-4

Chemical Structure

CA Index Name   Spiro[isobenzofuran-1(3H),9′-[9H] xanthene]-5-carboxylic acid, 3′,6′-bis(acetyloxy)-3-oxo-

Other Names   5-Carboxyfluorescein diacetate; 5-CFDA

Merck Index Number   Not listed

Chemical/Dye Class   Xanthene

Molecular Formula $C_{25}H_{16}O_9$

Molecular Weight   460.39

Physical Form   Colorless solid

Solubility   Soluble in acetone, acetonitrile, chloroform, N,N-dimethylformamide, dimethyl sulfoxide, methanol

Boiling Point (Calcd.)   693.1 ± 55.0 °C, pressure: 760 Torr

pK$_a$ (Calcd.)   3.77 ± 0.20, most acidic, temperature: 25 °C

Absorption (λ_{max})   290 nm (MeOH)

Emission (λ_{max})   None

Molar Extinction Coefficient   5,500 cm^{-1} M^{-1} (MeOH)

Synthesis   Synthetic methods[1–4]

Imaging/Labeling Applications   Bacteria;[5–9,11,12,19] bone marrow stromal cells;[10] cells;[11,12,15,16,18–21] chloroplasts;[13] endospores;[14] eosinophils;[22] fungi;[15,16] low-density lipoprotein (LDL);[17] microorganisms;[18–21] neutrophils;[22] proteins;[2] protein kinase C;[4] sperms;[23] tumors;[24–26] yeasts[19,27,28]

Biological/Medical Applications   Analyzing intracellular glutathione levels in cells;[29] assessing gap junctional intercellular communication in cells;[29] characterizing antifungal activity of micafungin,[15] caspofungin;[16] counting/detecting/evaluating microorganisms;[18–21] detecting cells;[11,12,15,16,18–21] detecting/measuring intracellular pH in yeast;[27] determining extracellular/intracellular pH in bacteria under CO_2 treatment;[8] evaluating/measuring keratinocytes growth rate;[30] identifying cancerous cells[24–26] measuring gas (carbon dioxide or ammonia) concentration in body fluids;[31] screening xenobiotic phloem mobility in plants;[32] studying low-density lipoprotein (LDL)/cholesterol metabolism;[17] as a substrate for measuring esterase activity[18,19,24,28]

Industrial Applications   Not reported

Safety/Toxicity   No data available

REFERENCES

1. Sabnis, R. W. *Handbook of Acid–base Indicators*; CRC Press: Boca Raton, **2008**; pp 68–69.

2. Yao, S. Q.; Yeo, S. D. Site-specific labelling of proteins. U.S. Pat. Appl. Publ. US 2005158804, 2005.

3. Mattingly, P. G. 5(6)-Methyl substituted fluorescein derivatives. PCT Int. Appl. WO 9320060, 1993.

4. Chen, C.; Poenie, M. New fluorescent probes for protein kinase C. Synthesis, characterization, and application. *J. Biol. Chem.* **1993**, *268*, 15812–15822.

5. Ashida, N.; Ishii, S.; Hayano, S.; Tago, K.; Tsuji, T.; Yoshimura, Y.; Otsuka, S.; Senoo, K. Isolation of functional single cells from environments using a micromanipulator: application to study denitrifying bacteria. *Appl. Microbiol. Biotechnol.* **2010**, *85*, 1211–1217.

6. Motoyama, Y.; Yamaguchi, N.; Matsumoto, M.; Kagami, N.; Tani, Y.; Satake, M.; Nasu, M. Rapid and sensitive detection of viable bacteria in contaminated platelet concentrates using a newly developed bioimaging system. *Transfusion* **2008**, *48*, 2364–2369.

7. Belkova, N. L.; Tazaki, K.; Zakharova, J. R.; Parfenova, V. V. Activity of bacteria in water of hot springs from Southern and Central Kamchatskaya geothermal provinces, Kamchatka Peninsula, Russia. *Microbiol. Res.* **2007**, *162*, 99–107.

8. Spilimbergo, S.; Bertucco, A.; Basso, G.; Bertoloni, G. Determination of extracellular and intracellular pH of *Bacillus subtilis* suspension under CO_2 treatment. *Biotechnol. Bioeng.* **2005**, *92*, 447–451.

9. Hitomi, M. Method for measuring number of bacteria in methane fermentation tank, and method for processing methane fermentation. Jpn. Kokai Tokkyo Koho JP 2004008078, 2004.

10. Ouyang, H. W.; Goh, J. C. H.; Lee, E. H. Viability of allogeneic bone marrow stromal cells following local delivery into patella tendon in rabbit model. *Cell Transplant.* **2004**, *13*, 649–657.

11. Yoshimura, Y.; Kawasaki, Y.; Tsuji, A.; Kurane, R. Fluorometirc method for detecting living cells. Jpn. Kokai Tokkyo Koho JP 2002034595, 2002.

12. Kawasaki, Y.; Tsuji, T.; Tanaka, S. Detection of living cells and reagent kits containing fluorescent enzyme substrates used for it. Jpn. Kokai Tokkyo Koho JP 10179191, 1998.

13. Schulz, A.; Knoetzel, J.; Scheller, H. V.; Mant, A. Uptake of a fluorescent dye as a swift and simple indicator of organelle intactness: Import-competent chloroplasts from soil-grown *Arabidopsis*. *J. Histochem. Cytochem.* **2004**, *52*, 701–704.

14. Cronin, U. P.; Wilkinson, M. G. The use of flow cytometry to study the germination of *Bacillus cereus* endospores. *Cytometry, Part A* **2007**, *71*, 143–153.

15. Watabe, E.; Nakai, T.; Matsumoto, S.; Ikeda, F.; Hatano, K. Killing activity of micafungin against *Aspergillus fumigatus* hyphae assessed by specific fluorescent staining for cell viability. *Antimicrob. Agents Chemother.* **2003**, *47*, 1995–1998.

16. Bowman, J. C.; Hicks, P. Scott; Kurtz, M. B.; Rosen, H.; Schmatz, D. M.; Liberator, P. A.; Douglas, C. M. The antifungal echinocandin caspofungin acetate kills growing cells of *Aspergillus fumigatus in vitro*. *Antimicrob. Agents Chemother.* **2002**, *46*, 3001–3012.

17. Falck, J. R.; Krieger, M.; Goldstein, J. L.; Brown, M. S. Preparation and spectral properties of lipophilic fluorescein derivatives: application to plasma low-density lipoprotein. *J. Am. Chem. Soc.* **1981**, *103*, 7396–7398.

18. Horikiri, S. Microorganism cell detection method using multiple fluorescent indicators. Jpn. Kokai Tokkyo Koho JP 2006238779, 2006.

19. Fukutome, K. Method for evaluating microorganism cell activity by flow cytometry analysis. Jpn. Kokai Tokkyo Koho JP 2006238771, 2006.

20. Breeuwer, P.; Drocourt, J. Method for assessing microorganism viability. PCT Int. Appl. WO 9500660, 1995.

21. Sugata, K.; Ueda, R.; Doi, T.; Onishi, T.; Matsumoto, K. Method for counting living cells of microbes and apparatus therefore. U.S. Patent 5389544, 1995.

22. Davenpeck, K. L.; Chrest, F. J.; Sterbinsky, S. A.; Bickel, C. A.; Bochner, B. S. Carboxyfluorescein diacetate labeling does not affect adhesion molecule expression or function in human neutrophils or eosinophils. *J. Immunol. Methods* **1995**, *188*, 79–89.

23. Chaveiro, A.; Liu, J.; Engel, B.; Critser, J. K.; Woelders, H. Significant variability among bulls in the sperm membrane permeability for water and glycerol: Possible implications for semen freezing protocols for individual males. *Cryobiology* **2006**, *53*, 349–359.

24. Thompson, J. A.; Huang, W. Use of fluorescent reagents in identification of cancerous cells and activated lymphocytes. PCT Int. Appl. WO 2000003033, 2000.

25. Connors, K. M.; Monosov, A. Native-state method and system for determining viability and proliferative capacity of tissues *in vitro*. U.S. Patent 5726009, 1998.

26. Connors, K. M.; Monosov, A. Native-state method and system for determining viability and proliferative capacity of tissue *in vitro*. PCT Int. Appl. WO 9501455, 1995.

27. Liu, S.; Li, C.; Xie, L.; Cao, Z. Intracellular pH and metabolic activity of long-chain dicarboxylic acid-producing yeast *Candida tropicalis*. *J. Biosci. Bioeng.* **2003**, *96*, 349–353.

28. Breeuwer, P.; Drocourt, J. L.; Bunschoten, N.; Zwietering, M. H.; Rombouts, F. M.; Abee, T. Characterization of uptake and hydrolysis of fluorescein diacetate and carboxyfluorescein diacetate by intracellular esterases in *Saccharomyces cerevisiae*, which result in accumulation of fluorescent product. *Appl. Environ. Microbiol.* **1995**, *61*, 1614–1619.

29. Barhoumi, R.; Bowen, J. A.; Stein, L. S.; Echols, J.; Burghardt, R. C. Concurrent analysis of intracellular glutathione content and gap junctional intercellular communication. *Cytometry* **1993**, *14*, 747–756.

30. Hanthamrongwit, M.; Reid, W. H.; Courtney, J. M.; Grant, M. H. 5-carboxyfluorescein diacetate as a probe for measuring the growth of keratinocytes. *Hum. Exp. Toxicol.* **1994**, *13*, 423–427.

31. Bentsen, J. G.; Wood, K. B. Sensor with improved drift stability. U.S. Patent 5403746, 1995.

32. Wang, N.; Lau, S. C.; Rogers, G.; Ray, T. A new method for rapid screening of xenobiotic phloem mobility in plants. *Aust. J. Plant Physiol.* **2000**, *27*, 835–843.

6-CARBOXYFLUORESCEIN DIACETATE

CAS Registry Number   3348-03-6

Chemical Structure

CA Index Name   Spiro[isobenzofuran-1(3*H*),9′-[9*H*] xanthene]-6-carboxylic acid, 3′,6′-bis(acetyloxy)-3-oxo-

Other Names   Spiro[phthalan-1,9′-xanthene]-6-carboxylic acid, 3′,6′-dihydroxy-3-oxo-, diacetate; Terephthalic acid, (3,6,9-trihydroxyxanthen-9-yl)-, γ-lactone, diacetate; 6-Carboxyfluorescein diacetate; 6-CFDA

Merck Index Number   Not listed

Chemical/Dye Class   Xanthene

Molecular Formula $C_{25}H_{16}O_9$

Molecular Weight   460.39

Physical Form   Light yellow or off-white solid

Solubility   Soluble in acetonitrile, chloroform, *N,N*-dimethylformamide, dimethyl sulfoxide, methanol

Melting Point   152–153 °C[4]

Boiling Point (Calcd.)   701.6 ± 60.0 °C, pressure: 760 Torr

pK$_a$ (Calcd.)   3.55 ± 0.20, most acidic, temperature: 25 °C

Absorption (λ$_{max}$)   290 nm (MeOH)

Emission (λ$_{max}$)   None

Molar Extinction Coefficient   6,150 cm^{-1} M^{-1} (MeOH)

Synthesis   Synthetic methods[1–4]

Imaging/Labeling Applications   Bacteria;[5–24,28,29,42] cells;[5,7,13,14,17,19,20,25–30,41,42,45,46,52,58,59] chloroplasts;[31] eosinophils;[48] fungi;[32–38] low-density lipoprotein (LDL);[39] microorganisms;[40–47] neutrophils;[48] protein kinase C;[3] sperms;[49] tumors;[50,51] yeasts[42,47,52–58]

Biological/Medical Applications   Analyzing/detecting cells;[5,7,13,14,17,19,20,25–30,41,42,45,46,52,58,59] counting/detecting/evaluating microorganisms;[40–47] detecting/measuring bacteria;[5–24,28,29,42] quantifying adhesion of tumor cells to endothelial monolayers;[50] measuring gas (carbon dioxide or ammonia) concentration in body fluids;[61] monitoring intracellular nitric oxide production;[62] screening xenobiotic phloem mobility in plants;[63,64] studying low-density lipoprotein (LDL)/cholesterol metabolism;[39] apoptosis assay;[59] method for antifungal susceptibility testing;[35,37] as a substrate for measuring esterase activity[5,14,17,22,24,33,41,42,51,55,57,58,60]

Industrial Applications   Not reported

Safety/Toxicity   Hepatotoxicity;[65] ploem mobility of xenobiotics;[64] retinal toxicity[66]

REFERENCES

1. Sabnis, R. W. *Handbook of Acid–base Indicators*; CRC Press: Boca Raton, **2008**; pp 70–71.

2. Mattingly, P. G. 5(6)-Methyl substituted fluorescein derivatives. PCT Int. Appl. WO 9320060, 1993.

3. Chen, C.; Poenie, M. New fluorescent probes for protein kinase C. Synthesis, characterization, and application. *J. Biol. Chem.* **1993**, *268*, 15812–15822.

4. Drechsler, G.; Smagin, S. Preparation of trimelliteins and 5′-carboxyfluorescein. *J. Prakt. Chem.* **1965**, *28*, 315–324.

5. Cronin, U. P.; Wilkinson, M. G. Physiological response of *Bacillus cereus* vegetative cells to simulated food processing treatments. *J. Food Prot.* **2008**, *71*, 2168–2176.

6. Motoyama, Y.; Yamaguchi, N.; Matsumoto, M.; Kagami, N.; Tani, Y.; Satake, M.; Nasu, M. Rapid and sensitive detection of viable bacteria in contaminated platelet concentrates using a newly developed bioimaging system. *Transfusion* **2008**, *48*, 2364–2369.

7. Miyanaga, K.; Takano, S.; Morono, Y.; Hori, K.; Unno, H.; Tanji, Y. Optimization of distinction between viable and dead cells by fluorescent staining method and its application to bacterial consortia. *Biochem. Eng. J.* **2007**, *37*, 56–61.

8. Rault, A.; Beal, C.; Ghorbal, S.; Ogier, J.; Bouix, M. Multiparametric flow cytometry allows rapid assessment and comparison of lactic acid bacteria viability after freezing and during frozen storage. *Cryobiology* **2007**, *55*, 35–43.

9. Papadimitriou, K.; Pratsinis, H.; Nebe-von-Caron, G.; Kletsas, D.; Tsakalidou, E. Acid tolerance of *Streptococcus macedonicus* as assessed by flow cytometry and single-cell sorting. *Appl. Environ. Microbiol.* **2007**, *73*, 465–476.

10. Papadimitriou, K.; Pratsinis, H.; Nebe-von-Caron, G.; Kletsas, D.; Tsakalidou, E. Rapid assessment of the physiological status of *Streptococcus macedonicus* by flow cytometry and fluorescence probes. *Int. J. Food Microbiol.* **2006**, *111*, 197–205.

11. Hoefel, D.; Monis, P. T.; Grooby, W. L.; Andrews, S.; Saint, C. P. Profiling bacterial survival through a water treatment process and subsequent distribution system. *J. Appl. Microbiol.* **2005**, *99*, 175–186.

12. Bianchi, M. A.; del Rio, D.; Pellegrini, N.; Sansebastiano, G.; Neviani, E.; Brighenti, F. A fluorescence-based method for the detection of adhesive properties of lactic acid bacteria to Caco-2 cells. *Lett. Appl. Microbiol.* **2004**, *39*, 301–305.

13. Ananta, E.; Heinz, V.; Knorr, D. Assessment of high pressure induced damage on *Lactobacillus rhamnosus* GG by flow cytometry. *Food Microbiol.* **2004**, *21*, 567–577.

14. Morono, Y.; Takano, S.; Miyanaga, K.; Tanji, Y.; Unno, H.; Hori, K. Application of glutaraldehyde for the staining of esterase-active cells with carboxyfluorescein diacetate. *Biotechnol. Lett.* **2004**, *26*, 379–383.

15. Hitomi, M. Method for measuring number of bacteria in methane fermentation tank, and method for processing methane fermentation. Jpn. Kokai Tokkyo Koho JP 2004008078, 2004.

16. Hoefel, D.; Grooby, W. L.; Monis, P. T.; Andrews, S.; Saint, C. P. A comparative study of carboxyfluorescein diacetate and carboxyfluorescein diacetate succinimidyl ester as indicators of bacterial activity. *J. Microbiol. Methods* **2003**, *52*, 379–388.

17. Ben Amor, K.; Breeuwer, P.; Verbaarschot, P.; Rombouts, F. M.; Akkermans, A. D. L.; De Vos, W. M.; Abee, T. Multiparametric flow cytometry and cell sorting for the assessment of viable, injured, and dead Bifidobacterium cells during bile salt stress. *Appl. Environ. Microbiol.* **2002**, *68*, 5209–5216.

18. Forster, S.; Snape, J. R.; Lappin-Scott, H. M.; Porter, J. Simultaneous fluorescent gram staining and activity assessment of activated sludge bacteria. *Appl. Environ. Microbiol.* **2002**, *68*, 4772–4779.

19. Bunthof, C. J.; Abee, T. Development of a flow cytometric method to analyze subpopulations of bacteria in probiotic products and dairy starters. *Appl. Environ. Microbiol.* **2002**, *68*, 2934–2942.

20. Bunthof, C. J.; Bloemen, K.; Breeuwer, P.; Rombouts, F. M.; Abee, T. Flow cytometric assessment of viability of lactic acid bacteria. *Appl. Environ. Microbiol.* **2001**, *67*, 2326–2335.

21. Budde, B. B.; Rasch, M. A comparative study on the use of flow cytometry and colony forming units for assessment of the antibacterial effect of bacteriocins. *Int. J. Food Microbiol.* **2001**, *63*, 65–72.

22. Yamaguchi, N.; Nasu, M. Flow cytometric analysis of bacterial respiratory and enzymic activity in the natural aquatic environment. *J. Appl. Microbiol.* **1997**, *83*, 43–52.

23. Porter, J.; Diaper, J.; Edwards, C.; Pickup, R. Direct measurements of natural planktonic bacterial community viability by flow cytometry. *Appl. Environ. Microbiol.* **1995**, *61*, 2783–2786.

24. Shechter, E.; Letellier, L.; Simons, E. R. Fluorescence dye as monitor of internal pH in *Escherichia coli* cells. *FEBS Lett.* **1982**, *139*, 121–124.

25. Kemna, E. W. M.; Wolbers, F.; Vermes, I.; van den Berg, A. On chip electrofusion of single human B cells and mouse myeloma cells for efficient hybridoma generation. *Electrophoresis* **2011**, *32*, 3138–3146.

26. Shirai, M.; Azuma, Y. Method for determining existence ratio of viable cell, dead-cell, and false viable cell. Jpn. Kokai Tokkyo Koho JP 2008054509, 2008.

27. McClain, M. A.; Culbertson, C. T.; Jacobson, S. C.; Allbritton, N. L.; Sims, C. E.; Ramsey, J. M. Microfluidic devices for the high-throughput chemical analysis of cells. *Anal. Chem.* **2003**, *75*, 5646–5655.

28. Yoshimura, Y.; Kawasaki, Y.; Tsuji, A.; Kurane, R. Fluorometirc method for detecting living cells. Jpn. Kokai Tokkyo Koho JP 2002034595, 2002.

29. Kawasaki, Y.; Tsuji, T.; Tanaka, S. Detection of living cells and reagent kits containing fluorescent enzyme substrates used for it. Jpn. Kokai Tokkyo Koho JP 10179191, 1998.

30. Niwa, M.; Yasui, T. Fluorescent dye solutions for cell staining and detection. Jpn. Kokai Tokkyo Koho JP 04102060, 1992.

31. Schulz, A.; Knoetzel, J.; Scheller, H. V.; Mant, A. Uptake of a fluorescent dye as a swift and simple

indicator of organelle intactness: Import-competent chloroplasts from soil-grown Arabidopsis. *J. Histochem. Cytochem.* **2004**, *52*, 701–704.

32. De Vos, M. M.; Sanders, N. N.; Nelis, H. J. Detection of *Aspergillus fumigatus* hyphae in respiratory secretions by membrane filtration, fluorescent labelling and laser scanning. *J. Microbiol. Methods* **2006**, *64*, 420–423.

33. Peter, J.; Armstrong, D.; Lyman, C. A.; Walsh, T. J. Use of fluorescent probes to determine MICs of amphotericin B and caspofungin against *Candida* spp. and *Aspergillus* spp. *J. Clin. Microbiol.* **2005**, *43*, 3788–3792.

34. Watabe, E.; Nakai, T.; Matsumoto, S.; Ikeda, F.; Hatano, K. Killing activity of micafungin against *Aspergillus fumigatus* hyphae assessed by specific fluorescent staining for cell viability. *Antimicrob. Agents Chemother.* **2003**, *47*, 1995–1998.

35. Liao, R. S.; Rennie, R. P.; Talbot, J. A. Comparative evaluation of a new fluorescent carboxyfluorescein diacetate-modified microdilution method for antifungal susceptibility testing of *Candida albicans* isolates. *Antimicrob. Agents Chemother.* **2002**, *46*, 3236–3242.

36. Bowman, J. C.; Hicks, P. Scott; Kurtz, M. B.; Rosen, H.; Schmatz, D. M.; Liberator, P. A.; Douglas, C. M. The antifungal echinocandin caspofungin acetate kills growing cells of *Aspergillus fumigatus in vitro*. *Antimicrob. Agents Chemother.* **2002**, *46*, 3001–3012.

37. Liao, R. S.; Rennie, R. P.; Talbot, J. A. Novel fluorescent broth microdilution method for fluconazole susceptibility testing of *Candida albicans*. *J. Clin. Microbiol.* **2001**, *39*, 2708–2712.

38. Cole, L.; Hyde, G. J.; Ashford, A. E. Uptake and compartmentalization of fluorescent probes by *Pisolithus tinctorius* hyphae: evidence for an anion transport mechanism at the tonoplast but not for fluid-phase endocytosis. *Protoplasma* **1997**, *199*, 18–29.

39. Falck, J. R.; Krieger, M.; Goldstein, J. L.; Brown, M. S. Preparation and spectral properties of lipophilic fluorescein derivatives: application to plasma low-density lipoprotein. *J. Am. Chem. Soc.* **1981**, *103*, 7396–7398.

40. Ribault, S.; Chollet, R. Fluorogenic culture medium for detection of microorganisms including a dye to mask residual fluorescence. Fr. Demande FR 2955121, 2011.

41. Horikiri, S. Microorganism cell detection method using multiple fluorescent indicators. Jpn. Kokai Tokkyo Koho JP 2006238779, 2006.

42. Fukutome, K. Method for evaluating microorganism cell activity by flow cytometry analysis. Jpn. Kokai Tokkyo Koho JP 2006238771, 2006.

43. Maruyama, K.; Saiga, T.; Asai, R. Microorganism test method and kit using adhesive sheet and fluorecent labeling. Jpn. Kokai Tokkyo Koho JP 2005341870, 2005.

44. Maruyama, K.; Matsunawa, A.; Saiga, T. Method and kit for measuring microorganism quantity. Jpn. Kokai Tokkyo Koho JP 2005065624, 2005.

45. Matsunawa, A.; Saiga, T.; Maruyama, K. Method and kit for detecting and counting microorganism or cells by fluorescent staining. Jpn. Kokai Tokkyo Koho JP 2005006567, 2005.

46. Tanaka, Y.; Maruyama, K.; Saiga, T. Method and kit for evaluating physiological activity of microorganism or cell. Jpn. Kokai Tokkyo Koho JP 2004187534, 2004.

47. Abe, F.; Horikoshi, H. The microorganism detection by the flow cytometry method. Jpn. Kokai Tokkyo Koho JP 2000069995, 2000.

48. Davenpeck, K. L.; Chrest, F. J.; Sterbinsky, S. A.; Bickel, C. A.; Bochner, B. S. Carboxyfluorescein diacetate labeling does not affect adhesion molecule expression or function in human neutrophils or eosinophils. *J. Immunol. Methods* **1995**, *188*, 79–89.

49. Marti, J. I.; Cebrian-Perez, J. A.; Muino-Blanco, T. Assessment of the acrosomal status of ram spermatozoa by RCA lectin-binding and partition in an aqueous two-phase system. *J. Androl.* **2000**, *21*, 541–548.

50. Price, E. A.; Coombe, D. R.; Murray, J. C. A simple fluorometric assay for quantifying the adhesion of tumor cells to endothelial monolayers. *Clin. Exp. Metastasis* **1995**, *13*, 155–164.

51. Thomas, J. A.; Buchsbaum, R. N.; Zimniak, A.; Racker, E. Intracellular pH measurements in Ehrlich ascites tumor cells utilizing spectroscopic probes generated *in situ*. *Biochemistry* **1979**, *18*, 2210–2218.

52. Garcia, M. T.; Sanz, R.; Galceran, M. T.; Puignou, L. Use of fluorescent probes for determination of yeast cell viability by gravitational field-flow fractionation. *Biotechnol. Prog.* **2006**, *22*, 847–852.

53. Bouchez, J. C.; Cornu, M.; Danzart, M.; Leveau, J. Y.; Duchiron, F.; Bouix, M. Physiological significance of the cytometric distribution of fluorescent yeasts after viability staining. *Biotechnol. Bioeng.* **2004**, *86*, 520–530.

54. Bouix, M.; Leveau, J. Rapid assessment of yeast viability and yeast vitality during alcoholic fermentation. *J. Inst. Brew.* **2001**, *107*, 217–225.

55. Degrassi, G.; Uotila, L.; Klima, R.; Venturi, V. Purification and properties of an esterase from the yeast *Saccharomyces cerevisiae* and identification of the encoding gene. *Appl. Environ. Microbiol.* **1999**, *65*, 3470–3472.

56. Abe, F. Hydrostatic pressure enhances vital staining with carboxyfluorescein or carboxydichlorofluorescein in *Saccharomyces cerevisiae*: efficient detection of labeled yeasts by flow cytometry. *Appl. Environ. Microbiol.* **1998**, *64*, 1139–1142.

57. Breeuwer, P.; Drocourt, J. L.; Bunschoten, N.; Zwietering, M. H.; Rombouts, F. M.; Abee, T. Characterization of uptake and hydrolysis of fluorescein diacetate and carboxyfluorescein diacetate by intracellular esterases in *Saccharomyces cerevisiae*, which result in accumulation of fluorescent product. *Appl. Environ. Microbiol.* **1995**, *61*, 1614–1619.

58. Breeuwer, P.; Drocourt, J. L.; Rombouts, F. M.; Abee, T. Energy-dependent, carrier-mediated extrusion of carboxyfluorescein from *Saccharomyces cerevisiae* allows rapid assessment of cell viability by flow cytometry. *Appl. Environ. Microbiol.* **1994**, *60*, 1467–1472.

59. Farinacci, M. Improved apoptosis detection in ovine neutrophils by annexin V and carboxyfluorescein diacetate staining. *Cytotechnology* **2007**, *54*, 149–155.

60. Dive, C.; Cox, H.; Watson, J. V.; Workman, P. Polar fluorescein derivatives as improved substrate probes for flow cytoenzymological assay of cellular esterases. *Mol. Cell. Probes* **1988**, *2*, 131–145.

61. Bentsen, J. G.; Wood, K. B. Sensor with improved drift stability. U.S. Patent 5403746, 1995.

62. Mainz, E. R.; Gunasekara, D. B.; Caruso, G.; Jensen, D. T.; Hulvey, M. K.; Fracassi da Silva, J. A.; Metto, E. C.; Culbertson, A. H.; Culbertson, C. T.; Lunte, S. M. Monitoring intracellular nitric oxide production using microchip electrophoresis and laser-induced fluorescence detection. *Anal. Methods* **2012**, *4*, 414–420.

63. Wang, N.; Lau, S. C.; Rogers, G.; Ray, T. A new method for rapid screening of xenobiotic phloem mobility in plants. *Aust. J. Plant Physiol.* **2000**, *27*, 835–843.

64. Wright, K. M.; Horobin, R. W.; Oparka, K. J. Phloem mobility of fluorescent xenobiotics in Arabidopsis in relation to their physicochemical properties. *J. Exp. Bot.* **1996**, *47*, 1779–1787.

65. Wang, P. Y. T.; Boccanfuso, M.; Lemay, A.; Devries, H.; Sui, J.; She, Y.; Hill, C. E. Sex-specific extraction of organic anions by the rat liver. *Life Sci.* **2008**, *82*, 436–443.

66. Kato A.; Kimura H.; Okabe K.; Okabe J.; Kunou N.; Ogura Y. Feasibility of drug delivery to the posterior pole of the rabbit eye with an episcleral implant. *Invest. Ophthalmol. Vis. Sci.* **2004**, *45*, 238–244.

5-CARBOXYFLUORESCEIN SUCCINIMIDYL ESTER

CAS Registry Number 92557-80-7

Chemical Structure

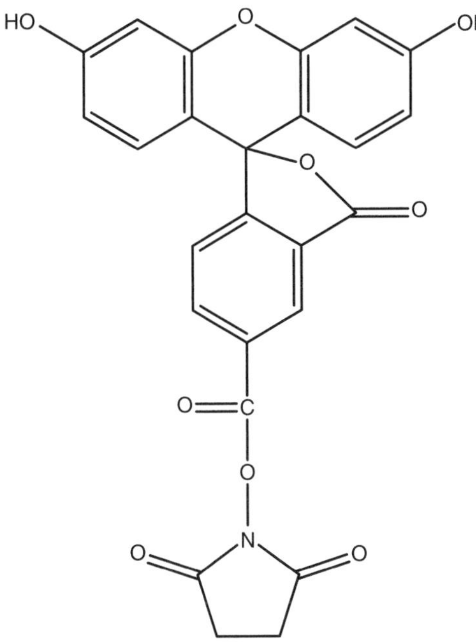

CA Index Name Spiro[isobenzofuran-1(3*H*),9′-[9*H*] xanthene]-5-carboxylic acid, 3′,6′-dihydroxy-3-oxo-, 2,5-dioxo-1-pyrrolidinyl ester

Other Names 2,5-Pyrrolidinedione, 1-[[(3′,6′-dihydroxy-3-oxospiro[isobenzofuran-1(3*H*),9′-[9*H*]xanthen]-5-yl)carbonyl]oxy]-; Spiro[isobenzofuran-1(3*H*),9′-[9*H*]xanthene], 2,5-pyrrolidinedione deriv.; 5-Carboxyfluorescein *N*-hydroxysuccinimide ester; 5-Carboxyfluorescein *N*-succinimidyl ester; 5-Carboxyfluorescein NHS ester; 5-Carboxyfluorescein succinimidyl ester; 5-CFSE; CFSE; 5-FAM, SE

Merck Index Number Not listed

Chemical/Dye Class Xanthene

Molecular Formula $C_{25}H_{15}NO_9$

Molecular Weight 473.39

Physical Form Orange solid

Solubility Soluble in *N,N*-dimethylformamide, dimethyl sulfoxide, methanol

Boiling Point (Calcd.) 778.7 ± 70.0 °C, pressure: 760 Torr

pK$_a$ (Calcd.) 9.31 ± 0.20, most acidic, temperature: 25 °C

Absorption (λ_{max}) 494 nm (Buffer pH 9)

Emission (λ_{max}) 520 nm (Buffer pH 9)

Molar Extinction Coefficient 78,000 cm^{-1} M^{-1} (Buffer pH 9)

Synthesis Synthetic methods[1–5]

Imaging/Labeling Applications Actin;[6] amino acids;[7,8] antibodies;[9] bacteria;[10] peptide chloromethyl ketones;[11] herbicides;[12] histamine H1/H2 receptor antagonists;[13,14] lymphocytes;[15,16] metal ions (Fe(III);[17] Pt(IV);[18]) nanoparticles;[19–24] nucleic acids/nucleotides;[25–34] pheomelanin;[35] polystyrene latex particles;[36] proteins/peptides;[37–41] risedronate;[42] serotonin;[43] stem cells[44,45]

Biological/Medical Applications Analyzing/assessing T cell function proliferation;[15,16] detecting degree of cross-linked allogeneic tissue;[46] determining phosphorus-containing amino acid herbicides;[12] evaluating/studying mesenchymal stem cells subpopulations proliferative activity;[45] locating/investigating hemopoietic stem cell niche *in situ*;[44] monitoring bacterial pathogens;[23] multiplexed cancer cell monitoring;[21] quantifying serotonin;[43] for synthesizing vasopressin analogs;[47] as a substrate for measuring histone acetyltransferases (HATs) activity,[48] kinase activity,[49,50] phosphatase activity[50]

Industrial Applications Abrasives;[19] platings[19]

Safety/Toxicity No data available

REFERENCES

1. Adamczyk, M.; Johnson, D. D.; Mattingly, P. G.; Moore, J. A.; Pan, Y. Immunoassay reagents for thyroid testing. 3. Determination of the solution binding affinities of a T4 monoclonal antibody fab fragment for a library of thyroxine analogs using surface plasmon resonance. *Bioconjugate Chem.* **1998**, *9*, 23–32.

2. Adamczyk, M.; Fishpaugh, J. R.; Heuser, K. J. Preparation of succinimidyl and pentafluorophenyl active esters of 5- and 6-carboxyfluorescein. *Bioconjugate Chem.* **1997**, *8*, 253–255.

3. Vanderbilt, A. S.; Osikowicz, E. W.; Fino, J. R.; Shipchandler, M. T. Total estriol fluorescence polarization immunoassay. Eur. Pat. Appl. EP 200960, 1986.

4. Flentge, C. A.; Kirkemo, C. L. Substituted carboxyfluoresceins. Eur. Pat. Appl. EP 108403, 1984.

5. Fino, J. R.; Kirkemo, C. L. Substituted carboxyfluoresceins. Eur. Pat. Appl. EP 108399, 1984.

6. Milroy, L.; Rizzo, S.; Calderon, A.; Ellinger, B.; Erdmann, S.; Mondry, J.; Verveer, P.; Bastiaens, P.; Waldmann, H.; Dehmelt, L.; Arndt, H. Selective chemical imaging of static actin in live cells. *J. Am. Chem. Soc.* **2012**, *134*, 8480–8486.

7. Danger, G.; Ross, D. Chiral separation with gradient elution isotachophoresis for future *in situ* extraterrestrial analysis. *Electrophoresis* **2008**, *29*, 4036–4044.

8. Kei Lau, S.; Zaccardo, F.; Little, M.; Banks, P. Nanomolar derivatizations with 5-carboxyfluorescein succinimidyl ester for fluorescence detection in capillary electrophoresis. *J. Chromatogr., A* **1998**, *809*, 203–210.

9. Wei, A.; Blumenthal, D. K.; Herron, J. N. Antibody-mediated fluorescence enhancement based on shifting the intramolecular dimer .dblarw. monomer equilibrium of fluorescent dyes. *Anal. Chem.* **1994**, *66*, 1500–1506.

10. Breeuwer, P.; Drocourt, J.; Rombouts, F. M.; Abee, T. A novel method for continuous determination of the intracellular pH in bacteria with the internally conjugated fluorescent probe 5 (and 6-)-carboxyfluorescein succinimidyl ester. *Appl. Environ. Microbiol.* **1996**, *62*, 178–183.

11. Williams, E. B.; Krishnaswamy, S.; Mann, K. G. Zymogen/enzyme discrimination using peptide chloromethyl ketones. *J. Biol. Chem.* **1989**, *264*, 7536–7545.

12. Molina, M.; Silva, M. Analytical potential of fluorescein analogues for ultrasensitive determinations of phosphorus-containing amino acid herbicides by micellar electrokinetic chromatography with laser-induced fluorescence detection. *Electrophoresis* **2002**, *23*, 1096–1103.

13. Li, L.; Kracht, J.; Peng, S.; Bernhardt, G.; Buschauer, A. Synthesis and pharmacological activity of fluorescent histamine H1 receptor antagonists related to mepyramine. *Bioorg. Med. Chem. Lett.* **2003**, *13*, 1245–1248.

14. Li, L.; Kracht, J.; Peng, S.; Bernhardt, G.; Elz, S.; Buschauer, A. Synthesis and pharmacological activity of fluorescent histamine H2 receptor antagonists related to potentidine. *Bioorg. Med. Chem. Lett.* **2003**, *13*, 1717–1720.

15. Tsuge, I.; Okumura, A.; Kondo, Y.; Itomi, S.; Kakami, M.; Kawamura, M.; Nakajima, Y.; Komatsubara, R.; Urisu, A. Allergen-specific T-cell response in patients with phenytoin hypersensitivity; simultaneous analysis of proliferation and cytokine production by carboxyfluorescein succinimidyl ester (CFSE) dilution assay. *Allergol. Int.* **2007**, *56*, 149–155.

16. Fulcher, D. A.; Wong, S. W. J. Carboxyfluorescein succinimidyl ester-based proliferative assays for assessment of T cell function in the diagnostic laboratory. *Immunol. Cell Biol.* **1999**, *77*, 559–564.

17. Su, B.; Moniotte, N.; Nivarlet, N.; Tian, G.; Desmet, J. Design and synthesis of fluorescence-based siderophore sensor molecules for Fe(III) ion determination. *Pure Appl. Chem.* **2010**, *82*, 2199–2216.

18. Song, Y.; Suntharalingam, K.; Yeung, J. S.; Royzen, M.; Lippard, S. J. Synthesis and Characterization of Pt(IV) fluorescein conjugates to investigate Pt(IV) intracellular transformations. *Bioconjugate Chem.* **2013**, *24*, 1733–1740.

19. Komatsu, N.; Ito, M. Surface modified nanodiamond, method for its manufacture, and nanodiamond which is surface modified for fluorescent labeling. Jpn. Kokai Tokkyo Koho JP 2010202458, 2010.

20. Takimoto, T.; Chano, T.; Shimizu, S.; Okabe, H.; Ito, M.; Morita, M.; Kimura, T.; Inubushi, T.; Komatsu, N. Preparation of fluorescent diamond nanoparticles stably dispersed under a physiological environment through multistep organic transformations. *Chem. Mater.* **2010**, *22*, 3462–3471.

21. Chen, X.; Estevez, M. C.; Zhu, Z.; Huang, Y.; Chen, Y.; Wang, L.; Tan, W. Using aptamer-conjugated fluorescence resonance energy transfer nanoparticles for multiplexed cancer cell monitoring. *Anal. Chem.* **2009**, *81*, 7009–7014.

22. Takagi, T.; Okubo, N. Method to produce silica nanoparticle having a laminated structure and containing functional molecules. Jpn. Kokai Tokkyo Koho JP 2009221059, 2009.

23. Wang, L.; Zhao, W.; O'Donoghue, M. B.; Tan, W. Fluorescent nanoparticles for multiplexed bacteria monitoring. *Bioconjugate Chem.* **2007**, *18*, 297–301.

24. Wang, L.; Tan, W. Multicolor FRET silica nanoparticles by single wavelength excitation. *Nano Lett.* **2006**, *6*, 84–88.

25. Kimoto, M.; Kawai, R.; Mitsui, T.; Yokoyama, S.; Hirao, I. An unnatural base pair system for efficient PCR amplification and functionalization of DNA molecules. *Nucleic Acids Res.* **2009**, *37*, e14/1–e14/9.

26. Kawai, R.; Kimoto, M.; Ikeda, S.; Mitsui, T.; Endo, M.; Yokoyama, S.; Hirao, I. Site-specific fluorescent labeling of RNA molecules by specific transcription using unnatural base pairs. *J. Am. Chem. Soc.* **2005**, *127*, 17286–17295.

27. Finn, P. J.; Sun, L.; Nampalli, S.; Xiao, H.; Nelson, J. R.; Mamone, J. A.; Grossmann, G.; Flick, P. K.; Fuller, C. W.; Kumar, S. Synthesis and application of charge-modified dye-labeled dideoxynucleoside-5′-triphosphates to "direct-load" DNA sequencing. *Nucleic Acids Res.* **2002**, *30*, 2877–2885.

28. Nampalli, S.; McDougall, M. G.; Lavrenov, K.; Xiao, H.; Kumar, S. Utility of thiol-cross-linked fluorescent dye labeled terminators for DNA sequencing. *Bioconjugate Chem.* **2002**, *13*, 468–473.

29. Cummins, W. J.; Hamilton, A. L.; Smith, C. L.; Briggs, M. S. J. Synthesis and study of the fluorescein conjugate of the nucleotide dPTP. *Nucleosides, Nucleotides Nucleic Acids* **2001**, *20*, 1049–1051.

30. Nampalli, S.; Khot, M.; Nelson, J. R.; Flick, P. K.; Fuller, C. W.; Kumar, S. Fluorescence resonance energy transfer dye nucleotide terminators: a new synthetic approach for high-throuhout DNA sequencing. *Nucleosides, Nucleotides Nucleic Acids* **2001**, *20*, 361–367.

31. Nampalli, S.; Khot, M.; Kumar, S. Fluorescence resonance energy transfer terminators for DNA sequencing. *Tetrahedron Lett.* **2000**, *41*, 8867–8871.

32. Igloi, G. L. Nonradioactive labeling of RNA. *Anal. Biochem.* **1996**, *233*, 124–129.

33. Shoukry, S.; Anderson, M. W.; Glickman, B. W. Use of fluorescently tagged DNA and an automated DNA sequencer for the comparison of the sequence selectivity of SN1 and SN2 alkylating agents. *Carcinogenesis* **1993**, *14*, 155–157.

34. Folsom, V.; Hunkeler, M. J.; Haces, A.; Harding, J. D. Detection of DNA targets with biotinylated and fluoresceinated RNA probes. Effects of the extent of derivatization on detection sensitivity. *Anal. Biochem.* **1989**, *182*, 309–314.

35. Zhang, X.; Yang, Q.; Jiao, B.; Dai, J.; Zhang, J.; Ni, W. Sensitive determination of pheomelanin after 5-carboxyfluorescein succinimidyl ester precapillary derivatization and micellar electrokinetic capillary chromatography with laser-induced fluorescence detection. *J. Chromatogr., B* **2008**, *861*, 136–139.

36. Charreyre, M.; Yekta, A.; Winnik, M. A.; Delair, T.; Pichot, C. Fluorescence energy transfer from fluorescein to tetramethylrhodamine covalently bound to the surface of polystyrene latex particles. *Langmuir* **1995**, *11*, 2423–2428.

37. Mizusawa, K.; Takaoka, Y.; Hamachi, I. Specific cell surface protein imaging by extended self-assembling fluorescent turn-on nanoprobes. *J. Am. Chem. Soc.* **2012**, *134*, 13386–13395.

38. Hofmann, F. T.; Szostak, J. W.; Seebeck, F. P. *In vitro* selection of functional lantipeptides. *J. Am. Chem. Soc.* **2012**, *134*, 8038–8041.

39. Ye, G.; Nam, N.; Kumar, A.; Saleh, A.; Shenoy, D. B.; Amiji, M. M.; Lin, X.; Sun, G.; Parang, K. Synthesis and evaluation of tripodal peptide analogues for cellular delivery of phosphopeptides. *J. Med. Chem.* **2007**, *50*, 3604–3617.

40. Geoghegan, K. F.; Rosner, P. J.; Hoth, L. R. Dye-pair reporter systems for protein-peptide molecular interactions. *Bioconjugate Chem.* **2000**, *11*, 71–77.

41. Hoogerhout, P.; Stittelaar, K. J.; Brugghe, H. F.; Timmermans, J. A. M.; Hove, G. J. ten; Jiskoot, W.; Hoekman, J. H. G.; Roholl, P. J. M. Solid-phase synthesis and application of double-fluorescent-labeled lipopeptides, containing a CTL-epitope from the measles fusion protein. *J. Pept. Res.* **1999**, *54*, 436–443.

42. Kashemirov, B. A.; Bala, J. L. F.; Chen, X.; Ebetino, F. H.; Xia, Z.; Russell, R. G. G.; Coxon, F. P.; Roelofs, A. J.; Rogers, M. J.; McKenna, C. E. Fluorescently labeled risedronate and related analogues: "magic linker" synthesis. *Bioconjugate Chem.* **2008**, *19*, 2308–2310.

43. Qi, S.; Tian, S.; Xu, H.; Sung, J. J. Y.; Bian, Z. Quantification of luminally released serotonin in rat proximal colon by capillary electrophoresis with laser-induced fluorescence detection. *Anal. Bioanal. Chem.* **2009**, *393*, 2059–2066.

44. Ellis, S. L.; Williams, B.; Asquith, S.; Bertoncello, I.; Nilsson, S. K. An innovative triple immunogold labeling method to investigate the hemopoietic stem cell niche *in situ*. *Microsc. Microanal.* **2009**, *15*, 403–414.

45. Saccardi, R.; Urbani, S.; Caporale, R.; Lombaradini, L.; Bosi, A. Use of CFDA-SE for evaluating the *in vitro* proliferation pattern of human mesenchymal stem cells. *Cytotherapy* **2006**, *8*, 243–253.

46. Zhang, J.; Liu, W.; Liu, L. 5-Carboxyfluorescein N-succinimidyl ester in detecting the degree of cross-linked allogeneic tissue. *Shengwu Yixue Gongchengxue Zazhi* **2010**, *27*, 816–819.

47. Lutz, W. H.; Londowski, J. M.; Kumar, R. The synthesis and biological activity of four novel

fluorescent vasopressin analogs. *J. Biol. Chem.* **1990**, *265*, 4657–4663.

48. Wu, J.; Zheng, Y. G. Fluorescent reporters of the histone acetyltransferase. *Anal. Biochem.* **2008**, *380*, 106–110.

49. Katayama, Y.; Koga, H.; Toita, R.; Niidome, T.; Mori, T. A novel fluorometric assay method for the detection of protein kinase activities. Jpn. Kokai Tokkyo Koho JP 2012254938, 2012.

50. Klink, T. A.; Beebe, J. A.; Lasky, D. A.; Kleman-Leyer, K. M.; Somberg, R. Kinase and phosphatase assays. U.S. Pat. Appl. Publ. US 20070059787, 2007.

6-CARBOXYFLUORESCEIN SUCCINIMIDYL ESTER

CAS Registry Number 92557-81-8

Chemical Structure

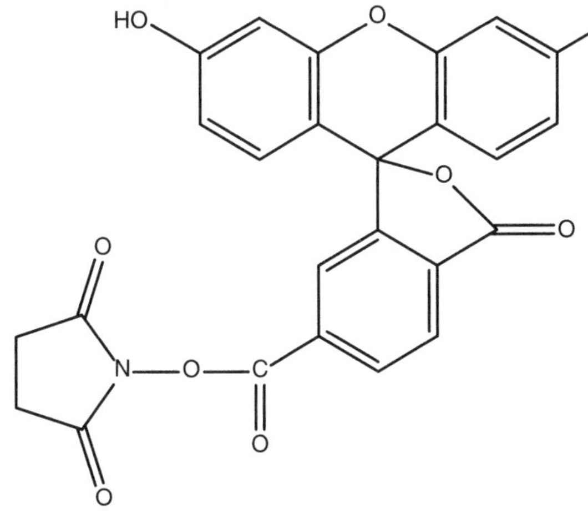

CA Index Name Spiro[isobenzofuran-1(3H),9'-[9H] xanthene]-6-carboxylic acid, 3',6'-dihydroxy-3-oxo-, 2,5-dioxo-1-pyrrolidinyl ester

Other Names 2,5-Pyrrolidinedione, 1-[[(3',6'-dihydroxy-3-oxospiro[isobenzofuran-1(3H),9'-[9H]xanthen]-6-yl)carbonyl]oxy]-; Spiro[isobenzofuran-1(3H),9'-[9H]xanthene], 2,5-pyrrolidinedione deriv.; 6-Carboxyfluorescein N-hydroxysuccinimide ester; 6-Carboxyfluorescein succinimidyl ester; Fluorescein-6-carboxylic acid N-succinimidyl ester; 6-CFSE; CFSE; 6-FAM, SE

Merck Index Number Not listed

Chemical/Dye Class Xanthene

Molecular Formula $C_{25}H_{15}NO_9$

Molecular Weight 473.39

Physical Form Orange solid

Solubility Soluble in N,N-dimethylformamide, dimethyl sulfoxide

Boiling Point (Calcd.) 788.1 ± 70.0 °C, pressure: 760 Torr

pK$_a$ (Calcd.) 9.31 ± 0.20, most acidic, temperature: 25 °C

Absorption (λ_{max}) 496 nm (Buffer pH 9)

Emission (λ_{max}) 516 nm (Buffer pH 9)

Molar Extinction Coefficient 83,000 cm^{-1} M^{-1} (Buffer pH 9)

Synthesis Synthetic methods[1-6]

Imaging/Labeling Applications Amantadine;[16] aminoglycosides;[7] anti-digoxin;[8] baclofen;[9] bacteria;[10] peptide chloromethyl ketones;[11] gabapentin;[12] herbicides;[13] histamine H1/H2 receptor antagonists;[14,15] memantine[16] metal ions (Fe(III));[17] nanoparticles;[18] nucleic acids/ nucleotides;[19-23] phosphatidylethanolamine;[24] proteins/ peptides;[5,25-35] risedronate;[36] amino functionalized rhodamine spirolactams[37]

Biological/Medical Applications Analyzing aminoglycosides;[7] analyzing/determining gabapentin;[12] detecting/analyzing baclofen in human plasma at micromolar level;[9] determining phosphorus-containing amino acid herbicides;[13] multiplex genetic analysis;[19] visualizing/detecting proteins;[5,25-35] for synthesizing vancomycin analogs,[38] vasopressin analogs;[39] as a substrate for measuring kinase activity,[40,41] phosphatase activity[41]

Industrial Applications Not reported

Safety/Toxicity No data available

REFERENCES

1. Adamczyk, M.; Fishpaugh, J. R.; Heuser, K. J. Preparation of succinimidyl and pentafluorophenyl active esters of 5- and 6-carboxyfluorescein. *Bioconjugate Chem.* **1997**, *8*, 253–255.

2. Vanderbilt, A. S.; Osikowicz, E. W.; Fino, J. R.; Shipchandler, M. T. Total estriol fluorescence polarization immunoassay. Eur. Pat. Appl. EP 0200960, 1986.

3. Flentge, C. A.; Kirkemo, C. L. Substituted carboxyfluoresceins. Eur. Pat. Appl. EP 0108403, 1984.

4. Fino, J. R.; Kirkemo, C. L. Substituted carboxyfluoresceins. Eur. Pat. Appl. EP 0108399, 1984.

5. Huchzermeier, R. F.; Spring, T. G. Determination of unsaturated thyroxine binding protein sites using fluorescence polarization techniques. Eur. Pat. Appl. EP 0108400, 1984.

6. Yoshida, R. A. Support-ligand analog-fluorescer conjugate and serum assay method involving such conjugate. Eur. Pat. Appl. EP 0015695, 1980.

7. Lin, Y.; Wang, Y.; Chang, S. Y. Capillary electrophoresis of aminoglycosides with argon-ion laser-induced fluorescence detection. *J. Chromatogr., A* **2008**, *1188*, 331–333.

8. Ghazarossian, V.; Shanafelt, A. B.; Skold, C. N.; Ullman, E. F. Methods and devices for (immuno)chromatographic analysis and their use. Eur. Pat. Appl. EP 342913, 1989.

9. Gu, Y.; Whang, C. Capillary electrophoresis of baclofen with argon-ion laser-induced fluorescence detection. *J. Chromatogr., A* **2002**, *972*, 289–293.

10. Breeuwer, P.; Drocourt, J.; Rombouts, F. M.; Abee, T. A novel method for continuous determination of the intracellular pH in bacteria with the internally conjugated fluorescent probe 5 (and 6-)-carboxyfluorescein succinimidyl ester. *Appl. Environ. Microbiol.* **1996**, *62*, 178–183.

11. Williams, E. B.; Krishnaswamy, S.; Mann, K. G. Zymogen/enzyme discrimination using peptide chloromethyl ketones. *J. Biol. Chem.* **1989**, *264*, 7536–7545.

12. Chang, S. Y.; Wang, F. Determination of gabapentin in human plasma by capillary electrophoresis with laser-induced fluorescence detection and acetonitrile stacking technique. *J. Chromatogr., B* **2004**, *799*, 265–270.

13. Molina, M.; Silva, M. Analytical potential of fluorescein analogues for ultrasensitive determinations of phosphorus-containing amino acid herbicides by micellar electrokinetic chromatography with laser-induced fluorescence detection. *Electrophoresis* **2002**, *23*, 1096–1103.

14. Li, L.; Kracht, J.; Peng, S.; Bernhardt, G.; Buschauer, A. Synthesis and pharmacological activity of fluorescent histamine H1 receptor antagonists related to mepyramine. *Bioorg. Med. Chem. Lett.* **2003**, *13*, 1245–1248.

15. Li, L.; Kracht, J.; Peng, S.; Bernhardt, G.; Elz, S.; Buschauer, A. Synthesis and pharmacological activity of fluorescent histamine H2 receptor antagonists related to potentidine. *Bioorg. Med. Chem. Lett.* **2003**, *13*, 1717–1720.

16. Yeh, H.; Yang, Y.; Chen, S. Simultaneous determination of memantine and amantadine in human plasma as fluorescein derivatives by micellar electrokinetic chromatography with laser-induced fluorescence detection and its clinical application. *Electrophoresis* **2010**, *31*, 1903–1911.

17. Su, B.; Moniotte, N.; Nivarlet, N.; Tian, G.; Desmet, J. Design and synthesis of fluorescence-based siderophore sensor molecules for FeIII ion

determination. *Pure Appl. Chem.* **2010**, *82*, 2199–2216.

18. Wang, L.; Tan, W. Multicolor FRET silica nanoparticles by single wavelength excitation. *Nano Lett.* **2006**, *6*, 84–88.

19. Richardson, J. A.; Gerowska, M.; Shelbourne, M.; French, D.; Brown, T. Six-colour HyBeacon probes for multiplex genetic analysis. *ChemBioChem* **2010**, *11*, 2530–2533.

20. Tanaka, K.; Tainaka, K.; Kamei, T.; Okamoto, A. Direct labeling of 5-methylcytosine and its applications. *J. Am. Chem. Soc.* **2007**, *129*, 5612–5620.

21. Seo, T. S.; Li, Z.; Ruparel, H.; Ju, J. Click chemistry to construct fluorescent oligonucleotides for DNA sequencing. *J. Org. Chem.* **2003**, *68*, 609–612.

22. Cummins, W. J.; Hamilton, A. L.; Smith, C. L.; Briggs, M. S. J. Synthesis and study of the fluorescein conjugate of the nucleotide dPTP. *Nucleosides, Nucleotides Nucleic Acids* **2001**, *20*, 1049–1051.

23. Albarella, J. P.; Anderson Deriemer, L. H.; Carrico, R. J. Hybridization assay employing labeled probe and anti-hybrid. Eur. Pat. Appl. EP 0144913, 1985.

24. Rzepecki, P. W.; Prestwich, G. D. Synthesis of hybrid lipid probes: Derivatives of phosphatidylethanolamine-extended phosphatidylinositol 4,5-bisphosphate (Pea-PIP2). *J. Org. Chem.* **2002**, *67*, 5454–5460.

25. Hori, Y.; Nakaki, K.; Sato, M.; Mizukami, S.; Kikuchi, K. Development of protein-labeling probes with a redesigned fluorogenic switch based on intramolecular association for no-wash live-cell imaging. *Angew. Chem., Int. Ed.* **2012**, *51*, 5611–5614.

26. Mizukami, S.; Watanabe, S.; Akimoto, Y.; Kikuchi, K. No-wash protein labeling with designed fluorogenic probes and application to real-time pulse-chase analysis. *J. Am. Chem. Soc.* **2012**, *134*, 1623–1629.

27. Watanabe, S.; Mizukami, S.; Akimoto, Y.; Hori, Y.; Kikuchi, K. Intracellular protein labeling with prodrug-Like probes using a mutant β-lactamase tag. *Chem. Eur. J.* **2011**, *17*, 8342–8349

28. Watanabe, S.; Mizukami, S.; Hori, Y.; Kikuchi, K. Multicolor protein labeling in living cells using mutant β-lactamase-tag technology. *Bioconjugate Chem.* **2010**, *21*, 2320–2326.

29. Sadhu, K. K.; Mizukami, S.; Watanabe, S.; Kikuchi, K. Turn-on fluorescence switch involving aggregation and elimination processes

for β-lactamase-tag. *Chem. Commun.* **2010**, *46*, 7403–7405.

30. Hori, Y.; Egashira, Y.; Kamiura, R.; Kikuchi, K. Noncovalent-interaction-promoted ligation for protein labeling. *ChemBioChem* **2010**, *11*, 646–648.

31. Takiyama, K.; Kinoshita, E.; Kinoshita-Kikuta, E.; Fujioka, Y.; Kubo, Y.; Koike, T. A Phos-tag-based fluorescence resonance energy transfer system for the analysis of the dephosphorylation of phosphopeptides. *Anal. Biochem.* **2009**, *388*, 235–241.

32. Los, G. V.; Encell, L. P.; McDougall, M. G.; Hartzell, D. D.; Karassina, N.; Zimprich, C.; Wood, M. G.; Learish, R.; Ohana, R. F.; Urh, M.; Simpson, D.; Mendez, J.; Zimmerman, K.; Otto, P.; Vidugiris, G.; Zhu, J.; Darzins, A.; Klaubert, D. H.; Bulleit, R. F.; Wood, K. V. HaloTag: A novel protein labeling technology for cell imaging and protein analysis. *ACS Chem. Biol.* **2008**, *3*, 373–382.

33. Calloway, N. T.; Choob, M.; Sanz, A.; Sheetz, M. P.; Miller, L. W.; Cornish, V. W. Optimized fluorescent trimethoprim derivatives for *in vivo* protein labeling. *ChemBioChem* **2007**, *8*, 767–774.

34. Lata, S.; Reichel, A.; Brock, R.; Tampe, R.; Piehler, J. High-affinity adaptors for switchable recognition of histidine-tagged proteins. *J. Am. Chem. Soc.* **2005**, *127*, 10205–10215.

35. Pellois, J.; Hahn, M. E.; Muir, T. W. Simultaneous triggering of protein activity and fluorescence. *J. Am. Chem. Soc.* **2004**, *126*, 7170–7171.

36. Kashemirov, B. A.; Bala, J. L. F.; Chen, X.; Ebetino, F. H.; Xia, Z.; Russell, R. G. G.; Coxon, F. P.; Roelofs, A. J.; Rogers, M. J.; McKenna, C. E. Fluorescently labeled risedronate and related analogues: "magic linker" synthesis. *Bioconjugate Chem.* **2008**, *19*, 2308–2310.

37. Adamczyk, M.; Grote, J. Synthesis of probes with broad pH range fluorescence. *Bioorg. Med. Chem. Lett.* **2003**, *13*, 2327–2330.

38. Adamczyk, M.; Grote, J.; Moore, J. A.; Rege, S. D.; Yu, Z. Structure-binding relationships for the interaction between a vancomycin monoclonal antibody Fab fragment and a library of vancomycin analogs and tracers. *Bioconjugate Chem.* **1999**, *10*, 176–185.

39. Lutz, W. H.; Londowski, J. M.; Kumar, R. The synthesis and biological activity of four novel fluorescent vasopressin analogs. *J. Biol. Chem.* **1990**, *265*, 4657–4663.

40. Katayama, Y.; Koga, H.; Toita, R.; Niidome, T.; Mori, T. A novel fluorometric assay method for the detection of protein kinase activities. Jpn. Kokai Tokkyo Koho JP 2012254938, 2012.

41. Klink, T. A.; Beebe, J. A.; Lasky, D. A.; Kleman-Leyer, K. M.; Somberg, R. Kinase and phosphatase assays. U.S. Pat. Appl. Publ. US 20070059787, 2007.

5-CARBOXYNAPHTHO-FLUORESCEIN

CAS Registry Number   145103-60-2

Chemical Structure

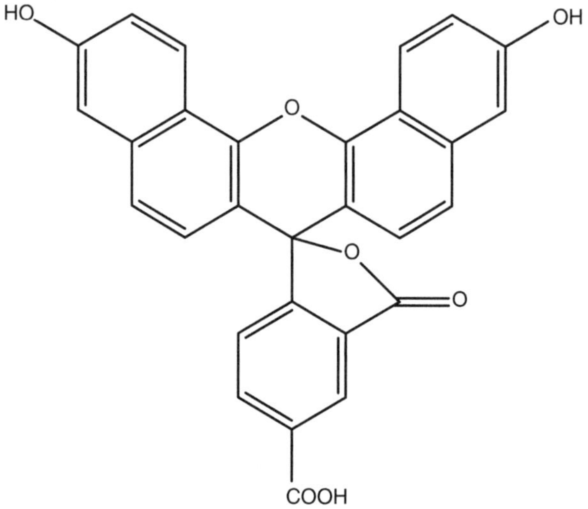

CA Index Name   Spiro[7H-dibenzo[c,h]xanthene-7,1′(3′H)-isobenzofuran]-5′-carboxylic acid, 3,11-dihydroxy-3′-oxo-

Other Names   Spiro[7H-dibenzo[c,h]xanthene-7,1′(3′H)-isobenzofuran]-5′-carboxylic acid, 3,6a-dihydro-11-hydroxy-3,3′-dioxo-; 5-Carboxynaphthofluorescein; CNF

Merck Index Number   Not listed

Chemical/Dye Class   Xanthene

Molecular Formula $C_{29}H_{16}O_7$

Molecular Weight   476.44

Physical Form   Purple solid or dark blue crystalline powder

Solubility   Insoluble in water; soluble in N,N-dimethylformamide, dimethyl sulfoxide, methanol

Boiling Point (Calcd.)   845.9 ± 65.0 °C, pressure: 760 Torr

pK$_a$ 7.6, temperature: 22 °C

pK$_a$ (Calcd.)   3.68 ± 0.20, most acidic, temperature: 25 °C

Absorption (λ_{max})   512 nm (Buffer pH 6.0); 598 nm (Buffer pH 10.0)

Emission (λ_{max})   563 nm (Buffer pH 6.0); 668 (Buffer pH 10.0)

Molar Extinction Coefficient   11,000 $cm^{-1} M^{-1}$ (Buffer pH 6.0); 49,000 $cm^{-1} M^{-1}$ (Buffer pH 10.0)

Synthesis   Synthetic methods[1,2]

Imaging/Labeling Applications   Live cells;[3,4] oligonucleotides[5]

Biological/Medical Applications   Evaluating multimeric complexes (proteins, peptides, antibody molecules, small/large molecules);[6] fluorosensor for measurement of near-neutral pH values;[7] organic electroluminescence (OEL)-based biochips;[8,9] optical sensors for bio-medical applications;[10-12] substrates for assaying esterase;[2] studying protein crystal nucleation;[13] as temperature sensor[14,15]

Industrial Applications   Adhesive;[16] electronic device manufacturing;[16] fabricating multiplexed arrays of chemosensors[17]

Safety/Toxicity   No data available

REFERENCES

1. Sabnis, R. W. *Handbook of Acid–base Indicators*; CRC Press: Boca Raton, **2008**; pp 74–75.

2. Coleman, D. J.; Naleway, J. J. Fluorogenic enzyme substrates for visualizing acidic organelles. U.S. Pat. Appl. Publ. US 20100233744, 2010.

3. Brasuel, M.; Kopelman, R.; Aylott, J. W.; Clark, H.; Xu, H.; Hoyer, M.; Miller, T. J.; Tjalkens, R.; Philbert, M. A. Production, characteristics and applications of fluorescent PEBBLE nanosensors: potassium, oxygen, calcium and pH imaging inside live cells. *Sens. Mater.* **2002**, *14*, 309–338.

4. Clark, H. A.; Hoyer, M.; Parus, S.; Philbert, M. A.; Kopelman, R. Optochemical nanosensors and subcellular applications in living cells. *Mikrochim. Acta* **1999**, *131*, 121–128.

5. Vu, H.; Majlessi, M.; Adelpour, D.; Russell, J. The use of 6-carboxynaphthofluorescein phosphoramidite in the automated synthesis of quencher-dye oligonucleotide probes (QDOPs). *Tetrahedron Lett.* **2009**, *50*, 737–740.

6. Yang, X.; Fitz, L. J.; Lee, J. M.; Brooks, J.; Wolf, S. F. Assays and methods using homogeneous proximity-based detection methods for evaluating multimeric complexes. PCT Int. Appl. WO 2009061911, 2009.

7. Wolfbeis, O. S.; Rodriguez, N. V.; Werner, T. LED-compatible fluorosensor for measurement of

near-neutral pH values. *Mikrochim. Acta* **1992**, *108*, 133–141.

8. Jang, S.; Suen, S.; Chen, S.; Lin, C.; Yang, J. Organic electroluminescence (OEL)-based biochips. Taiwan TW 504945, 2002.

9. Chen, S.; Yang, C.; Chang, S.; Lin, C.; Sun, S. Organic electroluminescence (OEL)-based biochips. U.S. Pat. Appl. Publ. US 20030035755, 2003.

10. Chen-Esterlit, Z.; Aylott, J. W.; Kopelman, R. Development of oxygen and pH optical sensors using phase modulation technique. *Proc. SPIE-Int. Soc. Opt. Eng.* **1999**, *3540*, 19–27.

11. Kopelman, R.; Clark, H.; Monson, E.; Parus, S.; Philbert, M.; Thorsrud, B. Optical fiberless sensors. PCT Int. Appl. WO 9902651, 1999.

12. Bentsen, J. G.; Wood, K. B. Sensor with improved drift stability. U.S. Patent 5403746, 1995.

13. Pusey, M. L.; Sumida, J. Fluorescence studies of protein crystal nucleation. *Proc. SPIE-Int. Soc. Opt. Eng.* **2000**, *4098*, 1–10.

14. Bousseksou, A.; Salmon, L.; Molnar, G.; Cobo, S. Heat-sensitive spin-transition materials doped with one or more fluorescent agents for use as temperature sensor. PCT Int. Appl. WO 2011058277, 2011.

15. Bousseksou, A.; Salmon, L.; Molnar, G.; Cobo, S. Materials with thermochromic spin transition doped with one or more fluorescent agents for use as temperature sensor. Fr. Demande FR 2952371, 2011.

16. Lu, D.; Li, E. J. Adhesive with differential optical properties and its application for substrate processing and detection of fluorescent adhesive residue. U.S. Pat. Appl. Publ. US 20060189093, 2006.

17. Martinez-Otero, A.; Gonzalez-Monje, P.; Maspoch, D.; Hernando, J.; Ruiz-Molina, D. Multiplexed arrays of chemosensors by parallel dip-pen nanolithography. *Chem. Commun.* **2011**, *47*, 6864–6866.

6-CARBOXYNAPHTHO-FLUORESCEIN

CAS Registry Number   145103-61-3

Chemical Structure

CA Index Name   Spiro[7H-dibenzo[c,h]xanthene-7,1′(3′H)-isobenzofuran]-6′-carboxylic acid, 3,11-dihydroxy-3′-oxo-

Other Names   Spiro[7H-dibenzo[c,h]xanthene-7,1′(3′H)-isobenzofuran]-6′-carboxylic acid, 3,11-dihydro-3′-oxo-; 6-Carboxynaphthofluorescein; CNF

Merck Index Number   Not listed

Chemical/Dye Class   Xanthene

Molecular Formula $C_{29}H_{16}O_7$

Molecular Weight   476.44

Physical Form   Purple solid or dark blue crystalline powder

Solubility   Insoluble in water; soluble in N,N-dimethylformamide, dimethyl sulfoxide, methanol

Boiling Point (Calcd.)   855.5 ± 65.0 °C, pressure: 760 Torr

pK$_a$ 7.6, temperature: 22 °C

pK$_a$ (Calcd.) 3.48 ± 0.20, most acidic, temperature: 25 °C

Absorption (λ_{max})   512 nm (Buffer pH 6.0); 598 nm (Buffer pH 10.0)

Emission (λ_{max})   563 nm (Buffer pH 6.0); 668 (Buffer pH 10.0)

Molar Extinction Coefficient   11,000 cm^{-1} M^{-1} (Buffer pH 6.0); 49,000 cm^{-1} M^{-1} (Buffer pH 10.0)

Synthesis   Synthetic methods[1–4]

Imaging/Labeling Applications   Live cells;[5,6] oligonucleotides[3]

Biological/Medical Applications   Fluorosensor for measurement of near-neutral pH values;[4] organic electroluminescence (OEL)-based biochips;[7,8] optical sensors for bio-medical applications;[9–11] substrates for assaying esterase;[2] studying protein crystal nucleation[12]

Industrial Applications   Not reported

Safety/Toxicity   No data available

REFERENCES

1. Sabnis, R. W. *Handbook of Acid–base Indicators*; CRC Press: Boca Raton, **2008**; p 76.

2. Coleman, D. J.; Naleway, J. J. Fluorogenic enzyme substrates for visualizing acidic organelles. U.S. Pat. Appl. Publ. US 20100233744, 2010.

3. Vu, H.; Majlessi, M.; Adelpour, D.; Russell, J. The use of 6-carboxynaphthofluorescein phosphoramidite in the automated synthesis of quencher-dye oligonucleotide probes (QDOPs). *Tetrahedron Lett.* **2009**, *50*, 737–740.

4. Wolfbeis, O. S.; Rodriguez, N. V.; Werner, T. LED-compatible fluorosensor for measurement of near-neutral pH values. *Mikrochim. Acta* **1992**, *108*, 133–141.

5. Brasuel, M.; Kopelman, R.; Aylott, J. W.; Clark, H.; Xu, H.; Hoyer, M.; Miller, T. J.; Tjalkens, R.; Philbert, M. A. Production, characteristics and applications of fluorescent PEBBLE nanosensors: potassium, oxygen, calcium and pH imaging inside live cells. *Sens. Mater.* **2002**, *14*, 309–338.

6. Clark, H. A.; Hoyer, M.; Parus, S.; Philbert, M. A.; Kopelman, R. Optochemical nanosensors and subcellular applications in living cells. *Mikrochim. Acta* **1999**, *131*, 121–128.

7. Jang, S.; Suen, S.; Chen, S.; Lin, C.; Yang, J. Organic electroluminescence (OEL)-based biochips. Taiwan TW 504945, 2002.

8. Chen, S.; Yang, C.; Chang, S.; Lin, C.; Sun, S. Organic electroluminescence (OEL)-based biochips. U.S. Pat. Appl. Publ. US 20030035755, 2003.

9. Chen-Esterlit, Z.; Aylott, J. W.; Kopelman, R. Development of oxygen and pH optical sensors using

phase modulation technique. *Proc. SPIE-Int. Soc. Opt. Eng.* **1999**, *3540*, 19–27.

10. Kopelman, R.; Clark, H.; Monson, E.; Parus, S.; Philbert, M.; Thorsrud, B. Optical fiberless sensors. PCT Int. Appl. WO 9902651, 1999.

11. Bentsen, J. G.; Wood, K. B. Sensor with improved drift stability. U.S. Patent 5403746, 1995.

12. Pusey, M. L.; Sumida, J. Fluorescence studies of protein crystal nucleation. *Proc. SPIE-Int. Soc. Opt. Eng.* **2000**, *4098*, 1–10.

CASCADE BLUE ACETYL AZIDE TRISODIUM SALT

CAS Registry Number 138039-58-4

Chemical Structure

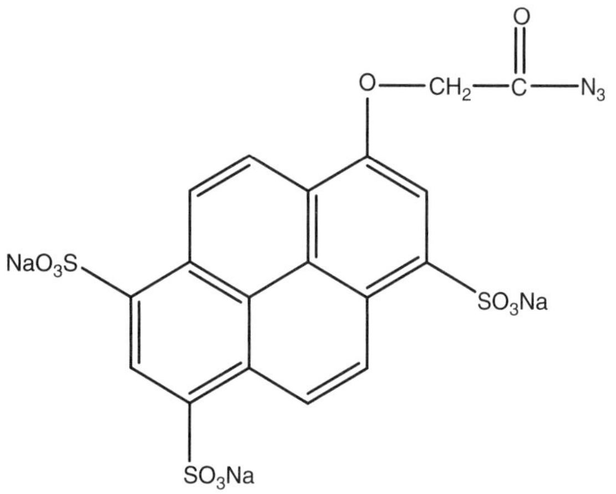

CA Index Name 1,3,6-Pyrenetrisulfonic acid, 8-(2-azido-2-oxoethoxy)-, sodium salt (1:3)

Other Names 1,3,6-Pyrenetrisulfonic acid, 8-(2-azido-2-oxoethoxy)-, trisodium salt; Cascade Blue Acetyl Azide; Cascade Blue acetyl azide trisodium salt

Merck Index Number Not listed

Chemical/Dye Class Pyrene

Molecular Formula $C_{18}H_8N_3Na_3O_{11}S_3$

Molecular Weight 607.42

Physical Form Tan powder;[1,2] Yellow solid

Solubility Soluble in water, methanol

Absorption (λ_{max}) 396 nm, 376 nm (MeOH)

Emission (λ_{max}) 410 nm (MeOH)

Molar Extinction Coefficient 29,000 cm^{-1} M^{-1} (MeOH)

Synthesis Synthetic methods[1,2]

Imaging/Labeling Applications Bacteria;[3] cells;[4,5] cyclodextrin;[6] nuclei;[1] nucleic acids;[7–9] proteins[3–5,10–14]

Biological/Medical Applications Detecting/identifying proteins;[3–5,10–14] detecting/quantifying bacteria;[3] diagnosing/treating chronic lymphocytic leukemia (CLL)[7]

Industrial Applications Not reported

Safety/Toxicity No data available

REFERENCES

1. Whitaker, J. E.; Haugland, R. P.; Moore, P. L.; Hewitt, P. C.; Reese, M.; Haugland, R. P. Cascade Blue derivatives: water soluble, reactive, blue emission dyes evaluated as fluorescent labels and tracers. *Anal. Biochem.* **1991**, *198*, 119–130.

2. Haugland, R. P.; Whitaker, J. E. Chemically reactive pyrenesulfonic acid dyes. U.S. Patent 5132432, 1992.

3. Semprevivo, L. H.; Stuart, E. S. Detection and quantification of intracellular pathogens. PCT Int. Appl. WO 2005116234, 2005.

4. Mao, F.; Leung, W.; Cheung, C. Fluorescent pyrene compounds. U.S. Pat. Appl. Publ. US 20110097735, 2011.

5. Mao, F.; Leung, W.; Cheung, C.; Hoover, H. E. Fluorescent compounds. PCT Int. Appl. WO 2009078970, 2009.

6. Huff, J. B.; Bieniarz, C.; Horng, W. J. Fluorescent polymer labeled conjugates and intermediates. U.S. Patent 5661040, 1997.

7. Drandi, D.; Gribben, J. G.; Lee, C.; Dal Cin, P. Human chromosome 14 specific DNA probes Compositions and methods for diagnosing and treating chronic lymphocytic leukemia. PCT Int. Appl. WO 2005046606, 2005.

8. Lee, L. G. Method for detecting oligonucleotides using UV light source. U.S. Patent 6218124, 2001.

9. Lee, L. G. UV-excitable fluorescent energy transfer dyes. PCT Int. Appl. WO 2001016369, 2001.

10. Nguyen, T.; Joshi, N. S.; Francis, M. B. An affinity-based method for the purification of fluorescently-labeled biomolecules. *Bioconjugate Chem.* **2006**, *17*, 869–872.

11. Taylor, R. M.; Lin, B.; Foubert, T. R.; Burritt, J. B.; Sunner, J.; Jesaitis, A. J. Cascade blue as a donor for resonance energy transfer studies of heme-containing proteins. *Anal. Biochem.* **2002**, *302*, 19–27.

12. Yefimov, S.; Sjomeling, C.; Yergey, A. L.; Li, T.; Chrambach, A. Recovery of sodium dodecyl sulfate-proteins from gel electrophoretic bands in a single electroelution step for mass spectrometric analysis. *Anal. Biochem.* **2000**, *284*, 288–295.

13. Gombocz, E.; Chrambach, A.; Yefimov, S.; Yergey, A. L. Electroelution of nonfluorescent stacked

proteins detected by fluorescence optics from gel electrophoretic bands for transfer into mass spectrometry. *Electrophoresis* **2000**, *21*, 846–849.

14. Yefimov, S.; Yergey, A. L.; Chrambach, A. Transfer of SDS-proteins from gel electrophoretic zones into mass spectrometry, using electroelution of the band into buffer without sectioning of the gel. *J. Biochem. Biophys. Methods* **2000**, *42*, 65–78.

CASCADE BLUE HYDRAZIDE TRISODIUM SALT

CAS Registry Number   137182-38-8

Chemical Structure

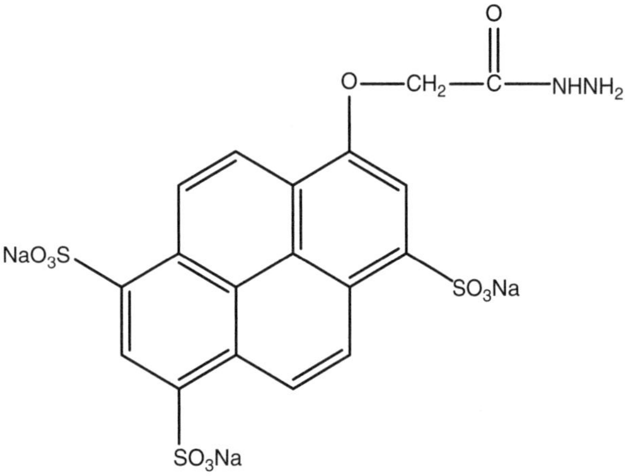

CA Index Name   Acetic acid, 2-[(3,6,8-trisulfo-1-pyrenyl)oxy]-, hydrazide, sodium salt (1 : 3)

Other Names   Acetic acid, [(3,6,8-trisulfo-1-pyrenyl) oxy]-, 1-hydrazide, trisodium salt; Cascade Blue hydrazide trisodium salt

Merck Index Number   Not listed

Chemical/Dye Class   Pyrene

Molecular Formula $C_{18}H_{11}N_2Na_3O_{11}S_3$

Molecular Weight   596.44

Physical Form   Yellow powder[1,2]

Solubility   Soluble in water

Absorption (λ_{max})   399 nm, 376 nm (H_2O)

Emission (λ_{max})   421 nm (H_2O)

Molar Extinction Coefficient   30,000 $cm^{-1} M^{-1}$ (H_2O)

Synthesis   Synthetic methods[1,2]

Imaging/Labeling Applications   Cytoplasm;[3] nuclei;[1,3] nucleic acid[4]

Biological/Medical Applications   Cosmetics;[5,6] reagents for modifying aldehydes/ketones[1,2,7]

Industrial Applications   Solid-state materials[8]

Safety/Toxicity   No data available

REFERENCES

1. Whitaker, J. E.; Haugland, R. P.; Moore, P. L.; Hewitt, P. C.; Reese, M.; Haugland, R. P. Cascade Blue derivatives: water soluble, reactive, blue emission dyes evaluated as fluorescent labels and tracers. *Anal. Biochem.* **1991**, *198*, 119–130.

2. Haugland, R. P.; Whitaker, J. E. Chemically reactive pyrenesulfonic acid dyes. U.S. Patent 5132432, 1992.

3. Oparka, K. J.; Murant, E. A.; Wright, K. M.; Prior, D. A. M.; Harris, N. The drug probenecid inhibits the vacuolar accumulation of fluorescent anions in onion epidermal cells. *J. Cell Sci.* **1991**, *99*, 557–563.

4. McCormick, M. R. Method for precipitating nucleic acid with visible carrier. PCT Int. Appl. WO 9712994, 1997.

5. Vidal, L.; Huguet, E.; Barbarat, P. A cosmetic composition including at least one fluorophore compound. PCT Int. Appl. WO 2011089571, 2011.

6. Vidal, L.; Huguet, E.; Barbarat, P. A cosmetic composition including at least one fluorophore compound. Fr. Demande FR 2956028, 2010.

7. Som, A.; Matile, S. Rigid-rod β-barrel ion channels with internal "Cascade Blue" cofactors - catalysis of amide, carbonate, and ester hydrolysis. *Eur. J. Org. Chem.* **2002**, 3874–3883.

8. Sun, Y.; Chan, B. C.; Ramnarayanan, R.; Leventry, W. M.; Mallouk, T. E.; Bare, S. R.; Willis, R. R. Split-pool method for synthesis of solid-state material combinatorial libraries. *J. Comb. Chem.* **2002**, *4*, 569–575.

3-CYANO-7-ETHOXYCOUMARIN (7-ETHOXYCOUMARIN-3-CARBONITRILE)

CAS Registry Number   117620-77-6

Chemical Structure

CA Index Name   2H-1-Benzopyran-3-carbonitrile, 7-ethoxy-2-oxo-

Other Names   3-Cyano-7-ethoxycoumarin; 7-Ethoxycoumarin-3-carbonitrile; 7-Ethoxy-2-oxo-2H-chromene-3-carbonitrile

Merck Index Number   Not listed

Chemical/Dye Class   Coumarin

Molecular Formula $C_{12}H_9NO_3$

Molecular Weight   215.20

Physical Form   Off-white solid

Solubility   Soluble in acetone, acetonitrile, chloroform, N,N-dimethylformamide, dimethyl sulfoxide, methanol

Melting Point   212 °C[1,2]

Boiling Point (Calcd.)   396.8 ± 42.0 °C, pressure: 760 Torr

Absorption (λ_{max})   356 nm (Buffer pH 7); 355 nm (MeOH)

Emission (λ_{max})   411 nm (Buffer pH 7); 411 nm (MeOH)

Molar Extinction Coefficient   20,000 cm^{-1} M^{-1} (Buffer pH 7); 22,500 cm^{-1} M^{-1} (MeOH)

Synthesis   Synthetic methods[1,2]

Imaging/Labeling Applications   Cytochrome P 450 enzymes[1–23]

Biological/Medical Applications   As a substrate for measuring cytochrome P 450 enzymes activity;[1–11] cytochrome P450 inhibition assays;[12–16] drug design;[17,18] drug metabolism;[17–23] treating viral or parasite infections[24]

Industrial Applications   Not reported

Safety/Toxicity   No data available

REFERENCES

1. White, I. N. H. Preparation of 7-alkoxy-3-cyanocoumarins used as fluorometric assays for enzymes. Brit. UK Pat. Appl. GB 2211500, 1989.

2. White, I. N. H. A continuous fluorometric assay for cytochrome P-450-dependent mixed function oxidases using 3-cyano-7-ethoxycoumarin. *Anal. Biochem.* **1988**, *172*, 304–310.

3. Ho, S. H. Y.; Singh, M.; Holloway, A. C.; Crankshaw, D. J. The effects of commercial preparations of herbal supplements commonly used by women on the biotransformation of fluorogenic substrates by human cytochromes P450. *Phytother. Res.* **2011**, *25*, 983–989.

4. Makaji, E.; Ho, S. H. Y.; Holloway, A. C.; Crankshaw, D. J. Effects in rats of maternal exposure to raspberry leaf and its constituents on the activity of cytochrome P450 enzymes in the offspring. *Int. J. Toxicol.* **2011**, *30*, 216–224.

5. Scornaienchi, M. L.; Thornton, C.; Willett, K. L.; Wilson, J. Y. Functional differences in the cytochrome P450 1 family enzymes from Zebrafish (*Danio rerio*) using heterologously expressed proteins. *Arch. Biochem. Biophys.* **2010**, *502*, 17–22.

6. Smith, E. M.; Wilson, J. Y. Assessment of cytochrome P450 fluorometric substrates with rainbow trout and killifish exposed to dexamethasone, pregnenolone-16α-carbonitrile, rifampicin, and β-naphthoflavone. *Aquat. Toxicol.* **2010**, *97*, 324–333.

7. Ekins, S.; Iyer, M.; Krasowski, M. D.; Kharasch, E. D. Molecular characterization of CYP2B6 substrates. *Curr. Drug Metab.* **2008**, *9*, 363–373.

8. Donato, M. T.; Jimenez, N.; Castell, J. V.; Gomez-Lechon, M. J. Fluorescence-based assays for screening nine cytochrome P450 (P450) activities in intact cells expressing individual human P450 enzymes. *Drug Metab. Dispos.* **2004**, *32*, 699–706.

9. Stresser, D. M.; Turner, S. D.; Blanchard, A. P.; Miller, V. P.; Crespi, C. L. Cytochrome P450 fluorometric substrates: identification of isoform-selective probes for rat CYP2D2 and human CYP3A4. *Drug Metab. Dispos.* **2002**, *30*, 845–852.

10. Ekins, S.; Bravi, G.; Ring, B. J.; Gillespie, T. A.; Gillespie, J. S.; Vandenbranden, M.; Wrighton, S. A.; Wikel, J. H. Three-dimensional quantitative structure activity relationship analyses of substrates

for CYP2B6. *J. Pharmacol. Exp. Ther.* **1999**, *288*, 21–29.

11. Ekins, S.; Vandenbranden, M.; Ring, B. J.; Gillespie, J. S.; Yang, T. J.; Gelboin, H. V.; Wrighton, S. A. Further characterization of the expression in liver and catalytic activity of CYP2B6. *J. Pharmacol. Exp. Ther.* **1998**, *286*, 1253–1259.

12. Makaji, E.; Trambitas, C. S.; Shen, P.; Holloway, A. C.; Crankshaw, D. J. Effects of cytochrome P450 inhibitors on the biotransformation of fluorogenic substrates by adult male rat liver microsomes and cDNA-expressed rat cytochrome P450 isoforms. *Toxicol. Sci.* **2010**, *113*, 293–304.

13. Wang, H.; Kim, R. A.; Sun, D.; Gao, Y.; Wang, H.; Zhu, J.; Chen, C. Evaluation of the effects of 18 non-synonymous single-nucleotide polymorphisms of CYP450 2C19 on *in vitro* drug inhibition potential by a fluorescence-based high-throughput assay. *Xenobiotica* **2011**, *41*, 826–835.

14. Persiani, S.; Canciani, L.; Larger, P.; Rotini, R.; Trisolino, G.; Antonioli, D.; Rovati, L. C. *In vitro* study of the inhibition and induction of human cytochromes P450 by crystalline glucosamine sulfate. *Drug Metab. Drug Interact.* **2009**, *24*, 195–209.

15. Makings, L. R.; Zlokarnik, G. Optical molecular sensors cytochrome P450 activity. U.S. Patent 6514687, 2003.

16. Miller, V. P.; Stresser, D.; Crespi, C. L. The use of fluorescein aryl ethers in high throughput cytochrome P450 inhibition assays. PCT Int. Appl. WO 2001014361, 2001.

17. Wang, J.; Wei, D.; Chen, C.; Li, Y.; Chou, K. Molecular modeling of two CYP2C19 SNPs and its implications for personalized drug design. *Protein Pept. Lett.* **2008**, *15*, 27–32.

18. Wang, J.; Wei, D.; Li, L.; Zheng, S.; Li, Y.; Chou, K. 3D structure modeling of cytochrome P450 2C19 and its implication for personalized drug design. *Biochem. Biophys. Res. Commun.* **2007**, *355*, 513–519.

19. Rydberg, P.; Vasanthanathan, P.; Oostenbrink, C.; Olsen, L. Fast prediction of cytochrome P450 mediated drug metabolism. *ChemMedChem* **2009**, *4*, 2070–2079.

20. Ghosal, A.; Hapangama, N.; Yuan, Y.; Lu, X.; Horne, D.; Patrick, J. E.; Zbaida, S. Rapid determination of enzyme activities of recombinant human cytochromes P450, human liver microsomes and hepatocytes. *Biopharm. Drug Dispos.* **2003**, *24*, 375–384.

21. Wang, Q.; Halpert, J. R. Combined three-dimensional quantitative structure-activity relationship analysis of cytochrome P450 2B6 substrates and protein homology modeling. *Drug Metab. Dispos.* **2002**, *30*, 86–95.

22. Kariv, I.; Fereshteh, M. P.; Oldenburg, K. R. Development of a miniaturized 384-well high throughput screen for the detection of substrates of cytochrome P450 2D6 and 3A4 metabolism. *J. Biomol. Screen.* **2001**, *6*, 91–99.

23. Crespi, C. L.; Miller, V. P.; Penman, B. W. Microtiter plate assays for inhibition of human, drug-metabolizing cytochromes P450. *Anal. Biochem.* **1997**, *248*, 188–190.

24. Prendergast, P. T. Use of flavones, coumarins and related compounds to treat infections. PCT Int. Appl. WO 2001003681, 2001.

3-CYANO-7-HYDROXYCOUMARIN (3-CYANOUMBELLIFERONE)

CAS Registry Number   19088-73-4

Chemical Structure

CA Index Name   2H-1-Benzopyran-3-carbonitrile, 7-hydroxy-2-oxo-

Other Names   Coumarin, 3-cyano-7-hydroxy-; 3-Cyano-7-hydroxycoumarin; 3-Cyanoumbelliferone; 7-Hydroxy-3-cyanocoumarin; 7-Hydroxycoumarin-3-carbonitrile

Merck Index Number   Not listed

Chemical/Dye Class   Coumarin

Molecular Formula $C_{10}H_5NO_3$

Molecular Weight   187.15

Physical Form   Yellow solid[1] or powder

Solubility   Soluble in acetonitrile, N,N-dimethylformamide, dimethyl sulfoxide, ethanol, methanol

Melting Point   273–275 °C;[7] 270–272 °C;[10] 262 °C;[14] 252–254 °C;[2] 250–252 °C;[4] 249–251 °C;[5,8] 246–248 °C[1]

Boiling Point (Calcd.)   436.6 ± 45.0 °C, pressure: 760 Torr

pK$_a$ (Calcd.)   7.03 ± 0.20, most acidic, temperature: 25 °C

Absorption (λ_{max})   408 nm (Buffer pH 9); 408 nm (MeOH)

Emission (λ_{max})   450 nm (Buffer pH 9); 450 nm (MeOH)

Molar Extinction Coefficient   43,000 cm^{-1} M^{-1} (Buffer pH 9)

Synthesis   Synthetic methods[1–14]

Imaging/Labeling Applications   Esters;[15] ethers;[16–25] phosphates;[26–32] microorganisms;[33] proteins[34]

Biological/Medical Applications   Detecting microorganisms,[33] nucleic acids;[26–32] inhibiting cell proliferation/tumor growth;[10] reference standard for esters;[15] ethers,[16–25] phosphates substrates;[26–32] as a substrate for measuring nucleic acid polymerases activity;[26–32] treating viral or parasite infections[35]

Industrial Applications   Electroluminescent device;[36] laser dyes;[37,38] modified polyester for food containers/beverage bottles;[11] monitoring of cationic photopolymerization processes;[39] textiles;[12] thin films[40]

Safety/Toxicity   No data available

REFERENCES

1. Valizadeh, H.; Mahmoodian, M.; Gholipour, H. ZrCl$_4$/[bmim]BF$_4$-Catalyzed condensation of salicylaldehydes and malononitrile: Single-step synthesis of 3-cyanocoumarin derivatives. *J. Heterocycl. Chem.* **2011**, *48*, 799–802.

2. Rajesha; Kiran Kumar, H. C.; Bhojya Naik, H. S.; Mahadevan, K. M. Aqueous synthesis of coumarins using tetramethylammonium hydroxide as surfactant. *Org. Chem.: Indian J.* **2011**, *7*, 365–368.

3. Valizadeh, H.; Fakhari, A. Facile, efficient, and eco-friendly synthesis of benzo[b]pyran-2-imines over MgO and transformation to the coumarin derivatives. *J. Heterocycl. Chem.* **2009**, *46*, 1392–1395.

4. Valizadeh, H.; Shockravi, A.; Gholipur, H. Microwave assisted synthesis of coumarins via potassium carbonate catalyzed Knoevenagel condensation in 1-n-butyl-3-methylimidazolium bromide ionic liquid. *J. Heterocycl. Chem.* **2007**, *44*, 867–870.

5. Valizadeh, H.; Mamaghani, M.; Badrian, A. Effect of microwave irradiation on reaction of arylaldehyde derivatives with some active methylene compounds in aqueous media. *Synth. Commun.* **2005**, *35*, 785–790.

6. Williams, L.; Sjovik, R.; Falck-Pedersen, M. L. ChemScreen: Planar synthesis, separation and screening of antioxidants. *J. Planar Chromatogr.– Mod. TLC* **2004**, *17*, 244–249.

7. Fringuelli, F.; Piermatti, O.; Pizzo, F. One-pot synthesis of 3-carboxycoumarins via consecutive Knoevenagel and Pinner reactions in water. *Synthesis* **2003**, 2331–2334.

8. Shockravi, A.; Shargi, H.; Valizadeh, H.; Heravi, M. M. Solvent free synthesis of coumarins. *Phosphorus, Sulfur Silicon Relat. Elem.* **2002**, *177*, 2555–2559.

9. Brufola, G.; Fringuelli, F.; Piermatti, O.; Pizzo, F. Simple and efficient one-pot preparation of 3-substituted coumarins in water. *Heterocycles* **1996**, *43*, 1257–1266.

10. Keri, G.; Bajor, T.; Orfi, L.; Tihanyi, E.; Balogh, A.; Bokonyi, G.; Horvath, A.; Idei, M. Coumarin derivatives and their analogues inhibiting cell proliferation and tumor growth, pharmaceutical compositions containing them and process for preparing same. PCT Int. Appl. WO 9316064, 1993.

11. Weaver, M. A.; Coates, C. A., Jr.,; Pruett, W. P.; Hilbert, S. D. Polyester polymer containing the residue of the UV-absorbing benzopyran compound and shaped articles produced therefrom. U.S. Patent 4882412, 1989.

12. Lokhande, S. B.; Rangnekar, D. W. Synthesis of 3-substituted 7-hydroxy-8-arylazocoumarins. *Indian J. Chem., Sect. B* **1986**, *25B*, 638–639.

13. Wolfbeis, O. S. pH-Dependent fluorescence spectra of 3-substituted umbelliferones. *Z. Naturforsch., Teil A* **1977**, *32A*, 1065–1067.

14. Balaiah, V.; Seshadri, T. R.; Venkateswarlu, V. Visible fluorescence and chemical constitution of compounds of the benzopyrone group. III. Further study of structural influences in coumarins. *Proc. Indian Acad. Sci., Sect. A* **1942**, *16A*, 68–82.

15. Gaydukov, L.; Khersonsky, O.; Tawfik, D. S. Method of determining total PON1 level. PCT Int. Appl. WO 2008111062, 2008.

16. Traylor, M. J.; Chai, J.; Clark, D. S. Simultaneous measurement of CYP1A2 activity, regioselectivity, and coupling: implications for environmental sensitivity of enzyme-substrate binding. *Arch. Biochem. Biophys.* **2011**, *505*, 186–193.

17. Persiani, S.; Canciani, L.; Larger, P.; Rotini, R.; Trisolino, G.; Antonioli, D.; Rovati, L. C. *In vitro* study of the inhibition and induction of human cytochromes P450 by crystalline glucosamine sulfate. *Drug Metab. Drug Interact.* **2009**, *24*, 195–209.

18. Sereemaspun, A.; Hongpiticharoen, P.; Rojanathanes, R.; Maneewattanapinyo, P.; Ekgasit, S.; Warisnoicharoen, W. Inhibition of human cytochrome P450 enzymes by metallic nanoparticles: a preliminary to nanogenomics. *Int. J. Pharmacol.* **2008**, *4*, 492–495.

19. Aydinli, M.; Tutas, M.; Atasoy, B.; Bozdemir, O. A. Synthesis and characterization of poly(aryl ether) dendritic structures functionalized with coumarin derivatives. *React. Funct. Polym.* **2005**, *65*, 317–327.

20. Mistry, S.; Desai, S.; Rao, S. S. M.; Shah, A. Synthesis of furocoumarins: Claisen rearrangement of 7-allyloxycoumarins. *Indian J. Heterocycl. Chem.* **2004**, *13*, 301–306.

21. Makings, L. R.; Zlokarnik, G. Optical molecular sensors cytochrome P450 activity. U.S. Patent 6514687, 2003.

22. Miller, V. P.; Stresser, D.; Crespi, C. L. The use of fluorescein aryl ethers in high throughput cytochrome P450 inhibition assays. PCT Int. Appl. WO 2001014361, 2001.

23. Makings, L.; Zlokarnik, G. Optical molecular sensors for cytochrome P450 activity. PCT Int. Appl. WO 2000035900, 2000.

24. Ekins, S.; Vandenbranden, M.; Ring, B. J.; Gillespie, J. S.; Yang, T. J.; Gelboin, H. V.; Wrighton, S. A. Further characterization of the expression in liver and catalytic activity of CYP2B6. *J. Pharmacol. Exp. Ther.* **1998**, *286*, 1253–1259.

25. White, I. N. H. Preparation of 7-alkoxy-3-cyanocoumarins used as fluorometric assays for enzymes. Brit. UK Pat. Appl. GB 2211500, 1989.

26. Sood, A.; Kumar, S.; Nampalli, S.; Nelson, J. R.; Macklin, J.; Fuller, C. W. Terminal phosphate-labeled nucleotides with improved substrate properties for homogeneous nucleic acid assays. *J. Am. Chem. Soc.* **2005**, *127*, 2394–2395.

27. Sood, A.; Kumar, S.; Fuller, C.; Nelson, J. Analytes detection. U.S. Pat. Appl. Publ. US 20040224319, 2004.

28. Sood, A.; Kumar, S.; Nelson, J.; Fuller, C. Solid phase sequencing. U.S. Pat. Appl. Publ. US 20040152119, 2004.

29. Kumar, S.; Mcdougall, M.; Sood, A.; Nelson, J.; Fuller, C.; Macklin, J.; Mitsis, P. Terminal phosphate-labeled nucleotides with new linkers. PCT Int. Appl. WO 2004072238, 2004.

30. Fuller, C.; Kumar, S.; Sood, A.; Nelson, J. Terminal-phosphate-labeled nucleotides and methods of use. U.S. Pat. Appl. Publ. US 20030162213, 2003.

31. Nelson, J.; Fuller, C.; Sood, A.; Kumar, S. Terminal-phosphate-labeled nucleotides and methods of use. PCT Int. Appl. WO 2003020984, 2003.

32. Kumar, S.; Sood, A. Labeled nucleoside polyphosphates. PCT Int. Appl. WO 2003020734, 2003.

33. Roscoe, S. B.; Bolea, P. A.; Moeller, S. J. Methods of detecting microorganisms and kits therefore. PCT Int. Appl. WO 2011087711, 2011.

34. Chen, Y.; Muller, J. D.; Tetin, S. Y.; Tyner, J. D.; Gratton, E. Probing ligand protein binding equilibria with fluorescence fluctuation spectroscopy. *Biophys. J.* **2000**, *79*, 1074–1084.

35. Prendergast, P. T. Use of flavones, coumarins and related compounds to treat infections. PCT Int. Appl. WO 2001003681, 2001.

36. Kathirgamanathan, P.; Ganeshamurugan, S.; Kumaraverl, M.; Ravichandran, S. Electroluminescent device. Brit. UK Pat. Appl. GB 2440367, 2008.

37. Takakusa, M.; Itoh, U.; Anzai, H.; Masuko, H.; Sato, T. Laser emission from new hydroxycoumarin dyes. *Jpn. J. Appl. Phys.* **1978**, *17*, 1461–1462.

38. Wolfbeis, O. S.; Rapp, W.; Lippert, E. Luminescent heterocycles. Part 6. 3-Substituted umbelliferones: a class of blue emitting, photostable and easily pumpable laser dyes. *Monatsh. Chem.* **1978**, *109*, 899–903.

39. Ortyl, J.; Popielarz, R. The performance of 7-hydroxycoumarin-3-carbonitrile and 7-hydroxycoumarin-3-carboxylic acid as fluorescent probes for monitoring of cationic photopolymerization processes by FPT. *J. Appl. Polym. Sci.* **2013**, *128*, 1974–1978.

40. Frenette, M.; Coenjarts, C.; Scaiano, J. C. Mapping acid-catalyzed deprotection in thin polymer films: Fluorescence imaging using prefluorescent 7-hydroxycoumarin probes. *Macromol. Rapid Commun.* **2004**, *25*, 1628–1631.

4′,6-DIAMIDINO-2-PHENYLINDOLE DIHYDROCHLORIDE (DAPI)

CAS Registry Number   28718-90-3

Chemical Structure

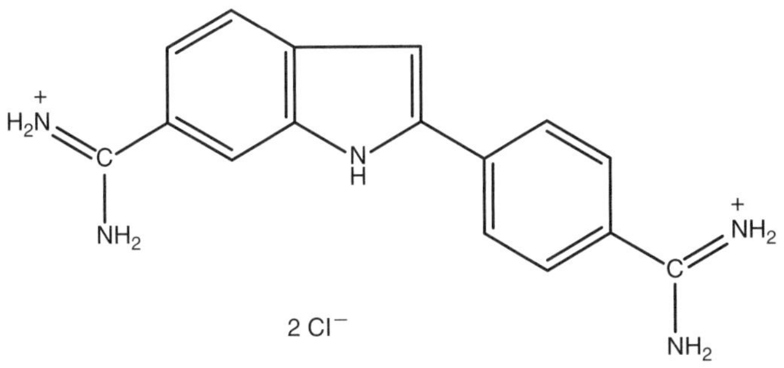

2 Cl⁻

CA Index Name   1H-Indole-6-carboximidamide, 2-[4-(aminoiminomethyl)-phenyl]-, hydrochloride (1 : 2)

Other Names   1H-Indole-6-carboximidamide, 2-[4-(aminoiminomethyl)-phenyl]-, dihydrochloride; Indole-6-carboxamidine, 2-(p-amidinophenyl)-, dihydrochloride; 4′,6-Diamidino-2-phenylindole dihydrochloride; 6-Amidino-2-(4-amidinophenyl)-indole dihydrochloride; FxCycle Violet; DAPI

Merck Index Number   Not listed

Chemical/Dye Class   Heterocycle; Indole

Molecular Formula $C_{16}H_{17}Cl_2N_5$

Molecular Weight   350.25

Physical Form   Yellow powder[4,5] or solid

Solubility   Soluble in water, N,N-dimethylformamide, dimethyl sulfoxide, ethanol, methanol

Melting Point   360–362 °C (decompose)[4,5]

Absorption (λ_{max})   358 nm (H_2O/DNA); 342 nm (H_2O); 348 nm (MeOH)

Emission (λ_{max})   461 nm (H_2O/DNA); 450 nm (H_2O); 457 nm (MeOH)

Molar Extinction Coefficient   24,000 cm^{-1} M^{-1} (H_2O/DNA); 28,000 cm^{-1} M^{-1} (H_2O); 31,000 cm^{-1} M^{-1} (MeOH)

Quantum Yield   0.34 (H_2O/DNA)

Synthesis   Synthetic methods[1–6]

Imaging/Labeling Applications   Nucleic acids;[1,7–18,29] cells;[1,19–22] nuclei;[23] antigens;[24] bacteria;[25–28] chromosomes;[29] endotoxins;[30] microorganisms;[31] neuron-specific nuclear protein NeuN;[32] polynucleotides;[33] proteins;[34] mesenchymal stem cells[35]

Biological/Medical Applications   Detecting nucleic acids,[1,7–18,29] cells,[1,19–22] neuron-specific nuclear protein NeuN;[28] identifying/quantifying bacteria;[25–28] monitoring microorganisms;[31] quantifying proteins;[34] nuclear apoptosis assay;[36] treating Alzheimer's disease,[37,38] Down's syndrome,[37] type II diabetes,[37,38] cerebral amyloid angiopathy,[38] ischemic disease;[39] monitoring atmospheric/indoor bioaerosols[40,41]

Industrial Applications   Not reported

Safety/Toxicity   Neurotoxicity[42]

REFERENCES

1. Sabnis, R. W. *Handbook of Biological Dyes and Stains*; John Wiley & Sons Inc.: Hoboken, **2010**; pp 127–128.

2. Dann, O. Trypanocidal diamidine compounds. Fr. Demande FR 1586113, 1970.

3. Dann, O.; Bergen, G.; Demant, E.; Vol, G. Trypanosomicidal diamidines of 2-phenylbenzofuran, 2-phenylindene, and 2-phenylindole. *Justus Liebigs Ann. Chem.* **1971**, *749*, 68–89.

4. Dann, O. Diamidine compounds. U.S. Patent 3652591, 1972.

5. Dann, O. Benzofuran diamidine compounds. U.S. Patent 3689506, 1972.

Handbook of Fluorescent Dyes and Probes, First Edition. R. W. Sabnis.
© 2015 John Wiley & Sons, Inc. Published 2015 by John Wiley & Sons, Inc.

6. Zhang, X.; Chen, G.; Li, G.; Chen, L.; Dai, Z. Antimalarial agents. Part 14. Synthesis of 4′,6-diamidino-2-phenylindole and its antimalarial effect. *Yiyao Gongye* **1983**, 2–4.

7. Kjaerulff, S.; Glensbjerg, M. Method for analysis of cellular DNA content. PCT Int. Appl. WO 2011098085, 2011.

8. Del Castillo, P.; Horobin, R. W.; Blazquez-Castro, A.; Stockert, J. C. Binding of cationic dyes to DNA: distinguishing intercalation and groove binding mechanisms using simple experimental and numerical models. *Biotech. Histochem.* **2010**, *85*, 247–256.

9. Park, H. O.; Kim, H. B.; Chi, S. M. Detection method of DNA amplification using probes labeled with intercalating dye. PCT Int. Appl. WO 2006004267, 2006.

10. Noirot, M.; Barre, P.; Louarn, J.; Duperray, C.; Hamon, S. Consequences of stoichiometric error on nuclear DNA content evaluation in *Coffea liberica* var. dewevrei using DAPI and propidium iodide. *Ann. Bot.* **2002**, *89*, 385–389.

11. Buel, E.; Schwartz, M. The use of DAPI as a replacement for ethidium bromide in forensic DNA analysis. *J. Forensic Sci.* **1995**, *40*, 275–278.

12. Vollenweider, I.; Groscurth, P. Comparison of four DNA staining fluorescence dyes for measuring cell proliferation of lymphokine-activated killer (LAK) cells. *J. Immunol. Methods* **1992**, *149*, 133–135.

13. Mabuchi, T.; Nishikawa, S. Selective staining with two fluorochromes of DNA fragments on gels depending on their AT content. *Nucleic Acids Res.* **1990**, *18*, 7461–7462.

14. Leusch, H. G.; Hoffmann, M.; Emeis, C. C. Fluorometric determination of the total DNA content of different yeast species using 4′,6-diamidino-2-phenyl-indol-dihydrochloride. *Can. J. Microbiol.* **1985**, *31*, 1164–1166.

15. Legros, M.; Kepes, A. One-step fluorometric microassay of DNA in prokaryotes. *Anal. Biochem.* **1985**, *147*, 497–502.

16. Hamada, S.; Fujita, S. Fluorescence enhancement in DNA-DAPI complexes. *Acta Histochem. Cytochem.* **1983**, *16*, 606–609.

17. Kapuscinski, J.; Skoczylas, B. Fluorescent complexes of DNA with DAPI (4′,6-diamidine-2-phenyl indole dihydrochloride) or DCI (4′,6-dicarboxyamide-2-phenyl indole). *Nucleic Acids Res.* **1978**, *5*, 3775–3799.

18. Kapuscinski, J.; Skoczylas, B. Simple and rapid fluorimetric method for DNA microassay. *Anal. Biochem.* **1977**, *83*, 252–257.

19. Shirai, M.; Azuma, Y. Method for determining existence ratio of viable cell, dead-cell, and false viable cell. Jpn. Kokai Tokkyo Koho JP 2008054509, 2008.

20. Skyggebjerg, O.; Glensbjerg, M. A method and a system for counting cells from a plurality of species. PCT Int. Appl. WO 2002101087, 2002.

21. Cao, F.; Eckert, R.; Elfgang, C.; Nitsche, J. M.; Snyder, S. A.; Hulser, D. F.; Willecke, K.; Nicholson, B. J. A quantitative analysis of connexin-specific permeability differences of gap junctions expressed in HeLa transfectants and Xenopus oocytes. *J. Cell Sci.* **1998**, *111*, 31–43.

22. Takahama, M. Reagents and method for determination of protein/DNA ratios in cells. Jpn. Kokai Tokkyo Koho JP 62135769, 1987.

23. Thornthwaite, J. T. Nuclear isolation medium and procedure for separating cell nuclei. U.S. Patent 4668618, 1987.

24. Krenacs, T.; Krenacs, L.; Raffeld, M. Multiple antigen immunostaining procedures. *Methods Mol. Biol.* **2010**, *588*, 281–300.

25. Ando, M.; Kamei, R.; Komagoe, K.; Inoue, T.; Yamada, K.; Katsu, T. *In situ* potentiometric method to evaluate bacterial outer membrane-permeabilizing ability of drugs: Example using antiprotozoal diamidines. *J. Microbiol. Methods* **2012**, *91*, 497–500.

26. Hu, B.; Zheng, P.; Tang, C.; Chen, J.; van der Biezen, E.; Zhang, L.; Ni, B.; Jetten, M. S. M.; Yan, J.; Yu, H. Identification and quantification of anammox bacteria in eight nitrogen removal reactors. *Water Res.* **2010**, *44*, 5014–5020.

27. Kawaharasaki, M.; Tanaka, H.; Kanagawa, T.; Nakamura, K. *In situ* identification of polyphosphate-accumulating bacteria in activated sludge by dual staining with rRNA-targeted oligonucleotide probes and 4′,6-diamidino-2-phenylindol (DAPI) at a polyphosphate-probing concentration. *Water Res.* **1998**, *33*, 257–265.

28. Kawaharasaki, M.; Kanagawa, T.; Tanaka, H.; Nakamura, K. Development and application of 16S rRNA-targeted oligonucleotide probe for detection of phosphate-accumulating bacterium, *Microlunatus phosphovorus* in an enhanced biological phosphorus

removal process. *Water Sci. Technol.* **1998**, *37*, 481–484.

29. Gosden, J. R.; Spowart, G.; Lawrie, S. S. Satellite DNA and cytological staining patterns in heterochromatic inversions of human chromosome 9. *Hum. Genet.* **1981**, *58*, 276–278.

30. Evans, D.; Jindal, S. Methods and compositions for binding endotoxins. PCT Int. Appl. WO 9641185, 1996.

31. Li, C. S.; Chia, W. C.; Chen, P. S. Fluorochrome and flow cytometry to monitor microorganisms in treated hospital wastewater. *J. Environ. Sci. Health, Part A* **2007**, *42*, 195–203.

32. Gill, S. K.; Ishak, M.; Rylett, R. J. Exposure of nuclear antigens in formalin-fixed, paraffin-embedded necropsy human spinal cord tissue: Detection of NeuN. *J. Neurosci. Methods* **2005**, *148*, 26–35.

33. Cavatorta, P.; Masotti, L.; Szabo, A. G. A time-resolved fluorescence study of 4′,6′-diamidine-2-phenylindole dihydrochloride binding to polynucleotides. *Biophys. Chem.* **1985**, *22*, 11–16.

34. Kai, K.; Kitajima, Y.; Miyazaki, K.; Tokunaga, O. Protein quantitation method by fluorescent multiple staining. Jpn. Kokai Tokkyo Koho JP 2008268167, 2008.

35. Ocarino, N. M.; Bozzi, A.; Pereira, R. D. O.; Breyner, N. M.; Silva, V. L.; Castanheira, P.; Goes, A. M.; Serakides, R. Behavior of mesenchymal stem cells stained with 4′,6-diamidino-2-phenylindole dihydrochloride (DAPI) in osteogenic and non osteogenic cultures. *Biocell* **2008**, *32*, 175–183

36. Susin, S. A.; Zamzami, N.; Larochette, N.; Dallaporta, B.; Marzo, I.; Brenner, C.; Hirsch, T.; Petit, P. X.; Geuskens, M.; Kroemer, G. A cytofluorometric assay of nuclear apoptosis induced in a cell-free system: application to ceramide-induced apoptosis. *Exp. Cell Res.* **1997**, *236*, 397–403.

37. Chalifour, R. J.; Kong, X.; Wu, X.; Lu, W. Amidine derivatives for treating amyloidosis. PCT Int. Appl. WO 2003017994, 2003.

38. Chalifour, R. J.; Kong, X.; Wu, X.; Lu, W.; Tidwell, R. R.; Boykin, D. Amidine derivatives for treating amyloidosis. PCT Int. Appl. WO 2003103598, 2003.

39. Wakita, H.; Igarashi, K.; Oie, K. Diagnosis and treatment of ischemic disease. PCT Int. Appl. WO 2009022756, 2009.

40. Chi, M.; Li, C. Fluorochrome in monitoring atmospheric bioaerosols and correlations with meteorological factors and air pollutants. *Aerosol Sci. Technol.* **2007**, *41*, 672–678.

41. Li, C.; Huang, T. Fluorochrome in monitoring indoor bioaerosols. *Aerosol Sci. Technol.* **2006**, *40*, 237–241.

42. Chouliaras, L.; van den Hove, D. L. A.; Kenis, G.; Keitel, S.; Hof, P. R.; van Os, J.; Steinbusch, H. W. M.; Schmitz, C.; Rutten, B. P. F. Prevention of age-related changes in hippocampal levels of 5-methylcytidine by caloric restriction. *Neurobiol. Aging* **2012**, *33*, 1672–1681.

7-DIETHYLAMINOCOUMARIN-3-CARBONYL AZIDE

CAS Registry Number   157673-16-0

Chemical Structure

CA Index Name   2*H*-1-Benzopyran-3-carbonyl azide, 7-(diethylamino)-2-oxo-

Other Names   7-Diethylaminocoumarin-3-carbonyl azide

Merck Index Number   Not listed

Chemical/Dye Class   Coumarin

Molecular Formula $C_{14}H_{14}N_4O_3$

Molecular Weight   286.29

Physical Form   Orange solid

Solubility   Soluble in acetonitrile, *N,N*-dimethylformamide, methanol

Absorption (λ_{max})   436 nm (MeOH)

Emission (λ_{max})   478 nm (MeOH)

Molar Extinction Coefficient   57,000 cm^{-1} M^{-1} (MeOH)

Synthesis   Synthetic method[1]

Imaging/Labeling Applications   Alcohols;[2] cyclic ketones;[3] epolactaene derivatives;[4] iejimalide carbamate derivatives;[5] lysophospholipids;[6] macrolide analogs;[7] platelet-activating factor acetylhydrolase (PAF-AH);[6,8] phosphonic acid-based NAALADase inhibitors;[9] uronium salts;[10] vinblastine analog[11]

Biological/Medical Applications   Detecting lysophospholipids,[6] uronium salts;[10] identifying macrolide analogs;[7] measuring platelet-activating factor acetylhydrolase (PAF-AH) activity;[6,8] reagents for alcohols/hydroxyl groups[1,2,12]

Industrial Applications   Polyelectrolytes[12]

Safety/Toxicity   No data available

REFERENCES

1. Haugland, R. P. *Handbook of Fluorescent Probes and Research Chemicals*; Molecular Probes Inc.: Eugene, **1996**; pp 64–67.

2. Nelson, T. J. Fluorescent high-performance liquid chromatography assay for lipophilic alcohols. *Anal. Biochem.* **2011**, *419*, 40–45.

3. Moersel, J.; Schmiedl, D. Determination of 2-alkylcyclobutanone using fluorescent labeling. *Fresenius' J. Anal. Chem.* **1994**, *349*, 538–541.

4. Kuramochi, K.; Yukizawa, S.; Ikeda, S.; Sunoki, T.; Arai, S.; Matsui, R.; Morita, A.; Mizushina, Y.; Sakaguchi, K.; Sugawara, F. Syntheses and applications of fluorescent and biotinylated epolactaene derivatives: Epolactaene and its derivative induce disulfide formation. *Bioorg. Med. Chem.* **2008**, *16*, 5039–5049.

5. Schweitzer, D.; Zhu, J.; Jarori, G.; Tanaka, J.; Higa, T.; Davisson, V. J.; Helquist, P. Synthesis of carbamate derivatives of iejimalides. Retention of normal antiproliferative activity and localization of binding in cancer cells. *Bioorg. Med. Chem.* **2007**, *15*, 3208–3216.

6. Servillo, L.; Iorio, E. L.; Quagliuolo, L.; Camussi, G.; Balestrieri, C.; Giovane, A. Simultaneous determination of lysophospholipids by high-performance liquid chromatography with fluorescence detection. *J. Chromatogr., B: Biomed. Appl.* **1997**, *689*, 281–286.

7. Marriott, G. Macrolide analogs and methods for identifying same. U.S. Pat. Appl. Publ. US 20040259185, 2004.

8. Balestrieri, C.; Camussi, G.; Giovane, A.; Lorio, E. L.; Quangliuolo, L.; Servillo, L. Measurement of platelet-activating factor acetylhydrolase activity by quantitative high-performance liquid chromatography determination of coumarin-derivatized 1-O-alkyl-2-sn-lysoglyceryl-3-phosphorylcholine. *Anal. Biochem.* **1996**, *233*, 145–150.

9. Ding, P.; Helquist, P.; Miller, M. J. Design, synthesis and pharmacological activity of novel enantiomerically pure phosphonic acid-based NAALADase inhibitors. *Org. Biomol. Chem.* **2007**, *5*, 826–831.

10. Almog, J. Detection of uronium salts. U.S. Pat. Appl. Publ. US 20060084176, 2006.

11. Chatterjee, S. K.; Laffray, J.; Patel, P.; Ravindra, R.; Qin, Y.; Kuehne, M. E.; Bane, S. L. Interaction of tubulin with a new fluorescent analogue of vinblastine. *Biochemistry* **2002**, *41*, 14010–14018.

12. Southard, G. E.; Woo, J. T. K.; Massingill, J. L. Synthesis of fluorescently tagged polyelectrolytes. *Prog. Org. Coat.* **2004**, *49*, 160–164.

7-DIETHYLAMINOCOUMARIN-3-CARBOXYLIC ACID (DAC) (DEAC)

CAS Registry Number   50995-74-9

Chemical Structure

CA Index Name   2H-1-Benzopyran-3-carboxylic acid, 7-(diethylamino)-2-oxo-

Other Names   3-Carboxy-7-(diethylamino)coumarin; 7-(Diethylamino)-3-(carboxy)coumarin; 7-Diethylaminocoumarin-3-carboxylic acid; Coumarin D 1421; D 1421; DAC; DCCA; 7-DCCA; DEAC

Merck Index Number   Not listed

Chemical/Dye Class   Coumarin

Molecular Formula $C_{14}H_{15}NO_4$

Molecular Weight   261.27

Physical Form   Bright orange solid;[11] Orange crystals;[13] Orange crystalline solid;[27] Orange laths;[37,38] Orange powder;[4] Orange solid[2,3,6,9,12,16,20,28,35]

Solubility   Soluble in chloroform, N,N-dimethylformamide, dimethyl sulfoxide,

Melting Point   243–248 °C;[16]   231–232 °C;[37,38] 230–232 °C;[13]   230 °C;[2]   229–231 °C;[27]   228 °C;[35] 227–229 °C;[40]   224–225.5 °C;[30]   224–225 °C;[10,18]

222.4–223.3 °C;[7] 222–224 °C;[33] 220 °C;[39] 217–218 °C;[32] 215–217 °C;[36] 214–215 °C;[17]

Boiling Point (Calcd.)   464.7 ± 45.0 °C, pressure: 760 Torr

pK$_a$ (Calcd.)   −98.36 ± 0.20, most acidic, temperature: 25 °C; 6.06 ± 0.20, most basic, temperature: 25 °C

Absorption (λ_{max})   409 nm (Buffer pH 9.0)

Emission (λ_{max})   473 nm (Buffer pH 9.0)

Molar Extinction Coefficient $34{,}000\,cm^{-1}\,M^{-1}$ (Buffer pH 9.0)

Synthesis   Synthetic methods[1–41]

Imaging/Labeling Applications   Alcohols;[42] ATP;[43] copper ions;[22,44–46] iron ions;[3,30] fluoride ions;[7] cyclodextrins (CDs)/dextrans;[47–49] hydrogen peroxide;[50] hydrogen sulfide;[51] hypochlorite;[52] lysine;[28,29] nitric oxide;[53] nucleotides/nucleic acids;[35,54–56] proteins/amino acids;[9,37,38,57–60] sulfide ions;[61] thiols[62]

Biological/Medical Applications   Detecting hydrogen peroxide;[50] evaluating iron ions;[3,30] cysteine cathepsins inhibitors;[8] β-glucosidase inhibitors;[63,64] human monoamine oxidase inhibitors;[15] N-Methyl-D-aspartate receptors (NMDARs) inhibitors;[65] phosphate assay;[37,60] probing actomyosin mechanism;[55] treating cancer/antitumor activity;[5] treating cognitive deficits or schizophrenia;[65] as a substrate for measuring protein kinase activity[66,67]

Industrial Applications   Epoxy resins photoluminescent devices;[68] lithographic plate materials;[69] oil-repellent copolymers/coatings;[19,20,70,71] polymer matrices;[72] silica nanoparticles for imaging, security, and sensing applications[73,74]

Safety/Toxicity   No data available

REFERENCES

1. Maldonado-Dominguez, M.; Arcos-Ramos, R.; Romero, M.; Flores-Perez, B.; Farfan, N.; Santillan, R.; Lacroix, P. G.; Malfant, I. The amide bridge in donor-acceptor systems: delocalization depends on push-pull stress. *New J. Chem.* **2014**, *38*, 260–268.

2. Inal, S.; Koelsch, J. D.; Sellrie, F.; Schenk, J. A.; Wischerhoff, E.; Laschewsky, A.; Neher, D. A water soluble fluorescent polymer as a dual colour sensor for temperature and a specific protein. *J. Mater. Chem. B* **2013**, *1*, 6373–6381.

3. Ge, F.; Ye, H.; Zhang, H.; Zhao, B. A novel ratiometric probe based on rhodamine B and coumarin for selective recognition of Fe(III) in aqueous solution. *Dyes Pigm.* **2013**, *99*, 661–665.

4. Li, Y.; Schaffer, P.; Perrin, D. M. Dual isotope labeling: Conjugation of 32P-oligonucleotides with 18 F-aryltrifluoroborate via copper(I) catalyzed cycloaddition. *Bioorg. Med. Chem. Lett.* **2013**, *23*, 6313–6316.

5. Draoui, N.; Schicke, O.; Fernandes, A.; Drozak, X.; Nahra, F.; Dumont, A.; Douxfils, J.; Hermans, E.; Dogne, J.; Corbau, R.; Marchand, A.; Chaltin, P.; Sonveaux, P.; Feron, O.; Riant, O. Synthesis and pharmacological evaluation of carboxycoumarins as a new antitumor treatment targeting lactate transport in cancer cells. *Bioorg. Med. Chem.* **2013**, *21*, 7107–7117.

6. Liu, W. Incorporation of two different noncanonical amino acids into a single protein.

U.S. Pat. Appl. Publ. US 20130078671, 2013.

7. Li, J.; Zhou, X.; Zhou, Y.; Fang, Y.; Yao, C. A highly specific tetrazole-based chemosensor for fluoride ion: A new sensing functional group based on intramolecular proton transfer. *Spectrochim. Acta, Part A* **2013**, *102*, 66–70.

8. Frizler, M.; Mertens, M. D.; Guetschow, M. Fluorescent nitrile-based inhibitors of cysteine cathepsins. *Bioorg. Med. Chem. Lett.* **2012**, *22*, 7715–7718.

9. Wu, B.; Wang, Z.; Huang, Y.; Liu, W. R. Catalyst-free and site-specific one-pot dual-labeling of a protein directed by two genetically incorporated noncanonical amino acids. *ChemBioChem* **2012**, *13*, 1405–1408.

10. Bardajee, G. R.; Moallem, S. A. Synthesis of 3-carboxycoumarins catalyzed by $CuSO_4.5H_2O$ under ultrasound irradiation in aqueous media. *Asian J. Biochem. Pharm. Res.* **2012**, *2*, 410–414.

11. Gustafson, T. P.; Metzel, G. A.; Kutateladze, A. G. Externally sensitized deprotection of PPG-masked carbonyls as a spatial proximity probe in photoamplified detection of binding events. *Photochem. Photobiol. Sci.* **2012**, *11*, 564–577.

12. Simonin, J.; Vernekar, S. V.; Thompson, A. J.; Hothersall, J. D.; Connolly, C. N.; Lummis, S. C. R.; Lochner, M. High-affinity fluorescent ligands for the 5-HT3 receptor. *Bioorg. Med. Chem. Lett.* **2012**, *22*, 1151–1155.

13. Danko, M.; Szabo, E.; Hrdlovic, P. Synthesis and spectral characteristics of fluorescent dyes based on coumarin fluorophore and hindered amine stabilizer in solution and polymer matrices. *Dyes Pigm.* **2011**, *90*, 129–138.

14. Shahinian, E. G. H.; Sebe, I. Synthesis of some new benzocoumarin heterocyclic fluorescent dyes. *Rev. Chim.* **2011**, *62*, 1098–1101.

15. Secci, D.; Carradori, S.; Bolasco, A.; Chimenti, P.; Yanez, M.; Ortuso, F.; Alcaro, S. Synthesis and selective human monoamine oxidase inhibition of 3-carbonyl, 3-acyl, and 3-carboxyhydrazido coumarin derivatives. *Eur. J. Med. Chem.* **2011**, *46*, 4846–4852.

16. Harishkumar, H. N.; Mahadevan, K. M.; Kiran Kumar, H. C.; Satyanarayan, N. D. A facile, choline chloride/urea catalyzed solid phase synthesis of coumarins via Knoevenagel condensation. *Org. Commun.* **2011**, *4*, 26–32.

17. Yao, C.; Zeng, H. Synthesis and spectroscopic study of coumarin derivatives. *Chem. Res. Chin. Univ.* **2011**, *27*, 599–603.

18. Bardajee, G. R.; Jafarpour, F.; Afsari, H. S. $ZrOCl_2.8H_2O$, an efficient catalyst for rapid one-pot synthesis of 3-carboxycoumarins under ultrasound irradiation in water. *Cent. Eur. J. Chem.* **2010**, *8*, 370–374.

19. Hoshino, T.; Shirota, N.; Asakawa, A. Compound having fluorescent functional group and method for producing polymer of the same. PCT Int. Appl. WO 2009011427, 2009.

20. Hoshino, T.; Shirota, N.; Otozawa, N.; Asakawa, A. Oil repellent copolymer, method for its production and oil-repellent treatment solution. U.S. Pat. Appl. Publ. US 20090281239, 2009.

21. He, G.; Guo, D.; He, C.; Zhang, X.; Zhao, X.; Duan, C. A color-tunable europium complex emitting three primary colors and white light. *Angew. Chem., Int. Ed.* **2009**, *48*, 6132–6135.

22. Jung, H. S.; Kwon, P. S.; Lee, J. W.; Kim, J. Il; Hong, C. S.; Kim, J. W.; Yan, S.; Lee, J. Y.; Lee, J. H.; Joo, T.; K., Jong S. Coumarin-derived Cu2+−selective fluorescence sensor: Synthesis, mechanisms, and applications in living cells. *J. Am. Chem. Soc.* **2009**, *131*, 2008–2012.

23. Chen, K.; Liao, J.; Chan, H.; Fang, J. A fluorescence sensor for detection of geranyl pyrophosphate by the chemo-ensemble method. *J. Org. Chem.* **2009**, *74*, 895–898.

24. Wells, G.; Suggitt, M.; Coffils, M.; Baig, M. A. H.; Howard, P. W.; Loadman, P. M.; Hartley, J. A.; Jenkins, T. C.; Thurston, D. E. Fluorescent 7-diethylaminocoumarin pyrrolobenzodiazepine conjugates: Synthesis, DNA interaction, cytotoxicity and differential cellular localization. *Bioorg. Med. Chem. Lett.* **2008**, *18*, 2147–2151.

25. Xu, Q.; Wei, J.; Ma, W. Synthesis of 7-N,N-diethylamino coumarin derivatives under microwave irradiation. *Ranliao Yu Ranse* **2007**, *44*, 20–23.

26. Aujard, I.; Benbrahim, C.; Gouget, M.; Ruel, O.; Baudin, J.; Neveu, P.; Jullien, L. o-Nitrobenzyl photolabile protecting groups with red-shifted absorption: syntheses and uncaging cross-sections for one- and two-photon excitation. *Chem. Eur. J.* **2006**, *12*, 6865–6879.

27. Bardajee, G. R.; Winnik, M. A.; Lough, A. J. 7-Diethylamino-2-oxo-2H-chromene-3-carboxylic acid. *Acta Crystallogr., Sect. E* **2006**, *E62*, o3076–o3078.

28. Berthelot, T.; Talbot, J.; Lain, G.; Deleris, G.; Latxague, L. Synthesis of Nε-(7-diethyl-aminocoumarin-3-carboxyl)- and Nε-(7-methoxy-coumarin-3-carboxyl)-L-Fmoc lysine as tools for protease cleavage detection by fluorescence. *J. Pept. Sci.* **2005**, *11*, 153–160.

29. Berthelot, T.; Lain, G.; Latxague, L.; Deleris, G. Synthesis of novel fluorogenic L-fmoc lysine derivatives as potential tools for imaging cells. *J. Fluoresc.* **2004**, *14*, 671–675.

30. Ma, Y.; Luo, W.; Quinn, P. J.; Liu, Z.; Hider, R. C. Design, synthesis, physicochemical properties, and evaluation of novel iron chelators with fluorescent sensors. *J. Med. Chem.* **2004**, *47*, 6349–6362.

31. Jia, J.; Sheng, W.; Gao, J.; Chen, B. Study on the synthesis of solvent yellow X16. *Huagong Shikan* **2003**, *17*, 25–27.

32. Bogdal, D.; Stepien, I.; Sanetra, J.; Gondek, E. Microwave-assisted synthesis and copolymerization of photo- and electroluminescent methacrylates containing carbazole and coumarin pendant groups. *Polimery* **2003**, *48*, 111–115.

33. Song, A.; Wang, X.; Lam, K. S. A convenient synthesis of coumarin-3-carboxylic acids via Knoevenagel condensation of Meldrum's acid with ortho-hydroxyaryl aldehydes or ketones. *Tetrahedron Lett.* **2003**, *44*, 1755–1758.

34. Brunet, E.; Garcia-Losada, P.; Rodriguez-Ubis, J.; Rodriguez-Ubis, O.; Juanes, O. Synthesis of new fluorophores derived from monoazacrown ethers and coumarin nucleus. *Can. J. Chem.* **2002**, *80*, 169–174.

35. Muller, C.; Even, P.; Viriot, M.; Carre, M. Protection and labelling of thymidine by a fluorescent photolabile group. *Helv. Chim. Acta* **2001**, *84*, 3735–3741.

36. Bonsignore, L.; Cottiglia, F.; Lavagna, S. M.; Loy, G.; Secci, D. Synthesis of coumarin-3-O-acylisoureas by different carbodiimides. *Heterocycles* **1999**, *50*, 469–478.

37. Webb, M. R.; Brune, M. H.; Corrie, J. E. T. Phosphate assay. PCT Int. Appl. WO 9502825, 1995.

38. Corrie, J. E. T. Thiol-reactive fluorescent probes for protein labeling. *J. Chem. Soc., Perkin Trans. 1* **1994**, 2975–2982.

39. Ayyangar, N. R.; Srinivasan, K. V.; Daniel, T. Polycyclic compounds. Part VI. Structural features of C.I. Disperse Yellow 232. *Dyes Pigm.* **1990**, *13*, 301–310.

40. Sassiver, M. L.; Boothe, J. H. 6-[D-α-(Coumarin-3-carboxamido)arylacetamido]-penicillanic acids or salts. U.S. Patent 4317774, 1982.

41. Schwander, H. Oxadiazol-5-yl-coumarin derivatives. Ger. Offen. DE 2319230, 1973.

42. Lo, L.; Liao, Y.; Kuo, C.; Chen, C. A novel coumarin-type derivatizing reagent of alcohols: Application in the CD exciton chirality method for microscale structural determination. *Org. Lett.* **2000**, *2*, 683–685.

43. Schneider, S. E.; O'Neil, S. N.; Anslyn, E. V. Coupling rational design with libraries leads to the production of an ATP selective chemosensor. *J. Am. Chem. Soc.* **2000**, *122*, 542–543.

44. Chen, Y.; Zhu, C.; Cen, J.; Li, J.; He, W.; Jiao, Y.; Guo, Z. A reversible ratiometric sensor for intracellular Cu2+ imaging: metal coordination-altered FRET in a dual fluorophore hybrid. *Chem. Commun.* **2013**, *49*, 7632–7634.

45. Jiang, Z.; Lv, H.; Zhu, J.; Zhao, B. New fluorescent chemosensor based on quinoline and coumarine for Cu^{2+}. *Synth. Met.* **2012**, *162*, 2112–2116.

46. Ciesienski, K. L.; Hyman, L. M.; Derisavifard, S.; Franz, K. J. Toward the detection of cellular copper(II) by a light-activated fluorescence increase. *Inorg. Chem.* **2010**, *49*, 6808–6810.

47. Chatterjee, A.; Seth, D. Photophysical properties of 7-(diethylamino)coumarin-3-carboxylic acid in the nanocage of cyclodextrins and in different solvents and solvent mixtures. *Photochem. Photobiol.* **2013**, *89*, 280–293.

48. Tablet, C.; Matei, I.; Pincu, E.; Meltzer, V.; Hillebrand, M. Spectroscopic and thermodynamic studies of 7-diethylamino-coumarin-3-carboxylic acid in interaction with β- and 2-hydroxypropyl-β-cyclodextrins. *J. Mol. Liq.* **2012**, *168*, 47–53.

49. Even, P.; Chaubet, F.; Letourneur, D.; Viriot, M. L.; Carre, M. C. Coumarin-like fluorescent molecular rotors for bioactive polymers probing. *Biorheology* **2003**, *40*, 261–263.

50. Albers, A. E.; Okreglak, V. S.; Chang, C. J. A FRET-based approach to ratiometric fluorescence detection of hydrogen peroxide. *J. Am. Chem. Soc.* **2006**, *128*, 9640–9641.

51. Wei, C.; Wei, L.; Xi, Z.; Yi, L. A FRET-based fluorescent probe for imaging H2S in living cells. *Tetrahedron Lett.* **2013**, *54*, 6937–6939.

52. Long, L.; Zhang, D.; Li, X.; Zhang, J.; Zhang, C.; Zhou, L. A fluorescence ratiometric sensor for hypochlorite based on a novel dual-fluorophore

response approach. *Anal. Chim. Acta* **2013**, *775*, 100–105.

53. Yuan, L.; Lin, W.; Xie, Y.; Chen, B.; Song, J. Development of a ratiometric fluorescent sensor for ratiometric imaging of endogenously produced nitric oxide in macrophage cells. *Chem. Commun.* **2011**, *47*, 9372–9374.

54. Xie, Y.; Dix, A. V.; Tor, Y. FRET enabled real time detection of RNA-small molecule binding. *J. Am. Chem. Soc.* **2009**, *131*, 17605–17614.

55. Webb, M. R.; Reid, G. P.; Munasinghe, V. R. N.; Corrie, J. E. T. A series of related Nucleotide analogues that aids optimization of fluorescence signals in probing the mechanism of P-loop ATPases, such as actomyosin. *Biochemistry* **2004**, *43*, 14463–14471.

56. Webb, M. R.; Corrie, J. E. T. Fluorescent coumarin-labeled nucleotides to measure ADP release from actomyosin. *Biophys. J.* **2001**, *81*, 1562–1569.

57. Yi, L.; Sun, H.; Itzen, A.; Triola, G.; Waldmann, H.; Goody, R. S.; Wu, Y. One-pot dual-labeling of a protein by two chemoselective reactions. *Angew. Chem., Int. Ed.* **2011**, *50*, 8287–8290.

58. Kitamatsu, M.; Yamamoto, T.; Futami, M.; Sisido, M. Quantitative screening of EGF receptor-binding peptides by using a peptide library with multiple fluorescent amino acids as fluorescent tags. *Bioorg. Med. Chem. Lett.* **2010**, *20*, 5976–5978.

59. Tsukiji, S.; Miyagawa, M.; Takaoka, Y.; Tamura, T.; Hamachi, I. Ligand-directed tosyl chemistry for protein labeling *in vivo. Nat. Chem. Biol.* **2009**, *5*, 341–343.

60. Riechers, A.; Schmidt, F.; Stadlbauer, S.; Koenig, B. Detection of protein phosphorylation on SDS-PAGE using probes with a phosphate-sensitive emission response. *Bioconjugate Chem.* **2009**, *20*, 804–807.

61. Wu, X.; Li, H.; Kan, Y.; Yin, B. A regeneratable and highly selective fluorescent probe for sulfide detection in aqueous solution. *Dalton Trans.* **2013**, *42*, 16302–16310.

62. Katritzky, A. R.; Ibrahim, T. S.; Tala, S. R.; Abo-Dya, N. E.; Abdel-Samii, Z. K.; El-Feky, S. A. Synthesis of coumarin conjugates of biological thiols for fluorescent detection and estimation. *Synthesis* **2011**, 1494–1500.

63. Aguilar-Moncayo, M.; Garcia-Moreno, M. I.; Stuetz, A. E.; Garcia Fernandez, J. M.; Wrodnigg, T. M.; Mellet, C. O. Fluorescent-tagged sp2-iminosugars with potent β-glucosidase inhibitory activity. *Bioorg. Med. Chem.* **2010**, *18*, 7439–7445.

64. Wrodnigg, T. M.; Withers, S. G.; Stutz, A. E. Novel, lipophilic derivatives of 2,5-dideoxy-2,5-imino-D-mannitol (DMDP) are powerful β-glucosidase inhibitors. *Bioorg. Med. Chem. Lett.* **2001**, *11*, 1063–1064.

65. Irvine, M. W.; Costa, B. M.; Volianskis, A.; Fang, G.; Ceolin, L.; Collingridge, G. L.; Monaghan, D. T.; Jane, D. E. Coumarin-3-carboxylic acid derivatives as potentiators and inhibitors of recombinant and native N-methyl-D-aspartate receptors. *Neurochem. Int.* **2012**, *61*, 593–600.

66. Nomura, W.; Ohashi, N.; Okuda, Y.; Narumi, T.; Ikura, T.; Ito, N.; Tamamura, H. Fluorescence-quenching screening of protein kinase C ligands with an environmentally sensitive fluorophore. *Bioconjugate Chem.* **2011**, *22*, 923–930.

67. Agnes, R. S.; Jernigan, F.; Shell, J. R.; Sharma, V.; Lawrence, D. S. Suborganelle sensing of mitochondrial cAMP-dependent protein kinase activity. *J. Am. Chem. Soc.* **2010**, *132*, 6075–6080.

68. Bogdal, D.; Pielichowski, J.; Gorczyk, J. Microwave-assisted synthesis of high molecular weight epoxy resins. *Nonlinear Opt., Quantum Opt.* **2004**, *32*, 111–116.

69. Ishidai, K.; Okubo, K. Photopolymerizable resin composition, presensitized lithographic plate material, platemaking method, and coumarin derivative sensitizing dye. Jpn. Kokai Tokkyo Koho JP 2006106675, 2006.

70. Hoshino, Y. Visible oil-repellent copolymers, their manufacture, and solutions for oil-repellent treatment. Jpn. Kokai Tokkyo Koho JP 2009029902, 2009.

71. Hoshino, T.; Shirota, N.; Otozawa, N.; Asakawa, A. Oil-repellent copolymer, method for producing the same and oil-repellent processing liquid. PCT Int. Appl. WO 2008087915, 2008.

72. Felorzabihi, N.; Haley, J. C.; Bardajee, G. R.; Winnik, M. A. Systematic study of the fluorescence decays of amino-coumarin dyes in polymer matrices. *J. Polym. Sci., Part B: Polym. Phys.* **2007**, *45*, 2333–2343.

73. Genovese, D.; Bonacchi, S.; Juris, R.; Montalti, M.; Prodi, L.; Rampazzo, E.; Zaccheroni, N. Prevention of self-quenching in fluorescent silica nanoparticles by efficient energy transfer. *Angew. Chem., Int. Ed.* **2013**, *52*, 5965–5968.

74. Herz, E.; Marchincin, T.; Connelly, L.; Bonner, D.; Burns, A.; Switalski, S.; Wiesner, U. Relative quantum yield measurements of coumarin encapsulated in core-shell silica nanoparticles. *J. Fluoresc.* **2010**, *20*, 67–72.

7-DIETHYLAMINOCOUMARIN-3-CARBOXYLIC ACID HYDRAZIDE (DCCH)

CAS Registry Number   100343-98-4

Chemical Structure

CA Index Name   2H-1-Benzopyran-3-carboxylic acid, 7-(diethylamino)-2-oxo-, hydrazide;

Other Names   7-Diethylaminocoumarin-3-carboxylic acid hydrazide; DCCH; 7-Diethylaminocoumarin-3-carbohydrazide; CHH; 7-Diethylamino-2-oxo-2H-chromene-3-carbohydrazide

Merck Index Number   Not listed

Chemical/Dye Class   Coumarin

Molecular Formula $C_{14}H_{17}N_3O_3$

Molecular Weight   275.30

Physical Form   Orange needles;[1] Yellow crystals[2]

Solubility   Soluble in acetonitrile, chloroform, N,N-dimethylformamide, methanol

Melting Point   162–165 °C;[2] 160–165 °C[1]

pK$_a$ (Calcd.)   10.95 ± 0.20, most acidic, temperature: 25 °C; 3.01 ± 0.20, most basic, temperature: 25 °C

Absorption (λ_{max})   420 nm (MeOH)

Emission (λ_{max})   468 nm (MeOH)

Molar Extinction Coefficient   46,000 cm^{-1} M^{-1} (MeOH)

Synthesis   Synthetic methods[1,2]

Imaging/Labeling Applications   Cellulose fibre interfaces;[3,4] copper;[5,6] iron;[6] sugar chains;[7–11] skin;[12] hair;[12] tetrahydroprogesterones[13]

Biological/Medical Applications   Analyzing/detecting/quantifying sugar chains;[7–11] detecting copper,[5,6] iron,[6] hydrogen peroxide;[6] analyzing skin/hair;[12] detecting/labeling/quantifying biological carbonyl compounds (aldehydes and/or ketones)[13–15]

Industrial Applications   Cellulosic materials/paper;[3,4] detecting carboxylic acids[16]

Safety/Toxicity   No data available

REFERENCES

1. Takechi, H.; Oda, Y.; Nishizono, N.; Oda, K.; Machida, M. Screening search for organic fluorophores: syntheses and fluorescence properties of 3-azolyl-7-diethylaminocoumarin derivatives. *Chem. Pharm. Bull.* **2000**, *48*, 1702–1710.

2. Munasinghe, V. R. N.; Corrie, J. E. T.; Kelly, G.; Martin, S. R. Fluorescent ligands for the hemagglutinin of influenza A: Synthesis and ligand binding assays. *Bioconjugate Chem.* **2007**, *18*, 231–237.

3. Thomson, C. I.; Lowe, R. M.; Ragauskas, A. J. First characterization of the development of bleached kraft softwood pulp fiber interfaces during drying and rewetting using FRET microscopy. *Holzforschung* **2008**, *62*, 383–388.

4. Thomson, C. I.; Lowe, R. M.; Ragauskas, A. J. Imaging cellulose fibre interfaces with fluorescence microscopy and resonance energy transfer. *Carbohydr. Polym.* **2007**, *69*, 799–804.

5. Zhao, X.; Zhang, Y.; He, G.; Zhou, P. Highly sensitive fluorescent coumarin-based probes for selective detection of copper ion. *Faguang Xuebao* **2010**, *31*, 433–438.

6. Franz, K. J.; Hyman, L. M. Fluorescent prochelators for cellular iron and copper detection. U.S. Pat. Appl. Publ. US 20090253161, 2009.

7. Abe, K.; Shimaoka, H.; Abe, A.; Aihara, O. Method for determining sugar chains using substrate having sugar chains immobilized thereon. Jpn. Kokai Tokkyo Koho JP 2012211817, 2012.

8. Shimaoka, H.; Abe, M. Method for preparation of sugar chain sample, sugar chain sample, and method for analysis of sugar chain. PCT Int. Appl. WO 2009150834, 2009.

9. Shimaoka, H.; Abe, A. Glycoprotein sugar chain analysis method. Jpn. Kokai Tokkyo Koho JP 2009156587, 2009.

10. Shimaoka, H.; Nishimura, S.; Shinohara, Y.; Miura, Y.; Furukawa, J. Sugar chain-capturing substance and use thereof. PCT Int. Appl. WO 2008018170, 2008.

11. Hamaji, I. Protein for fluorescent sugar detection and its production method. Jpn. Kokai Tokkyo Koho JP 2000338044, 2000.

12. Hanes, R. E.; Windsor, J. B.; Borich, D. V.; Neeser, J. A.; Rasoulian, M. B.; Escamilla, P. R. Device, method and apparatus for analyzing skin and hair. PCT Int. Appl. WO 2010048431, 2010.

13. Touchstone, J. C.; Fang, X. Fluorometric labeling of tetrahydroprogesterones. *Steroids* **1991**, *56*, 601–605.

14. Anderson, J. M. Fluorescent hydrazides for the high-performance liquid chromatographic determination of biological carbonyls. *Anal. Biochem.* **1986**, *152*, 146–153.

15. Milic, I.; Hoffmann, R.; Fedorova, M. Simultaneous detection of low and high molecular weight carbonylated compounds derived from lipid peroxidation by electrospray ionization-tandem mass spectrometry. *Anal. Chem.* **2013**, *85*, 156–162.

16. Grayeski, M. L.; DeVasto, J. K. Coumarin derivatizing agents for carboxylic acid detection using peroxyoxalate chemiluminescence with liquid chromatography. *Anal. Chem.* **1987**, *59*, 1203–1206.

7-DIETHYLAMINOCOUMARIN-3-CARBOXYLIC ACID SUCCINIMIDYL ESTER (DEAC SE)

CAS Registry Number   139346-57-9

Chemical Structure

CA Index Name   2*H*-1-Benzopyran-3-carboxylic acid, 7-(diethylamino)-2-oxo-, 2,5-dioxo-1-pyrrolidinyl ester

Other Names   2,5-Pyrrolidinedione, 1-[[[7-(diethylamino)-2-oxo-2*H*-1-benzopyran-3-yl]carbonyl]oxy]-; 7-(Diethylamino)-coumarin-3-carboxylic acid *N*-succinimidyl ester; 7-Diethylaminocoumarin-3-carboxylic acid succinimidyl ester; 7-Ethylaminocoumarin-3-carboxylic acid succinimidyl ester; DCCS; DEAC SE

Merck Index Number   Not listed

Chemical/Dye Class   Coumarin

Molecular Formula $C_{18}H_{18}N_2O_6$

Molecular Weight   358.35

Physical Form   Yellow solid;[4–6] Bright yellow solid;[1] Organge solid;[3] Orange powder[2]

Solubility   Soluble in acetonitrile, *N,N*-dimethylformamide, dimethyl sulfoxide, methanol

Boiling Point (Calcd.)   559.3 $\pm$ 60.0 °C, pressure: 760 Torr

pK$_a$ (Calcd.)   2.60 $\pm$ 0.20, most basic, temperature: 25 °C

Absorption (λmax)   442 nm (Buffer pH 9)

Emission (λmax)   483 nm (Buffer pH 9)

Molar Extinction Coefficient   64,000 cm^{-1} M^{-1} (Buffer pH 9)

Synthesis   Synthetic methods[1–7]

Imaging/Labeling Applications   Amino acids;[2,6–9] analytes;[10] azobenzene photoswitching *in vivo*;[11] leucine;[2] lysines;[6–8] small molecules;[12,13] metal ions;[13] transition metal complexes;[14] nucleic acids;[15,16] peptide analog;[11] proteins;[4,17–20] silane coupling agent;[24] silica particles/beads;[24] uronium salts[21]

Biological/Medical Applications   Detecting amino acids,[2,6–9] hydrogen peroxide,[5] metal ions/small molecules;[12,13] detecting/identifying uronium salts;[21] identifying chromosome,[15] genome;[16] monitoring proteolysis;[6] quantifying analytes;[10] as a substrate for measuring extracellular matrix metalloprotease (MMP-1) activity,[6] proteases activity;[22] measuring lysosomal membrane potential *in situ*;[23] colloidal diagnostic devices;[24] useful for synthesizing peptides[24]

Industrial Applications   Silica particles/beads[24]

Safety/Toxicity   No data available

REFERENCES

1. Gustafson, T. P.; Metzel, G. A.; Kutateladze, A. G. Externally sensitized deprotection of PPG-masked carbonyls as a spatial proximity probe in photoamplified detection of binding events. *Photochem. Photobiol. Sci.* **2012**, *11*, 564–577.

2. Tsutsumi, H.; Λbe, S.; Mino, T.; Nomura, W.; Tamamura, H. Intense blue fluorescence in a leucine zipper assembly. *ChemBioChem* **2011**, *12*, 691–694.

3. Wada, A.; Tamaru, S.; Ikeda, M.; Hamachi, I. MCM-enzyme-supramolecular hydrogel hybrid as a fluorescence sensing material for polyanions of biological significance. *J. Am. Chem. Soc.* **2009**, *131*, 5321–5330.

4. Wakabayashi, H.; Miyagawa, M.; Koshi, Y.; Takaoka, Y.; Tsukiji, S.; Hamachi, I. Affinity-labeling-based introduction of a reactive handle for natural protein modification. *Chem.-Asian J.* **2008**, *3*, 1134–1139.

5. Albers, A. E.; Okreglak, V. S.; Chang, C. J. A FRET-based approach to ratiometric fluorescence detection of hydrogen peroxide. *J. Am. Chem. Soc.* **2006**, *128*, 9640–9641.

6. Berthelot, T.; Talbot, J.; Lain, G.; Deleris, G.; Latxague, L. Synthesis of Nε-(7-diethylaminocoumarin-3-carboxyl)- and Nε-(7-methoxycoumarin-3-carboxyl)-L-Fmoc lysine as tools for protease cleavage detection by fluorescence. *J. Pept. Sci.* **2005**, *11*, 153–160.

7. Berthelot, T.; Lain, G.; Latxague, L.; Deleris, G. Synthesis of novel fluorogenic L-fmoc lysine derivatives as potential tools for imaging cells. *J. Fluoresc.* **2004**, *14*, 671–675.

8. Tanaka, K.; Masuyama, T.; Hasegawa, K.; Tahara, T.; Mizuma, H.; Wada, Y.; Watanabe, Y.; Fukase, K. A submicrogram-scale protocol for biomolecule-based PET imaging by rapid 6π-azaelectrocyclization: visualization of sialic acid dependent circulatory residence of glycoproteins. *Angew. Chem., Int. Ed.* **2008**, *47*, 102–105.

9. Higashijima, T.; Fuchigami, T.; Imasaka, T.; Ishibashi, N. Determination of amino acids by capillary zone electrophoresis based on semiconductor laser fluorescence detection. *Anal. Chem.* **1992**, *64*, 711–714

10. Lakowicz, J. R.; Maliwal, B. P.; Thompson, R.; Ozinskas, A. Fluorescent energy transfer immunoassay. U.S. Patent 5631169, 1997.

11. Beharry, A. A.; Wong, L.; Tropepe, V.; Woolley, G. Andrew fluorescence imaging of azobenzene photoswitching *in vivo*. *Angew. Chem., Int. Ed.* **2011**, *50*, 1325–1327.

12. Devaraj, N. K.; Hilderbrand, S.; Upadhyay, R.; Mazitschek, R.; Weissleder, R. Bioorthogonal turn-on probes for imaging small molecules inside living cells. *Angew. Chem., Int. Ed.* **2010**, *49*, 2869–2872.

13. Thompson, R. B.; Patchan, M. W.; Ge, Z. Enzyme-based fluorescence biosensor for chemical analysis. U.S. Patent 5952236, 1999.

14. Lacombe, M.; Opdam, F. J. M.; Talman, E. G.; Veuskens, J. T. M. Labeled transition metal complexes. PCT Int. Appl. WO 2006062391, 2006.

15. Bittner, M. L.; Morrison, L. E.; Legator, M. S. Probe compositions for chromosome identification and methods. PCT Int. Appl. WO 9205185, 1992.

16. Morrison, L. E.; Legator, M. S.; Bittner, M. L. Probe composition for genome identification and methods. PCT Int. Appl. WO 9306245, 1993.

17. Ribbert, M.; Barth, S.; Kampmeier, F. Self coupling recombinant antibody fusion proteins. PCT Int. Appl. WO 2009013359, 2009.

18. Gautier, A.; Johnsson, K.; Kindermann, M.; Juillerat, A.; Beaufils, F. Labelling of fusion proteins with synthetic probes. PCT Int. Appl. WO 2008012296, 2008.

19. Tanaka, K.; Fujii, Y.; Fukase, K. Site-selective and nondestructive protein labeling through azaelectrocyclization-induced cascade reactions. *ChemBioChem* **2008**, *9*, 2392–2397.

20. Bhunia, A. K.; Miller, S. C. Labeling tetracysteine-tagged proteins with a SplAsH of color: a modular approach to bis-arsenical fluorophores. *ChemBioChem* **2007**, *8*, 1642–1645.

21. Almog, J. Detection of uronium salts. U.S. Pat. Appl. Publ. US 2006084176, 2006.

22. Alouini, M. A.; Berthelot, T.; Moustoifa, E. F.; Albenque-Rubio, S.; Deleris, G. Biosensor for detecting the presence of proteases and optionally quantifying the enzymatic activity thereof. PCT Int. Appl. WO 2012093162, 2012.

23. Koivusalo, M.; Steinberg, B. E.; Mason, D.; Grinstein, S. *In situ* measurement of the electrical potential across the lysosomal membrane using FRET. *Traffic* **2011**, *12*, 972–982.

24. Lawrie, G.; Grondahl, L.; Battersby, B.; Keen, I.; Lorentzen, M.; Surawski, P.; Trau, M. Tailoring surface properties to build colloidal diagnostic devices: Controlling interparticle associations. *Langmuir* **2006**, *22*, 497–505.

4-[4-(DIETHYLAMINO)STYRYL]-*N*-METHYLPYRIDINIUM IODIDE (4-Di-2-ASP)

CAS Registry Number   105802-46-8

Chemical Structure

CA Index Name   Pyridinium, 4-[2-[4-(diethylamino) phenyl]ethenyl]-1-methyl-, iodide (1 : 1)

Other Names   Pyridinium, 4-[2-[4-(diethylamino) phenyl]ethenyl]-1-methyl-, iodide; 4-Di-2-ASP; 4-[4-(Diethylamino)styryl]-1-methylpyridinium iodide; 4-[4-Diethylamino)styryl]-*N*-methylpyridinium iodide; D289

Merck Index Number   Not listed

Chemical/Dye Class   Styryl

Molecular Formula $C_{18}H_{23}IN_2$

Molecular Weight   394.30

Physical Form   Red solid

Solubility   Soluble in water, chloroform, *N,N*-dimethylformamide, dimethyl sulfoxide, methanol

Melting Point   231.3–232 °C;[2] 226 °C;[1] 221–222 °C[4]

Absorption (λ_{max}) 488 nm (MeOH); 484.7 nm (DMSO); 471 nm (H_2O)

Emission (λ_{max}) 607 nm (MeOH)

Molar Extinction Coefficient   48,000 cm^{-1} M^{-1} (MeOH)

Synthesis   Synthetic methods[1–5]

Imaging/Labeling Applications   Amyloid plaques;[6] axons;[17,27,28,38] blood cells (erythrocytes, reticulocytes, blood platelets);[7] blood vessels;[13] ganglia;[16,20,22] Merkel cells;[28] mitochondria;[27,38] myelinated fibers;[10] nerve fibers;[16,22,28] nerve plexuses;[20,22] nerve terminals;[10,14,23,24,26,27,30,38] neuroepithelial bodies (NEBs);[8,9] neuromuscular junctions (NMJs);[14,19,21,24–27,29] neurons;[11–13,15,18,22] polylactic acid (PLA) particles[31]

Biological/Medical Applications   Assessing nerve ultrastructure;[15] characterizing Merkel cells and mechanosensory axons;[28] diagnosing of Hirschsprung's disease;[20] identifying neuroepithelial bodies (NEBs);[8,9] investigating stability and release properties of biodegradable polylactic acid (PLA) particles;[31] measuring blood cells (erythrocytes, reticulocytes, blood platelets);[7] visualizing nerve terminals and myelinated fibers[10]

Industrial Applications   Bentonite clay;[32] characterizing supramolecular materials;[33] visualizing mesopores and defects in porous molecular sieves;[33] nonlinear optical devices/materials;[34,35] photographic materials/systems;[36] photoresists[37]

Safety/Toxicity   Neurotoxicity[27]

REFERENCES

1. Okada, S.; Nogi, K.; Anwar; Tsuji, K.; Duan, X.; Oikawa, H.; Matsuda, H.; Nakanishi, H. Ethyl-substituted stilbazolium derivatives for second-order nonlinear optics. *Jpn. J. Appl. Phys., Part 1* **2003**, *42*, 668–671.

2. Shah, S. S.; Ahmad, R.; Shah, S. W. H.; Asif, K. M.; Naeem, K. Synthesis of cationic hemicyanine dyes and their interactions with ionic surfactants. *Colloids Surf., A* **1998**, *137*, 301–305.

3. Finkelstein, J.; Lee, J. Piperidine derivatives. U.S. Patent 2686784, 1954.

4. Phillips, A. P. Condensation of aromatic aldehydes with 4-picoline methiodide. *J. Org. Chem.* **1949**, *14*, 302–305.

5. Doja, M. Q.; Prasad, K. B. Cyanine dyes of the pyridine series. V. *J. Indian Chem. Soc.* **1947**, *24*, 301–306.

6. Kudo, Y.; Suzuki, M.; Suemoto, T.; Shimazu, H. Image diagnosis probe based on substituted azobenzene or analogue thereof for disease attributable to amyloid accumulation and composition for image diagnosis containing the same. PCT Int. Appl. WO 2001070667, 2001.

7. Sakata, T.; Takami, T. Reagent for the measurement of hemocytes. Jpn. Kokai Tokkyo Koho JP 61079163, 1986.

8. De Proost, I.; Pintelon, I.; Brouns, I.; Kroese, A. B. A.; Riccardi, D.; Kemp, P. J.; Timmermans, J.; Adriaensen, D. Functional live cell imaging of the pulmonary neuroepithelial body microenvironment. *Am. J. Respir. Cell Mol. Biol.* **2008**, *39*, 180–189.

9. Pintelon, I.; De Proost, I.; Brouns, I.; Van Herck, H.; Van Genechten, J.; Van Meir, F.; Timmermans, J.; Adriaensen, D. Selective visualisation of neuroepithelial bodies in vibratome slices of living lung by 4-Di-2-ASP in various animal species. *Cell Tissue Res.* **2005**, *321*, 21–33.

10. De Proost, I.; Pintelon, I.; Brouns, I.; Timmermans, J.; Adriaensen, D. Selective visualization of sensory receptors in the smooth muscle layer of ex-vivo airway whole-mounts by styryl pyridinium dyes. *Cell Tissue Res.* **2007**, *329*, 421–431.

11. Campanucci, V. A.; Nurse, C. A. Biophysical characterization of whole-cell currents in O2-sensitive neurons from the rat glossopharyngeal nerve. *Neuroscience* **2005**, *132*, 437–451

12. Schrodl, F.; De Laet, A.; Tassignon, M.; Van Bogaert, P.; Brehmer, A.; Neuhuber, W. L; Timmermans, J. Intrinsic choroidal neurons in the human eye: projections, targets, and basic electrophysiological data. *Invest. Ophthalmol. Vis. Sci.* **2003**, *44*, 3705–3712

13. Papworth, G. D.; Delaney, P. M.; Bussau, L. J.; Vo, L. T.; King, R. G. *In vivo* fibre optic confocal imaging of microvasculature and nerves in the rat vas deferens and colon. *J. Anat.* **1998**, *192*, 489–495.

14. Marques, M. J.; Santo N. H. Imaging neuromuscular junctions by confocal fluorescence microscopy: individual endplates seen in whole muscles with vital intracellular staining of the nerve terminals. *J. Anat.* **1998**, *192*, 425–430.

15. Cushway, T. R.; Lanzetta, M.; Cox, G.; Trickett, R.; Owen, E. R. Assessment of nerve ultrastructure by fibre-optic confocal microscopy. *Microsurgery* **1996**, *17*, 444–448.

16. Bergua, A.; Neuhuber, W. L.; Naumann, G. O. H. Visualization of human choroidal ganglion cells with the supravital fluorescent dye 4-(4-diethylaminostyryl)-N-methylpyridinium iodide. *Ophthalmic Res.* **1994**, *26*, 290–295.

17. Kurdyak, P.; Atwood, H. L.; Stewart, B. A.; Wu, C. F. Differential physiology and morphology of motor axons to ventral longitudinal muscles in larval *Drosophila. J. Comp. Neurol.* **1994**, *350*, 463–472.

18. Hilbig, H.; Schierwagen, A. Interlayer neurones in the rat superior colliculus: a tracer study using DiI/Di-ASP. *Neuroreport* **1994**, *5*, 477–480.

19. Chen, L.; Ko, C. P. Extension of synaptic extracellular matrix during nerve terminal sprouting in living frog neuromuscular junctions. *J. Neurosci.* **1994**, *14*, 796–808.

20. Hanani, M.; Udassin, R.; Ariel, I.; Freund, H. R. A simple and rapid method for staining the enteric ganglia: application for Hirschsprung's disease. *J. Pediatr. Surg.* **1993**, *28*, 939–941.

21. Langenfeld-Oster, B.; Dorlochter, M.; Wernig, A. Regular and photodamage-enhanced remodelling in vitally stained frog and mouse neuromuscular junctions. *J. Neurocytol.* **1993**, *22*, 517–530.

22. Hanani, M. Visualization of enteric and gallbladder ganglia with a vital fluorescent dye. *J. Auton. Nerv. Syst.* **1992**, *38*, 77–84.

23. Melamed, N.; Rahamimoff, R. Confocal microscopy of the lizard motor nerve terminals. *J. Basic Clin. Physiol. Pharmacol.* **1991**, *2*, 63–85.

24. Chen, L. L.; Folsom, D. B.; Ko, C. P. The remodeling of synaptic extracellular matrix and its dynamic relationship with nerve terminals at living frog neuromuscular junctions. *J. Neurosci.* **1991**, *11*, 2920–2930.

25. Wigston, D. J. Repeated *in vivo* visualization of neuromuscular junctions in adult mouse lateral gastrocnemius. *J. Neurosci.* **1990**, *10*, 1753–1761.

26. Herrera, A. A.; Banner, L. R.; Nagaya, N. Repeated, *in vivo* observation of frog neuromuscular junctions: remodelling involves concurrent growth and retraction. *J. Neurocytol.* **1990**, *19*, 85–99.

27. Herrera, A. A.; Banner, L. R. The use and effects of vital fluorescent dyes: Observation of motor nerve terminals and satellite cells in living frog muscles. *J. Neurocytol.* **1990**, *19*, 67–83.

28. Nurse, C. A.; Farraway, L. Characterization of Merkel cells and mechanosensory axons of the rat by styryl pyridinium dyes. *Cell Tissue Res.* **1989**, *255*, 125–128.

29. Lichtman, J. W.; Magrassi, L.; Purves, D. Visualization of neuromuscular junctions over periods of several months in living mice. *J. Neurosci.* **1987**, *7*, 1215–1222.

30. Magrassi, L.; Purves, D.; Lichtman, J. W. Fluorescent probes that stain living nerve terminals. *J. Neurosci.* **1987**, *7*, 1207–1214.

31. Rancan, F.; Todorova, A.; Hadam, S.; Papakostas, D.; Luciani, E.; Graf, C.; Gernert, U.; Ruehl, E.; Verrier, B.; Sterry, W. Stability of polylactic acid particles and release of fluorochromes upon topical application on human skin explants. *Eur. J. Pharm. Biopharm.* **2012**, *80*, 76–84.

32. Coradin, T.; Nakatani, K.; Ledoux, I.; Zyss, J.; Clement, R. Second harmonic generation of dye aggregates in bentonite clay. *J. Mater. Chem.* **1997**, *7*, 853–854.

33. Seebacher, C.; Rau, J.; Deeg, F.; Brauchle, C.; Altmaier, S.; Jager, R.; Behrens, P. Visualization of mesostructures and organic guest inclusion in molecular sieves with confocal microscopy. *Adv. Mater.* **2001**, *13*, 1374–1377.

34. Kubodera, K.; Kanbara, H.; Kurihara, T.; Kaino, T. Nonlinear optical device. Jpn. Kokai Tokkyo Koho JP 02195330, 1990.

35. Kaino, T.; Amano, M.; Kurihara, T.; Shudo, Y.; Kimura, T.; Kitagawa, T. Nonlinear optical device. Jpn. Kokai Tokkyo Koho JP 02199432, 1990.

36. Farid, S. Y.; Haley, N. F.; Moody, R. E.; Specht, D. P. Dye-sensitized photographic imaging system. U.S. Patent 4743531, 1988.

37. Farid, S. Y.; Haley, N. F.; Moody, R. E.; Specht, D. P. Negative working photoresists responsive to shorter visible wavelengths and novel coated articles. U.S. Patent 4743529, 1988.

3,3′-DIETHYLOXACARBOCYANINE IODIDE (DiOC₂(3))

CAS Registry Number   905-96-4

Chemical Structure

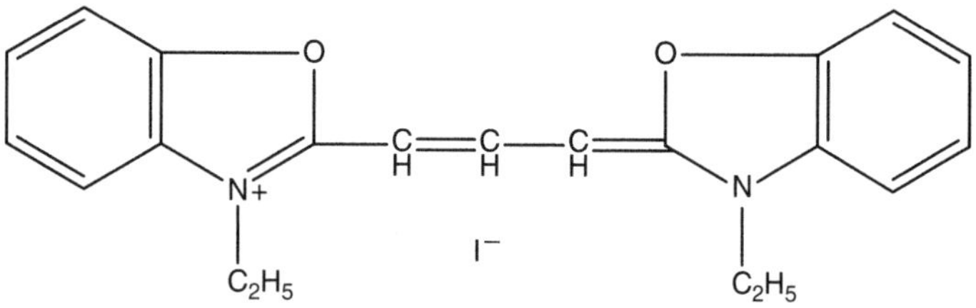

CA Index Name   Benzoxazolium, 3-ethyl-2-[3-(3-ethyl-2(3*H*)-benzoxazolylidene)-1-propen-1-yl]-, iodide (1:1)

Other Names   3-Ethyl-2-[3-(3-ethyl-2-benzoxazolinylidene)propenyl]-benzoxazolium iodide; Benzoxazolium, 3-ethyl-2-[3-(3-ethyl-2(3*H*)-benzoxazolylidene)-1-propenyl]-, iodide; Benzoxazolium, 3-ethyl-2-[3-(3-ethyl-2-benzoxazolinylidene)propenyl]-, iodide; 3,3′-Diethyl-oxadicarbocyanine iodide; 3,3′-Diethyl-2,2′-oxacarbocyanine iodide; 3,3′-Diethyloxacarbocyanine iodide; 3-Ethyl-2-[3-ethyl-2(3*H*)-benzoxazolylidene)-1-propenyl] benzoxazolium iodide); 3-Ethyl-2-[3-(3-ethyl-3H-benzoxazol-2-ylidene)propenyl]benzoxazolium iodide; DOC; DOC (dye); DOCCI; DOCI; DiOC 2; DiOC 2(3); DODCI; G 1745; NK 85

Merck Index Number   Not listed

Chemical/Dye Class   Cyanine

Molecular Formula $C_{21}H_{21}IN_2O_2$

Molecular Weight   460.31

Physical Form   Dark red crystals or powder; Red-violet crystalline powder;[2] Orange solid[9]

Solubility   Soluble in acetone, acetonitrile, chloroform, *N,N*-dimethylformamide, dimethyl sulfoxide, ethanol, methanol

Melting Point   287 °C;[8] 279 °C;[2] 272–273 °C;[4] 210 °C (decompose)[9]

Absorption (λ_{max})   482 nm (MeOH); 478 nm (H₂O)

Emission (λ_{max})   497 nm (MeOH); 496 nm (H₂O)

Molar Extinction Coefficient   165,000 cm⁻¹ M⁻¹ (MeOH)

Quantum Yield   0.04 (MeOH)

Synthesis   Synthetic methods[1–9]

Imaging/Labeling Applications   Bacteria;[10,11] cells;[10,12,13] hairs;[14] hepatocytes;[15] malaria infected cells;[16] mitochondria;[17–19,25] nucleic acids;[20–22] P-glycoprotein;[23,24] sperms[25]

Biological/Medical Applications   Analyzing sperm motility;[25] assessing/determining P-glycoprotein expression and/or function;[23,24] detecting malaria infected cells,[16] prostate cancer;[17] inhibiting mitosis and cell growth;[12,13] measuring/monitoring membrane potential;[4,10,11,25,26] treating cancer;[24] cytotoxicity assay;[26] as temperature sensor[27,28]

Industrial Applications   Displays;[29] dye laser;[30–34] electroluminescent devices;[35,36] electrophotographic material;[37] imaging materials;[38] light emitting diode lens;[39] micelles;[40] microemulsions;[40] mortar composition;[41] nonlinear optical materials;[42] photographic materials;[6,9,43–45] polymer nanocomposites (PNCs);[46] recording materials;[47–50] solar panels;[51] thin films[52]

Safety/Toxicity   As substrate to evaluate efflux pump activity in leukemias;[53] toxicity[54]

REFERENCES

1. Sabnis, R. W. *Handbook of Biological Dyes and Stains*; John Wiley & Sons Inc.: Hoboken, **2010**; pp 158–159.

2. Mustroph, H.; Reiner, K.; Mistol, J.; Ernst, S.; Keil, D.; Hennig, L. Relationship between the molecular structure of cyanine dyes and the vibrational fine structure of their electronic absorption spectra. *ChemPhysChem* **2009**, *10*, 835–840.

3. Glazer, A. N.; Mathies, R. A.; Hung, S.; Jue, J. Cyanine dyes with high-absorbance cross section as donor chromophores in energy transfer labels. PCT Int. Appl. WO 9814612, 1998.

4. Sims, P. J.; Waggoner, A. S.; Wang, C.; Hoffman, J. F. Mechanism by which cyanine dyes measure membrane potential in red blood cells and

phosphatidylcholine vesicles. *Biochemistry* **1974**, *13*, 3315–3330.

5. Ciernik, J. V. Formylation of nitrogen-containing heterocycles and their quaternary salts. *Collect. Czech. Chem. Commun.* **1972**, *37*, 2273–2281.

6. Jones, J. E.; Kalenda, N. W. Silver halide emulsions containing supersensitizing dye combinations. FR 1480762, 1967.

7. Hishiki, Y. J. Trinuclear cyanine dyes. V. Decomposition of oxacyanine. *J. Sci. Res. Inst., Tokyo* **1954**, *48*, 130–142.

8. Dent, S. G., Jr.,; Brooker, L. G. S. Process for preparing polymethine dyes. U.S. Patent 2537880, 1951.

9. Hamer, F. M.; Rathbone, R. J. Improvements in and relating to photographic layers, and the manufacture of dyes and intermediates for use therein. GB 541330, 1941.

10. David, F.; Hebeisen, M.; Schade, G.; Franco-Lara, E.; Di Berardino, M. Viability and membrane potential analysis of *Bacillus megaterium* cells by impedance flow cytometry. *Biotechnol. Bioeng.* **2012**, *109*, 483–492.

11. Gentry, D. R.; Wilding, I.; Johnson, J. M.; Chen, D.; Remlinger, K.; Richards, C.; Neill, S.; Zalacain, M.; Rittenhouse, S. F.; Gwynn, M. N. A rapid microtiter plate assay for measuring the effect of compounds on Staphylococcus aureus membrane potential. *J. Microbiol. Methods* **2010**, *83*, 254–256.

12. Zigman, S.; Gilman, P. B., Jr., Process for inhibiting the growth of sea urchin eggs. U.S. Patent 4226868, 1980.

13. Zigman, S.; Gilman, P., Jr., Inhibition of cell division and growth by a redox series of cyanine dyes. *Science* **1980**, *208*, 188–191.

14. Ohashi, Y.; Miyabe, H.; Matsunaga, K. Hair dye composition. Eur. Pat. Appl. EP 1166753, 2002.

15. Li, M.; Yuan, H.; Li, N.; Song, G.; Zheng, Y.; Baratta, M.; Hua, F.; Thurston, A.; Wang, J.; Lai, Y. Identification of interspecies difference in efflux transporters of hepatocytes from dog, rat, monkey and human. *Eur. J. Pharm. Sci.* **2008**, *35*, 114–126.

16. Sakata, T.; Matsumoto, H. Reagent for detecting malaria infected cells and a detecting method for malaria infected cells using the same. U.S. Patent 5470751, 1995.

17. Dickman, D. Methods of detecting prostate cancer. PCT Int. Appl. WO 2006054296, 2006.

18. Bunting, J. R. Influx and efflux kinetics of cationic dye binding to respiring mitochondria. *Biophys. Chem.* **1992**, *42*, 163–175.

19. Bunting, J. R.; Phan, T. V.; Kamali, E.; Dowben, R. M. Fluorescent cationic probes of mitochondria. Metrics and mechanism of interaction. *Biophys. J.* **1989**, *56*, 979–993.

20. Kawabe, Y.; Kato, S.; Honda, M.; Yoshida, J. Effects of DNA on the optical properties of cyanine dyes. *Proc. SPIE* **2010**, *7765*, 776503/1–776503/10.

21. Kerwin, S. M.; Sun, D.; Kern, J. T.; Rangan, A.; Thomas, P. W. G-Quadruplex DNA binding by a series of carbocyanine dyes. *Bioorg. Med. Chem. Lett.* **2001**, *11*, 2411–2414.

22. Mikheikin, A. L.; Zhuze, A. L.; Zasedatelev, A. S. Binding of symmetrical cyanine dyes into the DNA minor groove. *J. Biomol. Struct. Dyn.* **2000**, *18*, 59–72.

23. Marcelletti, J. F.; Multani, P. S.; Lancet, J. E.; Baer, M. R.; Sikic, B. I. Leukemic blast and natural killer cell P-glycoprotein function and inhibition in a clinical trial of zosuquidar infusion in acute myeloid leukemia. *Leuk. Res.* **2009**, *33*, 769–774.

24. Sikic, B.; Hoth, D.; Socks, D.; Glenn, S.; Marcelletti, J.; Walsh, M. J.; Multani, P. S. Treatment of cancer patients exhibiting activation of the P-glycoprotein efflux pump mechanism. PCT Int. Appl. WO 2007008499, 2007.

25. Nascimento, J. M.; Shi, L. Z.; Chandsawangbhuwana, C.; Tam, J.; Durrant, B.; Botvinick, E. L.; Berns, M. W. Use of laser tweezers to analyze sperm motility and mitochondrial membrane potential. *J. Biomed. Opt.* **2008**, *13*, 014002/1–014002/7.

26. Murakami, T. Cytotoxicity test method by measuring membrane electric potential. Jpn. Kokai Tokkyo Koho JP 2000300290, 2000.

27. Bousseksou, A.; Salmon, L.; Molnar, G.; Cobo, S. Materials with thermochromic spin transition doped with one or more fluorescent agents for use as temperature sensor. Fr. Demande FR 2952371, 2011.

28. Bousseksou, A.; Salmon, L.; Molnar, G.; Cobo, S. Heat-sensitive spin-transition materials doped with one or more fluorescent agents for use as temperature sensor. PCT Int. Appl. WO 2011058277, 2011.

29. Hajto, J.; Hindle, C.; Graham, A. Displays. PCT Int. Appl. WO 2000007039, 2000.

30. Kessler, W. J.; Davis, S. J.; Ferguson, D. R.; Pugh, E. R. Solid-state dye laser host. U.S. Patent 5610932, 1997.

31. Sharp, T. E.; Dane, C. B.; Barber, D.; Tittel, F. K.; Wisoff, P. J.; Szabo, G. Tunable, high-power,

subpicosecond blue-green dye laser system with a two-stage dye amplifier. *IEEE J. Quantum Electron.* **1991**, *27*, 1121–1127.

32. French, P. M. W.; Taylor, J. R. The passively mode locked Coumarin 6H ring dye laser. *Opt. Commun.* **1988**, *67*, 51–52.

33. Wyatt, R. Broadly tunable blue-green picosecond pulses from a flashlamp-pumped dye laser. *Opt. Commun.* **1981**, *38*, 64–66.

34. Ammann, E. O.; Decker, C. D.; Falk, J. High peak-power 532-nm pumped dye laser. *IEEE J. Quantum Electron.* **1974**, *10*, 463–465.

35. Pan, J.; Stoessel, P. Organic electroluminescent device. PCT Int. Appl. WO 2012084115, 2012.

36. Pan, J.; Heun, S. Organic electroluminescent device comprising an integrated layer for colour conversion. PCT Int. Appl. WO 2011091946, 2011.

37. Kamota, H. Electrophotographic material. Jpn. Kokai Tokkyo Koho JP 54121741, 1979.

38. Ishida, E.; Takashima, Y.; Shimoda, H.; Oda, F. Imaging materials based on semiconductors and silver salts. Jpn. Kokai Tokkyo Koho JP 50067643, 1975.

39. Butterworth, M. M.; Helbing, R. P. Fluorescent dye added to epoxy of light emitting diode lens. U.S. Patent 5847507, 1998.

40. Basu, S.; Mondal, S.; Mandal, D. 3,3'-Diethyloxacarbocyanine iodide: a new microviscosity probe for micelles and microemulsions. *Colloids Surf., A* **2010**, *363*, 41–48.

41. Shi, X.; Guo, H.; Wang, Y. Faming Method for production of photosensitive UV-resistant protective agent for cement-based waterproof mortar. Zhuanli Shenqing Gongkai Shuomingshu CN 101648787, 2010.

42. Morita, K.; Suehiro, T.; Yokoh, Y.; Ashitaka, H. The development of organic third-order nonlinear optical materials. *J. Photopolym. Sci. Technol.* **1993**, *6*, 229–238.

43. Zaleski, A.; Nowak, P.; Weglinska-Flis, J.; Mora, C.; Latacz, L.; Rajkowski, B.; Sicinska, P. Silver halide photographic materials for holography. *Chemik* **2000**, *53*, 321–324.

44. Hermans, T.; Vrielynck, M. Static pressure effects on photographic materials. Part III. The influence of dyes. *J. Photogr. Sci.* **1984**, *32*, 153–157.

45. Kitova, S.; Malinovski, I. Dye sensitized photoprocess in silver bromide evaporated layers. *Photogr. Sci. Eng.* **1980**, *24*, 50–54.

46. Jee, A.; Kwon, H.; Lee, M. Molecular rotor dynamics influenced by the elastic modulus of polyethylene nanocomposites. *J. Chem. Phys.* **2009**, *131*, 171104/1–171104/4.

47. Nomura, T.; Takizawa, H. Recording materials and method for hologram. Jpn. Kokai Tokkyo Koho JP 2007086179, 2007.

48. Takano, H.; Yamashita, N. Manufacture of optical recording material. Jpn. Kokai Tokkyo Koho JP 2007199125, 2007.

49. Takizawa, H.; Inoue, N. Hologram recording material and method, and optical recording medium. Jpn. Kokai Tokkyo Koho JP 2006078877, 2006.

50. Murofushi, K.; Hosoda, K. Photodecolorizing recording material. Jpn. Kokai Tokkyo Koho JP 06236000, 1994.

51. Krokoszinski, H.; Mayer, O.; Stromberger, J.; Korman, C. S. Solar panels, methods of manufacture thereof and articles comprising the same. U.S. Pat. Appl. Publ. US 20070137696, 2007.

52. O'Regan, B.; Schwartz, D. T. Efficient photo-hole injection from adsorbed cyanine dyes into electrodeposited copper(I) thiocyanate thin films. *Chem. Mater.* **1995**, *7*, 1349–1354.

53. Vasconcelos, F. C.; Cavalcanti, G. B.; Silva, K. L.; de Meis, E.; Kwee, J. K.; Rumjanek, V. M.; Maia, R. C. Contrasting features of MDR phenotype in leukemias by using two fluorochromes: implications for clinical practice. *Leuk. Res.* **2007**, *31*, 445–454.

54. Kues, H. A.; Lutty, G. A. Dyes can be deadly. *Laser Focus* **1975**, *11*, 59–61.

3,3′-DIHEPTYLOXACARBOCYANINE IODIDE (DiOC$_7$(3))

CAS Registry Number   53213-83-5

Chemical Structure

CA Index Name   Benzoxazolium, 3-heptyl-2-[3-(3-heptyl-2(3H)-benzoxazolylidene)-1-propen-1-yl]-, iodide (1:1)

Other Names   Benzoxazolium, 3-heptyl-2-[3-(3-heptyl-2(3H)-benzoxazolylidene)-1-propenyl]-, iodide; 3,3′-Diheptyloxacarbocyanine iodide; D 378; DiOC$_7$(3)

Merck Index Number   Not listed

Chemical/Dye Class   Cyanine

Molecular Formula   C$_{31}$H$_{41}$IN$_2$O$_2$

Molecular Weight   600.58

Physical Form   Red powder or crystals

Solubility   Soluble in chloroform, N,N-dimethylformamide, dimethyl sulfoxide, methanol

Melting Point   194–197 °C[2]

Absorption (λ_{max})   482 nm (MeOH)

Emission (λ_{max})   504 nm (MeOH)

Molar Extinction Coefficient   148,000 cm^{-1} M^{-1} (MeOH)

Synthesis   Synthetic methods[1,2]

Imaging/Labeling Applications   Mitochondria;[1,3,4] bacteria;[5] cells;[6,7] fungi;[8] tumors[9–11]

Biological/Medical Applications   Analyzing cells;[6,7] distinguishing between viable and non-viable bacteria;[5] identifying/quantifying transient perfusion in tumors;[10] marking murine tumor vasculature;[11] measuring/monitoring membrane potential;[1,2,12] mitochondria stain;[1,3,4] cytotoxicity assays[12,13]

Industrial Applications   Not reported

Safety/Toxicity   No data available

REFERENCES

1. Sabnis, R. W. *Handbook of Biological Dyes and Stains*; John Wiley & Sons Inc.: Hoboken, **2010**; p 166.

2. Sims, P. J.; Waggoner, A. S.; Wang, C.; Hoffman, J. F. Mechanism by which cyanine dyes measure membrane potential in red blood cells and phosphatidylcholine vesicles. *Biochemistry* **1974**, *13*, 3315–3330.

3. Hattori, F.; Fukuda, K. Method of selecting myocardial cells by using intracellular mitochondria as indication. PCT Int. Appl. WO 2006022377, 2006.

4. Liu, Z.; Bushnell, W. R.; Brambl, R. Potentiometric cyanine dyes are sensitive probes for mitochondria in intact plant cells. *Plant Physiol.* **1987**, *84*, 1385.

5. Mason, D. J.; Lopez-Amoros, R.; Allman, R.; Stark, J. M.; Lloyd, D. The ability of membrane potential dyes and calcafluor white to distinguish between viable and non-viable bacteria. *J. Appl. Bacteriol.* **1995**, *78*, 309–315.

6. Garini, Y.; Mcnamara, G.; Soenksen, D. G.; Cabib, D.; Buckwald, R. A. *In situ* method of analyzing cells. PCT Int. Appl. WO 2000031534, 2000.

7. Olive, P. L.; Durand, R. E. Characterization of a carbocyanine derivative as a fluorescent penetration probe. *Cytometry* **1987**, *8*, 571–575.

8. Thrane, C.; Olsson, S.; Harder Nielsen, T.; Sorensen, J. Vital fluorescent stains for detection of stress in *Pythium ultimum* and *Rhizoctonia solani* challenged with viscosinamide from *Pseudomonas fluorescens* DR54. *FEMS Microbiol. Ecol.* **1999**, *30*, 11–23.

9. Fenton, B. M. Influence of hydralazine administration on oxygenation in spontaneous and transploanted

tumor models. *Int. J. Radiat. Oncol., Biol., Phys.* **2001**, *49*, 799–808.

10. Trotter, M. J.; Chaplin, D. J.; Durand, R. E.; Olive, P. L. The use of fluorescent probes to identify regions of transient perfusion in murine tumors. *Int. J. Radiat. Oncol., Biol., Phys.* **1989**, *16*, 931–934.

11. Trotter, M. J.; Chaplin, D. J.; Olive, P. L. Use of a carbocyanine dye as a marker of functional vasculature in murine tumors. *Br. J. Cancer* **1989**, *59*, 706–709.

12. Murakami, T. Cytotoxicity test method by measuring membrane electric potential. Jpn. Kokai Tokkyo Koho JP 2000300290, 2000.

13. Wadkins, R. M.; Houghton, P. J. Kinetics of transport of dialkyloxacarbocyanines in multidrug-resistant cell lines overexpressing p-glycoprotein: Interrelationship of dye alkyl chain length, cellular flux, and drug resistance. *Biochemistry* **1995**, *34*, 3858–3872.

3,3′-DIHEXYLOXACARBOCYANINE IODIDE (DiOC$_6$(3))

CAS Registry Number 53213-82-4

Chemical Structure

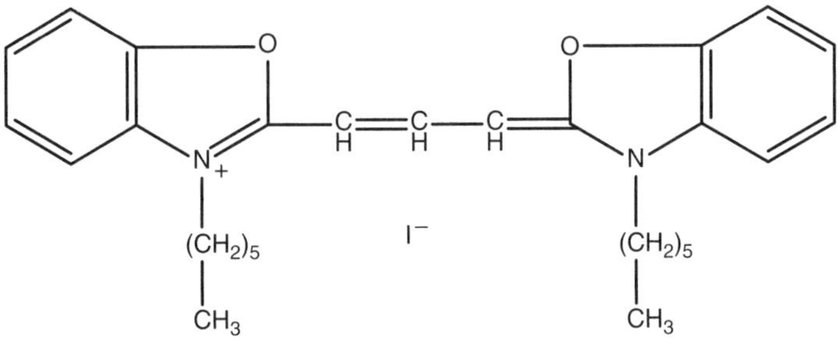

CA Index Name Benzoxazolium, 3-hexyl-2-[3-(3-hexyl-2(3H)-benzoxazolylidene)-1-propen-1-yl]-, iodide (1:1)

Other Names Benzoxazolium, 3-hexyl-2-[3-(3-hexyl-2(3H)-benzoxazolylidene)-1-propenyl]-, iodide; 3,3′-Dihexyloxacarbocyanine iodide; D 273; DiOC$_6$(3); Dioc6; NK 2280

Merck Index Number Not listed

Chemical/Dye Class Cyanine

Molecular Formula C$_{29}$H$_{37}$IN$_2$O$_2$

Molecular Weight 572.53

Physical Form Red powder or crystals

Solubility Soluble in chloroform, N,N-dimethylformamide, dimethyl sulfoxide, ethanol, methanol

Melting Point 222–224 °C[3]

Absorption (λ_{max}) 484 nm (MeOH)

Emission (λ_{max}) 501 nm (MeOH)

Molar Extinction Coefficient 154,000 cm^{-1} M^{-1} (MeOH)

Synthesis Synthetic methods[1–5]

Imaging/Labeling Applications

Mitochondria;[1,2,6–24,59,60,70] endoplasmic reticulum;[1,2,20,25–45] astrocytes;[46] bacteria;[47–50] Bacillus spores;[51] cells;[4] fungi;[48,52] lymphocytes;[53,54] monocytes;[55] myelin sheath;[56] nucleic acids;[57] presynaptic nerve terminals;[58] sperms[59–61]

Biological/Medical Applications

Analyzing germinating Bacillus spores;[51] assessing bacterial viability;[50] detecting prostate cancer,[9] diagnosing bipolar/unipolar disorders;[62] identifying myelin sheath;[56] measuring/monitoring membrane potential;[1–3,6,10–18,21,24,51,53,54,59–66,70] monitoring bacterial batch cultures;[47] apoptosis assays;[67–70] autophagy assays;[71] in photodynamic therapy[72]

Industrial Applications

Carbon nanotubes;[73,74] Copper electrodeposition;[75] dye sensitized photographic imaging system;[76,77] photoresists[78]

Safety/Toxicity

Phototoxicity[79]

REFERENCES

1. Sabnis, R. W. *Handbook of Biological Dyes and Stains*; John Wiley & Sons Inc.: Hoboken, **2010**; pp 162–165.

2. Sabnis, R. W.; Deligeorgiev, T. G.; M. N. Jachak, M. N.; Dalvi, T. S. DiOC$_6$(3): A useful dye for staining the endoplasmic reticulum. *Biotech. Histochem.* **1997**, *72*, 253–258.

3. Sims, P. J.; Waggoner, A. S.; Wang, C.; Hoffman, J. F. Mechanism by which cyanine dyes measure membrane potential in red blood cells and phosphatidylcholine vesicles. *Biochemistry* **1974**, *13*, 3315–3330.

4. Adier, C.; Thomas, M.; Turpin, F.; Brigand, C.; Cenatiempo, Y.; Belgsir, E. M. Method for obtaining water-soluble fluorescent complexes from lipophilic dyes and cyclodextrin derivatives. Fr. Demande FR 2878853, 2006.

5. Akins, D. L.; Kumar, V. T. High-performance liquid chromatography of cyanine dyes: Multiphase separation, purification, and substitution of the counter ion. *J. Chromatogr., A* **1995**, *689*, 269–273.

6. Danilovich, G. V.; Danilovich, Yu. V.; Gorchev, V. F. Comparative study of plasma and inner mitochondrial membrane polarization in smooth muscle cell by spectrofluorimetry and flow cytometry using

potential-sensitive probe DiOC$_6$(3). *Ukr. Biokhim. Zh.* **2011**, *83*, 99–105.

7. Swayne, T. C.; Gay, A. C.; Pon, L. A. Visualization of mitochondria in budding yeast. *Methods Cell Biol.* **2007**, *80*, 591–626.

8. Hattori, F.; Fukuda, K. Method for selecting myocardial cells by using intracellular mitochondria as indication. PCT Int. Appl. WO 2006022377, 2006.

9. Dickman, D. Methods of detecting prostate cancer. PCT Int. Appl. WO 2006054296, 2006.

10. Le, S. B.; Holmuhamedov, E. L.; Narayanan, V. L.; Sausville, E. A.; Kaufmann, S. H. Adaphostin and other anticancer drugs quench the fluorescence of mitochondrial potential probes. *Cell Death Differ.* **2006**, *13*, 151–159.

11. Kataoka, M.; Fukura, Y.; Shinohara, Y.; Baba, Y. Analysis of mitochondrial membrane potential in the cells by microchip flow cytometry. *Electrophoresis* **2005**, *26*, 3025–3031.

12. Kalbacova, M.; Vrbacky, M.; Drahota, Z.; Melkova, Z. Comparison of the effect of mitochondrial inhibitors on mitochondrial membrane potential in 2 different cell lines using flow cytometry and spectrofluorometry. *Cytometry* **2003**, *52A*, 110–116.

13. Zamzami, N.; Metivier, D.; Kroemer, G. Quantitation of mitochondrial transmembrane potential in cells and in isolated mitochondria. *Methods Enzymol.* **2000**, *322*, 208–213.

14. Mathur, A.; Hong, Y.; Kemp, B. K.; Barrientos, A. A.; Erusalimsky, J. D. Evaluation of fluorescent dyes for the detection of mitochondrial membrane potential changes in cultured cardiomyocytes. *Cardiovasc. Res.* **2000**, *46*, 126–138.

15. Rottenberg, H.; Wu, S. Quantitative assay by flow cytometry of the mitochondrial membrane potential in intact cells. *Biochim. Biophys. Acta* **1998**, *1404*, 393–404.

16. Metivier, D.; Dallaporta, B.; Zamzami, N.; Larochette, N.; Susin, S. A.; Marzo, I.; Kroemer, G. Cytofluorometric detection of mitochondrial alterations in early CD95/Fas/APO-1 triggered apoptosis of Jurkat T lymphoma cells. Comparison of seven mitochondrion-specific fluorochromes. *Immunol. Lett.* **1998**, *61*, 157–163.

17. Salvioli, S.; Ardizzoni, A.; Franceschi, C.; Cossarizza, A. JC-1, but not DiOC6(3) or rhodamine 123, is a reliable fluorescent probe to assess $\Delta\Psi$ changes in intact cells: Implications for studies on mitochondrial functionality during apoptosis. *FEBS Lett.* **1997**, *411*, 77–82.

18. Yang, H. C.; Taguchi, H.; Nishimura, K.; Miyaji, M. Effect of miconazole on diO-C6-(3) accumulation in mitochondria of *Candida albicans*. *Mycoscience* **1996**, *37*, 243–248.

19. Miyakawa, I.; Higo, K.; Osaki, F.; Sando, N. Double staining of mitochondria and mitochondrial nucleoids in the living yeast during the life cycle. *J. Gen. Appl. Microbiol.* **1994**, *40*, 1–14.

20. Soltys, B. J.; Gupta, R. S. Interrelationships of endoplasmic reticulum, mitochondria, intermediate filaments, and microtubules: A quadruple fluorescence labeling study. *Biochem. Cell Biol.* **1992**, *70*, 1174–1186.

21. Petit, P. X.; O'Connor, J. E.; Grunwald, D.; Brown, S. C. Analysis of the membrane potential of rat- and mouse-liver mitochondria by flow cytometry and possible applications. *Eur. J. Biochem.* **1990**, *194*, 389–397.

22. Hayashi, Y. New method for observation of mitochondria. *Cell (Tokyo, Japan)* **1990**, *22*, 192–195.

23. Korchak, H. M.; Rich, A. M.; Wilkenfeld, C.; Rutherford, L. E.; Weissmann, G. A carbocyanine dye, DiOC6(3), acts as a mitochondrial probe in human neutrophils. *Biochem. Biophys. Res. Commun.* **1982**, *108*, 1495–1501.

24. Johnson, L. V.; Wash, M. L.; Bockus, B. J.; Chen, L. B. Monitoring of relative mitochondrial membrane potential in living cells by fluorescence microscopy. *J. Cell Biol.* **1981**, *88*, 526–535.

25. Foissner, I. Fluorescent phosphocholine-a specific marker for the endoplasmic reticulum and for lipid droplets in Chara internodal cells. *Protoplasma* **2009**, *238*, 47–58

26. Ramoino, P.; Diaspro, A.; Fato, M.; Beltrame, F.; Robello, M. Changes in the endoplasmic reticulum structure of *Paramecium primaurelia* in relation to different cellular physiological states. *J. Photochem. Photobiol., B: Biol.* **2000**, *54*, 35–42.

27. Terasaki, M.; Reese, T. S. Interaction among endoplasmic reticulum, microtubules and retrograde movements of the cell surface. *Cell Motil. Cytoskeleton* **1994**, *29*, 291–300.

28. Terasaki, M. Probes for the endoplasmic reticulum. In *Fluorescent and Luminescent Probes for Biological Activity*; Mason, W. T., Ed.; Academic Press: London, **1993**; pp 120–123.

29. Terasaki, M.; Reese, T. S. Characterization of endoplasmic reticulum by co-localization of BiP and dicarbocyanine dyes. *J. Cell Sci.* **1992**, *101*, 315–322.

30. Toyoshima, I.; Yu, H.; Steuer, E. R.; Sheetz, M. P. Kinectin, a major kinesin-binding protein on ER. *J. Cell Biol.* **1992**, *118*, 1121–1131.

31. Chen, L. B.; Lee, C. Probing endoplasmic reticulum in living cells by epifluorescence and digitized video microscopy. In *Optical Microscopy for Biology*; Herman, B.; Jacobson, E., Eds.; Wiley-Liss: New York, **1990**; pp 409–418.

32. Terasaki, M. Recent progress on structural interactions of the endoplasmic reticulum. *Cell Motil. Cytoskeleton* **1990**, *15*, 71–75.

33. Quader, H. Formation and disintegration of cisternae of the endoplasmic reticulum visualized in live cells by conventional fluorescence and confocal laser scanning microscopy: Evidence for the involvement of calcium and the cytoskeleton. *Protoplasma* **1990**, *155*, 166–175.

34. Quader, H.; Fast, H. Influence of cytosolic pH changes on the organization of the endoplasmic reticulum in epidermal cells of onion bulb scales: Acidification by loading with weak organic acids. *Protoplasma* **1990**, *157*, 216–224.

35. Dailey, M. E.; Bridgman, P. C. Dynamics of the endoplasmic reticulum and other membranous organelles in growth cones of cultured neurons. *J. Neurosci.* **1989**, *9*, 1897–1909.

36. Terasaki, M. Fluorescent labeling of endoplasmic reticulum. *Methods Cell Biol.* **1989**, *29*, 125–135.

37. Lee, C.; Ferguson, M.; Chen, L. B. Construction of the endoplasmic reticulum. *J. Cell Biol.* **1989**, *109*, 2045–2055.

38. Quader, H.; Hofmann, A.; Schnepf, E. Reorganization of the endoplasmic reticulum in epidermis cells of onion bulb scales after cold stress: involvement of cytoskeletal elements. *Planta* **1989**, *177*, 273–280.

39. Sanger, J. M.; Dome, J. S.; Mittal, B.; Somlyo, A. V.; Sanger, J. W. Dynamics of the endoplasmic reticulum in living non-muscle and muscle cells. *Cell Motil. Cytoskeleton* **1989**, *13*, 301–319.

40. Lee, C.; Chen, L. B. Dynamic behavior of endoplasmic reticulum in living cells. *Cell* **1988**, *54*, 37–46.

41. Allen, N. S.; Brown, D. T. Dynamics of the endoplasmic reticulum in living onion epidermal cells in relation to microtubules, microfilaments and intracellular particle movement. *Cell Motil. Cytoskeleton* **1988**, *10*, 153–163.

42. Quader, H.; Hofmann, A.; Schnepf, E. Shape and movement of the endoplasmic reticulum in onion bulb epidermis cells: Possible involvement of actin. *Eur. J. Cell Biol.* **1987**, *44*, 17–26.

43. Terasaki, M.; Chen, L. B.; Fujiwara, K. Microtubules and the endoplasmic reticulum are highly interdependent structures. *J. Cell Biol*, **1986**, *103*, 1557–1568.

44. Quader, H.; Schnepf, E. Endoplasmic reticulum and cytoplasmic streaming: Fluorescence microscopical observations in adaxial epidermis cells of onion bulb scales. *Protoplasma* **1986**, *131*, 250–252.

45. Terasaki, M.; Song, J.; Wong, J. R.; Weiss, M. J.; Chen, L. B. Localization of endoplasmic reticulum in living and glutaraldehyde-fixed cells with fluorescent dyes. *Cell* **1984**, *38*, 101–108.

46. Chen, X.; Schluesener, H. J. Mode of dye loading affects staining outcomes of fluorescent dyes in astrocytes exposed to multiwalled carbon nanotubes. *Carbon* **2010**, *48*, 730–743.

47. Lopes da Silva, T.; Piekova, L.; Mileu, J.; Roseiro, J. C. A comparative study using the dual staining flow cytometric protocol applied to *Lactobacillus rhamnosus* and *Bacillus licheniformis* batch cultures. *Enzyme Microb. Technol.* **2009**, *45*, 134–138.

48. Little, R. G., II,; Abrahamson, S.; Wong, P. Identification of novel antimicrobial agents using membrane potential indicator dyes. PCT Int. Appl. WO 2000018951, 2000.

49. Mason, D. J.; Lopez-Amoros, R.; Allman, R.; Stark, J. M.; Lloyd, D. The ability of membrane potential dyes and calcofluor white to distinguish between viable and non-viable bacteria. *J. Appl. Bacteriol.* **1995**, *78*, 309–315.

50. Diaper, J. P.; Tither, K.; Edwards, C. Rapid assessment of bacterial viability by flow cytometry. *Appl. Microbiol. Biotechnol.* **1992**, *38*, 268–272.

51. Laflamme, C.; Ho, J.; Veillette, M.; Latremoille, M. C.; Verreault, D.; Meriaux, A.; Duchaine, C. Flow cytometry analysis of germinating *Bacillus* spores, using membrane potential dye. *Arch. Microbiol.* **2005**, *183*, 107–112.

52. Duckett, J. G.; Read, D. J. The use of the fluorescent dye, 3,3′-dihexyloxacarbocyanine iodide, for selective staining of ascomycete fungi associated with liverwort rhizoids and ericoid mycorrhizal roots. *New Phytol.* **1991**, *118*, 259–272.

53. Cercek, B.; Cercek, L. Differential binding of membrane-potential sensitive materials to lymphocytes. U.S. Patent 4835103, 1989.

54. Shapiro, H. M.; Natale, P. J.; Kamentsky, L. A. Estimation of membrane potentials of individual lymphocytes by flow cytometry. *Proc. Natl. Acad. Sci. U.S.A.* **1979**, *76*, 5728–5730.

55. Sakata, T.; Kuroda, T. Reagent for classifying leukocytes by flow cytometry. U.S. Patent 5175109, 1992.

56. Micu, I.; Ridsdale, A.; Zhang, L.; Woulfe, J.; McClintock, J.; Brantner, C. A.; Andrews, S. B.; Stys, P. K. Real-time measurement of free Ca2+ changes in CNS myelin by two-photon microscopy. *Nat. Med.* **2007**, *13*, 874–879.

57. Honda, M.; Nakai, N.; Fukuda, M.; Kawabe, Y. Optical amplification and laser action in cyanine dyes doped in DNA complex. *Proc. SPIE-Int. Soc. Opt. Eng.* **2007**, *6646*, 664609/1–664609/8.

58. Yoshikami, D.; Okun, L. M. Staining of living presynaptic nerve terminals with selective fluorescent dyes. *Nature* **1984**, *310*, 53–56.

59. Chen, T.; Shi, L. Z.; Zhu, Q.; Chandsawangbhuwana, C.; Berns, M. W. Optical tweezers and non-ratiometric fluorescent-dye-based studies of respiration in sperm mitochondria. *J. Opt.* **2011**, *13*, 044010/1-044010/6.

60. Kumaresan, A.; Kadirvel, G.; Bujarbaruah, K. M.; Bardoloi, R. K.; Das, A.; Kumar, S.; Naskar, S. Preservation of boar semen at 18° induces lipid peroxidation and apoptosis like changes in spermatozoa. *Anim. Reprod. Sci.* **2009**, *110*, 162–171.

61. Brewis, I. A.; Morton, I. E.; Mohammad, S. N.; Browes, C. E.; Moore, H. D. M. Measurement of intracellular calcium concentration and plasma membrane potential in human spermatozoa using flow cytometry. *J. Androl.* **2000**, *21*, 238–249.

62. Thiruvengadam, A. P.; Chandrasekaran, K.; Regenold, W. T. Methods for diagnosing bipolar and unipolar disorder. U.S. Pat. Appl. Publ. US 20050095579, 2005.

63. Tanner, M. K.; Wellhausen, S. R. Flow cytometric detection of fluorescent redistributional dyes for measurement of cell transmembrane potential. *Methods Mol. Biol.* **1998**, *91*, 85–95.

64. Kovac, L.; Poliachova, V. Membrane potential monitoring cyanine dyes uncouple respiration and induce respiration-deficient mutants in intact yeast cells. *Biochem. Int.* **1981**, *2*, 503–507.

65. Miller, J. B.; Koshland, D. E., Jr., Effects of cyanine dye membrane probes on cellular properties. *Nature* **1978**, *272*, 83–84.

66. Burckhardt, G. Non-linear relationship between fluorescence and membrane potential. *Biochim. Biophys. Acta, Biomembr.* **1977**, *468*, 227–237.

67. Buenz, E. J.; Limburg, P. J.; Howe, C. L. A high-throughput 3-parameter flow cytometry-based cell death assay. *Cytometry* **2007**, *71A*, 170–173.

68. Ozgen, U.; Savasan, S.; Buck, S.; Ravindranath, Y. Comparison of DiOC6(3) uptake and annexin V labeling for quantification of apoptosis in leukemia cells and non-malignant T lymphocytes from children. *Cytometry* **2000**, *42*, 74–78.

69. Li, X.; Darzynkiewicz, Z. The Schrodinger's cat quandary in cell biology: Integration of live cell functional assays with measurements of fixed cells in analysis of apoptosis. *Exp. Cell Res.* **1999**, *249*, 404–412.

70. Castedo, M.; Macho, A.; Zamzami, N.; Hirsch, T.; Marchetti, P.; Uriel, J.; Kroemer, G. Mitochondrial perturbations define lymphocytes undergoing apoptotic depletion *in vivo*. *Eur. J. Immunol.* **1995**, *25*, 3277–3284.

71. Tettamanti, G.; Malagoli, D. *In vitro* methods to monitor autophagy in *Lepidoptera*. *Methods Enzymol.* **2008**, *451*, 685–709.

72. Lipshutz, G. S.; Castro, D. J.; Saxton, R. E.; Haugland, R. P.; Soudant, J. Evaluation of four new carbocyanine dyes for photodynamic therapy with lasers. *Laryngoscope* **1994**, *104*, 996–1002.

73. Mureau, N.; Mendoza, E.; Silva, S. R. P. Dielectrophoretic manipulation of fluorescing single-walled carbon nanotubes. *Electrophoresis* **2007**, *28*, 1495–1498.

74. Prakash, R.; Washburn, S.; Superfine, R.; Cheney, R. E.; Falvo, M. R. Visualization of individual carbon nanotubes with fluorescence microscopy using conventional fluorophores. *Appl. Phys. Lett.* **2003**, *83*, 1219–1221.

75. Chung, D. S.; Alkire, R. C. Confocal microscopy for simultaneous imaging of Cu electrodeposit morphology and adsorbate fluorescence. *J. Electrochem. Soc.* **1997**, *144*, 1529–1536.

76. Farid, S. Y.; Haley, N. F.; Moody, R. E.; Specht, D. P. Dye sensitized photographic imaging system. U.S. Patent 4743531, 1988.

77. Simpson, S. M.; Boon, J. R. Supersensitization of silver halide emulsions. Eur. Pat. Appl. EP 271260, 1988.

78. Farid, S. Y.; Haley, N. F.; Moody, R. E.; Specht, D. P. Negative working photoresists responsive to shorter wavelength visible light and novel coated articles. U.S. Patent 4743529, 1988.

79. Lee, C.; Wu, S. S.; Chen, L. B. Photosensitization by 3,3′-dihexyloxacarbocyanine iodide: specific disruption of microtubules and inactivation of organelle motility. *Cancer Res.* **1995**, *55*, 2063–2069.

DIHYDRORHODAMINE 123 (DHR 123)

CAS Registry Number   109244-58-8

Chemical Structure

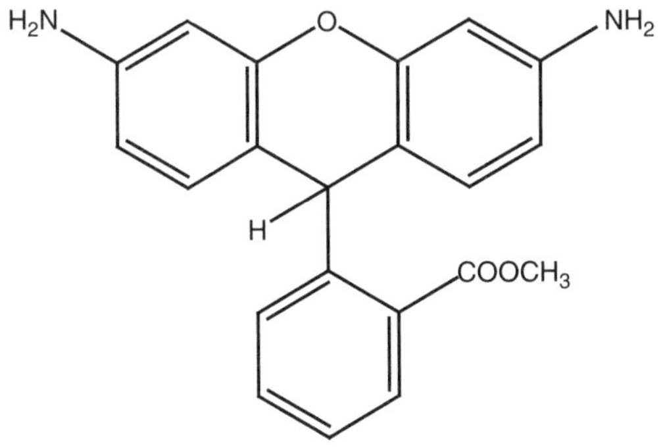

CA Index Name   Benzoic acid, 2-(3,6-diamino-9H-xanthen-9-yl)-, methyl ester

Other Names   D 23806; D 632; DHR; DHR 123; Dihydrorhodamine 123

Merck Index Number   Not listed

Chemical/Dye Class   Xanthene

Molecular Formula $C_{21}H_{18}N_2O_3$

Molecular Weight   346.38

Physical Form   Pinkish-white solid[2,3]

Solubility   Insoluble in water; soluble in acetonitrile, chloroform, N,N-dimethylformamide, dimethyl sulfoxide, ethanol, methanol

Melting Point   163–165 °C[2,3]

Boiling Point (Calcd.)   526.9 ± 50.0 °C, pressure: 760 Torr

pK$_a$ (Calcd.)   4.70 ± 0.40, most basic, temperature: 25 °C

Absorption (λ_{max})   289 nm (MeOH)

Emission (λ_{max})   None

Molar Extinction Coefficient   7100 cm^{-1} M^{-1} (MeOH)

Synthesis   Synthetic methods[1–3]

Imaging/Labeling Applications   Cells;[1–4] high density lipoprotein (HDL);[5] leukocytes (neutrophils, lymphocytes, monocytes);[6–24] mitochondria;[25–29] sperms[30,31]

Biological/Medical Applications   Analyzing/measuring oxidative/respiratory burst activity;[1,6–24] assessing the oxidative properties of high density lipoprotein (HDL);[5] generating/detecting/measuring reactive oxygen species;[1,4,6–26,30–48,56,66] generating/detecting/measuring reactive nitrogen species;[1,10,31,46–50] producing/detecting/measuring peroxynitrite;[51–54] apoptosis assays;[8] autophagy assays;[34] chronic granulomatous disease (CGD) assay;[55] screening antioxidant activity[56]

Industrial Applications   Not reported

Safety/Toxicity   Benzene-mediated toxicity;[57] cytotoxicity;[58] hepatotoxicity;[59] lung toxicity;[60,61] nephrotoxicity;[62] neurotoxicity;[63–65] skin toxicity[66]

REFERENCES

1. Sabnis, R. W. *Handbook of Biological Dyes and Stains*; John Wiley & Sons Inc.: Hoboken, **2010**; pp 151–152.

2. Kinsey, B. M.; Kassis, A. I.; Fayad, F.; Layne, W. W.; Adelstein, S. J. Synthesis and biological studies of iodinated ($^{127/125}$I) derivatives of rhodamine 123. *J. Med. Chem.* **1987**, *30*, 1757–1761.

3. Kinsey, B. M.; Kassis, A. I.; Adelstein, S. J. Dihydrorhodamines and halogenated derivatives theirof. PCT Int. Appl. WO 8706138, 1987.

4. Zeigler, F. C. Determination of cell viability and phenotype. U.S. Patent 7018804, 2006.

5. Kelesidis, T.; Currier, J. S.; Huynh, D.; Meriwether, D.; Charles-Schoeman, C.; Reddy, S. T.; Fogelman, A. M.; Navab, M.; Yang, O. O. A biochemical fluorometric method for assessing the oxidative properties of HDL. *J. Lipid Res.* **2011**, *52*, 2341–2351.

6. Peluso, I.; Morabito, G.; Riondino, S.; La Farina, F.; Serafini, M. Lymphocytes as internal standard in oxidative burst analysis by cytometry: A new data analysis approach. *J. Immunol. Methods* **2012**, *379*, 61–65.

7. Chen, Y.; Junger, W. G. Measurement of oxidative burst in neutrophils. *Methods Mol. Biol.* **2012**, *844*, 115–124.

8. Honda, F.; Kano, H.; Kanegane, H.; Nonoyama, S.; Kim, E.; Lee, S.; Takagi, M.; Mizutani, S.; Morio, T. The kinase Btk negatively regulates the production of reactive oxygen species and stimulation-induced apoptosis in human neutrophils. *Nat. Immunol.* **2012**, *13*, 369–378.

9. Kalgraff, C. A. K.; Wergeland, H. I.; Pettersen, E. F. Flow cytometry assays of respiratory burst in Atlantic salmon (*Salmo salar* L.) and in Atlantic cod (*Gadus morhua* L.) leucocytes. *Fish Shellfish Immunol.* **2011**, *31*, 381–388.

10. Kumar, S.; Patel, S.; Jyoti, A.; Keshari, R. S.; Verma, A.; Barthwal, M. K.; Dikshit, M. Nitric oxide-mediated augmentation of neutrophil reactive oxygen and nitrogen species formation: critical use of probes. *Cytometry, Part A* **2010**, *77A*, 1038–1048.

11. Freitas, M.; Porto, G.; Lima, J. L. F. C.; Fernandes, E. Optimization of experimental settings for the analysis of human neutrophils oxidative burst *in vitro*. *Talanta* **2009**, *78*, 1476–1483.

12. Rinaldi, M.; Moroni, P.; Paape, M. J.; Bannerman, D. D. Evaluation of assays for the measurement of bovine neutrophil reactive oxygen species. *Vet. Immunol. Immunopathol.* **2007**, *115*, 107–125.

13. Walrand, S.; Valeix, S.; Rodriguez, C.; Ligot, P.; Chassagne, J.; Vasson, M. Flow cytometry study of polymorphonuclear neutrophil oxidative burst: a comparison of three fluorescent probes. *Clin. Chim. Acta* **2003**, *331*, 103–110.

14. Ito, Y.; Lipschitz, D. A. Assay of intracellular hydrogen peroxide generation in activated individual neutrophils by flow cytometry. *Methods Mol. Biol.* **2002**, *196*, 111–116.

15. Dambaeva, S. V.; Mazurov, D. V.; Pinegin, B. V. Assessment of active oxygen form production in human peripheral blood using laser flow cytometry. *Immunologiya* **2001**, 58–61.

16. Bassoe, C.; Smith, I.; Sornes, S.; Halstensen, A.; Lehmann, A. K. Concurrent measurement of antigen- and antibody-dependent oxidative burst and phagocytosis in monocytes and neutrophils. *Methods* **2000**, *21*, 203–220.

17. Richardson, M. P.; Ayliffe, M. J.; Helbert, M.; Davies, E. G. A simple flow cytometry assay using dihydrorhodamine for the measurement of the neutrophil respiratory burst in whole blood: Comparison with the quantitative nitrobluetetrazolium test. *J. Immunol. Methods* **1998**, *219*, 187–193.

18. Szucs, S.; Vamosi, G.; Poka, R.; Sarvary, A.; Bardos, H.; Balazs, M.; Kappelmayer, J.; Toth, L.; Szollosi, J.; Adany, R. Single-cell measurement of superoxide anion and hydrogen peroxide production by human neutrophils with digital imaging fluorescence microscopy. *Cytometry* **1998**, *33*, 19–31.

19. van Pelt, L. J.; van Zwieten, R.; Weening, R. S.; Roos, D.; Verhoeven, A. J.; Bolscher, B. G. J. M. Limitations on the use of dihydrorhodamine 123 for flow cytometric analysis of the neutrophil respiratory burst. *J. Immunol. Methods* **1996**, *191*, 187–196.

20. Prince, H. E.; Lape-Nixon, M. Influence of specimen age and anticoagulant on flow cytometric evaluation of granulocyte oxidative burst generation. *J. Immunol. Methods* **1995**, *188*, 129–138.

21. Vowells, S. J.; Sekhsaria, S.; Malech, H. L.; Shalit, M.; Fleisher, T. A. Flow cytometric analysis of the granulocyte respiratory burst: a comparison study of fluorescent probes. *J. Immunol. Methods* **1995**, *178*, 89–97.

22. Smith, J. A.; Weidemann, M. J. Further characterization of the neutrophil oxidative burst by flow cytometry. *J. Immunol. Methods* **1993**, *162*, 261–268.

23. Rothe, G.; Emmendoerffer, A.; Oser, A.; Roesler, J.; Valet, G. Flow cytometric measurement of the respiratory burst activity of phagocytes using dihydrorhodamine 123. *J. Immunol. Methods* **1991**, *138*, 133–135.

24. Rothe, G.; Oser, A.; Valet, G. Dihydrorhodamine-123: a new flow-cytometric indicator for respiratory burst activity in neutrophil granulocytes. *Naturwissenschaften* **1988**, *75*, 354–355.

25. Matteucci, E.; Manzini, S.; Ghimenti, M.; Consani, C.; Giampietro, O. Rapid flow cytometric method for measuring mitochondrial membrane potential, respiratory burst activity, and intracellular thiols of human whole blood leukocytes. *Open Chem. Biomed. Methods J.* **2009**, *2*, 65–68.

26. Qin, Y.; Lu, M.; Gong, X. Dihydrorhodamine 123 is superior to 2,7-dichlorodihydrofluorescein diacetate and dihydrorhodamine 6G in detecting intracellular hydrogen peroxide in tumor cells. *Cell Biol. Int.* **2008**, *32*, 224–228.

27. Hattori, F.; Fukuda, K. Method of selecting myocardial cells by using intracellular mitochondria as indication. PCT Int. Appl. WO 2006022377, 2006.

28. Sobreira, C.; Davidson, M.; King, M. P.; Miranda, A. F. Dihydrorhodamine 123 identifies impaired mitochondrial respiratory chain function in cultured cells harboring mitochondrial DNA mutations. *J. Histochem. Cytochem.* **1996**, *44*, 571–579.

29. Ronot, X.; Benel, L. Mitochondrial probes in analytical and quantitative cytology. *Appl. Fluoresc. Technol.* **1989**, *1*, 11–12.

30. Kiani-Esfahani, A.; Tavalaee, M.; Deemeh, M. R.; Hamiditabar, M.; Nasr-Esfahani, M. H. DHR123: an alternative probe for assessment of ROS in human

spermatozoa. *Syst. Biol. Reprod. Med.* **2012**, *58*, 168–174.

31. Aziz, N.; Novotny, J.; Oborna, I.; Fingerova, H.; Brezinova, J.; Svobodova, M. Comparison of chemiluminescence and flow cytometry in the estimation of reactive oxygen and nitrogen species in human semen. *Fertil. Steril.* **2010**, *94*, 2604–2608.

32. Wang, S. T.; Zhegalova, N. G.; Gustafson, T. P.; Zhou, A.; Sher, J.; Achilefu, S.; Berezin, O. Y.; Berezin, M. Y. Sensitivity of activatable reactive oxygen species probes by fluorescence spectroelectrochemistry. *Analyst* **2013**, *138*, 4363–4369.

33. Boulton, S.; Anderson, A.; Swalwell, H.; Henderson, J. R.; Manning, P.; Birch-Machin, M. A. Implications of using the fluorescent probes, dihydrorhodamine 123 and 2′,7′-dichlorodihydrofluorescein diacetate, for the detection of UVA-induced reactive oxygen species. *Free Radical Res.* **2011**, *45*, 115–122.

34. Rouschop, K. M. A.; Ramaekers, C. H. M. A.; Schaaf, M. B. E.; Keulers, T. G. H.; Savelkouls, K. G. M.; Lambin, P.; Koritzinsky, M.; Wouters, B. G. Autophagy is required during cycling hypoxia to lower production of reactive oxygen species. *Radiother. Oncol.* **2009**, *92*, 411–416.

35. Dimitrijevic, N. M.; Rozhkova, E.; Rajh, T. Dynamics of localized charges in dopamine-modified TiO_2 and their effect on the formation of reactive oxygen species. *J. Am. Chem. Soc.* **2009**, *131*, 2893–2899.

36. Chang, S.; Rodrigues, N. P.; Zurgil, N.; Henderson, J. R.; Bedioui, F.; McNeil, C. J.; Deutsch, M. Simultaneous intra- and extracellular superoxide monitoring using an integrated optical and electrochemical sensor system. *Biochem. Biophys. Res. Commun.* **2005**, *327*, 979–984.

37. Sun, Y.; Yin, X. F.; Ling, Y. Y.; Fang, Z. L. Determination of reactive oxygen species in single human erythrocytes using microfluidic chip electrophoresis. *Anal. Bioanal. Chem.* **2005**, *382*, 1472–1476.

38. Ling, Y. Y.; Yin, X. F.; Fang, Z. L. Simultaneous determination of glutathione and reactive oxygen species in individual cells by microchip electrophoresis. *Electrophoresis* **2005**, *26*, 4759–4766.

39. Hanson, K. M.; Clegg, R. M. Two-photon fluorescence imaging and reactive oxygen species detection within the epidermis. *Methods Mol. Biol.* **2004**, *289*, 413–421.

40. Grzelak, A.; Rychlik, B.; Bartosz, G. Light-dependent generation of reactive oxygen species in cell culture media. *Free Radical Biol. Med.* **2001**, *30*, 1418–1425.

41. Canbolat, O.; Fandrey, J.; Jelkmann, W. Effects of modulators of the production and degradation of hydrogen peroxide on erythropoietin synthesis. *Respir. Physiol.* **1998**, *114*, 175–183.

42. Buxser, S. E.; Sawada, G.; Raub, T. J. Analytical and numerical techniques for evaluation of free radical damage in cultured cells using imaging cytometry and fluorescent indicators. *Methods Enzymol.* **1999**, *300*, 256–275.

43. Henderson, L. M.; Chappell, J. B. Dihydrorhodamine 123: A fluorescent probe for superoxide generation? *Eur. J. Biochem.* **1993**, *217*, 973–980.

44. Royall, J. A.; Ischiropoulos, H. Evaluation of 2′,7′-dichlorofluorescin and dihydrorhodamine 123 as fluorescent probes for intracellular hydrogen peroxide in cultured endothelial cells. *Arch. Biochem. Biophys.* **1993**, *302*, 348–355.

45. Sakurada, H.; Koizumi, H.; Ohkawara, A.; Ueda, T.; Kamo, N. Use of dihydrorhodamine 123 for detecting intracellular generation of peroxides upon UV irradiation in epidermal keratinocytes. *Arch. Dermatol. Res.* **1992**, *284*, 114–116.

46. Kalyanaraman, B.; Darley-Usmar, V.; Davies, K. J. A.; Dennery, P. A.; Forman, H. J.; Grisham, M. B.; Mann, G. E.; Moore, K.; Roberts, L. J., II,; Ischiropoulos, H. Measuring reactive oxygen and nitrogen species with fluorescent probes: challenges and limitations. *Free Radical Biol. Med.* **2012**, *52*, 1–6.

47. Kalyanaraman, B. Oxidative chemistry of fluorescent dyes: implications in the detection of reactive oxygen and nitrogen species. *Biochem. Soc. Trans.* **2011**, *39*, 1221–1225.

48. Price, M.; Kessel, D. On the use of fluorescence probes for detecting reactive oxygen and nitrogen species associated with photodynamic therapy. *J. Biomed. Opt.* **2010**, *15*, 051605/1-051605/3.

49. Crow, J. P. Dichlorodihydrofluorescein and dihydrorhodamine 123 are sensitive indicators of peroxynitrite *in vitro*: Implications for intracellular measurement of reactive nitrogen and oxygen species. *Nitric Oxide* **1997**, *1*, 145–157.

50. Ischiropoulos, H.; Gow, A.; Thom, S. R.; Kooy, N. W.; Royall, J. A.; Crow, J. P. Detection of reactive nitrogen species using 2,7-dichlorodihydrofluorescein and dihydrorhodamine 123. *Methods Enzymol.* **1999**, *301*, 367–373.

51. Santos, M. R.; Mira, L. Protection by flavonoids against the peroxynitrite-mediated oxidation of dihydrorhodamine. *Free Radical Res.* **2004**, *38*, 1011–1018.

52. Martin-Romero, F. J.; Gutierrez-Martin, Y.; Henao, F.; Gutierrez-Merino, C. Fluorescence measurements of steady state peroxynitrite production upon SIN-1 decomposition: NADH versus dihydrodichlorofluorescein and dihydrorhodamine 123. *J. Fluoresc.* **2004**, *14*, 17–23.

53. Malcolm, S.; Foust, R.; Hertkom, C.; Ischiropoulos, H. Detection of peroxynitrite in biological fluids. *Methods Mol. Biol.* **2000**, *36*, 171–177.

54. Kooy, N. W.; Royall, J. A.; Ischiropoulos, H.; Beckman, J. S. Peroxynitrite-mediated oxidation of dihydrorhodamine 123. *Free Radical Biol. Med.* **1994**, *16*, 149–156.

55. Jirapongsananuruk, O.; Malech, H. L.; Kuhns, D. B.; Niemela, J. E.; Brown, M. R.; Anderson-Cohen, M.; Fleisher, T. A. Diagnostic paradigm for evaluation of male patients with chronic granulomatous disease, based on the dihydrorhodamine 123 assay. *J. Allergy Clin. Immunol.* **2003**, *111*, 374–379.

56. Dunlap, W.; Llewellyn, L.; Doyle, J.; Yamamoto, Y. A microtiter plate assay for screening antioxidant activity in extracts of marine organisms. *Mar. Biotechnol.* **2003**, *5*, 294–301.

57. Wan, J.; Winn, L. M. Benzene's metabolites alter c-MYB activity via reactive oxygen species in HD3 cells. *Toxicol. Appl. Pharmacol.* **2007**, *222*, 180–189.

58. Choi, J. J.; Kong, M. Y.; Lee, S. J.; Kim, H.C.; Ko, K. H.; Kim, W. K. Ciclopirox prevents peroxynitrite toxicity in astrocytes by maintaining their mitochondrial function: A novel mechanism for cytoprotection by ciclopirox. *Neuropharmacology* **2002**, *43*, 408–417.

59. Weng, D.; Lu, Y.; Wei, Y.; Liu, Y.; Shen, P. The role of ROS in microcystin-LR-induced hepatocyte apoptosis and liver injury in mice. *Toxicology* **2007**, *232*, 15–23.

60. Hulo, S.; Tiesset, H.; Lancel, S.; Edme, J. L.; Viollet, B.; Sobaszek, A.; Neviere, R. AMP-activated protein kinase deficiency reduces ozone-induced lung injury and oxidative stress in mice. *Respir. Res.* **2011**, *12*, 64.

61. Nichol, A. D.; O'Cronin, D. F.; Naughton, F.; Hopkins, N.; Boylan, J.; McLoughlin, P. Hypercapnic acidosis reduces oxidative reactions in endotoxin-induced lung injury. *Anesthesiology* **2010**, *113*, 116–125.

62. Khand, F. D.; Gordge, M. P.; Robertson, W. G.; Noronha-Dutra, A. A.; Hothersall, J. S. Mitochondrial superoxide production during oxalate-mediated oxidative stress in renal epithelial cells. *Free Radical Biol. Med.* **2002**, *32*, 1339–1350.

63. Vanderveldt, G. M.; Regan, R. F. The neurotoxic effect of sickle cell hemoglobin. *Free Radical Res.* **2004**, *38*, 431–437.

64. Rogers, B.; Yakopson, V.; Teng, Z.; Guo, Y.; Regan, R. F. Heme oxygenase-2 knockout neurons are less vulnerable to hemoglobin toxicity. *Free Radical Biol. Med.* **2003**, *35*, 872–881.

65. Velasco, I.; Tapia, R. Alterations of intracellular calcium homeostasis and mitochondrial function are involved in ruthenium red neurotoxicity in primary cortical cultures. *J. Neurosci. Res.* **2000**, *60*, 543–551.

66. Hanson, K. M.; Clegg, R. M. Observation and quantification of ultraviolet-induced reactive oxygen species in *ex vivo* human skin. *Photochem. Photobiol.* **2002**, *76*, 57–63.

DIHYDRORHODAMINE 6G (DHR 6G)

CAS Registry Number   217176-83-5

Chemical Structure

CA Index Name   Benzoic acid, 2-[3,6-bis(ethylamino)-2,7-dimethyl-9H-xanthen-9-yl]-, ethyl ester

Other Names   D 633; DHR 6G; Dihydrorhodamine 6G; d-R 6G

Merck Index Number   Not listed

Chemical/Dye Class   Xanthene

Molecular Formula $C_{28}H_{32}N_2O_3$

Molecular Weight   444.57

Physical Form   Light pink solid

Solubility   Soluble in acetonitrile, chloroform, N,N-dimethylformamide, dimethyl sulfoxide, methanol

Boiling Point (Calcd.) $572.4 \pm 50.0\,°C$, pressure: 760 Torr

pK_a (Calcd.) 4.97 ± 0.40, most basic, temperature: $25\,°C$

Absorption (λ_{max})   296 nm (MeOH)

Emission (λ_{max})   None

Molar Extinction Coefficient $11,000\,cm^{-1}\,M^{-1}$ (MeOH)

Synthesis   Synthetic methods[1,2]

Imaging/Labeling Applications   Fungi;[3] mitochondria;[4,5] nucleic acids[6–9]

Biological/Medical Applications   Detecting intracellular hydrogen peroxide in tumor cells,[4] 5-methylcytosine,[10] nucleic acids;[7,8] detecting/identifying gene duplications;[8] detecting/quantifying fungi;[3] generating/detecting/monitoring reactive oxygen species (ROS);[11–15] measuring hydrogen peroxide production in eukaryotic cells;[16] measuring/monitoring membrane potential;[5] nucleic acid ligation assays;[6] nucleic acids sequencing;[9] nanocomposite particles for biomedical applications;[17] skin care products[18]

Industrial Applications   Light-transmitting materials/construction boards[19]

Safety/Toxicity   No data available

REFERENCES

1. Sabnis, R. W. *Handbook of Biological Dyes and Stains*; John Wiley & Sons Inc.: Hoboken, **2010**; pp 149–150.

2. Haugland, R. P. *Handbook of Fluorescent Probes and Research Chemicals*; Molecular Probes Inc.: Eugene, **1996**; pp 266–274.

3. Liu, C.; Kachur, S.; Price, L. Methods and kits used in the detection of fungus. PCT Int. Appl. WO 2011094356, 2011.

4. Qin, Y.; Lu, M.; Gong, X. Dihydrorhodamine 123 is superior to 2,7-dichlorodihydrofluorescein diacetate and dihydrorhodamine 6G in detecting intracellular hydrogen peroxide in tumor cells. *Cell Biol. Int.* **2008**, *32*, 224–228.

5. Hattori, F.; Fukuda, K. Method of selecting myocardial cells by using intracellular mitochondria as indication. PCT Int. Appl. WO 2006022377, 2006.

6. Wenz, H. M.; Day, J. P. Controls for determining reaction performance in polynucleotide sequence detection assays. U.S. Pat. Appl. Publ. US 20060014189, 2006.

7. Brandis, J.; Bolchakova, E. V.; Karger, A. E. Detection of small RNAs. U.S. Pat. Appl. Publ. US 20060003337, 2006.

8. Livak, K. J.; Stevens, J.; Lazaruk, K. D.; Ziegle, J. S.; Wong, L. Y. Detection of gene duplications. U.S. Pat. Appl. Publ. US 20050255485, 2005.

9. Kao, P. Method and device for rapid color detection. U.S. Pat. Appl. Publ. US 20020155485, 2002.

10. Zon, G. 5-Methylcytosine detection, compositions and methods thereof. U.S. Patent 7399614, 2008.

11. Liu, H.; Sun, S.; Zong, Y.; Li, P.; Xie, J. Fluorescence evaluation of scavenging efficiency of antioxidants against reactive oxygen species (ROS) in cigarette smoke. *Anal. Lett.* **2013**, *46*, 682–693.

12. Liu, L.; Xu, S.; Li, S. Determination of reactive oxygen species in cigarette smoke by DR6G fluorescent probe. *Adv. Mater. Res.* **2012**, *361–363*, 1863–1867.

13. Liu, L.; Xu, S.; Li, S. Detection of reactive oxygen species in mainstream cigarette smoke by a fluorescent probe. *Proc. SPIE* **2009**, *7382*, 738236/1–738236/6.

14. Xu, S.; Liu, L.; Tang, J.; Li, S. Detection of reactive oxygen species (ROS) in mainstream cigarette smoke using fluorescent probe. *Yingyong Guangxue* **2009**, *30*, 291–295.

15. Ou, B.; Huang, D. Fluorescent approach to quantitation of reactive oxygen species in mainstream cigarette smoke. *Anal. Chem.* **2006**, *78*, 3097–3103.

16. Bortolami, S.; Cavallini, L. Measurement of hydrogen peroxide production in eukaryotic cells using horseradish peroxidase synthesized *in situ*. Ital. Appl. IT 2007MI2165, 2008.

17. Adair, J. H.; Rouse, S. M.; Wang, J.; Kester, M.; Siedlecki, C.; White, W. B.; Vogler, E.; Snyder, A.; Pantano, C. G.; Sinoway, L.; Luck, J. Unagglomerated core/shell nanocomposite particles. PCT Int. Appl. WO 2005118702, 2005.

18. Ou, B.; Zhang, L.; Kondo, M.; Ji, H.; Kou, Y. Method for assaying the antioxidant capacity of a skin care product. U.S. Pat. Appl. Publ. US 20100159613, 2010.

19. He, Z.; Xu, Z. Fabrication of energy-saving light-transmitting construction boards integrated with solar cells. Faming Zhuanli Shenqing Gongkai Shuomingshu CN 101294435, 2008.

6,7-DIMETHOXYCOUMARIN-4-ACETIC ACID

CAS Registry Number 88404-26-6

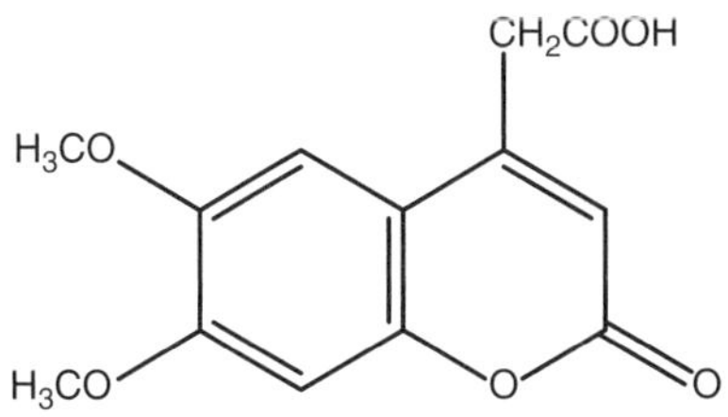

CA Index Name 2*H*-1-Benzopyran-4-acetic acid, 6,7-dimethoxy-2-oxo-

Other Names 6,7-Dimethoxy-4-coumarinyl acetic acid; 6,7-Dimethoxycoumarin-4-acetic acid; (6,7-Dimethoxy-2-oxo-2*H*-chromen-4-yl)acetic acid

Merck Index Number Not listed

Chemical/Dye Class Coumarin

Molecular Formula $C_{13}H_{12}O_6$

Molecular Weight 264.23

Physical Form Light tan solid

Solubility Soluble in dimethyl sulfoxide, ethanol, methanol, tetrahydrofuran

Melting Point 218 °C;[1] 174.5–175 °C[2]

Boiling Point (Calcd.) 498.0 ± 45.0 °C, pressure: 760 Torr

pK$_a$ (Calcd.) 4.12 ± 0.10, most acidic, temperature: 25 °C

Absorption (λ_{max}) 340 nm (MeOH)

Emission (λ_{max}) 415 nm (MeOH)

Molar Extinction Coefficient 12,800 cm^{-1} M^{-1} (MeOH)

Quantum Yield 0.36 (MeOH)

Synthesis Synthetic methods[1,2]

Imaging/Labeling Applications Cyclic peptides[3–5]

Biological/Medical Applications Reagent for carboxylic acids;[1] as a substrate for measuring proteinases activity[3–5]

Industrial Applications Not reported

Safety/Toxicity No data available

REFERENCES

1. Farinotti, R.; Siard, P.; Bourson, J.; Kirkiacharian, S.; Valeur, B.; Mahuzier, G. 4-Bromomethyl-6,7-dimethoxycoumarin as a fluorescent label for carboxylic acids in chromatographic detection. *J. Chromatogr.* **1983**, *269*, 81–90.

2. Ma, Y.; Luo, W.; Quinn, P. J.; Liu, Z.; Hider, R. C. Design, synthesis, physicochemical properties, and evaluation of novel iron chelators with fluorescent sensors. *J. Med. Chem.* **2004**, *47*, 6349–6362.

3. Berthelot, T.; Deleris, G. Fluorescent cyclic peptides, methods for the preparation thereof, and use of said peptides for measuring the enzymatic activity of a proteinase enzyme. PCT Int. Appl. WO 2010000591, 2010.

4. Berthelot, T.; Deleris, G. Microarrays of assay substrates for the detection of proteinases. Fr. Demande FR 2932189, 2009.

5. Berthelot, T.; Deleris, G. Cyclic peptides labeled with FRET pairs of dyes as assay substrates for proteinases and their preparation. Fr. Demande FR 2932190, 2009.

7-DIMETHYLAMINOCOUMARIN-4-ACETIC ACID (DMACA)

CAS Registry Number 80883-54-1

Chemical Structure

CA Index Name 2*H*-1-Benzopyran-4-acetic acid, 7-(dimethylamino)-2-oxo-

Other Names 7-(Dimethylamino)coumarin-4-acetic acid; Coumarin D 126; D 126; DMACA

Merck Index Number Not listed

Chemical/Dye Class Coumarin

Molecular Formula $C_{13}H_{13}NO_4$

Molecular Weight 247.25

Physical Form Yellow crystals;[5] Yellow needles;[7] Yellow precipitate;[3] Yellow solid;[1] Light yellow solid[2]

Solubility Soluble in *N,N*-dimethylformamide, methanol

Melting Point 168 °C;[7] 166–167 °C;[5] 163.5–164.5 °C[4]

Boiling Point (Calcd.) 497.6 ± 45.0 °C, pressure: 760 Torr

pK$_a$ (Calcd.) 4.27 ± 0.10, most acidic, temperature: 25 °C; 1.70 ± 0.20, most basic, temperature: 25 °C

Absorption (λ_{max}) 370 nm (MeOH)

Emission (λ_{max}) 459 nm (MeOH)

Molar Extinction Coefficient 22,000 $cm^{-1} M^{-1}$ (MeOH)

Synthesis Synthetic methods[1–7]

Imaging/Labeling Applications Antibodies;[1] bovine pancreatic trypsin inhibitor (BPTI);[13,14] iron chelators;[4] natural products;[2,8] oligo[(R)-3-hydroxybutanoic acids] (OHB);[9] proteins;[1,3,10–14] toxins[8]

Biological/Medical Applications Distinguishing producer and nonproducer strains;[8] evaluating iron chelators;[4] synthesizing didemnin,[5] neovibsanin derivatives;[15] studying natural product biology;[2] as a substrate for measuring/monitoring β-secretase activity[16]

Industrial Applications Not reported

Safety/Toxicity No data available

REFERENCES

1. Song, H. Y.; Ngai, M. H.; Song, Z. Y.; MacAry, P. A.; Hobley, J.; Lear, M. J. Practical synthesis of maleimides and coumarin-linked probes for protein and antibody labelling via reduction of native disulfides. *Org. Biomol. Chem.* **2009**, *7*, 3400–3406.

2. Alexander, M. D.; Burkart, M. D.; Leonard, M. S.; Portonovo, P.; Liang, B.; Ding, X.; Joullie, M. M.; Gulledge, B. M.; Aggen, J. B.; Chamberlin, A. R.; Sandler, J.; Fenical, W.; Cui, J.; Gharpure, S. J.; Polosukhin, A.; Zhang, H.; Evans, P. A.; Richardson, A. D.; Harper, M. K.; Ireland, C. M.; Vong, B. G.; Brady, T. P.; Theodorakis, E. A.; La Clair, J. J. A central strategy for converting natural products into fluorescent probes. *ChemBioChem* **2006**, *7*, 409–416.

3. Clarke, K. M.; Mercer, A. C.; La Clair, J. J.; Burkart, M. D. *In vivo* reporter labeling of proteins via metabolic delivery of coenzyme A analogues. *J. Am. Chem. Soc.* **2005**, *127*, 11234–11235.

4. Ma, Y.; Luo, W.; Quinn, P. J.; Liu, Z.; Hider, R. C. Design, synthesis, physicochemical properties, and evaluation of novel iron chelators with fluorescent sensors. *J. Med. Chem.* **2004**, *47*, 6349–6362.

5. Portonovo, P.; Ding, X.; Leonard, M. S.; Joullie, M. M. First total synthesis of a fluorescent didemnin. *Tetrahedron* **2000**, *56*, 3687–3690.

6. Tolmachev, A. I.; Sribnaya, V. P.; Shcheglova, L. V. Chlorobenzopyran monomethinecyanines. *Zh. Obshch. Khim.* **1963**, *33*, 440–447.

7. Dey, B. B. A study in the coumarin condensation. *J. Chem. Soc., Trans.* **1915**, *107*, 1606–1651.

8. Sandler, J. S.; Fenical, W.; Gulledge, B. M.; Chamberlin, A. R.; La Clair, J. J. Fluorescent profiling of natural product producers. *J. Am. Chem. Soc.* **2005**, *127*, 9320–9321.

9. Fritz, M. G.; Seebach, D. Synthesis of amino acid-, carbohydrate-, coumarin-, and biotin-labeled oligo[(R)-3-hydroxybutanoic acids] (OHB). *Helv. Chim. Acta* **1998**, *81*, 2414–2429.

10. Hughes, C. C.; MacMillan, J. B.; Gaudencio, S. P.; Fenical, W.; La Clair, J. J. Ammosamides A and

B target myosin. *Angew. Chem., Int. Ed.* **2009**, *48*, 728–732.

11. Burkart, M. D.; Clarke, K.; Mercer, A.; Laclair, J. J.; Meier, J. Metabolic delivery of coenzyme A analogs. U.S. Pat. Appl. Publ. US 20070128683, 2007.

12. Meier, J. L.; Mercer, A. C.; Rivera, II., Jr.,; Burkart, M. D. Synthesis and evaluation of bioorthogonal pantetheine analogues for *in vivo* protein modification. *J. Am. Chem. Soc.* **2006**, *128*, 12174–12184.

13. Amir, D.; Haas, E. Determination of intramolecular distance distributions in a globular protein by nonradiative excitation energy transfer measurements. *Biopolymers* **1986**, *25*, 235–240.

14. Amir, D.; Haas, E. A series of site-specific fluorescently labeled BPTI derivatives prepared by nonselective acylation and chromatographic separations. *Int. J. Pept. Protein Res.* **1986**, *27*, 7–17.

15. Imagawa, H.; Saijo, H.; Yamaguchi, H.; Maekawa, K.; Kurisaki, T.; Yamamoto, H.; Nishizawa, M.; Oda, M.; Kabura, M.; Nagahama, M. Syntheses of structurally-simplified and fluorescently-labeled neovibsanin derivatives and analysis of their neurite outgrowth activity in PC12 cells. *Bioorg. Med. Chem. Lett.* **2012**, *22*, 2089–2093.

16. Folk, D. S.; Torosian, J. C.; Hwang, S.; McCafferty, D. G.; Franz, K. J. Monitoring β-secretase activity in living cells with a membrane-anchored FRET probe. *Angew. Chem., Int. Ed.* **2012**, *51*, 10795–10799.

7-DIMETHYLAMINOCOUMARIN-4-ACETIC ACID SUCCINIMIDYL ESTER (DMACA SE)

CAS Registry Number 96686-59-8

Chemical Structure

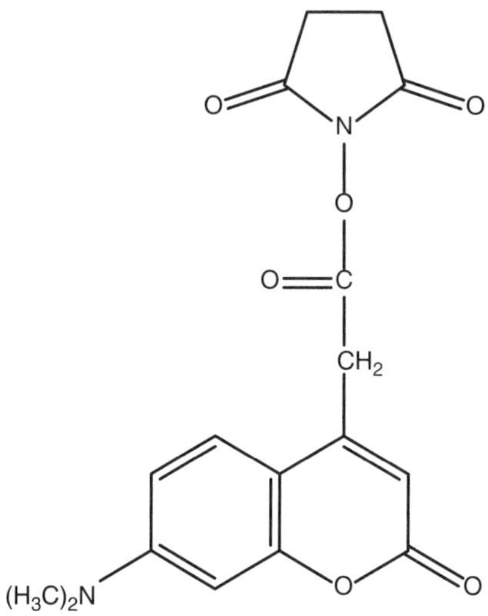

CA Index Name 2H-1-Benzopyran-4-acetic acid, 7-(dimethylamino)-2-oxo-, 2,5-dioxo-1-pyrrolidinyl ester

Other Names 2,5-Pyrrolidinedione, 1-[[[7-(dimethylamino)-2-oxo-2H-1-benzopyran-4-yl]acetyl]oxy]-; 7-Dimethylaminocoumarin-4-acetic acid, succinimidyl ester

Merck Index Number Not listed

Chemical/Dye Class Coumarin

Molecular Formula $C_{17}H_{16}N_2O_6$

Molecular Weight 344.32

Physical Form Yellow solid

Solubility Soluble in acetonitrile, chloroform, *N,N*-dimethylformamide, dimethyl sulfoxide, methanol

Boiling Point (Calcd.) 552.3 $\pm$ 60.0 °C, pressure: 760 Torr

pK$_a$ (Calcd.) 1.60 $\pm$ 0.20, most basic, temperature: 25 °C

Absorption (λ_{max}) 376 nm (MeOH)

Emission (λ_{max}) 468 nm (MeOH)

Molar Extinction Coefficient 22,000 cm^{-1} M^{-1} (MeOH)

Synthesis Synthetic method[1]

Imaging/Labeling Applications Analytes;[2] bovine pancreatic trypsin inhibitor (BPTI) derivatives;[1,10] oligonucleotides;[3,4] proteins/ fusion proteins;[1,5–10] silane coupling agent;[19] silica particles/beads;[19] taxol[11]

Biological/Medical Applications Amine reactive probe;[1,6,10,17,18] detecting oligonucleotides;[3,4] detecting/isolating proteins/fusion proteins;[1,5–10] quantifying analytes;[2] nucleic acid sequencing;[3,4] as a substrate for measuring angiotensin I-converting enzyme (ACE) activity,[12] glutathione and/or glutathione transferase activity,[13] chloramphenicol acetyltransferase activity,[14,15] proteases activity;[16] visualizing biomolecular dynamics in living cells,[5] taxol-microtubule system in cultured cells;[11] colloidal diagnostic devices;[19] useful for synthesizing peptides[19]

Industrial Applications Cations/metal ion sensing by self-assembled monolayers on glass;[17,18] silica particles/beads[19]

Safety/Toxicity No data available

REFERENCES

1. Amir, D.; Haas, E. A series of site-specific fluorescently labeled BPTI derivatives prepared by nonselective acylation and chromatographic separations. *Int. J. Pept. Protein Res.* **1986**, *27*, 7–17.

2. Lakowicz, J. R.; Maliwal, B. P.; Thompson, R.; Ozinskas, A. Fluorescent energy transfer immunoassay. U.S. Patent 5631169, 1997.

3. Lee, L. G. Method for detecting oligonucleotides using UV light source. U.S. Patent 6218124, 2001.

4. Lee, L. G. UV-excitable fluorescent energy transfer dyes. PCT Int. Appl. WO 2001016369, 2001.

5. Beatty, K. E.; Fisk, J. D.; Smart, B. P.; Lu, Y. Y.; Szychowski, J.; Hangauer, M. J.; Baskin, J. M.; Bertozzi, C. R.; Tirrell, D. A. Live-cell imaging of cellular proteins by a strain-promoted azide-alkyne cycloaddition. *ChemBioChem* **2010**, *11*, 2092–2095.

6. Beatty, K. E.; Tirrell, D. A. Two-color labeling of temporally defined protein populations in mammalian cells. *Bioorg. Med. Chem. Lett.* **2008**, *18*, 5995–5999.

7. Gee, K. R.; Hart, C. R.; Haugland, R.; Patton, W. F.; Whitney, S. Site-specific labeling of affinity tags in fusion proteins. U.S. Pat. Appl. Publ. US 2006141554, 2006.

8. Gee, K.; Hart, C.; Haugland, R.; Patton, W.; Whitney, S. Site-specific labeling of affinity tags in fusion proteins. PCT Int. Appl. WO 2005038460, 2005.

9. Diwu, Z.; Gee, K.; Hart, C.; Haugland, R.; Leung, W.; Patton, W.; Rukavishnikov, A. Site-specific fluorescent labeling of affinity tags in fusion proteins. PCT Int. Appl. WO 2004025259, 2004.

10. Amir, D.; Haas, E. Estimation of intramolecular distance distributions in bovine pancreatic trypsin inhibitor by site-specific labeling and nonradiative excitation energy-transfer measurements. *Biochemistry* **1987**, *26*, 2162–2175.

11. Souto, A. A.; Acuna, A. U.; Andreu, J. M.; Barasoain, I.; Abal, M.; Amat-Guerri, F. New fluorescent water-soluble taxol derivatives. *Angew. Chem., Int. Ed. Engl.* **1996**, *34*, 2710–2712.

12. Cheviron, N.; Rousseau-Plasse, A.; Lenfant, M.; Adeline, M.; Potier, P.; Thierry, J. Coumarin-Ser-Asp-Lys-Pro-OH, a fluorescent substrate for determination of angiotensin-converting enzyme activity via high-performance liquid chromatography. *Anal. Biochem.* **2000**, *280*, 58–64.

13. Diwu, Z.; Haugland, R. P. Assay for glutathione transferase using polyhaloaryl-substituted reporter molecules. U.S. Patent 5773236, 1998.

14. Haughland, R. P.; Kang, H. C.; Young, S. L.; Melner, M. H. Fluorescent chloramphenicol derivatives for determination of chloramphenicol acetyltransferase activity. U.S. Patent 5364764, 1994.

15. Haughland, R. P.; Kang, H. C.; Young, S. L.; Melner, M. H. Fluorescent chloramphenicol derivatives for determination of chloramphenicol acetyltransferase activity. U.S. Patent 5262545, 1993.

16. Wendt, K.; Kindermann, M.; Miniejew, C.; Globisch, A. Covalently binding imaging probes. PCT Int. Appl. WO 2009103432, 2009.

17. Zimmerman, R.; Basabe-Desmonts, L.; van der Baan, F.; Reinhoudt, D. N.; Crego-Calama, M. A combinatorial approach to surface-confined cation sensors in water. *J. Mater. Chem.* **2005**, *15*, 2772–2777.

18. Crego-Calama, M.; Reinhoudt, D. N. New materials for metal ion sensing by self-assembled monolayers on glass. *Adv. Mater.* **2001**, *13*, 1171–1174.

19. Lawrie, G.; Grondahl, L.; Battersby, B.; Keen, I.; Lorentzen, M.; Surawski, P.; Trau, M. Tailoring surface properties to build colloidal diagnostic devices: Controlling interparticle associations. *Langmuir* **2006**, *22*, 497–505.

7-DIMETHYLAMINO-4-METHYLCOUMARIN-3-ISOTHIOCYANATE (DACITC)

CAS Registry Number 74802-04-3

Chemical Structure

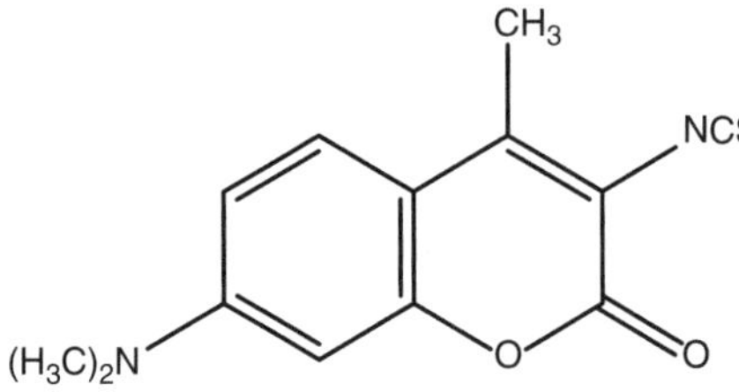

CA Index Name 2*H*-1-Benzopyran-2-one, 7-(dimethylamino)-3-isothiocyanato-4-methyl-

Other Names 7-(Dimethylamino)-4-methylcoumarin-3-isothiocyanate; DACITC

Merck Index Number Not listed

Chemical/Dye Class Coumarin

Molecular Formula $C_{13}H_{12}N_2O_2S$

Molecular Weight 260.31

Physical Form Yellow needles[1]

Solubility Soluble in acetonitrile, chloroform, *N,N*-dimethylformamide, methanol

Melting Point 200–201 °C[1]

Boiling Point (Calcd.) 475.6 ± 45.0 °C, pressure: 760 Torr

pK$_a$ (Calcd.) 1.17 ± 0.20, most basic, temperature: 25 °C

Absorption (λ_{max}) 400 nm (MeOH)

Emission (λ_{max}) 476 nm (MeOH)

Molar Extinction Coefficient 36,000 cm^{-1} M^{-1} (MeOH)

Synthesis Synthetic method[1]

Imaging/Labeling Applications Alcohols;[1] amines;[1] silica shells[2]

Biological/Medical Applications Fluorescent acylating reagent for alcohols and primary amines[1]

Industrial Applications Superparamagnetic iron oxide-silica nanoparticles coating[2]

Safety/Toxicity No data available

REFERENCES

1. Goya, S.; Takadate, A.; Tanaka, T.; Nakashima, F. Synthesis and fluorescent properties of coumarin derivatives as analytical reagents. *Yakugaku Zasshi* **1980**, *100*, 289–294.

2. Lu, Y.; Yin, Y.; Mayers, B. T.; Xia, Y. Modifying the surface properties of superparamagnetic iron oxide nanoparticles through a sol–gel approach. *Nano Lett.* **2002**, *2*, 183–186.

2-[4-(DIMETHYLAMINO)STYRYL]-N-ETHYLPYRIDINIUM IODIDE (DASPEI)

CAS Registry Number   3785-01-1

Chemical Structure

CA Index Name   Pyridinium, 2-[2-[4-(dimethylamino)phenyl]ethenyl]-1-ethyl-, iodide (1:1)

Other Names   2-[p-(Dimethylamino)styryl]-1-ethylpyridinium iodide; Pyridinium, 2-[2-[4-(dimethylamino)phenyl]ethenyl]-1-ethyl-, iodide; Pyridinium, 2-[p-(dimethylamino)styryl]-1-ethyl-, iodide; 2-[4-(Dimethylamino)styryl]-1-ethylpyridinium iodide; 2-[4-(Dimethylamino)styryl]-N-ethylpyridinium iodide; D 426; DASPEI; NK 557; Pinaflavol

Merck Index Number   Not listed

Chemical/Dye Class   Styryl

Molecular Formula $C_{17}H_{21}IN_2$

Molecular Weight   380.27

Physical Form   Bright red crystals[4]

Solubility   Soluble in chloroform, N,N-dimethylformamide, dimethyl sulfoxide, methanol

Melting Point   258–259 °C;[4] 255 °C[5]

Absorption (λ_{max})   461 nm (MeOH)

Emission (λ_{max})   589 nm (MeOH)

Molar Extinction Coefficient   39,000 cm^{-1} M^{-1} (MeOH)

Synthesis   Synthetic methods[1–8]

Imaging/Labeling Applications   Amyloid plaques;[9–11] axons;[25] bacteria;[12] epidermal cells;[13] hairs;[14–17] mitochondria;[1,18–25,34] nerve terminals;[25,26] neuromuscular junctions;[25] phospholipid vesicles/liposomes[27]

Biological/Medical Applications   Analyzing mitochondria;[21] detecting prostate cancer;[21] diagnosing/treating amyloidosis disorder;[9–11] measuring/monitoring membrane potential[1,18–25,27]

Industrial Applications   Color proofing materials;[32] holographic materials;[32] monitoring photopolymerization processes;[28] platemaking materials;[32] optoelectronic devices;[29] photographic materials/systems;[30–32] photoresists;[32] thermal cure monitoring of phenolic resole resins[33]

Safety/Toxicity   Mitochondrial toxicity[34]

REFERENCES

1. Sabnis, R. W. *Handbook of Biological Dyes and Stains*; John Wiley & Sons Inc.: Hoboken, **2010**; pp 130–131.

2. Antonious, M. S.; Mahmoud, M. R.; Guirguis, D. B. Solvent polarity indicators: Mass spectral properties of some styryl pyridinium and quinolinium salts. *Ann. Chim.* **1993**, *83*, 457–460.

3. Zhmurova, I. N.; Yurchenko, R. I.; Kirsanov, A. V. Auxochromic action of a phosphazo group. III. *Zh. Obshch. Khim.* **1970**, *40*, 982–986.

4. Phillips, A. P. Synthetic curare substitutes from stilbazoline bisquaternary ammonium salts. *J. Am. Chem. Soc.* **1952**, *74*, 3683–3685.

5. Takahashi, T.; Satake, K.; Nomura, N.; Yoshikawa, K.; Sawata, M. Syntheses of heterocyclic compounds of nitrogen. LXII. Photosensitizing dyes. *Yakugaku Zasshi* **1952**, *72*, 42–45.

6. Brooker, L. G. S.; Keyes, G. H.; Sprague, R. H.; VanDyke, R. H.; VanLare, E.; VanZandt, G.; White, F. L.; Cressman, H. W. J.; Dent, S. G., Jr., Color and constitution. X. Absorption of the merocyanines. *J. Am. Chem. Soc.* **1951**, *73*, 5332–5350.

7. Crippa, G. B.; Maffei, S. The reactivity of picoline alkiodides. *Gazz. Chim. Ital.* **1947**, *77*, 416–421.

8. Doja, M. Q.; Prasad, D. J. The cyanine dyes of the pyridine series. II. *J. Indian Chem. Soc.* **1942**, *19*, 125–129.

9. Gervais, F.; Kong, X.; Chalifour, R.; Migneault, D. Amyloid targeting imaging agents and uses thereof. U.S. Pat. Appl. Publ. US 20050048000, 2005.

10. Gervais, F.; Kong, X.; Chalifour, R.; Migneault, D. Amyloid targeting imaging agents and uses thereof. PCT Int. Appl. WO 2002007781, 2002.

11. Kudo, Y.; Suzuki, M.; Suemoto, T.; Shimazu, H. Image diagnosis probe based on substituted azobenzene or analogue thereof for disease attributable to amyloid accumulation and

composition for image diagnosis containing the same. PCT Int. Appl. WO 2001070667, 2001.

12. Sedgwick, E. G.; Bragg, P. D. The role of efflux systems and the cell envelope in fluorescence changes of the lipophilic cation 2-(4-dimethylaminostyryl)-1-ethylpyridinium in *Escherichia coli*. *Biochim. Biophys. Acta* **1996**, *1278*, 205–212.

13. Leise, E. M. Selective retention of the fluorescent dye DASPEI in a larval gastropod mollusc after paraformaldehyde fixation. *Microsc. Res. Tech.* **1996**, *33*, 496–500.

14. Chaisy, M.; Gourlaouen, L. Hair dye composition containing a soluble fluorescent compound, an electrophilic monomer, and at least a liquid organic solvent. Fr. Demande FR 2899778, 2007.

15. Rollat-Corvol, I.; Samain, H. Hair cosmetic compositions comprising a conducting polymer and an optical brightener and/or a fluorescent compound. Eur. Pat. Appl. EP 1498110, 2005.

16. Pastore, F.; Gourlaouen, L.; Lagrange, A. Hair dye brightener composition for human keratinic fibers. Fr. Demande FR 2830189, 2003.

17. Moeller, H.; Oberkobusch, D.; Hoeffkes, H. Hair dye compositions containing aromatic aldehydes or ketones. Ger. Offen. DE 19936911, 2001.

18. Peng, E. Method for screening chemical compounds targeting mitochondria with *Danio rerio*. Faming Zhuanli Shenqing CN 101968484, 2011.

19. Jensen, K. H. R.; Rekling, J. C. Development of a no-wash assay for mitochondrial membrane potential using the styryl dye DASPEI. *J. Biomol. Screen.* **2010**, *15*, 1071–1081.

20. Hattori, F.; Fukuda, K. Method of selecting myocardial cells by using intracellular mitochondria as indication. PCT Int. Appl. WO 2006022377, 2006.

21. Dickman, D. Methods of detecting prostate cancer. PCT Int. Appl. WO 2006054296, 2006.

22. Dykens, J. A.; Velicelebi, G.; Ghosh, S. S. Compositions and methods for assaying subcellular conditions and processes using energy transfer. PCT Int. Appl. WO 2000079274, 2000.

23. Sakamoto, T.; Yokota, S.; Ando, M. Rapid morphological oscillation of mitochondrion-rich cell in estuarine mudskipper following salinity changes. *J. Exp. Zool.* **2000**, *286*, 666–669.

24. Jezek, P.; Borecký, J. Mitochondrial uncoupling protein may participate in futile cycling of pyruvate and other monocarboxylates. *Am. J. Physiol.* **1998**, *275*, C496-C504.

25. Herrera, A. A.; Banner, L. R. The use and effects of vital fluorescent dyes: Observation of motor nerve terminals and satellite cells in living frog muscles. *J. Neurocytol.* **1990**, *19*, 67–83.

26. Magrassi, L.; Purves, D.; Lichtman, J. W. Fluorescent probes that stain living nerve terminals. *J. Neurosci.* **1987**, *7*, 1207–1214.

27. Sedgwick, E. G.; Bragg, P. D. Mechanism of uptake of the fluorescent dye 2-(4-dimethylaminostyryl)-1-ethylpyridinium cation (DMP+) by phospholipid vesicles. *Biochim. Biophys. Acta* **1993**, *1146*, 113–120.

28. Jager, W. F.; Kudasheva, D.; Neckers, D. C. Organic donor-π-acceptor salts: A new type of probe for monitoring photopolymerization processes. *Macromolecules* **1996**, *29*, 7351–7355.

29. Cynora, G. Compositions for singlet harvesting for optoelectronic devices and their use and devices including them and methods for reducing emission lifetimes and raising quantum yields. Ger. Offen. DE 102010025547, 2011.

30. Thiry, H. Preparation and properties of ultrafine grain silver bromide emulsions. *J. Photogr. Sci.* **1987**, *35*, 150–154.

31. Robinson, I. D.; Gerlach, J. C. Nonsilver, direct-positive dyes bleachout system using polymethine dyes and colorless activators. U.S. Patent 3595655, 1971.

32. Toba, Y.; Yamaguchi, T.; Yasuike, M. Photopolymerizable composition. Jpn. Kokai Tokkyo Koho JP 06348011, 1994.

33. Vatanparast, R.; Li, S.; Hakala, K.; Lemmetyinen, H. Thermal cure monitoring of phenolic resole resins based on in situ fluorescence technique. *J. Appl. Polym. Sci.* **2002**, *83*, 1773–1780.

34. Zhang, H.; Chen, Q.; Xiang, M.; Ma, C.; Huang, Q.; Yang, S. In silico prediction of mitochondrial toxicity by using GA-CG-SVM approach. *Toxicol. in Vitro* **2009**, *23*, 134–140.

4-[4-(DIMETHYLAMINO)STYRYL]-*N*-METHYLPYRIDINIUM IODIDE (4-Di-1-ASP) (DASPMI)

CAS Registry Number 959-81-9

Chemical Structure

CA Index Name Pyridinium, 4-[2-[4-(dimethylamino) phenyl]ethenyl]-1-methyl-, iodide (1 : 1)

Other Names 4-[*p*-(Dimethylamino)styryl]-1-methylpyridinium iodide; Pyridinium, 4-[2-[4-(dimethylamino)phenyl]ethenyl]-1-methyl-, iodide; Pyridinium, 4-[*p*-(dimethylamino)styryl]-1-methyl-, iodide; 1-Methyl 4-(4-dimethylamino styryl) pyridinium iodide; 4-[2-[4-(Dimethylamino)phenyl]-ethenyl]-1-methylpyridinium iodide; 4-[4-(Dimethylamino)-α-styryl]-1-methylpyridinium iodide; 4-[4-(Dimethylamino)styryl]-1-methylpyridinium iodide; 4-[4-(Dimethylamino) styryl]-*N*-methylpyridinium iodide; 4′-Dimethylamino-1-methylstilbazolium iodide; 4′-Dimethylamino-*N*-methyl-4-stilbazolium iodide; D 288; DASPMI; 4-Di-1-ASP; ω-(*N*′-Methylpyridyl-4′)-4-dimethylaminostyrene iodide

Merck Index Number Not listed

Chemical/Dye Class Styryl

Molecular Formula $C_{16}H_{19}IN_2$

Molecular Weight 366.24

Physical Form Greenish purple solid;[2] Purple solid; Red needles[16]

Solubility Soluble in chloroform, *N,N*-dimethylformamide, dimethyl sulfoxide, methanol

Melting Point 262–263 °C;[6] 261–262 °C;[17] 261 °C;[7,8] 259 °C;[12] 258–259 °C;[20] 256–258 °C;[16] 255.5–256.2 °C;[11] 240–242 °C[2]

Absorption (λ_{max}) 475 nm (MeOH); 472 nm (DMSO); 505 nm (CHCl$_3$); 450 nm (H$_2$O)

Emission (λ_{max}) 605 nm (MeOH)

Molar Extinction Coefficient 45,000 cm^{-1} M^{-1} (MeOH); 42,000 cm^{-1} M^{-1} (DMSO); 58,000 cm^{-1} M^{-1} (CHCl$_3$)

Synthesis Synthetic methods[1–21]

Imaging/Labeling Applications Amyloid plaques[22–24] epithelial cells;[25] mitochondria;[1,26–32] nerve terminals;[33] nucleic acids;[4,34–36] proteins[37]

Biological/Medical Applications Analyzing/monitoring protein aggregation;[37] analyzing/visualizing mitochondria;[1,26–32] detecting nucleic acids,[4,34–36] prostate cancer,[29] diagnosing/treating amyloidosis disorder;[22–24] measuring/monitoring membrane potential;[1,30,32] treating bacterial infection[38]

Industrial Applications Electrochromic materials;[39] glass;[40] nonlinear optical (NLO) materials;[1,2,7,8,14,41–43] photoconductors;[44,45] thin films[46]

Safety/Toxicity No data available

REFERENCES

1. Sabnis, R. W. *Handbook of Biological Dyes and Stains*; John Wiley & Sons Inc.: Hoboken, **2010**; pp 132–133.

2. Teshome, A.; Bhuiyan, M. D. H.; Gainsford, G. J.; Ashraf, M.; Asselberghs, I.; Williams, G. V. M.; Kay, A. J.; Clays, K. Synthesis, linear and quadratic nonlinear optical properties of ionic indoline and N,N-dimethylaniline based chromophores. *Opt. Mater.* **2011**, *33*, 336–345.

3. Vasilev, A.; Deligeorgiev, T.; Gadjev, N.; Kaloyanova, S.; Vaquero, J. J.; Alvarez-Builla, J.; Baeza, A. G. Novel environmentally benign procedures for the synthesis of styryl dyes. *Dyes Pigm.* **2008**, *77*, 550–555.

4. Balanda, A. O.; Volkova, K. D.; Kovalska, V. B.; Losytskyy, M. Yu.; Lukashov, S. S.; Yarmoluk, S. M. Novel styryl cyanines and their dimers as fluorescent dyes for nucleic acids detection: Synthesis and spectral-luminescent studies. *Ukr. Bioorg. Acta* **2006**, *4*, 17–29.

5. Chang, Y.; Rosania, G. Combinatorial fluorescent library based on the styryl scaffold. U.S. Pat. Appl. Publ. US 2005054006, 2005.

6. Wang, L.; Zhang, X.; Shi, Y.; Zhang, Z. Microwave-assisted solvent-free synthesis of some hemicyanine dyes. *Dyes Pigm.* **2004**, *62*, 21–25.

7. Okada, S.; Nogi, K.; Anwar; Tsuji, K.; Duan, X.; Oikawa, H.; Matsuda, H.; Nakanishi, H. Ethyl-substituted stilbazolium derivatives for second-order nonlinear optics. *Jpn. J. Appl. Phys., Part 1* **2003**, *42*, 668–671.

8. Umezawa, H.; Tsuji, K.; Okada, S.; Oikawa, H.; Matsuda, H.; Nakanishi, H. Molecular design on substituted DAST derivatives for second-order nonlinear optics. *Opt. Mater.* **2003**, *21*, 75–78.

9. Berneth, H.; Bruder, F.; Haese, W.; Hagen, R.; Hassenrueck, K.; Kostromine, S.; Landenberger, P.; Oser, R.; Sommermann, T.; Stawitz, J.; Bieringer, T. Optical data support comprising a hemicyanine dye in the information layer as light-absorbing compound. PCT Int. Appl. WO 2002086879, 2002.

10. Zhang, J.; Wu, Y.; Wang, Z.; Li, F.; Jin, L. Electrochemical properties of 4-[2-4(dimethylamino)phenyl]ethenylpyridine N-oxide. *Wuli Huaxue Xuebao* **2000**, *16*, 362–365.

11. Shah, S. S.; Ahmad, R.; Shah, S. W. H.; Asif, K. M.; Naeem, K. Synthesis of cationic hemicyanine dyes and their interactions with ionic surfactants. *Colloids Surf., A* **1998**, *137*, 301–305.

12. Stewart, K. R. Preparation of high melting point stilbazolium salts as second harmonic generators. U.S. Patent 5292888, 1994.

13. Matsui, M.; Kawamura, S.; Shibata, K.; Muramatsu, H. Synthesis and characterization of mono-, bis-, and tris-substituted pyridinium and pyrylium dyes. *Bull. Chem. Soc. Jpn.* **1992**, *65*, 71–74.

14. Okada, S.; Matsuda, H.; Nakanishi, H.; Kato, M.; Muramatsu, R. Organic nonlinear optical materials. Jpn. Kokai Tokkyo Koho JP 63048265, 1988.

15. Barni, E.; Savarino, P.; Larovere, R.; Viscardi, G.; Pelizzetti, E. Long-chain heterocyclic dyes. Part I. Hydrophobic structures. *J. Heterocycl. Chem.* **1986**, *23*, 209–221.

16. Kramer, D. N.; Bisauta, L. P.; Bato, R. Kinetics of the condensation of N-methyl-4-picolinium iodide with p-dimethylaminobenzaldehyde in aqueous ethanol. *J. Org. Chem.* **1974**, *39*, 3132–3136.

17. Sheinkman, A. K.; Prilepskaya, A. N.; Kolomoitsev, L. R.; Kost, A. N. 4-p-Dialkylaminophenylpyridinium salts: A new group of bactericides, fungicides, and herbicides. *Vestnik Moskov. Univ., Ser. 2, Khim.* **1964**, *19*, 74–82.

18. Sheinkman, A. K.; Rudenko, N. Z.; Kazarinova, N. F.; Lysenko, V. B. Structure of quaternary salts of 4-(p-dimethylaminophenyl) and 4-(p-dimethylaminostyryl)pyridines. *Zh. Obshch. Khim.* **1963**, *33*, 1964–1969.

19. Finkelstein, J.; Lee, J. Piperidine derivatives. U.S. Patent 2686784, 1954.

20. Phillips, A. P. Condensation of aromatic aldehydes with 4-picoline methiodide. *J. Org. Chem.* **1949**, *14*, 302–305.

21. Doja, M. Q.; Prasad, K. B. Cyanine dyes of the pyridine series. V. *J. Indian Chem. Soc.* **1947**, *24*, 301–306.

22. Gervais, F.; Kong, X.; Chalifour, R.; Migneault, D. Amyloid targeting imaging agents and uses thereof. U.S. Pat. Appl. Publ. US 20050048000, 2005.

23. Gervais, F.; Kong, X.; Chalifour, R.; Migneault, D. Amyloid targeting imaging agents and uses thereof. U.S. Pat. Appl. Publ. US 20020115717, 2002.

24. Gervais, F.; Kong, X.; Chalifour, R.; Migneault, D. Amyloid targeting imaging agents and uses thereof. PCT Int. Appl. WO 2002007781, 2002.

25. Salomon, J. J.; Endter, S.; Tachon, G.; Falson, F.; Buckley, S. T.; Ehrhardt, C. Transport of the fluorescent organic cation 4-(4-(dimethylamino)styryl)-N-methylpyridinium iodide (ASP+) in human respiratory epithelial cells. *Eur. J. Pharm. Biopharm.* **2012**, *81*, 351–359.

26. Brickley, M. R.; Lawrie, E.; Weise, V.; Hawes, C.; Cobb, A. H. Use of a potentiometric vital dye to determine the effect of the herbicide bromoxynil octanoate on mitochondrial bioenenergetics in *Chlamydomonas reinhardtii*. *Pest Manag. Sci.* **2012**, *68*, 580–586.

27. Brickley, M. R.; Weise, V.; Hawes, C.; Cobb, A. H. Morphology and dynamics of mitochondria in *Mougeotia* sp. *Eur. J. Phycol.* **2010**, *45*, 258–266.

28. Swayne, T. C.; Gay, A. C.; Pon, L. A. Visualization of mitochondria in budding yeast. *Methods Cell Biol.* **2007**, *80*, 591–626.

29. Dickman, D. Methods of detecting prostate cancer. PCT Int. Appl. WO 2006054296, 2006.

30. Hattori, F.; Fukuda, K. Method of selecting myocardial cells by using intracellular mitochondria as indication. PCT Int. Appl. WO 2006022377, 2006.

31. Villa, A. M.; Doglia, S. M. Mitochondria in tumor cells studied by laser scanning confocal microscopy. *J. Biomed. Opt.* **2004**, *9*, 385–394.

32. Dykens, J. A.; Velicelebi, G.; Ghosh, S. S. Compositions and methods for assaying subcellular

conditions and processes using energy transfer for drug screening. PCT Int. Appl. WO 2000079274, 2006.

33. Magrassi, L.; Purves, D.; Lichtman, J. W. Fluorescent probes that stain living nerve terminals. *J. Neurosci.* **1987**, *7*, 1207–1214.

34. Yashchuk, V. M.; Kudrya, V. Y.; Losytskyy, M. Y.; Tokar, V. P.; Yarmoluk, S. M.; Dmytruk, I. M.; Prokopets, V. M.; Kovalska, V. B.; Balanda, A. O.; Kryvorotenko, D. V.; Ogul'chansky, T. Y. The optical biomedical sensors for the DNA detection and imaging based on two-photon excited luminescent styryl dyes: Phototoxic influence on the DNA. *Proc. SPIE-Int. Soc. Opt. Eng.* **2007**, *6796*, 67960M/1–67960M/14.

35. Valyukh, I. V.; Vyshnyak, V. V.; Slobodyanyuk, A. V.; Yarmoluk, S. M. Spectroscopic study of the fluorescent dyes interaction with DNA. *Funct. Mater.* **2003**, *10*, 528–533.

36. Kumar, C. V.; Turner, R. S.; Asuncion, E. H. Groove binding of a styrylcyanine dye to the DNA double helix: the salt effect. *J. Photochem. Photobiol., A: Chem.* **1993**, *74*, 231–238.

37. Patton, W. F.; Yarmoluk, S. M.; Pande, P.; Kovalska, V.; Dai, L.; Volkova, K.; Coleman, J.; Losytskyy, M.; Ludlum, A.; Balanda, A. Novel dyes and compositions, and processes for using same in analysis of protein aggregation and other applications. U.S. Pat. Appl. Publ. US 20110130305, 2011.

38. Mendz, G. L.; Hazell, S. L. Method of treating *Helicobacter pylori* infection. PCT Int. Appl. WO 9427606, 1994.

39. Heller, W. R.; Kumamoto, J.; Luzzi, J. J. R.; Powers, J. C., Jr., Electrochromic light valve. U.S. Patent 3317266, 1967.

40. Imashita, K.; Yokokura, S. Glass containing photofunctional organic compound and its manufacture. Jpn. Kokai Tokkyo Koho JP 07064135, 1995.

41. Umezawa, H.; Okada, S.; Oikawa, H.; Matsuda, H.; Nakanishi, H. Synthesis and crystal structures of phenylethynylpyridinium derivatives for second-order nonlinear optics. *Bull. Chem. Soc. Jpn.* **2005**, *78*, 344–348.

42. Di Bella, S.; Fragala, I.; Ratner, M. A.; Marks, T. J. Chromophore environmental effects in saltlike nonlinear optical materials. A computational study of architecture/anion second-order response relationships in high-β stilbazolium self-assembled films. *Chem. Mater.* **1995**, *7*, 400–404.

43. Matsuda, H.; Okada, S.; Nakanishi, H.; Kato, M. Nonlinear optical polymeric materials. Jpn. Kokai Tokkyo Koho JP 02012133, 1990.

44. Yadav, H. O. Relation between the thermal activation energy of conduction and the first excited singlet state energy-a case of photo-conducting organic materials. *Thin Solid Films* **2005**, *477*, 222–226.

45. Narasimharaghavan, P. K.; Yadav, H. O.; Varadarajan, T. S.; Patnaik, L. N.; Das, S. Organic photoconductors: Dark and photoconduction studies in two p-dimethylamino styryl dyes derived from pyridine-2 and pyridine-4. *J. Mater. Sci.* **1991**, *26*, 4774–4786.

46. Forrest, S. R.; Burrows, P.; Ban, V. S. Low pressure vapor phase deposition of organic thin films. PCT Int. Appl. WO 9925894, 1999.

3,3′-DIMETHYL-α-NAPHTHOXACARBOCYANINE IODIDE (JC-9)

CAS Registry Number   522592-13-8

Chemical Structure

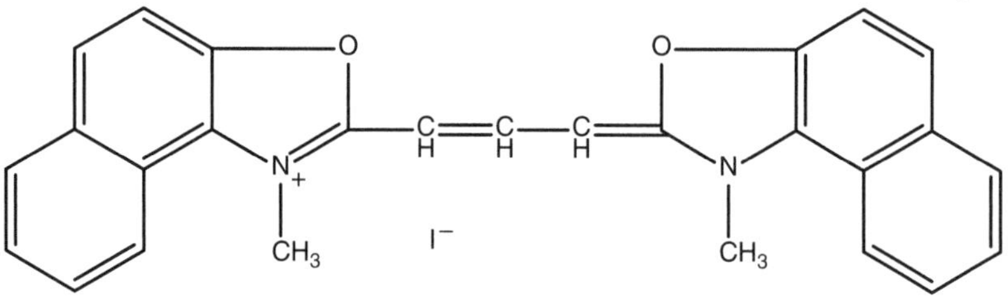

CA Index Name   Naphth[1,2-*d*]oxazolium, 1-methyl-2-[3-(1-methylnaphth-[1,2-*d*]oxazol-2(1*H*)-ylidene)-1-propen-1-yl]-, iodide (1 : 1)

Other Names   Naphth[1,2-*d*]oxazolium, 1-methyl-2-[3-(1-methylnaphth-[1,2-*d*]oxazol-2(1*H*)-ylidene)-1-propenyl]-, iodide; D 22421; DiNOC₁(3); 3,3′-Dimethyl-α-naphthoxacarbocyanine iodide; JC 9

Merck Index Number   Not listed

Chemical/Dye Class   Cyanine

Molecular Formula $C_{27}H_{21}IN_2O_2$

Molecular Weight   532.38

Physical Form   Red solid

Solubility   Soluble in chloroform, *N,N*-dimethylformamide, dimethyl sulfoxide

Absorption (λ_{max})   522 nm (CHCl₃)

Emission (λ_{max})   535 nm (CHCl₃)

Molar Extinction Coefficient   143,000 cm⁻¹ M⁻¹ (CHCl₃)

Synthesis   Synthetic methods[1,2]

Imaging/Labeling Applications   Mitochondria;[1–6] proteins;[7] sperms;[3] yeast cells[7]

Biological/Medical Applications   Assessing sperm quality;[3] detecting ligated products;[8] identifying proteins;[7] measuring *in vivo* hematotoxicity;[5] measuring/monitoring membrane potential;[1–6] apoptosis assay;[5,9] as temperature sensor[10,11]

Industrial Applications   Not reported

Safety/Toxicity   No data available

REFERENCES

1. Sabnis, R. W. *Handbook of Biological Dyes and Stains*; John Wiley & Sons Inc.: Hoboken, **2010**; p 256.

2. Haugland, R. P. *Handbook of Fluorescent Probes and Research Products*; Molecular Probes Inc.: Eugene, **2002**; pp 473–488.

3. Hayrabedyan, S.; Georgiev, B.; Kacheva, D.; Chervenkov, M.; Shumkov, K.; Taushanova, P.; Kistanova, E. Flow cytometry as a method for advanced evaluation of boar semen. *Comptes Rendus Acad. Bulg. Sci.* **2012**, *65*, 541–548.

4. Michalowicz, J.; Sicinska, P. Chlorophenols and chlorocatechols induce apoptosis in human lymphocytes (*in vitro*). *Toxicol. Lett.* **2009**, *191*, 246–252.

5. Dertinger, S. D.; Bemis, J. C.; Bryce, S. M. Method for measuring *in vivo* hematotoxicity with an emphasis on radiation exposure assessment. U.S. Pat. Appl. Publ. US 20080311586, 2008.

6. Hattori, F.; Fukuda, K. Method for selecting myocardial cells using intracellular mitochondria

as indication. PCT Int. Appl. WO 2006022377, 2006.

7. Kim, E. J.; Barker, L.; Burnet, M.; Guse, J.; Luyten, K.; Tsotsou, G. Method for identifying transport proteins. PCT Int. Appl. WO 2003038092, 2003.

8. Mancebo, R. Reagents and methods for autolitigation chain reaction. PCT Int. Appl. WO 2013102150, 2013.

9. Smyth, P. G.; Berman, S. A. Markers of apoptosis: Methods for elucidating the mechanism of apoptotic cell death from the nervous system. *Biotechniques* **2002**, *32*, 648–650, 652, 654.

10. Bousseksou, A.; Salmon, L.; Molnar, G.; Cobo, S. Materials with thermochromic spin transition doped with one or more fluorescent agents for use as temperature sensor. Fr. Demande FR 2952371, 2011.

11. Bousseksou, A.; Salmon, L.; Molnar, G.; Cobo, S. Heat-sensitive spin-transition materials doped with one or more fluorescent agents for use as temperature sensor. PCT Int. Appl. WO 2011058277, 2011.

3,3′-DIPENTYLOXACARBOCYANINE IODIDE (DiOC$_5$(3))

CAS Registry Number   53213-81-3

Chemical Structure

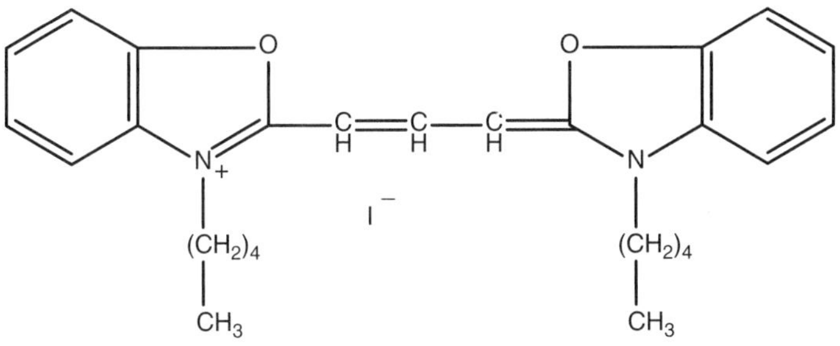

CA Index Name   Benzoxazolium, 3-pentyl-2-[3-(3-pentyl-2(3H)-benzoxazolylidene)-1-propen-1-yl]-, iodide (1 : 1)

Other Names   Benzoxazolium, 3-pentyl-2-[3-(3-pentyl-2(3H)-benzoxazolylidene)-1-propenyl]-, iodide; 3,3′-Dipentyloxacarbocyanine iodide; DiOC$_5$(3); NK 2453

Merck Index Number   Not listed

Chemical/Dye Class   Cyanine

Molecular Formula   C$_{27}$H$_{33}$IN$_2$O$_2$

Molecular Weight   544.47

Physical Form   Red powder or solid

Solubility   Soluble in chloroform, N,N-dimethylformamide, dimethyl sulfoxide, methanol

Melting Point   214–215 °C[2]

Absorption (λ_{max})   484 nm (MeOH)

Emission (λ_{max})   500 nm (MeOH)

Molar Extinction Coefficient   155,000 cm^{-1} M^{-1} (MeOH)

Synthesis   Synthetic methods[1,2]

Imaging/Labeling Applications   Mitochondria;[3–7] bacteria;[8,15,16,28] cells;[9] endoplasmic reticulum;[10] fungi;[27] hairs;[11] intraocular tissue;[12] lymphocytes;[13,14] leukocytes;[15] squamous epithelial cells;[15] microorganisms;[15,16,25–28] parasites;[26] platelets;[13] monocytes;[17] retina tissue;[18] sperms;[19] yeasts[25]

Biological/Medical Applications   Detecting prostate cancer;[3] detecting/quantitating microorganisms,[15] leukocytes,[15] squamous epithelial cells;[15] measuring/monitoring membrane potential;[1,2,4,7,8,13,14,19–23,25–28] mitochondria stain;[3–7] cytotoxicity assay;[23,24] susceptibility assay[25–28]

Industrial Applications   Photoresists[29]

Safety/Toxicity   No data available

REFERENCES

1. Sabnis, R. W. *Handbook of Biological Dyes and Stains*; John Wiley & Sons Inc.: Hoboken, **2010**; pp 160–161.

2. Sims, P. J.; Waggoner, A. S.; Wang, C.; Hoffman, J. F. Mechanism by which cyanine dyes measure membrane potential in red blood cells and phosphatidylcholine vesicles. *Biochemistry* **1974**, *13*, 3315–3330.

3. Dickman, D. Methods of detecting prostate cancer. PCT Int. Appl. WO 2006054296, 2006.

4. Shapiro, H. M. Cell membrane potential analysis. *Methods Cell Biol.* **1994**, *41*, 121–133.

5. Liu, Z.; Bushnell, W. R.; Brambl, R. Potentiometric cyanine dyes are sensitive probes for mitochondria in intact plant cells. *Plant Physiol.* **1987**, *84*, 1385.

6. Haanen, C.; Muus, P.; Pennings, A. The effect of cytosine arabinoside upon mitochondrial staining kinetics in human hematopoietic cells. *Histochemistry* **1986**, *84*, 609–613.

7. Johnson, L. V.; Wash, M. L.; Bockus, B. J.; Chen, L. B. Monitoring of relative mitochondrial membrane potential in living cells by fluorescence microscopy. *J. Cell Biol.* **1981**, *88*, 526–535.

8. Miller, J. B.; Koshland, D. E., Jr., Effects of cyanine dye membrane probes on cellular properties. *Nature* **1978**, *272*, 83–84.

9. Crissman, H. A.; Hofland, M. H.; Stevenson, A. P.; Wilder, M. E.; Tobey, R. A. Supravital cell staining with Hoechst 33342 and DiOC5(3). *Methods Cell Biol.* **1990**, *33*, 89–95.

10. Terasaki, M.; Song, J.; Wong, J. R.; Weiss, M. J.; Chen, L. B. Localization of endoplasmic reticulum in living and glutaraldehyde-fixed cells with fluorescent dyes. *Cell* **1984**, *38*, 101–108.

11. Ohashi, Y.; Miyabe, H.; Matsunaga, K. Hair dye composition. Eur. Pat. Appl. EP 1166753, 2002.

12. Watanabe, K.; Shinto, T.; Nomoto, T.; Miyazaki, T.; Tanaka, T.; Nishimura, Y.; Shimada, Y.; Nishimura, N. Labeling composition for intraocular tissue, labeling method of intraocular tissue, and screening method. PCT Int. Appl. WO 2010074325, 2010.

13. Bramhall, J. S.; Morgan, J. I.; Perris, A. D.; Britten, A. Z. The use of a fluorescent probe to monitor alterations in trans-membrane potential in single cell suspensions. *Biochem. Biophys. Res. Commun.* **1976**, *72*, 654–662.

14. Morgan, J. I.; Bramhall, J. S.; Britten, A. Z.; Perris, A. D. The use of a fluorescent probe to demonstrate the interaction of calcium and estradiol-17β in the thymic lymphocyte plasma membrane. *J. Endocrinol.* **1976**, *69*, 29P–30P.

15. Mansour, J. D.; Schulte, T. H.; Sage, B. H. Detection and quantitation of microorganisms, leukocytes and squamous epithelial cells in urine. U.S. Patent 4622298, 1986.

16. Mansour, J. D. Fluorescent gram stain. U.S. Patent 4665024, 1987.

17. Sakata, T.; Kuroda, T. Reagent for classifying leukocytes by flow cytometry. U.S. Patent 5175109, 1992.

18. Watanabe, K.; Shindo, T.; Nomoto, T.; Miyazaki, T.; Tanaka, T.; Nishimura, A.; Shimada, Y.; Nishimura, K. Retina tissue staining composition, retina tissue staining method. Jpn. Kokai Tokkyo Koho JP 2010148447, 2010.

19. Peterson, R. N.; Bundman, D.; Freund, M. Use of a fluorescent dye to measure drug-induced changes in the membrane potential of boar spermatozoa. *Life Sci.* **1978**, *22*, 659–666.

20. Remani, P.; Ostapenko, V. V.; Akagi, K.; Bhattathiri, V. N.; Nair, M. K.; Tanaka, Y. Relation of transmembrane potential to cell survival following hyperthermia in HeLa cells. *Cancer Lett.* **1999**, *144*, 117–123.

21. Waggoner, A. S. Dye indicators for membrane potential. *Annu. Rev. Biophys. Bioeng.* **1979**, *8*, 47–68.

22. Pick, U.; Avron, M. Measurement of transmembrane potentials in *Rhodhospirillium rubrum* chromatophores with an oxacarbocyanine dye. *Biochim. Biophys. Acta* **1976**, *440*, 189–204.

23. Murakami, T. Cytotoxicity test method by measuring membrane electric potential. Jpn. Kokai Tokkyo Koho JP 2000300290, 2000.

24. Wadkins, R. M.; Houghton, P. J. Kinetics of transport of dialkyloxacarbocyanines in multidrug-resistant cell lines overexpressing p-glycoprotein: Interrelationship of dye alkyl chain length, cellular flux, and drug resistance. *Biochemistry* **1995**, *34*, 3858–3872.

25. Peyron, F.; Favel, A.; Guiraud-Dauriac, H.; El Mzibri, M.; Chastin, C.; Dumenil, G.; Regli, P. Evaluation of a flow cytofluorometric method for rapid determination of amphotericin B susceptibility of yeast isolates. *Antimicrob. Agents Chemther.* **1997**, *41*, 1537–1540.

26. Azas, N.; Di Giorgio, C.; Delmas, F.; Gasquet, M.; Timon-David, P. Assessment of amphotericin B susceptibility in *Leishmania infantum* promastigotes by flow cytometric membrane potential assay. *Cytometry* **1997**, *28*, 165–169.

27. Ordonez, J. V.; Wehman, N. M. Amphotericin B susceptibility of *Candida* species assessed by rapid flow cytometric membrane potential assay. *Cytometry* **1995**, *22*, 154–157.

28. Ordonez, J. V.; Wehman, N. M. Rapid flow cytometric antibiotic susceptibility assay for *Staphylococcus aureus*. *Cytometry* **1993**, *14*, 811–818.

29. Noppakundilograt, S.; Miyagawa, N.; Takahara, S.; Yamaoka, T. Visible light-sensitive positive-working photopolymer based on poly(p-hydroxystyrene) and vinyl ether crosslinker. *J. Photopolym. Sci. Technol.* **2000**, *13*, 719–721.

3,3′-DIPROPYLTHIADICARBO-CYANINE IODIDE (DiSC$_3$(5))

CAS Registry Number 53213-94-8

Chemical Structure

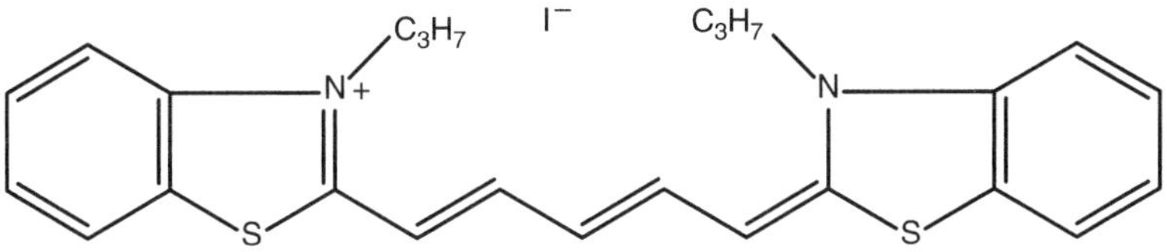

CA Index Name Benzothiazolium, 3-propyl-2-[5-(3-propyl-2(3H)-benzothiazolylidene)-1,3-pentadien-1-yl]-, iodide (1 : 1)

Other Names Benzothiazolium, 3-propyl-2-[5-(3-propyl-2(3H)-benzothiazolylidene)-1,3-pentadienyl]-, iodide; 3,3′-Dipropyl-2,2′-thiadicarbocyanine iodide; 3,3′-Dipropylthiadicarbocyanine iodide; Di-S-C$_3$-5; DiSC$_3$(5); NK 2251

Merck Index Number Not listed

Chemical/Dye Class Cyanine

Molecular Formula C$_{25}$H$_{27}$IN$_2$S$_2$

Molecular Weight 546.53

Physical Form Green powder or solid

Solubility Soluble in N,N-dimethylformamide, dimethyl sulfoxide, methanol

Melting Point 248–249 °C[2]

Absorption (λ_{max}) 651 nm (MeOH)

Emission (λ_{max}) 675 nm (MeOH)

Molar Extinction Coefficient 258,000 cm^{-1} M^{-1} (MeOH)

Synthesis Synthetic methods[1,2]

Imaging/Labeling Applications Adipocytes;[3] bacteria;[4–7] brush border membrane vesicles (BBMV);[8–12] cells;[13] Ehrlich ascites tumor cells;[57–59] epithelial cells;[14] erythrocytes;[15–21] granulocytes;[22,23] leukocytes;[24] lipid membranes/lipid bilayer membranes;[25–27] liposomes;[26,28,29] lymphocytes;[30,31] lysosomes;[32,33] microorganisms;[34] mitochondria;[30,35–41] mycoplasma membranes;[42] neuroblastoma cells;[43,44] neutrophils;[45,46] platelets;[47,48] proteins;[49] sperms;[50,51] yeast[52–55]

Biological/Medical Applications Detecting prostate cancer,[35] calcium-binding proteins;[49] measuring/monitoring membrane potential;[1–7,10–24,27–33,36–61] studying conformational changes,[49] neutrophil activation;[45] as anticancer agents;[56–59] cytotoxicity assays;[60] for photodynamic therapy[62]

Industrial Applications Not reported

Safety/Toxicity No data available

REFERENCES

1. Sabnis, R. W. *Handbook of Biological Dyes and Stains*; John Wiley & Sons Inc.: Hoboken, **2010**; pp 169–170.

2. Sims, P. J.; Waggoner, A. S.; Wang, C.; Hoffman, J. F. Mechanism by which cyanine dyes measure membrane potential in red blood cells and phosphatidylcholine vesicles. *Biochemistry* **1974**, *13*, 3315–3330.

3. Bailey, F.; Hill, N.; Malinski, T.; Kiechle, F. Changes in membrane potential of intact adipocytes measured with fluorescent dyes. *Bioelectrochem. Bioenerg.* **1989**, *21*, 333–342.

4. Chehimi, S.; Pons, A.; Sable, S.; Hajlaoui, M.; Limam, F. Mode of action of thuricin S, a new class

IId bacteriocin from *Bacillus thuringiensis*. *Can. J. Microbiol.* **2010**, *56*, 162–167.

5. Kirouac, M.; Vachon, V.; Rivest, S.; Schwartz, J. L.; Laprade, R. Analysis of the properties of *Bacillus thuringiensis* insecticidal toxins using a potential-sensitive fluorescent probe. *J. Membr. Biol.* **2003**, *196*, 51–59.

6. Suzuki, H.; Wang, Z.; Yamakoshi, M.; Kobayashi, M.; Nozawa, T. Probing the transmembrane potential of bacterial cells by voltage-sensitive dyes. *Anal. Sci.* **2003**, *19*, 1239–1242.

7. Goulbourne, E. A., Jr.,; Greenberg, E. P. Chemotaxis of *Spirochaeta aurantia*: Involvement of membrane

potential in chemosensory signal transduction. *J. Bacteriol.* **1981**, *148*, 837–844.

8. Faria, J. L.; Berberan-Santos, M.; Prieto, M. J. E. A comment on the localization of cyanine dye binding to brush-border membranes by the fluorescence quenching of n-(9-anthroyloxy) fatty acid probes. *Biochim. Biophys. Acta, Biomembr.* **1990**, *1026*, 133–134.

9. Lipkowitz, M. S.; Abramson, R. G. Ionic permeabilities of rat renal cortical brush-border membrane vesicles. *Am. J. Physiol.* **1987**, *252*, F700–F711.

10. Cabrini, G.; Verkman, A. S. Potential-sensitive response mechanism of 3,3′-dipropylthiodicarbocyanine iodide (DiS-C3-(5)) in biological membranes. *J. Membr. Biol.* **1986**, *92*, 171–182.

11. Cabrini, G.; Verkman, A. S. Mechanism of interaction of the cyanine dye DiS-C3-(5) with renal brush-border vesicles. *J. Membr. Biol.* **1986**, *90*, 163–175.

12. Ganapathy, V.; Burckhardt, G.; Leibach, F. H. Peptide transport in rabbit intestinal brush-border membrane vesicles studied with a potential-sensitive dye. *Biochim. Biophys. Acta* **1985**, *816*, 234–240.

13. Plasek, J.; Dale, R. E.; Sigler, K.; Laskay, G. Transmembrane potentials in cells: A diS-C3(3) assay for relative potentials as an indicator of real changes. *Biochim. Biophys. Acta, Biomembr.* **1994**, *1196*, 181–190.

14. Kaunitz, J. D. Preparation and characterization of viable epithelial cells from rabbit distal colon. *Am. J. Physiol.* **1988**, *254*, G502-G512.

15. Waczulikova, I.; Sikurova, L.; Bryszewska, M. R; kawiecka, K.; Carsky, J.; Ulicna, O. Impaired erythrocyte transmembrane potential in diabetes mellitus and its possible improvement by resorcylidene aminoguanidine. *Bioelectrochemistry* **2000**, *52*, 251–256.

16. Kaji, D. M. Effect of membrane potential on K-Cl transport in human erythrocytes. *Am. J. Physiol.* **1993**, *264*, C376-C382.

17. Glaser, R.; Gengnagel, C.; Donath, J. The influence of valinomycin induced membrane potential on erythrocyte shape. *Biomed. Biochim. Acta* **1991**, *50*, 869–877.

18. Pape, L. Effect of extracellular Ca2+, K+ and OH- on erythrocyte membrane potential as monitored by the fluorescent probe 3,3′-dipropylthiodicarbocyanine. *Biochim. Biophys. Acta* **1982**, *686*, 225–232.

19. Freedman, J. C.; Hoffman, J. F. The relation between dicarbocyanine dye fluorescence and the membrane potential of human red blood cells set at varying *Donnan equilibriums. J. Gen. Physiol.* **1979**, *74*, 187–212.

20. Tsien, R. Y.; Hladky, S. B. A quantitative resolution of the spectra of a membrane potential indicator, diS-C3-(5), bound to cell components and to red blood cells. *J. Membr. Biol.* **1978**, *38*, 73–97.

21. Hladky, S. B.; Rink, T. J. pH changes in human erythrocytes reported by 3,3′-dipropylthiadicarbocyanine. *J. Physiol.* **1976**, *263*, 213P-214P.

22. Castranova, V.; Van Dyke, K. Analysis of oxidation of the membrane potential probe, Di-S-C3(5), during granulocyte activation. *Microchem. J.* **1984**, *29*, 151–161.

23. Whitin, J. C.; Clark, R. A.; Simons, E. R.; Cohen, H. J. Effects of the myeloperoxidase system on fluorescent probes of granulocyte membrane potential. *J. Biol. Chem.* **1981**, *256*, 8904–8906.

24. Kuroki, M.; Kamo, N.; Kobatake, Y.; Okimasu, E.; Utsumi, K. Measurement of membrane potential in polymorphonuclear leukocytes and its changes during surface stimulation. *Biochim. Biophys. Acta* **1982**, *693*, 326–334.

25. Krishna, M. M. G.; Periasamy, N. Fluorescence of organic dyes in lipid membranes: Site of solubilization and effects of viscosity and refractive index on lifetimes. *J. Fluoresc.* **1998**, *8*, 81–91.

26. Matylevich, N. P. Fluctuation accumulations of cholesterol molecules in lipid bilayer determine substance distribution between the membrane and water phases. *Biofizika* **1986**, *31*, 714–716.

27. Waggoner, A. S.; Wang, C. H.; Tolles, R. L. Mechanism of potential-dependent light absorption changes of lipid bilayer membranes in the presence of cyanine and oxonol dyes. *J. Membr. Biol.* **1977**, *33*, 109–140.

28. Koyano, T.; Saito, M.; Myamoto, H.; Umibe, K.; Kato, M. Biochip with liposome-encapsulated with light-responsive chemical and liposome-encapsulated with fluorescent dye. Jpn. Kokai Tokkyo Koho JP 05175574, 1993.

29. Kumazawa, T.; Nomura, T.; Kurihara, K. Liposomes as model for taste cells: receptor sites for bitter substances including N-C=S substances and mechanism of membrane potential changes. *Biochemistry* **1988**, *27*, 1239–1244.

30. Gulyaeva, N. V.; Konoshenko, G. I.; Mokhova, E. N. Mitochondrial membrane potential in lymphocytes as monitored by fluorescent cation diS-C3-(5). *Membr. Biochem.* **1985**, *6*, 19–32.

31. Montecucco, C.; Pozzan, T.; Rink, T. Dicarbocyanine fluorescent probes of membrane potential block lymphocyte capping, deplete cellular ATP and inhibit respiration of isolated mitochondria. *Biochim. Biophys. Acta, Biomembr.* **1979**, *552*, 552–557.

32. Harikumar, P.; Reeves, J. P. The lysosomal proton pump is electrogenic. *J. Biol. Chem.* **1983**, *258*, 10403–10410.

33. Ohkuma, S.; Moriyama, Y.; Takano, T. Electrogenic nature of lysosomal proton pump as revealed with a cyanine dye. *J. Biochem.* **1983**, *94*, 1935–1943.

34. Lykov, V. P.; Khovrychev, M. P.; Polin, A. N. Determination of the sensitivity of microorganisms to antibiotics. U.S.S.R. SU 1337411, 1987.

35. Dickman, D. Methods of detecting prostate cancer. PCT Int. Appl. WO 2006054296, 2006.

36. Yamamoto, T.; Tachikawa, A.; Terauchi, S.; Yamashita, K.; Kataoka, M.; Terada, H.; Shinohara, Y. Multiple effects of DiS-C3(5) on mitochondrial structure and function. *Eur. J. Biochem.* **2004**, *271*, 3573–3579.

37. Farrelly, E.; Amaral, M. C.; Marshall, L.; Huang, S. G. A high-throughput assay for mitochondrial membrane potential in permeabilized yeast cells. *Anal. Biochem.* **2001**, *293*, 269–276.

38. Huang, S.; Chen, J. High-throughput screening assays for modulators of mitochondrial membrane potential. PCT Int. Appl. WO 2000068686, 2000.

39. Bammel, B. P.; Brand, J. A.; Germon, W.; Smith, J. C. Interaction of the extrinsic potential-sensitive molecular probe diS-C3-(5) with pigeon heart mitochondria under equilibrium and time-resolved conditions. *Arch. Biochem. Biophys.* **1986**, *244*, 67–84.

40. Bakeeva, L. E.; Derevyanchenko, I. G.; Konoshenko, G. I.; Mokhova, E. N. Interaction of diS-C3-(5) and ethylrhodamine with lymphocyte mitochondria. *Biokhimiya* **1983**, *48*, 1463–1470.

41. Kovac, L.; Poliachova, V. Membrane potential monitoring cyanine dyes uncouple respiration and induce respiration-deficient mutants in intact yeast cells. *Biochem. Int.* **1981**, *2*, 503–507.

42. Schummer, U.; Schiefer, H. G. Ion diffusion potentials across mycoplasma membranes determined by a novel method using a carbocyanine dye. *Arch. Biochem. Biophys.* **1986**, *244*, 553–562.

43. Milligan, G.; Strange, P. G. Reduction in accumulation of [3H]triphenylmethylphosphonium cation in neuroblastoma cells caused by optical probes of membrane potential. Evidence for interactions between carbocyanine dyes and lipophilic anions. *Biochim. Biophys. Acta, Mol. Cell Res.* **1983**, *762*, 585–592.

44. Milligan, G.; Strange, P. G. Biochemical estimation of membrane potential in neuroblastoma cells. *Biochem. Soc. Trans.* **1981**, *9*, 414–415.

45. Seligmann, B. E.; Gallin, J. I. Comparison of indirect probes of membrane potential utilized in studies of human neutrophils. *J. Cell. Physiol.* **1983**, *115*, 105–115.

46. Tatham, P. E. R.; Delves, P. J.; Shen, L.; Roitt, I. M. Chemotactic factor-induced membrane potential changes in rabbit neutrophils monitored by the fluorescent dye 3,3′-dipropylthiadicarbocyanine iodide. *Biochim. Biophys. Acta, Biomembr.* **1980**, *602*, 285–298.

47. Pipili, E. Platelet membrane potential: simultaneous measurement of diSC3(5) fluorescence and optical density. *Thromb. Haemost.* **1985**, *54*, 645–649.

48. Friedhoff, L. T.; Sonenberg, M. The membrane potential of human platelets. *Blood* **1983**, *61*, 180–185.

49. Orlov, S. N.; Pokudin, N. I.; Ryazhskii, G. G.; Kravtsov, G. M. Use of 3,3′-dipropylthio carbocyanine iodide [diS-C3-(5)] for the study of conformation changes and for detection of calcium-binding proteins. *Biokhimiya* **1984**, *49*, 51–59.

50. Boitano, S.; Omoto, C. K. Membrane hyperpolarization activates trout sperm without an increase in intracellular pH. *J. Cell Sci.* **1991**, *98*, 343–349.

51. Chou, K.; Chen, J.; Yuan, S. X.; Haug, A. The membrane potential changes polarity during capacitation of murine epididymal sperm. *Biochem. Biophys. Res. Commun.* **1989**, *165*, 58–64.

52. Pena, A.; Sanchez, N. S.; Calahorra, M. Estimation of the electric plasma membrane potential difference in yeast with fluorescent dyes: comparative study of methods. *J. Bioenerg. Biomembr.* **2010**, *42*, 419–432.

53. Pena, A.; Uribe, S.; Pardo, J. P.; Borbolla, M. The use of a cyanine dye in measuring membrane potential in yeast. *Arch. Biochem. Biophys.* **1984**, *231*, 217–225.

54. Van den Broek, P. J.; Christianse, K.; Van Steveninck, J. The energetics of D-fucose transport in *Saccharomyces fragilis*. The influence of the protonmotive force on sugar accumulation. *Biochim. Biophys. Acta* **1982**, *692*, 231–237.

55. Kovac, L.; Varecka, L. Membrane potentials in respiring and respiration-deficient yeasts monitored by a fluorescent dye. *Biochim. Biophys. Acta* **1981**, *637*, 209–216.

56. Iwagaki, H.; Fuchimoto, S.; Shiiki, S.; Miyake, M.; Orita, K. Monitoring the effect of an anticancer drug on RPMI 4788 cells by a membrane potential probe, dis-C3-(5). *J. Med.* **1989**, *20*, 135–141.

57. Smith, T. C.; Herlihy, J. T.; Robinson, S. C. The effect of the fluorescent probe, 3,3′-dipropylthiadicarbocyanine iodide, on the energy metabolism of Ehrlich ascites tumor cells. *J. Biol. Chem.* **1981**, *256*, 1108–1110.

58. Smith, T. C.; Robinson, S. C. The effect of the fluorescent probe, 3,3′-dipropylthiodicarbocyanine iodide, on the membrane potential of Ehrlich ascites tumor cells. *Biochem. Biophys. Res. Commun.* **1980**, *95*, 722–727.

59. Uchiumi, K.; Yasui, S.; Hara, H. Cyanine heterocycles as anticancer agents. Jpn. Kokai Tokkyo Koho JP 54157839, 1979.

60. Murakami, T. Cytotoxicity test method by measuring membrane electric potential. Jpn. Kokai Tokkyo Koho JP 2000300290, 2000.

61. Plasek, J.; Hrouda, V. Assessment of membrane potential changes using the carbocyanine dye diS-C3-(5): Synchronous excitation spectroscopy studies. *Eur. Biophys. J.* **1991**, *19*, 183–188.

62. Lipshutz, G. S.; Castro, D. J.; Saxton, R. E.; Haugland, R. P.; Soudant, J. Evaluation of four new carbocyanine dyes for photodynamic therapy with lasers. *Laryngoscope* **1994**, *104*, 996–1002.

ETHIDIUM BROMIDE (HOMIDIUM BROMIDE)

CAS Registry Number 1239-45-8

Chemical Structure

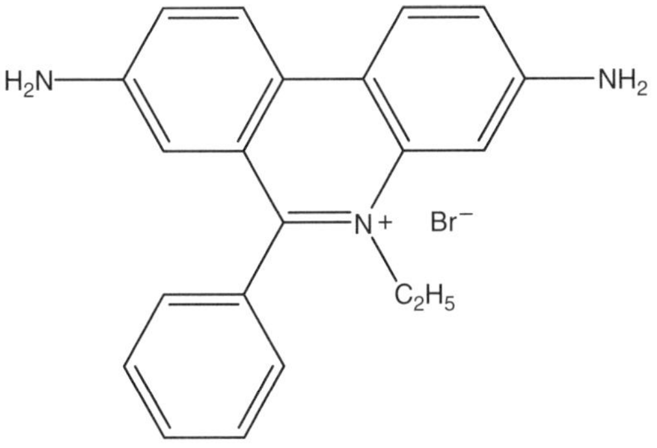

CA Index Name Phenanthridinium, 3,8-diamino-5-ethyl-6-phenyl-, bromide (1 : 1)

Other Names 3,8-Diamino-5-ethyl-6-phenyl-phenanthridinium bromide; Phenanthridinium, 3,8-diamino-5-ethyl-6-phenyl-, bromide; 2,7-Diamino-10-ethyl-9-phenylphenanthridinium bromide; 2,7-Diamino-9-phenyl-10-ethylphenanthridinium bromide; 2,7-Diamino-9-phenylphenanthridine ethobromide; Dromilac; Ethidium bromide; Homidium bromide

Merck Index Number 4769

Chemical/Dye Class Heterocycle; Phenanthridine

Molecular Formula $C_{21}H_{20}BrN_3$

Molecular Weight 394.31

Physical Form Dark red crystals;[2,4,5] Dark red elongated plates[6]

Solubility Soluble in water, chloroform, N,N-dimethylformamide, dimethyl sulfoxide, methanol

Melting Point 261–264 °C (decompose);[2] 248–249 °C (decompose);[6] 238–240 °C[4,5]

Absorption (λmax) 518 nm (H$_2$O/DNA); 480 nm (H$_2$O)

Emission (λmax) 605 nm (H$_2$O/DNA); 620 nm (H$_2$O)

Molar Extinction Coefficient 5,200 cm^{-1} M^{-1} (H$_2$O/DNA); 5,600 cm^{-1} M^{-1} (H$_2$O)

Synthesis Synthetic methods[1–6]

Imaging/Labeling Applications Nucleic acids;[1,2,7–46] bacteria;[47–50] cells;[51,52] chromatin;[53–56] chromosomes;[57–59] leukocytes;[60] microorganisms;[61–64] nerve terminals;[65] ribosomes;[66,67] sperms;[68–71] tumors;[72–74] yeast[75]

Biological/Medical Applications Assessing sperms quality;[70,71] detecting nucleic acids,[1,2,7–46] classifying/counting leukocytes;[60] counting sperms;[69] apoptosis assay;[76,77] cytotoxicity assay;[78,79] mitochondria toxicity assay;[80] as a substrate for measuring ribonuclease activity;[81,82] studying multidrug resistance;[83] as temperature sensor;[84,85] treating horses with pyroplasmosis;[86] treating trypanosomiasis[12,87]

Industrial Applications Electroluminescent displays;[88] photoresists[89]

Safety/Toxicity Carcinogenicity;[90–92] chromosome aberration;[93] genotoxicity;[94–97] DNA damage;[98] embryotoxicity;[99] genotoxicity;[100–104] mitochondrial toxicity;[105] mutagenicity;[106–113] neurotoxicity;[114,115] phototoxicity[116]

REFERENCES

1. Sabnis, R. W. *Handbook of Biological Dyes and Stains*; John Wiley & Sons Inc.: Hoboken, **2010**; pp 183–185.

2. Osadchii, S. A.; Shubin, V. G.; Kozlova, L. P.; Varlamenko, V. S.; Filipenko, M. L.; Boyarskikh, U. A. Improvement of ways to obtain ethidium bromide and synthesis of ethidium ethyl sulfate, a new fluorescent dye for detection of nucleic acids. *Russ. J. Appl. Chem.* **2011**, *84*, 1541–1548.

3. Osadchii, S. A.; Kozlova, L. P.; Varlamenko, V. S. Method of producing ethidium bromide. Russ. RU 2396260, 2010.

4. Short, W. F.; Peak, D. A.; Watkins, T. I. Phenanthridine compounds. U.S. Patent 2662082, 1953.

5. Watkins, T. I. New phenanthridine compounds. GB 697296, 1953.

6. Watkins, T. I. Trypanosides of the phenanthridine series. Part I. The effect of changing the quarternary grouping in dimidium bromide. *J. Chem. Soc.* **1952**, 3059–3064.

7. Ogino, M.; Makino, Y. System for detecting nucleic acid by fluorescent substance using energy transfer. Jpn. Kokai Tokkyo Koho JP 2012147722, 2012.

Handbook of Fluorescent Dyes and Probes, First Edition. R. W. Sabnis.
© 2015 John Wiley & Sons, Inc. Published 2015 by John Wiley & Sons, Inc.

8. Nandi, S.; Routh, P.; Layek, R. K.; Nandi, A. K. Ethidium bromide-adsorbed graphene templates as a platform for preferential sensing of DNA. *Biomacromolecules* **2012**, *13*, 3181–3188.

9. Chen, X.; Liu, G.; Liang, S.; Qian, S.; Liu, J.; Chen, Z. An assay of DNA by resonance light scattering technique and its application in screening anticancer drugs. *Anal. Methods* **2012**, *4*, 1546–1551.

10. Villa, A. M.; Fusi, P.; Pastori, V.; Amicarelli, G.; Pozzi, C.; Adlerstein, D.; Doglia, S. M. Ethidium bromide as a marker of mtDNA replication in living cells. *J. Biomed. Opt.* **2012**, *17*, 046001/1-046001/9.

11. Kjaerulff, S.; Glensbjerg, M. Method for analysis of cellular DNA content. PCT Int. Appl. WO 2011098085, 2011.

12. Wainwright, M. Dyes, trypanosomiasis and DNA: a historical and critical review. *Biotech. Histochem.* **2011**, *85*, 341–354.

13. Miyamoto, S.; Kato, T.; Tomono, J. Nucleic acid identification method. Jpn. Kokai Tokkyo Koho JP 2010233530, 2010.

14. Del Castillo, P.; Horobin, R. W.; Blazquez-Castro, A.; Stockert, J. C. Binding of cationic dyes to DNA: distinguishing intercalation and groove binding mechanisms using simple experimental and numerical models. *Biotech. Histochem.* **2010**, *85*, 247–256.

15. Villa, A. M.; Doglia, S. M. Ethidium bromide as a vital probe of mitochondrial DNA in carcinoma cells. *Eur. J. Cancer* **2009**, *45*, 2588–2597.

16. Rajendran, A.; Magesh, C. J.; Perumal, P. T. DNA-DNA cross-linking mediated by bifunctional [SalenAlIII]+ complex. *Biochim. Biophys. Acta* **2008**, *1780*, 282–288.

17. Exner, M.; Rogers, A. Methods for detecting nucleic acids using multiple signals. U.S. Pat. Appl. Publ. US 2007172836, 2007.

18. Bonasera, V.; Alberti, S.; Sacchetti, A. Protocol for high-sensitivity/long linear-range spectrofluorimetric DNA quantification using ethidium bromide. *BioTechniques* **2007**, *43*, 173–174, 176.

19. Ragazzon, P. A.; Garbett, N. C.; Chaires, J. B. Competition dialysis: a method for the study of structural selective nucleic acid binding. *Methods* **2007**, *42*, 173–182.

20. Dolezel, J.; Greilhuber, J.; Suda, J. Estimation of nuclear DNA content in plants using flow cytometry. *Nat. Protoc.* **2007**, *2*, 2233–2244.

21. Hilal, H.; Taylor, J. A. Determination of the stoichiometry of DNA-dye interaction and application to the study of a bis-cyanine dye-DNA complex. *Dyes Pigm.* **2007**, *75*, 483–490.

22. Kim, K.; Min, J.; Lee, I.; Kim, A. Method for highly sensitive nucleic acid detection using nanopore and non-specific nucleic acid-binding agent. U.S. Pat. Appl. Publ. US 2006292605, 2006.

23. Hooper-McGrevy, K. E.; MacDonald, B.; Whitcombe, L. Quick, simple, and sensitive RNA quantitation. *Anal. Biochem.* **2003**, *318*, 318–320.

24. Vardevanyan, P. O.; Antonyan, A. P.; Parsadanyan, M. A.; Davtyan, H. G.; Karapetyan, A. T. The binding of ethidium bromide with DNA: Interaction with single- and double-stranded structures. *Exp. Mol. Med.* **2003**, *35*, 527–533.

25. Tomita, N.; Mori, Y. Method for efficiently detecting double-stranded nucleic acid. PCT Int. Appl. WO 2002103053, 2002.

26. Deka, C.; Gordon, K. M.; Gupta, R.; Horton, A. Methods and compositions for rapid staining of nucleic acids in whole cells. U.S. Patent 6271035, 2001.

27. Okamoto, H.; Suzuki, T.; Yamamoto, N. Method for detecting/quantitating target nucleic acid by dry fluorometry. Jpn. Kokai Tokkyo Koho JP 2001033439, 2001.

28. Yonekura, N.; Mutoh, M.; Miyagi, Y.; Takushi, E. Ethidium bromide binding sites in DNA gel. *Chem. Lett.* **2000**, 954–955.

29. King, M. P. Use of ethidium bromide to manipulate ratio of mutated and wild-type mitochondrial DNA in cultured cells. *Methods Enzymol.* **1996**, *264*, 339–344.

30. Bruno, J. G.; Sincock, S. A.; Stopa, P. J. Highly selective acridine and ethidium staining of bacterial DNA and RNA. *Biotech. Histochem.* **1996**, *71*, 130–136.

31. Harriman, W. D.; Wabl, M. A video technique for the quantification of DNA in gels stained with ethidium bromide. *Anal. Biochem.* **1995**, *228*, 336–342.

32. Haugland, R. P. Phenanthridium dye staining of nucleic acids in living cells. U.S. Patent 5437980, 1995.

33. Giache, V.; Pirami, L.; Becciolini, A. A method for measuring microquantities of DNA. *J. Biolumin. Chemilum.* **1994**, *9*, 229–232.

34. Biggiogera, M.; Biggiogera, F. F. Ethidium bromide- and propidium iodide-PTA staining of nucleic acids at the electron microscopic level. *J. Histochem. Cytochem.* **1989**, *37*, 1161–1166.

35. Villanueva, A.; Stockert, J. C.; Armas-Portela, R. A simple method for the fluorescence analysis of nucleic acid-dye complexes in cytological preparations. *Histochemistry* **1984**, *81*, 103–104.

36. Taylor, I. W.; Milthorpe, B. K. An evaluation of DNA fluorochromes, staining techniques, and analysis for flow cytometry. I. Unperturbed cell populations. *J. Histochem. Cytochem.* **1980**, *28*, 1224–1232.

37. Taylor, I. W. A rapid single step staining technique for DNA analysis by flow microfluorimetry. *J. Histochem. Cytochem.* **1980**, *28*, 1021–1124.

38. Jones, C. R.; Bolton, P. H.; Kearns, D. R. Ethidium bromide binding to transfer RNA: transfer RNA as a model system for studying drug-RNA interactions. *Biochemistry* **1978**, *17*, 601–607.

39. Boer, G. J. Simplified microassay of DNA and RNA using ethidium bromide. *Anal. Biochem.* **1975**, *65*, 225–231.

40. Urbanke, C.; Roemer, R.; Maass, G. Binding of ethidium bromide to different conformations of tRNA. Unfolding of tertiary structure. *Eur. J. Biochem.* **1973**, *33*, 511–516.

41. Douthart, R. J.; Burnett, J. P.; Beasley, F. W.; Frank, B. H. Binding of ethidium bromide to double-stranded ribonucleic acid. *Biochemistry* **1973**, *12*, 214–220.

42. Haag, D.; Tschahargane, C.; Goerttler, K. Does ethidium bromide bind selectively and stoichiometrically to nucleic acids in histological tissues. *Histochemie* **1971**, *27*, 119–124.

43. Bittman, R. Binding of ethidium bromide to transfer ribonucleic acid: absorption, fluorescence, ultracentrifugation, and kinetic investigations. *J. Mol. Biol.* **1969**, *46*, 251–268.

44. Le Pecq, J. B.; Paoletti, C. A fluorescent complex between ethidium bromide and nucleic acids: Physical-chemical characterization. *J. Mol. Biol.* **1967**, *27*, 87–106.

45. Le Pecq, J. B.; Paoletti, C. A new fluorometric method for RNA and DNA determination. *Anal. Biochem.* **1966**, *17*, 100–107.

46. Waring, M. J. Complex formation between ethidium bromide and nucleic acids. *J. Mol. Biol.* **1965**, *13*, 269–282.

47. Hannig, C.; Hannig, M.; Rehmer, O.; Braun, G.; Hellwig, E.; Al-Ahmad, A. Fluorescence microscopic visualization and quantification of initial bacterial colonization on enamel *in situ*. *Arch. Oral Biol.* **2007**, *52*, 1048–1056.

48. Li, Y.; Dick, W. A.; Tuovinen, O. H. Evaluation of fluorochromes for imaging bacteria in soil. *Soil Biol. Biochem.* **2003**, *35*, 737–744.

49. Tikhonravov, V. V.; Ageev, V. A.; Kovtun, V. P.; Kovtun, A. L.; Shvedov, V. V. Method of quantitative determination of bacteria in biopreparations by fluorescence staining with ethidium bromide. Russ. RU 2117291, 1998.

50. Nasu, M.; Kawase, K.; Aoki, N.; Yamada, K. A method of monitoring filamentous bacteria in activated sludge by image analysis. Jpn. Kokai Tokkyo Koho JP 10070999, 1998.

51. Guda, K.; Natale, L.; Markowitz, S. D. An improved method for staining cell colonies in clonogenic assays. *Cytotechnology* **2007**, *54*, 85–88.

52. Stockert, J. C. Ethidium bromide fluorescence in living and fixed cells. *Naturwissenschaften* **1974**, *61*, 363–364.

53. Vergani, L.; Gavazzo, P.; Mascetti, G.; Nicolini, C. Ethidium bromide intercalation and chromatin structure: A spectropolarimetric analysis. *Biochemistry* **1994**, *33*, 6578–6585.

54. Kinouchi, Y.; Yamada, M. Ethidium bromide fluorescence of the chromatin structure in relation to sodium chloride and EDTA concentrations. *Acta Histochem. Cytochem.* **1977**, *10*, 426–442

55. Doenecke, D. Binding of ethidium bromide to fractionated chromatin. *Exp. Cell Res.* **1976**, *100*, 223–227.

56. Lawrence, J. J.; Louis, M. Ethidium bromide as a probe of chromatin structure. *FEBS Lett.* **1974**, *40*, 9–12.

57. Lee, S. J. Kit for identifying chromosome x 16 loci by multiplex-PCR. Repub. Korean Kongkae Taeho Kongbo KR 2013008486, 2013.

58. Langlois, R. G.; Carrano, A. V.; Gray, J. W.; Van Dilla, M. A. Cytochemical studies of metaphase chromosomes by flow cytometry. *Chromosoma* **1980**, *77*, 229–251.

59. McGill, M.; Pathak, S.; Hsu, T. C. Effects of ethidium bromide on mitosis and chromosomes. Possible material basis for chromosome stickiness. *Chromosoma* **1974**, *47*, 157–166.

60. Sakata, T.; Mizukami, T.; Hatanaka, K. Method for classifying and counting immature leukocytes. Eur. Pat. Appl. EP 844481, 1998.

61. Horikiri, S. Microorganism cell detection method using multiple fluorescent indicators. Jpn. Kokai Tokkyo Koho JP 2006238779, 2006.

62. Besson, F. I.; Hermet, J. P.; Ribault, S. Reaction medium and process for universal detection of

microorganisms. Fr. Demande FR 2847589, 2004.

63. Scholefield, J. Staining microorganisms. Ger. Offen. DE 2728077, 1978.

64. Tomchick, R.; Mandel, H. G. Biochemical effects of ethidium bromide in microorganisms. *J. Gen. Microbiol.* **1964**, *36*, 225–236.

65. Magrassi, L.; Purves, D.; Lichtman, J. W. Fluorescent probes that stain living nerve terminals. *J. Neurosci.* **1987**, *7*, 1207–1214.

66. Yamada, M.; Kinouchi, Y.; Fujimori, K.; Yamamoto, K. Fluorometric studies on the ribosomal structure by ethidium bromide. *Cell. Mol. Biol.* **1978**, *23*, 183–190.

67. Suryanarayana, T.; Burma, D. P. Effects of intercalating agents on the structure of the ribosome. *Biochem. Biophys. Res. Commun.* **1975**, *65*, 708–713.

68. Shi, H.; Wang, J.; Yuan, Y.; Zhao, H. Method for measuring sperm with fluorescent staining. Faming Zhuanli Shenqing CN 101308131, 2008.

69. Matsumoto, T.; Okada, H.; Hamaguchi, Y. Method and reagent for counting sperm by flow cytometry. Jpn. Kokai Tokkyo Koho JP 2001242168, 2001.

70. Bondzio, A.; Schuelke, B.; Risse, S. Sperm quality assessment by biochemical parameters. *Monatsh. Vet.* **1992**, *47*, 301–306.

71. Tereshchenko, A. V.; Sakhatskii, N. I.; Artemenko, A. B. Determination of the quality of spermia. U.S.S.R. SU 1329780, 1987.

72. Pivonkova, H.; Sebest, P.; Pecinka, P.; Ticha, O.; Nemcova, K.; Brazdova, M.; Jagelska, E. B.; Brazda, V.; Fojta, M. Selective binding of tumor suppressor p53 protein to topologically constrained DNA: Modulation by intercalative drugs. *Biochem. Biophys. Res. Commun.* **2010**, *393*, 894–899.

73. Richardson, V. J.; Vodinelich, L.; Wilson, A.; Potter, C. W. Effect of ethidium bromide on transplanted virus-induced tumor cells. *J. Natl. Cancer Inst.* **1976**, *57*, 815–819.

74. Kramer, M. J.; Grunberg, E. Effect of ethidium bromide against transplantable tumors in mice and rats. *Chemotherapy* **1973**, *19*, 254–258.

75. Chan, L. L. Yeast concentration and viability measurement. PCT Int. Appl. WO 2011156249, 2011.

76. Sailer, B. L.; Valdez, J. G.; Steinkamp, J. A.; Crissman, H. A. Apoptosis induced with different cycle-perturbing agents produces differential changes in the fluorescence lifetime of DNA-bound ethidium bromide. *Cytometry* **1998**, *31*, 208–216.

77. Hueber, A.; Pieres, M.; He, H. Quantitating apoptosis by a nonradioactive DNA dot blot assay. *Anal. Biochem.* **1994**, *221*, 431–433.

78. Gedda, L.; Silvander, M.; Sjoberg, S.; Tjarks, W.; Carlsson, J. Cytotoxicity and subcellular localization of boronated phenanthridinium analogs. *Anti-Cancer Drug Des.* **1997**, *12*, 671–685.

79. Kemp, R. B. Cytotoxicity in an anchorage-independent fibroblast cell line measured by a combination of fluorescent dyes. *Methods Mol. Biol.* **1995**, *43*, 211–218.

80. Deguchi, J. Method for assaying mitochondria toxicity possessed by substance. Jpn. Kokai Tokkyo Koho JP 2005160392, 2005.

81. Tripathy, D. R.; Dinda, A. K.; Dasgupta, S. A simple assay for the ribonuclease activity of ribonucleases in the presence of ethidium bromide. *Anal. Biochem.* **2013**, *437*, 126–129.

82. Lee, J.; Choi, S. Improved fluorometric assay method for ribonuclease activity. *J. Biochem. Mol. Biol.* **1997**, *30*, 258–261.

83. Neyfakh, A. A. Use of fluorescent dyes as molecular probes for the study of multidrug resistance. *Exp. Cell Res.* **1988**, *174*, 168–176.

84. Bousseksou, A.; Salmon, L.; Molnar, G.; Cobo, S. Materials with thermochromic spin transition doped with one or more fluorescent agents for use as temperature sensor. Fr. Demande FR 2952371, 2011.

85. Bousseksou, A.; Salmon, L.; Molnar, G.; Cobo, S. Heat-sensitive spin-transition materials doped with one or more fluorescent agents for use as temperature sensor. PCT Int. Appl. WO 2011058277, 2011.

86. Shaikin, V. I.; Deineko, G. I.; Efremova, E. A.; Yuzhakov, A. Y. Method for applying ethidium bromide for treatment of horses with pyroplasmosis. Russ. RU 2258501, 2005.

87. Woolfe, G. Trypanocidal action of phenanthridine compounds: effect of changing the quaternary groups of known trypanocides. *Br. J. Pharmacol. Chemother.* **1956**, *11*, 330–333.

88. Kinlen, P. J. Light-emitting phosphor particles and electroluminescent devices employing same. U.S. Pat. Appl. Publ. US 20040018379, 2004.

89. Garza, C. M.; Cho, S. Metrology of bilayer photoresist processes. U.S. Pat. Appl. Publ. US 20090220895, 2009.

90. Sakai, A.; Sasaki, K.; Muramatsu, D.; Arai, S.; Endou, N.; Kuroda, S.; Hayashi, K.; Lim, Y.; Yamazaki, S.; Umeda, M. A Bhas 42

cell transformation assay on 98 chemicals: The characteristics and performance for the prediction of chemical carcinogenicity. *Mutat. Res., Genet. Toxicol. Environ. Mutagen.* **2010**, *702*, 100–122.

91. Na, M. R.; Koo, S. K.; Kim, D. Y.; Park, S. D.; Rhee, S. K.; Kang, K. W.; Joe, C. O. *In vitro* inhibition of gap junctional intercellular communication by chemical carcinogens. *Toxicology* **1995**, *98*, 199–206.

92. Mamber, S. W.; Bryson, V.; Katz, S. E. The *Escherichia coli* WP2/WP100 rec assay for detection of potential chemical carcinogens. *Mutat. Res. Lett.* **1983**, *119*, 135–144.

93. Wu, Y.; Cai, J. Analysis of chromosome aberration due to ethidium bromide using AFM. *Zhongguo Bingli Shengli Zazhi* **2007**, *23*, 2451–2454.

94. Medley, C. D.; Smith, J. E.; Wigman, L. S.; Chetwyn, N. P. A DNA-conjugated magnetic nanoparticle assay for assessing genotoxicity. *Anal. Bioanal. Chem.* **2012**, *404*, 2233–2239.

95. Gonzalez, N. V.; Soloneski, S.; Larramendy, M. L. Dicamba-induced genotoxicity in Chinese hamster ovary (CHO) cells is prevented by vitamin E. *J. Hazard. Mater.* **2009**, *163*, 337–343.

96. Huang, R.; Southall, N.; Cho, M.; Xia, M.; Inglese, J.; Austin, C. P. Characterization of diversity in toxicity mechanism using *in vitro* cytotoxicity assays in quantitative high throughput screening. *Chem. Res. Toxicol.* **2008**, *21*, 659–667.

97. Scaife, M. C. An *in vitro* cytotoxicity test to predict the ocular irritation potential of detergents and detergent products. *Food Chem. Toxicol.* **1985**, *23*, 253–258.

98. Barclay, B. J.; DeHaan, C. L.; Hennig, U. G. G.; Iavorovska, O.; Von Borstel, R. W.; Von Borstel, R. C. A rapid assay for mitochondrial DNA damage and respiratory chain inhibition in the yeast *Saccharomyces cerevisiae*. *Environ. Mol. Mutagen.* **2001**, *38*, 153–158.

99. Kohler, M.; Kundig, A.; Reist, H. W.; Michel, C. Modification of *in vitro* mouse embryogenesis by x-rays and fluorochromes. *Radiat. Environ. Biophys.* **1994**, *33*, 341–351.

100. Knight, A. W.; Billinton, N.; Cahill, P. A.; Scott, A.; Harvey, J. S.; Roberts, K. J.; Tweats, D. J.; Keenan, P. O.; Walmsley, R. M. An analysis of results from 305 compounds tested with the yeast RAD54-GFP genotoxicity assay (GreenScreen GC) - including relative predictivity of regulatory tests and rodent carcinogenesis and performance

with autofluorescent and colored compounds. *Mutagenesis* **2007**, *22*, 409–416.

101. Ohno, K.; Tanaka-Azuma, Y.; Yoneda, Y.; Yamada, T. Genotoxicity test system based on p53R2 gene expression in human cells: Examination with 80 chemicals. *Mutat. Res., Genet. Toxicol. Environ. Mutagen.* **2005**, *588*, 47–57.

102. Cahill, P. A.; Knight, A. W.; Billinton, N.; Barker, M. G.; Walsh, L.; Keenan, P. O.; Williams, C. V.; Tweats, D. J.; Walmsley, R. M. The GreenScreen genotoxicity assay: A screening validation programme. *Mutagenesis* **2004**, *19*, 105–119.

103. Min, J.; Gu, M. B. Genotoxicity assay using chromosomally-integrated bacterial recA::lux. *J. Microbiol. Biotechnol.* **2003**, *13*, 99–103.

104. Fernandez, M.; Gauthier, L.; Jaylet, A. Use of newt larvae for *in vivo* genotoxicity testing of water: results on 19 compounds evaluated by the micronucleus test. *Mutagenesis* **1989**, *4*, 17–26.

105. Zhang, H.; Chen, Q.; Xiang, M.; Ma, C.; Huang, Q.; Yang, S. *In silico* prediction of mitochondrial toxicity by using GA-CG-SVM approach. *Toxicology in Vitro* **2009**, *23*, 134–140.

106. Nishigaki, K.; Nikami, M.; Gautum, S. G.; Kamiseki, M. Mutagenicity test method using mammalian cells. Jpn. Kokai Tokkyo Koho JP 2012139169, 2012.

107. Prabhu, N.; Sudha, E. G.; Anna, J. P.; Soumya, T. S. Effect of *Rosa multiflora* extract on chemical mutagens using Ames Assay. *Pharm. Chem.* **2010**, *2*, 91–97.

108. Singer, V. L.; Lawlor, T. E.; Yue, S. Comparison of SYBR Green I nucleic acid gel stain mutagenicity and ethidium bromide mutagenicity in the *Salmonella*/mammalian microsome reverse mutation assay (Ames test). *Mutat. Res., Genet. Toxicol. Environ. Mutagen.* **1999**, *439*, 37–47.

109. Buschmann, N. Does the two-phase titration of surfactants require a mutagenic indicator? *J. Am. Oil Chem. Soc.* **1995**, *72*, 1243.

110. Ferguson, L. R.; Baguley, B. C. Verapamil as a co-mutagen in the *Salmonella*/mammalian microsome mutagenicity test. *Mutat. Res. Lett.* **1988**, *209*, 57–62.

111. Yates, I. E. Differential sensitivity to mutagens by *Photobacterium phosphoreum*. *J. Microbiol. Methods* **1985**, *3*, 171–180.

112. Xamena, N.; Creus, A.; Marcos, R. Mutagenic activity of some intercalating compounds in the *Drosophila* zeste somatic eye mutation test.

Mutat. Res., Genet. Toxicol. Test. **1984**, *138*, 169–173.

113. Fukunaga, M.; Yielding, L. W. Structure-function characterization of phenanthridinium compounds as mutagens in *Salmonella. Mutat. Res. Lett.* **1983**, *121*, 89–94

114. Guazzo, E. P. A technique for producing demyelination of the rat optic nerves. *J. Clin. Neurosci.* **2005**, *12*, 54–58.

115. Riet-Correa, G.; Fernandes, C. G.; Pereira, L. A. V.; Graca, D. L. Ethidium bromide-induced demyelination of the sciatic nerve of adult Wistar rats. *Braz. J. Med. Biol. Res.* **2002**, *35*, 99–104.

116. Dobrucki, J. W.; Feret, D.; Noatynska, A. Scattering of exciting light by live cells in fluorescence confocal imaging: Phototoxic effects and relevance for FRAP studies. *Biophys. J.* **2007**, *93*, 1778–1786.

ETHIDIUM HOMODIMER-1 (EthD-1)

CAS Registry Number 61926-22-5

Chemical Structure

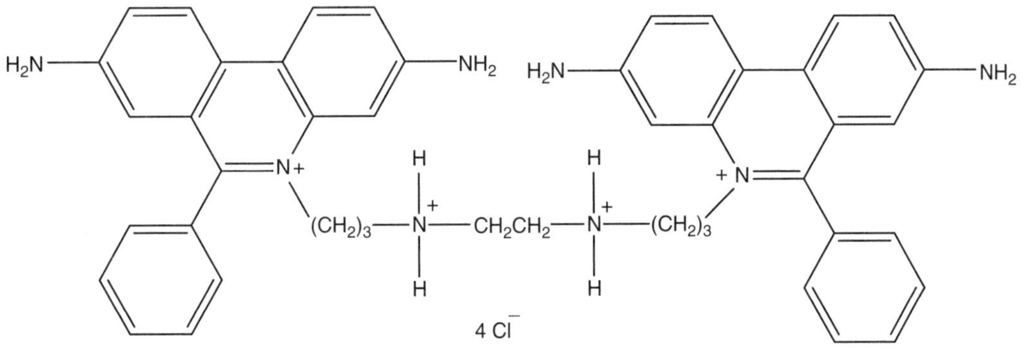

Absorption (λ_{max})   528 nm (H_2O/DNA); 493 nm (H_2O)

Emission (λ_{max})   617 nm (H_2O/DNA); 631 nm (H_2O)

Molar Extinction Coefficient   7000 cm^{-1} M^{-1} (H_2O/DNA); 9100 cm^{-1} M^{-1} (H_2O)

CA Index Name Phenanthridinium, 5,5'-[1,2-ethanediylbis(imino-3,1-propanediyl)]bis[3,8-diamino-6-phenyl-, chloride, hydrochloride (1 : 2 : 2)

Other Names Phenanthridinium, 5,5'-[1,2-ethanediyl-bis(imino-3,1-propanediyl)]bis[3,8-diamino-6-phenyl-, dichloride, dihydrochloride; 5,5'-(4,7-Diazadecamethyl-ene)bis(3,8-diamino-6-phenylphenanthridinium) dichloride dihydrochloride; EthD 1; Ethidium homodimer; Ethidium homodimer 1

Merck Index Number Not listed

Chemical/Dye Class Heterocycle; Phenanthridine

Molecular Formula $C_{46}H_{50}Cl_4N_8$

Molecular Weight 856.77

Physical Form Purple crystals;[2] Red solid

Solubility Soluble in water, *N,N*-dimethylformamide, dimethyl sulfoxide, methanol

Melting Point 275 °C[2,3]

Synthesis Synthetic methods[1-4]

Imaging/Labeling Applications Nucleic acids;[1-18] cells;[4,18-24] nuclei;[28-31] algae;[26] bacteria;[4] leukocytes;[32] megakaryocytes;[33] microorganisms;[34-39] oligonucleotides;[45] sperms;[18,40-43,58] Schwann cells (SCs) in whole nerves[44]

Biological/Medical Applications Detecting nucleic acids,[1-18] cell viability assay;[4,18-24] cytotoxicity assay;[24-27,52] classifying/counting leukocytes,[32] megakaryocyte;[33] counting erythroblasts;[30,31] detecting/counting sperms;[18,40-43,58] detecting/measuring microorganisms;[34-39] nucleic acid amplification;[45] evaluating freshness of a fish product;[46] as temperature sensor[47,48]

Industrial Applications For authentification[49]

Safety/Toxicity Cytotoxicity;[50-52] gastric toxicity;[53] neurotoxicity;[54,55] phototoxicity;[56,57] retinal toxicity;[56,57] reproductive toxicity;[58] tumor necrosis factor activity[59]

REFERENCES

1. Sabnis, R. W. *Handbook of Biological Dyes and Stains*; John Wiley & Sons Inc.: Hoboken, **2010**; pp 186–188.

2. Gaugain, B.; Barbet, J.; Oberlin, R.; Roques, B. P.; Le Pecq, J. B. DNA bifunctional intercalators. 1. Synthesis and conformational properties of an ethidium homodimer and of an acridine ethidium heterodimer. *Biochemistry* **1978**, *17*, 5071–5078.

3. Roques, B. P.; Barbet, J.; Oberlin, R.; Le Pecq, J. B. DNA intercalating drugs: Synthesis of phenanthridinium monomers and of one dimer with aminoalkyl chains. *Comptes Rendus Sean. Acad. Sci., Ser. D: Sci. Naturelles* **1976**, *283*, 1365–1367.

4. Millard, P. J.; Roth, B. L.; Yue, S. T.; Haugland, R. P. Fluorescent viability assay using cyclic-substituted unsymmetrical cyanine dyes. U.S. Patent 5534416, 1996.

5. Kjaerulff, S.; Glensbjerg, M. Method for analysis of cellular DNA content. PCT Int. Appl. WO 2011098085, 2011.

6. Exner, M.; Rogers, A. Methods for detecting nucleic acids using multiple signals. U.S. Pat. Appl. Publ. US 2007172836, 2007.

7. Erikson, G. H.; Daksis, J. I. Pre-incubation method to improve signal/noise ratio of nucleic acid assays. U.S. Pat. Appl. Publ. US 2004180345, 2004.

8. Cui, H. H.; Valdez, J. G.; Steinkamp, J. A.; Crissman, H. A. Fluorescence lifetime-based discrimination and quantification of cellular DNA and RNA with phase-sensitive flow cytometry. *Cytometry, Part A* **2003**, *52A*, 46–55.

9. Erikson, G. H. Method for modifying transcription and/or translation in an organism for therapeutic, prophylactic and/or analytic uses. U.S. Pat. Appl. Publ. US 2003181412, 2003.

10. Erikson, G. H.; Daksis, J. I.; Kandic, I.; Picard, P. Nucleic acid multiplex formation. PCT Int. Appl. WO 2002103051, 2002.

11. Scheinert, P. Electrophoresis system for nucleic acids and gel staining methods. *BioTec (Germany)* **1996**, *8*, 47–49.

12. Rye, H. S.; Glazer, A. N. Interaction of dimeric intercalating dyes with single-stranded DNA. *Nucleic Acids Res.* **1995**, *23*, 1215–1222.

13. Rye, H. S.; Quesada, M. A.; Peck, K.; Mathies, R. A.; Glazer, A. N. High-sensitivity two-color detection of double-stranded DNA with a confocal fluorescence gel scanner using ethidium homodimer and thiazole orange. *Nucleic Acids Res.* **1991**, *19*, 327–333.

14. Glazer, A. N.; Peck, K.; Mathies, R. A. A stable double-stranded DNA-ethidium homodimer complex: application to picogram fluorescence detection of DNA in agarose gels. *Proc. Natl. Acad. Sci. U.S.A.* **1990**, *87*, 3851–3855.

15. Reinhardt, C. G.; Roques, B. P.; Le Pecq, J. B. Binding of bifunctional ethidium intercalators to transfer RNA. *Biochem. Biophys. Res. Commun.* **1982**, *104*, 1376–1385.

16. Markovits, J.; Roques, B. P.; Le Pecq, J. B. Ethidium dimer: a new reagent for the fluorimetric determination of nucleic acids. *Anal. Biochem.* **1979**, *94*, 259–264.

17. Gaugain, B.; Barbet, J.; Capelle, N.; Roques, B. P.; Le Pecq, J. B.; Le Bret, M. DNA bifunctional intercalators. 2. Fluorescence properties and DNA binding interaction of an ethidium homodimer and an acridine ethidium heterodimer. Appendix: Numerical solution of McGhee and von Hippel equations for competing ligands. *Biochemistry* **1978**, *17*, 5078–5088.

18. Anderson, A. L.; Knutson, C. R.; Mueth, D.; Plewa, J.; Tanner, E. Methods for staining cells for identification and sorting. U.S. Pat. Appl. Publ. US 2006172315, 2006.

19. Wheeler, A. R.; Barbulovic-Nad, I. Droplet-based cell culture and cell assays using digital microfluidics. U.S. Pat. Appl. Publ. US 20090203063, 2009.

20. Gantenbein-Ritter, B.; Potier, E.; Zeiter, S.; van der Werf, M.; Sprecher, C. M.; Ito, K. Accuracy of three techniques to determine cell viability in 3D tissues or scaffolds. *Tissue Eng., Part C: Methods* **2008**, *14*, 353–358.

21. Oberhardt, B. J. Cell analysis methods. U.S. Patent 6251615, 2001.

22. Thompson, T. A. Viability assays for cells *in vitro*: the ethidium/calcein assay and the immunofluorescence combination assay. *Methods Mol. Med.* **1999**, *22*, 145–155.

23. Merrilees, M. J.; Beaumont, B. W.; Scott, L. J. Fluoroprobe quantification of viable and non-viable cells in human coronary and internal thoracic arteries sampled at autopsy. *J. Vasc. Res.* **1996**, *32*, 371–377.

24. Haugland, R. P.; MacCoubrey, I. C.; Moore, P. L. Dual-fluorescence cell viability assay using ethidium homodimer and calcein AM. U.S. Patent 5314805, 1994.

25. Guzman, E.; McCrae, M. A. A rapid and accurate assay for assessing the cytotoxicity of viral proteins. *J. Virol. Methods* **2005**, *127*, 119–125.

26. Faber, M. J.; Smith, L. M. J.; Boermans, H. J.; Stephenson, G. R.; Thompson, D. G.; Solomon, K. R. Cryopreservation of fluorescent marker-labeled algae (*Selenastrum capricornutum*) for toxicity testing using flow cytometry. *Environ. Toxicol. Chem.* **1997**, *16*, 1059–1067.

27. Papadopoulos, N. G.; Dedoussis, G. V. Z.; Spanakos, G.; Gritzapis, A. D.; Baxevanis, C. N.; Papamichail, M. An improved fluorescence assay for the determination of lymphocyte-mediated cytotoxicity using flow cytometry. *J. Immunol. Methods* **1994**, *177*, 101–111.

28. Gauer, C.; Mann, W.; Alunni-Fabbroni, M. Method for carrying out an enzymic reactions. Ger. DE 102006056694, 2010.

29. Gauer, C.; Mann, W.; Alunni-Fabbroni, M. Method for carrying out an enzymic reaction. PCT Int. Appl. WO 2008064730, 2008.

30. Heuven, B.; Wong, F.; Tsuji, T.; Sakata, T.; Hamaguchi, I. Method for classifying and counting

erythroblasts by flow cytometry. Jpn. Kokai Tokkyo Koho JP 11326323, 1999.

31. Heuven, B.; Wong, F.; Tsuji, T.; Sakata, T.; Hamaguchi, I. Process for discriminating and counting erythroblasts. U.S. Pat. Appl. Publ. US 20020006631, 2002.

32. Sakata, T.; Mizukami, T.; Hatanaka, K. Method for classifying and counting immature leukocytes. Eur. Pat. Appl. EP 844481, 1998.

33. Minakami, T.; Mori, Y.; Tsuji, T.; Ikeuchi, Y. Megakaryocyte classification/counting method by double fluorescent staining and flow cytometry. Jpn.Kokai Tokkyo Koho JP 2006275985, 2006.

34. Eckert, R. H.; Kaplan, C.; He, J.; Yarbrough, D. K.; Anderson, M.; Sim, J. Methods and devices for the selective detection of microorganisms. U.S. Pat. Appl. Publ. US 20120003661, 2012.

35. Noda, N.; Mizutani, T. Microorganism-measuring method using multiple staining. Jpn. Kokai Tokkyo Koho JP 2006340684, 2006.

36. Fukutome, K. Method for evaluating microorganism cell activity by flow cytometry analysis. Jpn. Kokai Tokkyo Koho JP 2006238771, 2006.

37. Horikiri, S. Microorganism cell detection method using fluorescent indicator. Jpn. Kokai Tokkyo Koho JP 2006262775, 2006.

38. Tokuda, Y.; Ishiyama, Y. Instant method for evaluating live and dead microorganisms by fluorescent staining. Jpn. Kokai Tokkyo Koho JP 2000232897, 2000.

39. Kaneshiro, E. S.; Wyder, M. A.; Wu, Y. P.; Cushion, M. T. Reliability of calcein acetoxy methyl ester and ethidium homodimer or propidium iodide for viability assessment of microbes. J. Microbiol. Methods 1993, 17, 1–16.

40. Sokolowska, A.; Garcia, B. M.; Fernandez, L. G.; Ortega-Ferrusola, C.; Tapia, J. A.; Pena, F. J. Activated caspases are present in frozen-thawed canine sperm and may be related to post thaw sperm quality. Zygote 2009, 17, 297–305.

41. Kato, M.; Makino, S.; Kimura, H.; Ota, T.; Furuhashi, T.; Nagamura, Y.; Hirano, K. In vitro evaluation of acrosomal status and motility in rat epididymal spermatozoa treated with α-chlorohydrin for predicting their fertilizing capacity. J. Reprod. Dev. 2002, 48, 461–468.

42. Matsumoto, T.; Okada, H.; Hamaguchi, Y. Method and reagent for counting sperm by flow cytometry. Jpn. Kokai Tokkyo Koho JP 2001242168, 2001.

43. Kimura, H.; Kato, M. Determination of rat sperm viability with fluorescence. Jpn. Kokai Tokkyo Koho JP 08332098, 1996.

44. Decherchi, P.; Cochard, P.; Gauthier, P. Dual staining assessment of Schwann cell viability within whole peripheral nerves using calcein-AM and ethidium homodimer. J. Neurosci. Methods 1997, 71, 205–213.

45. Park, H. O.; Kim, H. B.; Chi, S. M. Detection method of DNA amplification using probes labeled with intercalating dye. PCT Int. Appl. WO 2006004267, 2006.

46. Liberman, B. A method of evaluating freshness of a fish product. PCT Int. Appl. WO 2005089254, 2005.

47. Bousseksou, A.; Salmon, L.; Molnar, G.; Cobo, S. Materials with thermochromic spin transition doped with one or more fluorescent agents for use as temperature sensor. Fr. Demande FR 2952371, 2011.

48. Bousseksou, A.; Salmon, L.; Molnar, G.; Cobo, S. Heat-sensitive spin-transition materials doped with one or more fluorescent agents for use as temperature sensor. PCT Int. Appl. WO 2011058277, 2011.

49. Isler, U.; Hoehener, K.; Meier, W.; Poux, S. Procedure for the treatment of yarn with encapsulated marking substance for authentification and device for proof of authenticity. Ger. Offen. DE 102005047786, 2007.

50. Luo, Y.; Wang, C.; Hossain, M.; Qiao, Y.; Ma, L.; An, J.; Su, M. Three-dimensional microtissue assay for high-throughput cytotoxicity of nanoparticles. Anal. chem. 2012, 84, 6731–6738.

51. Ho, J.; Tsai, Ray J.; Chen, S.; Chen, H. Cytotoxicity of indocyanine green on retinal pigment epithelium: implications for macular hole surgery. Arch. Ophthalmol. 2003, 121, 1423–1429.

52. Dias, N.; Lima, N. A comparative study using a fluorescence-based and a direct-count assay to determine cytotoxicity in Tetrahymena pyriformis. Res. Microbiol. 2002, 153, 313–322.

53. Kokoska, E. R.; Smith, G. S.; Deshpande, Y.; Wolff, A. B.; Rieckenberg, C.; Miller, T. A. Calcium accentuates injury induced by ethanol in human gastric cells. J. Gastrointest. Surg. 1999, 3, 308–318.

54. Guzman-Lenis, M.; Vallejo, C.; Navarro, X.; Casas, C. Analysis of FK506-mediated protection in an organotypic model of spinal cord damage: heat shock protein 70 levels are modulated in microglial cells. Neuroscience 2008, 155, 104–113.

55. Nattie, E. E.; Erlichman, J. S.; Li, A. Brain stem lesion size determined by DEAD red or conjugation of neurotoxin to fluorescent beads. J. Appl. Physiol. 1998, 85, 2370–2375.

56. Jackson, T. L.; Vote, B.; Knight, B. C.; El-Amir, A.; Stanford, M. R.; Marshall, J. Safety testing of infracyanine green using retinal pigment epithelium and glial cell cultures. *Invest. Ophthalmol. Vis. Sci.* **2004**, *45*, 3697–3703.

57. Ho, J.; Chen, H.; Chen, S.; Tsai, R. Reduction of indocyanine green-associated photosensitizing toxicity in retinal pigment epithelium by sodium elimination. *Arch. Ophthalmol.* **2004**, *122*, 871–878.

58. Kato, M.; Makino, S.; Kimura, H.; Ota, T.; Furuhashi, T.; Nagamura, Y. Evaluation of mitochondrial function and membrane integrity by dual fluorescent staining for assessment of sperm status in rats. *J. Toxicol. Sci.* **2002**, *27*, 11–18.

59. Levesque, A.; Paquet, A.; Page, M. Measurement of tumor necrosis factor activity by flow cytometry. *Cytometry* **1995**, *20*, 181–184.

ETHIDIUM HOMODIMER-2 (EthD-2)

CAS Registry Number   180389-01-9

Chemical Structure

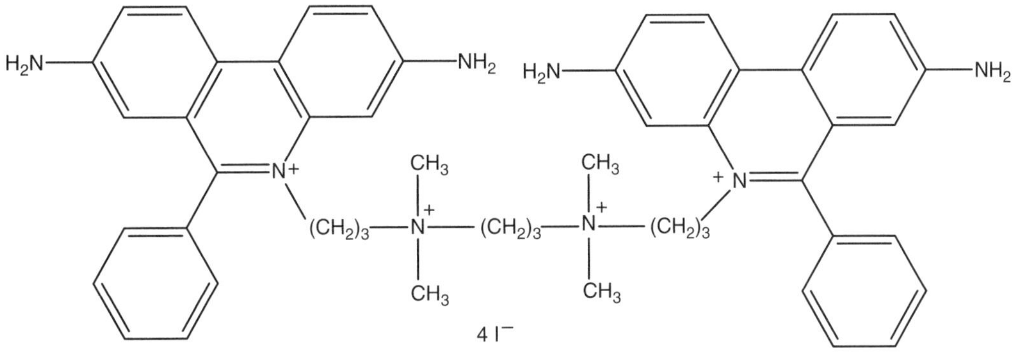

$4 \, I^-$

CA Index Name   Phenanthridinium, 5,5′-[1,3-propanediylbis-[(dimethyliminio)-3,1-propanediyl]]bis[3,8-diamino-6-phenyl-, iodide (1 : 4)

Other Names   Phenanthridinium, 5,5′-[1,3-propanediyl-bis-[(dimethyliminio)-3,1-propanediyl]]bis[3,8-diamino-6-phenyl-, tetraiodide; EthD 2; Ethidium homodimer 2

Merck Index Number   Not listed

Chemical/Dye Class   Heterocycle; Phenanthridine

Molecular Formula $C_{51}H_{60}I_4N_8$

Molecular Weight   1292.71

Physical Form   Purple crystals or red solid

Solubility   Soluble in water, *N,N*-dimethylformamide, dimethyl sulfoxide, methanol

Absorption (λ_{max})   535 nm (H_2O/DNA); 498 (H_2O)

Emission (λ_{max})   624 nm (H_2O/DNA); 631 (H_2O)

Molar Extinction Coefficient   8000 cm^{-1} M^{-1} (H_2O/DNA); 10,800 (H_2O)

Synthesis   Synthetic methods[1,2]

Imaging/Labeling Applications   Nucleic acids;[1-9] cells;[1,2,9] nuclei;[10,11] bacteria;[2] leukocytes;[12] megakaryocyte;[13] microorganisms;[14] oligonucleotides;[15] sperms[9,16]

Biological/Medical Applications   Detecting nucleic acids,[1-9] cells,[1,2,9] microorganisms;[14] classifying/counting leukocytes,[12] megakaryocyte;[13] counting erythroblasts,[10,11] sperms;[9,16] nucleic acids amplification[15]

Industrial Applications   Not reported

Safety/Toxicity   No data available

REFERENCES

1. Sabnis, R. W. *Handbook of Biological Dyes and Stains*; John Wiley & Sons Inc.: Hoboken, **2010**; pp 189–190.

2. Millard, P. J.; Roth, B. L.; Yue, S. T.; Haugland, R. P. Fluorescent viability assay using cyclic-substituted unsymmetrical cyanine dyes. U.S. Patent 5534416, 1996.

3. Kjaerulff, S.; Glensbjerg, M. Method for analysis of cellular DNA content. PCT Int. Appl. WO 2011098085, 2011.

4. Exner, M.; Rogers, A. Methods for detecting nucleic acids using multiple signals. U.S. Pat. Appl. Publ. US 2007172836, 2007.

5. Erikson, G. H.; Daksis, J. I. Pre-incubation method to improve signal/noise ratio of nucleic acid assays. U.S. Pat. Appl. Publ. US 2004180345, 2004.

6. Erikson, G. H. Method for modifying transcription and/or translation in an organism for therapeutic, prophylactic and/or analytic uses. U.S. Pat. Appl. Publ. US 2003181412, 2003.

7. Erikson, G. H.; Daksis, J. I.; Kandic, I.; Picard, P. Nucleic acid multiplex formation. PCT Int. Appl. WO 2002103051, 2002.

8. Leng, F.; Graves, D.; Chaires, J. B. Chemical crosslinking of ethidium to DNA by glyoxal. *Biochim. Biophys. Acta, Gene Struct. Expression* **1998**, *1442*, 71–81.

9. Anderson, A. L.; Knutson, C. R.; Mueth, D.; Plewa, J.; Tanner, E. Methods for staining cells for identification and sorting. U.S. Pat. Appl. Publ. US 2006172315, 2006.

10. Heuven, B.; Wong, F.; Tsuji, T.; Sakata, T.; Hamaguchi, I. Method for classifying and counting erythroblasts by flow cytometry. Jpn. Kokai Tokkyo Koho JP 11326323, 1999.

11. Heuven, B.; Wong, F.; Tsuji, T.; Sakata, T.; Hamaguchi, I. Process for discriminating and counting erythroblasts. U.S. Pat. Appl. Publ. US 20020006631, 2002.

12. Sakata, T.; Mizukami, T.; Hatanaka, K. Method for classifying and counting immature leukocytes. Eur. Pat. Appl. EP 844481, 1998.

13. Minakami, T.; Mori, Y.; Tsuji, T.; Ikeuchi, Y. Megakaryocyte classification/counting method by double fluorescent staining and flow cytometry. Jpn.Kokai Tokkyo Koho JP 2006275985, 2006.

14. Eckert, R. H.; Kaplan, C.; He, J.; Yarbrough, D. K.; Anderson, M.; Sim, J. Methods and devices for the selective detection of microorganisms. U.S. Pat. Appl. Publ. US 20120003661, 2012.

15. Park, H. O.; Kim, H. B.; Chi, S. M. Detection method of DNA amplification using probes labeled with intercalating dye. PCT Int. Appl. WO 2006004267, 2006.

16. Matsumoto, T.; Okada, H.; Hamaguchi, Y. Method and reagent for counting sperm by flow cytometry. Jpn. Kokai Tokkyo Koho JP 2001242168, 2001.

FLUORESCEIN

CAS Registry Number 2321-07-5

Chemical Structure

CA Index Name Spiro[isobenzofuran-1(3H),9'-[9H] xanthen]-3-one, 3',6'-dihydroxy-

Other Names Fluorescein; 3,6-Dihydroxyspiro [xanthene-9,3'-phthalide]; 3',6'-Dihydroxyfluoran; 3',6'-Fluorandiol; 9-(o-Carboxyphenyl)-6-hydroxy-3-isoxanthenone; Benzoic acid, 2-(6-hydroxy-3-oxo-3H-xanthen-9-yl)-; C.I. 45350:1; C.I. Solvent Yellow 94; D and C Yellow No. 7; D&C Yellow No. 7; Fluorescein acid; Japan Yellow 201; Japan Yellow No. 201; NSC 667256; Resorcinolphthalein; Solvent Yellow 94; Yellow fluorescein

Merck Index Number 4192

Chemical/Dye Class Xanthene

Molecular Formula $C_{20}H_{12}O_5$

Molecular Weight 332.31

Physical Form Red crystalline solid;[12] Reddish mass;[22] Red powder[3,21,24]

Solubility Insoluble in water, benzene, chloroform, ether; Soluble in acetone, N,N-dimethylformamide, dimethyl sulfoxide, hot ethanol, glacial acetic acid

Melting Point >350°C;[27] 314–316°C;[21,24] 312–315°C;[3] >300°C[12]

Boiling Point (Calcd.) 620.8 ± 55.0°C, pressure: 760 Torr

pK$_a$ 6.4, temperature: 22°C

pK$_a$ (Calcd.) 9.35 ± 0.20, most acidic, temperature: 25°C

Absorption (λ_{max}) 473 nm (Buffer pH 5.0); 490 nm (Buffer pH 9.0)

Emission (λ_{max}) 514 nm (Buffers pH 5.0, pH 9.0)

Molar Extinction Coefficient 34,000 cm^{-1} M^{-1} (Buffer pH 5.0); 93,000 cm^{-1} M^{-1} (Buffer pH 9.0)

Synthesis Synthetic methods[1–29]

Imaging/Labeling Applications Amyloid-β peptides;[30] antibodies;[31–35] bacteria;[36] brain;[37] chromosomes;[38,39] cornea;[40–43] cyanide;[44] dental plaque;[45,46] dextran;[47–49] fibrin;[50] hairs;[51–54] hydrogen peroxide;[55] insects;[56] liposomes;[57] lymphocytes;[58–60] metal ions (aluminum ions,[61] arsenic ions,[62] chromium ions,[61] copper ions,[63] iron ions,[61] lead ions,[64] mercury ions,[65,66] palladium ions,[67] silver ions,[68] zinc ions;[69]) nucleic acids/nucleotides;[70–79] plasma;[80,81] polylactic acid (PLA) microspheres;[82] proteins;[83–96] protoplasm;[97] silica nanoparticles;[98–100] steroids[101]

Biological/Medical Applications Detecting chromosomal aberration;[38,39] diagnosing corneal diseases;[40] marking/detecting insects;[56] as antitumor agent;[102] as antihistaminic agent;[103] use in agriculture and plant cultivation;[104] use in cosmetics;[105–109] use in ophthalmology;[40–43,110,111] as a substrate for measuring α-amylases activity,[112] elastases activity,[113] esterases activity,[114,115] β-glucuronidases activity,[116] horseradish peroxidases activity,[117] kinases activity,[118,119] monoamine oxidases (MAOs) activity,[120] phosphatases activity,[121] phospholipases activity,[122] proteases/proteinases activity,[123,124] telomerases activity activity[125]

Industrial Applications Aerosol products;[126] bar code systems;[127] cathode-ray tube;[128] color filters;[129,130] corrosion inhibitors;[131] detecting cracks on concrete;[132] dye lasers;[133–140] electrophotographic materials;[141,142] glass articles;[143] liquid-crystal displays;[144] light emitting diode;[145] metal oxide particles;[146] organic light-emitting device (OLED);[147] paints;[148,149] papers;[150–153] photoconductive materials;[154] photoresists;[155] photosensitive materials;[156,157] recording materials;[158–160] rubber;[161,162] textiles;[163,164] thin films;[165] toners;[166,167] for water tracing (tracers for hydrology)[168–170]

Safety/Toxicity Acute toxicity;[171] cytotoxicity;[172,173] mutagenicity;[172,174,175] ophthalmic toxicity;[176,177] phototoxicity;[178] skin toxicity[179,180]

Handbook of Fluorescent Dyes and Probes, First Edition. R. W. Sabnis.
© 2015 John Wiley & Sons, Inc. Published 2015 by John Wiley & Sons, Inc.

REFERENCES

1. Sabnis, R. W. *Handbook of Acid–base Indicators*; CRC Press: Boca Raton, **2008**; pp 155–157.

2. Chaudhari, A. S.; Parab, Y. S.; Patil, V.; Sekar, N.; Shukla, S. R. Intrinsic catalytic activity of Bronsted acid ionic liquids for the synthesis of triphenylmethane and phthalein under microwave irradiation. *RSC Adv.* **2012**, *2*, 12112–12117.

3. Eshghi, H.; Mirzaie, N. Ferric hydrogensulphate as a recyclable catalyst for the synthesis of fluorescein derivatives. *Chem. Papers* **2011**, *65*, 504–509.

4. Guan, Y.; Yan, D.; Liu, Z. Method for preparing high purity fluorescein sodium. Faming Zhuanli Shenqing CN 101792429, 2010.

5. Bydlinski, G.; Harris, G. R.; Scott, B. S. Substantially pure fluorescein. PCT Int. Appl. WO 2008073764, 2008.

6. Gao, Q.; Li, F.; Lv, R. Research on purification of fluorescein and fluorescein disodium. *Shandong Huagong* **2008**, *37*, 1–2.

7. Gao, Q. A process for the preparation and purification of fluorescein and its salts. Faming Zhuanli Shenqing Gongkai Shuomingshu CN 101270124, 2008.

8. Woodroofe, C. C.; Lim, M. H.; Bu, W.; Lippard, S. J. Synthesis of isomerically pure carboxylate- and sulfonate-substituted xanthene fluorophores. *Tetrahedron* **2005**, *61*, 3097–3105.

9. Tran, G. J.; Scherninski, F. High purity phthaleines and their preparation process using an organic ester as cyclocondensation solvent and a strong acid to precipitate them. Fr. Demande FR 2846331, 2004.

10. Cihelnik, S.; Stibor, I.; Lhotak, P. Solvent-free synthesis of sulfonephthaleins, sulfonefluoresceins and fluoresceins under microwave irradiation. *Collect. Czech. Chem. Commun.* **2002**, *67*, 1779–1789.

11. Tselinskii, I. V.; Brykov, A. S.; Astrat'ev, A. A. Effect of microwave heating on the course of different types of organic reactions. *Zh. Obshch. Khim.* **1996**, *66*, 1696–1698.

12. Tadic, D.; Brossi, A. Chiral prodyes. Ethers and esters of dihydrofluorescein. Part 1. Dibenzyldihydrofluorescein (DBDF) a new reagent. *Heterocycles* **1990**, *31*, 1975–1982.

13. Kroupa, J. Continuous manufacture of fluorescein. Czech. CS 250342, 1987.

14. Friedrich, F.; Kottke, K.; Kuehmstedt, H. Process for producing fluorescein for injection purposes. *Pharmazie* **1980**, *35*, 300–301.

15. Paucescu, S. D.; Visan, V. Xanthene staining solutions. I. Fluorescein synthesis. *Rev. Chim.* **1980**, *31*, 245–247.

16. Friedrich, F.; Kottke, K.; Kuehmstedt, H. Highly purified fluorescein for injection purposes. Ger. (East) DD 136498, 1979.

17. Ito, M.; Saito, Y.; Imaseki, I.; Hirano, S.; Amano, K. Manufacture of bath dyes. Jpn. Kokai Tokkyo Koho JP 53066940, 1978.

18. Andreyanov, V. V.; Klein, A. G.; Fomina, N. A. Fluorescein. U.S.S.R. SU 457710, 1975.

19. Unoda, T.; Murakami, N. Xanthene derivatives. JP 39020088, 1964.

20. Dolinsky, M.; Jones, J. H. Studies on coal-tar colors. IX. D&C Yellow No. 7, D&C Orange No. 5, D&C Red No. 21, tetrachlorofluorescein, D&C Red No. 27, and FD&C Red No. 3. *J. Assoc. Off. Agric. Chem.* **1951**, *34*, 114–126.

21. Welcher, F. J. *Organic Analytical Reagents;* D. Van Nostrand Company: New York, **1948**; Vol. *4*, pp 488–496.

22. McKenna, J. F.; Sowa, F. J. Organic reactions with boron fluoride: XVII. *J. Am. Chem. Soc.* **1938**, *60*, 124–125.

23. Woods, P. B.; Ogilvie, J.; Kranz, F. H. Process of making dihydroxyhydroxyfluorans. U.S. Patent 1931049, 1933.

24. Orndorff, W. R.; Hemmer, A. J. Fluorescein and some of its derivatives. *J. Am. Chem. Soc.* **1927**, *49*, 1272–1280.

25. George, E. Preparation of phthaleins directly from naphthalene. *J. S. Afr. Chem. Inst.* **1926**, *9*, 3–5.

26. Fischer, O.; Bollmann, M. Preparation and characteristics of fluorescein. *J. Prakt. Chem.* **1922**, *104*, 123–131.

27. von Liebig, H. Resorcinolbenzein and fluorescein. *J. Prakt. Chem.* **1912**, *85*, 241–284.

28. Ges. fur Teerverwertung. Fluorescein. DE 203312, 1908.

29. Baeyer, A. Ueber eine neue klasse von farbstoffen. *Ber. Dtsch. Chem. Ges.* **1871**, *4*, 555–558.

30. Hillen, H.; Striebinger, A.; Krantz, C.; Moeller, A.; Mueller, R. Preparation of

amyloid-β(1–42)-oligomers and their derivatives for use in diagnostics and therapy. Ger. Offen. DE 10303974, 2004.

31. Arcangeli, A.; Becchetti, A.; Pillozzi, S.; Masselli, M.; De Lorenzo, E. Method and kit for the prevention and/or the monitoring of chemoresistance of leukemia forms. PCT Int. Appl. WO 2011058509, 2011.

32. Carrero, J.; Voss, E. W., Jr., Temperature and pH dependence of fluorescein binding within the monoclonal antibody 9–40 active site as monitored by hydrostatic pressure. *J. Biol. Chem.* **1996**, *271*, 5332–5337.

33. Curtain, C. C. The chromatographic purification of fluorescein-antibody. *J. Histochem. Cytochem.* **1961**, *9*, 484–486.

34. Goldstein, G.; Slizys, I. S.; Chase, M. W. Studies on fluorescent antibody staining. I. Nonspecific fluorescence with fluorescein-coupled sheep antirabbit globulins. *J. Exp. Med.* **1961**, *114*, 89–110.

35. Coons, A. H. Histochemistry with labeled antibody. *Bull. N.Y. Acad. Med.* **1956**, *32*, 168.

36. Drevets, D. A.; Elliott, A. M. Fluorescence labeling of bacteria for studies of intracellular pathogenesis. *J. Immunol. Methods* **1995**, *187*, 69–79.

37. Martinez, J. L., Jr.; Koda, L. Penetration of fluorescein into the brain: a sex difference. *Brain Res.* **1988**, *450*, 81–85.

38. Hauke, S. Method for detecting a chromosomal aberration. PCT Int. Appl. WO 2012150022, 2012.

39. Poulsen, T. S.; Poulsen, S. M.; Petersen, K. H. Methods for detecting chromosome aberrations. PCT Int. Appl. WO 2005111235, 2005.

40. Soma, T. Vital staining of the cornea. *Atarashii Ganka* **2012**, *29*, 1593–1598.

41. Malkhanov, V. B.; Gajnutdinova, G. K.; Khafizov, G. G. Diagnosis of eye cornea disease and trauma. Russ. RU 2098089, 1997.

42. Araie, M. Carboxyfluorescein. A dye for evaluating the corneal endothelial barrier function *in vivo*. *Exp. Eye Res.* **1986**, *42*, 141–150.

43. Norn, M. S. Congo red vital staining of cornea and conjunctiva. *Acta Ophthalmol.* **1976**, *54*, 601–610.

44. Gong, G.; Zhao, H.; Wang, L. A fluorimetric method for the determination of cyanide with fluorescein and iodine. *Anal. Lett.* **1994**, *27*, 2797–2803.

45. Halberg, E. Dental treatment compositions containing plaque indicators. S. African ZA 9203194, 1993.

46. Gillings, B. R. D. Recent developments in dental plaque disclosants. *Aust. Dent. J.* **1977**, *22*, 260–266.

47. Shkurupii, V. A.; Luzgina, N. G.; Troitskii, A. V.; Gulyaeva, E. P.; Bystrova, T. N.; Medvedev, V. S. Method of obtaining fluorescent derivatives of dextran. Russ. RU 2426545, 2011.

48. Baker, G. A.; Neal Watkins, A.; Pandey, S.; Bright, F. V. Static and time-resolved fluorescence of fluorescein-labeled dextran dissolved in aqueous solution or sequestered within a sol–gel-derived hydrogel. *Analyst* **1999**, *124*, 373–379.

49. Nance, D. M.; Burns, J. Fluorescent dextrans as sensitive anterograde neuroanatomical tracers: applications and pitfalls. *Brain Res. Bull.* **1990**, *25*, 139–145.

50. Pappenhagen, A. R.; Koppel, J. L.; Olwin, J. H. Use of fluorescein-labeled fibrin for the determination of fibrinolytic activity. *J. Lab. Clin. Med.* **1962**, *59*, 1039–1046.

51. Greaves, A. Hair dyeing compositions comprising a dye/pigment, a photoactive compound and a light source. Fr. Demande FR 2990944, 2013.

52. Moeller, H.; Meinigke, B. Hair dye compositions containing xanthenes. Ger. Offen. DE 19926377, 2000.

53. Mizumaki, K. Sensitization-free hair dyes. Jpn. Kokai Tokkyo Koho JP 02231414, 1990.

54. Watanabe, K.; Ono, T.; Ota, T.; Minei, T.; Horikoshi, T. Wave-setting hair dye. Jpn. Kokai Tokkyo Koho JP 02076807, 1990.

55. Daniel, K. B.; Agrawal, A.; Manchester, M.; Cohen, S. M. Readily accessible fluorescent probes for sensitive biological imaging of hydrogen peroxide. *ChemBioChem* **2013**, *14*, 593–598.

56. Musgrave, A. J. Use of a fluorescent material for marking and detecting insects. *Can. Entomologist* **1949**, *81*, 173.

57. Ishida, A.; Otsuka, C.; Tani, H.; Kamidate, T. Fluorescein chemiluminescence method for estimation of membrane permeability of liposomes. *Anal. Biochem.* **2005**, *342*, 338–340.

58. Kinoshita, S.; Fukami, T.; Ido, Y.; Kushida, T. Spectroscopic properties of fluorescein in living lymphocytes. *Cytometry* **1987**, *8*, 35–41.

59. Kinoshita, S.; Fukami, T.; Saito, A.; Hirata, K.; Kushida, T.; Kinoshita, Y.; Kimura, S. Fluorescence depolarization of fluorescein in living lymphocytes. *J. Luminesc.* **1984**, *31–32*, 902–904.

60. Djaldetti, M.; Bessler, H.; Sandbank, U.; De Vries, A. Fluorescein staining of human lymphocytes

induced by *Echis colorata* venom. *Experientia* **1968**, *24*, 295–296.

61. Barba-Bon, A.; Costero, A. M.; Gil, S.; Parra, M.; Soto, J.; Martinez-Manez, R.; Sancenon, F. A new selective fluorogenic probe for trivalent cations. *Chem. Commun.* **2012**, *48*, 3000–3002.

62. He, X.; Gong, G.; Zhao, H.; Li, H. Fluorometric determination of arsenic(III) with fluorescein. *Microchem. J.* **1997**, *56*, 327–331.

63. Lin, J.; Hobo, T. Chemiluminescence investigation of NH2OH-fluorescein-Cu^{2+} system and its application to copper analysis in serum. *Talanta* **1995**, *42*, 1619–1623.

64. Kim, J. H.; Han, S. H.; Chung, B. H. Improving Pb^{2+} detection using DNAzyme-based fluorescence sensors by pairing fluorescence donors with gold nanoparticles. *Biosens. Bioelectron.* **2011**, *26*, 2125–2129.

65. Wan, J.; Ma, X.; Xing, L. Highly sensitive fluorescence detection of mercury (II) ions based on DNA machine amplification. *Sens. Actuators, B: Chem.* **2013**, *178*, 615–620.

66. Yang, X.; Li, Y.; Bai, Q. A highly selective and sensitive fluorescein-based chemodosimeter for Hg^{2+} ions in aqueous media. *Anal. Chim. Acta* **2007**, *584*, 95–100.

67. Santra, M.; Ko, S.; Shin, I.; Ahn, K. H. Fluorescent detection of palladium species with an O-propargylated fluorescein. *Chem. Commun.* **2010**, *46*, 3964–3966.

68. Nishio, Y. Electrodes having silver ion adsorbents and capable of visual check of lifetime. Jpn. Kokai Tokkyo Koho JP 2004157015, 2004.

69. Godwin, H. A.; Berg, J. M. A fluorescent zinc probe based on metal-induced peptide folding. *J. Am. Chem. Soc.* **1996**, *118*, 6514–6515.

70. Carrocci, T. J.; Hoskins, A. A. Imaging of RNAs in live cells with spectrally diverse small molecule fluorophores. *Analyst* **2014**, *139*, 44–47.

71. Murata, A.; Sato, S.; Kawazoe, Y.; Uesugi, M. Small-molecule fluorescent probes for specific RNA targets. *Chem. Commun.* **2011**, *47*, 4712–4714.

72. Al Attar, H. A.; Monkman, A. P. FRET and competing processes between conjugated polymer and dye substituted DNA strands: A comparative study of probe selection in DNA detection. *Biomacromolecules* **2009**, *10*, 1077–1083.

73. Ichikawa, K.; Nakamura, K.; Kurata, S. Real-time PCR method using the novel type of oligonucleotide probes. Jpn. Kokai Tokkyo Koho JP 2008182974, 2008.

74. Ozaki, H.; Iwase, N.; Sawai, H.; Kodama, T.; Kyogoku, Y. Detection of DNA bending in a DNA-PAP1 protein complex by fluorescence resonance energy transfer. *Biochem. Biophys. Res. Commun.* **1997**, *231*, 553–556.

75. Patil, S. V.; Salunkhe, M. M. Fluorescent-labeled oligonucleotide probes with non-nucleotide linker: detection of hybrid formation by fluorescence anisotropy. *Nucleosides Nucleotides* **1996**, *15*, 1603–1610.

76. Walker, G. T.; Linn, C. P.; Nadeau, J. G. DNA detection by strand displacement amplification and fluorescence polarization with signal enhancement using a DNA binding protein. *Nucleic Acids Res.* **1996**, *24*, 348–353.

77. Hou, W.; Smith, L. M. Fluorescence-based DNA sequencing with hexamer primers. *Nucleic Acids Res.* **1993**, *21*, 3331–3332.

78. Hoeltke, H. J.; Ettl, I.; Finken, M.; West, S.; Kunz, W. Multiple nucleic acid labeling and rainbow detection. *Anal. Biochem.* **1992**, *207*, 24–31.

79. Schubert, F.; Cech, D.; Wagner, K. Preparation of fluorophore derivatives for fluorescent labeling of nucleosides, nucleotides, and oligonucleotides. Ger. Offen. DE 4023212, 1991.

80. Lund-Andersen, H.; Krogsaa, B.; Jensen, P. K. Fluorescein in human plasma *in vivo*. *Acta Ophthalmol.* **1982**, *60*, 709–716.

81. Lund-Andersen, H.; Krogsaa, B. Fluorescein in human plasma *in vitro*. *Acta Ophthalmol.* **1982**, *60*, 701–708.

82. Armstrong, D. J.; Elliott, P. N. C.; Ford, J. L.; Gadsdon, D.; McCarthy, G. P.; Rostron, C.; Worsley, M. D. Poly(DL-lactic acid) microspheres incorporating histological dyes for intra-pulmonary histopathological investigations. *J. Pharm. Pharmacol.* **1996**, *48*, 258–262.

83. Patra, S.; Santhosh, K.; Pabbathi, A.; Samanta, A. Diffusion of organic dyes in bovine serum albumin solution studied by fluorescence correlation spectroscopy. *RSC Adv.* **2012**, *2*, 6079–6086.

84. Yi, L.; Sun, H.; Itzen, A.; Triola, G.; Waldmann, H.; Goody, R. S.; Wu, Y. One-pot dual-labeling of a protein by two chemoselective reactions. *Angew. Chem., Int. Ed.* **2011**, *50*, 8287–8290.

85. Kaneko, E.; Isoe, J. Method and reagent for measuring protein in sample. Jpn. Kokai Tokkyo Koho JP 2007240163, 2007.

86. Machleidt, T.; Robers, M.; Hanson, G. T. Protein labeling with FlAsH and ReAsH. *Methods Mol. Biol.* **2007**, *356*, 209–220.

87. Hamasaki, K. Fluorescent protein and drug screening method. Jpn. Kokai Tokkyo Koho JP 2002241397, 2002.

88. Rockey, J. H.; Li, W.; Eccleston, J. F. Binding of fluorescein and carboxyfluorescein by human serum proteins: significance of kinetic and equilibrium parameters of association in ocular fluorometric studies. *Exp. Eye Res.* **1983**, *37*, 455–466.

89. Li, W.; Rockey, J. H. Fluorescein binding to normal human serum proteins demonstrated by equilibrium dialysis. *Arch. Ophthalmol.* **1982**, *100*, 484–487.

90. Ianacone, D. C.; Felberg, N. T.; Federman, J. L. Tritiated fluorescein binding to normal human plasma proteins. *Arch. Ophthalmol.* **1980**, *98*, 1643–1645.

91. Johnson, E. A.; Brighton, W. D. Properties of some [14C]-fluorescein-protein conjugates. *J. Immunol. Methods* **1971**, *1*, 91–99.

92. Kierszenbaum, F.; Levison, S. A.; Dandliker, W. B. Fractionation of fluorescent-labeled proteins according to the degree of labeling. *Anal. Biochem.* **1969**, *28*, 563–572.

93. Ditseherlein, G.; Bausdorf, B. Fixing of tissue after intravenous injection of fluorescein-labeled protein. *Histochemie* **1965**, *5*, 55–69.

94. Chadwick, C. S.; Nairn, R. C.; McEntegart, M. G. The unreacted fluorescent material in fluorescein-protein conjugates. *Biochem. J.* **1959**, *73*, 41P.

95. Vodrazka, Z. Protein interactions. XIX. Interaction of serum albumin with fluorescein dyes. *Collect. Czech. Chem. Commun.* **1960**, *25*, 410–419.

96. Schiller, A. A.; Schayer, R. W.; Hess, E. L. Fluorescein-conjugated bovine albumin. Physical and biological properties. *J. Gen. Physiol.* **1953**, *36*, 489–506.

97. Hoefler, K.; Ziegler, A.; Luhan, M. Thresholds for uranin staining of the protoplasm of some *Florideae. Protoplasma* **1962**, *55*, 357–371.

98. Saleh, S. M.; Mueller, R.; Mader, H. S.; Duerkop, A.; Wolfbeis, O. S. Novel multicolor fluorescently labeled silica nanoparticles for interface fluorescence resonance energy transfer to and from labeled avidin. *Anal. Bioanal. Chem.* **2010**, *398*, 1615–1623.

99. Aizawa, H.; Okubo, N. Labeled silica nanoparticles for immunochromatography method, immunochromatography method reagent, immunochromatography method test strip using it, and immunochromatography method fluorescence detection system. Jpn. Kokai Tokkyo Koho JP 2008304401, 2008.

100. Gao, F.; Tang, L.; Dai, L.; Wang, L. A fluorescence ratiometric nano-pH sensor based on dual-fluorophore-doped silica nanoparticles. *Spectrochim. Acta, Part A: Mol. Biomol. Spectrosc.* **2007**, *67A*, 517–521.

101. Daxenbichler, G.; Grill, H. J.; Domanig, R.; Moser, E.; Dapunt, O. Receptor binding of fluorescein-labeled steroids. *J. Steroid Biochem.* **1980**, *13*, 489–493.

102. Doughtery, T. J. Activated dyes as antitumor agents. *J. Natl. Cancer Inst.* **1974**, *52*, 1333–1336.

103. Bukantz, S. C.; Dammin, G. J. Fluorescein as an indicator of antihistaminic activity. Inhibition of histamine-induced fluorescence in the skin of human subjects. *Science* **1948**, *107*, 224–225.

104. Kuznetsov, Y. P.; Yasinskii, A. M. Luminescent greenhouse material for use in agriculture and plant cultivation. Russ. RU 2248386, 2005.

105. Schroeder, K.; Manderla, S. Cosmetic powder product comprising cosmetic dye. Ger. Offen. DE 102011001256, 2012.

106. Mann, T.; Wendel, V.; Buelow, R. Visualization of sun protective or suntanning agents on skin by using fluorescein or its derivatives. Ger. Offen. DE 102004002601, 2005.

107. Momose, S.; Hirai, K. Lipsticks containing fluorine-base oils and fluorescein-based dyes. Jpn. Kokai Tokkyo Koho JP 07082116, 1995.

108. Ohtsu, Y.; Matsumoto, I. Studies on analysis of cosmetic dyes. 1. Rapid analysis of xanthene group dyes by high-speed liquid chromatography. *Nippon Kagaku Kaishi* **1979**, 511–516.

109. Dutton, K. R.; Reinisch, W. B. Lipstick and cosmetic compositions containing dyes. Brit. GB 1173743, 1969.

110. Amsler, M.; Huber, A. Nine years of testing fluorescein. *Ophthalmologica* **1955**, *129*, 271–275.

111. Sasiain, M. R. Value of the fluorescein test in ophthalmology. *Clin. Lab.* **1952**, *54*, 368–370.

112. Hofman, M.; Shaffar, M. Fluorescence depolarization assay for quantifying α-amylase in serum and urine. *Clin. Chem.* **1985**, *31*, 1478–1480.

113. Rinderknecht, H.; Geokas, M. C.; Silverman, P.; Lillard, Y.; Haverback, B. J. New methods for the determination of elastase. *Clin. Chim. Acta* **1968**, *19*, 327–339.

114. Metcalf, R. L.; Maxon, M.; Fukuto, T. R.; March, R. B. Aromatic esterase in insects. *Ann. Entomol. Soc. Am.* **1956**, *49*, 274–279.

115. Burch, J. Purification and properties of horse-liver esterase. *Biochem. J.* **1954**, *58*, 415–426.

116. Starkey, D. E.; Han, A.; Bao, J. J.; Ahn, C. H.; Wehmeyer, K. R.; Prenger, M. C.; Halsall, H. B.; Heineman, W. R. Fluorogenic assay for β-glucuronidase using microchip-based capillary electrophoresis. *J. Chromatogr., B: Biomed. Sci. Appl.* **2001**, *762*, 33–41.

117. Schoenhuber, W.; Fuchs, B.; Juretschko, S.; Amann, R. Improved sensitivity of whole-cell hybridization by the combination of horseradish peroxidase-labeled oligonucleotides and tyramide signal amplification. *Appl. Environ. Microbiol.* **1997**, *63*, 3268–3273.

118. Katayama, Y.; Koga, H.; Toita, R.; Niidome, T.; Mori, T. A novel fluorometric assay method for the detection of protein kinase activities. Jpn. Kokai Tokkyo Koho JP 2012254938, 2012.

119. Simard, J. R.; Getlik, M.; Grutter, C.; Pawar, V.; Wulfert, S.; Rabiller, M.; Rauh, D. Development of a fluorescent-tagged kinase assay system for the detection and characterization of allosteric kinase inhibitors. *J. Am. Chem. Soc.* **2009**, *131*, 13286–13296.

120. Li, X.; Zhang, H.; Xie, Y.; Hu, Y.; Sun, H.; Zhu, Q. Fluorescent probes for detecting monoamine oxidase activity and cell imaging. *Org. Biomol. Chem.* **2014**, *12*, 2033–2036.

121. Wang, Q.; Scheigetz, J.; Gilbert, M.; Snider, J.; Ramachandran, C. Fluorescein monophosphates as fluorogenic substrates for protein tyrosine phosphatases. *Biochim. Biophys. Acta, Protein Struct. Mol. Enzymol.* **1999**, *1431*, 14–23.

122. Zaikova, T. O.; Rukavishnikov, A. V.; Birrell, G. B.; Griffith, O. H.; Keana, J. F. W. Synthesis of fluorogenic substrates for continuous assay of phosphatidylinositol-specific phospholipase C. *Bioconjugate Chem.* **2001**, *12*, 307–313.

123. Kim, G. M.; Ryu, J. H.; Kwon, I. C.; Choi, G. W.; Yoon, I. C. *In vitro* kit for detecting protease activity and its preparation method. Repub. Korean Kongkae Taeho Kongbo KR 2014024193, 2014.

124. Krafft, G. A.; Wang, G. T.; Matayoshi, E. D. Fluorogenic substrates for the detection of proteolytic enzymes. Eur. Pat. Appl. EP 428000, 1991.

125. Qian, R.; Ding, L.; Ju, H. Switchable fluorescent imaging of intracellular telomerase activity using telomerase-responsive mesoporous silica nanoparticle. *J. Am. Chem. Soc.* **2013**, *135*, 13282–13285.

126. Vorozhtsov, G. N.; Tambieva, O. A.; Feizulova, R. K. G.; Khan, I. G.; Khromov, A. V. Compound for making signal or masking aerosol product. Russ. RU 2305676, 2007.

127. Nibu, M. Bar code systems using phosphors and reading apparatuses therefor. Jpn. Kokai Tokkyo Koho 09212576, 1997.

128. Morishita, H.; Hayashi, N.; Miura, S.; Sasaya, O.; Ito, M. Manufacture of color cathode-ray tube. Jpn. Kokai Tokkyo Koho JP 08087962, 1996.

129. Kwon, J. Y.; Yoo, Y. J.; Han, G. S.; Jeon, H. S.; Cho, H. Y. Photosensitive resin composition for color filter. Repub. Korean Kongkae Taeho Kongbo KR 2013073539, 2013.

130. Yamazaki, M.; Hofmann, O.; Ryu, G.; Li, X.; Lee, T. K.; de Mello, A. J.; de Mello, J. C. Non-emissive colour filters for fluorescence detection. *Lab Chip* **2011**, *11*, 1228–1233.

131. Schapira, J.; Droniou, P.; Hilaire, P.; Vincent-Ozanne, F. Corrosion inhibitors for and corrosion prevention treatment of iron alloy parts. Fr. Demande FR 2633947, 1990.

132. Kajimoto, K.; Nakane, T. Method for detecting cracks on concrete. Jpn. Kokai Tokkyo Koho JP 01114740, 1989.

133. Fujii, T.; Matsui, M. Multiple wavelength-emitting dye laser. Jpn. Kokai Tokkyo Koho JP 04302488, 1992.

134. Pavlopoulos, T. G. A figure of merit for laser dyes. *Opt. Commun.* **1981**, *38*, 397–401.

135. Govindanunny, T.; Sivaram, B. M. Solvation effects on the tunability of a fluorescein dye laser. *Opt. Commun.* **1980**, *32*, 425–428.

136. Mau, A. W. H. Broadband tunability of dye lasers. *Opt. Commun.* **1974**, *11*, 356–359.

137. Meyer, J. A.; Johnson, C. L.; Kierstead, E.; Sharma, R. D.; Itzkan, I. Dye laser stimulation. *Laser J.* **1971**, *3*, 19–20.

138. Myer, J. A.; Sharma, R. D.; Kierstead, E. J. Device for producing excited-state radiation in dyes and other laser materials. Ger. Offen. DE 2057791, 1975.

139. Peterson, O. G.; Webb, J. P.; McColgin, W. C.; Eberly, J. H. Organic dye laser threshold. *J. Appl. Phys.* **1971**, *42*, 1917–1928.

140. Schaefer, F. P.; Schmidt, W.; Marth, K. New dye lasers covering the visible spectrum. *Phys. Lett. A* **1967**, *24*, 280–281.

141. Koshelev, K. K.; Vasilenko, N. A.; Bojko, I. I.; Kosheleva, G. A.; Bojko, T. N.; Kotov, B.

V. Electrophotographic materials. U.S.S.R. SU 1194178, 1996.

142. Fiedler, H.; Herms, W. Electrophotographic materials. Ger. (East) DD 67033, 1969.

143. Obara, K.; Kawamura, H.; Tomita, Y.; Kawamura, T.; Misumi, A. Coloring of glass articles. Jpn. Kokai Tokkyo Koho JP 63030346, 1988.

144. Baur, G.; Greubel, W. Fluorescence-activated liquid-crystal display. *Appl. Phys. Lett.* **1977**, *31*, 4–6.

145. Zheng, Y.; Zhao, C.; Jiang, D.; Shen, D.; Yin, C. Organic material for white light emitting diode. Faming Zhuanli Shenqing CN 1327027, 2001.

146. Ocana, M.; Levy, D.; Serna, C. J. Preparation and optical properties of spherical metal oxide particles containing fluorescent dyes. *J. Non-Cryst. Solids* **1992**, *147–148*, 621–626.

147. Dai, Q.; Liu, Z.; Sun, L. Organic light-emitting device (OLED) and its manufacture method. Faming Zhuanli Shenqing CN 103187434, 2013.

148. Oka, S.; Mizuguchi, H. Fluorescent paints. Jpn. Kokai Tokkyo Koho JP 62015268, 1987.

149. Heidrich, H. Organic pigment phosphorescent paints. DE 873294, 1953.

150. Thomas, J. L.; Clot, O.; Banks, R. H. Fluorometric method for monitoring surface additives in a papermaking process. U.S. Pat. Appl. Publ. US 20090126889, 2009.

151. Dettling, B.; Ahonen, H. Fluorescent coating for paper and polymer webs. Finn. FI 109217, 2002.

152. Ikenoue, S.; Masuda, T. Thermographic printing paper. Jpn. Kokai Tokkyo Koho JP 50002924, 1975.

153. Denisova, L. M.; Polishchuk, L. A.; Markevich, N. N. Photosemiconductor paper. U.S.S.R. SU 313925, 1971.

154. Schoustra, B.; Roncken, H. Indirect electrophotography using photoconductive material containing light-sensitive dyes. Ger. Offen. DE 2432388, 1975.

155. Garza, C. M.; Cho, S. Metrology of bilayer photoresist processes. U.S. Pat. Appl. Publ. US 20090220895, 2009.

156. Kimura, K.; Inaba, N. Electrophotographic photosensitive materials. Jpn. Kokai Tokkyo Koho JP 54004136, 1979.

157. Miyatsuka, H.; Honjo, S.; Takimoto, M. Photosensitive material for electrophotography. Jpn. Tokkyo Koho JP 50040016, 1975.

158. Arai, N. Thermal recording material. Jpn. Kokai Tokkyo Koho JP 61092892, 1986.

159. Morinaka, A.; Oikawa, S. Optical recording materials. Jpn. Kokai Tokkyo Koho JP 60219647, 1985.

160. Arneth, R.; Lorenz, B.; Kluepfel, K. W. Electrophotographic recording material. Ger. DE 1302986, 1971.

161. Otoyama, T.; Mori, N. Polyrotaxane-containing rubber compositions with good workability, and pneumatic tires with good driving stability and durability having their treads. Jpn. Kokai Tokkyo Koho JP 2009062485, 2009.

162. Fujino, K.; Mori, N. Polyrotaxane-containing rubber compositions, and pneumatic tires with good driving stability and durability having their treads. Jpn. Kokai Tokkyo Koho JP 2009062484, 2009.

163. Spogli, R.; Sisani, M.; Latterini, L.; Nocchetti, M. Method for coloring natural textile fibers. Eur. Pat. Appl. EP 2628849, 2013.

164. Kaynak, A.; Foitzik, R. C.; Pfeffer, F. M. Fluorescence and conductivity studies on wool. *Mater. Chem. Phys.* **2009**, *113*, 480–484.

165. Qian, H. Q.; Mao, H. Y.; Chen, Q.; Song, F.; Hu, Y. W.; Huang, H.; Zhang, H. J.; Li, H. Y.; He, P. M.; Bao, S. N. The growth of thin fluorescein films on Ag (110). *Appl. Surf. Sci.* **2006**, *253*, 2336–2339.

166. Murayama, H.; Kawabata, T. Photoconductive toner. Jpn. Kokai Tokkyo Koho JP 62028770, 1987.

167. Watanabe, H.; Shoji, H.; Kawakado, K.; Seto, N. Photoconductive toners. Jpn. Kokai Tokkyo Koho JP 60153053, 1985.

168. Lyons, R. G. Identification and separation of water tracing dyes using pH response characteristics. *J. Hydrol.* **1993**, *152*, 13–29.

169. Viriot, M. L.; Andre, J. C. Fluorescent dyes: a search for new tracers for hydrology. *Analusis* **1989**, *17*, 97–111.

170. Smart, P. L.; Laidlaw, I. M. S. An evaluation of some fluorescent dyes for water tracing. *Water Resour. Res.* **1977**, *13*, 15–33.

171. Pouliquen, H.; Algoet, M.; Buchet, V.; Le Bris, H. Acute toxicity of fluorescein to turbot (*Scophthalmus maximus*). *Vet. Hum. Toxicol.* **1995**, *37*, 527–529.

172. Alford, R.; Simpson, H. M.; Duberman, J.; Hill, G. C.; Ogawa, M.; Regino, C.; Kobayashi, H.; Choyke, P. L. Toxicity of organic fluorophores used in molecular imaging: literature review. *Mol. Imaging* **2009**, *8*, 341–354.

173. Chang, Y.; Tseng, S.; Tseng, Sung-H.; Chen, Y.; Hsiao, J. Comparison of dyes for cataract surgery.

Part 1: cytotoxicity to corneal endothelial cells in a rabbit model. *J. Cataract Refract. Surg.* **2005**, *31*, 792–798.

174. Valencia, R.; Mason, J. M.; Woodruff, R. C.; Zimmering, S. Chemical mutagenesis testing in *Drosophila*. III. Results of 48 coded compounds tested for the National Toxicology Program. *Environ. Mutagen.* **1985**, *7*, 325–348.

175. Muzzall, J. M.; Cook, W. L. Mutagenicity test of dyes used in cosmetics with the *Salmonella*/mammalian-microsome test. *Mutat. Res., Genet. Toxicol. Test.* **1979**, *67*, 1–8.

176. Plange, N.; Bienert, M.; Remky, A.; Arend, K. O. Optic disc fluorescein leakage and intraocular pressure in primary open-angle glaucoma. *Curr. Eye Res.* **2012**, *37*, 508–512.

177. Yannuzzi, L. A.; Rohrer, K. T.; Tindel, L. J.; Sobel, R. S.; Costanza, M. A.; Shields, W.; Zang E. Fluorescein angiography complication survey. *Ophthalmology* **1986**, *93*, 611–617.

178. Danis, R. P.; Wolverton, S.; Steffens, T. Phototoxicity from systemic sodium fluorescein. *Retina* **2000**, *20*, 370–373.

179. Saleh, T. A.; Chidgey, C.; Wong, K. K. Fluorescein angiography and patchy skin discoloration: a case report. *Eye* **2004**, *18*, 553–554.

180. Carson, S. Skin painting studies in mice on 11 colors. *J. Toxicol., Cutan. Ocul. Toxicol.* **1984**, *3*, 309–331.

FLUORESCEIN DIACETATE (FDA)

CAS Registry Number   596-09-8

Chemical Structure

CA Index Name   Spiro[isobenzofuran-1(3H),9'-[9H]-xanthen]-3-one, 3',6'-bis(acetyloxy)-

Other Names   Fluorescein, diacetate; 3',6'-Diacetyl-fluorescein; F 1303; FDA; Fluorescein 3',6'-diacetate; MFCD 5062; NSC 4726; NSC 667259

Merck Index Number   Not listed

Chemical/Dye Class   Xanthene

Molecular Formula $C_{24}H_{16}O_7$

Molecular Weight   416.38

Physical Form   White crystals[5,9,10] or colorless solid; Beige crystals[11]

Solubility   Soluble in acetone, chloroform, dimethyl sulfoxide, ethanol, methanol

Melting Point   205–206 °C;[17]   205 °C;[11,16] 203–205.5 °C;[5]   203–205 °C;[15]   202–203 °C;[2]   202 °C;[8] 199–203 °C[9,10]

Boiling Point (Calcd.)   604.7 ± 55.0 °C, pressure: 760 Torr

Absorption (λ_{max})   289 nm (MeOH)

Emission (λ_{max})   None

Molar Extinction Coefficient   5300 cm^{-1} M^{-1} (MeOH)

Synthesis   Synthetic methods[1–17]

Imaging/Labeling Applications   Algae;[18,115–117] bacteria;[19–25] cells;[26–33] corneal endothelium;[34] cysts;[35,36] embryo;[37,128] hepatocytes;[38,39] hyphae;[40,41] marine plankton;[42] membrane integrity/permeability;[43–50] micro-organisms;[51–57] microspores;[58] oocytes;[59] perborate;[60,61] pollen;[62–65] protoplasts;[66–69] seeds;[70] sperms;[43,45] spores;[71] teliospores;[72] tumors;[73,74] yeasts[75]

Biological/Medical Applications   Assessing membrane integrity/permeability;[43–50] detecting/counting micro-organisms;[51–57] measuring microbial activity;[76–86] as a substrate for measuring acylases activity,[101] amidases activity,[87] aminopeptidases activity,[97] arylsulfatases activity,[87] chymotrypsin activity,[101] esterases activity;[88–100] galactosidases activity,[87] glucosidases activity,[87,97] lipases activity,[2,8,89,101] phosphatases activity,[97] proteases activity;[89] cell viability assay;[26–33] cytotoxicity assay;[46,117,120–125,130] implantable medical devices;[102] as temperature sensor;[103,104] treating ischemic disease[105]

Industrial Applications   Display devices;[106] electro-photography;[107,108] laser dyes;[109,110] papers;[111] petroleum markers,[3,9,10] recording materials[112,113]

Safety/Toxicity   Acute toxicity;[114] algal toxicity;[115–117] aluminum toxicity;[118] arsenic toxicity;[119] cytotoxicity;[46,117,120–125,130] ecotoxicity;[126,127] embryotoxicity;[128] marine toxicity;[129] nephrotoxicity;[130] neurotoxicity;[131–133] phytotoxicity[134]

REFERENCES

1. Sabnis, R. W. *Handbook of Acid–base Indicators*; CRC Press: Boca Raton, **2008**; pp 158–159.

2. Eshghi, H.; Mirzaie, N.; Asoodeh, A. Synthesis of fluorescein aromatic esters in the presence of P_2O_5/SiO_2 as catalyst and their hydrolysis studies in the presence of lipase. *Dyes Pigm.* **2011**, *89*, 120–126.

3. Raduly, M.; Raditoiu, V.; Raditoiu, A.; Wagner, L.; Amariutei, V. Synthesis and characterization of some xanthene fluorophore-markers for petroleum products. *Rev. Chim.* **2010**, *61*, 372–376.

4. Guan, Y.; Yan, D.; Liu, Z. Method for preparing high purity fluorescein sodium. Faming Zhuanli Shenqing CN 101792429, 2010.

5. Bydlinski, G.; Harris, G. R.; Scott, B. S. Substantially pure fluorescein. PCT Int. Appl. WO 2008073764, 2008.

6. Gao, Q.; Li, F.; Lv, R. Research on purification of fluorescein and fluorescein disodium. *Shandong Huagong* **2008**, *37*, 1–2.

7. Gao, Q. A process for the preparation and purification of fluorescein and its salts. Faming Zhuanli Shenqing Gongkai Shuomingshu CN 101270124, 2008.

8. Ge, F.; Chen, L.; Zhou, X.; Pan, H.; Yan, F.; Bai, G.; Yan, X. Synthesis and study on hydrolytic properties of fluorescein esters. *Dyes Pigm.* **2007**, *72*, 322–326.

9. Smith, M. J. Developer system for base reactable petroleum fuel markers. PCT Int. Appl. WO 9632461, 1996.

10. Smith, M. J. Fluorescent petroleum markers. U.S. Patent 5498808, 1996.

11. Tadic, D.; Brossi, A. Chiral prodyes. Ethers and esters of dihydrofluorescein. Part 1. Dibenzyldihydrofluorescein (DBDF) a new reagent. *Heterocycles* **1990**, *31*, 1975–1982.

12. Rypacek, F.; Drobnik, J.; Kalal, J. Fluorescence labeling method for estimation of soluble polymers in the living material. *Anal. Biochem.* **1980**, *104*, 141–149.

13. Friedrich, F.; Kottke, K.; Kuehmstedt, H. Process for producing fluorescein for injection purposes. *Pharmazie* **1980**, *35*, 300–301.

14. Friedrich, F.; Kottke, K.; Kuehmstedt, H. Highly purified fluorescein for injection purposes. Ger. (East) DD 136498, 1979.

15. McKenna, J. F.; Sowa, F. J. Organic reactions with boron fluoride. XVII. *J. Am. Chem. Soc.* **1938**, *60*, 124–125.

16. Hurd, C. D.; Schmerling, L. Alkenyl derivatives of fluorescein. *J. Am. Chem. Soc.* **1937**, *59*, 112–117.

17. von Liebig, H. Resorcinolbenzein and Fluorescein. *J. Prakt. Chem.* **1912**, *85*, 241–284.

18. Sato, S. Reagent and method for evaluating live/dead state of algae. Jpn. Kokai Tokkyo Koho JP 2002323490, 2002.

19. Hannig, C.; Hannig, M.; Rehmer, O.; Braun, G.; Hellwig, E.; Al-Ahmad, A. Fluorescence microscopic visualization and quantification of initial bacterial colonization on enamel *in situ*. *Arch. Oral Biol.* **2007**, *52*, 1048–1056.

20. Nakajima, H. Viable bacteria detection method, and viable bacteria counting apparatus. Jpn. Kokai Tokkyo Koho JP 2006029793, 2006.

21. Bruheim, P.; Eimhjellen, K. Effects of non-ionic surfactants on the uptake and hydrolysis of fluorescein diacetate by alkane-oxidizing bacteria. *Can. J. Microbiol.* **2000**, *46*, 387–390.

22. Diaper, J. P.; Tither, K.; Edwards, C. Rapid assessment of bacterial viability by flow cytometry. *Appl. Microbiol. Biotechnol.* **1992**, *38*, 268–272.

23. Bercovier, H.; Resnick, M.; Kornitzer, D.; Levy, L. Rapid method for testing drug-susceptibility of *Mycobacteria* spp. and gram-positive bacteria using rhodamine 123 and fluorescein diacetate. *J. Microbiol. Methods* **1987**, *7*, 139–142.

24. Ikeda, M. Apparatus for counting of bacteria in water. Jpn. Kokai Tokkyo Koho JP 61186854, 1986.

25. Brunius, G. Technical aspects of the use of 3′,6′-diacetyl fluorescein for vital fluorescent staining of bacteria. *Curr. Microbiol.* **1980**, *4*, 321–323.

26. Castro-Concha, L. A.; Escobedo, R. M.; de Lourdes Miranda-Ham, M. Measurement of cell viability. *Methods Mol. Biol.* **2012**, *877*, 49–56.

27. Powell, H. M.; Armour, A. D.; Boyce, S. T. Fluorescein diacetate for determination of cell viability in 3D fibroblast-collagen-gag constructs. *Methods Mol. Biol.* **2011**, *740*, 115–126.

28. Armour, A. D.; Powell, H. M.; Boyce, S. T. Fluorescein diacetate for determination of cell viability in tissue-engineered skin. *Tissue Eng., Part C: Methods* **2008**, *14*, 89–96.

29. Amano, T.; Hirasawa, K.; O'Donohue, M. J.; Pernolle, J.; Shioi, Y. A versatile assay for the accurate, time-resolved determination of cellular viability. *Anal. Biochem.* **2003**, *314*, 1–7.

30. Yoshimura, Y.; Kawasaki, Y.; Tsuji, A.; Kurane, R. Fluorometirc method for detecting living cells. Jpn. Kokai Tokkyo Koho JP 2002034595, 2002.

31. Niwa, M.; Yasui, T. Fluorescent dye solutions for cell staining and detection. Jpn. Kokai Tokkyo Koho JP 04102060, 1992.

32. Nadel, B. L. The use of fluorescein diacetate as a viability probe for plant cells. *Appl. Fluoresc. Technol.* **1989**, *1*, 13–16.

33. Szydlowska, H.; Zaporowska, E.; Kuszlik-Jochym, K.; Korohoda, W.; Branny, J. Membranolytic activity of detergents as studied with cell viability tests. *Folia Histochem. Cytochem.* **1978**, *16*, 69–77.

34. Wilhelm, F.; Melzig, M.; Franke, G. Staining time of fluorescein diacetate in corneal endothelium. *Ophthalmologe* **1993**, *90*, 171–173.

35. Jarmey-Swan, C.; Gibbs, R. A.; Ho, G. E.; Bailey, I. W.; Howgrave-Graham, A. R. A novel method for detection of viable *Giardia* cysts in water samples. *Water Res.* **2000**, *34*, 1948–1951.

36. Smith, A. L.; Smith, H. V. A comparison of fluorescein diacetate and propidium iodide staining and in vitro excystation for determining *Giardia intestinalis* cyst viability. *Parasitology* **1989**, *99*, 329–331.

37. Mohr, L. R.; Trounson, A. O. The use of fluorescein diacetate to assess embryo viability in the mouse. *J. Reprod. Fertil.* **1980**, *58*, 189–196.

38. Barth, C. A.; Schwarz, L. R. Transcellular transport of fluorescein in hepatocyte monolayers: evidence for functional polarity of cells in culture. *Proc. Natl. Acad. Sci. U.S.A.* **1982**, *79*, 4985–4987.

39. Gumucio, J. J.; Miller, D. L.; Krauss, M. D.; Zanolli, C. C. Transport of fluorescent compounds into hepatocytes and the resultant zonal labeling of the hepatic acinus in the rat. *Gastroenterology* **1981**, *80*, 639–646.

40. Ingham, E. R.; Klein, D. A. Soil fungi: relationships between hyphal activity and staining with fluorescein diacetate. *Soil Biol. Biochem.* **1984**, *16*, 273–278.

41. Ingham, E. R.; Klein, D. A. Relationship between fluorescein diacetate-stained hyphae and oxygen utilization, glucose utilization, and biomass of submerged fungal batch cultures. *Appl. Environ. Microbiol.* **1982**, *44*, 363–370.

42. Baek, S. H.; Shin, K. Applicability of fluorescein diacetate (FDA) and calcein-AM to determine the viability of marine plankton. *Ocean Polar Res.* **2009**, *31*, 349–357.

43. Dou, W.; Wang, M.; Zheng, J.; Yu, X. Sperm viability rate and membrane integrity of *Pinctada martensii* with a combination of FDA and PI. *Guangdong Haiyang Daxue Xuebao* **2012**, *32*, 59–63.

44. Umebayashi, Y.; Miyamoto, Y.; Wakita, M.; Kobayashi, A.; Nishisaka, T. Elevation of plasma membrane permeability on laser irradiation of extracellular latex particles. *J. Biochem.* **2003**, *134*, 219–224.

45. Reyes, R.; Martinez, J. C.; Delgado, N. M.; Merchant-Larios, H. Heparin-glutathione III: Study with fluorescent probes as indicators of membrane status of bull sperm. *Arch. Androl.* **2002**, *48*, 209–219.

46. Aeschbacher, M.; Reinhardt, C. A.; Zbinden, G. A rapid cell membrane permeability test using fluorescent dyes and flow cytometry. *Cell Biol. Toxicol.* **1986**, *2*, 247–255.

47. Prosperi, E.; Croce, A. C.; Bottiroli, G.; Supino, R. Flow cytometric analysis of membrane permeability properties influencing intracellular accumulation and efflux of fluorescein. *Cytometry* **1986**, *7*, 70–75.

48. Persidsky, M. D.; Baillie, G. S. Fluorometric test of cell membrane integrity. *Cryobiology* **1977**, *14*, 322–331.

49. Augsten, K.; Guettner, J. FDA [fluoresceindiacetate] hydrolysis for the fluorometrical evidence of a cell membrane alteration in peritoneal macrophages. *Acta Histochem.* **1975**, *52*, 79–87.

50. Rotman, B.; Papermaster, B. W. Membrane properties of living mammalian cells as studied by enzymic hydrolysis of fluorogenic esters. *Proc. Natl. Acad. Sci. U.S.A.* **1966**, *55*, 134–141.

51. Ribault, S.; Chollet, R. Fluorogenic culture medium for detection of microorganisms including a dye to mask residual fluorescence. Fr. Demande FR 2955121, 2011.

52. Horikiri, S. Microorganism cell detection method using multiple fluorescent indicators. Jpn. Kokai Tokkyo Koho JP 2006238779, 2006.

53. Riis, V.; Lorbeer, H.; Babel, W. Extraction of microorganisms from soil: evaluation of the efficiency by counting methods and activity measurements. *Soil Biol. Biochem.* **1998**, *30*, 1573–1581.

54. Buenning, G.; Hempel, D. C. Vital-fluorochromization of microorganisms using 3′,6′-diacetylfluorescein to determine damages of cell membranes and loss of metabolic activity by ozonation. *Ozone: Sci. Eng.* **1996**, *18*, 173–181.

55. Sugata, K.; Ueda, R. Fluorometric microorganism counting in food or pharmaceuticals. Jpn. Kokai Tokkyo Koho JP 03043069, 1991.

56. Desrosier, J. P.; Ward, N. N., Jr., Apparatus for the detection of microorganisms, based on the gas produced by the microorganism. Braz. Pedido PI BR 8705260, 1988.

57. Scholefield, J. Staining microorganisms. Ger. Offen. DE 2728077, 1978.

58. Deslauriers, C.; Powell, A. D.; Fuchs, K.; Pauls, K. P. Flow cytometric characterization and sorting of cultured *Brassica napus* microspores. *Biochim. Biophys. Acta, Mol. Cell Res.* **1991**, *1091*, 165–172.

59. Boender, J. Fluorescein-diacetate, a fluorescent dye compound stain for rapid evaluation of the viability of mammalian oocytes prior to *in vitro* studies. *Vet. Q.* **1984**, *6*, 236–240.

60. Jang, S. G.; Choi, M. G.; Lee, H. G.; Jeon, H. R.; Park, J. E. Sensor containing fluorescein compounds having selectivity for perborate and method for detecting perborate. Repub. Korea KR 1099508, 2011.

61. Choi, M. G.; Cha, S.; Park, J. E.; Lee, H.; Jeon, H. L.; Chang, S. Selective perborate signaling by deprotection of fluorescein and resorufin acetates. *Org. Lett.* **2010**, *12*, 1468–1471.

62. Pinillos, V.; Cuevas, J. Standardization of the fluorochromatic reaction test to assess pollen viability. *Biotech. Histochem.* **2008**, *83*, 15–21.

63. Class, S.; Braun, P.; Siegert, A. Fluorochromatic determination of pollen viability of *Triticum aestivum* L. evaluation of chemical hybridizing agents. *Angew. Bot.* **1989**, *63*, 1–6.

64. Yang, H. Y. Fluorescein diacetate used as a vital stain for labeling living pollen tubes. *Plant Sci.* **1986**, *44*, 59–63.

65. Heslop-Harrison, J.; Heslop-Harrison, Y. Evaluation of pollen viability by enzymically induced fluorescence; intracellular hydrolysis of fluorescein diacetate. *Stain Technol.* **1970**, *45*, 115–120.

66. Li, Y.; Shi, S.; He, Y. Effects of different fluorescent dyes on staining alfalfa protoplast. *Caodi Xuebao* **2012**, *20*, 348–351.

67. Saito, N. Double staining method and enzyme reaction chamber for protoplasts preparation. Jpn. Kokai Tokkyo Koho JP 06102158, 1994.

68. Galbraith, D. W. Isolation and flow cytometric characterization of plant protoplasts. *Methods Cell Biol.* **1990**, *33*, 527–547.

69. Uchimiya, H. Parameters influencing the liposome-mediated insertion of fluorescein diacetate into plant protoplasts. *Plant Physiol.* **1981**, *67*, 629–632.

70. Thorogood, C. J.; Rumsey, F. J.; Hiscock, S. J. Seed viability determination in parasitic broomrapes (*Orobanche* and *Phelipanche*) using fluorescein diacetate staining. *Weed Res.* **2009**, *49*, 461–468.

71. Yang, H.; Nemoto, Y.; Homma, T.; Matsuoka, H.; Yamada, S.; Sumita, O.; Takatori, K.; Kurata, H. Rapid viability assessment of spores of several fungi by an ionic intensified fluorescein diacetate method. *Curr. Microbiol.* **1995**, *30*, 173–176.

72. Chastain, T. G.; King, B. A biochemical method for estimating viability of teliospores of *Tilletia controversa*. *Phytopathology* **1990**, *80*, 474–476.

73. Nagano, T.; Urano, Y.; Kojima, H.; Fujikawa, Y.; Takamatsu, T.; Harada, Y. Tumor-selective fluorescent dye. PCT Int. Appl. WO 2007099924, 2007.

74. Thomas, J. A.; Buchsbaum, R. N.; Zimniak, A.; Racker, E. Intracellular pH measurements in Ehrlich ascites tumor cells utilizing spectroscopic probes generated *in situ*. *Biochemistry* **1979**, *18*, 2210–2218.

75. Chilver, M. J.; Harrison, J.; Webb, T. J. B. Use of immunofluorescence and viability stains in quality control. *J. Am. Soc. Brew. Chem.* **1978**, *36*, 13–18.

76. Swiontek Brzezinska, M.; Burkowska, A.; Walczak, M. Microbial activity in the landfill soil. *Appl. Biochem. Microbiol.* **2012**, *48*, 371–376.

77. Gomez, M. A.; Baldini, M.; Marcos, M.; Martinez, A.; Fernandez, S.; Reyes, S. Aerobic microbial activity and solid waste biodegradation in a landfill located in a semi-arid region of Argentina. *Ann. Microbiol.* **2012**, *62*, 745–752.

78. Gravel, V.; Dorais, M.; Menard, C. Organic fertilization and its effect on development of sweet pepper transplants. *HortScience* **2012**, *47*, 198–204.

79. Nunes, J. S.; Araujo, A. S. F.; Nunes, L. A. P. L.; Lima, L. M.; Carneiro, R. F. V.; Salviano, A. A. C.; Tsai, S. M. Impact of land degradation on soil microbial biomass and activity in northeast Brazil. *Pedosphere* **2012**, *22*, 88–95.

80. Alarcon-Gutierrez, E.; Floch, C.; Ruaudel, F.; Criquet, S. Non-enzymatic hydrolysis of fluorescein diacetate (FDA) in a Mediterranean oak (*Quercus ilex* L.) litter. *Eur. J. Soil Sci.* **2008**, *59*, 139–146.

81. Rietz, D. N.; Haynes, R. J. Effects of irrigation-induced salinity and sodicity on soil microbial activity. *Soil Biol. Biochem.* **2003**, *35*, 845–854.

82. Adam, G.; Duncan, H. Development of a sensitive and rapid method for the measurement of total microbial activity using fluorescein diacetate (FDA) in a range of soils. *Soil Biol. Biochem.* **2001**, *33*, 943–951.

83. Fontvieille, D. A.; Outagaguerouine, A.; Thevenot, D. R. Fluorescein diacetate hydrolysis as a measure of microbial activity in aquatic systems: application to activated sludges. *Environ. Technol.* **1992**, *13*, 531–540.

84. Zelles, L.; Adrian, P.; Bai, Q. Y.; Stepper, K.; Adrian, M. V.; Fischer, K.; Maier, A.; Ziegler, A. Microbial activity measured in soils stored under different temperature and humidity conditions. *Soil Biol. Biochem.* **1991**, *23*, 955–962.

85. Stubberfield, L. C. F.; Shaw, P. J. A. A comparison of tetrazolium reduction and FDA hydrolysis with other measures of microbial activity. *J. Microbiol. Methods* **1990**, *12*, 151–162.

86. Schnuerer, J.; Rosswall, T. Fluorescein diacetate hydrolysis as a measure of total microbial activity in soil and litter. *Appl. Environ. Microbiol.* **1982**, *43*, 1256–1261.

87. Bandick, A. K.; Dick, R. P. Field management effects on soil enzyme activities. *Soil Biol. Biochem.* **1999**, *31*, 1471–1479.

88. Ye, L.; Shi, X.; Wu, X.; Yu, Y.; Zhang, M.; Kong, F. Assessing phytoplankton cell lysis rate in lake Taihu by esterase assay. *Hupo Kexue* **2012**, *24*, 712–716.

89. Prosser, J. A.; Speir, T. W.; Stott, D. E. Soil oxidoreductases and FDA hydrolysis. *Soil Sci. Soc. Am. Book Ser.* **2011**, *9*, 103–124.

90. Kwolek-Mirek, M.; Bednarska, S.; Zadrag-Tecza, R.; Bartosz, G. The hydrolytic activity of esterases in the yeast *Saccharomyces cerevisiae* is strain dependent. *Cell Biol. Int.* **2011**, *35*, 1111–1119.

91. Zhan, R.; Tan, A. J. H.; Liu, B. Conjugated polyelectrolyte as signal amplifier for fluorogenic probe based enzyme activity study. *Polym. Chem.* **2011**, *2*, 417–421.

92. Vitecek, J.; Adam, V.; Petrek, J.; Babula, P.; Novotna, R.; Kizek, R.; Havel, L. Application of fluorimetric determination of esterases in plant material. *Chem. Listy* **2005**, *99*, 496–501.

93. Chrost, R. J.; Gajewski, A.; Siuda, W. Fluorescein-diacetate (FDA) assay for determining microbial esterase activity in lake water. *Adv. Limnol.* **1999**, *54*, 167–183.

94. Breeuwer, P.; Drocourt, J.; Bunschoten, N.; Zwietering, M. H.; Rombouts, F. M.; Abee, T. Characterization of uptake and hydrolysis of fluorescein diacetate and carboxyfluorescein diacetate by intracellular esterases in *Saccharomyces cerevisiae*, which result in accumulation of fluorescent product. *Appl. Environ. Microbiol.* **1995**, *61*, 1614–1619.

95. Cohen, J. D.; Husic, H. D. Carbonic anhydrase catalyzed hydrolysis of fluorogenic esterase substrates. *Phytochem. Anal.* **1991**, *2*, 60–64.

96. Dive, C.; Cox, H.; Watson, J. V.; Workman, P. Polar fluorescein derivatives as improved substrate probes for flow cytoenzymological assay of cellular esterases. *Mol. Cell. Probes* **1988**, *2*, 131–145.

97. Holzapfel-Pschorn, A.; Obst, U.; Haberer, K. Sensitive methods for the determination of microbial activities in water samples using fluorigenic substrates. *Fresenius' Z. Anal. Chem.* **1987**, *327*, 521–523.

98. Hofmann, J.; Sernetz, M. A kinetic study on the enzymatic hydrolysis of fluorescein diacetate and fluorescein di-β-D-galactopyranoside. *Anal. Biochem.* **1983**, *131*, 180–186.

99. Szollosi, J.; Kertai, P.; Somogyi, B.; Damjanovich, S. Characterization of living normal and leukemic mouse lymphocytes by fluorescein diacetate. *J. Histochem. Cytochem.* **1981**, *29*, 503–510.

100. Sernetz, M.; Puchinger, H.; Couwenbergs, C.; Ostwald, M. A new method for the evaluation of reaction kinetics of immobilized enzymes investigated in single enzyme-Sepharose beads by microfluorometry. *Anal. Biochem.* **1976**, *72*, 24–37.

101. Guilbault, G. G.; Kramer, D. N. Fluorometric determination of lipase, acylase, α- and γ-chymotrypsin and inhibitors of these enzymes. *Anal. Chem.* **1964**, *36*, 409–412.

102. Steele, T. W. J.; Huang, C. L.; Kumar, S.; Widjaja, E.; Boey, F. Y. C.; Loo, J. S. C.; Venkatraman, S. S. High-throughput screening of PLGA thin films utilizing hydrophobic fluorescent dyes for hydrophobic drug compounds. *J. Pharm. Sci.* **2011**, *100*, 4317–4329.

103. Bousseksou, A.; Salmon, L.; Molnar, G.; Cobo, S. Materials with thermochromic spin transition doped with one or more fluorescent agents for use as temperature sensor. Fr. Demande FR 2952371, 2011.

104. Bousseksou, A.; Salmon, L.; Molnar, G.; Cobo, S. Heat-sensitive spin-transition materials doped with one or more fluorescent agents for use as temperature sensor. PCT Int. Appl. WO 2011058277, 2011.

105. Wakita, H.; Igarashi, K.; Oie, K. Diagnosis and treatment of ischemic disease. PCT Int. Appl. WO 2009022756, 2009.

106. Jung, H. G.; Kang, C. S.; Park, H. J. Method for making flexible plastic substrate for display devices. Repub. Korean Kongkae Taeho Kongbo KR 2012078316, 2012.

107. Inoue, S.; Katagiri, N. Multicolor electrophotographic process. Jpn. Kokai Tokkyo Koho JP 54009929, 1979.

108. Hojo, K.; Futaki, K.; Suzuki, S.; Komura, I. Color electrophotography. Jpn. Kokai Tokkyo Koho JP 51113630, 1976.

109. Marling, J. B.; Gregg, D. W.; Thomas, S. J. Effect of oxygen on flashlamp-pumped organic-dye lasers. *IEEE J. Quantum Electron.* **1970**, *6*, 570–572.

110. Gregg, D. W.; Querry, M. R.; Marling, J. B.; Thomas, S. J.; Dobler, C. V.; Davies, N. J.; Belew, J. F. Wavelength tunability of new flashlamp-pumped laser dyes. *IEEE J. Quantum Electron.* **1970**, *6*, 270.

111. Komura, I.; Futaki, K.; Haino, K. Multicolor heat-sensitive copying papers. Jpn. Kokai Tokkyo Koho JP 50147736, 1975.

112. Inaba, N.; Iiyama, K. Heat-sensitive two-color recording material. Ger. Offen. DE 3540627, 1986.

113. Mitsubishi Paper Mills, Ltd. Multicolor thermal recording materials. Jpn. Tokkyo Koho JP 55036519, 1980.

114. Leszczynska, M.; Oleszkiewic, J. A. Application of the fluorescein diacetate hydrolysis as an acute toxicity test. *Environ. Technol.* **1996**, *17*, 79–85.

115. Cronin, M. T. D.; Netzeva, T. I.; Dearden, J. C.; Edwards, R.; Worgan, A. D. P. Assessment and modeling of the toxicity of organic chemicals to *Chlorella vulgaris*: development of a novel database. *Chem. Res. Toxicol.* **2004**, *17*, 545–554.

116. Franklin, N. M.; Adams, M. S.; Stauber, J. L.; Lim, R. P. Development of an improved rapid enzyme inhibition bioassay with marine and freshwater microalgae using flow cytometry. *Arch. Environ. Contam. Toxicol.* **2001**, *40*, 469–480.

117. Faber, M. J.; Smith, L. M. J.; Boermans, H. J.; Stephenson, G. R.; Thompson, D. G.; Solomon, K. R. Cryopreservation of fluorescent marker-labeled algae (*Selenastrum capricornutum*) for toxicity testing using flow cytometry. *Environ. Toxicol. Chem.* **1997**, *16*, 1059–1067.

118. Gunse, B.; Garzon, T.; Barcelo, J. Study of aluminum toxicity by means of vital staining profiles in four cultivars of *Phaseolus vulgaris* L. *J. Plant Physiol.* **2003**, *160*, 1447–1450.

119. Ghosh, A. K.; Bhattacharyya, P.; Pal, R. Effect of arsenic contamination on microbial biomass and its activities in arsenic contaminated soils of Gangetic West Bengal, India. *Environ. Int.* **2004**, *30*, 491–499.

120. Kleszczynski, K.; Skladanowski, A. C. Mechanism of cytotoxic action of perfluorinated acids. I. alteration in plasma membrane potential and intracellular pH level. *Toxicol. Appl. Pharmacol.* **2009**, *234*, 300–305.

121. Lindhagen, E.; Nygren, P.; Larsson, R. The fluorometric microculture cytotoxicity assay. *Nat. Protoc.* **2008**, *3*, 1364–1369.

122. Keshelava, N.; Frgala, T.; Krejsa, J.; Kalous, O.; Reynolds, C. P. DIMSCAN a microcomputer fluorescence-based cytotoxicity assay for preclinical testing of combination chemotherapy. *Methods Mol. Med.* **2005**, *110*, 139–153.

123. Diaz, T. M.; Pertega, S.; Ortega, D.; Lopez, E.; Centeno, A.; Manez, R.; Domenech, N. FDA/PI flow cytometry assay of complement-mediated cytotoxicity of antibodies generated during xenotransplantation. *Cytometry, Part A* **2004**, *62A*, 54–60.

124. Larsson, R.; Nygren, P. Pharmacological modification of multi-drug resistance (MDR) in vitro detected by a novel fluorometric microculture cytotoxicity assay. Reversal of resistance and selective cytotoxic actions of cyclosporin A and verapamil on MDR leukemia T-cells. *Int. J. Cancer* **1990**, *46*, 67–72.

125. Scaife, M. C. An *in vitro* cytotoxicity test to predict the ocular irritation potential of detergents and detergent products. *Food Chem. Toxicol.* **1985**, *23*, 253–258.

126. Nancharaiah Y. V.; Rajadurai, M.; Venugopalan, V. P. Single cell level microalgal ecotoxicity assessment by confocal microscopy and digital image analysis. *Environ. Sci. Technol.* **2007**, *41*, 2617–2621.

127. Stauber, J. L.; Franklin, N. M.; Adams, M. S. Applications of flow cytometry to ecotoxicity testing using microalgae. *Trends Biotechnol.* **2002**, *20*, 141–143.

128. Kola, I.; Folb, P. I. An assessment of the effects of cyclophosphamide and sodium valproate on the viability of preimplantation mouse embryos using the fluorescein diacetate test. *Teratogen., Carcinogen., Mutagen.* **1986**, *6*, 23–31.

129. Moffat, B. D.; Snell, T. W. Rapid toxicity assessment using an *in vivo* enzyme test for *Brachionus plicatilis* (Rotifera). *Ecotoxicol. Environ. Saf.* **1995**, *30*, 47–53.

130. Tsai, C. Y.; Wu, T. H.; Yu, C. L.; Chou, C. T. The *in vitro* immunomodulatory effects of sulfasalazine on human polymorphonuclear leukocytes, mononuclear cells, and cultured glomerular mesangial cells. *Life Sci.* **2000**, *67*, 1149–1161.

131. Noelker, C.; Bacher, M.; Gocke, P.; Wei, X.; Klockgether, T.; Du, Y.; Dodel, R. The flavanoide caffeic acid phenethyl ester blocks 6-hydroxydopamine-induced neurotoxicity. *Neurosci. Lett.* **2005**, *383*, 39–43.

132. Hu, J.; el-Fakahany, E. E. An artifact associated with using trypan blue exclusion to measure effects of amyloid beta on neuron viability. *Life Sci.* **1994**, *55*, 1009–1016.

133. Keilhoff, G.; Wolf, G. Comparison of double fluorescence staining and LDH-test for monitoring cell viability *in vitro*. *Neuroreport* **1993**, *5*, 129–132.

134. Gigliotti, G.; Onofri, A.; Pannacci, E.; Businelli, D.; Trevisan, M. Influence of dissolved organic matter from waste material on the phytotoxicity and environmental fate of triflusulfuron methyl. *Environ. Sci. Technol.* **2005**, *39*, 7446–7451.

FLUOROSCEIN-5-ISOTHIOCYANATE (FITC)

CAS Registry Number 3326-32-7
Chemical Structure

CA Index Name Spiro[isobenzofuran-1(3*H*),9'-[9*H*]xanthen]-3-one, 3',6'-dihydroxy-5-isothiocyanato-

Other Names Fluorescein, 5-isothiocyanato-; 5-Isothiocyanatofluorescein; FITC isomer I; Fitc isomer 1; Fluorescein 5-isothiocyanate; Fluorescein isothiocyanate isomer 1; Fluorescein isothiocyanate isomer I; Fluoroscein-5-isothiocyanate

Merck Index Number 4120

Chemical/Dye Class Xanthene

Molecular Formula $C_{21}H_{11}NO_5S$

Molecular Weight 389.38

Physical Form Orange-yellow crystal;[2,6] Orange solid[9]

Solubility Soluble in acetone, *N,N*-dimethylformamide, dimethyl sulfoxide, ethanol, methanol

Melting Point >193 °C (decompose);[2] >160 °C (decompose)[6]

Boiling Point (Calcd.) 708.6 ± 60.0 °C, pressure: 760 Torr

pK_a (Calcd.) 9.31 ± 0.20, most acidic, temperature: 25 °C

Absorption (λ_{max}) 494 nm (Buffer pH 9.0)

Emission (λ_{max}) 519 nm (Buffer pH 9.0)

Molar Extinction Coefficient 77,000 $cm^{-1} M^{-1}$ (Buffer pH 9.0)

Synthesis Synthetic methods[1–9]

Imaging/Labeling Applications Amino groups;[10] antibodies;[11–15] biotin;[16] cells;[17,18] cellulose nanocrystals;[19,20] α-cobratoxin;[21] dendrimers;[22–25,80] dextrans;[26,27] digoxin;[28] enzymes: ATPase,[29–33] soybean lipoxygenase;[34] heparins;[35,36] Hodgkin and Reed Sternberg (HRS) cells;[37] insulin;[38] nanoparticles (gold nanoparticles,[39] silica nanoparticles;[40,41]) nucleotides/nucleic acids;[12,42–47] paclitaxel;[48,49] PEO-PCL-PEO triblock copolymers;[50] polymersomes;[51] proteins/peptides;[5,7,9,12,52–66] strychnine;[67] sugar chains;[68] thiol groups[69,70]

Biological/Medical Applications Analyzing sugar chain;[68] detecting iodide/iodate or total iodine;[39] detecting/imaging thiols;[69,70] detecting/quantifying biotin derivatives;[16] diagnosing classical Hodgkin lymphoma (CHL) in lymph nodes;[37] cellulose nanocrystals for bioimaging applications;[20] treating neuroinflammation;[23] treating/diagnosing cancer;[24] PEO-PCL-PEO triblock copolymers for topical delivery;[50] polymersomes in biomedical or cosmetic applications;[51] apoptosis assay;[71] as a substrate for measuring heparanase activity,[72] human immunodeficiency virus 1 (HIV-1) proteinase activity,[73] kinase activity,[74] phosphatase activity,[74] β-lactamase activity,[75] transglutaminase (TG2) activity;[76] implantable drug-delivery devicess[77]

Industrial Applications Gel coating/gel monolith for fiber-optic sensor applications;[78] ultrathin films[79]

Safety/Toxicity Cytotoxicity;[80,81] genotoxicity;[82] membrane toxicity;[83] neurotoxicity;[84] ophthalmic toxicity;[85] phototoxicity;[86] pulmonary toxicity;[87] skin toxicity[88–92]

REFERENCES

1. Sabnis, R. W. *Handbook of Acid–base Indicators*; CRC Press: Boca Raton, **2008**; pp 162–163.

2. Muramoto, K.; Kamiya, H.; Kawauchi, H.; Meguro, H. High performance liquid chromatography of fluorescein isothiocyanate isomers and intermediate products. *Anal. Sci.* **1985**, *1*, 447–450.

3. Pitra, J.; Zoula, V. Fluoresceinyl isothiocyanates. Czech. CS 201892, 1980.

4. Pitra, J.; Perina, Z.; Brda, M. Aryl isothiocyanates. Czech. CS 190225, 1979.

5. Steinbach, G. Characterization of fluorescein isothiocyanate. Synthesis and testing methods for fluorescein isothiocyanate isomers. *Acta Histochem.* **1974**, *50*, 19–34.

6. Sinsheimer, J. E.; Jagodic, V.; Burckhalter, J. H. Fluorescein isothiocyanates: Improved synthesis and

purity, Spectral studies. *Anal. Biochem.* **1974**, *57*, 227–231.

7. Tsou, K. C. Isothiocyanates of phthalein. U.S. Patent 3028397. 1962.

8. University of Kansas Research Foundation, Xanthene compounds and dyestuffs and means of producing the same. GB 846674, 1960.

9. Riggs, J. L.; Seiwald, R. J.; Burckhalter, J. H.; Downs, C. M.; Metcalf, T. G. Isothiocyanate compounds as fluorescent labeling agents for immune serum. *Am. J. Pathol.* **1958**, *34*, 1081–1097.

10. Amir, R. J.; Albertazzi, L.; Willis, J.; Khan, A.; Kang, T.; Hawker, C. J. Multifunctional trackable dendritic scaffolds and delivery agents. *Angew. Chem., Int. Ed.* **2011**, *50*, 3425–3429.

11. Arcangeli, A.; Becchetti, A.; Pillozzi, S.; Masselli, M.; De Lorenzo, E. Method and kit for the prevention and/or the monitoring of chemoresistance of leukemia forms. PCT Int. Appl. WO 2011058509, 2011.

12. Green, D. P. L.; Rawle, C. B. Analysis system and method. PCT Int. Appl. WO 2009082242, 2009.

13. Pashanina, T. P.; Napalkova, G. M.; Manankov, V. V. Method for producing immunofluorescent antibodies for detecting toxoplasmosis pathogen. Russ. RU 2174233, 2001.

14. Forni, L.; De Petris, S. Use of fluorescent antibodies in the study of lymphoid cell membrane molecules. *Methods Enzymol.* **1984**, *108*, 413–425.

15. Goldstein, G.; Spalding, B. H.; Hunt, W. B., Jr., Fluorescent antibody staining. II. Inhibition by suboptimally conjugated antibody globulins. *Proc. Soc. Exp. Biol. Med.* **1962**, *111*, 416–420.

16. Wilbur, D. S.; Pathare, P. M.; Hamlin, D. K.; Frownfelter, M. B.; Kegley, B. B.; Leung, W.; Gee, K. R. Evaluation of biotin-dye conjugates for use in an HPLC assay to assess relative binding of biotin derivatives with avidin and streptavidin. *Bioconjugate Chem.* **2000,** *11*, 584–598.

17. Elliott, J. T.; Tona, A.; Plant, A. L. Comparison of reagents for shape analysis of fixed cells by automated fluorescence microscopy. *Cytometry, Part A* **2003**, *52A*, 90–100.

18. Bestvater, F.; Spiess, E.; Stobrawa, G.; Hacker, M.; Feurer, T.; Porwol, T.; Berchner-Pfannschmidt, U.; Wotzlaw, C.; Acker, H. Two-photon fluorescence absorption and emission spectra of dyes relevant for cell imaging. *J. Microsc.* **2002**, *208*, 108–115.

19. Nielsen, L. J.; Eyley, S.; Thielemans, W.; Aylott, J. W. Dual fluorescent labelling of cellulose nanocrystals for pH sensing. *Chem. Commun.* **2010**, *46*, 8929–8931.

20. Dong, S.; Roman, M. Fluorescently labeled cellulose nanocrystals for bioimaging applications. *J. Am. Chem. Soc.* **2007**, *129*, 13810–13811.

21. Kang, S.; Maelicke, A. Fluorescein isothiocyanate-labeled α-cobratoxin. Biochemical characterization and interaction with acetylcholine receptor from *Electrophorus electricus. J. Biol. Chem.* **1980**, *255*, 7326–7332.

22. Menjoge, A. R.; Navath, R. S.; Asad, A.; Kannan, S.; Kim, C. J.; Romero, R.; Kannan, R. M. Transport and biodistribution of dendrimers across human fetal membranes: Implications for intravaginal administration of dendrimer-drug conjugates. *Biomaterials* **2010**, *31*, 5007–5021.

23. Kannan, R. M.; Kannan, S.; Romero, R. Dendrimer based nanodevices for therapeutic and imaging purposes. PCT Int. Appl. WO 2010147831, 2010.

24. Shi, X.; Wang, S.; Baker, J. R. Dendrimer based compositions and methods of using the same. PCT Int. Appl. WO 2008008483, 2008.

25. Khopade, A. J.; Moehwald, H. Statistical megamer morphologies and materials from PAMAM dendrimers. *Macromol. Rapid Commun.* **2005**, *26*, 445–449.

26. del Mercato, L. L.; Abbasi, A. Z.; Parak, W. J. Synthesis and characterization of ratiometric ion-sensitive polyelectrolyte capsules. *Small* **2011**, *7*, 351–363.

27. De Belder, A. N.; Granath, K. Preparation and properties of fluorescein-labeled dextrans. *Carbohydr. Res.* **1973**, *30*, 375–378.

28. Lingwood, C. A.; Soldin, S.; Nutikka, A. Synthesis of a fluorescent, cardioactive analog of digoxin. *Anal. Lett.* **1988**, *21*, 813–826.

29. Lin, S.; Faller, L. D. Preparation of Na,K-ATPase specifically modified on the anti-fluorescein antibody-inaccessible site by fluorescein 5′-isothiocyanate. *Anal. Biochem.* **2000**, *287*, 303–312.

30. Friedman, M. L.; Ball, W. J., Jr., Determination of monoclonal antibody-induced alterations in sodium-potassium ATPase conformations using fluorescein-labeled enzyme. *Biochim. Biophys. Acta, Protein Struct. Mol. Enzymol.* **1989**, *995*, 42–53.

31. Karlish, S. J. D. Use of formycin nucleotides, intrinsic protein fluorescence, and fluorescein isothiocyanate-labeled enzymes for measurement of conformational states of sodium-potassium ATPase. *Methods Enzymol.* **1988**, *156*, 271–277.

32. Kirley, T. L.; Wallick, E. T.; Lane, L. K. The amino acid sequence of the fluorescein isothiocyanate reactive site of lamb and rat kidney sodium and potassium dependent ATPase. *Biochem. Biophys. Res. Commun.* **1984**, *125*, 767–773.

33. Pick, U. Interaction of fluorescein isothiocyanate with nucleotide-binding sites of the calcium ATPase from sarcoplasmic reticulum. *Eur. J. Biochem.* **1981**, *121*, 187–195.

34. Garcia-Barrado, J. A.; Gata, J. L.; Santano, E.; Solis, J. I.; Pinto, M. C.; Macias, P. The use of fluorescein 5′-isothiocyanate for studies of structural and molecular mechanisms of soybean lipoxygenase. *Biochem. Biophys. Res. Commun.* **1999**, *265*, 489–493.

35. Harenberg, J.; Malsch, R.; Piazolo, L.; Huhle, G.; Heene, D.L. Analysis of heparin binding to human leukocytes using a fluorescein-5-isothiocyanate labeled heparin fragment. *Cytometry* **1996**, *23*, 59–66.

36. Nagasawa, K.; Uchiyama, H. Preparation and properties of biologically active fluorescent heparins. *Biochim. Biophys. Acta, Gen. Sub.* **1978**, *544*, 430–440.

37. Fromm, J. R.; Thomas, A.; Wood, B. L. Flow cytometry can diagnose classical Hodgkin lymphoma in lymph nodes with high sensitivity and specificity. *Am. J. Clin. Pathol.* **2009**, *131*, 322–332.

38. Liu, F.; Kohn, W. D.; Mayer, J. P. Site-specific fluorescein labeling of human insulin. *J. Pept. Sci.* **2012**, *18*, 336–341.

39. Chen, Y.; Cheng, T.; Tseng, W. Fluorescence turn-on detection of iodide, iodate and total iodine using fluorescein-5-isothiocyanate-modified gold nanoparticles. *Analyst* **2009**, *134*, 2106–2112.

40. Estevez, M. C.; O'Donoghue, M. B.; Chen, X.; Tan, W. Highly fluorescent dye-doped silica nanoparticles increase flow cytometry sensitivity for cancer cell monitoring. *Nano Res.* **2009**, *2*, 448–461.

41. Nyffenegger, R.; Quellet, C.; Ricka, J. Synthesis of fluorescent, monodisperse, colloidal silica particles. *J. Colloid Interface Sci.* **1993**, *159*, 150–157.

42. Ichikawa, K.; Nakamura, K.; Kurata, S. Real-time PCR method using the novel type of oligonucleotide probes. Jpn. Kokai Tokkyo Koho JP 2008182974, 2008.

43. Gordon, K. M.; Duckett, L.; Daul, B.; Petrie, H. T. A simple method for detecting up to five immunofluorescent parameters together with DNA staining for cell cycle or viability on a benchtop flow cytometer. *J. Immunol. Methods* **2003**, *275*, 113–121.

44. Reeves, R. H.; Bennison, B. W.; LaRock, P. A. Species-specific DNA probes for *Vibrio vulnificus* methods and kits. U.S. Patent 5426025, 1995.

45. Lee, S. P.; Porter, D.; Chirikjian, J. G.; Knutson, J. R.; Han, M. K. A fluorometric assay for DNA cleavage reactions characterized with BamHI restriction endonuclease. *Anal. Biochem.* **1994**, *220*, 377–383.

46. Murakami, A.; Nakaura, M.; Nakatsuji, Y.; Nagahara, S.; Tran, C. Q.; Makino, K. Fluorescent-labeled oligonucleotide probes: detection of hybrid formation in solution by fluorescence polarization spectroscopy. *Nucleic Acids Res.* **1991**, *19*, 4097–4102.

47. Smith, L. M.; Sanders, J. Z.; Kaiser, R. J.; Hughes, P.; Dodd, C.; Connell, C. R.; Heiner, C.; Kent, S. B. H.; Hood, L. E. Fluorescence detection in automated DNA sequence analysis. *Nature* **1986**, *321*, 674–679.

48. Rao, C. S.; Chu, J.; Liu, R.; Lai, Y. Synthesis and evaluation of [^{14}C]-labeled and fluorescent-tagged paclitaxel derivatives as new biological probes. *Bioorg. Med. Chem.* **1998**, *6*, 2193–2204.

49. Bicamumpaka, C.; Page, M. *In situ* localization of paclitaxel binding structures: labeling with a paclitaxel fluorescent analog. *Int. J. Mol. Med.* **1998**, *2*, 161–165.

50. Cho, H. K.; Lone, S.; Kim, D. D.; Choi, J. H.; Choi, S. W.; Cho, J. H.; Kim, J. H.; Cheong, I. W. Synthesis and characterization of fluorescein isothiocyanate (FITC)-labeled PEO-PCL-PEO triblock copolymers for topical delivery. *Polymer* **2009**, *50*, 2357–2364.

51. Marguet, M.; Edembe, L.; Lecommandoux, S. Polymersomes in polymersomes: Multiple loading and permeability control. *Angew. Chem., Int. Ed.* **2012**, *51*, 1173–1176.

52. Borra, R.; Dong, D.; Elnagar, A. Y.; Woldemariam, G. A.; Camarero, J. A. In-cell fluorescence activation and labeling of proteins mediated by FRET-quenched split inteins. *J. Am. Chem. Soc.* **2012**, *134*, 6344–6353.

53. Murikipudi, S.; Lancaster, T. M.; Zion, T. C. Uses of macrophage mannose receptor to screen compounds and uses of these compounds. PCT Int. Appl. WO 2012050822, 2012.

54. Sugawara, S.; Osumi, K. Labeled glycopeptide derivatives, their manufacture, and kits for detection of influenza viruses. Jpn. Kokai Tokkyo Koho JP 2011231293, 2011.

55. Mishra, G.; Easton, C. D.; Fowler, G. J. S.; McArthur, S. L. Spontaneously reactive plasma polymer micropatterns. *Polymer* **2011**, *52*, 1882–1890.

56. Jullian, M.; Hernandez, A.; Maurras, A.; Puget, K.; Amblard, M.; Martinez, J.; Subra, G. N-terminus FITC labeling of peptides on solid support: the truth behind the spacer. *Tetrahedron Lett.* **2009**, *50*, 260–263.

57. Chen, A. K.; Cheng, Z.; Behlke, M. A.; Tsourkas, A. Assessing the sensitivity of commercially available fluorophores to the intracellular environment. *Anal. Chem.* **2008**, *80*, 7437–7444.

58. Wu, S.; Lu, J. J.; Wang, S.; Peck, K. L.; Li, G.; Liu, S. Staining method for protein analysis by capillary gel electrophoresis. *Anal. Chem.* **2007**, *79*, 7727–7733.

59. Clayton, A. H.; Walker, F.; Orchard, S. G.; Henderson, C.; Fuch, D.; Rothacker, J.; Nice, E. C.; Burgess, A. W. Ligand-induced dimer-tetramer transition during the activation of the cell surface epidermal growth factor receptor-A multidimentional microscopy analysis. *J. Biol. Chem.* **2005**, *280*, 30392–30399.

60. Wu, X.; Pawliszyn, J. Whole-column fluorescence-imaged capillary electrophoresis. *Electrophoresis* **2004**, *25*, 3820–3824.

61. Hellweg, C. E.; Baumstark-Khan, C.; Rettberg, P.; Horneck, G. Suitability of enhanced green fluorescent protein as a reporter component for bioassays. *Anal. Chim. Acta* **2001**, *426*, 175–184.

62. Schnaible, V.; Przybylski, M. Identification of fluorescein-5'-isothiocyanate-modification sites in proteins by electrospray-ionization mass spectrometry. *Bioconjugate Chem.* **1999**, *10*, 861–866.

63. Chersi, A.; di Modugno, F.; Rosano, L. Selective 'in synthesis' labeling of peptides by fluorochromes. *Biochim. Biophys. Acta, Gen. Sub.* **1997**, *1336*, 83–88.

64. Kwon, G.; Remmers, A. E.; Datta, S.; Neubig, R. R. Synthesis and characterization of fluorescently labeled bovine brain G protein subunits. *Biochemistry* **1993**, *32*, 2401–2408.

65. Corin, A. F.; Blatt, E.; Jovin, T. M. Triplet-state detection of labeled proteins using fluorescence recovery spectroscopy. *Biochemistry* **1987**, *26*, 2207–2217.

66. Rinderknecht, H. Ultrarapid fluorescent labeling of proteins. *Nature* **1962**, *193*, 167–168.

67. Bhattacharyya, P. K.; Bhattacharyya, A. Fluorescent labeling of strychnine: a novel approach for recognition of strychnine binding sites on neuronal membrane. *Biochem. Biophys. Res. Commun.* **1981**, *101*, 273–280.

68. Toyooka, T.; Bin, T.; Onosawa, T.; Kumata, J.; Ishikawa, K. Novel fluorescent agents for sugar chain analysis. Jpn. Kokai Tokkyo Koho JP 2006321775, 2006.

69. Wu, Y. W.; Tsai, Y. H. Method for modifying thiol group. Taiwan. TW 375666, 2012.

70. Lou, Z.; Li, P.; Sun, X.; Yang, S.; Wang, B.; Han, K. A fluorescent probe for rapid detection of thiols and imaging of thiols reducing repair and H_2O_2 oxidative stress cycles in living cells. *Chem. Commun.* **2013**, *49*, 391–393.

71. Vermes, I.; Haanen, C.; Steffens-Nakken, H.; Reutelingsperger, C. A novel assay for apoptosis. Flow cytometric detection of phosphatidylserine expression on early apoptotic cells using fluorescein labeled Annexin V. *J. Immunol. Methods* **1995**, *184*, 39–51.

72. Huang, K.; Holmgren, J.; Reik, L.; Lucas-McGady, D.; Roberts, J.; Liu, C.; Levin, W. High-throughput methods for measuring heparanase activity and screening potential antimetastatic and anti-inflammatory agents. *Anal. Biochem.* **2004**, *333*, 389–398.

73. Anjuere, F.; Monsigny, M.; Lelievre, Y.; Mayer, R. Sensitive, hydrosoluble, macromolecular fluorogenic substrates for human immunodeficiency virus 1 proteinase. *Biochem. J.* **1993**, *291*, 869–873.

74. Klink, T. A.; Beebe, J. A.; Lasky, D. A.; Kleman-Leyer, K. M.; Somberg, R. Kinase and phosphatase assays. U.S. Pat. Appl. Publ. US 20070059787, 2007.

75. Shao, Q.; Zheng, Y.; Dong, X.; Tang, K.; Yan, X.; Xing, B. A covalent reporter of β-lactamase activity for fluorescent imaging and rapid Screening of antibiotic-resistant bacteria. *Chem. Eur. J.* **2013**, *19*, 10903–10910.

76. Gnaccarini, C.; Ben-Tahar, W.; Lubell, W. D.; Pelletier, J. N.; Keillor, J. W. Fluorometric assay for tissue transglutaminase-mediated transamidation activity. *Bioorg. Med. Chem.* **2009**, *17*, 6354–6359.

77. Meng, E.; Humayun, M.; Lo, R.; Li, P.; Saati, S. Implantable drug-delivery devices, and apparatus and methods for refilling the devices. U.S. Pat. Appl. Publ. US 20090192493, 2009.

78. Grattan, K. T. V.; Badini, G. E.; Palmer, A. W.; Tseung, A. C. C. Use of sol–gel techniques for fiber-optic sensor applications. *Sens. Actuators, A* **1991**, *26*, 483–487.

79. Wong, E. H. H.; Guntari, S. N.; Blencowe, A.; van Koeverden, M. P.; Caruso, F.; Qiao, G. G. Phototriggered, metal-free continuous assembly of

polymers for the fabrication of ultrathin films. *ACS Macro Lett.* **2012**, *1*, 1020–1023.

80. Scutaru, A. M.; Kruger, M.; Wenzel, M.; Richter, J.; Gust, R. Investigations on the use of fluorescence dyes for labeling dendrimers: cytotoxicity, accumulation kinetics, and intracellular distribution. *Bioconjugate Chem.* **2010**, *21*, 2222–2226.

81. Armeni, T.; Damiani, E.; Battino, M.; Greci, L.; Principato, G. Lack of *in vitro* protection by a common sunscreen ingredient on UVA-induced cytotoxicity in keratinocytes. *Toxicology* **2004**, *203*, 165–178.

82. Schmitt, E.; Lehmann, L.; Metzler, M.; Stopper, H. Hormonal and genotoxic activity of resveratrol. *Toxicol. Lett.* **2002**, *136*, 133–142.

83. Hamid K. A.; Katsumi H.; Sakane T.; Yamamoto A. The effects of common solubilizing agents on the intestinal membrane barrier functions and membrane toxicity in rats. *Int. J. Pharm.* **2009**, *379*, 100–108.

84. Vaslin, A.; Puyal, J.; Borsello, T.; Clarke, P. G. H. Excitotoxicity-related endocytosis in cortical neurons. *J. Neurochem.* **2007**, *102*, 789–800.

85. Vehanen, K.; Hornof, M.; Urtti, A.; Uusitalo, H. Peribulbar poloxamer for ocular drug delivery. *Acta Ophthalmol.* **2008**, *86*, 91–96.

86. Rumbaut, R. E.; Sial, A. J. Differential phototoxicity of fluorescent dye-labeled albumin conjugates. *Microcirculation* **1999**, *6*, 205–213.

87. He, L.; Gao, Y.; Lin, Y.; Katsumi, H.; Fujita, T.; Yamamoto, A. Improvement of pulmonary absorption of insulin and other water-soluble compounds by polyamines in rats. *J. Controlled Release* **2007**, *122*, 94–101.

88. Natsch, A.; Emter, R.; Ellis, G. Filling the concept with data: Integrating data from different *in vitro* and *in silico* assays on skin sensitizers to explore the battery approach for animal-free skin sensitization testing. *Toxicol. Sci.* **2009**, *107*, 106–121.

89. Natsch, A.; Gfeller, H. LC-MS-based characterization of the peptide reactivity of chemicals to improve the *in vitro* prediction of the skin sensitization potential. *Toxicol. Sci.* **2008**, *106*, 464–478.

90. Natsch, A.; Emter, R. Skin sensitizers induce antioxidant response element dependent genes: application to the in vitro testing of the sensitization potential of chemicals. *Toxicol. Sci.* **2008**, *102*, 110–119.

91. Patlewicz, G.; Roberts, D. W.; Uriarte, E. A comparison of reactivity schemes for the prediction skin sensitization potential. *Chem. Res. Toxicol.* **2008**, *21*, 521–541.

92. Ashby, J.; Basketter, D. A.; Paton, D.; Kimber, I. Structure activity relationships in skin sensitization using the murine local lymph node assay. *Toxicology* **1995**, *103*, 177–194.

FLUORESCEIN-5-THIOSEMICARBAZIDE

CAS Registry Number   76863-28-0

Chemical Structure

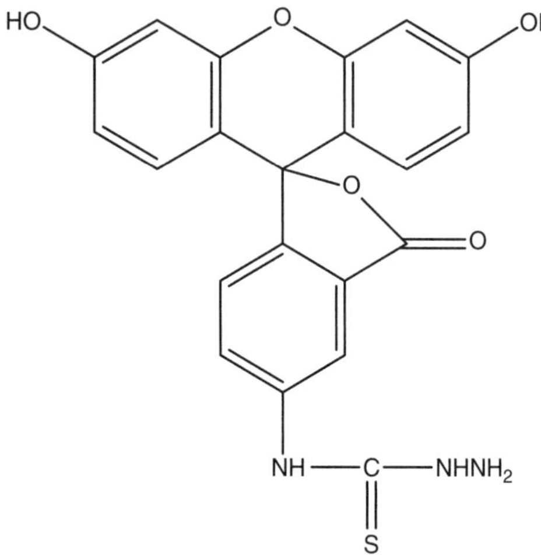

CA Index Name   Hydrazinecarbothioamide, N-(3′,6′-dihydroxy-3-oxospiro[isobenzofuran-1(3H),9′-[9H]xanthen]-5-yl)-

Other Names   Spiro[isobenzofuran-1(3H),9′-[9H]xanthene], hydrazinecarbothioamide deriv.; Fluorescein-5-thiosemicarbazide; 5-FTSC; FTSC; FTZ;

Merck Index Number   Not listed

Chemical/Dye Class   Xanthene

Molecular Formula $C_{21}H_{15}N_3O_5S$

Molecular Weight   421.43

Physical Form   Yellow solid[1]

Solubility   Soluble in water, N,N-dimethylformamide, dimethyl sulfoxide

Boiling Point (Calcd.)   723.8 ± 70.0 °C, pressure: 760 Torr

pK$_a$ (Calcd.)   9.33 ± 0.20, most acidic, temperature: 25 °C; 1.76 ± 0.19, most basic, temperature: 25 °C

Absorption (λ_{max})   492 nm (Buffer pH 9.0)

Emission (λ_{max})   516 nm (Buffer pH 9.0)

Molar Extinction Coefficient   85,000 cm^{-1} M^{-1} (Buffer pH 9.0)

Synthesis   Synthetic methods[1,2]

Imaging/Labeling Applications   Carbonyl groups (aldehydes/ketones/carboxylic acids);[1,3,4,7,9–12,] cellulose fibre interfaces;[5,6] dendrimer surface;[7] glycosaminoglycan;[8] hairs;[9–12] hyaluronic acid;[8] lipobeads;[13] nanoparticles;[14] nucleic acids;[2,15–20,] polysaccharides;[21] proteins;[9–12,22–31] sterols[32]

Biological/Medical Applications   Analyzing low-molecular mass aldehydes in urine,[3] drinking waters;[4] assaying glycosaminoglycan-degrading enzyme activity;[8] detecting sterols;[32] measuring/quantifying degree of hair damage[9–12]

Industrial Applications   Characterizing soft material interfaces;[6] understanding fibrous network structures;[6] in nanodevice preparation;[7] functionalizing/ encapsulatiing particles by initiated chemical vapor deposition (iCVD);[33,34] functional polymer coatings;[33] high refractive index coatings;[34] papers[5]

Safety/Toxicity   No data available

REFERENCES

1. Haugland, R. P. Fluorescent labeling reagents containing the fluorescein and eosin chromophores. U.S. Patent 4213904, 1980.

2. Bauman, J. G. J.; Wiegant, J.; Van Duijn, P. Cytochemical hybridization with fluorochrome-labeled RNA. I. Development of a method using nucleic acids bound to agarose beads as a model. *J. Histochem. Cytochem.* **1981**, *29*, 227–237.

3. Banos, C. E.; Silva, M. A novel clean-up method for urine analysis of low-molecular mass aldehydes by capillary electrophoresis with laser-induced fluorescence detection. *J. Chromatogr., B* **2011**, *879*, 1412–1418.

4. Banos, C. E.; Silva, M. Analysis of low-molecular mass aldehydes in drinking waters through capillary electrophoresis with laser-induced fluorescence detection. *Electrophoresis* **2010**, *31*, 2028–2036.

5. Thomson, C. I.; Lowe, R. M.; Ragauskas, A. J. First characterization of the development of bleached kraft softwood pulp fiber interfaces during drying and rewetting using FRET microscopy. *Holzforschung* **2008**, *62*, 383–388.

6. Thomson, C. I.; Lowe, R. M.; Ragauskas, A. J. Imaging cellulose fibre interfaces with fluorescence microscopy and resonance energy transfer. *Carbohydr. Polym.* **2007**, *69*, 799–804.

7. Lee, C. Y.; Ki, D. W.; Sharma, A. Method for amine-amine attachment in nanodevice preparation. *J. Mater. Sci.* **2009**, *44*, 3179–3185.

8. Tanaka, M.; Ishimaru, T. Glycosaminoglycan-degrading enzyme activity assay method. Jpn. Kokai Tokkyo Koho JP 2006025625, 2006.

9. Kawazoe, T.; Fujii, T. Method for measuring degree of hair damage caused by heat using keratin film. Jpn. Kokai Tokkyo Koho JP 2012242151, 2012.

10. Kawazoe, T.; Watanabe, T.; Fujii, T. A novel method using a keratin film for quantifying the photo-modification of hair proteins. *Nippon Keshohin Gijutsusha Kaishi* **2011**, *45*, 100–107.

11. Kawazoe, T.; Watanabe, T.; Fujii, T. Visualization of modified human hair by artificial sunlight with carbonylated proteins as an indicator of hair damage. *Nippon Koshohin Gakkaishi* **2010**, *34*, 287–291.

12. Kawazoe, T.; Watanabe, T.; Fujii, T. UV ray hair damage degree-measuring method using keratin film. Jpn. Kokai Tokkyo Koho JP 2008180709, 2008.

13. Ma, A.; Rosenzweig, Z. Synthesis and analytical properties of micrometric biosensing lipobeads. *Anal. Bioanal. Chem.* **2005**, *382*, 28–36.

14. Sun, G.; Berezin, M. Y.; Fan, J.; Lee, H.; Ma, J.; Zhang, K.; Wooley, K. L.; Achilefu, S. Bright fluorescent nanoparticles for developing potential optical imaging contrast agents. *Nanoscale* **2010**, *2*, 548–558.

15. Morse, D. P. Direct selection of RNA beacon aptamers. *Biochem. Biophys. Res. Commun.* **2007**, *359*, 94–101.

16. Wu, T. P.; Ruan, K. C.; Liu, W. Y. A fluorescence-labeling method for sequencing small RNA on polyacrylamide gel. *Nucleic Acids Res.* **1996**, *24*, 3472–3473.

17. Rychlik, W.; Odom, O. W.; Hardesty, B. Localization of the elongation factor Tu binding site on *Escherichia coli* ribosomes. *Biochemistry* **1983**, *22*, 85–93.

18. Stoffler-Meilicke, M.; Stoffler, G.; Odom, O. W.; Zinn, A.; Kramer, G.; Hardesty, B. Localization of 3′ ends of 5S and 23S rRNAs in reconstituted subunits of *Escherichia coli* ribosomes. *Proc. Natl. Acad. Sci. U.S.A.* **1981**, *78*, 5538–5542.

19. Odom, O. W.; Robbins, D. J.; Lynch, J. Dottavio-Martin, D.; Kramer, G.; Hardesty, B. Distance between 3′ ends of ribosomal ribonucleic acids reassembled into *Escherichia coli* ribosomes. *Biochemistry* **1980**, *19*, 5947–5954.

20. Wells, B. D.; Cantor, C. R. Ribosome binding by tRNAs with fluorescent labeled 3′ termini. *Nucleic Acids Res.* **1980**, *8*, 3229–3246.

21. Zhang, Y.; Wang, Z.; Zhang, X.; Zhou, W.; Huang, L. One-pot fluorescent labeling of saccharides with fluorescein-5-thiosemicarbazide for imaging polysaccharides transported in living cells. *Carbohydr. Res.* **2011**, *346*, 2156–2164.

22. Pazos, M.; Pereira da Rocha, A.; Roepstorff, P.; Rogowska-Wrzesinska, A. Fish proteins as targets of ferrous-catalyzed oxidation: Identification of protein carbonyls by fluorescent labeling on two-dimensional gels and MALDI-TOF/TOF mass spectrometry. *J. Agric. Food Chem.* **2011**, *59*, 7962–7977.

23. Fujita, H.; Hirao, T.; Takahashi, M. A simple and non-invasive visualization for assessment of carbonylated protein in the stratum corneum. *Skin Res. Technol.* **2007**, *13*, 84–90.

24. Iwai, I.; Hirao, T. Method for evaluating flexibility and elasticity of skin using oxidized proteins in horny cell layer as index. Jpn. Kokai Tokkyo Koho JP 2006349372, 2006.

25. Chaudhuri, A. R.; de Waal, E. M.; Pierce, A.; Van Remmen, H.; Ward, W. F.; Richardson, A. Detection of protein carbonyls in aging liver tissue: a fluorescence-based proteomic approach. *Mech. Ageing Dev.* **2006**, *127*, 849–861.

26. Hamaji, I. Protein for fluorescent sugar detection and its production method. Jpn. Kokai Tokkyo Koho JP 2000338044, 2000.

27. Jung, Y.; Chay, K.; Song, D.; Yang, S.; Lee, M.; Ahn, B. Protein carbonyl formation in blood plasma by cephalosporins. *Arch. Biochem. Biophys.* **1997**, *345*, 311–317.

28. Gao, X.; Liu, W.; Ruan, K. Improved method for detecting fluorescent-labeled glycoprotein on sodium dodecyl sulfate-polyacrylamide gel. *Bioorg. Chem.* **1997**, *25*, 163–168.

29. Corbett, J. D.; Cho, M. R.; Golan, D. E. Deoxygenation affects fluorescence photobleaching recovery measurements of red cell membrane protein lateral mobility. *Biophys. J.* **1994**, *66*, 25–30.

30. Ahn, B.; Rhee, S. G.; Stadtman, E. R. Use of fluorescein hydrazide and fluorescein thiosemicarbazide reagents for the fluorometric determination of protein carbonyl groups and for the detection of oxidized protein on polyacrylamide gels. *Anal. Biochem.* **1987**, *161*, 245–257.

31. Bar-Noy, S.; Darmon, A.; Ginsburg, H.; Cabantchik, Z. I. Orientation of transmembrane polypeptides as revealed by antibody quenching of fluorescence.

Biochim. Biophys. Acta, Biomembr. **1984**, *778*, 612–614.

32. Yoshida, M.; Nishimura, S.; Kobayashi, T.; Iwamoto, K. Sterol detection method, and probe for it. Jpn. Kokai Tokkyo Koho JP 2009109361, 2009.

33. Gleason, K. K.; Lau, K. K. S. Initiated chemical vapor deposition of vinyl polymers for the encapsulation of particles. U.S. Pat. Appl. Publ. US 20070104860, 2007.

34. Lau, K. K. S.; Gleason, K. K. Particle functionalization and encapsulation by initiated chemical vapor deposition (iCVD). *Surf. Coat. Technol.* **2007**, *201*, 9189–9194.

4-HEPTADECYL-7-HYDROXYCOUMARIN (4-HEPTADECYLUMBELLIFERONE)

CAS Registry Number   26038-83-5

Chemical Structure

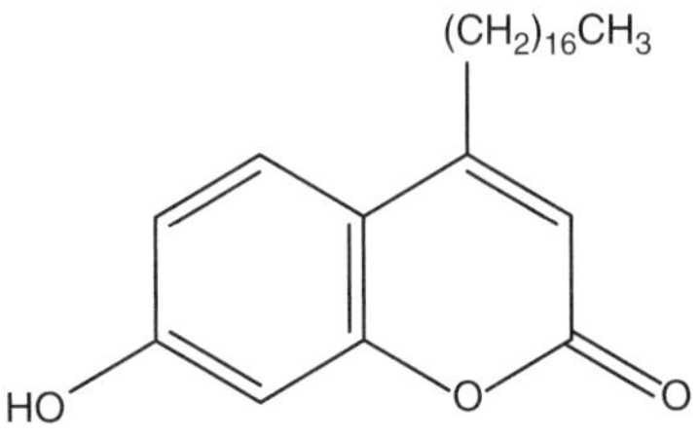

CA Index Name   2H-1-Benzopyran-2-one, 4-heptadecyl-7-hydroxy-

Other Names   Coumarin, 4-heptadecyl-7-hydroxy-; 4-Heptadecyl-7-hydroxycoumarin; 4-heptadecyl-hydroxycoumarin; heptadecyl-hydroxycoumarin; 4-Heptadecylumbelliferone; 7-hydroxy-4-heptadecyl coumarin; C17HC; C17-HC; HC; HHC; HUF

Merck Index Number   Not listed

Chemical/Dye Class   Coumarin

Molecular Formula $C_{26}H_{40}O_3$

Molecular Weight   400.59

Physical Form   Off-white solid

Solubility   Soluble in *N,N*-dimethylformamide, dimethyl sulfoxide, ethanol, methanol

Melting Point   95 °C

Boiling Point (Calcd.) 544.7 ± 45.0 °C, pressure: 760 Torr

pK$_a$ 8.9, temperature: 22 °C

pK$_a$ (Calcd.)   8.02 ± 0.20, most acidic, tempeature: 25 °C

Absorption (λ_{max})   325 nm (MeOH/H$^+$); 366 nm (MeOH/OH$^-$)

Emission (λ_{max})   385 nm (MeOH/H$^+$); 453 nm (MeOH/OH$^-$)

Molar Extinction Coefficient   15,000 cm^{-1} M^{-1} (MeOH/H$^+$); 20,000 cm^{-1} M^{-1} (MeOH/OH$^-$)

Synthesis   Synthetic methods[1–3]

Imaging/Labeling Applications   Erythrocytes;[4,5] lipids;[6–10] liposomes;[21–24] proteins[24]

Biological/Medical Applications   Analyzing protein orientations in proteoliposomes;[24] characterizing interactions of cationic lipids with nucleic acids;[6] investigating lipid assemblies;[10] measuring surface potential in biomembranes;[7,8,13–23,] studying/evaluating human erythrocyte exovesiculation;[4,5] studying lipid bilayers properties;[7] Langmuir-Blodgett films;[2,11,12,] as a biological pH indicator;[1,7–20,22,23] as a probe of membrane surfaces[7,8,13–23,]

Industrial Applications   Not reported

Safety/Toxicity   No data available

REFERENCES

1. Sabnis, R. W. *Handbook of Acid–base Indicators*; CRC Press: Boca Raton, **2008**; p 183.

2. Yonezawa, Y.; Wataya, K.; Hada, H. Preparation and spectroscopic properties of LB (Langmuir-Blodgett) films composed of polar matrix molecules and a xanthene or coumarin dye. *Nippon Shashin Gakkaishi* **1989**, *52*, 550–560.

3. Moebius, D.; Buecher, H.; Kuhn, H.; Sondermann, J. Reversible change of surface and surface potential of monomolecular films of a photochromic system. *Ber. Bunsen-Ges.* **1969**, *73*, 845–850.

4. Santos, N. C.; Martins-Silva, J.; Saldanha, C. Gramicidin D and dithiothreitol effects on erythrocyte exovesiculation. *Cell Biochem. Biophys.* **2005**, *43*, 419–430.

5. Saldanha, C.; Santos, N. C.; Martins-Silva, J. Fluorescent probes DPH, TMA-DPH and C17-HC induce erythrocyte exovesiculation. *J. Membr. Biol.* **2002**, *190*, 75–82.

6. Hirsch-Lerner, D.; Zhang, M.; Eliyahu, H.; Ferrari, M. E.; Wheeler, C. J.; Barenholz, Y. Effect of "helper lipid" on lipoplex electrostatics. *Biochim. Biophys. Acta, Biomembr.* **2005**, *1714*, 71–84.

7. Pal, R.; Petri, W. A., Jr.,; Ben-Yashar, V.; Wagner, R. R.; Barenholz, Y. Characterization of the fluorophore 4-heptadecyl-7-hydroxycoumarin: a probe for the

Handbook of Fluorescent Dyes and Probes, First Edition. R. W. Sabnis.
© 2015 John Wiley & Sons, Inc. Published 2015 by John Wiley & Sons, Inc.

head-group region of lipid bilayers and biological membranes. *Biochemistry* **1985**, *24*, 573–581.

8. Schummer, U.; Schiefer, H. G. Surface potential of mycoplasma membranes. *FEMS Microbiol. Lett.* **1984**, *23*, 143–145.

9. Fromherz, P. Acid–base sensitization of fluorescence by energy transfer in an ordered dye-lipid-electrolyte lamella. *Chem. Phys. Lett.* **1974**, *26*, 221–224.

10. Fromherz, P. Investigation of lipid assemblies with a lipid pH indicator in monomolecular films. *Biochim. Biophys. Acta, Biomembr.* **1973**, *323*, 326–334.

11. Petrov, J. G.; Mobius, D. Interfacial acid–base equilibrium and electrostatic potentials of model Langmuir-Blodgett membranes in contact with phosphate buffer. *Colloids Surf., A* **2000**, *171*, 207–215.

12. Murray, B. S.; Godfrey, J. S.; Grieser, F.; Healy, T. W.; Lovelock, B.; Scales, P. J. Spectroscopic and electrokinetic study of pH-dependent ionization of Langmuir-Blodgett films. *Langmuir* **1991**, *7*, 3057–3064.

13. Warren, D. B.; Dyson, G.; Grieser, F.; Perera, J. M.; Stevens, G. W.; Rizzacasa, M. A. Characterisation of nickel(II) extraction by 2-hydroxy-5-nonylacetophenone oxime (LIX 84) in a micellar phase. *Colloids Surf., A* **2003**, *227*, 49–61.

14. Petrov, J. G.; Polymeropoulos, E. E.; Moehwald, H. On the three-capacitor model for surface potential of insoluble monolayers. *J. Phys. Chem.* **1996**, *100*, 9860–9869.

15. Hobson, R. A.; Grieser, F.; Healy, T. W. Surface potential measurements in mixed micelle systems. *J. Phys. Chem.* **1994**, *98*, 274–278.

16. Kraayenhof, R.; Sterk, G. J.; Wong F. S., Harro W. Probing biomembrane interfacial potential and pH profiles with a new type of float-like fluorophores positioned at varying distance from the membrane surface. *Biochemistry* **1993**, *32*, 10057–10066.

17. Fromherz, P. Lipid coumarin dye as a probe of interfacial electrical potential in biomembranes. *Methods Enzymol.* **1989**, *171*, 376–387.

18. Petrov, J. G.; Moebius, D. Determination of the electrostatic potential of positively charged monolayers at the air/water interface by means of fluorometric titration of 4-heptadecyl-7-hydroxycoumarin. *Langmuir* **1990**, *6*, 746–751.

19. Drummond, C. J.; Grieser, F. Absorption spectra and acid–base dissociation of the 4-alkyl derivatives of 7-hydroxycoumarin in self-assembled surfactant solution: comments on their use as electrostatic surface potential probes. *Photochem. Photobiol.* **1987**, *45*, 19–34.

20. Hartland, G. V.; Grieser, F.; White, L. R. Surface potential measurements in pentanol-sodium dodecyl sulfate micelles. *J. Chem. Soc., Faraday Trans. 1* **1987**, *83*, 591–613.

21. Yonezawa, Y.; Iwabuchi, H.; Sato, T.; Terashima, S.; Katoh, S. Excitation energy transfer in liposomes. Fluorescence quenching of coumarin dye by retinal. *Nippon Shashin Gakkaishi* **1992**, *55*, 325–336.

22. Pal, R.; Petri, W. A., Jr.,; Barenholz, Y.; Wagner, R. R. Lipid and protein contributions to the membrane surface potential of vesicular stomatitis virus probed by a fluorescent pH indicator, 4-heptadecyl-7-hydroxycoumarin. *Biochim. Biophys. Acta, Biomembr.* **1983**, *729*, 185–192.

23. Fernandez, M. S. Determination of surface potential in liposomes. *Biochim. Biophys. Acta, Biomembr.* **1981**, *646*, 23–26.

24. Nicholls, P.; Tattrie, B.; Butko, P.; Tihova, M. Probe and protein orientations in proteoliposomes: electron microscopy and topobiochemistry. *Biochem. Soc. Trans.* **1992**, *20*, 115S.

HEXIDIUM IODIDE

CAS Registry Number   211566-66-4

Chemical Structure

CA Index Name   Phenanthridinium, 3,8-diamino-5-hexyl-6-phenyl-, iodide (1:1)

Other Names   Phenanthridinium, 3,8-diamino-5-hexyl-6-phenyl-, iodide; Hexidium iodide

Merck Index Number   Not listed

Chemical/Dye Class   Heterocycle; Phenanthridine

Molecular Formula $C_{25}H_{28}IN_3$

Molecular Weight   497.42

Physical Form   Red solid

Solubility   Soluble in water, N,N-dimethylformamide, dimethyl sulfoxide, methanol

Absorption (λ_{max})   518 nm (H_2O/DNA); 482 nm (H_2O)

Emission (λ_{max})   600 nm (H_2O/DNA); 625 nm (H_2O)

Molar Extinction Coefficient   3900 $cm^{-1} M^{-1}$ (H_2O/DNA); 5500 $cm^{-1} M^{-1}$ (H_2O)

Synthesis   Synthetic methods[1,2]

Imaging/Labeling Applications   Nucleic acids;[1,3,4] nuclei;[5–7] bacteria;[8–11] fungi;[12] microorganisms;[13,14] mold;[8,15] oligonucleotides;[16] polynucleotides;[17] yeast[8]

Biological/Medical Applications   Detecting nucleic acids,[1,3,4] bacteria,[8–11] microorganisms;[13,14] molds,[8,15] nucleic acid binding proteins,[18] yeast;[8] analyzing polynucleotides;[17] nucleic acids amplification;[16] measuring membrane/transmembrane potential[19,20]

Industrial Applications   Not reported

Safety/Toxicity   No data available

REFERENCES

1. Sabnis, R. W. *Handbook of Biological Dyes and Stains*; John Wiley & Sons Inc.: Hoboken, **2010**; 227–228.

2. Schneider, K.; Naujok, A.; Zimmermann, H. W. Influence of trans-membrane potential and of hydrophobic interactions on dye accumulation in mitochondria of living cells: Photoaffinity labeling of mitochondrial proteins, action of potential dissipating drugs, and competitive staining. *Histochemistry* **1994**, *101*, 455–461.

3. Weber, J.; Brendler, A.; Bley, T. Method and kit for the analysis of the DNA content in cells, especially plant cells using barium hydroxide treatment prior staining. Ger. Offen. DE 102006046079, 2008.

4. McNally, A. J.; Wu, R. S.; Li, Z. Immunoassay based on DNA replication using labeled primer. U.S. Pat. Appl. Publ. US 2002072053, 2002.

5. Gauer, C.; Mann, W.; Alunni-Fabbroni, M. Method for carrying out an enzymic reactions. Ger. DE 102006056694, 2010.

6. Gauer, C.; Mann, W.; Alunni-Fabbroni, M. Method for carrying out an enzymic reaction. PCT Int. Appl. WO 2008064730, 2008.

7. Gunasekera, N.; Musier-Forsyth, K.; Arriaga, E. Electrophoretic behavior of individual nuclear species as determined by capillary electrophoresis with laser-induced fluorescence detection. *Electrophoresis* **2002**, *23*, 2110–2116.

8. Oppedahl, A. M.; Lasky, S. J.; Baker, D. D.; Buttry, D. A.; Steger, A. M. Simultaneous rapid detection of microbes. U.S. Pat. Appl. Publ. US 2008261229, 2008.

9. Holm, C.; Jespersen, L. A flow-cytometric gram-staining technique for milk-associated bacteria. *Appl. Environ. Microbiol.* **2003**, *69*, 2857–2863.

10. Forster, S.; Snape, J. R.; Lappin-Scott, H. M.; Porter, J. Simultaneous fluorescent gram staining and activity assessment of activated sludge bacteria. *Appl. Environ. Microbiol.* **2002**, *68*, 4772–4779.

11. Mason, D. J.; Shanmuganathan, S.; Mortimer, F. C.; Gant, V. A. A fluorescent Gram stain for flow cytometry and epifluorescence microscopy. *Appl. Environ. Microbiol.* **1998**, *64*, 2681–2685.

12. Bradner, J. R.; Nevalainen, K. M. H. Metabolic activity in filamentous fungi can be analyzed by flow cytometry. *J. Microbiol. Methods* **2003**, *54*, 193–201.

13. Clauss, M.; Springorum, A. C.; Hartung, J. Comparison of different fluorescence and non-fluorescence staining techniques for rapid detection of airborne micro-organisms collected on room temperature vulcanizing (RTV) silicones from generated aerosols and from ambient air. *Aerosol Sci. Technol.* **2012**, *46*, 818–827.

14. Horikiri, S. Microorganism cell detection method using fluorescent indicator. Jpn. Kokai Tokkyo Koho JP 2006262775, 2006.

15. Oppedahl, A. M.; Lasky, S. J.; Baker, D. D.; Buttry, D.; Steger, A. M. Rapid detection of mold by accelerated growth and detection. U.S. Pat. Appl. Publ. US 2007231852, 2007.

16. Park, H. O.; Kim, H. B.; Chi, S. M. Detection method of DNA amplification using probes labeled with intercalating dye. PCT Int. Appl. WO 2006004267, 2006.

17. Hyldig-Nielsen, J. J.; Fiandaca, M. J.; Coull, J. M. Compositions and methods for multiplex analysis of polynucleotides. PCT Int. Appl. WO 2004074447, 2004.

18. Loewy, Z.; Chaung, W.; Pottathil, R. Methods for high throughput screening and characterization of nucleic acid binding proteins. PCT Int. Appl. WO 2004011606, 2004.

19. Deutsch, M.; Namer, Y. A method and system for measuring membrane potential based on fluorescence polarization. PCT Int. Appl. WO 2007122602, 2007.

20. Farinas, J. A.; Wada, H. G. Use of Nernstein voltage sensitive dyes in measuring transmembrane voltage. PCT Int. Appl. WO 2001027253, 2001.

HOECHST 33258

CAS Registry Number 23491-45-4
Chemical Structure

CA Index Name Phenol, 4-[6-(4-methyl-1-piperazinyl) [2,6′-bi-1*H*-benzimidazol]-2′-yl]-, hydrochloride (1:3)
Other Names Phenol, 4-[5-(4-methyl-1-piperazinyl)[2, 5′-bi-1*H*-benzimidazol]-2′-yl]-, trihydrochloride; Phenol, *p*-[5-[5-(4-methyl-1-piperazinyl)-2-benzimidazolyl]-2-benzimidazolyl]-, trihydrochloride; 2-[2-(4-Hydroxyphenyl)-6-benzimidazolyl]-6-(1-methyl-4-piperazinyl)-benzimidazole trihydrochloride; 2-[2-(4-Hydroxyphenyl)-6-benzimidazolyl]-6-(1-methyl-4-piperazyl)-benzimidazole trihydrochloride; 2-[2-(4-Hydroxyphenyl)-6-benzimidazolyl]-6-(1-methyl-4-piperazyl)benzimidazole trichloride; 2-[2-(4-Hydroxyphenyl)-6-benzimidazolyl]-6-(1-methyl-4-piperazyl)benzimidazole-3HCl; 2′-(4-Hydroxyphenyl)-5-(4-methyl-1-piperazinyl)-2,5′-bi-1*H*-benzimidazole; 4-[5-[5-(4-Methyl-1-piperazinyl)-2-benzimidazolyl]-2-benzimidazolyl]phenol trihydrochloride; Bisbenzimide; Bisbenzimide (quenchant); Bisbenzimide trihydrochloride; H 33258; HOE 33258; Ho 33258; Hoechst 33258; NSC 322921; Pibenzimol hydrochloride
Merck Index Number 1250
Chemical/Dye Class Heterocycle; Benzimidazole
Molecular Formula $C_{25}H_{27}Cl_3N_6O$
Molecular Weight 533.88
Physical Form Yellow to green powder or solid

Solubility Soluble in water, *N,N*-dimethylformamide, ethanol, methanol
Melting Point 280 °C (decompose)[4,5]

Absorption (λ_{max}) 352 nm (H_2O/DNA); 357 nm (MeOH/H[+])
Emission (λ_{max}) 461 nm (H_2O/DNA); 536 nm (MeOH/H[+])
Molar Extinction Coefficient 40,000 cm[-1] M[-1] (H_2O/DNA); 47,000 cm[-1] M[-1] (MeOH/H[+])
Quantum Yield 0.42 (H_2O/DNA)
Synthesis Synthetic methods[1–5]

Imaging/Labeling Applications Nucleic acids;[1,6–55] cells;[1,55–59] chromatin;[60–63] chromosomes;[64–83] embryos;[84] lymphocytes;[85] microorganisms;[86,87] myoglobin;[88] nuclei;[89–95] parasites;[96,97] protoplasts;[98,99] sperms[100–105]

Biological/Medical Applications Analyzing cell division;[106] as anticancer agent;[1,2,107–110] as antimicrobial agent;[110,111] counting/identifying cells;[1,55–59] detecting/measuring microorganisms;[86,87] detecting/quantifying nucleic acids;[1,6–55] measuring membrane potential;[112,113] studying multidrug resistance;[114] apoptosis assay;[115,116] cell cycle assay;[117–123] cell proliferation assay[124]

Industrial Applications Photoresists;[125] thin films[126]
Safety/Toxicity Cytotoxicity;[127,128] DNA damage;[129] genotoxicity;[130] radiotoxicity[131]

REFERENCES

1. Sabnis, R. W. *Handbook of Biological Dyes and Stains*; John Wiley & Sons Inc.: Hoboken, **2010**; pp 229–232.

2. Loewe, H.; Urbanietz, J. Basic substituted 2,6-bisbenzimidazole derivatives, a novel series of substances with chemotherapeutic activity. *Arzneim.-Forsch.* **1974**, *24*, 1927–1933.

3. Farbwerke Hoechst AG. Anthelmintic 2-(2-phenyl-6-benzimidazolyl)benzimidazoles. FR 6681, 1969.

4. Loewe, H.; Taunus, K.; Urbanietz, J.; Taunus, S.; Lammler, G. Substituted piperazino-bis-benzimidazoles having anthelmintic and bacteriostatic activity. U.S. Patent 3538097, 1970.

5. Farbwerke Hoechst AG. Basically substituted bis-benzimidazole derivatives and process for their manufacture. FR 1519964, 1968.

6. Zhang, F.; Tsunoda, M.; Suzuki, K.; Kikuchi, Y.; Wilkinson, O.; Millington, C. L.; Margison, G. P.; Williams, D. M.; Czarina Morishita, E.; Takenaka, A. Structures of DNA duplexes containing O6-carboxymethylguanine, a lesion associated with gastrointestinal cancer, reveal a mechanism for inducing pyrimidine transition mutations. *Nucleic Acids Res.* **2013**, *41*, 5524–5532.

7. Silva, E. F.; Ramos, E. B.; Rocha, M. S. DNA interaction with Hoechst 33258: Stretching experiments decouple the different binding modes. *J. Phys. Chem. B* **2013**, *117*, 7292–7296.

8. Garg, D.; Beribisky, A. V.; Ponterini, G.; Ligabue, A.; Marverti, G.; Martello, A.; Costi, M. P.; Sattler, M.; Wade, R. C. Translational repression of thymidylate synthase by targeting its mRNA. *Nucleic Acids Res.* **2013**, *41*, 4159–4170.

9. Nakamura, N.; Ito, K.; Hashimoto, K.; Genma, N.; Nikaido, M. A novel quantitative nucleic acid assay by an improved LAMP method. Jpn. Kokai Tokkyo Koho JP 2012200223, 2012.

10. Kashanian, S.; Shariati, Z.; Roshanfekr, H.; Ghobadi, S. DNA binding studies of 3, 5, 6-trichloro-2-pyridinol pesticide metabolite. *DNA Cell Biol.* **2012**, *31*, 1341–1348.

11. Zeng, G.; Xiang, D.; Cai, J. A new trace analytic method for the determination of sequence-specific DNA with fluorescent probes. *Anal. Lett.* **2012**, *45*, 2334–2343.

12. Kjaerulff, S.; Glensbjerg, M. Method for analysis of cellular DNA content. PCT Int. Appl. WO 2011098085, 2011.

13. Miyamoto, S.; Kato, T.; Tomono, J. Nucleic acid identification method. Jpn. Kokai Tokkyo Koho JP 2010233530, 2010.

14. Harashima, H.; Akita, H. Development of the fluorescent imaging method for quantifying nucleic acid incorporated into target organelle. Jpn. Tokkyo Koho JP 4493939, 2010.

15. Aoyama, Y.; Yamahigashi, S.; Narita, A. Preparation of nucleic acid staining fluorescent probe reagents with an environmentally-responsive motif. Jpn. Kokai Tokkyo Koho JP 2009005594, 2009.

16. Vargiu, A. V.; Ruggerone, P.; Magistrato, A.; Carloni, P. Dissociation of minor groove binders from DNA: insights from metadynamics simulations. *Nucleic Acids Res.* **2008**, *36*, 5910–5921.

17. Guan, Y.; Zhou, W.; Yao, X.; Zhao, M.; Li, Y. Determination of nucleic acids based on the fluorescence quenching of Hoechst 33258 at pH 4.5. *Anal. Chim. Acta* **2006**, *570*, 21–28.

18. Horstkotte, B.; Rehbein, H. Determination of DNA content of whole fish. *Fish. Sci.* **2006**, *72*, 429–436.

19. Martin, R. M.; Leonhardt, H.; Cardoso, M. C. DNA labeling in living cells. *Cytometry, Part A* **2005**, *67A*, 45–52.

20. Gromyko, A. V.; Streltsov, S. A.; Zhuze, A. L. A DNA-specific dimeric bisbenzimidazole. *Russ. J. Bioorg. Chem.* **2004**, *30*, 400–402.

21. Morozkin, E. S.; Laktionov, P. P.; Rykova, E. Y.; Vlassov, V. V. Fluorometric quantification of RNA and DNA in solutions containing both nucleic acids. *Anal. Biochem.* **2003**, *322*, 48–50.

22. Cho, J; Rando, R R. Specific binding of Hoechst 33258 to site 1 thymidylate synthase mRNA. *Nucleic Acids Res.* **2000**, *28*, 2158–2163.

23. Wiederholt, K.; McLaughlin, L. W. Duplex stabilization by DNA-Hoechst 33258 conjugates: effects of base pair mismatches. *Nucleosides Nucleotides* **1998**, *17*, 1895–1904.

24. Vega, M. C.; Coll, M.; Aleman, C. Intrinsic conformational preferences of the Hoechst dye family and their influence on DNA binding. *Eur. J. Biochem.* **1996**, *239*, 376–383.

25. Durand, M.; Thuong, N. T.; Maurizot, J. C. Interaction of Hoechst 33258 with a DNA triple helix. *Biochimie* **1994**, *76*, 181–186.

26. Kapuscinski, J. Interactions of nucleic acids with fluorescent dyes: spectral properties of condensed

complexes. *J. Histochem. Cytochem.* **1990**, *38*, 1323–1329.

27. Churikov, N. A.; Chernov, B. K.; Golova, Y. B.; Nechipurenko, Y. D. Parallel DNA: generation of a duplex between two *Drosophila* sequences *in vitro*. *FEBS Lett.* **1989**, *257*, 415–418.

28. Lipman, J. M. Fluorophotometric quantitation of DNA in articular cartilage utilizing Hoechst 33258. *Anal. Biochem.* **1989**, *176*, 128–131.

29. Van de Sande, J. H.; Ramsing, N. B.; Germann, M. W.; Elhorst, W.; Kalisch, B. W.; Kitzing, E. V.; Pon, R. T.; Clegg, R. C.; Jovin, T. M. Parallel stranded DNA. *Science* **1988**, *241*, 551–557.

30. Bonaly, J.; Bre, M. H.; Lefort-Tran, M.; Mestre, J. C. A flow cytometric study of DNA staining in situ in exponentially growing and stationary *Euglena gracilis*. *Cytometry* **1987**, *8*, 42–45.

31. Mocharla, R.; Mocharla, H.; Hodes, M. E. A novel, sensitive fluorometric staining technique for the detection of DNA in RNA preparations. *Nucleic Acids Res.* **1987**, *15*, 10589.

32. Murray, M. G.; Paaren, H. E. Nucleic acid quantitation by continuous flow fluorometry. *Anal. Biochem.* **1986**, *154*, 638–642.

33. Sterzel, W.; Bedford, P.; Eisenbrand, G. Automated determination of DNA using the fluorochrome Hoechst 33258. *Anal. Biochem.* **1985**, *147*, 462–467.

34. Harshman, K. D.; Dervan, P. B. Molecular recognition of B-DNA by Hoechst 33258. *Nucleic Acids Res.* **1985**, *13*, 4825–4835.

35. Stokke, T.; Steen, H. B. Multiple binding modes for Hoechst 33258 to DNA. *J. Histochem. Cytochem.* **1985**, *33*, 333–338.

36. Villanueva, A.; Stockert, J. C.; Armas-Portela, R. A simple method for the fluorescence analysis of nucleic acid-dye complexes in cytological preparations. *Histochemistry* **1984**, *81*, 103–104.

37. Bjelkenkrantz, K. An evaluation of Feulgen-acriflavine-sulfur dioxide and Hoechst 33258 for DNA cytofluorometry in tumor pathology. *Histochemistry* **1983**, *79*, 177–191.

38. Hyman, B. C.; James, T. W. Detection of A + T-rich DNA in gels by differential fluorescence. *Anal. Biochem.* **1983**, *131*, 205–210.

39. Dean, P. N.; Gray, J. W.; Dolbeare, F. A. The analysis and interpretation of DNA distributions measured by flow cytometry. *Cytometry* **1982**, *3*, 188–195.

40. Stout, D. L.; Becker, F. F. Fluorometric quantitation of single-stranded DNA: a method applicable to the

technique of alkaline elution. *Anal. Biochem.* **1982**, *127*, 302–307.

41. Noguchi, P. D.; Johnson, J. B.; Browne, W. Measurement of DNA synthesis by flow cytometry. *Cytometry* **1981**, *1*, 390–393.

42. Curtis, S. K.; Cowden, R. R. Four fluorochromes for the demonstration and microfluorometric estimation of RNA. *Histochemistry* **1981**, *72*, 39–48.

43. Labarca, C.; Paigen, K. A simple, rapid, and sensitive DNA assay procedure. *Anal. Biochem.* **1980**, *102*, 344–352.

44. Lydon, M. J.; Keeler, K. D.; Thomas, D. B. Vital DNA staining and cell sorting by flow microfluorometry. *J. Cell. Physiol.* **1980**, *102*, 175–181.

45. Cesarone, C. F.; Bolognesi, C.; Santi, L. Improved microfluorometric DNA determination in biological material using 33258 Hoechst. *Anal. Biochem.* **1979**, *100*, 188–197.

46. Steiner, R. F.; Sternberg, H. The interaction of Hoechst 33258 with natural and biosynthetic nucleic acids. *Arch. Biochem. Biophys.* **1979**, *197*, 580–588.

47. Brunk, C. F.; Jones, K. C.; James, T. W. Assay for nanogram quantities of DNA in cellular homogenates. *Anal. Biochem.* **1979**, *92*, 497–500.

48. Arndt-Jovin, D. J.; Jovin, T. M. Analysis and sorting of living cells according to deoxyribonucleic acid content. *J. Histochem. Cytochem.* **1977**, *25*, 585–589.

49. Manuelidis, L. A simplified method for preparation of mouse satellite DNA. *Anal. Biochem.* **1977**, *78*, 561–568.

50. Latt, S. A.; Stetten, G. Spectral studies on 33258 Hoechst and related bisbenzimidazole dyes useful for fluorescent detection of deoxyribonucleic acid synthesis. *J. Histochem. Cytochem.* **1976**, *24*, 24–33.

51. Bontemps, J.; Houssier, C.; Fredericq, E. Physicochemical study of the complexes of 33258 Hoechst with DNA and nucleohistone. *Nucleic Acids Res.* **1975**, *2*, 971–984.

52. Latt, S. A.; Stetten, G.; Juergens, L. A.; Willard, H. F.; Scher, C. D. Detection of deoxyribonucleic acid synthesis by 33258 Hoechst fluorescence. *J. Histochem. Cytochem.* **1975**, *23*, 493–505.

53. Russell, W. C.; Newman, C.; Williamson, D. H. A simple cytochemical technique for demonstration of DNA in cells infected with mycoplasmas and viruses. *Nature* **1975**, *253*, 461–462.

54. Mueller, W.; Gautier, F. Interactions of heteroaromatic compounds with nucleic acids.

A.T-specific nonintercalating DNA ligands. *Eur. J. Biochem.* **1975**, *54*, 385–394.

55. Anderson, A. L.; Knutson, C. R.; Mueth, D.; Plewa, J.; Tanner, E. Methods for staining cells for identification and sorting. U.S. Pat. Appl. Publ. US 2006172315, 2006.

56. Skyggebjerg, O.; Glensbjerg, M. A method and a system for counting cells from a plurality of species. PCT Int. Appl. WO 2002101087, 2002.

57. Winterbourne, D. J.; Thomas, S.; Stebbing, A.; Hermon-Taylor, J. A procedure for selection of cells with suppressed growth in suspension after gene transfection. *Biochem. Soc. Trans.* **1988**, *16*, 1018.

58. Steen, H. B.; Stokke, T. Fluorescence spectra of cells stained with a DNA-specific dye, measured by flow cytometry. *Cytometry* **1986**, *7*, 104–106.

59. Langlois, R. G.; Jensen, R. H. Interactions between pairs of DNA-specific fluorescent stains bound to mammalian cells. *J. Histochem. Cytochem.* **1979**, *27*, 72–79.

60. Miyata, K.; Takasugi, M.; Yamada, M. Heterogeneous bindings of fluorescent antibiotics to chromatin. *Cell. Mol. Biol.* **1986**, *32*, 577–585.

61. Stokke, T.; Steen, H. B. Binding of Hoechst 33258 to chromatin *in situ*. *Cytometry* **1986**, *7*, 227–234.

62. Cowden, R. R.; Curtis, S. K. Microfluorometric investigations of chromatin structure. I. Evaluation of nine DNA-specific fluorochromes as probes of chromatin organization. *Histochemistry* **1981**, *72*, 11–23.

63. Di Castro, M.; Prantera, G.; Marchetti, E.; Rocchi, A. Characterization of the chromatin of *Asellus aquaticus* (Crust. Isop.) by treatment *in vivo* with BrdU and Hoechst 33258. *Caryologia* **1979**, *32*, 81–88.

64. Iannuzzi, L. Methodologies applied to domestic animal chromosomes. *Methods Mol. Biol.* **2004**, *240*, 15–34.

65. Vanderbyl, S.; MacDonald, N.; De Jong, G. A flow cytometry technique for measuring chromosome-mediated gene transfer. *Cytometry* **2001**, *44*, 100–105.

66. Aten, J. A.; Buys, C. H. C. M.; Van der Veen, A. Y.; Mesa, J. R.; Yu, L. C.; Gray, J. W.; Osinga, J.; Stap, J. Stabilization of chromosomes by DNA intercalators for flow karyotyping and identification by banding of isolated chromosomes. *Histochemistry* **1987**, *87*, 359–366.

67. Trask, B.; Van den Engh, G.; Gray, J.; Vanderlaan, M.; Turner, B. Immunofluorescent detection of

histone 2B on metaphase chromosomes using flow cytometry. *Chromosoma* **1984**, *90*, 295–302.

68. Hauser-Urfer, I.; Leemann, U.; Ruch, F. Cytofluorometric determination of the DNA base content in human chromosomes with quinacrine mustard, Hoechst 33258, DAPI, and mithramycin. *Exp. Cell Res.* **1982**, *142*, 455–459.

69. Latt, S. A.; Sahar, E.; Eisenhard, M. E.; Juergens, L. A. Interactions between pairs of DNA-binding dyes: results and implications for chromosome analysis. *Cytometry* **1980**, *1*, 2–12.

70. Langlois, R. G.; Carrano, A. V.; Gray, J. W.; Van Dilla, M. A. Cytochemical studies of metaphase chromosomes by flow cytometry. *Chromosoma* **1980**, *77*, 229–251.

71. Carrano, A. V.; Gray, J. W.; Langlois, R. G.; Burkhart-Schultz, K. J.; Van Dilla, M. A. Measurement and purification of human chromosomes by flow cytometry and sorting. *Proc. Natl. Acad. Sci. U.S.A.* **1979**, *76*, 1382–1384.

72. Jorgenson, K. F.; Van de Sande, J. H.; Lin, C. C. The use of base pair specific DNA binding agents as affinity labels for the study of mammalian chromosomes. *Chromosoma* **1978**, *68*, 287–302.

73. Scheid, W.; Traupe, H. Further studies on the mechanism involved in differential staining of BUdR-substituted *Vicia faba* chromosomes. *Exp. Cell Res.* **1977**, *108*, 440–444.

74. Jensen, R. H.; Langlois, R. G.; Mayall, B. H. Strategies for choosing a deoxyribonucleic acid stain for flow cytometry of metaphase chromosomes. *J. Histochem. Cytochem.* **1977**, *25*, 954–964.

75. Comings, D. E.; Drets, M. E. Mechanisms of chromosome banding. IX. Are variations in DNA base composition adequate to account for quinacrine, Hoechst 33258 and daunomycin banding? *Chromosoma* **1976**, *56*, 199–211.

76. Rocchi, A.; Prantera, G.; Pimpinelli, S.; Di Castro, M. Effect of Hoechst 33258 on Chinese hamster chromosomes. *Chromosoma* **1976**, *56*, 41–46.

77. Comings, D. E. Mechanisms of chromosome banding. VIII. Hoechst 33258-DNA interaction. *Chromosoma* **1975**, *52*, 229–243.

78. Jalal, S. M.; Markvong, A.; Hsu, T. C. Differential chromosomal fluorescence with 33258 Hoechst. *Exp. Cell Res.* **1975**, *90*, 443–444.

79. Latt, S. A. Microfluorometric analysis of DNA replication in human X chromosomes. *Exp. Cell Res.* **1974**, *86*, 412–415.

80. Lin, M. S.; Latt, S. A.; Davidson, R. L. Microfluorometric detection of asymmetry in the centromeric region of mouse chromosomes. *Exp. Cell Res.* **1974**, *86*, 392–395.

81. Latt, S. A. Microfluorometric analysis of deoxyribonucleic acid replication kinetics and sister chromatid exchanges in human chromosomes. *J. Histochem. Cytochem.* **1974**, *22*, 478–491.

82. Hilwig, I.; Gropp, A. Decondensation of constitutive heterochromatin in L cell chromosomes by a benzimidazole compound (33258 Hoechst). *Exp. Cell Res.* **1973**, *81*, 474–482.

83. Hilwig, I.; Gropp, A. Staining of constitutive heterochromatin in mammalian chromosomes with a new fluorochrome. *Exp. Cell Res.* **1972**, *75*, 122–126.

84. Masci, J.; Monteiro, A. Visualization of early embryos of the butterfly *Bicyclus anynana*. *Zygote* **2005**, *13*, 139–144.

85. De Braekeleer, M.; Keushnig, M.; Lin, C. C. Synchronization of human lymphocyte cultures by fluorodeoxyuridine. *Can. J. Genet. Cytol.* **1985**, *27*, 622–625.

86. Noda, N.; Mizutani, T. Microorganism-measuring method using multiple staining. Jpn. Kokai Tokkyo Koho JP 2006340684, 2006.

87. Mizuki, H. *In situ* staining with DNA-binding fluorescent dye, Hoechst 33258, to detect microorganisms in the epithelial cells of oral leukoplakia. *Oral Oncol.* **2001**, *37*, 521–526.

88. Chaudhury, N. K.; Murari, B. M.; Gohil, N. K.; Anand, S. Fluorescence based sensors and sensing method for *in vitro* detection and quantitation of myoglobin. Indian Pat. Appl. IN 2007DE02294, 2009.

89. Gayoso, R. M. Fluorescent dye mixture and procedure for staining the nucleus and cytoplasm of the cell to permit study of the morphology of cells on any type of support. Span. ES 2326061, 2009.

90. Horobin, R. W.; Stockert, J. C.; Rashid-Doubell, F. Fluorescent cationic probes for nuclei of living cells: Why are they selective? A quantitative structure-activity relations analysis. *Histochem. Cell Biol.* **2006**, *126*, 165–175.

91. Pellicciari, C.; Marchese, G.; Bottone, M. G.; Manfredi Romanini, M. G. Kinetics of DNase I digestion of interphase chromatin in differentiated cell nuclei of the mouse: a flow cytometric study. *Basic Appl. Histochem.* **1987**, *31*, 501–509.

92. Araki, T.; Yamamoto, A.; Yamada, M. Accurate determination of DNA content in single cell nuclei stained with Hoechst 33258 fluorochrome at high salt concentration. *Histochemistry* **1987**, *87*, 331–338.

93. Kubbies, M.; Friedl, R. Flow cytometric correlation between BrdU/Hoechst quench effect and base pair composition in mammalian cell nuclei. *Histochemistry* **1985**, *83*, 133–137.

94. Laloue, M.; Courtois, D.; Manigault, P. Convenient and rapid fluorescent staining of plant cell nuclei with 33258 Hoechst. *Plant Sci. Lett.* **1980**, *17*, 175–179.

95. Latt, S. A. Detection of DNA synthesis in interphase nuclei by fluorescence microscopy. *J. Cell Biol.* **1974**, *62*, 546–550.

96. Smeijsters, L. J. J. W.; Zijlstra, N. M.; Franssen, F. F. J.; Overdulve, J. P. Simple, fast, and accurate fluorometric method to determine drug susceptibility of *Plasmodium falciparum* in 24-well suspension cultures. *Antimicrob. Agents Chemother.* **1996**, *40*, 835–838.

97. Dame, J. B.; McCutchan, T. F. *Plasmodium falciparum*: Hoechst dye 33258-cesium chloride ultracentrifugation for separating parasite and host DNAs. *Exp. Parasitol.* **1987**, *64*, 264–266.

98. Ulrich, I.; Fritz, B.; Ulrich, W. Application of DNA fluorochromes for flow cytometric DNA analysis of plant protoplasts. *Plant Sci.* **1988**, *55*, 151–158.

99. Reich, T. J.; Iyer, V. N.; Haffner, M.; Holbrook, L. A.; Miki, B. L. The use of fluorescent dyes in the microinjection of alfalfa protoplasts. *Can. J. Bot.* **1986**, *64*, 1259–1267.

100. Marco-Jimenez, F.; Garzon, D. L.; Penaranda, D. S.; Perez, L.; Viudes-de-Castro, M. P.; Vicente, J. S.; Jover, M.; Asturiano, J. F. Cryopreservation of European eel (*Anguilla anguilla*) spermatozoa: Effect of dilution ratio, fetal bovine serum supplementation, and cryoprotectants. *Cryobiology* **2006**, *53*, 51–57.

101. Vadnais, M. L.; Kirkwood, R. N.; Tempelman, R. J.; Sprecher, D. J.; Chou, K. Effect of cooling and seminal plasma on the capacitation status of fresh boar sperm as determined using chlortetracycline assay. *Anim. Reprod. Sci.* **2005**, *87*, 121–132.

102. Valcarcel, A.; de las Heras, M. A.; Perez, L.; Moses, D. F.; Baldassarre, H. Assessment of the acrosomal status of membrane-intact ram spermatozoa after freezing and thawing, by simultaneous lectin/Hoechst 33258 staining. *Anim. Reprod. Sci.* **1997**, *45*, 299–309.

103. Tao, J.; Critser, E. S.; Critser, J. K. Evaluation of mouse sperm acrosomal status and viability by flow cytometry. *Mol. Reprod. Dev.* **1993**, *36*, 183–194.

104. Yelian, F. D.; Dukelow, W. R. Effects of a DNA-specific fluorochrome, Hoechst 33258, on mouse sperm motility and fertilizing capacity. *Andrologia* **1992**, *24*, 167–170.

105. Holden, C. A.; Hyne, R. V.; Sathananthan, A. H.; Trounson, A. O. Assessment of the human sperm acrosome reaction using concanavalin A lectin. *Mol. Reprod. Dev.* **1990**, *25*, 247–257.

106. Boehmer, R. M. Cell division analysis using bromodeoxyuridine-induced suppression of Hoechst 33258 fluorescence. *Methods Cell Biol.* **1990**, *33*, 173–184.

107. Zeldis, J. B.; Zeitlin, A.; Barer, S. Compositions for the treatment of cancer comprising a topoisomerase inhibitor and thalidomide. Eur. Pat. Appl. EP 1676577, 2006.

108. Gabelica, V.; De Pauw, E.; Rosu, F. Interaction between antitumor drugs and a double-stranded oligonucleotide studied by electrospray ionization mass spectrometry. *J. Mass Spectrom.* **1999**, *34*, 1328–1337.

109. Soderlind, K.; Gorodetsky, B.; Singh, A. K.; Bachur, N. R.; Miller, G. G.; Lown, J. W. Bis-benzimidazole anticancer agents: targeting human tumour helicases. *Anti-Cancer Drug Des.* **1999**, *14*, 19–36.

110. Baraldi, P. G.; Bovero, A.; Fruttarolo, F.; Preti, D.; Tabrizi, M. A.; Pavani, M. G.; Romagnoli, R. DNA minor groove binders as potential antitumor and antimicrobial agents. *Med. Res. Rev.* **2004**, *24*, 475–528.

111. Disney, M. D.; Stephenson, R.; Wright, T. W.; Haidaris, C. G.; Turner, D. H.; Gigliotti, F. Activity of Hoechst 33258 against *Pneumocystis carinii* f. sp. muris, *Candida albicans*, and *Candida dubliniensis. Antimicrob. Agents Chemother.* **2005**, *49*, 1326–1330.

112. Deutsch, M.; Namer, Y. A method and system for measuring membrane potential based on fluorescence polarization. PCT Int. Appl. WO 2007122602, 2007.

113. Farinas, J. A.; Wada, H. G. Use of Nernstein voltage sensitive dyes in measuring transmembrane voltage. PCT Int. Appl. WO 2001027253, 2001.

114. Neyfakh, A. A. Use of fluorescent dyes as molecular probes for the study of multidrug resistance. *Exp. Cell Res.* **1988**, *174*, 168–176.

115. Telford, W. G.; Komoriya, A.; Packard, B. Z.; Bagwell, C. B. Multiparametric analysis of apoptosis by flow cytometry. *Methods Mol. Biol.* **2011**, *699*, 203–227.

116. Matsuo, A.; Watanabe, A.; Takahashi, T.; Futamura, M.; Mori, S.; Sugiyama, Y.; Takahashi, Y.; Saji, S. A simple method for classification of cell death by use of thin layer collagen gel for the detection of apoptosis and/or necrosis after cancer chemotherapy. *Jpn. J. Cancer Res.* **2001**, *92*, 813–819.

117. Ormerod, M. G. Cell cycle analysis of asynchronous populations. *Methods Mol. Biol.* **1998**, *91*, 157–165.

118. Ormerod, M. G.; Kubbies, M. Cell cycle analysis of asynchronous cell populations by flow cytometry using bromodeoxyuridine label and Hoechst-propidium iodide stain. *Cytometry* **1992**, *13*, 678–685.

119. Poot, M.; Hoehn, H.; Kubbies, M.; Grossmann, A.; Chen, Y.; Rabinovitch, P. S. Cell cycle analysis using continuous bromodeoxyuridine labeling and Hoechst 33258-ethidium bromide bivariate flow cytometry. *Methods Cell Biol.* **1990**, *33*, 185–198.

120. Rabinovitch, P. S.; Kubbies, M.; Chen, Y. C.; Schindler, D.; Hoehn, H. BrdU-Hoechst flow cytometry: a unique tool for quantitative cell cycle analysis. *Exp. Cell Res.* **1988**, *174*, 309–318.

121. Dubey, D. D.; Raman, R. Effects of Hoechst 33258 on different cell cycle events. I. Inhibition of synthetic activities in bone marrow cells of the mole rat *Bandicota bengalensis. Exp. Cell Res.* **1983**, *149*, 419–432.

122. Boehmer, R. M.; Ellwart, J. Cell cycle analysis by combining the 5-bromodeoxyuridine/33258 Hoechst technique with DNA-specific ethidium bromide staining. *Cytometry* **1981**, *2*, 31–34.

123. Boehmer, R. M. Flow cytometric cell cycle analysis using the quenching of Hoechst 33258 fluorescence by bromodeoxyuridine incorporation. *Cell Tissue Kinet.* **1979**, *12*, 101–110.

124. Ng, K. W.; Leong, D. T. W.; Hutmacher, D. W. The challenge to measure cell proliferation in two and three dimensions. *Tissue Eng.* **2005**, *11*, 182–191.

125. Garza, C. M.; Cho, S. Metrology of bilayer photoresist processes. U.S. Pat. Appl. Publ. US 20090220895, 2009.

126. Gupta, R.; Chaudhury, N. K. Probing internal environment of sol–gel bulk and thin films using multiple fluorescent probes. *J. Sol–gel Sci. Technol.* **2009**, *49*, 78–87.

127. Bielawski, K.; Wolczynski, S.; Bielawska, A. DNA-binding properties and cytotoxicity of extended aromatic bisamidines in breast cancer MCF-7 cells. *Pol. J. Pharmacol.* **2001**, *53*, 143–147.

128. Finlay, G. J.; Baguley, B. C. Potentiation by phenylbisbenzimidazoles of cytotoxicity of anticancer drugs directed against topoisomerase II. *Eur. J. Cancer* **1990**, *26*, 586–589.

129. Limoli, C. L.; Ward, J. F. DNA damage in bromodeoxyuridine substituted SV40 DNA and minichromosomes following UVA irradiation in the presence of Hoechst dye 33258. *Int. J. Radiat. Biol.* **1994**, *66*, 717–728.

130. He, L.; Jurs, P. C.; Custer, L. L.; Durham, S. K.; Pearl, G. M. Predicting the genotoxicity of polycyclic aromatic compounds from molecular structure with different classifiers. *Chem. Res. Toxicol.* **2003**, *16*, 1567–1580.

131. Singh, S. P.; Jayanth, V. R.; Chandna, S.; Dwarakanath, B. S.; Singh, S.; Adhikari, J. S.; Jain, V. Radioprotective effects of DNA ligands Hoechst-33342 and 33258 in whole body irradiated mice. *Indian J. Exp. Biol.* **1998**, *36*, 375–384.

HOECHST 33342

CAS Registry Number   875756-97-1
Chemical Structure

pK$_a$ (Calcd.)   11.08 ± 0.69, most acidic, temperature: 25 °C; 7.66 ± 0.42, most basic, temperature: 25 °C
Absorption (λ$_{max}$)   350 nm (H$_2$O/DNA); 357 nm (MeOH/H$^+$)

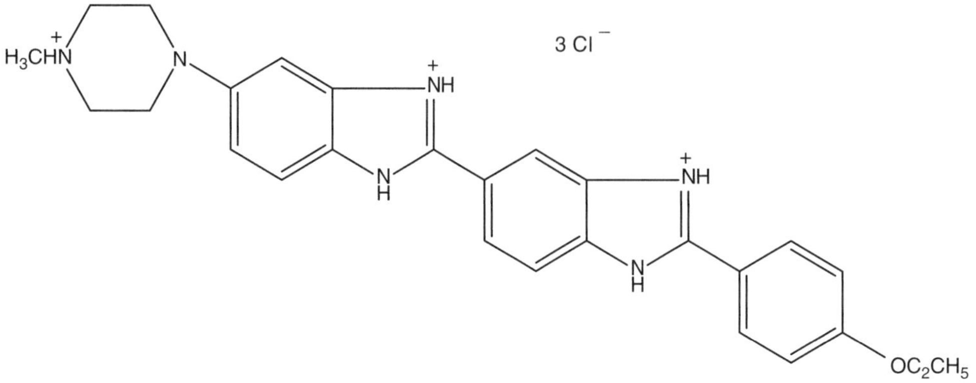

CA Index Name   2,5′-Bi-1*H*-benzimidazole, 2′-(4-ethoxyphenyl)-5-(4-methyl-1-piperazinyl)-, hydrochloride (1:3)

Other Names   2,5′-Bi-1*H*-benzimidazole, 2′-(4-ethoxy-phenyl)-5-(4-methyl-1-piperazinyl)-, trihydrochloride; 2,5′-Bibenzimidazole, 2′-(*p*-ethoxyphenyl)-5-(4-methyl-1-piperazinyl)-, trihydrochloride; 2′-(4-Ethoxyphenyl)-5-(4-methyl-1-piperazinyl)-2,5′-bi-1H-benzimidazole trihydrochloride; 2-[2-(4-Ethoxyphenyl)-6-benzimidazo-lyl]-6-(1-methyl-4-piperazinyl)benzimidazole trihydrochloride; Bisbenzimide; HOE 33342; Ho 342; Hoechst 33342; Hoechst 33342 trihydrochloride; NSC 334072

Merck Index Number   1250

Chemical/Dye Class   Heterocycle; Benzimidazole

Molecular Formula   C$_{27}$H$_{31}$Cl$_3$N$_6$O

Molecular Weight   561.94

Physical Form   Yellow to green powder or solid

Solubility   Soluble in water, *N,N*-dimethylformamide, methanol

Melting Point   >300 °C

Boiling Point (Calcd.)   725.9 ± 70.0 °C, pressure: 760 Torr

Emission (λ$_{max}$)   461 nm (H$_2$O/DNA); 538 nm (MeOH/H$^+$)

Molar Extinction Coefficient   45,000 cm^{-1} M^{-1} (H$_2$O/DNA); 43,000 cm^{-1} M^{-1} (MeOH/H$^+$)

Quantum Yield   0.38 (H$_2$O/DNA)

Synthesis   Synthetic methods[1–5]

Imaging/Labeling Applications   Nucleic acids;[6–21] β-amyloid (Aβ) plaques;[22] bacteria;[23,24] cells;[25–31] chromatin;[32,33] lymphocytes;[34–36] microorganisms;[37,38] nucleus;[39–41] neurons;[42] parasites;[43,44] sperms;[45–52] stem cells;[53–62] tumors[63–71]

Biological/Medical Applications   Detecting β-amyloid (Aβ) plaques;[22] detecting/measuring microorganisms;[37,38] identifying/counting cells;[25–31] studying multidrug resistance;[72] apoptosis assay;[73–83] cell cycle assay;[83–85] cell proliferation assay;[86] measuring membrane potential;[87–89] treating disc degenerative disease,[90] erectile dysfunction[91]

Industrial Applications   Photoresists[92]

Safety/Toxicity   Carcinogenicity;[93] cardiotoxicity;[94] cytotoxicity;[95,99,100] DNA damage;[96–100] genotoxicity;[101–103] metabolic toxicity;[104] mutagenicity;[100] neurotoxicity;[105–107] phototoxicity;[108,109] reproductive toxicity[110–112]

REFERENCES

1. Sabnis, R. W. *Handbook of Biological Dyes and Stains*; John Wiley & Sons Inc.: Hoboken, **2010**; pp 233–236.

2. Loewe, H.; Urbanietz, J. Basic substituted 2,6-bisbenzimidazole derivatives, a novel series of substances with chemotherapeutic activity. *Arzneim.-Forsch.* **1974**, *24*, 1927–1933.

3. Farbwerke Hoechst AG. Anthelmintic 2-(2-phenyl-6-benzimidazolyl)benzimidazoles. FR 6681, 1969.

4. Loewe, H.; Taunus, K.; Urbanietz, J.; Taunus, S.; Lammler, G. Substituted piperazino-bis-benzimidazoles having anthelmintic and bacteriostatic activity. U.S. Patent 3538097, 1970.

5. Farbwerke Hoechst AG. Basically substituted bis-benzimidazole derivatives and process for their manufacture. FR 1519964, 1968.

6. Vekshin, N. Binding of hoechst with nucleic acids using fluorescence spectroscopy. *J. Biophys. Chem.* **2011**, *2*, 443–447.

7. Kjaerulff, S.; Glensbjerg, M. Method for analysis of cellular DNA content. PCT Int. Appl. WO 2011098085, 2011.

8. Del Castillo, P.; Horobin, R. W.; Blazquez-Castro, A.; Stockert, J. C. Binding of cationic dyes to DNA: distinguishing intercalation and groove binding mechanisms using simple experimental and numerical models. *Biotech. Histochem.* **2010**, *85*, 247–256.

9. Wu, J.; Apontes, P.; Song, L.; Liang, P.; Yang, L.; Li, F. Molecular mechanism of upregulation of survivin transcription by the AT-rich DNA-binding ligand, Hoechst33342: evidence for survivin involvement in drug resistance. *Nucleic Acids Res.* **2007**, *35*, 2390–2402.

10. Evans, D. A.; Neidle, S. Virtual screening of DNA minor groove binders. *J. Med. Chem.* **2006**, *49*, 4232–4238.

11. Darzynkiewicz, Z. Flow cytometric methods for RNA content analysis. *Methods* **1991**, *2*, 200–206.

12. Kapuscinski, J. Interactions of nucleic acids with fluorescent dyes: spectral properties of condensed complexes. *J. Histochem. Cytochem.* **1990**, *38*, 1323–1329.

13. Van der Valk, H. C. P. M.; Blaas, J.; Van Eck, J. W.; Verhoeven, H. A. Vital DNA staining of agarose-embedded protoplasts and cell suspensions of *Nicotiana plumbaginifolia*. *Plant Cell Rep.* **1988**, *7*, 489–492.

14. Evenson, D.; Darzynkiewicz, Z.; Jost, L.; Janca, F.; Ballachey, B. Changes in accessibility of DNA to various fluorochromes during spermatogenesis. *Cytometry* **1986**, *7*, 45–53.

15. Puite, K. J.; Ten Broeke, W. R. R. DNA staining of fixed and nonfixed plant protoplasts for flow cytometry with Hoechst 33342. *Plant Sci. Lett.* **1983**, *32*, 79–88.

16. Shapiro, H. M. Flow cytometric estimation of DNA and RNA content in intact cells stained with Hoechst 33342 and pyronin Y. *Cytometry* **1981**, *2*, 143–150.

17. Grogan, W. M.; Farnham, W. F.; Sabau, J. M. DNA analysis and sorting of viable mouse testis cells. *J. Histochem. Cytochem.* **1981**, *29*, 738–746.

18. Taylor, I. W.; Milthorpe, B. K. An evaluation of DNA fluorochromes, staining techniques, and analysis for flow cytometry. I. Unperturbed cell populations. *J. Histochem. Cytochem.* **1980**, *28*, 1224–1232.

19. Lydon, M. J.; Keeler, K. D.; Thomas, D. B. Vital DNA staining and cell sorting by flow microfluorometry. *J. Cell. Physiol.* **1980**, *102*, 175–181.

20. Taylor, I. W. A rapid single step staining technique for DNA analysis by flow microfluorimetry. *J. Histochem. Cytochem.* **1980**, *28*, 1021–1024.

21. Arndt-Jovin, D. J.; Jovin, T. M. Analysis and sorting of living cells according to deoxyribonucleic acid content. *J. Histochem. Cytochem.* **1977**, *25*, 585–589.

22. Uchida, Y.; Takahashi, H. Rapid detection of Aβ deposits in APP transgenic mice by Hoechst 33342. *Neurosci. Lett.* **2008**, *448*, 279–281.

23. Webber, M. A.; Coldham, N. G. Measuring the activity of active efflux in gram-negative bacteria. *Methods Mol. Biol.* **2010**, *642*, 173–180.

24. Monger, B. C.; Landry, M. R. Flow cytometric analysis of marine bacteria with Hoechst 33342. *Appl. Environ. Microbiol.* **1993**, *59*, 905–911.

25. Adamski, D.; Mayol, J. F.; Platet, N.; Berger, F.; Herodin, F.; Wion, D. Effects of Hoechst 33342 on C2C12 and PC12 cell differentiation. *FEBS Lett.* **2007**, *581*, 3076–3080.

26. Anderson, A. L.; Knutson, C. R.; Mueth, D.; Plewa, J.; Tanner, E. Methods for staining cells for identification and sorting. U.S. Pat. Appl. Publ. US 2006172315, 2006.

27. Bestvater, F.; Spiess, E.; Stobrawa, G.; Hacker, M.; Feurer, T.; Porwol, T.; Berchner-Pfannschmidt, U.; Wotzlaw, C.; Acker, H. Two-photon fluorescence

absorption and emission spectra of dyes relevant for cell imaging. *J. Microsc.* **2002**, *208*, 108–115.

28. Skyggebjerg, O.; Glensbjerg, M. A method and a system for counting cells from a plurality of species. PCT Int. Appl. WO 2002101087, 2002.

29. Foglieni, C.; Meoni, C.; Davalli, A. M. Fluorescent dyes for cell viability: an application on prefixed conditions. *Histochem. Cell Biol.* **2001**, *115*, 223–229.

30. Pollack, A.; Ciancio, G. Cell cycle phase-specific analysis of cell viability using Hoechst 33342 and propidium iodide after ethanol preservation. *Methods Cell Biol.* **1990**, *33*, 19–24.

31. Crissman, H. A.; Hofland, M. H.; Stevenson, A. P.; Wilder, M. E.; Tobey, R. A. Supravital cell staining with Hoechst 33342 and DiOC5(3). *Methods Cell Biol.* **1990**, *33*, 89–95.

32. Rousselle, C.; Paillasson, S.; Robert-Nicoud, M.; Ronot, X. Chromatin texture analysis in living cells. *Histochem. J.* **1999**, *31*, 63–70.

33. Gregoire, M.; Hernandez-Verdun, D.; Bouteille, M. Visualization of chromatin distribution in living PTO cells by Hoechst 33342 fluorescent staining. *Exp. Cell Res.* **1984**, *152*, 38–46.

34. Samlowski, W. E.; Robertson, B. A.; Draper, B.; Prystas, E.; McGregor, J. R. Effects of supravital fluorochromes used to analyze the *in vivo* homing of murine lymphocytes on cellular function. *J. Immunol. Methods* **1991**, *144*, 101–115.

35. Bialek, R.; Abken, H. A fluorometric assay to monitor mitogenic stimulation of human lymphocytes. *J. Immunol. Methods* **1991**, *144*, 223–229.

36. Loeffler, D.; Ratner, S. *In vivo* localization of lymphocytes labeled with low concentrations of Hoechst 33342. *J. Immunol. Methods* **1989**, *119*, 95–101.

37. Noda, N.; Mizutani, T. Microorganism-measuring method using multiple staining. Jpn. Kokai Tokkyo Koho JP 2006340684, 2006.

38. Horikiri, S. Microorganism cell detection method using fluorescent indicator. Jpn. Kokai Tokkyo Koho JP 2006262775, 2006.

39. Gayoso, R. M. Fluorescent dye mixture and procedure for staining the nucleus and cytoplasm of the cell to permit study of the morphology of cells on any type of support. Span. ES 2326061, 2009.

40. Yasui, L. S.; Chen, K.; Wang, K.; Jones, T. P.; Caldwell, J.; Guse, D.; Kassis, A. I. Using Hoechst 33342 to target radioactivity to the cell nucleus. *Radiat. Res.* **2007**, *167*, 167–175.

41. Horobin, R. W.; Stockert, J. C.; Rashid-Doubell, F. Fluorescent cationic probes for nuclei of living cells: why are they selective? A quantitative structure-activity relations analysis. *Histochem. Cell Biol.* **2006**, *126*, 165–175.

42. Casella, G. T. B.; Bunge, M. B.; Wood, P. M. Improved immunocytochemical identification of neural, endothelial, and inflammatory cell types in paraffin-embedded injured adult rat spinal cord. *J. Neurosci. Methods* **2004**, *139*, 1–11.

43. Joanny, F.; Held, J.; Mordmueller, B. *In vitro* activity of fluorescent dyes against asexual blood stages of *Plasmodium falciparum*. *Antimicrob. Agents Chemother.* **2012**, *56*, 5982–5985.

44. Jean-Moreno, V.; Rojas, R.; Goyeneche, D.; Coombs, G. H.; Walker, J. *Leishmania donovani*: Differential activities of classical topoisomerase inhibitors and antileishmanials against parasite and host cells at the level of DNA topoisomerase I and in cytotoxicity assays. *Exp. Parasitol.* **2006**, *112*, 21–30.

45. Yaniz, Y. L.; Palacin, I.; Vicente-Fiel, S.; Gosalvez, J.; Lopez-Fernandez, C.; Santolaria, P. Comparison of membrane-permeant fluorescent probes for sperm viability assessment in the ram. *Reprod. Dom. Anim.* **2013**, *48*, 598–603.

46. Balao da Silva, C.; Macias-Garcia, B.; Morillo Rodriguez, A.; Gallardo Bolanos, J. M.; Tapia, J. A.; Aparicio, I. M.; Morrell, J. M.; Rodriguez-Martinez, H.; Ortega-Ferrusola, C.; Pena, F. J. Effect of Hoechst 33342 on stallion spermatozoa incubated in KMT or Tyrodes modified INRA96. *Anim. Reprod. Sci.* **2012**, *131*, 165–171.

47. Seo, T. G. Method for separating sperm of mammal. Repub. Korean Kongkae Taeho Kongbo KR 2012025128, 2012.

48. Lu, Y. Q.; Zhang, M.; Meng, B.; Lu, S. S.; Wei, Y. M.; Lu, K. H. Identification of X- and Y-chromosome bearing buffalo (*Bubalus bubalis*) sperm. *Reprod. Sci.* **2006**, *95*, 158–164.

49. Hallap, T.; Nagy, S.; Jaakma, U.; Johannisson, A.; Rodriguez-Martinez, H. Usefulness of a triple fluorochrome combination merocyanine 540/YO-PRO 1/Hoechst 33342 in assessing membrane stability of viable frozen-thawed spermatozoa from Estonian Holstein AI bulls. *Theriogenology* **2006**, *65*, 1122–1136.

50. Green, D. P. L. Comparison of Hoechst 33342 and propidium iodide as fluorescent markers for sperm fusion with hamster oocytes. *J. Reprod. Fertil.* **1992**, *96*, 581–591.

51. Conover, J. C.; Gwatkin, R. B. L. Pre-loading of mouse oocytes with DNA-specific fluorochrome (Hoechst 33342) permits rapid detection of sperm-oocyte fusion. *J. Reprod. Fertil.* **1988**, *82*, 681–690.

52. Hinkley, R. E.; Wright, B. D.; Lynn, J. W. Rapid visual detection of sperm-egg fusion using the DNA-specific fluorochrome Hoechst 33342. *Dev. Biol.* **1986**, *118*, 148–154.

53. Teramura, T.; Fukuda, K.; Asada, S.; Kato, K. Cell culture method of concentrating stem cells or cell line derived from synovial membrane. Jpn. Kokai Tokkyo Koho JP 2009055866, 2009.

54. Allen, J. E.; Hart, L. S.; Dicker, D. T.; Wang, W.; El-Deiry, W. S. Visualization and enrichment of live putative cancer stem cell populations following p53 inactivation or Bax deletion using non-toxic fluorescent dyes. *Cancer Biol. Ther.* **2009**, *8*, 2193–2204.

55. Donnenberg, V. S.; Meyer, E. M.; Donnenberg, A. D. Measurement of multiple drug resistance transporter activity in putative cancer stem/progenitor cells. *Methods Mol. Biol.* **2009**, *568*, 261–279.

56. Zheng, X.; Shen, G.; Yang, X.; Liu, W. Most C6 cells are cancer stem cells: Evidence from clonal and population analyses. *Cancer Res.* **2007**, *67*, 3691–3697.

57. Grezina, N. M.; Zinov'eva, N. A. *In vitro* microscopic determination of animal mammary gland stem cells. Russ. RU 2295565, 2007.

58. Lin, K. K.; Goodell, M. A. Purification of hematopoietic stem cells using the side population. *Methods Enzymol.* **2006**, *420*, 255–264.

59. Uchida, N.; Dykstra, B.; Lyons, K.; Leung, F.; Kristiansen, M.; Eaves, C. ABC transporter activities of murine hematopoietic stem cells vary according to their developmental and activation status. *Blood* **2004**, *103*, 4487–4495.

60. Uryvaeva, I. V.; Tsitrin, E. B.; Gorodetsky, S. I.; Tsvetkova, I. A.; Delone, G. V.; Gulyaev, D. V.; Khrushchov, N. G. The phenotypic characters of the stem cells in hepatocytes during liver regeneration: The expression of the Bcrp1/Abcg2 membrane transporter and the Hoechst 33342 export. *Dokl. Biol. Sci.* **2004**, *398*, 413–416.

61. Bertoncello, I.; Williams, B. Hematopoietic stem cell characterization by Hoechst 33342 and rhodamine 123 staining. *Methods Mol. Biol.* **2004**, *263*, 181–200.

62. Huttmann, A.; Liu, S. L.; Boyd, A. W.; Li, C. L. Functional heterogeneity within rhodamine123[lo]

Hoechst 33342[lo/sp] primitive hemopoietic stem cells revealed by pyronin Y. *Exp. Hematol.* **2001**, *29*, 1109–1116.

63. Giorgini, G.; Manaresi, N.; Medoro, G. Method for identification, selection and analysis of tumour cells. Ital. IT 1391619, 2012.

64. Adhikari, J. S.; Divya, K.; Arya, M. B.; Dwarakanath, B. S. Heterogeneity in the radiosensitizing effects of the DNA ligand Hoechst-33342 in human tumor cell lines. *J. Cancer Res. Ther.* **2005**, *1*, 151–161.

65. Loeffler, D. A.; KuKuruga, M. A.; Juneau, P. L.; Heppner, G. H. Analysis of distribution of tumor- and preneoplasia-infiltrating lymphocytes using simultaneous Hoechst 33342 labeling and immunophenotyping. *Cytometry* **1992**, *13*, 169–174.

66. Trotter, M. J.; Olive, P. L.; Chaplin, D. J. Effect of vascular marker Hoechst 33342 on tumor perfusion and cardiovascular function in the mouse. *Br. J. Cancer* **1990**, *62*, 903–908.

67. Young, S. D.; Hill, R. P. Radiation sensitivity of tumor cells stained *in vitro* or *in vivo* with the bisbenzimide fluorochrome Hoechst 33342. *Br. J. Cancer* **1989**, *60*, 715–721.

68. Trotter, M. J.; Chaplin, D. J.; Durand, R. E.; Olive, P. L. The use of fluorescent probes to identify regions of transient perfusion in murine tumors. *Int. J. Radiat. Oncol., Biol., Phys.* **1989**, *16*, 931–934.

69. Smith, K. A.; Hill, S. A.; Begg, A. C.; Denekamp, J. Validation of the fluorescent dye Hoechst 33342 as a vascular space marker in tumors. *Br. J. Cancer* **1988**, *57*, 247–253.

70. Olive, P. L.; Chaplin, D. J.; Durand, R. E. Pharmacokinetics, binding and distribution of Hoechst 33342 in spheroids and murine tumors. *Br. J. Cancer* **1985**, *52*, 739–746.

71. Smith, P. J.; Nakeff, A.; Watson, J. V. Flow-cytometric detection of changes in the fluorescence emission spectrum of a vital DNA-specific dye in human tumor cells. *Exp. Cell Res.* **1985**, *159*, 37–46.

72. Neyfakh, A. A. Use of fluorescent dyes as molecular probes for the study of multidrug resistance. *Exp. Cell Res.* **1988**, *174*, 168–176.

73. Giordano, G.; Costa, L. G. Measurements of neuronal apoptosis. *Methods Mol. Biol.* **2011**, *758*, 179–193.

74. Boivin, W. A.; Granville, D. J. Detection and quantification of apoptosis in the vasculature. *Methods Mol. Med.* **2007**, *139*, 181–195.

75. Koeppel, F.; Jaiswal, J. K.; Simon, S. M. Quantum dot-based sensor for improved detection of apoptotic cells. *Nanomedicine* **2007**, *2*, 71–78.

76. Wlodkowic, D.; Skommer, J.; Pelkonen, J. Towards an understanding of apoptosis detection by SYTO dyes. *Cytometry* **2007**, *71A*, 61–72.

77. Boffa, D. J.; Waka, J.; Thomas, D.; Suh, S.; Curran, K.; Sharma, V. K.; Besada, M.; Muthukumar, T.; Yang, H.; Suthanthiran, M. Measurement of apoptosis of intact human islets by confocal optical sectioning and stereologic analysis of YO-PRO-1-stained islets. *Transplantation* **2005**, *79*, 842–845.

78. Welsh, N. Assessment of apoptosis and necrosis in isolated islets of Langerhans: Methodological considerations. *Curr. Top. Biochem. Res.* **2000**, *3*, 189–200.

79. Bossy-Wetzel, E.; Green, D. R. Detection of apoptosis by annexin V labeling. *Methods Enzymol.* **2000**, *322*, 15–18.

80. Weber, G. F.; Daley, J.; Kraeft, S.; Chen, L. B.; Cantor, H. Measurement of apoptosis in heterogeneous cell populations. *Cytometry* **1997**, *27*, 136–144.

81. Reynolds, J. E.; Li, J.; Eastman, A. Detection of apoptosis by flow cytometry of cells simultaneously stained for intracellular pH (carboxy SNARF-1) and membrane permeability (Hoechst 33342). *Cytometry* **1996**, *25*, 349–357.

82. Frey, T. Nucleic acid dyes for detection of apoptosis in live cells. *Cytometry* **1995**, *21*, 265–274.

83. Reid, S.; Cross, R.; Snow, E. C. Combined Hoechst 33342 and merocyanine 540 staining to examine murine B cell cycle stage, viability and apoptosis. *J. Immunol. Methods* **1996**, *192*, 43–54.

84. Boltz, R. C.; Fischer, P. A.; Wicker, L. S.; Peterson, L. B. Single UV excitation of Hoechst 33342 and ethidium bromide for simultaneous cell cycle analysis and viability determinations on *in vitro* cultures of murine B lymphocytes. *Cytometry* **1994**, *15*, 28–34.

85. Ciancio, G.; Pollack, A.; Taupier, M. A.; Block, N. L.; Irvin, G. L., III, Measurement of cell-cycle phase-specific cell death using Hoechst 33342 and propidium iodide: preservation by ethanol fixation. *J. Histochem. Cytochem.* **1988**, *36*, 1147–1152.

86. Vollenweider, I.; Groscurth, P. Comparison of four DNA staining fluorescence dyes for measuring cell proliferation of lymphokine-activated killer (LAK) cells. *J. Immunol. Methods* **1992**, *149*, 133–135.

87. Deutsch, M.; Namer, Y. A method and system for measuring membrane potential based on fluorescence polarization. PCT Int. Appl. WO 2007122602, 2007.

88. Chen, J. C.; Zhang, X.; Singleton, T. P.; Kiechle, F. L. Mitochondrial membrane potential change induced by Hoechst 33342 in myelogenous leukemia cell line HL-60. *Ann. Clin. Lab. Sci.* **2004**, *34*, 458–466.

89. Farinas, J. A.; Wada, H. G. Use of Nernstein voltage sensitive dyes in measuring transmembrane voltage. PCT Int. Appl. WO 2001027253, 2001.

90. Ichim, T. E. Treatment of disc degenerative disease and compositions for same. PCT Int. Appl. WO 2007136673, 2007.

91. Ichim, T. E. Treatment of erectile dysfunction by stem cell therapy. PCT Int. Appl. WO 2007149548, 2007.

92. Garza, C. M.; Cho, S. Metrology of bilayer photoresist processes. U.S. Pat. Appl. Publ. US 20090220895, 2009.

93. Heil, J.; Reifferscheid, G. Detection of mammalian carcinogens with an immunological DNA synthesis-inhibition test. *Carcinogenesis* **1992**, *13*, 2389–2394.

94. Jiang, B.; Zhang, L.; Wang, Y.; Li, M.; Wu, W.; Guan, S.; Liu, X.; Yang, M.; Wang, J.; Guo, D. Tanshinone IIA sodium sulfonate protects against cardiotoxicity induced by doxorubicin *in vitro* and *in vivo*. *Food Chem. Toxicol.* **2009**, *47*, 1538–1544.

95. Wiezorek, C. Cell cycle dependence of Hoechst 33342 dye cytotoxicity on sorted living cells. *Histochemistry* **1984**, *81*, 493–495.

96. Jeong, J. B.; Seo, E. W.; Jeong, H. J. Effect of extracts from pine needle against oxidative DNA damage and apoptosis induced by hydroxyl radical via antioxidant activity. *Food Chem. Toxicol.* **2009**, *47*, 2135–2141.

97. Zhao, H.; Traganos, F.; Dobrucki, J.; Wlodkowic, D.; Darzynkiewicz, Z. Induction of DNA damage response by the supravital probes of nucleic acids. *Cytometry Part A* **2009**, *75*, 510–519.

98. Dinant, C.; de Jager, M.; Essers, J.; van Cappellen, W. A.; Kanaar, R.; Houtsmuller, A. B.; Vermeulen, W. Activation of multiple DNA repair pathways by subnuclear damage induction methods. *J. Cell Sci.* **2007**, *120*, 2731–2740.

99. Erba, E.; Ubezio, P.; Broggini, M.; Ponti, M.; D'Incalci, M. DNA damage, cytotoxic effect and cell-cycle perturbation of Hoechst 33342 on L1210 cells *in vitro*. *Cytometry* **1988**, *9*, 1–6.

100. Durand, R. E.; Olive, P. L. Cytotoxicity, mutagenicity and DNA damage by Hoechst 33342. *J. Histochem. Cytochem.* **1982**, *30*, 111–116.

101. Argentin, G.; Cicchetti, R. Genotoxic and antiapoptotic effect of nicotine on human gingival fibroblasts. *Toxicol. Sci.* **2004**, *79*, 75–81.

102. He, L.; Jurs, P. C.; Custer, L. L.; Durham, S. K.; Pearl, G. M. Predicting the genotoxicity of polycyclic aromatic compounds from molecular structure with different classifiers. *Chem. Res. Toxicol.* **2003**, *16*, 1567–1580.

103. Mueller, S. O.; Stopper, H. Characterization of the genotoxicity of anthraquinones in mammalian cells. *Biochim. Biophys. Acta* **1999**, *1428*, 406–414.

104. Downing, T. W.; Garner, D. L.; Ericsson, S. A.; Redelman, D. Metabolic toxicity of fluorescent stains on thawed cryopreserved bovine sperm cells. *J. Histochem. Cytochem.* **1991**, *39*, 485–489.

105. Pin, S.; Chen, H.; Lein, P. J.; Wang, M. M. Nucleic acid binding agents exert local toxic effects on neurites via a non-nuclear mechanism. *J. Neurochem.* **2006**, *96*, 1253–1266.

106. Shimazawa, M.; Yamashima, T.; Agarwal, N.; Hara, H. Neuroprotective effects of minocycline against *in vitro* and *in vivo* retinal ganglion cell damage. *Brain Res.* **2005**, *1053*, 185–194.

107. Celsi, F.; Ferri, A.; Casciati, A.; D'Ambrosi, N.; Rotilio, G.; Costa, A.; Volonte, C.; Carri, M. T. Overexpression of superoxide dismutase 1 protects against beta-amyloid peptide toxicity: Effect of estrogen and copper chelators. *Neurochem. Int.* **2004**, *44*, 25–33.

108. Purschke, M.; Rubio, N.; Held, K. D.; Redmond, R. W. Phototoxicity of Hoechst 33342 in time-lapse fluorescence microscopy. *Photochem. Photobiol. Sci.* **2010**, *9*, 1634–1639.

109. Kessel, D.; Luo, Y. Mitochondrial photodamage and PDT-induced apoptosis. *J. Photochem. Photobiol., B: Biol.* **1998**, *42*, 89–95.

110. Garner, D. L. Hoechst 33342: the dye that enabled differentiation of living X-and Y-chromosome bearing mammalian sperm. *Theriogenology* **2009**, *71*, 11–21.

111. Velilla, E.; Lopez-Bejar, M.; Rodriguez-Gonzalez, E.; Vidal, F.; Paramio, M. Effect of Hoechst 33342 staining on developmental competence of prepubertal goat oocytes. *Zygote* **2002**, *10*, 201–208.

112. Watkins, A. M.; Chan, P. J.; Kalugdan, T. H.; Patton, W. C.; Jacobson, J. D.; King, A. Analysis of the flow cytometer stain Hoechst 33342 on human spermatozoa. *Mol. Hum. Reprod.* **1996**, *2*, 709–712.

HOECHST 34580

CAS Registry Number   23555-00-2

Chemical Structure

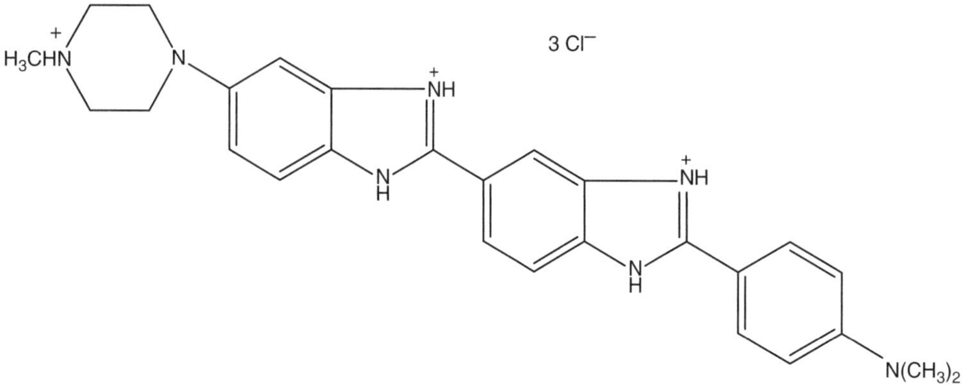

Boiling Point (Calcd.)   733.2 ± 70.0 °C, pressure: 760 Torr

pK$_a$ (Calcd.)   10.97±0.10, most acidic, temperature: 25 °C; 7.66 ± 0.42, most basic, temperature: 25 °C

Absorption (λ$_{max}$)   392 nm (H$_2$O/DNA)

CA Index Name   Benzenamine, *N,N*-dimethyl-4-[5-(4-methyl-1-piperazinyl)[2,5′-bi-1*H*-benzimidazol]-2′-yl]-, hydrochloride (1:3)

Other Names   2,5′-Bibenzimidazole, 2′-[*p*-(dimethylamino)-phenyl]-5-(4-methyl-1-piperazinyl)-, trihydrochloride; HOE 34580; Hoechst 34580; Proamine

Merck Index Number   Not listed

Chemical/Dye Class   Heterocycle; Benzimidazole

Molecular Formula   C$_{27}$H$_{32}$Cl$_3$N$_7$

Molecular Weight   560.96

Physical Form   Light brown powder;[5] Yellow powder or solid

Solubility   Soluble in dimethyl sulfoxide

Melting Point   225–226 °C (decompose);[2–5] 210 °C[6,7]

Emission (λ$_{max}$)   440 nm (H$_2$O/DNA)

Molar Extinction Coefficient   47,000 cm^{-1} M^{-1} (H$_2$O/DNA)

Synthesis   Synthetic methods[1–7]

Imaging/Labeling Applications   Nucleic acids;[1–13] cells;[13] nuclei;[14,15] microorganisms;[16] oligonucleotides;[17] polynucleotides;[18] sperms[13]

Biological/Medical Applications   Detecting nucleic acids,[1–13] cells,[13] microorganisms;[16] analyzing polynucleotides;[18] nucleic acid amplification;[17] nucleic acid sequencing;[19] radioprotectors;[2–4] as temperature sensor[20,21]

Industrial Applications   Not reported

Safety/Toxicity   No data available

REFERENCES

1. Sabnis, R. W. *Handbook of Biological Dyes and Stains*; John Wiley & Sons Inc.: Hoboken, **2010**; pp 237–238.

2. Martin, R. F.; Kelly, D. P.; White, J. M. Radioprotectors. U.S. Patent 6548505, 2003.

3. Martin, R. F.; Kelly, D. P.; White, J. M. Radioprotectors. U.S. Patent 6194414, 2001.

4. Martin, R. F.; Kelly, D. P.; White, J. M. Radioprotectors. PCT Int. Appl. WO 9704776, 1997.

5. Kelly, D. P.; Bateman, S. A.; Hook, R. J.; Martin, R. F.; Reum, M. E.; Rose, M.; Whittaker, A. R. D. DNA binding compounds. VI. Synthesis and characterization of 2,5′-disubstituted bibenzimidazoles related to the DNA minor groove binder Hoechst 33258. *Aust. J. Chem.* **1994**, *47*, 1751–1769.

6. Loewe, H.; Urbanietz, J. Basic substituted 2,6-bisben-zimidazole derivatives, a novel series of substances with chemotherapeutic activity. *Arzneim.-Forsch.* **1974**, *24*, 1927–1933.

7. Farbwerke Hoechst AG. Piperazino bisbenzimidazoles. Fr. Demande FR 1519964, 1968.

8. Kjaerulff, S.; Glensbjerg, M. Method for analysis of cellular DNA content. PCT Int. Appl. WO 2011098085, 2011.

9. Evans, D. A.; Neidle, S. Virtual screening of DNA minor groove binders. *J. Med. Chem.* **2006**, *49*, 4232–4238.

10. Martin, R. F.; Broadhurst, S.; D'Abrew, S.; Budd, R.; Sephton, R.; Reum, M.; Kelly, D. P. Radioprotection by DNA ligands. *Br. J. Cancer, Suppl.* **1996**, *74*, S99–S101.

11. Kogo, K.; Kubota, Y. The interaction of DNA with Hoechst 33258 analogs. *Nucleic Acids Symp. Ser.* **1994**, *31*, 81–82.

12. Latt, S. A.; Stetten, G. Spectral studies on 33258 Hoechst and related bisbenzimidazole dyes useful for fluorescent detection of deoxyribonucleic acid synthesis. *J. Histochem. Cytochem.* **1976**, *24*, 24–33.

13. Anderson, A. L.; Knutson, C. R.; Mueth, D.; Plewa, J.; Tanner, E. Methods for staining cells for identification and sorting. U.S. Pat. Appl. Publ. US 2006172315, 2006.

14. Gauer, C.; Mann, W.; Alunni-Fabbroni, M. Method for carrying out an enzymic reactions. Ger. DE 102006056694, 2010.

15. Gauer, C.; Mann, W.; Alunni-Fabbroni, M. Method for carrying out an enzymic reaction. PCT Int. Appl. WO 2008064730, 2008.

16. Horikiri, S. Microorganism cell detection method using fluorescent indicator. Jpn. Kokai Tokkyo Koho JP 2006262775, 2006.

17. Park, H. O.; Kim, H. B.; Chi, S. M. Detection method of DNA amplification using probes labeled with intercalating dye. PCT Int. Appl. WO 2006004267, 2006.

18. Hyldig-Nielsen, J. J.; Fiandaca, M. J.; Coull, J. M. Compositions and methods for multiplex analysis of polynucleotides. PCT Int. Appl. WO 2004074447, 2004.

19. Hoser, M. J. Nucleic acid sequencing methods, kits and reagents. PCT Int. Appl. WO 2004074503, 2004.

20. Bousseksou, A.; Salmon, L.; Molnar, G.; Cobo, S. Materials with thermochromic spin transition doped with one or more fluorescent agents for use as temperature sensor. Fr. Demande FR 2952371, 2011.

21. Bousseksou, A.; Salmon, L.; Molnar, G.; Cobo, S. Heat-sensitive spin-transition materials doped with one or more fluorescent agents for use as temperature sensor. PCT Int. Appl. WO 2011058277, 2011.

7-HYDROXYCOUMARIN-3-CARBOXYLIC ACID (UMBELLIFERONE-3-CARBOXYLIC ACID)

CAS Registry Number   779-27-1

Chemical Structure

CA Index Name   2*H*-1-Benzopyran-3-carboxylic acid, 7-hydroxy-2-oxo-

Other Names   3-Carboxy-7-hydroxycoumarin; 3-Carboxyumbelliferone; 7-Hydroxycoumarin-3-carboxylic acid; NSC 115545; Umbelliferone-3-carboxylic acid; HCC; 7-OHCCA; 7-OH-C-3-COOH

Merck Index Number   Not listed

Chemical/Dye Class   Coumarin

Molecular Formula $C_{10}H_6O_5$

Molecular Weight   206.15

Physical Form   Yellow solid;[1,16,17] Orange solid;[14] Off-white powder[2]

Solubility   Soluble in *N,N*-dimethylformamide, dimethyl sulfoxide, methanol

Melting Point   283–284 °C;[16] 271 °C (decompose);[27] 264–265 °C;[26]   263.4–265.3 °C;[1]   262–266 °C;[14] 262–263 °C;[8,24] 262 °C (decompose);[25] 261–263 °C;[20] 260–262 °C;[3] 259–262 °C;[7] 259–260 °C;[4] 258–259 °C;[13] 248–250 °C (decompose)[19]

Boiling Point (Calcd.)   474.8 ± 45.0 °C, pressure: 760 Torr

pK_a   7.04 ± 0.02, temperature: 23 °C

pK_a (Calcd.)   1.98 ± 0.20, most acidic, temperature: 25 °C

Absorption (λ_{max})   386 nm (Buffer pH 10.0); 339 nm (Buffer pH 4.0)

Emission (λ_{max})   448 nm (Buffer pH 10.0, 4.0)

Molar Extinction Coefficient   29,000 $cm^{-1} M^{-1}$ (Buffer pH 10.0); 19,000 $cm^{-1} M^{-1}$ (Buffer pH 4.0)

Synthesis   Synthetic methods[1–27]

Imaging/Labeling Applications   Amines;[28,29] antibodies;[30] bacteria;[31,32] hairs/keratinous fibers;[33,34] hydroxyl radicals;[10,12,15,35–44] nucleic acids;[44,45] palladium;[46] photoactive yellow protein[47–55]

Biological/Medical Applications   Chemical dosimeter for radiation therapy;[56] detecting bacteria;[31,32] detecting/sensing hydroxyl radicals;[10,12,15,35–44] studying nucleic acids;[44,45] as a substrate for measuring carboxylesterases activity,[57] cellulases activity,[58] β-galactosidases activity,[21,59,60] glycosidases activity,[61] glycosyltransferases activity,[62–64] hydrolases activity,[65] 4-hydroxycinnamate decarboxylases activity,[66] β-lactamases activity,[9,67] phosphatases activity,[57] sulfatases activity;[57] treating cancer/antitumor activity[68–70]

Industrial Applications   Carbon-dioxide sensor;[71,72] chemical sensors based on non-linear optics;[73] monitoring of cationic photopolymerization processes;[74] nanomaterials[75]

Safety/Toxicity   No data available

REFERENCES

1. Liu, J.; Wu, F.; Chen, L.; Zhao, L.; Zhao, Z.; Wang, M.; Lei, S. Biological evaluation of coumarin derivatives as mushroom tyrosinase inhibitors. *Food Chem.* **2012**, *135*, 2872–2878.

2. Gnaccarini, C.; Ben-Tahar, W.; Mulani, A.; Roy, I.; Lubell, W. D.; Pelletier, J. N.; Keillor, J. W. Site-specific protein propargylation using tissue transglutaminase. *Org. Biomol. Chem.* **2012**, *10*, 5258–5265.

3. Bardajee, G. R.; Moallem, S. A. Synthesis of 3-carboxycoumarins catalyzed by $CuSO_4.5H_2O$ under ultrasound irradiation in aqueous media. *Asian J. Biochem. Pharm. Res.* **2012**, *2*, 410–414.

4. Karami, B.; Farahi, M.; Khodabakhshi, S. Rapid synthesis of novel and known coumarin-3-carboxylic acids using stannous chloride dihydrate under solvent-free conditions. *Helv. Chim. Acta* **2012**, *95*, 455–460.

5. Shahinian, E. G. H.; Sebe, I. Synthesis of some new benzocoumarin heterocyclic fluorescent dyes. *Rev. Chim.* **2011**, *62*, 1098–1101.

6. Davison, H. R.; Taylor, S.; Drake, C.; Phuan, P.; Derichs, N.; Yao, C.; Jones, E. F.; Sutcliffe, J.; Verkman, A. S.; Kurth, M. J. Functional fluorescently labeled bithiazole ΔF508-CFTR corrector imaged in whole body slices in mice. *Bioconjugate Chem.* **2011**, *22*, 2593–2599.

7. Harishkumar, H. N.; Mahadevan, K. M.; Kiran Kumar, H. C.; Satyanarayan, N. D. A facile, choline chloride/urea catalyzed solid phase synthesis of coumarins via Knoevenagel condensation. *Org. Commun.* **2011**, *4*, 26–32.

8. Bardajee, G. R.; Jafarpour, F.; Afsari, H. S. ZrOCl₂.8H₂O: an efficient catalyst for rapid one-pot synthesis of 3-carboxycoumarins under ultrasound irradiation in water. *Cent. Eur. J. Chem.* **2010**, *8*, 370–374.

9. Mizukami, S.; Watanabe, S.; Hori, Y.; Kikuchi, K. Covalent protein labeling based on noncatalytic β-lactamase and a designed FRET substrate. *J. Am. Chem. Soc.* **2009**, *131*, 5016–5017.

10. Singh, A.; Yang, Y.; Adelstein, S. J.; Kassis, A. I. Synthesis and application of molecular probe for detection of hydroxyl radicals produced by Na125I and γ-rays in aqueous solution. *Int. J. Radiat. Biol.* **2008**, *84*, 1001–1010.

11. Darvatkar, N. B.; Deorukhkar, A. R.; Bhilare, S. V.; Raut, D. G.; Salunkhe, M. M. Ionic liquid-mediated synthesis of coumarin-3-carboxylic acids via Knoevenagel condensation of Meldrum's acid with ortho-hydroxyaryl aldehydes. *Synth. Commun.* **2008**, *38*, 3508–3513.

12. Singh, A.; Chen, K.; Adelstein, S. J.; Kassis, A. I. Synthesis of coumarin-polyamine-based molecular probe for the detection of hydroxyl radicals generated by gamma radiation. *Radiat. Res.* **2007**, *168*, 233–242.

13. Du, J.; Li, L.; Zhang, D. Ultrasound promoted synthesis of 3-carboxycoumarins in aqueous media. *E-J. Chem.* **2006**, *3*, 1–4.

14. Creaven, B. S.; Egan, D. A.; Kavanagh, K.; McCann, M.; Noble, A.; Thati, B.; Walsh, M. Synthesis, characterization and antimicrobial activity of a series of substituted coumarin-3-carboxylato silver(I) complexes. *Inorg. Chim. Acta* **2006**, *359*, 3976–3984.

15. Newton, G. L.; Milligan, J. R. Fluorescence detection of hydroxyl radicals. *Radiat. Phys. Chem.* **2006**, *75*, 473–478.

16. Alvim, J. Jr.,; Dias, R. L. A.; Castilho, M. S.; Oliva, G.; Correa, A. G. Preparation and evaluation of a coumarin library towards the inhibitory activity of the enzyme gGAPDH from *Trypanosoma cruzi*. *J. Braz. Chem. Soc.* **2005**, *16*, 763–773.

17. Zhou, M.; Ghosh, I. Noncovalent multivalent assembly of jun peptides on a leucine zipper dendrimer displaying fos peptides. *Org. Lett.* **2004**, *6*, 3561–3564.

18. Fringuelli, F.; Piermatti, O.; Pizzo, F. One-pot synthesis of 7-hydroxy-3-carboxycoumarin in water. *J. Chem. Educ.* **2004**, *81*, 874–876.

19. Fringuelli, F.; Piermatti, O.; Pizzo, F. One-pot synthesis of 3-carboxycoumarins via consecutive Knoevenagel and Pinner reactions in water. *Synthesis* **2003**, 2331–2334.

20. Song, A.; Wang, X.; Lam, K. S. A convenient synthesis of coumarin-3-carboxylic acids via Knoevenagel condensation of Meldrum's acid with ortho-hydroxyaryl aldehydes or ketones. *Tetrahedron Lett.* **2003**, *44*, 1755–1758.

21. Chilvers, K. F.; Perry, J. D.; James, A. L.; Reed, R. H. Synthesis and evaluation of novel fluorogenic substrates for the detection of bacterial β-galactosidase. *J. Appl. Microbiol.* **2001**, *91*, 1118–1130.

22. Bandgar, B. P.; Uppalla, L. S.; Kurule, D. S. Solvent-free one-pot rapid synthesis of 3-carboxycoumarins. *Green Chem.* **1999**, *1*, 243–245.

23. Watson, B. T.; Christiansen, G. E. Solid phase synthesis of substituted coumarin-3-carboxylic acids via the Knoevenagel condensation. *Tetrahedron Lett.* **1998**, *39*, 6087–6090.

24. Shirokova, E. A.; Segal, G. M.; Torgov, I. V. Use of Meldrum's acid in the syntheses of low molecular weight bioregulators. 3. Synthesis of coumarin-3-carboxylic acids and their derivatives. *Bioorg. Khim.* **1988**, *14*, 236–242.

25. Lokhande, S. B.; Rangnekar, D. W. Synthesis of 3-substituted 7-hydroxy-8-arylazocoumarins. *Indian J. Chem., Sect. B* **1986**, *25B*, 638–639.

26. Woods, L. L.; Sapp, J. Coumarin-3-carboxylic acids. *J. Org. Chem.* **1965**, *30*, 312–313.

27. Shah, D. N.; Shah, N. M. The Fries isomerization of acetyl and benzoyl esters of umbelliferones. *J. Org. Chem.* **1954**, *19*, 1681–1685.

28. Stewart, J. T.; Lotti, D. M. Fluorometric determination of amphetamines with 3-carboxy-7-hydroxycoumarin. *J. Pharm. Sci.* **1971**, *60*, 461–463.

29. Stewart, J. T.; Lotti, D. M. Fluorimetric determination of some aliphatic and cyclic amines with 3-carboxy-7-hydroxycoumarin. *Anal. Chim. Acta* **1970**, *52*, 390–393.

30. Sakai, Y.; Shimada, K.; Ooe, K.; Yano, H. Endoscope containing immobilized antibody for cancer diagnosis. Jpn. Kokai Tokkyo Koho JP 06027110, 1994.

31. Nijak, G. M., Jr.,; Geary, J. R.; Sawtelle, M. B. A biochemical sensor for quantitative simultaneous multi-species bacteria detection *in situ*. PCT Int. Appl. WO 2013134689, 2013.

32. Jadamec, J. R.; Bauman, R.; Anderson, C. P.; Jakubielski, S. A.; Sutton, N. D.; Kovacs, M. J. Method for detecting bacteria in a sample. U.S. Patent 5968762, 1999.

33. Morel, O.; Christie, R. M.; Greaves, A.; Morgan, K. M. Enhanced model for the diffusivity of a dye molecule into human hair fibre based on molecular modeling techniques. *Color. Technol.* **2008**, *124*, 301–309.

34. Nagase, S.; Okada, T.; Ueda, S.; Sato, N.; Itoh, T. Keratinous fiber treatment compositions. Eur. Pat. Appl. EP 0629395, 1994.

35. Soh, N.; Makihara, K.; Ariyoshi, T.; Seto, D.; Maki, T.; Nakajima, H.; Nakano, K.; Imato, T. Phospholipid-linked coumarin: a fluorescent probe for sensing hydroxyl radicals in lipid membranes. *Anal. Sci.* **2008**, *24*, 293–296.

36. Iakovlev, A.; Broberg, A.; Stenlid, J. Fungal modification of the hydroxyl radical detector coumarin-3-carboxylic acid. *FEMS Microbiol. Ecol.* **2003**, *46*, 197–202.

37. King, M.; Kopelman, R. Development of a hydroxyl radical ratiometric nanoprobe. *Sens. Actuators, B* **2003**, *B90*, 76–81.

38. Tornberg, K.; Olsson, S. Detection of hydroxyl radicals produced by wood-decomposing fungi. *FEMS Microbiol. Ecol.* **2002**, *40*, 13–20.

39. Lindqvist, C.; Nordstrom, T. Generation of hydroxyl radicals by the antiviral compound phosphonoformic acid (foscarnet). *Pharmacol. Toxicol.* **2001**, *89*, 49–55.

40. Ali, M. A.; Yasui, F.; Matsugo, S.; Konishi, T. The lactate-dependent enhancement of hydroxyl radical generation by the Fenton reaction. *Free Radical Res.* **2000**, *32*, 429–438.

41. Manevich, Y.; Held, K. D.; Biaglow, J. E. Coumarin-3-carboxylic acid as a detector for hydroxyl radicals generated chemically and by gamma radiation. *Radiat. Res.* **1997**, *148*, 580–591.

42. Kachur, A. V.; Manevich, Y.; Biaglow, J. E. Effect of purine nucleoside phosphates on OH-radical generation by reaction of Fe2+ with oxygen. *Free Radical Res.* **1997**, *26*, 399–408.

43. Kachur, A. V.; Held, K. D.; Koch, C. J.; Biaglow, J. E. Mechanism of production of hydroxyl radicals in the copper-catalyzed oxidation of dithiothreitol. *Radiat. Res.* **1997**, *147*, 409–415.

44. Tang, V. J.; Konigsfeld, K. M.; Aguilera, J. A.; Milligan, J. R. DNA binding hydroxyl radical probes. *Radiat. Phys. Chem.* **2012**, *81*, 46–51.

45. Mitsui, T.; Nakano, H.; Yamana, K. Coumarin-fluorescein pair as a new donor-acceptor set for fluorescence energy transfer study of DNA. *Tetrahedron Lett.* **2000**, *41*, 2605–2608.

46. Garner, A. L.; Song, F.; Koide, K. Enhancement of a catalysis-based fluorometric detection method for palladium through rational fine-tuning of the palladium species. *J. Am. Chem. Soc.* **2009**, *131*, 5163–5171.

47. Kubota, K.; Shingae, T.; Foster, N. D.; Kumauchi, M.; Hoff, W. D.; Unno, M. Active site structure of photoactive yellow protein with a locked chromophore analogue revealed by near-infrared Raman optical activity. *J. Phys. Chem. Lett.* **2013**, *4*, 3031–3038.

48. van der Horst, M. A.; Stalcup, T. P.; Kaledhonkar, S.; Kumauchi, M.; Hara, M.; Xie, A.; Hellingwerf, K. J.; Hoff, Wouter D. Locked chromophore analogs reveal that photoactive yellow protein regulates biofilm formation in the deep sea bacterium *Idiomarina ioihiensis*. *J. Am. Chem. Soc.* **2009**, *131*, 17443–17451.

49. Vengris, M.; Larsen, D. S.; Van der Horst, M. A.; Larsen, O. F. A.; Hellingwerf, K. J.; Van Grondelle, R. Ultrafast dynamics of isolated model photoactive yellow protein chromophores: "Chemical perturbation theory" in the laboratory. *J. Phys. Chem. B* **2005**, *109*, 4197–4208.

50. Vengris, M.; van der Horst, M. A.; Zgrablic, G.; van Stokkum, I. H. M.; Haacke, S.; Chergui, M.; Hellingwerf, K. J.; van Grondelle, R.; Larsen, D. S. Contrasting the excited-state dynamics of the photoactive yellow protein chromophore: Protein versus solvent environments. *Biophys. J.* **2004**, *87*, 1848–1857.

51. Mataga, N.; Chosrowjan, H.; Taniguchi, S.; Hamada, N.; Tokunaga, F.; Imamoto, Y.; Kataoka, M. Ultrafast photoreactions in protein nanospaces as revealed by fs fluorescence dynamics measurements on photoactive yellow protein and related systems. *Phys. Chem. Chem. Phys.* **2003**, *5*, 2454–2460.

52. Chosrowjan, H.; Taniguchi, S.; Mataga, N.; Hamada, N.; Tokunaga, F.; Imamoto, Y.; Kataoka, M. Coherent oscillations in photoisomerization reaction dynamics of photoactive yellow protein (PYP) and related systems. *Trends Opt. Photonics* **2002**, *72*, 380–381.

53. Larsen, D. S.; van Stokkum, I.; Vengris, M.; van Grondelle, R.; van der Horst, M.; Hellingwerf, K. J. Polarized transient absorption studies of

the initial isomerization dynamics of photoactive yellow protein, its E46Q derivative, and model chromophore compounds. *Trends Opt. Photonics* **2002**, *72*, 304–305.

54. Cordfunke, R.; Kort, R.; Pierik, A.; Gobets, B.; Koomen, G.; Verhoeven, J. W.; Hellingwerf, K. J. trans / cis (Z / E) Photoisomerization of the chromophore of photoactive yellow protein is not a prerequisite for the initiation of the photocycle of this photoreceptor protein. *Proc. Natl. Acad. Sci. U.S.A.* **1998**, *95*, 7396–7401.

55. Changenet, P.; Zhang, H.; van der Meer, M. J.; Hellingwerf, K. J.; Glasbeek, M. Subpicosecond fluorescence upconversion measurements of primary events in yellow proteins. *Chem. Phys. Lett.* **1998**, *282*, 276–282.

56. Collins, A. K.; Makrigiorgos, G. M.; Svensson, G. K. Coumarin chemical dosimeter for radiation therapy. *Med. Phys.* **1994**, *21*, 1741–1747.

57. Koller, E.; Wolfbeis, O. S. Syntheses and spectral properties of longwave absorbing and fluorescing substrates for the direct and continuous kinetic assay of carboxylesterases, phosphatases, and sulfatases. *Monatsh. Chem.* **1985**, *116*, 65–75.

58. Ostafe, R.; Prodanovic, R.; Commandeur, U.; Fischer, R. Flow cytometry-based ultra-high-throughput screening assay for cellulase activity. *Anal. Biochem.* **2013**, *435*, 93–98.

59. Komatsu, T.; Kikuchi, K.; Takakusa, H.; Hanaoka, K.; Ueno, T.; Kamiya, M.; Urano, Y.; Nagano, T. Design and synthesis of an enzyme activity-based labeling molecule with fluorescence spectral change. *J. Am. Chem. Soc.* **2006**, *128*, 15946–15947.

60. Sparks, A. L. Membrane transportable fluorescent substrates. PCT Int. Appl. WO 2003025192, 2003.

61. Perry, J. D.; James, A. L.; Morris, K. A.; Oliver, M.; Chilvers, K. F.; Reed, R. H.; Gould, F. K. Evaluation of novel fluorogenic substrates for the detection of glycosidases in *Escherichia coli* and enterococci. *J. Appl. Microbiol.* **2006**, *101*, 977–985.

62. Williams, G. J.; Zhang, C.; Thorson, J. S. Expanding the promiscuity of a natural-product glycosyltransferase by directed evolution. *Nat. Chem. Biol.* **2007**, *3*, 657–662.

63. Yang, M.; Proctor, M. R.; Bolam, D. N.; Errey, J. C.; Field, R. A.; Gilbert, H. J.; Davis, B. G. Probing the breadth of macrolide glycosyltransferases: *In vitro* remodeling of a polyketide antibiotic creates active bacterial uptake and enhances potency. *J. Am. Chem. Soc.* **2005**, *127*, 9336–9337.

64. Higai, K.; Masuda, D.; Matsuzawa, Y.; Satoh, T.; Matsumoto, K. A fluorometric assay for

glycosyltransferase activities using sugars aminated and tagged with 7-hydroxycoumarin-3-carboxylic acid as substrates and high performance liquid chromatography. *Biol. Pharm. Bull.* **1999**, *22*, 333–338.

65. Despotovic, D.; Vojcic, L.; Prodanovic, R.; Martinez, R.; Maurer, K.; Schwaneberg, U. Fluorescent assay for directed evolution of perhydrolases. *J. Biomol. Screen.* **2012**, *17*, 796–805.

66. Hashidoko, Y.; Tanaka, T.; Tahara, S. Induction of 4-hydroxycinnamate decarboxylase in *Klebsiella oxytoca* cells exposed to substrates and non-substrate 4-hydroxycinnamate analogs. *Biosci., Biotechnol., Biochem.* **2001**, *65*, 2604–2612.

67. Minakami, S.; Kikuchi, K.; Watanabe, S.; Yamamoto, H. Novel fluorescent indicator labeling method for recombinant proteins by using β-lactamase fusion protein partner. Jpn. Kokai Tokkyo Koho JP 2010119382, 2010.

68. Yin, H.; Saludes, J. P.; Morton, L. A. Compositions, methods and uses for peptides in diagnosis, progression and treatment of cancers. PCT Int. Appl. WO 2013033459, 2013.

69. Zhang, J.; Wang, X.; Li, H.; Xu, G. Preparation of peptide containing coumarin derivatives for treatment of cancer. Faming Zhuanli Shenqing CN 102898502, 2013.

70. Thati, B.; Noble, A.; Creaven, B. S.; Walsh, M.; McCann, M.; Kavanagh, K.; Devereux, M.; Egan, D. A. *In vitro* antitumor and cyto-selective effects of coumarin-3-carboxylic acid and three of its hydroxylated derivatives, along with their silver-based complexes, using human epithelial carcinoma cell lines. *Cancer Lett.* **2007**, *248*, 321–331.

71. Leiner, M. J.; Tusa, J.; Klimant, I. Dry optical-chemical carbon-dioxide sensor. U.S. Pat. Appl. Publ. US 20090004751, 2009.

72. Mills, A.; Chang, Q. Carbon dioxide detector. PCT Int. Appl. WO 9314399, 1993.

73. Draxler, S.; Pflanzl, I.; Xiang, Z.; Lippitsch, M. E. Chemical sensors based on non-linear optics. *Sens. Actuators, B* **1993**, *B11*, 129–131.

74. Ortyl, J.; Popielarz, R. The performance of 7-hydroxycoumarin-3-carbonitrile and 7-hydroxycoumarin-3-carboxylic acid as fluorescent probes for monitoring of cationic photopolymerization processes by FPT. *J. Appl. Polym. Sci.* **2013**, *128*, 1974–1978.

75. Guo, T. Chemical enhancement by nanomaterials under x-ray irradiation. PCT Int. Appl. WO 2013106219, 2013.

7-HYDROXYCOUMARIN-3-CARBOXYLIC ACID SUCCINIMIDYL ESTER

CAS Registry Number   134471-24-2

Chemical Structure

CA Index Name   2*H*-1-Benzopyran-3-carboxylic acid, 7-hydroxy-2-oxo-, 2,5-dioxo-1-pyrrolidinyl ester

Other Names   2,5-Pyrrolidinedione, 1-[[(7-hydroxy-2-oxo-2*H*-1-benzopyran-3-yl)carbonyl]oxy]-; 7-Hydroxycoumarin-3-carboxylic acid *N*-succinimidyl ester; 7-Hydroxycoumarin-3-carboxylic acid succinimidyl ester; HCC-NHS; HCCS

Merck Index Number   Not listed

Chemical/Dye Class   Coumarin

Molecular Formula $C_{14}H_9NO_7$

Molecular Weight   303.22

Physical Form   Yellow solid[3]

Solubility   Soluble in acetonitrile, *N,N*-dimethylformamide, dimethyl sulfoxide, methanol

Boiling Point (Calcd.)   558.9 ± 60.0 °C, pressure: 760 Torr

pK$_a$ (Calcd.)   7.09 ± 0.20, most acidic, temperature: 25 °C

Absorption (λ_{max})   419 nm (MeOH)

Emission (λ_{max})   447 nm (MeOH)

Molar Extinction Coefficient   36,000 cm^{-1} M^{-1} (MeOH)

Synthesis   Synthetic methods[1–4]

Imaging/Labeling Applications   Analytes;[5] cells;[3] NPTX-594 toxin analogs;[6–8] nucleic acids;[9] peptides;[10] proteins[1,11–14]

Biological/Medical Applications   Analyzing peptides;[10] detecting/measuring/sensing hydroxyl radicals;[2,4,15–21] identifying genome;[9] quantifying analytes;[5] as a substrate for measuring glutathione and/or glutathione transferase activity,[22] glycosyltransferases activity;[23] viscosity sensors[24]

Industrial Applications   Not reported

Safety/Toxicity   No data available

REFERENCES

1. Mizukami, S.; Watanabe, S.; Hori, Y.; Kikuchi, K. Covalent protein labeling based on noncatalytic β-lactamase and a designed FRET substrate. *J. Am. Chem. Soc.* **2009**, *131*, 5016–5017.

2. Singh, A.; Chen, K.; Adelstein, S. J.; Kassis, A. I. Synthesis of coumarin-polyamine-based molecular probe for the detection of hydroxyl radicals generated by gamma radiation. *Radiat. Res.* **2007**, *168*, 233–242

3. Li, W.; Zhao, Y. Photo-caged fluorescent molecules. U.S. Pat. Appl. Publ. US 2005042662, 2005.

4. Makrigiorgos, G. M.; Baranowska-Kortylewicz, J.; Bump, E.; Sahu, S. K.; Berman, R. M.; Kassis, A. I. A method for detection of hydroxyl radicals in the vicinity of biomolecules using radiation-induced fluorescence of coumarin. *Int. J. Radiat. Biol.* **1993**, *63*, 445–458.

5. Lakowicz, J. R.; Maliwal, B. P.; Thompson, R.; Ozinskas, A. Fluorescent energy transfer immunoassay. U.S. Patent 5631169, 1997.

6. Nishimaru, T.; Sano, M.; Yamaguchi, Y.; Wakamiya, T. Syntheses and biological activities of fluorescent-labeled analogs of acylpolyamine toxin NPTX-594 isolated from the venom of *Madagascar Joro* spider. *Bioorg. Med. Chem.* **2009**, *17*, 57–63.

7. Nishimaru, T.; Sano, M.; Shimamoto, K.; Nakajima, T.; Yamaguchi, Y.; Wakamiya, T. Syntheses and biological activities of fluorescent labeled analogs of acylpolyaminetoxin NPTX-594 isolated from the venom of *Madagascar Joro* spider. *Pept. Sci.* **2006**, *43*, 153–154.

8. Sano, M.; Nakagawa, K.; Yamaguchi, Y.; Nakajima, T.; Wakamiya, T. Study on the synthesis and biological activities of fluorescent labeled analogs of spider toxin NPTX-594. *Kinki Daigaku Rikogakubu Kenkyu Hokoku* **2006**, *42*, 23–28.

9. Morrison, L. E.; Legator, M. S.; Bittner, M. L. Probe composition for genome identification and methods PCT Int. Appl. WO 9306245, 1993.

10. Pashkova, A.; Moskovets, E.; Karger, B. L. Coumarin tags for improved analysis of peptides by MALDI-TOF MS and MS/MS. 1. Enhancement in MALDI MS signal intensities. *Anal. Chem.* **2004**, *76*, 4550–4557.

11. Kikuchi, K.; Mizukami, S.; Watanabe, S.; Akimoto, Y. Method for fluorescent labeling of protein. PCT Int. Appl. WO 2012105596, 2012.

12. Sadhu, K. K.; Mizukami, S.; Lanam, C. R.; Kikuchi, K. Fluorogenic protein labeling through photoinduced electron transfer-based BL-Tag technology. *Chem.-Asian J.* **2012**, *7*, 272–276.

13. Jin, X.; Uttamapinant, C.; Ting, A. Y. Synthesis of 7-aminocoumarin by Buchwald-Hartwig cross coupling for specific protein labeling in living cells. *ChemBioChem* **2011**, *12*, 65–70.

14. Minakami, S.; Kikuchi, K.; Watanabe, S.; Yamamoto, H. Novel fluorescent indicator labeling method for recombinant proteins by using β-lactamase fusion protein partner. Jpn. Kokai Tokkyo Koho JP 2010119382, 2010.

15. Soh, N.; Makihara, K.; Ariyoshi, T.; Seto, D.; Maki, T.; Nakajima, H.; Nakano, K.; Imato, T. Phospholipid-linked coumarin: a fluorescent probe for sensing hydroxyl radicals in lipid membranes. *Anal. Sci.* **2008**, *24*, 293–296.

16. Singh, A.; Yang, Y.; Adelstein, S. J.; Kassis, A. I. Synthesis and application of molecular probe for detection of hydroxyl radicals produced by Na125I and γ-rays in aqueous solution. *Int. J. Radiat. Biol.* **2008**, *84*, 1001–1010.

17. Soh, N.; Sakawaki, O.; Makihara, K.; Odo, Y.; Fukaminato, T.; Kawai, T.; Irie, M.; Imato, T. Design and development of a fluorescent probe for monitoring hydrogen peroxide using photoinduced electron transfer. *Bioorg. Med. Chem.* **2005**, *13*, 1131–1139.

18. Chakrabarti, S.; Kassis, A. I.; Slayter, H. S.; Bump, E. A.; Sahu, S. K.; Makrigiorgos, G. M. Continuous detection of radiation or metal generated hydroxyl radicals within core chromatin particles. *Int. J. Radiat. Biol.* **1998**, *73*, 53–63.

19. Chakrabarti, S.; Mahmood, A.; Kassis, A. I.; Bump, E. A.; Jones, A. G.; Makrigiorgos, G. M. Generation of hydroxyl radicals by nucleohistone-bound metal-adriamycin complexes. *Free Radical Res.* **1996**, *25*, 207–220.

20. Chakrabarti, S.; Makrigiorgos, G. M.; O'Brien, K.; Bump, E.; Kassis, A. I. Measurement of hydroxyl radicals catalyzed in the immediate vicinity of DNA by metal-bleomycin complexes. *Free Radical Biol. Med.* **1996**, *20*, 777–783.

21. Makrigiorgos, G. M.; Folkard, M.; Huang, C.; Bump, E.; Baranowska-Kortylewicz, J.; Sahu, S. K.; Michael, B. D.; Kassis, A. I. Quantification of radiation-induced hydroxyl radicals within nucleohistones using a molecular fluorescent probe. *Radiat. Res.* **1994**, *138*, 177–185.

22. Diwu, Z.; Haugland, R. P. Assay for glutathione transferase using polyhaloaryl-substituted reporter molecules. U.S. Patent 5773236, 1998.

23. Yang, G.; Rich, J. R.; Gilbert, M.; Wakarchuk, W. W.; Feng, Y.; Withers, S. G. Fluorescence activated cell sorting as a general ultra-high-throughput screening method for directed evolution of glycosyltransferases. *J. Am. Chem. Soc.* **2010**, *132*, 10570–10577.

24. Yoon, H.; Dakanali, M.; Lichlyter, D.; Chang, W. M.; Nguyen, K. A.; Nipper, M. E.; Haidekker, M. A.; Theodorakis, E. A. Synthesis and evaluation of self-calibrating ratiometric viscosity sensors. *Org. Biomol. Chem.* **2011**, *9*, 3530–3540.

7-HYDROXY-4-METHYLCOUMARIN-3-ACETIC ACID

CAS Registry Number   5852-10-8

Chemical Structure

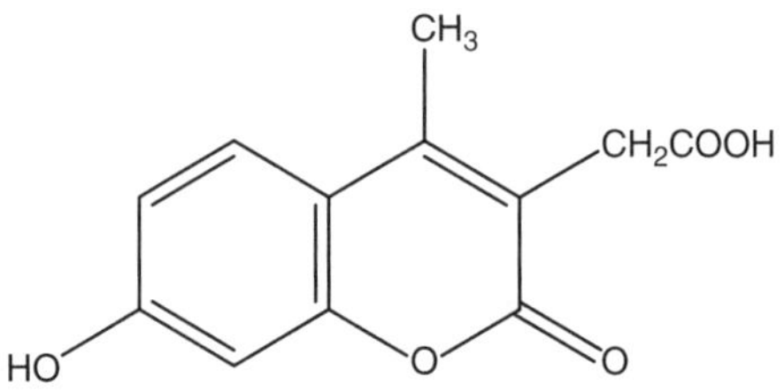

CA Index Name   2*H*-1-Benzopyran-3-acetic acid, 7-hydroxy-4-methyl-2-oxo-

Other Names   7-Hydroxy-4-methyl-2-oxo-2*H*-1-benzopyran-3-acetic acid; 7-Hydroxy-4-methylcoumarin-3-acetic acid; 7-Hydroxy-4-methyl-3-coumarinylacetic acid; 4-Methylumbelliferone-3-acetic acid

Merck Index Number   Not listed

Chemical/Dye Class   Coumarin

Molecular Formula $C_{12}H_{10}O_5$

Molecular Weight   234.20

Physical Form   Yellow solid

Solubility   Soluble in *N,N*-dimethylformamide, dimethyl sulfoxide

Melting Point   268–269 °C;[1] 268 °C;[5] 265–268 °C;[3] 265–266 °C;[4] 265 °C;[6]

Boiling Point (Calcd.)   512.9 ± 50.0 °C, pressure: 760 Torr

pK$_a$ (Calcd.)   3.78 ± 0.10, most acidic, temperature: 25 °C

Absorption (λ_{max})   360 nm (Buffer pH 10.0)

Emission (λ_{max})   455 nm (Buffer pH 10.0)

Molar Extinction Coefficient   19,000 cm^{-1} M^{-1} (Buffer pH 10.0)

Synthesis   Synthetic methods[1–6]

Imaging/Labeling Applications   Botulinum neurotoxins (BoNTs);[7] nucleic acids;[8] steroids;[9,10]

Biological/Medical Applications   Detecting nucleic acids;[8] detecting/monitoring botulinum neurotoxins (BoNTs);[7] assaying steroids;[9,10] as a substrate for measuring proteinases activity[11]

Industrial Applications   Not reported

Safety/Toxicity   No data available

REFERENCES

1. Nagorichna, I. V.; Ogorodniichuk, A. S.; Garazd, M. M.; Vinogradova, V. I.; Khilya, V. P. Modified coumarins. 25. N-acyl cytisine derivatives containing a coumarin fragment. *Chem. Nat. Compd.* **2007**, *43*, 10–14.

2. Woods, L. L.; Johnson, D. Some highly substituted coumarins. *Texas J. Sci.* **1965**, *17*, 340–344.

3. Laskowski, S. C.; Clinton, R. O. Coumarin II. Derivatives of coumarin-3- and -4-acetic acids. *J. Am. Chem. Soc.* **1950**, *72*, 3987–3991.

4. Shah, R. H.; Shah, N. M. Condensation of α-substituted acetoacetates with phenols. VI. The condensation of phenols with ethyl acetosuccinate. *J. Indian Chem. Soc.* **1942**, *19*, 481–485.

5. Banerjee, S. K. Condensation of ethyl acetosuccinate with phenols. *J. Indian Chem. Soc.* **1931**, *8*, 777–782.

6. Dey, B. B.; Sankaranarayanan, Y. Coumarin-3-acetic acids. *J. Indian Chem. Soc.* **1931**, *8*, 817–827.

7. Bagramyan, K.; Kaplan, B. E.; Cheng, L. W.; Strotmeier, J.; Rummel, A.; Kalkum, M. Substrates and controls for the quantitative detection of active botulinum neurotoxin in protease-containing samples. *Anal. Chem.* **2013**, *85*, 5569–5576.

8. Franzini, R. M.; Kool, E. T. 7-azidomethoxy-coumarins as profluorophores for templated nucleic acid detection. *ChemBioChem* **2008**, *9*, 2981–2988.

9. Exley, D.; Ekeke, G. I. Fluoroimmunoassay of 5-alpha-dihydrotestosterone. *J. Steroid Biochem.* **1981**, *14*, 1297–1302.

10. Ekeke, G. I.; Exley, D. The assay of steroids by fluoroimmunoassay. *Enzyme Labelled Immunoassay Horm. Drugs, Proc. Int. Symp.* **1978**, 195–205.

11. Khalfan, H.; Abuknesha, R.; Robinson, D. Fluorigenic method for the assay of proteinase activity with the use of 4-methylumbelliferyl-casein. *Biochem. J.* **1983**, *209*, 265–267.

8-HYDROXYPYRENE-1,3,6-TRISULFONIC ACID TRISODIUM SALT (HPTS) (PYRANINE)

CAS Registry Number 6358-69-6

Chemical Structure

CA Index Name 1,3,6-Pyrenetrisulfonic acid, 8-hydroxy-, sodium salt (1:3)

Other Names 1,3,6-Pyrenetrisulfonic acid, 8-hydroxy-, trisodium salt; 11389 Green; 8-Hydroxy-1,3,6-pyrenetrisulfonic acid trisodium salt; C.I. 59040; C.I. Solvent Green 7; D and C Green No. 8; D&C Green 8; D&C Green No. 8; Fluka 56360; Green 204; Green No. 204; HPTS; Japan Green 204; Japan Green No. 204; Keyacid Pyranine 10G; PY; Pyranine; Pyranine 10G; Pyranine 120; Pyranine Concentrated; Pyranine conk; Pyrene 1; Sanolin Pyranine Green; Solvent Green 7; Trisodium 1-hydroxy-3,6,8-pyrenetrisulfonate; Trisodium 8-hydroxypyrene-1,3,6-trisulfonate; Yellow Pyracide G

Merck Index Number 8065

Chemical/Dye Class Pyrene

Molecular Formula $C_{16}H_7Na_3O_{10}S_3$

Molecular Weight 524.37

Physical Form Yellow solid

Solubility Soluble in water, N,N-dimethylformamide, dimethyl sulfoxide; slightly soluble in glacial acetic acid

Melting Point >300 °C

pK$_a$ 7.3, temperature: 22 °C

Absorption (λ_{max}) 403 nm (Buffer pH 4.0); 454 nm (Buffer pH 9.0)

Emission (λ_{max}) 511 nm (Buffers pH 4.0, 9.0)

Molar Extinction Coefficient 20,000 cm^{-1} M^{-1} (Buffer pH 4.0); 24,000 cm^{-1} M^{-1} (Buffer pH 9.0)

Synthesis Synthetic method[1]

Imaging/Labeling Applications Bile acids;[2] living cells/tissues;[3–5] chitosan;[6] cornea;[7] cyclen 1;[8] cytochrome c/cytochrome o;[9–11] domes;[12] eye shadows;[13] fungi;[14] ferric/ferrous derivatives of hemoglobin;[15] lipids;[16,40] liposomes;[17–31] proteoliposomes;[32,33] microorganism;[34] mitochondria;[35,36] myoglobin;[37] plant cells;[38,39] prostaglandins;[40] yeast[41–44]

Biological/Medical Applications Antiperspirants;[45] bath preparations;[46–51] cleansing products;[52,53] cosmetics;[54–61] contact lens;[62] hair dyes;[63–75,201] hygiene products/soaps;[76,77] food products;[78] controlled-release system for proteins;[79] drug delivery;[80] in ophthalmology/retinal surgery;[81] sugar sensors (glucose, fructose, lactose, sucrose);[82–90] as a substrate for measuring acetylcholinesterase (AChE) activity,[21] carbonic anhydrase activity,[91] esterases activity,[92,93] β-glucuronidase activity,[94] lipase activity,[95] acidic/alkaline phosphatases activity,[96–98] pyrophosphatase activity,[99] sucrose phosphorylase (SPO) activity,[100] phosphoglucomutase (PGM) activity;[100] treating cancer,[101] infection,[101] immune/autoimmune disease,[101] mastitis,[76] tonsillitis,[76] tuberculosis[102]

Industrial Applications Adhesives;[103,104] agrochemicals;[61] ammonia sensor;[105] carbon dioxide sensor;[106–115] oxygen sensor;[115,116] anionic surfactants sensor;[117] measuring the size of the organoclay aggregates;[118] detecting cracks/flaws in metallic surfaces,[119] refrigerant leaking;[120] detergents;[121–123] correction fluid;[124] highlighters;[125–127] inks;[127–139,144,145] markers;[140–142] paints;[142,144,145] pencils;[143] toners;[146] chemically amplified resist;[147] color filters;[148] chemiluminescent materials;[149] electrophoretic display (EPD) film materials;[150] electroluminescent displays/materials;[151,152] fuel cell;[153,154] lasers;[155–158] liquid crystal displays;[159] luminescent materials;[160–162] nanoparticles;[163] papers;[164–166] photoresist processes;[167] photonic ionic liquids;[168] soils;[169] solar cell;[170] sol–gel materials/processes/thin films;[171–181] sonophotoluminescence;[182–184] water tracers;[185–194,198] evaluating liquid water flow in wood during drying[195]

Safety/Toxicity Acute intravenous toxicity;[196] cytotoxicity;[197] ecotoxicity;[198,199] phototoxicity;[200] skin toxicity[201]

REFERENCES

1. Tietze, E.; Bayer, O. Sulfonic acids of pyrene and their derivatives. *Justus Liebigs Ann. Chem.* **1939**, *540*, 189–210.

2. Kellogg, T. F. Improved spray reagent for detection of bile acids on thin-layer chromatoplates. *J. Lipid Res.* **1970**, *11*, 498–499.

3. Hille, C.; Berg, M.; Bressel, L.; Munzke, D.; Primus, P.; Loehmannsroeben, H.; Dosche, C. Time-domain fluorescence lifetime imaging for intracellular pH sensing in living tissues. *Anal. Bioanal. Chem.* **2008**, *391*, 1871–1879.

4. Vecer, J.; Holoubek, A.; Herman, P. Manipulation of intracellular pH by electroporation: an alternative method for fast calibration of pH in living cells. *Anal. Biochem.* **2004**, *329*, 348–350.

5. Slavik, J.; Lanz, E.; Cimprich, P. Measurement of individual intracellular pH and membrane potential values in living cells. *Proc. SPIE* **1999**, *3600*, 76–83.

6. Barnadas-Rodriguez, R. Effect and mechanism of association of 8-Hydroxy-1,3,6-pyrenetrisulfonic acid to chitosan: Physicochemical properties of the complex. *Macromol. Chem. Phys.* **2013**, *214*, 99–106.

7. Thomas, J. V.; Brimijoin, M. R.; Neault, T. R.; Brubaker, R. F. The fluorescent indicator pyranine is suitable for measuring stromal and cameral pH *in vivo*. *Exp. Eye Res.* **1990**, *50*, 241–249.

8. Winschel, C. A.; Kalidindi, A.; Zgani, I.; Magruder, J. L.; Sidorov, V. Receptor for anionic pyrene derivatives provides the basis for new biomembrane assays. *J. Am. Chem. Soc.* **2005**, *127*, 14704–14713.

9. Marantz, Y.; Nachliel, E. Gauging of cytochrome c structural fluctuation by time-resolved proton pulse. *Isr. J. Chem.* **1999**, *39*, 439–445.

10. Kotlyar, A. B.; Borovok, N. Light-induced oxidation of cytochrome c. *Biochim. Biophys. Acta, Bioenerg.* **1995**, *1228*, 87–90.

11. Sedgwick, E. G.; Bragg, P. D. pH Probes respond to redox changes in cytochrome o. *Arch. Biochem. Biophys.* **1990**, *282*, 372–376.

12. Rotoli, B. M.; Orlandini, G.; Gatti, R.; Dall'Asta, V.; Gazzola, G. C.; Bussolati, O. Employment of confocal microscopy for the dynamic visualization of domes in intact epithelial cell cultures. *Cells Tissues Organs* **2002**, *170*, 237–245.

13. Bouchard, F. Eye shadow comprising least 50% colorants. Eur. Pat. Appl. EP 1757262, 2007.

14. Nishino, T.; Kususe, M.; Kishi, M.; Hasegawa, T. Oxygen bleach compositions for fungi containing oxidation-resistant dyes. Jpn. Kokai Tokkyo Koho JP 11029420, 1999.

15. Petcu, L. C.; Turcu, G.; Rosoiu, N. Comparative study of certain ferric and ferrous derivatives of hemoglobin labeled with HPT. *Rom. J. Biophys.* **2002**, *12*, 53–58.

16. Winschel, C. A.; Kaushik, V.; Abdrakhmanova, G.; Aris, S. M.; Sidorov, V. New noninvasive methodology for real-time monitoring of lipid flip. *Bioconjugate Chem.* **2007**, *18*, 1507–1515.

17. Nyren-Erickson, E. K.; Haldar, M. K.; Gu, Y.; Qian, S. Y.; Friesner, D. L.; Mallik, S. Fluorescent liposomes for differential interactions with glycosaminoglycans. *Anal. Chem.* **2011**, *83*, 5989–5995.

18. Katagiri, K.; Imai, Y.; Koumoto, K.; Kaiden, T.; Kono, K.; Aoshima, S. Magnetoresponsive on-demand release of hybrid liposomes formed from Fe_3O_4 nanoparticles and thermosensitive block copolymers. *Small* **2011**, *7*, 1683–1689.

19. Arai, M.; Kubo, K.; Yokoyama, S.; Kohno, K. Phospholipid derivative and pH-responsive liposomes containing it. PCT Int. Appl. WO 2011059073, 2011.

20. Ng, L.; Yuba, E.; Kono, K. Modification of liposome surface with pH-responsive polyampholytes for the controlled-release of drugs. *Res. Chem. Intermed.* **2009**, *35*, 1015–1025.

21. Vamvakaki, V.; Fournier, D.; Chaniotakis, N. A. Fluorescence detection of enzymatic activity within a liposome based nano-biosensor. *Biosens. Bioelectron.* **2005**, *21*, 384–388.

22. Roux, E.; Lafleur, M.; Lataste, E.; Moreau, P.; Leroux, J. On the characterization of pH-sensitive liposome/polymer complexes. *Biomacromolecules* **2003**, *4*, 240–248.

23. Papahadjopoulos, D.; Meyer, O.; Leroux, J. Polymeric, pH-sensitive, serum-stable liposomes. U.S. Patent 6426086, 2002.

24. Nicklin, S.; Clarke, D. J.; Lloyd, C. J.; Aojula, H. Singh; T., Marina; W., Michael T. Detection system using liposomes and signal modification. PCT Int. Appl. WO 9938009, 1999.

25. Kobayashi, Y.; Yoshino, K.; Takahata, H.; Matsuyama, T. Preparation of *N*-alkyl-*N*-acylamino acid-modified pharmaceutical liposomes with improved bioavailability. Jpn. Kokai Tokkyo Koho JP 07285878, 1995.

26. Venema, K.; Gibrat, R.; Grouzis, J.; Grignon, C. Quantitative measurement of cationic fluxes, selectivity and membrane potential using liposomes multilabeled with fluorescent probes. *Biochim. Biophys. Acta, Biomembr.* **1993**, *1146*, 87–96.

27. Straubinger, R. M.; Papahadjopoulos, D.; Hong, K. Endocytosis and intracellular fate of liposomes using pyranine as a probe. *Biochemistry* **1990**, *29*, 4929–4939.

28. Seigneuret, M.; Rigaud, J. L. Use of the fluorescent pH probe pyranine to detect heterogeneous directions of proton movement in bacteriorhodopsin reconstituted large liposomes. *FEBS Lett.* **1985**, *188*, 101–106.

29. Rossignol, M.; Thomas, P.; Grignon, C. Proton permeability of liposomes from natural phospholipid mixtures. *Biochim. Biophys. Acta, Biomembr.* **1982**, *684*, 195–199.

30. Nomura, T.; Kondo, H.; Sunamoto, J. Liposomal membranes. Part X. Adsorption of pyranine onto cationic liposomal membranes as evidenced by fluorescence polarization. *Bull. Chem. Soc. Jpn.* **1981**, *54*, 1239–1240.

31. Kano, K.; Fendler, J. H. Pyranine as a sensitive pH probe for liposome interiors and surfaces. pH gradients across phospholipid vesicles. *Biochim. Biophys. Acta, Biomembr.* **1978**, *509*, 289–299.

32. Picard, M.; Verchere, A.; Broutin, I. Monitoring the active transport of efflux pumps after their reconstitution into proteoliposomes: Caveats and keys. *Anal. Biochem.* **2012**, *420*, 194–196.

33. Nicholls, P.; Tattrie, B.; Butko, P.; Tihova, M. Probe and protein orientations in proteoliposomes: electron microscopy and topobiochemistry. *Biochem. Soc. Trans.* **1992**, *20*, 115S.

34. Noda, N.; Mizutani, T. Microorganism-measuring method using multiple staining. Jpn. Kokai Tokkyo Koho JP 2006340684, 2006.

35. Campo, M. L.; Tedeschi, H. Protonmotive force in giant mitochondria. *Eur. J. Biochemistry* **1984**, *141*, 5–7.

36. Ziegler, M.; Penefsky, H. S. The adenine nucleotide translocase modulates oligomycin-induced quenching of pyranine fluorescence in submitochondrial particles. *J. Biol. Chem.* **1993**, *268*, 25320–25328.

37. Chaudhury, N. K.; Murari, B. M.; Gohil, N. K.; Anand, S. Fluorescence based sensors and sensing method for *in vitro* detection and quantitation of myoglobin. Indian Pat. Appl. IN 2007DE02294, 2009.

38. Apostol, I.; Heinstein, P. F.; Low, P. S. Rapid stimulation of an oxidative burst during elicitation of cultured plant cells. Role in defense and signal transduction. *Plant Physiol.* **1989**, *90*, 109–116.

39. Low, P. S.; Heinstein, P. F. Elicitor stimulation of the defense response in cultured plant cells monitored by fluorescent dyes. *Arch. Biochem. Biophys.* **1986**, *249*, 472–479.

40. Goswami, S. K.; Kinsella, J. E. A nondestructive spray reagent for the detection of prostaglandins and other lipids on thin-layer chromatograms. *Lipids* **1981**, *16*, 759–760.

41. Herman, P.; Drapalova, H.; Muzikova, R.; Vecer, J. Electroporative adjustment of pH in living yeast cells: ratiometric fluorescence pH imaging. *J. Fluoresc.* **2005**, *15*, 763–768.

42. Aguedo, M.; Wache, Y.; Belin, J. M. Intracellular pH-dependent efflux of the fluorescent probe pyranine in the yeast *Yarrowia lipolytica*. *FEMS Microbiol. Lett.* **2001**, *200*, 185–189.

43. Calahorra, M.; Martinez, G. A.; Hernandez-Cruz, A.; Pena, A. Influence of monovalent cations on yeast cytoplasmic and vacuolar pH. *Yeast* **1998**, *14*, 501–515.

44. Pena, A.; Ramirez, J.; Rosas, G.; Calahorra, M. Proton pumping and the internal pH of yeast cells, measured with pyranine introduced by electroporation. *J. Bacteriol.* **1995**, *177*, 1017–1022.

45. Kasat, R. B. Colored anhydrous hydrophobic antiperspirants. Eur. Pat. Appl. EP 116406, 1984.

46. Kase, H.; Hirano, S. Stable bath compositions containing polyethylene glycol, auxiliary solubilizing-dispersing agents, and colorants. Jpn. Kokai Tokkyo Koho JP 2005263642, 2005.

47. Hashimoto, N. Bath preparations containing liquid-containing capsules coated with oils. Jpn. Kokai Tokkyo Koho JP 04026616, 1992.

48. Saimo, Y.; Shimada, A. Colorful bath preparations containing inorganic substances encapsulated with pigment-containing polymers. Jpn. Kokai Tokkyo Koho JP 01294618, 1989.

49. Osugi, T.; Tachibana, S.; Murayama, K. Bath preparations containing water-soluble and -insoluble coloring agents. Jpn. Kokai Tokkyo Koho JP 01175929, 1989.

50. Hashimoto, N.; Murakoshi, Y.; Uchiyama, I.; Tsunakawa, M. Bath preparations containing pigments and perfumes in microcapsules. Jpn. Kokai Tokkyo Koho JP 62252715, 1987.

51. Okubo, T. Bubble bath preparations containing organic acids, carbonates, pigments, and

polyethylene glycol. Jpn. Kokai Tokkyo Koho JP 62205018, 1987.

52. Krzysik, D. G.; Utschig, J. M.; Cole, D. B. Color changing liquid cleansing products containing surfactants, electrolytes and coloring agents. U.S. Pat. Appl. Publ. US 20050148490, 2005.

53. Colwell, D. J. Color-stabilized body cleansing composition. Ger. Offen. DE 4412235, 1994.

54. Marthaler, W. J.; Holmgren, C. D. Cosmetic compositions comprising blend or mixture of solid butter and a colorant. U.S. Pat. Appl. Publ. US 20120276030, 2012.

55. Cassin, G.; Simonnet, J. Cosmetic composition for lightening or unifying the complexion. PCT Int. Appl. WO 2008022834, 2008.

56. Mochizuki, M.; Utsuki, A. Cosmetics containing black light-fluorescing dyes. Jpn. Kokai Tokkyo Koho JP 2002087934, 2002.

57. Kamiya, M.; Kito, N.; Morita, K.; Kawaguchi, S. Cosmetics containing light-emitting substances. Jpn. Kokai Tokkyo Koho JP 2001233730, 2001.

58. Schrader, K.; Schrader, A. Use of catalase for stabilization of organic coloring materials in cosmetics. Ger. Offen. DE 19922025, 2000.

59. Peters, D. W.; Calvo, L. C. Fluorescent cosmetic compositions. Eur. Pat. Appl. EP 370470, 1990.

60. Someya, T.; Onaki, M. Cosmetics containing water- and oil-repelling dyes. Jpn. Kokai Tokkyo Koho JP 01294611, 1989.

61. Tanimura, E.; Horiuchi, K.; Komatsu, A. Discoloring concentrated solution for use in external medical supplies, cosmetics, and agrochemicals. Jpn. Kokai Tokkyo Koho JP 48072227, 1973.

62. Chen, F. S.; Maurice, D. M. The pH in the precorneal tear film and under a contact lens measured with a fluorescent probe. *Exp. Eye Res.* **1990**, *50*, 251–259.

63. Pratt, D.; Moehring, H. Bleaching and dyeing composition for hair. Eur. Pat. Appl. EP 2606875, 2013.

64. Molenda, M.; Lipinski, N. Hair dyeing compositions an anionic sugar surfactant. Eur. Pat. Appl. EP 2609905, 2013.

65. Wood, J.; Schaefer, S. Oxidative hair coloring composition containing magnesium salts. Eur. Pat. Appl. EP 2468241, 2012.

66. Lee, S. I. Hair dye composition for coating and dyeing hair. Repub. Korea KR 998694, 2010.

67. Trigg, D. L.; Jones, J. L. Multistep hair color revitalizing kit comprising hair dye and glossing agent. U.S. Pat. Appl. Publ. US 20080189876, 2008.

68. Tanaka, M. Acidic hair dyes containing (carboxymethyl) N-succinylchitosan. Jpn. Kokai Tokkyo Koho JP 2007284380, 2007.

69. Pollack, G. Hair dye compositions containing polymers. U.S. Pat. Appl. Publ. US 20040025264, 2004.

70. Takahashi, M.; Ota, T.; Iketa, H. Acidic hair dye bases containing polymers. Jpn. Kokai Tokkyo Koho JP 2003160452, 2003.

71. Tsuge, T.; Mori, K.; Mita, S.; Kato, S.; Arai, A. Hair dye compositions containing acidic dyes. Jpn. Kokai Tokkyo Koho JP 2003073238, 2003.

72. Yoshida, M.; Suzuki, K. Hair dye compositions. PCT Int. Appl. WO 9831330, 1998.

73. Imamura, T.; Shibata, Y.; Murai, M. Hair dye composition containing acid dyes and organic solvents. Brit. UK Pat. Appl. GB 2259717, 1993.

74. Watanabe, K.; Ono, T.; Ota, T.; Minei, T.; Horikoshi, T. Wave-setting hair dye. Jpn. Kokai Tokkyo Koho JP 02076807, 1990.

75. Fukunishi, A.; Tsunekawa, T.; Kawai, M. Hair dyeing preparations containing dialkyldi(meth)allylammonium salt polymers and acidic dyes and dyeing methods. Jpn. Kokai Tokkyo Koho JP 01279820, 1989.

76. Maruoka, T. Disposable oral hygiene product comprising waterproof container and porous drug-holding material. Jpn. Kokai Tokkyo Koho JP 11197217, 1999.

77. Sekine, K. Powdery cosmetic soaps containing pigments. Jpn. Kokai Tokkyo Koho JP 63145400, 1988.

78. Van Veen, J. J. F.; Koopal, C. G. J. Optical pH sensor for food products. Eur. Pat. Appl. EP 2506005, 2012.

79. Vogelhuber, W.; Magni, E.; Mouro, M.; Spruss, T.; Guse, C.; Gazzaniga, A.; Gopferich, A. Monolithic triglyceride matrixes: A controlled-release system for proteins. *Pharm. Dev. Technol.* **2003**, *8*, 71–79.

80. Roux, E.; Francis, M.; Winnik, F. M.; Leroux, J. Polymer based pH-sensitive carriers as a means to improve the cytoplasmic delivery of drugs. *Int. J. Pharm.* **2002**, *242*, 25–36.

81. Hagedorn, N.; Rizzo, S.; Rodrigues, E. Dye-containing preparation comprising perfluorinated alkane and semi-fluorinated alkane for use in ophthalmolecular and retinal surgery. PCT Int. Appl. WO 2011151079, 2011.

82. Yin, X.; Dong, J.; Wang, H.; Li, S.; Fan, L.; Cao, C. A simple chip free-flow electrophoresis

for monosaccharide sensing via supermolecule interaction of boronic acid functionalized quencher and fluorescent dye. *Electrophoresis* **2013**, *34*, 2185–2192.

83. Egawa, Y.; Seki, T.; Takahashi, S.; Anzai, J. Electrochemical and optical sugar sensors based on phenylboronic acid and its derivatives. *Mater. Sci. Eng., C* **2011**, *31*, 1257–1264.

84. Markle, D. R.; Suri, J. T.; Wessling, R. A.; Romey, M. A. Device for optical determination of two or more analytes, particularly pH and glucose, simultaneously with a single indicator. U.S. Pat. Appl. Publ. US 20080188722, 2008.

85. Sharrett, Z.; Gamsey, S.; Levine, P.; Cunningham-Bryant, D.; Vilozny, B.; Schiller, A.; Wessling, R. A.; Singaram, B. Boronic acid-appended bis-viologens as a new family of viologen quenchers for glucose sensing. *Tetrahedron Lett.* **2008**, *49*, 300–304.

86. Cordes, D. B.; Miller, A.; Gamsey, S.; Singaram, B. Simultaneous use of multiple fluorescent reporter dyes for glucose sensing in aqueous solution. *Anal. Bioanal. Chem.* **2007**, *387*, 2767–2773.

87. Gamsey, S.; Suri, J. T.; Wessling, R. A.; Singaram, B. Continuous glucose detection using boronic acid-substituted viologens in fluorescent hydrogels: Linker effects and extension to fiber optics. *Langmuir* **2006**, *22*, 9067–9074.

88. Pan, B.; Chakraborty, R.; Berglund, K. A. Time resolved fluorescence and anisotropy of 1-pyrene butyric acid and pyranine as probes of solvent organization in sucrose solutions. *J. Cryst. Growth* **1993**, *130*, 587–599.

89. Chakraborty, R.; Berglund, K. A. Steady state fluorescence spectroscopy of pyranine as a trace extrinsic probe to study structure in aqueous sugar solutions. *J. Cryst. Growth* **1992**, *125*, 81–96.

90. Chakraborty, R.; Berglund, K. A. The use of pyranine as a trace fluorescent probe to study structure in aqueous sucrose solutions. *AIChE Symp. Ser.* **1991**, *87*, 114–123.

91. Shingles, R.; Moroney, J. V. Measurement of carbonic anhydrase activity using a sensitive fluorometric assay. *Anal. Biochem.* **1997**, *252*, 190–197.

92. Clouet, A.; Darbre, T.; Revmond, J. Combinatorial synthesis, selection, and properties of esterase peptide dendrimers. *Biopolymers* **2006**, *84*, 114–123.

93. Som, A.; Sakai, N.; Matile, S. Complementary characteristics of homologous p-octiphenyl β-barrels with ion channel and esterase activity. *Bioorg. Med. Chem.* **2003**, *11*, 1363–1369.

94. Pope, M. R.; Bieniarz, C. Reagents and methods for the determination of glycohydrolytic enzymes. U.S. Patent 5272260, 1993.

95. Koller, E.; Wolfbeis, O. S. Pyrenesulfonic acids useful in fluorescent lipid probes. U.S. Patent 4844841, 1989.

96. Conrad, M. J.; He, L. Novel fluorogenic substrates for hydrolytic enzymes. PCT Int. Appl. WO 2000003034, 2000.

97. Sato, E.; Chiba, K.; Hoshi, M.; Kanaoka, Y. Organic fluorescent reagents. XXI. Pyranine phosphate as a new fluorogenic substrate for acidic and alkaline phosphatases. *Chem. Pharm. Bull.* **1992**, *40*, 786–788.

98. Harnisch, H. Fluorogenic phosphoric acid esters and their use in the detection and fluorimetric determination of phosphatases. Ger. Offen. DE 3303871, 1984.

99. Yaku, H.; Maeshima, M.; Nakanishi, Y.; Hirono, M.; Yukimasa, T.; Oka, H. Pyrophosphate assay with H+−pyrophosphatase and application to primer extension detection. PCT Int. Appl. WO 2005093088, 2005.

100. Vilozny, B.; Schiller, A.; Wessling, R. A.; Singaram, B. Enzyme assays with boronic acid appended bipyridinium salts. *Anal. Chim. Acta* **2009**, *649*, 246–251.

101. Brix, L.; Pedersen, H.; Jakobsen, T.; Schoeller, J.; Lohse, J.; Brunstedt, K.; Jacobsen, K. MHC multimers and conjugates for use in diagnosis, prognosis and therapy of cancer, infection, immune and autoimmune disease. PCT Int. Appl. WO 2009003492, 2009.

102. Scholler, J.; Brix, L.; Pedersen, H.; Jakobsen, T. Multimers of MHC complexed with Mycobacterium tuberculosis peptide as vaccine and for diagnosis, prognosis and therapy of tuberculosis. PCT Int. Appl. WO 2009039854, 2009.

103. Seki, K. Fluorescent solid stick adhesives showing good visibility and discoloring upon exposure to air. Jpn. Kokai Tokkyo Koho JP 2007238869, 2007.

104. Sumiya, M. Colored adhesives exhibiting loss of color after application to surface. Jpn. Kokai Tokkyo Koho JP 63243184, 1988.

105. Amali, A. J.; Awwad, N. H.; Rana, R. K.; Patra, D. Nanoparticle assembled microcapsules for application as pH and ammonia sensor. *Anal. Chim. Acta* **2011**, *708*, 75–83.

106. Lee, J. Il; Son, O. J. Method for manufacturing sensor film fixed with fluorescent dye and used for detecting carbon dioxide. Repub. Korean Kongkae Taeho Kongbo KR 2010069413, 2010.

107. Oter, O.; Ertekin, K.; Derinkuyu, S. Ratiometric sensing of CO_2 in ionic liquid modified ethyl cellulose matrix. *Talanta* **2008**, *76*, 557–563.

108. Chu, C.; Lo, Y. Fiber-optic carbon dioxide sensor based on fluorinated xerogels doped with HPTS. *Sens. Actuators, B* **2008**, *B129*, 120–125.

109. Ge, X.; Kostov, Y.; Rao, G. High-stability non-invasive autoclavable naked optical CO_2 sensor. *Biosens. Bioelectron.* **2003**, *18*, 857–865.

110. Waldner, A.; Barnard, S. M. Optical carbon dioxide sensors. PCT Int. Appl. WO 9909406, 1999.

111. He, X.; Rechnitz, G. A. Linear response function for fluorescence-based fiber-optic CO_2 sensors. *Anal. Chem.* **1995**, *67*, 2264–2268.

112. Goswami, K.; Kennedy, J. A.; Dandge, D. K.; Klainer, S. M.; Tokar, J. M. A fiber optic chemical sensor for carbon dioxide dissolved in sea water. *Proc. SPIE* **1990**, *1172*, 225–232.

113. Zhang, Z.; Seitz, W. Rudolf. A carbon dioxide sensor based on fluorescence. *Anal. Chim. Acta* **1984**, *160*, 305–309.

114. Marsoner, H.; Kroneis, H.; Wolfbeis, O. Measurement device for determining the carbon dioxide content of a sample. Eur. Pat. Appl. EP 105870, 1984.

115. Dixit, R.; Shen, L.; Ratterman, M.; Papautsky, I.; Klotzkin, D. A simple single-detector system for simultaneous monitoring of O_2 and CO_2 gas concentrations. *Proc. SPIE* **2012**, *8251*, 82510J/1–82510J/8.

116. Larsen, M.; Borisov, S. M.; Grunwald, B.; Klimant, I.; Glud, R. N. A simple and inexpensive high resolution color ratiometric planar optode imaging approach: application to oxygen and pH sensing. *Limnol. Oceanogr.: Methods* **2011**, *9*, 348–360.

117. An, Y.; Bai, H.; Li, C.; Shi, G. Disassembly-driven colorimetric and fluorescent sensor for anionic surfactants in water based on a conjugated polyelectrolyte/dye complex. *Soft Matter* **2011**, *7*, 6873–6877.

118. Yilmaz, Y.; Alemdar, A. Fluoro-surfactant as a tool for both controlling and measuring the size of the organoclay aggregates. *Appl. Clay Sci.* **2005**, *30*, 154–164.

119. Miles, S. J. Process of using an inspection dye for detecting cracks and flaws in metallic surfaces. U.S. Pat. Appl. Publ. US 20080259324, 2008.

120. Berton, G. Fluorescent liquid composition for detecting refrigerant leaking. Eur. Pat. Appl. EP 2380943, 2011.

121. Hsu, F. G.; Lee, K. H.; Puleo, A. M. Water-soluble package containing multiple colored layers of liquid laundry detergent. U.S. Patent 6521581, 2003.

122. Matsushita, T.; Kobayashi, H. Phase separation-type liquid detergent compositions. Jpn. Kokai Tokkyo Koho JP 62138595, 1987.

123. Rubin, F. K. Dye-stabilized detergent compositions. U.S. Patent 4526701, 1985.

124. Bethouart, C.; Duez, J.; Faure, B. Correction fluid with a drying indicator. Fr. Demande FR 2909680, 2008.

125. Kwan, W. S. V.; Duong, C. Highlighting marking compositions, highlighting kits, highlighted complexes, and application of eradicator. U.S. Pat. Appl. Publ. US 20070017413, 2007.

126. Duez, J.; Bethouart, C. Colored liquid composition for highlighting pen. PCT Int. Appl. WO 2002051950, 2002.

127. Miller, R. E.; Krieger, C. Color-changing ink compositions for highlighters. U.S. Patent 5492558, 1996.

128. Sang, Y. Aqueous washable writing ink for children uses. Faming Zhuanli Shenqing CN 102675982, 2012.

129. Takahashi, H.; Nishikawa, T. Aqueous ink with good discharging properties. Jpn. Kokai Tokkyo Koho JP 2001240777, 2001.

130. Onuki, Y. Coloring liquid compositions useful for ball pen inks. Jpn. Kokai Tokkyo Koho JP 2001131458, 2001.

131. Takahashi, H.; Nishikawa, T. Aqueous writing inks with smudge prevention. Jpn. Kokai Tokkyo Koho JP 11193361, 1999.

132. Eishima, S.; Teraoka, H. Ink-jet printing method using ink containing surfactant. Jpn. Kokai Tokkyo Koho JP 10100530, 1998.

133. Teraoka, H.; Nagashima, S. Inks for use in ink-jet printing cartridges of recording apparatus and method for their use. Jpn. Kokai Tokkyo Koho JP 09302293, 1997.

134. Cancellieri, J. Assembly or set of different color inks and an assembly of writing instruments. PCT Int. Appl. WO 9406872, 1994.

135. Nishimoto, T.; Takahashi, H.; Watanabe, K. Water-thinned fluorescent inks for ball-point pens. Jpn. Kokai Tokkyo Koho JP 06073324, 1994.

136. Inoue, S.; Tabayashi, I.; Yamada, Y.; Ametani, S. Water-thinned fluorescent jet-printing inks. Jpn. Kokai Tokkyo Koho JP 03081376, 1991.

137. Horvat, I.; Kramp, P.; Onczul, E. Water-thinned marking inks. Ger. DE 4020900, 1991.

138. Kunkel, E.; Pietsch, G. Improved fluorescent daylight marking inks. Ger. Offen. DE 2441823, 1976.

139. Jankewitz, A. Marking ink. Ger. DE 2315680, 1974.

140. Horvat, I.; Kramp, P.; Onczul, E. Marking fluids. Ger. Offen. DE 4020901, 1992.

141. Brachman, A.; Allison, K. Color-changing marking composition systems. U.S. Patent 5464470, 1995.

142. Miller, R. E.; Dereamus, R. C. Color changing systems using pan paint compositions and markers. U.S. Patent 5498282, 1996.

143. Lugert, G. Leads for colored pencils. Eur. Pat. Appl. EP 924273, 1999.

144. Miller, R. E.; Couch, C. R. Color-changing artistic paint and marking ink systems. Eur. Pat. Appl. EP 584571, 1994.

145. Miller, R. E. Color-changing compositions for inks and paints. U.S. Patent 5232494, 1993.

146. Ohno, Y.; Tadokoro, H.; Matsushima, K.; Yasukawa, H.; Koga, I. Electrophotographic green toners for forming green images with high brightness and high chroma, and full color electrophotographic image forming methods. U.S. Pat. Appl. Publ. US 20120219892, 2012.

147. Tarumoto, N.; Miyagawa, N.; Takahara, S.; Yamaoka, T. Diphenyliodonium salts with environment-friendly dye for a photoacid generator in chemically amplified resist. *Polym. J.* **2004**, *36*, 866–869.

148. Ito, H.; Hagiwara, H. Color filters with high contrast and good resistance to chemicals, heat and light and their manufacture. Jpn. Kokai Tokkyo Koho JP 2000047016, 2000.

149. Baretz, B. H.; Trzaskos, W. J.; Elliott, L. Use of water-soluble polymers in aqueous chemiluminescent materials. Eur. Pat. Appl. EP 175889, 1986.

150. Liu, Z. Method for manufacturing color microcapsule electrophoretic display (EPD) film material and method for manufacturing display device comprising the film material. Faming Zhuanli Shenqing CN 102629064, 2012.

151. Wu, X.; Nakua, A. M. Color electroluminescent displays. PCT Int. Appl. WO 2004036961, 2004.

152. Le Barny, P.; Soyer, F.; Facoetti, H. Electroluminescent materials containing electroluminescent polymers derived from polystyrene, their preparation and their use. Fr. Demande FR 2757531, 1998.

153. Wang, X.; Wu, J.; Fang, Y. Method for preparing high-electric conductivity proton exchange membrane used for direct methanol fuel cell. Faming Zhuanli Shenqing CN 102188913, 2011.

154. Lafitte, B.; Jannasch, P. Proton-conducting aromatic polymers carrying hypersulfonated side chains for fuel cell applications. *Adv. Funct. Mater.* **2007**, *17*, 2823–2834.

155. Ding, X.; Yasui, Y.; Kawaguchi, Y.; Niino, H.; Yabe, A. Laser-induced back-side wet etching of fused silica with an aqueous solution containing organic molecules. *Appl. Phys. A: Mater. Sci. Process.* **2002**, *75*, 437–440.

156. Fletcher, A. N.; Hollins, R. A.; Kubin, R. F.; Henry, R. A.; Atienza Moore, T. M.; Pietrak, M. E. Luminescent coolants for solid-state lasers. *Appl. Phys. B: Photophys. Laser Chem.* **1983**, *B30*, 195–202.

157. Denisov, L. K.; Kozlov, N. A.; Krasnov, I. V.; Uzhinov, B. M.; Rubeko, L. M. Active medium for liquid lasers. U.S.S.R. SU 764025, 1980.

158. Rubeko, L. M.; Krasnov, I. V.; Malyshev, G. A.; Krashakov, S. A.; Kozlov, N. A.; Denisov, L. K.; Uzhinov, B. M. Increase in the lasing efficiency of sodium 8-hydroxypyrene-1,3,6-trisulfonate in an aqueous solution. *Zh. Prikl. Spektros.* **1980**, *32*, 262–265.

159. Kuriyama, T.; Kawamura, S.; Funakura, S.; Shimada, K. Liquid crystal display. Jpn. Tokkyo Koho JP 5273494, 2013.

160. Sakaue, H.; Dan, R.; Shimizu, M.; Kazama, H. Note: *In vivo* pH imaging system using luminescent indicator and color camera. *Rev. Scient. Instrum.* **2012**, *83*, 076106/1–076106/3.

161. Dang, S.; Yan, D.; Lu, J. 8-Hydroxypyrene-1,3,6-trisulphonate and octanesulphonate co-assembled layered double hydroxide and its controllable solid-state luminescence by hydrothermal synthesis. *J. Solid State Chem.* **2012**, *185*, 219–224.

162. Lu, J.; Dang, S.; Wei, M.; Duan, X. 8-hydroxy-1,3,6-trisulfonyl pyrene anion and octane sulfonate anion co-intercalated compound luminescent material, and the preparation method thereof. Faming Zhuanli Shenqing CN 102127425, 2011.

163. Muto, S.; Oaki, Y.; Imai, H. Incorporation of dyes into silica-surfactant mesostructured nanoparticles as a nanoscale host material for organic molecules. *Chem. Lett.* **2006**, *35*, 880–881.

164. Caruso, O.; Patrucco, V. Color-changing paper. Ital. IT 1320150, 2003.

165. Raux, L. Method and indicator for identifying security paper, copying paper or heat-reactive paper. Fr. Demande FR 2539533, 1984.

166. Breunig, A. New perceptions during the qualitative assessment of the fiber composition of papers. *Papier* **1983**, *37*, 473–481.

167. Garza, C. M.; Cho, S. Metrology of bilayer photoresist processes. U.S. Pat. Appl. Publ. US 20090220895, 2009.

168. Yung, K. Y.; Schadock-Hewitt, A. J.; Hunter, N. P.; Bright, F. V.; Baker, G. A. Liquid litmus: chemosensory pH-responsive photonic ionic liquids. *Chem. Commun.* **2011**, *47*, 4775–4777.

169. Omoti, U.; Wild, A. Use of fluorescent dyes to mark the pathways of solute movement through soils under leaching conditions: 1. Laboratory experiments. *Soil Sci.* **1979**, *128*, 28–33.

170. Jiang, L.; Wen, L.; Zhai, J.; Nie, F.; Song, Y. Photoinduced protonation solar cell combining solar energy with photoacid (photosensitive) molecule. Faming Zhuanli Shenqing CN 101872866, 2010.

171. McDonagh, C.; Wencel, D. Sol–gel-derived materials for optical fluorescent pH sensing. PCT Int. Appl. WO 2009118271, 2009.

172. Gupta, R.; Chaudhury, N. K. Probing internal environment of sol–gel bulk and thin films using multiple fluorescent probes. *J. Sol–Gel Sci. Technol.* **2009**, *49*, 78–87.

173. Huang, M. H.; Soyez, H. M.; Dunn, B. S.; Zink, J. I.; Sellinger, A.; Brinker, C. J. *In situ* fluorescence probing of the chemical and structural changes during formation of hexagonal phase cetyltrimethylammonium bromide and lamellar phase CTAB/Poly-(dodecylmethacrylate) sol–gel silica thin films. *J. Sol–Gel Sci. Technol.* **2008**, *47*, 300–310.

174. Murari, B. M.; Anand, S.; Gohil, N. K.; Chaudhury, N. K. Fluorescence spectroscopic study of dip coated sol–gel thin film internal environment using fluorescent probes Hoechst33258 and Pyranine. *J. Sol–Gel Sci. Technol.* **2007**, *41*, 147–155.

175. Huang, M. H.; Soyez, H. M.; Dunn, B. S.; Zink, J. I. *In situ* fluorescence probing of molecular mobility and chemical changes during formation of dip-coated sol–gel silica thin films. *Chem. Mater.* **2000**, *12*, 231–235.

176. Flora, K. K.; Dabrowski, M. A.; Musson, S. P.; Brennan, J. D. The effect of preparation and aging conditions on the internal environment of sol–gel derived materials as probed by 7-azaindole and pyranine fluorescence. *Can. J. Chem.* **1999**, *77*, 1617–1625.

177. Wittouck, N.; De Schryver, F.; Snijkers-Hendrickx, I. Fluorescence of pyranine in sol–gel based silica and hybrid thin films. *J. Sol–Gel Sci. Technol.* **1997**, *8*, 895–899.

178. Pope, E. J. A. Fiber-optic microsensors using porous, dye-doped silica gel microspheres. *Mater. Res. Soc. Symp. Proc.* **1995**, *372*, 253–262.

179. Nishida, F.; McKiernan, J. M.; Dunn, B.; Zink, J. I.; Brinker, C. J.; Hurd, A. J. *In situ* fluorescence probing of the chemical changes during sol–gel thin film formation. *J. Am. Ceram. Soc.* **1995**, *78*, 1640–1648.

180. Samuel, J.; Strinkovski, A.; Shalom, S.; Lieberman, K.; Ottolenghi, M.; Avnir, D.; Lewis, A. Miniaturization of organically doped sol–gel materials: a microns-size fluorescent pH sensor. *Mater. Lett.* **1994**, *21*, 431–434.

181. Nishida, F.; Dunn, B.; McKiernan, J. M.; Zink, J. I.; Brinker, C. J.; Hurd, A. J. *In-situ* fluorescence imaging of sol–gel thin film deposition. *J. Sol–Gel Sci. Technol.* **1994**, *2*, 477–481.

182. Ashokkumar, M.; Grieser, F. Single-Bubble Sonophotoluminescence. *J. Am. Chem. Soc.* **2000**, *122*, 12001–12002.

183. Ashokkumar, M.; Grieser, F. Sonophotoluminescence from aqueous and non-aqueous solutions. *Ultrason. Sonochem.* **1999**, *6*, 1–5.

184. Ashokkumar, M; Grieser, F. Sonophotoluminescence: pyranine emission induced by ultrasound. *Chem. Commun.* **1998**, 561–562.

185. Xu, J.; Quan, C. Method for controlling concentration ratio of circulating cooling water based on fluorescent tracer concentration change, and concentration ratio device therefor. Faming Zhuanli Shenqing CN 101943919, 2011.

186. Magal, E.; Weisbrod, N.; Yakirevich, A.; Yechieli, Y. The use of fluorescent dyes as tracers in highly saline groundwater. *J. Hydrol.* **2008**, *358*, 124–133.

187. Otz, M. H.; Otz, H. K.; Otz, I.; Siegel, D. I. Surface water/groundwater interaction in the Piora

Aquifer, Switzerland: evidence from dye tracing tests. *Hydrogeol. J.* **2003**, *11*, 228–239.

188. Johnson, C. A.; Richner, G. A.; Vitvar, T.; Schittli, N.; Eberhard, M. Hydrological and geochemical factors affecting leachate composition in municipal solid waste incinerator bottom ash Part I: The hydrology of Landfill Lostorf, Switzerland. *J. Contam. Hydrol.* **1998**, *33*, 361–376.

189. Sivakumar, A.; Shah, J.; Rao, N. M.; Budd, S. S. Fluorescent tracer for monitoring wastewater treatment. U.S. Patent 5413719, 1995.

190. Lyons, R. G. Identification and separation of water tracing dyes using pH response characteristics. *J. Hydrol.* **1993**, *152*, 13–29.

191. Viriot, M. L.; Andre, J. C. Fluorescent dyes: a search for new tracers for hydrology. *Analusis* **1989**, *17*, 97–111.

192. Aldous, P. J.; Smart, P. L. Tracing ground-water movement in abandoned coal mined aquifers using fluorescent dyes. *Ground Water* **1988**, *26*, 172–178.

193. Launay, M.; Tripier, M.; Guizerix, J.; Viriot, M. L.; Andre, J. C. Pyranine used as a fluorescent tracer in hydrology: pH effects in determination of its concentration. *J. Hydrol.* **1980**, *46*, 377–383.

194. Smart, P. L.; Laidlaw, I. M. S. An evaluation of some fluorescent dyes for water tracing. *Water Resour. Res.* **1977**, *13*, 15–33.

195. Moettoenen, V.; Kaerki, T.; Martikka, O. Method for determining liquid water flow in wood during drying using a fluorescent dye tracer. *Eur. J. Wood Wood Prod.* **2011**, *69*, 287–293.

196. Lutty, G. A. The acute intravenous toxicity of biological stains, dyes, and other fluorescent substances. *Toxicol. Appl. Pharmacol.* **1978**, *44*, 225–249.

197. Lundberg, B. B.; Griffiths, G.; Hansen, H. J. Cellular association and cytotoxicity of anti-CD74-targeted lipid drug-carriers in B lymphoma cells. *J. Controlled Release* **2004**, *94*, 155–161.

198. Field, M. S.; Wilhelm, R. G.; Quinlan, J. F.; Aley, T. J. An assessment of the potential adverse properties of fluorescent tracer dyes used for groundwater tracing. *Environ. Monit. Assess.* **1995**, *38*, 75–96.

199. Behrens, H.; Beims, U.; Dieter, H.; Dietze, G.; Eikmann, T.; Grummt, T.; Hanisch, H.; Henseling, H.; Kass, W.; Kerndorff, H.; Leibundgut, C.; Muller-Wegener, U.; Ronnefahrt, I.; Scharenberg, B.; Schleyer, R.; Schloz, W.; Tilkes, F. Toxicological and ecotoxicological assessment of water tracers. *Hydrogeol. J.* **2001**, *9*, 321–325.

200. Kiskin, N. I.; Chillingworth, R.; McCray, J. A.; Piston, D.; Ogden, D. The efficiency of two-photon photolysis of a "caged" fluorophore, o-1-(2-nitrophenyl)ethylpyranine, in relation to photodamage of synaptic terminals. *Eur. Biophys. J.* **2002**, *30*, 588–604.

201. Sosted, H.; Basketter, D. A.; Estrada, E.; Johansen, J. D.; Patlewicz, G. Y. Ranking of hair dye substances according to predicted sensitization potency: quantitative structure-activity relationships. *Contact Dermatitis* **2004**, *51*, 241–254.

JOJO 1

CAS Registry Number   305801-87-0

Chemical Structure

Physical Form   Yellow-brown powder

Solubility   Soluble in dimethyl sulfoxide

Absorption (λ_{max})   529 nm (H_2O/DNA)

Emission (λ_{max})   545 nm (H_2O/DNA)

CA Index Name   Oxazolo[4,5-*b*]pyridinium, 2,2′-[1,3-propanediylbis-[(dimethyliminio)-3,1-propanediyl-1(4*H*)-quinolinyl-4-ylidenemethylidyne]]bis[4-methyl-, iodide (1 : 4)

Other Names   Oxazolo[4,5-*b*]pyridinium, 2,2′-[1,3-propanediylbis-[(dimethyliminio)-3,1-propanediyl-1(4*H*)-quinolinyl-4-ylidenemethylidyne]]bis[4-methyl-, tetraiodide; JOJO 1; JOJO 1 iodide

Merck Index Number   Not listed

Chemical/Dye Class   Cyanine

Molecular Formula $C_{47}H_{56}I_4N_8O_2$

Molecular Weight   1272.63

Molar Extinction Coefficient   171,400 $cm^{-1} M^{-1}$ (H_2O/DNA)

Quantum Yield   0.44 (H_2O/DNA)

Synthesis   Synthetic methods[1,2]

Imaging/Labeling Applications   Nucleic acids;[1-7] cells;[7] sperms;[7] hairs[8]

Biological/Medical Applications   Detecting nucleic acids,[1-7] cells,[7] oligonucleotides;[2] nucleic acid nanochannel device;[9,10] nucleic acid sequencing;[11,12] as temperature sensor[13,14]

Industrial Applications   Not reported

Safety/Toxicity   No data available

REFERENCES

1. Sabnis, R. W. *Handbook of Biological Dyes and Stains*; John Wiley & Sons Inc.: Hoboken, **2010**; pp 257–258.

2. Haugland, R. P.; Yue, S. T. Aza-benzazolium-containing cyanine dyes and their use in fluorescent biological stains. PCT Int. Appl. WO 2000066664, 2000.

3. Exner, M.; Rogers, A. Methods for detecting nucleic acids using multiple signals. U.S. Pat. Appl. Publ. US 2007172836, 2007.

4. Erikson, G. H.; Daksis, J. I.; Kandic, I.; Picard, P. Nucleic acid multiplex formation. PCT Int. Appl. WO 2002103051, 2002.

5. Erikson, G. H.; Daksis, J. I. Pre-incubation method to improve signal/noise ratio of nucleic acid assays. U.S. Pat. Appl. Publ. US 2004180345, 2004.

6. Erikson, G. H. Method for modifying transcription and/or translation in an organism for therapeutic, prophylactic and/or analytic uses. U.S. Pat. Appl. Publ. US 2003181412, 2003.

7. Anderson, A. L.; Knutson, C. R.; Mueth, D.; Plewa, J.; Tanner, E. Methods for staining cells for identification and sorting. U.S. Pat. Appl. Publ. US 2006172315, 2006.

8. Lagrange, A. Hair dye compositions containing a polycationic direct dye. Fr. Demande FR 2848840, 2004.

9. Mokkapati, V. R. S. S.; Di Virgilio, V.; Shen, C.; Mollinger, J.; Bastemeijer, J.; Bossche, A. DNA tracking within a nanochannel: device fabrication and experiments. *Lab Chip* **2011**, *11*, 2711–2719.

10. Tegenfeldt, J.; Reisner, W.; Flyvbjerg, H. Method for the mapping of the local AT/GC ratio along DNA. PCT Int. Appl. WO 2010042007, 2010.

11. Williams, J. G. K.; Anderson, J. P. Field-switch sequencing. U.S. Patent 7462452, 2008.

12. Williams, J. G. K.; Anderson, J. P. Field-switch sequencing. PCT Int. Appl. WO 2005111240, 2005.

13. Bousseksou, A.; Salmon, L.; Molnar, G.; Cobo, S.Materials with thermochromic spin transition doped with one or more fluorescent agents for use as temperature sensor. Fr. Demande FR 2952371, 2011.

14. Bousseksou, A.; Salmon, L.; Molnar, G.; Cobo, S. Heat-sensitive spin-transition materials doped with one or more fluorescent agents for use as temperature sensor. PCT Int. Appl. WO 2011058277, 2011.

JO-PRO 1

CAS Registry Number   305801-86-9

Chemical Structure

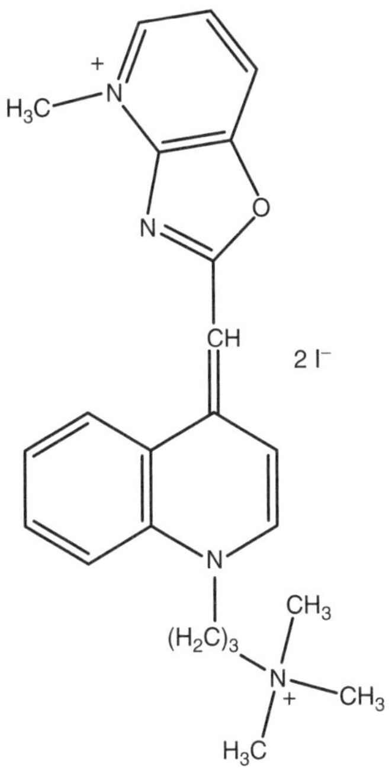

2 I⁻

CA Index Name   Quinolinium, 4-[(4-methyloxazolo[4,5-*b*]pyridin-2(4*H*)-ylidene)methyl]-1-[3-(trimethyl-ammonio)propyl]-, iodide (1 : 2)

Other Names   Quinolinium, 4-[(4-methyloxazolo[4,5-*b*]pyridin-2(4*H*)-ylidene)methyl]-1-[3-(trimethylammonio)propyl]-, diiodide; JO-PRO 1; JO-PRO 1 iodide

Merck Index Numbe   Not listed

Chemical/Dye Class   Cyanine

Molecular Formula $C_{23}H_{28}I_2N_4O$

Molecular Weight   630.31

Physical Form   Yellow-brown powder

Solubility   Soluble in dimethyl sulfoxide

Absorption (λ_{max})   530 nm (H_2O/DNA)

Emission (λ_{max})   546 nm (H_2O/DNA)

Molar Extinction Coefficient   94,400 cm^{-1} M^{-1} (H_2O/DNA)

Quantum Yield   0.38 (H_2O/DNA)

Synthesis   Synthetic methods[1,2]

Imaging/Labeling Applications   Nucleic acids;[1–9] cells;[9] sperms[9]

Biological/Medical Applications   Detecting nucleic acids,[1–9] cells;[9] apoptosis;[2] nucleic acid sequencing;[10] as temperature sensor[11,12]

Industrial Applications   Not reported

Safety/Toxicity   No data available

REFERENCES

1. Sabnis, R. W. *Handbook of Biological Dyes and Stains*; John Wiley & Sons Inc.: Hoboken, **2010**; p 259.

2. Haugland, R. P.; Yue, S. T. Aza-benzazolium-containing cyanine dyes and their use in fluorescent biological stains. PCT Int. Appl. WO 2000066664, 2000.

3. Sergeev, N. V.; Brevnov, M. G.; Furtado, M. R. Identification of nucleic acids. PCT Int. Appl. WO 2011143478, 2011.

4. Exner, M.; Rogers, A. Methods for detecting nucleic acids using multiple signals. U.S. Pat. Appl. Publ. US 2007172836, 2007.

5. Wittwer, C. T.; Dujols, V. E.; Reed, G.; Zhou, L. Amplicon melting analysis with saturation dyes. PCT Int. Appl. WO 2004038038, 2004.

6. Erikson, G. H.; Daksis, J. I.; Kandic, I.; Picard, P. Nucleic acid multiplex formation. PCT Int. Appl. WO 2002103051, 2002.

7. Erikson, G. H.; Daksis, J. I. Pre-incubation method to improve signal/noise ratio of nucleic acid

assays. U.S. Pat. Appl. Publ. US 2004180345, 2004.

8. Erikson, G. H. Method for modifying transcription and/or translation in an organism for therapeutic, prophylactic and/or analytic uses. U.S. Pat. Appl. Publ. US 2003181412, 2003.

9. Anderson, A. L.; Knutson, C. R.; Mueth, D.; Plewa, J.; Tanner, E. Methods for staining cells for identification and sorting. U.S. Pat. Appl. Publ. US 2006172315, 2006.

10. Hoser, M. J. Nucleic acid sequencing methods, kits and reagents. PCT Int. Appl. WO 2004074503, 2004.

11. Bousseksou, A.; Salmon, L.; Molnar, G.; Cobo, S. Materials with thermochromic spin transition doped with one or more fluorescent agents for use as temperature sensor. Fr. Demande FR 2952371, 2011.

12. Bousseksou, A.; Salmon, L.; Molnar, G.; Cobo, S. Heat-sensitive spin-transition materials doped with one or more fluorescent agents for use as temperature sensor. PCT Int. Appl. WO 2011058277, 2011.

LOLO 1

CAS Registry Number   305802-06-6
Chemical Structure

4 I⁻

CA Index Name   Thiazolo[4,5-b]pyridinium, 2,2′-[1,3-propanediylbis-[(dimethyliminio)-3,1-propanediyl-1(4H)-quinolinyl-4-ylidenemethylidyne]]bis[6-bromo-4-methyl-, iodide (1 : 4)

Other Names   Thiazolo[4,5-b]pyridinium, 2,2′-[1,3-propanediylbis-[(dimethyliminio)-3,1-propanediyl-1(4H)-quinolinyl-4-ylidenemethylidyne]]bis[6-bromo-4-methyl-, tetraiodide; LOLO 1; LOLO iodide

Merck Index Number   Not listed

Chemical/Dye Class   Cyanine

Molecular Formula $C_{47}H_{54}Br_2I_4N_8S_2$

Molecular Weight   1462.54

Physical Form   Yellow-brown powder

Solubility   Soluble in dimethyl sulfoxide

Absorption (λ_{max})   565 nm (H_2O/DNA)

Emission (λ_{max})   579 nm (H_2O/DNA)

Molar Extinction Coefficient   108,400 cm⁻¹ M⁻¹ (H_2O/DNA)

Quantum Yield   0.40 (H_2O/DNA)

Synthesis   Synthetic methods[1,2]

Imaging/Labeling Applications   Nucleic acids;[1–7] cells;[7] sperms;[7] microorganisms;[8] hairs[9]

Biological/Medical Applications   Detecting nucleic acids,[1–7] cells,[7] microorganisms;[8] nucleic acid nanochannel device;[10] nucleic acid sequencing;[11,12] as temperature sensor[13,14]

Industrial Applications   Not reported

Safety/Toxicity   No data available

REFERENCES

1. Sabnis, R. W. *Handbook of Biological Dyes and Stains*; John Wiley & Sons Inc.: Hoboken, **2010**; pp 264–265.

2. Haugland, R. P.; Yue, S. T. Aza-benzazolium-containing cyanine dyes and their use in fluorescent biological stains. PCT Int. Appl. WO 2000066664, 2000.

Handbook of Fluorescent Dyes and Probes, First Edition. R. W. Sabnis.
© 2015 John Wiley & Sons, Inc. Published 2015 by John Wiley & Sons, Inc.

3. Exner, M.; Rogers, A. Methods for detecting nucleic acids using multiple signals. U.S. Pat. Appl. Publ. US 2007172836, 2007.

4. Erikson, G. H.; Daksis, J. I. Pre-incubation method to improve signal/noise ratio of nucleic acid assays. U.S. Pat. Appl. Publ. US 2004180345, 2004.

5. Erikson, G. H. Method for modifying transcription and/or translation in an organism for therapeutic, prophylactic and/or analytic uses. U.S. Pat. Appl. Publ. US 2003181412, 2003.

6. Erikson, G. H.; Daksis, J. I.; Kandic, I.; Picard, P. Nucleic acid multiplex formation. PCT Int. Appl. WO 2002103051, 2002.

7. Anderson, A. L.; Knutson, C. R.; Mueth, D.; Plewa, J.; Tanner, E. Methods for staining cells for identification and sorting. U.S. Pat. Appl. Publ. US 2006172315, 2006.

8. Vannier, E. Methods for detection of pathogens in red blood cells. PCT Int. Appl. WO 2006031544, 2006.

9. Lagrange, A. Hair dye compositions containing a polycationic direct dye. Fr. Demande FR 2848840, 2004.

10. Tegenfeldt, J.; Reisner, W.; Flyvbjerg, H. Method for the mapping of the local AT/GC ratio along DNA. PCT Int. Appl. WO 2010042007, 2010.

11. Williams, J. G. K.; Anderson, J. P. Field-switch sequencing. U.S. Patent 7462452, 2008.

12. Williams, J. G. K.; Anderson, J. P. Field-switch sequencing. PCT Int. Appl. WO 2005111240, 2005.

13. Bousseksou, A.; Salmon, L.; Molnar, G.; Cobo, S. Materials with thermochromic spin transition doped with one or more fluorescent agents for use as temperature sensor. Fr. Demande FR 2952371, 2011.

14. Bousseksou, A.; Salmon, L.; Molnar, G.; Cobo, S. Heat-sensitive spin-transition materials doped with one or more fluorescent agents for use as temperature sensor. PCT Int. Appl. WO 2011058277, 2011.

LYSOSENSOR BLUE DND 167

CAS Registry Number   101821-61-8

Chemical Structure

CA Index Name   Morpholine, 4,4'-[9,10-anthra-cenediylbis(methylene)]bis-

Other Names   Morpholine, 4,4'-(9,10-anthrylene-dimethylene)di-; DND 167; LysoSensor Blue DND 167

Merck Index Number   Not listed

Chemical/Dye Class   Anthracene

Molecular Formula $C_{24}H_{28}N_2O_2$

Molecular Weight   376.49

Physical Form   Yellow plates[2]

Solubility   Insoluble in water; soluble in dimethyl sulfoxide

Melting Point   243 °C (decompose);[2] 229–230 °C[3]

Boiling Point (Calcd.)   546.5 ± 45.0 °C, pressure: 760 Torr

pK$_a$ 5.1, temperature: 22 °C

pK$_a$ (Calcd.)   6.97 ± 0.10, most basic, temperature: 25 °C

Absorption (λ_{max})   373 nm, 394 nm (Buffers pH 3.0, pH 7.0)

Emission (λ_{max})   425 nm, 401 nm (Buffers pH 3.0, pH 7.0)

Molar Extinction Coefficient   11,000 cm^{-1} M^{-1} (Buffers pH 3.0, pH 7.0)

Quantum Yield   0.80 (Buffer pH 3.0)

Synthesis   Synthetic methods[1–3]

Imaging/Labeling Applications   Lysosomes;[1,4–8] acidic vesicles[5]

Biological/Medical Applications   Analysizing of autophagic and necrotic cell death in *Dictyostelium discoideum*;[6] detecting nucleic acids;[9] measuring/controlling multidrug resistance;[7,8] sensing cholesterol composition of subcellular organelle membranes;[10] as temperature sensor[11,12]

Industrial Applications   Not reported

Safety/Toxicity   No data available

REFERENCES

1. Sabnis, R. W. *Handbook of Acid–base Indicators*; CRC Press: Boca Raton, **2008**; pp 206–207.

2. Beckett, A. H.; Walker, J. Steric interactions in substituted cyclohexadienes. II. Meso-substituted dihydroanthracenes; steric effects in the reactions of cis and trans isomers. *Tetrahedron* **1963**, *19*, 545–556.

3. Kaplan, S. Z.; Grad, N. M.; Zvontsova, A. S. N-Alkylated and N-aralkylated derivatives of morpholine. *Zh. Obshch. Khim.* **1958**, *28*, 3285–3289.

4. Galindo, F.; Burguete, M. I.; Vigara, L.; Luis, S. V.; Kabir, N.; Gavrilovic, J.; Russell, D. A.Synthetic macrocyclic peptidomimetics as tunable pH probes for the fluorescence imaging of acidic organelles in live cells. *Angew. Chem., Int. Ed.* **2005**, *44*, 6504–6508.

5. Lin, H.; Herman, P.; Kang, J. S.; Lakowicz, J. R. Fluorescence lifetime characterization of novel low-pH probes. *Anal. Biochem.* **2001**, *294*, 118–125.

6. Giusti, C.; Kosta, A.; Lam, D.; Tresse, E.; Luciani, M.; Golstein, P. Analysis of autophagic and necrotic cell death in *Dictyostelium. Methods Enzymol.* **2008**, *446*, 1–15.

7. Simon, S. M.; Schindler, M. S. Methods and agents for measuring and controlling multidrug resistance. U.S. Pat. Appl. Publ. US 20020042079, 2002.

8. Simon, S. I.; Schindler, M. S. Methods and agents for measuring and controlling multidrug resistance. PCT Int. Appl. WO 9960398, 1999.

9. Sergeev, N. V.; Brevnov, M. G.; Furtado, M. R. Identification of nucleic acids. PCT Int. Appl. WO 2011143478, 2011.

10. Wang, R.; Hosaka, M.; Han, L.; Yokota-Hashimoto, H.; Suda, M.; Mitsushima, D.; Torii, S.; Takeuchi, T. Molecular probes for sensing the cholesterol composition of subcellular organelle membranes. *Biochim.*

Biophys. Acta, Mol. Cell Biol. Lipids **2006**, *1761*, 1169–1181.

11. Bousseksou, A.; Salmon, L.; Molnar, G.; Cobo, S. Heat-sensitive spin-transition materials doped with one or more fluorescent agents for use as temperature sensor. PCT Int. Appl. WO 2011058277, 2011.

12. Bousseksou, A.; Salmon, L.; Molnar, G.; Cobo, S. Materials with thermochromic spin transition doped with one or more fluorescent agents for use as temperature sensor. Fr. Demande FR 2952371, 2011.

LYSOSENSOR BLUE DND 192

CAS Registry Number   166821-89-2

Chemical Structure

CA Index Name   9,10-Anthracenedimethanamine, N9,N9,N10,N10-tetramethyl-

Other Names   9,10-Anthracenedimethanamine, N,N,N′,N′-tetramethyl-; LysoSensor Blue DND 192; LysoSensor DND 192

Merck Index Number   Not listed

Chemical/Dye Class   Anthracene

Molecular Formula $C_{20}H_{24}N_2$

Molecular Weight   292.42

Physical Form   Yellow crystalline powder[2]

Solubility   Insoluble in water; soluble in dimethyl sulfoxide

Melting Point   >300 °C (decompose)[2]

Boiling Point (Calcd.)   418.0 ± 25.0 °C, pressure: 760 Torr

pK$_a$ 7.5, temperature: 22 °C

pK$_a$ (Calcd.)   9.11 ± 0.28 most basic, temperature: 25 °C

Absorption (λ_{max})   374 nm (Buffers pH 5.0, 9.0)

Emission (λ_{max})   424 nm (Buffers pH 5.0, 9.0)

Molar Extinction Coefficient $11{,}000\,cm^{-1}\,M^{-1}$ (Buffers pH 5.0, 9.0)

Quantum Yield   0.88 (Buffer pH 4.0)

Synthesis   Synthetic methods[1,2]

Imaging/Labeling Applications   Lysosomes;[1,3] acidic vesicles[1,3]

Biological/Medical Applications   As temperature sensor;[4,5] for ratiometric pH measurements[6]

Industrial Applications   Qualification of cooking oils;[7] quantitative temperature gradient focusing[8]

Safety/Toxicity   No data available

REFERENCES

1. Sabnis, R. W. *Handbook of Acid–base Indicators*; CRC Press: Boca Raton, **2008**; p 208.

2. Collins, C. J.; Lanz, M.; Goralski, C. T.; Singaram, B. Aminoborohydrides. 10. The synthesis of tertiary amine-boranes from various benzyl halides and lithium *N,N*-dialkylaminoborohydrides. *J. Org. Chem.* **1999**, *64*, 2574–2576.

3. Lin, H.; Herman, P.; Kang, J. S.; Lakowicz, J. R. Fluorescence lifetime characterization of novel low-pH probes. *Anal. Biochem.* **2001**, *294*, 118–125.

4. Bousseksou, A.; Salmon, L.; Molnar, G.; Cobo, S. Heat-sensitive spin-transition materials doped with one or more fluorescent agents for use as temperature sensor. PCT Int. Appl. WO 2011058277, 2011.

5. Bousseksou, A.; Salmon, L.; Molnar, G.; Cobo, S. Materials with thermochromic spin transition doped with one or more fluorescent agents for use as temperature sensor. Fr. Demande FR 2952371, 2011.

6. Kang, J. S.; Kostov, Y. Ratiometric pH measurements using LysoSensor DND-192. *J. Biochem. Mol. Biol.* **2002**, *35*, 384–388.

7. Wei, A.; Rajagopal, R.; Bineau, C. Device for the qualification of cooking oils, and methods. PCT Int. Appl. WO 2008086137, 2008.

8. Shackman, J. G.; Munson, M. S.; Kan, C.; Ross, D. Quantitative temperature gradient focusing performed using background electrolytes at various pH values. *Electrophoresis* **2006**, *27*, 3420–3427.

LYSOSENSOR GREEN DND 153

CAS Registry Number 231632-16-9
Chemical Structure

CA Index Name 7H-Benzimidazo[2,1-a]benz[de]
isoquinolin-7-one, 3-[[2-(dimethylamino)ethyl]amino]-
Other Names DND 153; LysoSensor Green DND 153
Merck Index Number Not listed

Chemical/Dye Class Heterocycle
Molecular Formula $C_{22}H_{20}N_4O$
Molecular Weight 356.42
Physical Form Orange solid
Solubility Insoluble in water; soluble in dimethyl sulfoxide, methanol
Boiling Point (Calcd.) $660.9 \pm 65.0\,°C$, pressure: 760 Torr
pK$_a$ 7.5, temperature: 22 °C
pK$_a$ (Calcd.) 9.10 ± 0.28, most basic, temperature: 25 °C
Absorption (λ_{max}) 442 nm (Buffers pH 5.0, pH 9.0)
Emission (λ_{max}) 505 nm (Buffers pH 5.0, pH 9.0)
Molar Extinction Coefficient $17,000\,cm^{-1}\,M^{-1}$ (Buffers pH 5.0, pH 9.0)
Quantum Yield 0.34 (Buffer pH 4.0)
Synthesis Synthetic method[1]
Imaging/Labeling Applications Lysosomes;[1,2] acidic vesicles;[1] living cells;[3] intestinal mucosa[4,5]
Biological/Medical Applications Detecting nucleic acids;[6] examining colitis;[2] test reagent for intestinal mucosa;[4,5] visualizing lysosomes[1]
Industrial Applications Not reported
Safety/Toxicity No data available

REFERENCES

1. Lin, H.; Herman, P.; Kang, J. S.; Lakowicz, J. R. Fluorescence lifetime characterization of novel low-pH probes. *Anal. Biochem.* **2001**, *294*, 118–125.

2. Ishiguro, K.; Ando, T.; Watanabe, O.; Goto, H. Novel application of low pH-dependent fluorescent dyes to examine colitis. *BMC Gastroenterol.* **2010**, *10*, 1–7.

3. Hille, C.; Berg, M.; Bressel, L.; Munzke, D.; Primus, P.; Loehmannsroeben, H.; Dosche, C. Time-domain fluorescence lifetime imaging for intracellular pH sensing in living tissues. *Anal. Bioanal. Chem.* **2008**, *391*, 1871–1879.

4. Ishiguro, K.; Goto, H.; Ando, T. Test reagent for intestinal mucosa fluorescence observation. Jpn. Kokai Tokkyo Koho JP 2010184866, 2010.

5. Ishiguro, K.; Goto, H.; Ando, T. Test reagent for intestinal mucosa fluorescence observation. PCT Int. Appl. WO 2010092917, 2010.

6. Sergeev, N. V.; Brevnov, M. G.; Furtado, M. R. Identification of nucleic acids. PCT Int. Appl. WO 2011143478, 2011.

LYSOSENSOR GREEN DND 189

CAS Registry Number   152584-38-8

Chemical Structure

CA Index Name   7H-Benzimidazo[2,1-a]benz[de] isoquinolin-7-one, 3-[[2-(4-morpholinyl)ethyl]amino]-

Other Names   DND 189; LysoSensor Green DND 189

Merck Index Number   Not listed

Chemical/Dye Class   Heterocycle

Molecular Formula $C_{24}H_{22}N_4O_2$

Molecular Weight   398.46

Physical Form   Red crystals[2]

Solubility   Insoluble in water; soluble in dimethyl sulfoxide

Melting Point   254–258 °C[2]

Boiling Point (Calcd.)   730.4 ± 70.0 °C, pressure: 760 Torr

pK$_a$ 5.2, temperature: 22 °C

pK$_a$ (Calcd.)   6.89 ± 0.10, most basic, temperature: 25 °C

Absorption (λ_{max})   443 nm (Buffers pH 3.0, 7.0)

Emission (λ_{max})   505 nm (Buffers pH 3.0, 7.0)

Molar Extinction Coefficient $16{,}000 \, cm^{-1} \, M^{-1}$ (Buffers pH 3.0, 7.0)

Quantum Yield   0.41 (Buffer pH 3.0); 0.48 (Buffer pH 4.4)

Synthesis   Synthetic methods[1,2]

Imaging/Labeling Applications   Lysosomes;[1–6] acidic vesicles;[3] amyloid oligomers;[7] endosomes;[5,6] intestinal mucosa[8,9]

Biological/Medical Applications   As an amyloid aggregation probe;[7] detecting nucleic acids;[10] examining colitis;[4] test reagent for intestinal mucosa;[8,9] visualizing lysosomes[1–6]

Industrial Applications   Not reported

Safety/Toxicity   No data available

REFERENCES

1. Sabnis, R. W. *Handbook of Acid–base Indicators*; CRC Press: Boca Raton, **2008**; p 209.

2. de Silva, A. P. Gunaratne, H. Q. N.; Lynch, P. L. M.; Patty, A. J.; Spence, G. L. Luminescence and charge transfer. Part 3. The use of chromophores with ICT (internal charge transfer) excited states in the construction of fluorescent PET (photoinduced electron transfer) pH sensors and related absorption pH sensors with aminoalkyl side chains. *J. Chem. Soc., Perkin Trans. 2* **1993**, 1611–1616.

3. Lin, H.; Herman, P.; Kang, J. S.; Lakowicz, J. R. Fluorescence lifetime characterization of novel low-pH probes. *Anal. Biochem.* **2001**, *294*, 118–125.

4. Ishiguro, K.; Ando, T.; Watanabe, O.; Goto, H. Novel application of low pH-dependent fluorescent dyes to examine colitis. *BMC Gastroenterol.* **2010**, *10*, 1–7.

5. Harashima, H.; Akita, H. Development of the fluorescent imaging method for quantifying nucleic acid incorporated into target organelle. Jpn. Tokkyo Koho JP 4493939, 2010.

6. Harashima, H.; Akita, H. Development of the fluorescent imaging method for quantifying nucleic acid incorporated into target organelle. PCT Int. Appl. WO 2004108962, 2004.

7. Kwon, H.; Jee, A. Y.; Lee, M. DND-189 as an amyloid aggregation probe. *Bull. Korean Chem. Soc.* **2009**, *30*, 1237–1238.

8. Ishiguro, K.; Goto, H.; Ando, T. Test reagent for intestinal mucosa fluorescence observation. Jpn. Kokai Tokkyo Koho JP 2010184866, 2010.

9. Ishiguro, K.; Goto, H.; Ando, T. Test reagent for intestinal mucosa fluorescence observation. PCT Int. Appl. WO 2010092917, 2010.

10. Sergeev, N. V.; Brevnov, M. G.; Furtado, M. R. Identification of nucleic acids. PCT Int. Appl. WO 2011143478, 2011.

LYSOSENSOR YELLOW/BLUE DND 160 (PDMPO)

CAS Registry Number   231632-18-1

Chemical Structure

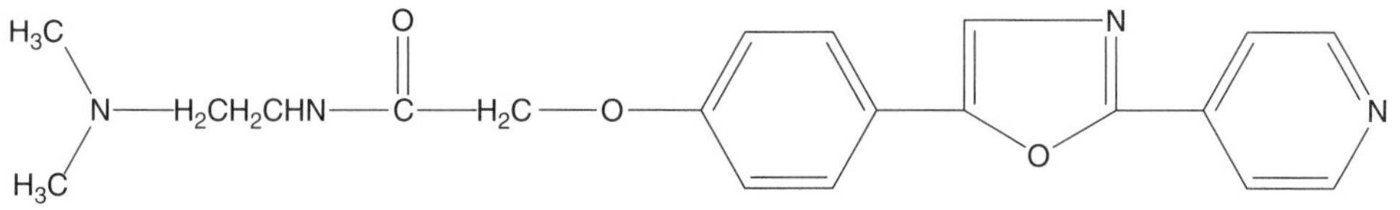

CA Index Name   Acetamide, N-[2-(dimethylamino) ethyl]-2-[4-[5-(4-pyridinyl)-2-oxazolyl]phenoxy]-

Other Names   DND 160; L 7545; Lyso Sensor DND 160; LysoSensor Yellow/Blue DND 160; 2-(4-Pyridyl)-5-{[4-(2-dimethylaminoethylaminocarbamoyl)-methoxy] phenyl}-oxazole; PDMPO

Merck Index Number   Not listed

Chemical/Dye Class   Heterocycle; Oxazole

Molecular Formula $C_{20}H_{22}N_4O_3$

Molecular Weight   366.41

Physical Form   Pale yellow solid[2]

Solubility   Insoluble in water; soluble in dimethyl sulfoxide, methanol

Melting Point   141–142 °C[2]

pK$_a$ 4.2, temperature: 22 °C

pK$_a$ (Calcd.)   14.25 ± 0.46, most acidic, temperature: 25 °C; 8.36 ± 0.10, most basic, temperature: 25 °C

Absorption (λ_{max})   384 nm (Buffer pH 3.0); 329 nm (Buffer pH 7.0)

Emission (λ_{max})   540 nm (Buffer pH 3.0); 440 nm (Buffer pH 7.0)

Molar Extinction Coefficient   21,000 cm^{-1} M^{-1} (Buffer pH 3.0); 23,000 cm^{-1} M^{-1} (Buffer pH 7.0)

Quantum Yield   0.31 (Buffer pH 3.0); 0.34 (Buffer pH 7.7)

Synthesis   Synthetic methods[1,2]

Imaging/Labeling Applications   Lysosomes;[1–4] acidic vesicles;[4–6] silica/silica deposits[7–12]

Biological/Medical Applications   Detecting nucleic acids;[13] biosensor system for liquid analytes;[14] regulating acidity within the lumen and tubulovesicle compartment of gastric parietal cells;[6] studying silica deposition;[7–12] treating age-related macular degeneration;[3] visualizing lysosomes[1–4]

Industrial Applications   Recording medium[15,16]

Safety/Toxicity   No data available

REFERENCES

1. Sabnis, R. W. *Handbook of Acid–base Indicators*; CRC Press: Boca Raton, **2008**; p 210.

2. Diwu, Z.; Chen, C.; Zhang, C.; Klaubert, D. H.; Haugland, R. P. A novel acidotrophic pH indicator and its potential application in labeling acidic organelles of live cells. *Chem. Biol.* **1999**, *6*, 411–418.

3. Mitchell, C.; Laties, A. M. Method for treatment of age-related macular degeneration by restoring lysosomal pH in retinal pigment epithelial cells. PCT Int. Appl. WO 2008042399, 2008.

4. Lin, H.; Herman, P.; Kang, J. S.; Lakowicz, J. R. Fluorescence lifetime characterization of novel low-pH probes. *Anal. Biochem.* **2001**, *294*, 118–125.

5. DePedro, H. M.; Urayama, P. Using LysoSensor Yellow/Blue DND-160 to sense acidic pH under high hydrostatic pressures. *Anal. Biochem.* **2009**, *384*, 359–361.

6. Gerbino, A.; Hofer, A. M.; Mckay, B.; Lau, B. W.; Soybel, D. I. Divalent cations regulate acidity within the lumen and tubulovesicle compartment of gastric parietal cells. *Gastroenterology* **2004**, *126*, 182–195.

7. Shimizu, K.; Del Amo, Y.; Brzezinski, M. A.; Stucky, G. D.; Morse, D. E. A novel fluorescent silica tracer for biological silicification studies. *Chem. Biol.* **2001**, *8*, 1051–1060.

8. Law, C.; Exley, C. New insight into silica deposition in horsetail (*Equisetum arvense*). *BMC Plant Biol.* **2011**, *11*, 112.

9. Ichinomiya, M.; Gomi, Y.; Nakamachi, M.; Ota, T.; Kobari, T. Temporal patterns in silica deposition among siliceous plankton during the spring bloom in the Oyashio region. *Deep-Sea Res., Part II: Top. Stud. Oceanogr.* **2010**, *57*, 1665–1670.

10. Znachor, P.; Nedoma, J. Application of the PDMPO technique in studying silica deposition in natural populations of *Fragilaria crotonensis* (Bacillariophyceae) at different depths in a eutrophic reservoir. *J. Phycol.* **2008**, *44*, 518–525.

11. Descles, J.; Vartanian, M.; El Harrak, A.; Quinet, M.; Bremond, N.; Sapriel, G.; Bibette, J.; Lopez, P. J. New tools for labeling silica in living diatoms. *New Phytol.* **2008**, *177*, 822–829.

12. Leblanc, K.; Hutchins, D. A. New applications of a biogenic silica deposition fluorophore in the study of oceanic diatoms. *Limnol. Oceanogr.: Methods* **2005**, *3*, 462–476.

13. Sergeev, N. V.; Brevnov, M. G.; Furtado, M. R. Identification of nucleic acids. PCT Int. Appl. WO 2011143478, 2011.

14. Walt, D. R.; Schauer, C. L.; Steemers, F. J. Cross-reactive sensors. PCT Int. Appl. WO 2001069245, 2001.

15. Mizuno, T.; Yamasaki, K.; Misawa, H. Three-dimensional optical memory in a photoacid-induced recording medium. *Jpn. J. Appl. Phys., Part 1* **2005**, *44*, 6593–6595.

16. Mizuno, T. Optic recording playback medium and optic recording playback method. Jpn. Kokai Tokkyo Koho JP 2005071538, 2005.

MARINA BLUE (6,8-DIFLUORO-7-HYDROXY-4-METHYLCOUMARIN) (DiFMU)

CAS Registry Number   215868-23-8

Chemical Structure

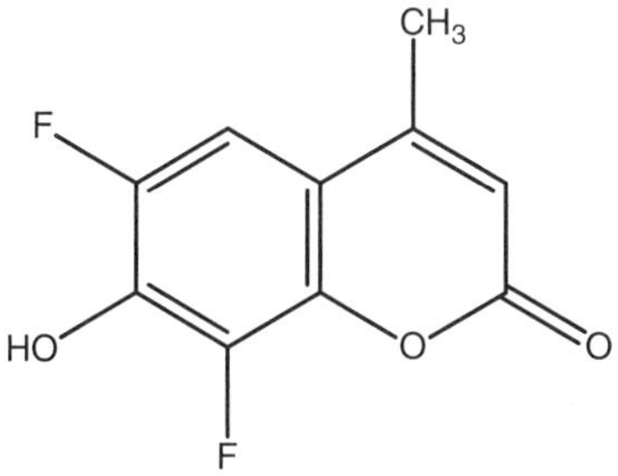

CA Index Name   2H-1-Benzopyran-2-one, 6,8-difluoro-7-hydroxy-4-methyl-

Other Names   6,8-Difluoro-7-hydroxy-4-methylcoumarin; 6,8-Difluoro-4-methylumbelliferon; DiFMU; Marina Blue

Merck Index Number   Not listed

Chemical/Dye Class   Coumarin

Molecular Formula $C_{10}H_6F_2O_3$

Molecular Weight   212.15

Physical Form   Colorless needles;[3] white solid[1]

Solubility   Insoluble in water; soluble in chloroform, *N,N*-dimethylformamide, dimethyl sulfoxide

Melting Point   154–155 °C[1]

Boiling Point (Calcd.)   332.3 ± 42.0 °C, pressure: 760 Torr

pK_a   4.7, temperature: 22 °C

pK_a (Calcd.)   5.50 ± 0.20, most acidic, temperature: 25 °C

Absorption (λ_{max})   358 nm (Buffers pH 10.0, pH 9.0)

Emission (λ_{max})   455 nm (Buffer pH 10.0); 452 nm (Buffer pH 9.0)

Molar Extinction Coefficient   17,500 cm^{-1} M^{-1} (Buffer pH 10.0); 18,000 cm^{-1} M^{-1} (Buffer pH 9.0)

Quantum Yield   0.89 (Buffer pH 10.0)

Synthesis   Synthetic methods[1–3]

Imaging/Labeling Applications   Aβ-proteins;[4,5] bacteria;[6] histone deacetylase;[7] nucleic acids;[14–20] peptides;[8] proteins;[9–12] sugar chains[13]

Biological/Medical Applications   Amplifying/detecting nucleic acid sequences;[14–18] analyzing peptides,[8] sugar chains;[13] detecting bacteria;[6] detecting/quantifying nucleic acids;[19,20] a substrate for measuring acyl protein thioesterase (APT1) activity,[1,30] α-amylase activity,[21] aryl sulfatases activity,[22] β-galactosidases activity,[3,23,29] guanidinobenzoatase activity,[3] β-lactamase activity,[24] lipase activity,[25] phosphatases activity,[3,26–29] phospholipase LYPLAL1 activity,[30] protein kinases activity,[31] xylanase activity;[32] diagnosing/treating traumatic brain injury;[4] treating amyloidogenic disease;[5] as temperature sensor[33,34]

Industrial Applications   Detecting/quantifying/measuring metals (beryllium, chromium, lead)[35]

Safety/Toxicity   No data available

REFERENCES

1. Hedberg, C.; Dekker, F. J.; Rusch, M.; Renner, S.; Wetzel, S.; Vartak, N.; Gerding-Reimers, C.; Bon, R. S.; Bastiaens, P. I. H.; Waldmann, H. Development of highly potent inhibitors of the ras-targeting human acyl protein thioesterases based on substrate similarity design. *Angew. Chem., Int. Ed.* **2011**, *50*, 9832–9837.

2. Sun, W.; Gee, K. R.; Haugland, R. P. Synthesis of novel fluorinated coumarins: excellent UV-light excitable fluorescent dyes. *Bioorg. Med. Chem. Lett.* **1998**, *8*, 3107-3110.

3. Gee, K. R.; Haugland, R. P.; Sun, W. C. Derivatives of 6,8-difluoro-7-hydroxycoumarin. U.S. Patent 5830912, 1998.

4. Duan, D. R.; Moll, J. R.; Rudolph, A. Diagnosis and treatment of traumatic brain injury using fluorescence labeled probes for Aβ-protein determination. PCT Int. Appl. WO 2011056958, 2011.

5. Duan, D. R.; Moll, J. R.; Rudolph, A.; Wegrzyn, R. Conformationally dynamic fluorescently labeled peptide probes and use for diagnosing and treating

an amyloidogenic disease. PCT Int. Appl. WO 2010088411, 2010.

6. Jadamec, J. R.; Bauman, R.; Anderson, C. P.; Jakubielski, S. A.; Sutton, N. D.; Kovacs, M. J. Method for detecting bacteria in a sample using a metabolizable fluorescent conjugate and measuring fluorescence emission intensity ratio. U.S. Patent 5968762, 1999.

7. Heidebrecht, R. W., Jr.,; Kral, A. M.; Miller, T. A. Fluorescent compounds that bind to histone deacetylase and their preparation. U.S. Pat. Appl. Publ. US 20090156825, 2009.

8. Pashkova, A.; Moskovets, E.; Karger, B. L. Coumarin tags for improved analysis of peptides by MALDI-TOF MS and MS/MS. 1. Enhancement in MALDI MS signal intensities. *Anal. Chem.* **2004**, *76*, 4550–4557.

9. Trowell, S. C.; Horne, I. M.; Dacres, H.; Leitch, V. Bioluminescent resonance energy transfer detection of compounds that bind or activate G protein-coupled receptors. PCT Int. Appl. WO 2010085844, 2010.

10. Kang, S.; Jun, S.; Kim, D. Fluorescent labeling of cell-free synthesized proteins by incorporation of fluorophore-conjugated nonnatural amino acids. *Anal. Biochem.* **2007**, *360*, 1–6.

11. Anderson, B. R.; Hall, D. L.; Groll, M. C.; Bingham, C. P.; Thompson, J. E.; Spendlove, R. S. Quality control and normalization methods for printed protein microarrays based on the use of fluorescent dye in printing buffer. U.S. Pat. Appl. Publ. US 20060063197, 2006.

12. Paysan, J.; Antz, C. Method for studying interactions of cellular molecules and their localization in cells using fluorescent-labeled fusion proteins. Eur. Pat. Appl. EP 969284, 2000.

13. Shimaoka, H.; Abe, A. Glycoprotein sugar chain analysis method. Jpn. Kokai Tokkyo Koho JP 2009156587, 2009.

14. Rohthmann, T.; Engel, H.; Himmelreich, R.; Wende, A.; Dahlke, R. Amplification and detection of multiple nucleic acid sequences in a reaction cartridge using probes differing in melting point. PCT Int. Appl. WO 2010128041, 2010.

15. Rohthmann, T.; Engel, H.; Himmelreich, R.; Wende, A.; Dahlke, R. Amplification and detection of multiple nucleic acid sequences in a reaction cartridge using probes differing in melting point. Eur. Pat. Appl. EP 2248915, 2010.

16. Rohthmann, T.; Engel, H. Simultaneous detection of multiple nucleic acid sequences in amplification reaction by probes containing multiple fluorescent labels and by different melting points for probes containing same label. PCT Int. Appl. WO 2009135832, 2009.

17. Rohthmann, T.; Engel, H.; Laue, T. Simultaneous detection of multiple nucleic acid sequences in amplification reaction by probes containing multiple fluorescent labels and by different melting points for probes containing same label. Eur. Pat. Appl. EP 2116614, 2009.

18. Barany, F.; Pingle, M.; Bergstrom, D. Methods for detection of target nucleic acid sequences using fluorescence resonance energy transfer. PCT Int. Appl. WO 2009126678, 2009.

19. Fang, N. Method for detecting and/or quantifying poly(A) RNA and mRNA by hybridization with poly(dT) and digestion with single strand-specific nuclease. Ger. Offen. DE 102009056729, 2011.

20. Fang, N. Method for detecting and/or quantifying poly(A) RNA and mRNA by hybridization with poly(dT) and digestion with single strand-specific nuclease. PCT Int. Appl. WO 2011067299, 2011.

21. Cohen, B. A. Determination of α-amylase in flour and other products by using labeled starch substrate. U.S. Pat. Appl. Publ. US 20040043116, 2004.

22. Ahmed, V.; Ispahany, M.; Ruttgaizer, S.; Guillemette, G.; Taylor, S. D. A fluorogenic substrate for the continuous assaying of aryl sulfatases. *Anal. Biochem.* **2005**, *340*, 80–88.

23. Szucs, J.; Pretsch, E.; Gyurcsanyi, R. E. Potentiometric enzyme immunoassay using miniaturized anion-selective electrodes for detection. *Analyst* **2009**, *134*, 1601–1607.

24. Corry, S.; Downey, W.; Filanoski, B.; Gee, K.; Greenfield, I. L.; Hirsch, J.; Johnson, I.; Rukavishnikov, A. Fluorogenic substrates for fluorometric determination of β-lactamase and use for detection of gene expression and in immunoassay. PCT Int. Appl. WO 2005071096, 2005.

25. Qian, Z.; Lutz, S. Improving the catalytic activity of *Candida antarctica* lipase B by circular permutation. *J. Am. Chem. Soc.* **2005**, *127*, 13466–13467.

26. Ni, L.; Swingle, M. S.; Bourgeois, A. C.; Honkanen, R. E. High yield expression of serine/threonine protein phosphatase Type 5, and a fluorescent assay suitable for use in the detection of catalytic inhibitors. *Assay Drug Dev. Technol.* **2007**, *5*, 645–654.

27. Dyhrman, S. T.; Ruttenberg, K. C. Presence and regulation of alkaline phosphatase activity in eukaryotic phytoplankton from the coastal ocean: implications for dissolved organic phosphorus

remineralization. *Limnol. Oceanogr.* **2006**, *51*, 1381–1390.

28. Kerby, M.; Chien, R. A fluorogenic assay using pressure-driven flow on a microchip. *Electrophoresis* **2001**, *22*, 3916–3923.

29. Gee, K. R.; Sun, W.; Bhalgat, M. K.; Upson, R. H.; Klaubert, D. H.; Latham, K. A.; Haugland, R. P. Fluorogenic substrates based on fluorinated umbelliferones for continuous assays of phosphatases and β-galactosidases. *Anal. Biochem.* **1999**, *273*, 41–48.

30. Buerger, M.; Zimmermann, T. J.; Kondoh, Y.; Stege, P.; Watanabe, N.; Osada, H.; Waldmann, H.; Vetter, I. R. Crystal structure of the predicted phospholipase LYPLAL1 reveals unexpected functional plasticity despite close relationship to acyl protein thioesterases. *J. Lipid Res.* **2012**, *53*, 43–50.

31. Lawrence, D. S. Fluorescent assays for screening for protein kinase inhibitors applicable in cancer treatment and diagnosis. PCT Int. Appl. WO 2004062475, 2004.

32. Ge, Y.; Antoulinakis, E. G.; Gee, K. R.; Johnson, I. An ultrasensitive, continuous assay for xylanase using the fluorogenic substrate 6,8-difluoro-4-methylumbelliferyl β-D-xylobioside. *Anal. Biochem.* **2007**, *362*, 63-68.

33. Bousseksou, A.; Salmon, L.; Molnar, G.; Cobo, S. Heat-sensitive spin-transition materials doped with one or more fluorescent agents for use as temperature sensor. PCT Int. Appl. WO 2011058277, 2011.

34. Bousseksou, A.; Salmon, L.; Molnar, G.; Cobo, S. Materials with thermochromic spin transition doped with one or more fluorescent agents for use as temperature sensor. Fr. Demande FR 2952371, 2011.

35. Agrawal, A.; Cronin, J. P.; Adams, L. L.; Agrawal, A.; Tonazzi, J. C. L. Methods for metal assays using optical techniques. U.S. Pat. Appl. Publ. US 20100035351, 2010.

7-METHOXYCOUMARIN-3-CARBONYL AZIDE

CAS Registry Number   97632-67-2

Chemical Structure

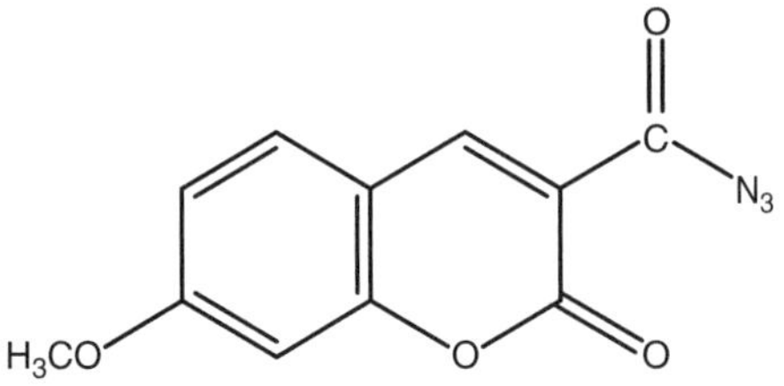

CA Index Name   2*H*-1-Benzopyran-3-carbonyl azide, 7-methoxy-2-oxo-

Other Names   7-Methoxycoumarin-3-carbonyl azide

Merck Index Number   Not listed

Chemical/Dye Class   Coumarin

Molecular Formula $C_{11}H_7N_3O_4$

Molecular Weight   245.19

Physical Form   Pale yellow needles[1]

Solubility   Soluble in acetonitrile, chloroform, *N,N*-dimethylformamide, methanol

Melting Point   170–171 °C[1]

Absorption (λ_{max})   360 nm (MeOH)

Emission (λ_{max})   415 nm (MeOH)

Molar Extinction Coefficient   25,000 cm^{-1} M^{-1} (MeOH)

Synthesis   Synthetic method[1]

Imaging/Labeling Applications   Alcohols;[1,2] biomacromolecules;[3,4] glycerophospholipids;[5] 7α-hydroxycholesterol[6]

Biological/Medical Applications   Analyzing molecular species of glycerophospholipids;[5] detecting 7α-hydroxycholesterol in dog plasma;[6] encapsulating/microencapsulating biomacromolecules;[3,4] reagent for alcohols[1,2]

Industrial Applications   Poly(propargyl acrylate) (pPA) polymer coatings/fims;[7] poly(propargyl methacrylate) (PPMA) polymer coatings/fims[8]

Safety/Toxicity   No data available

REFERENCES

1. Takadate, A.; Irikura, M.; Suehiro, T.; Fujino, H.; Goya, S. New labeling reagents for alcohols in fluorescence high-performance liquid chromatography. *Chem. Pharm. Bull.* **1985**, *33*, 1164–1169.

2. Nelson, T. J. Fluorescent high-performance liquid chromatography assay for lipophilic alcohols. *Anal. Biochem.* **2011**, *419*, 40–45.

3. Schwendeman, S. P. Methods for encapsulation of biomacromolecules in polymers. PCT Int. Appl. WO 2005117942, 2005.

4. Reinhold, S. E.; Desai, K. H.; Zhang, L.; Olsen, K. F.; Schwendeman, S. P. Self-healing microencapsulation of biomacromolecules without organic solvents. *Angew. Chem., Int. Ed.* **2012**, *51*, 10800–10803.

5. Mckeone, B. J.; Osmundsen, K.; Brauchi, D.; Pao, Q.; Payton-Ross, C.; Kilinc, C.; Kummerow, F. A.; Pownall, H. J. Alterations in serum phosphatidylcholine fatty acyl species by eicosapentenoic and docosahexaenoic ethyl esters in patients with severe hypertriglyceridemia *J. Lipid Res.* **1997**, *38*, 429–436.

6. Saisho, Y.; Shimada, C.; Umeda, T. Determination of 7α-hydroxycholesterol in dog plasma by high-performance liquid chromatography with fluorescence detection. *Anal. Biochem.* **1998**, *265*, 361–367.

7. Im, S. G.; Kim, B.; Tenhaeff, W. E.; Hammond, P. T.; Gleason, K. K. Single-step synthesis of polymer surfaces for "click" chemistry via initiative chemical vapor deposition (ICVD). *Polym. Prepr.* **2008**, *49*, 204–205.

8. Im, S. G.; Kim, B.; Lee, L. H.; Tenhaeff, W. E.; Hammond, P. T.; Gleason, K. K. A directly patternable, click-active polymer film via initiated chemical vapor deposition. *Macromol. Rapid Commun.* **2008**, *29*, 1648–1654.

7-METHOXYCOUMARIN-3-CARBOXYLIC ACID

CAS Registry Number   20300-59-8

Chemical Structure

CA Index Name $2H$-1-Benzopyran-3-carboxylic acid, 7-methoxy-2-oxo-

Other Names   3-Carboxy-7-methoxycoumarin; 7-methoxy-2-oxo-$2H$-chromene-3-carboxylic acid; 7-methoxy-$2H$-chromene-3-carboxylic acid; 7-methoxy-3-carboxycoumarin; 7-methoxycoumarin-3-carboxylic acid

Merck Index Number   Not listed

Chemical/Dye Class   Coumarin

Molecular Formula $C_{11}H_8O_5$

Molecular Weight   220.18

Physical Form   White crystals;[11] white crystalline solid;[10] white glistening flakes;[36] off-white solid[19,21] or colorless solid; Orange crystals;[17] pale greenish-yellow plates[35]

Solubility   Soluble in acetone, acetonitrile, N,N-dimethylformamide, dimethyl sulfoxide

Melting Point   231–233 °C;[17]   208–210 °C;[19] 203 °C;[27]   198–199 °C;[33]   197–199 °C;[21]   195 °C;[35,36] 193–195 °C;[3,13]   193–194 °C;[7,14,18]   192–194 °C;[25,34] 192–193 °C;[12,24,26,32]   192 °C;[28]   191.5–192.5 °C;[22] 190–191 °C;[6]   190–195 °C;[10]   178 °C;[2]   176–177 °C;[29,30] 169 °C[11]

Boiling Point (Calcd.) 424.4 ± 45.0 °C, pressure: 760 Torr

pK$_a$ (Calcd.) 1.92 ± 0.20, most acidic, temperature: 25 °C

Absorption (λ_{max})   336 nm (Buffer pH 9.0)

Emission (λ_{max})   402 nm (Buffer pH 9.0)

Molar Extinction Coefficient   20,000 cm^{-1} M^{-1} (Buffer pH 9.0)

Synthesis   Synthetic methods[1–36]

Imaging/Labeling Applications   Iron chelators;[22] lysine;[23,37] peptides;[38–41] proteins;[42,43] ribonucleopeptide (RNP);[44–48] sulfenic acids;[21,49] thiols[50]

Biological/Medical Applications   Assaying thiols;[50] evaluating iron chelators;[22] identifying/mapping sulfenic acid modifications;[21,49] as a substrate for measuring peptidases activity,[51] proteases/proteinases activity;[39–41] as *Mycobacterium bovis* (BCG) inhibitors;[5] human monoamine oxidase inhibitors;[9] *N*-Methyl-D-aspartate receptors (NMDARs) inhibitors;[52] dopamine sensors;[44] viscosity sensors;[53] treating cancer/antitumor activity;[1] treating cognitive deficits or schizophrenia[52]

Industrial Applications   For organic light-emitting device applications[54]

Safety/Toxicity   No data available

REFERENCES

1. Draoui, N.; Schicke, O.; Fernandes, A.; Drozak, X.; Nahra, F.; Dumont, A.; Douxfils, J.; Hermans, E.; Dogne, J.; Corbau, R.; Marchand, A.; Chaltin, P.; Sonveaux, P.; Feron, O.; Riant, O. Synthesis and pharmacological evaluation of carboxycoumarins as a new antitumor treatment targeting lactate transport in cancer cells. *Bioorg. Med. Chem.* **2013**, *21*, 7107–7117.

2. Lamba, M. S.; Makrandi, J. K.; Kumar, S. Synthesis of 3-carboxycoumarins using phase transfer catalysis. *Heterocycl. Lett.* **2013**, *3*, 177–181.

3. Chavan, H. V.; Bandgar, B. P. Aqueous extract of *Acacia concinna* pods: An efficient surfactant type catalyst for synthesis of 3-carboxycoumarins and cinnamic acids via Knoevenagel condensation. *ACS Sustainable Chem. Eng.* **2013**, *1*, 929–936.

4. Cui, Z. Preparation of 7-methoxycoumarin-3-carboxylic acid. Faming Zhuanli Shenqing CN 102924415, 2013.

5. Rezayan, A. H.; Azerang, P.; Sardari, S.; Sarvary, A. Synthesis and biological evaluation of coumarin derivatives as inhibitors of *Mycobacterium bovis* (BCG). *Chem. Biol. Drug Des.* **2012**, *80*, 929–936.

6. Bardajee, G. R.; Moallem, S. A. Synthesis of 3-carboxycoumarins catalyzed by CuSO$_4$.5H$_2$O under ultrasound irradiation in aqueous media. *Asian J. Biochem. Pharm. Res.* **2012**, *2*, 410–414.

7. Karami, B.; Farahi, M.; Khodabakhshi, S. Rapid synthesis of novel and known coumarin–3-carboxylic acids using stannous chloride dihydrate under solvent-free conditions. *Helv. Chim. Acta* **2012**, *95*, 455–460.

8. Shahinian, E. G. H.; Sebe, I. Synthesis of some new benzocoumarin heterocyclic fluorescent dyes. *Rev. Chim.* **2011**, *62*, 1098–1101.

9. Secci, D.; Carradori, S.; Bolasco, A.; Chimenti, P.; Yanez, M.; Ortuso, F.; Alcaro, S. Synthesis and selective human monoamine oxidase inhibition of 3-carbonyl, 3-acyl, and 3-carboxyhydrazido coumarin derivatives. *Eur. J. Med. Chem.* **2011**, *46*, 4846–4852.

10. Harishkumar, H. N.; Mahadevan, K. M.; Kiran Kumar, H. C.; Satyanarayan, N. D. A facile, choline chloride/urea catalyzed solid phase synthesis of coumarins via Knoevenagel condensation. *Org. Commun.* **2011**, *4*, 26–32.

11. Caron, K.; Lachapelle, V.; Keillor, J. W. Dramatic increase of quench efficiency in "spacerless" dimaleimide fluorogens. *Org. Biomol. Chem.* **2011**, *9*, 185–197.

12. Bardajee, G. R.; Jafarpour, F.; Afsari, H. S. ZrOCl$_2$.8H$_2$O, an efficient catalyst for rapid one-pot synthesis of 3-carboxycoumarins under ultrasound irradiation in water. *Cent. Eur. J. Chem.* **2010**, *8*, 370–374.

13. Shaabani, A.; Ghadari, R.; Rahmati, A.; Rezayan, A. H. Coumarin synthesis via Knoevenagel condensation reaction in 1,1,3,3,-N,N,N′,N′-tetramethylguanidinium trifluoroacetate ionic liquid. *J. Iran. Chem. Soc.* **2009**, *6*, 710–714.

14. Hekmatshoar, R.; Rezaei, A.; Beheshtiha, S. Y. S. Silica sulfuric acid, a versatile and reusable catalyst for synthesis of coumarin-3-carboxylic acids in a solventless system. *Phosphorus, Sulfur Silicon Relat. Elem.* **2009**, *184*, 2491–2496.

15. Li, J.; Shen, S.; Ma, W.; Xu, Q.; Sun, X.; Sun, X. Synthesis and spectrum properties of water-soluble coumarin derivative. *Huagong Shikan* **2007**, *21*, 7–9.

16. Moussaoui, Y.; Ben Salem, R. Catalyzed Knoevenagel reactions on inorganic solid supports: Application to the synthesis of coumarin compounds. *C. R. Chim.* **2007**, *10*, 1162–1169.

17. Bardajee, G. R.; Winnik, M. A.; Lough, A. J. 7-Methoxy-2-oxo-2H-chromene-3-carboxylic acid. *Acta Crystallogr., Sect. E* **2007**, *E63*, o1269–o1270.

18. Du, J.; Li, L.; Zhang, D. Ultrasound promoted synthesis of 3-carboxycoumarins in aqueous media. *E-J. Chem.* **2006**, *3*, 1–4.

19. Creaven, B. S.; Egan, D. A.; Kavanagh, K.; McCann, M.; Noble, A.; Thati, B.; Walsh, M. Synthesis, characterization and antimicrobial activity of a series of substituted coumarin-3-carboxylato silver(I) complexes. *Inorg. Chim. Acta* **2006**, *359*, 3976–3984.

20. Du, J.; Li, L.; Zhang, D. Solvent free synthesis of 3-carboxycoumarins. *Chem. Indian J.* **2006**, *3*, 6–7.

21. Poole, L. B.; Zeng, B.; Knaggs, S. A.; Yakubu, M.; King, S. B. Synthesis of chemical probes to map sulfenic acid modifications on proteins. *Bioconjugate Chem.* **2005**, *16*, 1624–1628.

22. Ma, Y.; Luo, W.; Quinn, P. J.; Liu, Z.; Hider, R. C. Design, synthesis, physicochemical properties, and evaluation of novel iron chelators with fluorescent sensors. *J. Med. Chem.* **2004**, *47*, 6349–6362.

23. Berthelot, T.; Lain, G.; Latxague, L.; Deleris, G. Synthesis of novel fluorogenic L-fmoc lysine derivatives as potential tools for imaging cells. *J. Fluoresc.* **2004**, *14*, 671–675.

24. Fringuelli, F.; Piermatti, O.; Pizzo, F. One-pot synthesis of 3-carboxycoumarins via consecutive Knoevenagel and Pinner reactions in water. *Synthesis* **2003**, 2331–2334.

25. Song, A.; Wang, X.; Lam, K. S. A convenient synthesis of coumarin-3-carboxylic acids via Knoevenagel condensation of Meldrum's acid with ortho-hydroxyaryl aldehydes or ketones. *Tetrahedron Lett.* **2003**, *44*, 1755–1758.

26. Tang, J.; Huang, X. An efficient solid-phase synthesis of 3-carboxycoumarins based on a scaffold-polymer-bound cyclic malonic ester. *J. Chem. Res. Synop.* **2003**, 354–355.

27. Deshmukh, M. N.; Burud, R.; Baldino, C.; Chan, P. C. M.; Liu, J. A practical and environmentally friendly preparation of 3-carboxycoumarins. *Synth. Commun.* **2003**, *33*, 3299–3303.

28. Bandgar, B. P.; Uppalla, L. S.; Kurule, D. S. Solvent-free one-pot rapid synthesis of 3-carboxycoumarins. *Green Chem.* **1999**, *1*, 243–245.

29. Bonsignore, L.; Cottiglia, F.; Maccioni, A. M.; Secci, D.; Lavagna, S. M. Synthesis of coumarin-3-O-acylisoureas by dicyclohexylcarbodiimide. *J. Heterocycl. Chem.* **1995**, *32*, 573–577.

30. Hormi, O. E. O.; Peltonen, C.; Bergstrom, R. A one-pot synthesis of coumarins from dipotassium O-methoxybenzylidenemalonates. *J. Chem. Soc., Perkin Trans. 1* **1991**, 219–221.

31. Takadate, A.; Yagashiro, I.; Irikura, M.; Fujino, H.; Goya, S. 3-(7-Methoxycoumarin-3-carbonyl)- and 3-(7-dimethylaminocoumarin-3-carbonyl)-2-oxazolones as new fluorescent labeling reagents for high-performance liquid chromatography. *Chem. Pharm. Bull.* **1989**, *37*, 373–376.

32. Armstrong, V.; Soto, O.; Valderrama, J. A.; Tapia, R. Synthesis of 3-carboxycoumarins from o-methoxybenzylidene Meldrum's acid derivatives. *Synth. Commun.* **1988**, *18*, 717–725.

33. Shirokova, E. A.; Segal, G. M.; Torgov, I. V. Use of Meldrum's acid in the syntheses of low molecular weight bioregulators. 3. Synthesis of coumarin-3-carboxylic acids and their derivatives. *Bioorg. Khim.* **1988**, *14*, 236–242.

34. Hoegberg, T.; Vora, M.; Drake, S. D.; Mitscher,L. A.; Chu, D. T. W. Structure-activity relationships among DNA-gyrase inhibitors. Synthesis and antimicrobial evaluation of chromones and coumarins related to oxolinic acid. *Acta Chem. Scand., Ser. B* **1984**, *B38*, 359–366.

35. Baker, W.; Collis, C. B. Fluorescent acylating agents derived from 7-hydroxycoumarin. *J. Chem. Soc. (Suppl.)* **1949**, S12–S15.

36. Rangaswami, S.; Seshadri, T. R.; Venkateswarlu, V. The remarkable fluorescence of certain coumarin derivatives. *Proc. Indian Acad. Sci., Sect. A* **1941**, *13A*, 316–321.

37. Berthelot, T.; Talbot, J.; Lain, G.; Deleris, G.; Latxague, L. Synthesis of Nε-(7-diethylaminocoumarin-3-carboxyl)- and Nε-(7-methoxycoumarin-3-carboxyl)-L-Fmoc lysine as tools for protease cleavage detection by fluorescence. *J. Pept. Sci.* **2005**, *11*, 153–160.

38. Katritzky, A. R.; Abdelmajeid, A.; Tala, S. R.; Amine, M. S.; Steel, P. J. Novel fluorescent aminoxy acids and aminoxy hybrid peptides. *Synthesis* **2011**, 83–90.

39. Berthelot, T.; Deleris, G. Biochips for detecting the enzymatic activity of a protease enzyme. PCT Int. Appl. WO 2010000593, 2010.

40. Berthelot, T.; Deleris, G. Fluorescent cyclic peptides, methods for the preparation thereof, and use of said peptides for measuring the enzymatic activity of a protease enzyme. PCT Int. Appl. WO 2010000591, 2010.

41. Berthelot, T.; Deleris, G. Cyclic peptides labeled with FRET pairs of dyes as assay substrates for proteinases and their preparation. Fr. Demande FR 2932190, 2009.

42. Keillor, J.; Caron, K. Fluorescent markers and use thereof for labelling specific protein targets. Can. Pat. Appl. CA 2746891, 2012.

43. Tegler, L. T.; Nonglaton, G.; Buettner, F.; Caldwell, K.; Christopeit, T.; Danielson, U. H.; Fromell, K.; Gossas, T.; Larsson, A.; Longati, P.; Norberg, T.; Ramapanicker, R.; Rydberg, J.; Baltzer, L. Powerful protein binders from designed polypeptides and small organic molecules-A general concept for protein recognition. *Angew. Chem., Int. Ed.* **2011**, *50*, 1823–1827.

44. Liew, F. F.; Hasegawa, T.; Fukuda, M.; Nakata, E.; Morii, T. Construction of dopamine sensors by using fluorescent ribonucleopeptide complexes. *Bioorg. Med. Chem.* **2011**, *19*, 4473–4481.

45. Fukuda, M.; Hayashi, H.; Hasegawa, T.; Morii, T. Development of a fluorescent ribonucleopeptide sensor for histamine. *Trans. Mater. Res. Soc. Jpn.* **2009**, *34*, 525–527.

46. Hasegawa, T.; Hagihara, M.; Fukuda, M.; Nakano, S.; Fujieda, N.; Morii, T. Context-dependent fluorescence detection of a phosphorylated tyrosine residue by a ribonucleopeptide. *J. Am. Chem. Soc.* **2008**, *130*, 8804–8812.

47. Hasegawa, T.; Hagihara, M.; Fukuda, M.; Morii, T. Stepwise functionalization of ribonucleopeptides: Optimization of the response of fluorescent ribonucleopeptide sensors for ATP. *Nucleosides, Nucleotides Nucleic Acids* **2007**, *26*, 1277–1281.

48. Hagihara, M.; Fukuda, M.; Hasegawa, T.; Morii, T. A modular strategy for tailoring fluorescent biosensors from ribonucleopeptide complexes. *J. Am. Chem. Soc.* **2006**, *128*, 12932–12940.

49. Poole, L. B.; King, S. B.; Fetrow, J. S. Sulfenic acid-reactive compounds and their methods of synthesis and use in detection or isolation of sulfenic acid-containing compounds. U.S. Pat. Appl. Publ. US 20060084173, 2006.

50. Katritzky, A. R.; Ibrahim, T. S.; Tala, S. R.; Abo-Dya, N. E.; Abdel-Samii, Z. K.; El-Feky, S. A. Synthesis of coumarin conjugates of biological thiols for fluorescent detection and estimation. *Synthesis* **2011**, 1494–1500.

51. Tisljar, U.; Knight, C. G.; Barrett, A. J. An alternative quenched fluorescence substrate for Pz-peptidase. *Anal. Biochem.* **1990**, *186*, 112–115.

52. Irvine, M. W.; Costa, B. M.; Volianskis, A.; Fang, G.; Ceolin, L.; Collingridge, G. L.; Monaghan, D. T.; Jane, D. E. Coumarin-3-carboxylic acid derivatives as potentiators and inhibitors of recombinant and native N-methyl-D-aspartate receptors. *Neurochem. Int.* **2012**, *61*, 593–600.

53. Yoon, H.; Dakanali, M.; Lichlyter, D.; Chang, W. M.; Nguyen, K. A.; Nipper, M. E.; Haidekker, M. A.; Theodorakis, E. A. Synthesis and evaluation of self-calibrating ratiometric viscosity sensors. *Org. Biomol. Chem.* **2011**, *9*, 3530–3540.

54. Karkkainen, A. H. O.; Hormi, O. E. O.; Rantala, J. T. Covalent bonding of coumarin molecules to sol-gel matrices for organic light-emitting device applications. *Proc. SPIE-Int. Soc. Opt. Eng.* **2000**, *3943*, 194–209.

7-METHOXYCOUMARIN-3-CARBOXYLIC ACID SUCCINIMIDYL ESTER

CAS Registry Number 150321-92-9

Chemical Structure

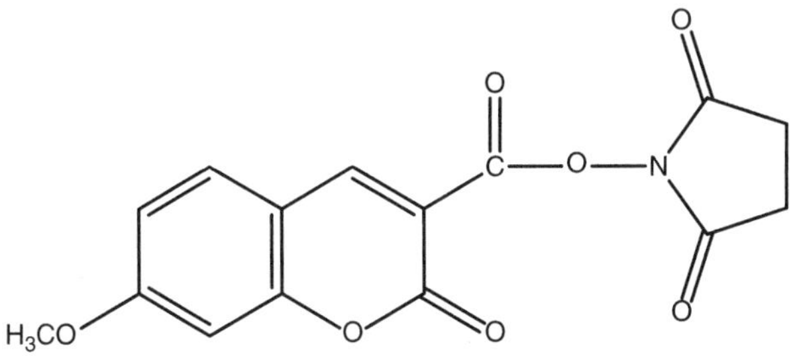

CA Index Name 2H-1-Benzopyran-3-carboxylic acid, 7-methoxy-2-oxo-, 2,5-dioxo-1-pyrrolidinyl ester

Other Names 2,5-Pyrrolidinedione, 1-[[(7-methoxy-2-oxo-2H-1-benzopyran-3-yl)carbonyl]oxy]-; 7-Methoxy-coumarin-3-carboxylic acid succinimidyl ester; Succinimidyl 7-methoxy-2H-chromene-3-carboxylate; MCCS

Merck Index Number Not listed

Chemical/Dye Class Coumarin

Molecular Formula $C_{15}H_{11}NO_7$

Molecular Weight 317.25

Physical Form White solid[1,2]

Solubility Soluble in acetonitrile, N,N-dimethylformamide, methanol

Boiling Point (Calcd.) $522.4 \pm 60.0\,°C$, pressure: 760 Torr

Absorption (λ_{max}) 358 nm (MeOH)

Emission (λ_{max}) 410 nm (MeOH)

Molar Extinction Coefficient $26,000\,cm^{-1}\,M^{-1}$ (MeOH)

Synthesis Synthetic methods[1–3]

Imaging/Labeling Applications Amino acids;[2,3] analytes;[4] lysines;[2,3] peptides;[5–8] silane coupling agent;[16] silica particles/beads[16]

Biological/Medical Applications Analyzing/screening peptides;[5–8] monitoring proteolysis;[2] quantifying analytes;[4] as ligands and/or substrates for transport proteins;[9] as a substrate for measuring extracellular matrix metalloprotease (MMP-1) activity,[2] reverse transcriptase (RT) polymerase activity,[10] proteases activity;[11] viscosity sensor;[12–15] colloidal diagnostic devices;[16] useful for synthesizing peptides[16]

Industrial Applications Silica particles/beads[16]

Safety/Toxicity No data available

REFERENCES

1. Bardajee, G. R.; Winnik, M. A.; Lough, A. J. Succinimidyl 7-methoxy-2H-chromene-3-carboxylate. *Acta Crystallogr., Sect. E: Struct. Rep.* **2007**, *E63*, o1513–o1514.

2. Berthelot, T.; Talbot, J.; Lain, G.; Deleris, G.; Latxague, L. Synthesis of Nε-(7-diethyl-aminocoumarin-3-carboxyl)- and Nε-(7-methoxy-coumarin-3-carboxyl)-L-Fmoc lysine as tools for protease cleavage detection by fluorescence. *J. Pept. Sci.* **2005**, *11*, 153–160.

3. Berthelot, T.; Lain, G.; Latxague, L.; Deleris, G. Synthesis of novel fluorogenic L-fmoc lysine derivatives as potential tools for imaging cells. *J. Fluoresc.* **2004**, *14*, 671–675.

4. Lakowicz, J. R.; Maliwal, B. P.; Thompson, R.; Ozinskas, A. Fluorescent energy transfer immunoassay. U.S. Patent 5631169, 1997.

5. Kitamatsu, M.; Futami, M.; Sisido, M. A novel method for screening peptides that bind to proteins by using multiple fluorescent amino acids as fluorescent tags. *Chem. Commun.* **2010**, *46*, 761–763.

6. Li, J.; Liu, X.; Zhao, R.; Xiong, S.; Ma, H.; Lu, Q.; Xu, L. Labeling analysis of neuropeptides with 7-methoxycoumarin-3-carboxylic acid N-succinimidyl ester. *Gaodeng Xuexiao Huaxue Xuebao* **2006**, *27*, 2297–2299.

7. Pashkova, A.; Chen, H.; Rejtar, T.; Zang, X.; Giese, R.; Andreev, V.; Moskovets, E.; Karger, B. L. Coumarin tags for analysis of peptides by MALDI-TOF MS and MS/MS. 2. Alexa Fluor 350 tag for increased peptide and protein identification by LC-MALDI-TOF/TOF MS. *Anal. Chem.* **2005**, *77*, 2085–2096.

8. Pashkova, A.; Moskovets, E.; Karger, B. L. Coumarin tags for improved analysis of peptides by

MALDI-TOF MS and MS/MS. 1. Enhancement in MALDI MS signal intensities. *Anal. Chem.* **2004**, *76*, 4550–4557.

9. Dower, W. J.; Gallop, M.; Barrett, R. W.; Cundy, K. C.; Chernov-Rogan, T. Substrates and screening methods for transport proteins. PCT Int. Appl. WO 2001020331, 2001.

10. Krebs, J. F.; Kore, A. R. Novel FRET-based assay to detect reverse transcriptase activity using modified dUTP analogues. *Bioconjugate Chem.* **2008**, *19*, 185–191.

11. Alouini, M. A.; Berthelot, T.; Moustoifa, E. F.; Albenque-Rubio, S.; Deleris, G. Biosensor for detecting the presence of proteases and optionally quantifying the enzymatic activity thereof. PCT Int. Appl. WO 2012093162, 2012.

12. Fischer, D.; Theodorakis, E. A.; Haidekker, M. A. Synthesis and use of an in-solution ratiometric fluorescent viscosity sensor. *Nat. Protoc.* **2007**, *2*, 227–236.

13. Yoon, H.; Dakanali, M.; Lichlyter, D.; Chang, W. M.; Nguyen, K. A.; Nipper, M. E.; Haidekker, M. A.; Theodorakis, E. A. Synthesis and evaluation of self-calibrating ratiometric viscosity sensors. *Org. Biomol. Chem.* **2011**, *9*, 3530–3540.

14. Nipper, M. E.; Dakanali, M.; Theodorakis, E.; Haidekker, M. A. Detection of liposome membrane viscosity perturbations with ratiometric molecular rotors. *Biochimie* **2011**, *93*, 988–994.

15. Dakanali, M.; Do, T. H.; Horn, A.; Chongchivivat, A.; Jarusreni, T.; Lichlyter, D.; Guizzunti, G.; Haidekker, M. A.; Theodorakis, E. A. Self-calibrating viscosity probes: Design and subcellular localization. *Bioorg. Med. Chem.* **2012**, *20*, 4443–4450.

16. Lawrie, G.; Grondahl, L.; Battersby, B.; Keen, I.; Lorentzen, M.; Surawski, P.; Trau, M. Tailoring surface properties to build colloidal diagnostic devices: Controlling interparticle associations. *Langmuir* **2006**, *22*, 497–505.

8-METHOXYPYRENE-1,3,6-TRISULFONIC ACID TRISODIUM SALT (MPTS)

CAS Registry Number   82962-86-5

Chemical Structure

CA Index Name   1,3,6-Pyrenetrisulfonic acid, 8-methoxy-, sodium salt (1:3)

Other Names   8-Methoxypyrene-1,3,6-trisulfonic acid, trisodium salt; MPTS; 1,3,6-Pyrenetrisulfonic acid, 8-methoxy-, trisodium salt; Trisodium 8-methoxy-1,3,6-pyrenetrisulfonate

Merck Index Number   Not listed

Chemical/Dye Class   Pyrene

Molecular Formula $C_{17}H_9Na_3O_{10}S_3$

Molecular Weight   538.41

Physical Form   Light brown to beige crystals or yellow solid

Solubility   Soluble in water, *N,N*-dimethylformamide, dimethyl sulfoxide, methanol

Melting Point   253–260 °C (decompose); 217 °C (decompose)[1]

Absorption (λ_{max})   404 nm (Buffer pH 8.0); 404 nm (H_2O)

Emission (λ_{max})   435 nm (Buffer pH 8.0); 431 nm (H_2O)

Molar Extinction Coefficient   29,000 $cm^{-1} M^{-1}$ (Buffer pH 8.0); 27,000 $cm^{-1} M^{-1}$ (H_2O)

Synthesis   Synthetic methods[1,2]

Imaging/Labeling Applications   Lung tissue;[3] reversed micelles[1,5,6]

Biological/Medical Applications   Detecting glucose;[4] measuring lipase activity[2]

Industrial Applications   Assessing the interior of the reversed micelles;[1] evaluating critical micelle concentrations (CMC) of non-ionic detergents;[5] investigating decarboxylation reaction of 6-nitro-1,2-benzisoxazole-3-carboxylate in cationic and anionic reversed micelles with particular attention to the microenvironmental effects[6]

Safety/Toxicity   No data available

REFERENCES

1. Kondo, H.; Miwa, I.; Sunamoto, J. Biphasic structure model for reversed micelles. Depressed acid dissociation of excited-state pyranine in the restricted reaction field. *J. Phys. Chem.* **1982**, *86*, 4826–4831.

2. Koller, E.; Wolfbeis, O. S. Preparation of pyrenesulfonic acid derivatives for photometric determination of enzyme activity. AT 385755, 1988.

3. Lahnstein, K.; Schmehl, T.; Ruesch, U.; Rieger, M.; Seeger, W.; Gessler, T. Pulmonary absorption of aerosolized fluorescent markers in the isolated rabbit lung. *Int. J. Pharm.* **2008**, *351*, 158–164.

4. Cordes, D. B.; Miller, A.; Gamsey, S.; Sharrett, Z.; Thoniyot, P.; Wessling, R.; Singaram, B. Optical glucose detection across the visible spectrum using anionic fluorescent dyes and a viologen quencher in a two-component saccharide sensing system. *Org. Biomol. Chem.* **2005**, *3*, 1708–1713.

5. Kriechbaum, M.; Wolfbeis, O. S.; Koller, E. Evaluation of critical micelle concentrations of non-ionic detergents using new superpolar lipid probes. *Chem. Phys. Lipids* **1987**, *44*, 19–29.

6. Sunamoto, J.; Iwamoto, K.; Nagamatsu, S.; Kondo, H. Effect of multiple field assistance with reversed micelles on the decarboxylation of 6-nitro-1,2-benzisoxazole-3-carboxylate. *Bull. Chem. Soc. Jpn.* **1983**, *56*, 2469–2472.

NBD C$_6$-CERAMIDE

CAS Registry Number 86701-10-2

Chemical Structure

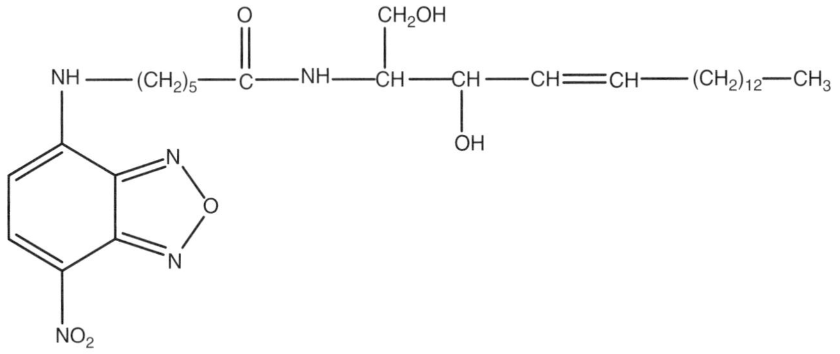

Absorption (λ_{max})   466 nm (MeOH)

Emission (λ_{max})   536 nm (MeOH)

CA Index Name   Hexanamide, *N*-[2-hydroxy-1-(hydroxymethyl)-3-heptadecen-1-yl]-6-[(7-nitro-2,1,3-benzoxadiazol-4-yl)amino]-

Other Names   Hexanamide, *N*-[2-hydroxy-1-(hydroxymethyl)-3-heptadecenyl]-6-[(7-nitro-2,1,3-benzoxadiazol-4-yl)amino]-; Hexanamide, *N*-[2-hydroxy-1-(hydroxymethyl)-3-heptadecenyl]-6-[(7-nitro-4-benzofurazanyl)amino]-; 2,1,3-Benzoxadiazole, hexanamide deriv.; 6-((N-(7-Nitrobenz-2-Oxa-1,3-Diazol-4-yl)amino)hexanoyl)Sphingosine; N-(6-[(7-nitrobenzo-2-oxa-1,3-diazol-4-yl)amino]caproyl)-sphingosine; NBD C$_6$-ceramide

Merck Index Number   Not listed

Chemical/Dye Class   Heterocycle; Benzoxadiazole

Molecular Formula   C$_{30}$H$_{49}$N$_5$O$_6$

Molecular Weight   575.74

Physical Form   Orange powder or solid

Solubility   Soluble in chloroform, *N,N*-dimethylformamide, dimethyl sulfoxide, methanol

pK$_a$ (Calcd.)   13.81 ± 0.20, most acidic, temperature: 25 °C; −0.77 ± 0.70, most basic, temperature: 25 °C

Molar Extinction Coefficient   22,000 cm^{-1} M^{-1} (MeOH)

Synthesis   Synthetic methods[1–3]

Imaging/Labeling Applications   Golgi apparatus;[1,2,4–28,31,44–46,48,55] cells;[18,28] erythrocytes;[29,30] fibroblasts;[3] lipids;[31–34] lipopolysaccharides;[35] lipoproteins;[36–40] lysosomes;[3] sphingolipids;[5,6,9,27,41–47] vacuoles[44,48]

Biological/Medical Applications   Selective stain for Golgi apparatus;[1,2,4–28,31,44–46,48,55] analyzing/evaluating lipoprotein profile;[36–40] examining the biogenesis of Golgi structure and function;[11] monitoring cholesterol at the Golgi apparatus of living cells;[13] studying sphingolipid transport and metabolism mechanisms;[5,6,9,27,41–47] as a substrate for measuring inositol phosphorylceramide synthase (IPC synthase) activity,[49–53] sphingomyelinase activity,[54] sphingomyelin synthase activity;[55] treating fungal infections[56] visualizing formation and dynamics of vacuoles[44,48]

Industrial Applications   Not reported

Safety/Toxicity   Fungal toxicity;[56] tumor necrosis[57]

REFERENCES

1. Sabnis, R. W. *Handbook of Biological Dyes and Stains*; John Wiley & Sons Inc.: Hoboken, **2010**; pp 317–319.

2. Pagano, R. E. A fluorescent derivative of ceramide: physical properties and use in studying the Golgi apparatus of animal cells. *Methods Cell Biol.* **1989**, *29*, 75–85.

3. Chen, W. C.; Moser, A. B.; Moser, H. W. Role of lysosomal acid ceramidase in the metabolism of ceramide in human skin fibroblasts. *Arch. Biochem. Biophys.* **1981**, *208*, 444–455.

4. Kaufmann, A. M.; Toro-Ramos, A. J.; Krise, J. P. Assessment of Golgi apparatus versus plasma membrane-localized multi-drug

Handbook of Fluorescent Dyes and Probes, First Edition. R. W. Sabnis.
© 2015 John Wiley & Sons, Inc. Published 2015 by John Wiley & Sons, Inc.

resistance-associated protein 1. *Mol. Pharm.* **2008**, *5*, 787–794.

5. Maier, O.; Hoekstra, D. Trans-Golgi network and subapical compartment of HepG2 cells display different properties in sorting and exiting of sphingolipids. *J. Biol. Chem.* **2003**, *278*, 164–173.

6. Wolf, K.; Hackstadt, T. Sphingomyelin trafficking in *Chlamydia pneumoniae*-infected cells. *Cell. Microbiol.* **2001**, *3*, 145–152.

7. Lanfredi-Rangel, A.; Kattenbach, W. M.; Diniz, J. A. Jr,; de Souza, W. Trophozoites of *Giardia lamblia* may have a Golgi-like structure. *FEMS Microbiol. Lett.* **1999**, *181*, 245–251.

8. Sandoval, R.; Leiser, J.; Molitoris, B. A. Aminoglycoside antibiotics traffic to the Golgi complex in LLC-PK1 cells. *J. Am. Soc. Nephrol.* **1998**, *9*, 167–174.

9. Hackstadt, T.; Rockey, D. D.; Heinzen, R. A.; Scidmore, M. A. *Chlamydia trachomatis* interrupts an exocytic pathway to acquire endogenously synthesized sphingomyelin in transit from the Golgi apparatus to the plasma membrane. *EMBO J.* **1996**, *15*, 964–977.

10. Seksek, O.; Biwersi, J.; Verkman, A. S. Direct measurement of trans-Golgi pH in living cells and regulation by second messengers. *J. Biol. Chem.* **1995**, *270*, 4967–4970.

11. Lujan, H. D.; Marotta, A.; Mowatt, M. R.; Sciaky, N.; Lippincott-Schwartz, J.; Nash, T. E. Developmental induction of Golgi structure and function in the primitive eukaryote *Giardia lamblia. J. Biol. Chem.* **1995**, *270*, 4612–4618.

12. Wu, Y. N.; Gadina, M.; Tao-Cheng, J. H.; Youle, R. J. Retinoic acid disrupts the Golgi apparatus and increases the cytosolic routing of specific protein toxins. *J. Cell Biol.* **1994**, *125*, 743–753.

13. Martin, O. C.; Comly, M. E.; Blanchette-Mackie, E. J.; Pentchev, P. G.; Pagano, R. E. Cholesterol deprivation affects the fluorescence properties of a ceramide analog at the Golgi apparatus of living cells. *Proc. Natl. Acad. Sci. U.S.A.* **1993**, *90*, 2661–2665.

14. Johnson, K. J.; Boekelheide, K. Visualization of Golgi complexes and spermatogonial cohorts of viable, intact seminiferous tubules. *J. Histochem. Cytochem.* **1993**, *41*, 299–306.

15. de Vries, H.; Schrage, C.; Hoekstra, K.; Kok, J. W.; van der Haar, M. E.; Kalicharan, D.; Liem, R. S.; Copray, J. C.; Hoekstra, D. Outstations of the Golgi complex are present in the processes of cultured rat oligodendrocytes. *J. Neurosci. Res.* **1993**, *36*, 336–343.

16. Sugai, M.; Chen, C. H.; Wu, H. C. Bacterial ADP-ribosyltransferase with a substrate specificity of the rho protein disassembles the Golgi apparatus in Vero cells and mimics the action of brefeldin A. *Proc. Natl. Acad. Sci. U.S.A.* **1992**, *89*, 8903–8907.

17. de Melo, E. J.; de Carvalho, T. U.; de Souza, W. Penetration of *Toxoplasma gondii* into host cells induces changes in the distribution of the mitochondria and the endoplasmic reticulum. *Cell Struct. Funct.* **1992**, *17*, 311–317.

18. Lippincott-Schwartz, J.; Glickman, J.; Donaldson, J. G.; Robbins, J.; Kreis, T. E.; Seamon, K. B.; Sheetz, M. P.; Klausner, R. D. Forskolin inhibits and reverses the effects of brefeldin A on Golgi morphology by a cAMP-independent mechanism. *J. Cell Biol.* **1991**, *112*, 567–577.

19. Chandra, S.; Kable, E. P.; Morrison, G. H.; Webb, W. W. Calcium sequestration in the Golgi apparatus of cultured mammalian cells revealed by laser scanning confocal microscopy and ion microscopy. *J. Cell Sci.* **1991**, *100*, 747–752.

20. Pagano, R. E.; Martin, O. C.; Kang, H. C.; Haugland, R. P. A novel fluorescent ceramide analogue for studying membrane traffic in animal cells: accumulation at the Golgi apparatus results in altered spectral properties of the sphingolipid precursor. *J. Cell Biol.* **1991**, *113*, 1267–1279.

21. Robenek, H.; Schmitz, G. Abnormal processing of Golgi elements and lysosomes in Tangier disease. *Arterioscler. Thromb.* **1991**, *11*, 1007–1020.

22. Crawford, J. M.; Vinter, D. W.; Gollan, J. L. Taurocholate induces pericanalicular localization of C$_6$-NBD-ceramide in isolated hepatocyte couplets. *Am. J. Physiol. Gastrointest. Liver Physiol.* **1991**, *260*, G119–G132.

23. Pagano, R. E. The Golgi apparatus: Insights from lipid biochemistry. *Biochem. Soc. Trans.* **1990**, *18*, 361–366.

24. Pagano, R. E.; Sepanski, M. A.; Martin, O. C. Molecular trapping of a fluorescent ceramide analogue at the Golgi apparatus of fixed cells: interaction with endogenous lipids provides a trans-Golgi marker for both light and electron microscopy. *J. Cell Biol.* **1989**, *109*, 2067–2079.

25. Ho, W. C.; Allan, V. J.; van Meer, G.; Berger, E. G.; Kreis, T. E. Reclustering of scattered Golgi elements occurs along microtubules. *Eur. J. Cell Biol.* **1989**, *48*, 250–263.

26. Lipsky, N. G.; Pagano, R. E. A vital stain for the Golgi apparatus. *Science* **1985**, *228*, 745–747.

27. Lipsky, N. G.; Pagano, R. E. Sphingolipid metabolism in cultured fibroblasts: Microscopic and biochemical studies employing a fluorescent ceramide analogue. *Proc. Natl. Acad. Sci. U.S.A.* **1983**, *80*, 2608–2612.

28. Alzhanov, D. T.; Suchland, R. J.; Bakke, A. C.; Stamm, W. E.; Rockey, D. D. Clonal isolation of chlamydia-infected cells using flow cytometry. *J. Microbiol. Methods* **2007**, *68*, 201–208.

29. Lauer, S. A.; Chatterjee, S.; Haldar, K. Uptake and hydrolysis of sphingomyelin analogues in *Plasmodium falciparum*-infected red cells. *Mol. Biochem. Parasitol.* **2001**, *115*, 275–281.

30. Haldar, K.; Uyetake, L.; Ghori, N.; Elmendorf, H. G.; Li, W. L. The accumulation and metabolism of a fluorescent ceramide derivative in *Plasmodium falciparum*-infected erythrocytes. *Mol. Biochem. Parasitol.* **1991**, *49*, 143–156.

31. Togo, T. Disruption of the plasma membrane stimulates rearrangement of microtubules and lipid traffic toward the wound site. *J. Cell Sci.* **2006**, *119*, 2780–2786.

32. Makino, A.; Ishii, K.; Murate, M.; Hayakawa, T.; Suzuki, Y.; Suzuki, M.; Ito, K.; Fujisawa, T.; Matsuo, H.; Ishitsuka, R.; Kobayashi, T. D-threo-1-phenyl-2-decanoylamino-3-morpholino-1-propanol alters cellular cholesterol homeostasis by modulating the endosome lipid domains. *Biochemistry* **2006**, *45*, 4530–4541.

33. Kuerschner, L.; Ejsing, C. S.; Ekroos, K.; Shevchenko, A.; Anderson, K. I.; Thiele, C. Polyene-lipids: a new tool to image lipids. *Nat. Methods* **2005**, *2*, 39–45.

34. Moffat, D.; Kusel, J. R. Fluorescent lipid uptake and transport in adult *Schistosoma mansoni. Parasitology* **1992**, *105*, 81–89.

35. Zimmermann, C.; Ginis, I.; Furuya, K.; Klimanis, D.; Ruetzler, C.; Spatz, M.; Hallenbeck, J. M. Lipopolysaccharide-induced ischemic tolerance is associated with increased levels of ceramide in brain and in plasma. *Brain Res.* **2001**, *895*, 59–65.

36. Larner, C. D.; Henriquez, R. R.; Johnson, J. D.; MacFarlane, R. D. Developing high performance lipoprotein density profiling for use in clinical studies relating to cardiovascular disease. *Anal. Chem.* **2011**, *83*, 8524–8530.

37. Wang, H.; Han, C.; Wang, H.; Jin, Q.; Wang, D.; Cao, L.; Wang, G. Simultaneous determination of high-density lipoprotein, very low-density lipoprotein and low-density lipoprotein subclass in human serum by microchip CE. *Chromatographia* **2011**, *74*, 799–805.

38. Troup, J. M. Method for analyzing blood for lipoprotein components. U.S. Pat. Appl. Publ. US 20080038762, 2008.

39. Troup, J. M. Method for analyzing blood for lipoprotein components. U.S. Pat. Appl. Publ. US 20080038763, 2008.

40. Boyanovsky, B.; Karakashian, A.; King, K.; Giltiay, N.; Nikolova-Karakashian, M. Uptake and metabolism of low density lipoproteins with elevated ceramide content by human microvascular endothelial cells: implications for the regulation of apoptosis. *J. Biol. Chem.* **2003**, *278*, 26992–26999.

41. Milis, D. G.; Moore, M. K.; Atshaves, B. P.; Schroeder, F.; Jefferson, J. R. Sterol carrier protein-2 expression alters sphingolipid metabolism in transfected mouse L-cell fibroblasts. *Mol. Cell. Biochem.* **2006**, *283*, 57–66.

42. Allan, D. Lipid metabolic changes caused by short-chain ceramides and the connection with apoptosis. *Biochem. J.* **2000**, *345*, 603–610.

43. Zegers, M. M.; Hoekstra, D. Sphingolipid transport to the apical plasma membrane domain in human hepatoma cells is controlled by PKC and PKA activity: A correlation with cell polarity in HepG2 cells. *J. Cell Biol.* **1997**, *138*, 307–321.

44. de Melo, E. J.; de Souza, W. Pathway of C$_6$-NBD-Ceramide on the host cell infected with *Toxoplasma gondii. Cell Struct. Funct.* **1996**, *21*, 47–52.

45. Scidmore, M. A.; Fischer, E. R.; Hackstadt, T. Sphingolipids and glycoproteins are differentially trafficked to the *Chlamydia trachomatis* inclusion. *J. Cell Biol.* **1996**, *134*, 363–374.

46. Hackstadt, T.; Scidmore, M. A.; Rockey, D. D. Lipid metabolism in *Chlamydia trachomatis*-infected cells: directed trafficking of Golgi-derived sphingolipids to the chlamydial inclusion. *Proc. Natl. Acad. Sci. U.S.A.* **1995**, *92*, 4877–4881.

47. Babia, T.; Kok, J. W.; Hulstaert, C.; de Weerd, H.; Hoekstra, D. Differential metabolism and trafficking of sphingolipids in differentiated versus undifferentiated HT29 cells. *Int. J. Cancer* **1993**, *54*, 839–845.

48. Voronina, S. G.; Sherwood, M. W.; Gerasimenko, O. V.; Petersen, O. H.; Tepikin, A. V. Visualizing formation and dynamics of vacuoles in living cells using contrasting dextran-bound indicator: Endocytic and nonendocytic vacuoles. *Am. J. Physiol. Gastrointest. Liver Physiol.* **2007**, *293*, G1333–G1338.

49. Elhammer, A. Novel assay for inositol phosphorylceramide synthase activity. U.S. Pat. Appl. Publ. US 20070269844, 2007.

50. Aeed, P. A.; Sperry, A. E.; Young, C. L.; Nagiec, M. M.; Elhammer, A. P. Effect of membrane perturbants on the activity and phase distribution of inositol phosphorylceramide synthase; development of a novel assay. *Biochemistry* **2004**, *43*, 8483–8493.

51. Radding, J.; Dickson, R. C.; Lester, R. L. Fungal IPC synthase assay. U.S. Patent 6022684, 2000.

52. Fischl, A. S.; Liu, Y.; Browdy, A.; Cremesti, A. E. Inositolphosphoryl ceramide synthase from yeast. *Methods Enzymol.* **2000**, *311*, 123–130.

53. Zhong, W.; Murphy, D. J.; Georgopapadakou, N. H. Inhibition of yeast inositol phosphorylceramide synthase by aureobasidin A measured by a fluorometric assay. *FEBS Lett.* **1999**, *463*, 241–244.

54. Loidl, A.; Claus, R.; Deigner, H. P.; Hermetter, A. High-precision fluorescence assay for sphingomyelinase activity of isolated enzymes and cell lysates. *J. Lipid Res.* **2002**, *43*, 815–823.

55. Nikolova-Karakashian, M. Assays for the biosynthesis of sphingomyelin and ceramide phosphoethanolamine. *Methods Enzymol.* **2000**, *311*, 31–42.

56. Poeta, M. D.; Luberto, C.; Kechichian, T. Methods for the diagnosis and treatment of fungal infections caused by microorganisms producing glucosylceramide. U.S. Pat. Appl. Publ. US 20080014192, 2008.

57. Bourteele, S.; Hausser, A.; Doppler, H.; Horn-Muller, J.; Ropke, C.; Schwarzmann, G.; Pfizenmaier, K.; Muller, G. Tumor necrosis factor induces ceramide oscillations and negatively controls sphingolipid synthases by caspases in apoptotic Kym-1 cells. *J. Biol. Chem.* **1998**, *273*, 31245–31251.

NBD METHYLHYDRAZINE (MNBDH)

CAS Registry Number   214147-22-5

Chemical Structure

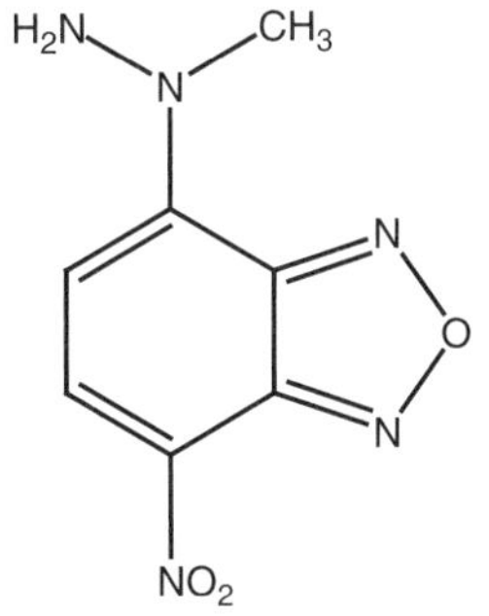

CA Index Name   2,1,3-Benzoxadiazole, 4-(1-methyl-hydrazinyl)-7-nitro-

Other Names   2,1,3-Benzoxadiazole, 4-(1-methyl-hydrazino)-7-nitro-; N-methyl-4-hydrazino-7-nitrobenzo-furazan; MNBDH; NBD methylhydrazine; 4-(1-Methylhydrazino)-7-nitrobenzofurazan; 4-(1-methylhydrazino)-7-nitro-benzooxadiazole; 4-(N-methylhydrazino)-7-nitro-1,2,3-benzooxadiazole; 4-(N-Methylhydrazino)-7-nitro-2,1,3-benzooxadiazole (MNBDH)

Merck Index Number   Not listed

Chemical/Dye Class   Heterocycle; Benzoxadiazole

Molecular Formula $C_7H_7N_5O_3$

Molecular Weight   209.16

Physical Form   Brown powder or solid

Solubility   Soluble in acetonitrile, chloroform, N,N-dimethylformamide, dimethyl sulfoxide, methanol

Melting Point   160 °C

Boiling Point (Calcd.) 425.7 ± 55.0 °C, pressure 760 Torr

pKa (Calcd.) 2.07 ± 0.30, most basic, temperature: 25 °C

Absorption (λ_{max})   487 nm (MeOH)

Emission (λ_{max})   None

Molar Extinction Coefficient   24,000 $cm^{-1} M^{-1}$ (MeOH)

Synthesis   Synthetic methods[1–3]

Imaging/Labeling Applications   Nitrite ions;[1,4] aldehyde-functionalized DNA[5]

Biological/Medical Applications   Detecting nitrite in water,[1,4] creatinine in body fluids,[12] telmisartan;[13] quantifying aldehyde-functionalized (carbonyl compound) DNA;[5] analyzing microperoxidases (microperoxidase substrate);[14] analyzing/detecting peroxidases (peroxidase substrate)[15,16]

Industrial Applications   Monitoring/detecting/measuring/quantifying aldehydes and/or ketones (carbonyl compounds) in air/smoke[1,2,6–11]

Safety/Toxicity   No data available

REFERENCES

1. Sabnis, R. W. *Handbook of Biological Dyes and Stains*; John Wiley & Sons Inc.: Hoboken, **2010**; pp. 320–321.

2. Bueldt, A.; Karst, U. N-Methyl-4-hydrazino-7-nitrobenzofurazan as a new reagent for air monitoring of aldehydes and ketones. *Anal. Chem.* **1999**, *71*, 1893–1898.

3. Bueldt, A.; Karst, U. New benzoxadiazoles. Ger. Offen. DE 19800537, 1998.

4. Bueldt, A.; Karst, U. Determination of nitrite in waters by microplate fluorescence spectroscopy and HPLC with fluorescence detection. *Anal. Chem.* **1999**, *71*, 3003–3007.

5. Raindlova, V.; Pohl, R.; Sanda, M.; Hocek, M. Direct polymerase synthesis of reactive aldehyde-functionalized DNA and its conjugation and staining with hydrazines. *Angew. Chem., Int. Ed.* **2010**, *49*, 1064–1066.

6. Olsen, R.; Thorud, S.; Hersson, M.; Ovrebo, S.; Lundanes, E.; Greibrokk, T.; Ellingsen, D. G.; Thomassen, Y.; Molander, P. Determination of the dialdehyde glyoxal in workroom air - development of personal sampling methodology. *J. Environ. Monit.* **2007**, *9*, 687–694.

7. Paolacci, H.; Tran, T. T. H. Nanoporous material for direct optical sensing of aldehydes. Fr. Demande FR 2890745, 2007.

8. Possanzini, M.; Di Palo, V.; Cecinato, A. Field evaluation of N-methyl-4-hydrazino-7-nitrobenzofurazan (MNBDH) coated silica gel cartridges for the measurement of lower carbonyls in air. *Chromatographia* **2004**, *60*, 715–719.

9. Schulte-Ladbeck, R.; Lindahl, R.; Levin, J.; Karst, U. Characterization of chemical interferences in the determination of unsaturated aldehydes using aromatic hydrazine reagents and liquid

chromatography. *J. Environ. Monit.* **2001**, *3*, 306-310.

10. Zurek, G.; Buldt, A.; Karst, U. Determination of acetaldehyde in tobacco smoke using *N*-methyl-4-hydrazino-7-nitrobenzofurazan and liquid chromatography/mass spectrometry. *Fresenius' J. Anal. Chem.* **2000**, *366*, 396–399.

11. Buldt, A.; Lindahl, R.; Levin, J.; Karst, U. A diffusive sampling device for the determination of formaldehyde in air using *N*-methyl-4-hydrazino-7-nitrobenzofurazan (MNBDH) as reagent. *J. Environ. Monit.* **1999**, *1*, 39–43.

12. Albarella, J. P.; Hatch, R. P. Fluorescent creatinine assay. U.S. Pat. Appl. Publ. US 2004132200, 2004.

13. Hempen, C.; Glaesle-Schwarz, L.; Kunz, U.; Karst, U. Determination of telmisartan in human blood plasma. *Anal. Chim. Acta* **2006**, *560*, 35–40.

14. Haselberg, R.; Hempen, C.; Van Leeuwen, S. M.; Vogel, M.; Karst, U. Analysis of microperoxidases using liquid chromatography, post-column substrate conversion and fluorescence detection. *J. Chromatogr., B* **2006**, *830*, 47–53.

15. Meyer, J.; Jachmann, N.; Bueldt, A.; Karst, U. Method for the determination of hydrogen peroxide and peroxides using peroxidase and *N*-methyl-4-hydrazino-7-nitrobenzofurazan (MNDBH). Ger. Offen. DE 19932380, 2000.

16. Meyer, J.; Buldt, A.; Vogel, M.; Karst, U. 4-(*N*-methylhydrazino)-7-nitro-1,2,3-benzooxadiazole (MNBDH): A novel fluorogenic peroxidase substrate. *Angew. Chem., Int. Ed.* **2000**, *39*, 1453–1455.

NUCLEAR YELLOW (HOECHST S 769121)

CAS Registry Number   74681-68-8
Chemical Structure

Molecular Weight   596.96
Physical Form   Yellow solid or powder
Solubility   Soluble in water, dimethyl sulfoxide
Melting Point   >300 °C
Absorption (λ_{max})   355 nm (H_2O/DNA)

CA Index Name   Benzenesulfonamide, 4-[5-(4-methyl-1-piperazinyl)[2,5′-bi-1*H*-benzimidazol]-2′-yl]-, hydrochloride (1:3)

Other Names   Benzenesulfonamide, 4-[5-(4-methyl-1-piperazinyl)[2,5′-bi-1*H*-benzimidazol]-2′-yl]-, trihydrochloride; 4-[5-(4-Methyl-1-piperazinyl)[2,5′-bi-1*H*-benzimidazol]-2′-yl]-benzenesulfonamide; Hoechst S 769121; Hoechst S 769121 trihydrochloride; Nuclear yellow; NY; 2-(4-sulfamylphenyl)-6-(6-(4-methylpiperazino)-2-benzimidazolyl)benzimidazoletrishydrochloride

Merck Index Number   Not listed

Chemical/Dye Class   Heterocycle; Benzimidazole

Molecular Formula $C_{25}H_{28}Cl_3N_7O_2S$

Emission (λ_{max})   495 nm (H_2O/DNA)

Molar Extinction Coefficient   36,000 cm^{-1} M^{-1} (H_2O/DNA)

Synthesis   Synthetic methods[1–3]

Imaging/Labeling Applications   Nucleic acids;[1,4–9] cells;[1,8] nuclei;[1,2,9–15] neurons;[1,3,12–19] chromosomes;[20] lysosomes;[21] sperms[8]

Biological/Medical Applications   Detecting/analyzing/quantifying nucleic acid;[1,4–9] cells;[1,8] retrograde neuronal tracer[1,3,12–19]

Industrial Applications   Not reported

Safety/Toxicity   No data available

REFERENCES

1. Sabnis, R. W. *Handbook of Biological Dyes and Stains*; John Wiley & Sons Inc.: Hoboken, **2010**; pp 342–343.

2. Bentivoglio, M.; Kuypers, H. G. J. M.; Catsman-Berrevoets, C. E.; Loewe, H.; Dann, O. Two new fluorescent retrograde neuronal tracers which are transported over long distances. *Neurosci. Lett.* **1980**, *18*, 25–30.

3. Bentivoglio, M.; Kuypers, H. G. J. M.; Catsman-Berrevoets, C. E. Two new fluorescent Retrograde neuronal labeling by means of bisbenzimide and Nuclear Yellow (Hoechst S 769121). Measures to prevent diffusion of the tracers out of retrogradely labeled neurons. *Neurosci. Lett.* **1980**, *18*, 19–24.

4. Del Castillo, P.; Horobin, R. W.; Blazquez-Castro, A.; Stockert, J. C. Binding of cationic dyes to DNA: distinguishing intercalation and groove binding mechanisms using simple experimental and numerical models. *Biotech. Histochem.* **2010**, *85*, 247–256.

5. Weber, J.; Brendler, A.; Bley, T. Method and kit for the analysis of the DNA content in cells, especially plant cells using barium hydroxide treatment prior staining. Ger. Offen. DE 102006046079, 2008.

6. Kobayashi, M.; Kaji, S.; Omi, M.; Tamiya, E. Electrochemical DNA quantification based on aggregation induced by phosphate group-binding substance. Jpn. Kokai Tokkyo Koho JP 2006145342, 2006.

7. Stockert, J. C.; Pinna-Senn, E.; Bella, J. L.; Lisanti, J. A. DNA-binding fluorochromes: correlation between C-banding of mouse metaphase chromosomes and hydrogen bonding to adenine-thymine base pairs. *Acta Histochem.* **2005**, *106*, 413–420.

8. Anderson, A. L.; Knutson, C. R.; Mueth, D.; Plewa, J.; Tanner, E. Methods for staining cells for identification and sorting. U.S. Pat. Appl. Publ. US 2006172315, 2006.

9. Curtis, S. K.; Cowden, R. R. Evaluation of five basic fluorochromes of potential use in microfluorometric studies of nucleic acids. *Histochemistry* **1983**, *78*, 503–511.

10. Gauer, C.; Mann, W.; Alunni-Fabbroni, M. Method for carrying out an enzymic reactions. Ger. DE 102006056694, 2010.

11. Gauer, C.; Mann, W.; Alunni-Fabbroni, M. Method for carrying out an enzymic reaction. PCT Int. Appl. WO 2008064730, 2008.

12. Taylor, D. C. M.; Pierau, F. K.; Schmid, H. The use of fluorescent tracers in the peripheral sensory nervous system. *J. Neurosci. Methods* **1983**, *8*, 211–224.

13. Pollin, B.; Laplante, S.; Cesaro, P.; Nguyen-Legros, J. Simultaneous visualization of Nuclear yellow and iron-dextran complex for demonstration of branched neurons by retrograde axonal transport. *J. Neurosci. Methods* **1983**, *8*, 205–209.

14. Keizer, K.; Kuypers, H. G. J. M.; Huisman, A. M.; Dann, O. Diamidino yellow dihydrochloride (DY·2HCl); a new fluorescent retrograde neuronal tracer, which migrates only very slowly out of the cell. *Exp. Brain Res.* **1983**, *51*, 179–191.

15. Katan, S.; Gottschall, J.; Neuhuber, W. Simultaneous visualization of horseradish peroxidase and nuclear yellow in tissue sections for neuronal double labeling. *Neurosci. Lett.* **1982**, *28*, 121–126.

16. Sripanidkulchai, K.; Wyss, J. M. Two rapid methods of counterstaining fluorescent dye tracer containing sections without reducing the fluorescence. *Brain Res.* **1986**, *397*, 117–129.

17. Skirboll, L.; Hoekfelt, T.; Norell, G.; Phillipson, O.; Kuypers, H. G. J. M.; Bentivoglio, M.; Catsman-Berrevoets, C. E.; Visser, T. J.; Steinbusch, H. A method for specific transmitter identification of retrogradely labeled neurons: immunofluorescence combined with fluorescence tracing. *Brain Res. Rev.* **1984**, *8*, 99–127.

18. Van der Krans, A.; Hoogland, P. V. Labeling of neurons following intravenous injections of fluorescent tracers in mice. *J. Neurosci. Methods* **1983**, *9*, 95–103.

19. Weidner, C.; Miceli, D.; Reperant, J. Orthograde axonal and transcellular transport of different fluorescent tracers in the primary visual system of the rat. *Brain Res.* **1983**, *272*, 129–136.

20. Pinna-Senn, E.; Lisanti, J. A.; Ortiz, M. I.; Dalmasso, G.; Bella, J. L.; Gosalvez, J.; Stockert, J. C. Specific heterochromatic banding of metaphase chromosomes using nuclear yellow. *Biotech. Histochem.* **2000**, *75*, 132–140.

21. Rashid, F.; Horobin, R. W.; Williams, M. A. Predicting the behavior and selectivity of fluorescent probes for lysosomes and related structures by means of structure-activity models. *Histochem. J.* **1991**, *23*, 450–459.

OREGON GREEN 488
(2′,7′-DIFLUOROFLUORESCEIN)

CAS Registry Number   195136-58-4

Chemical Structure

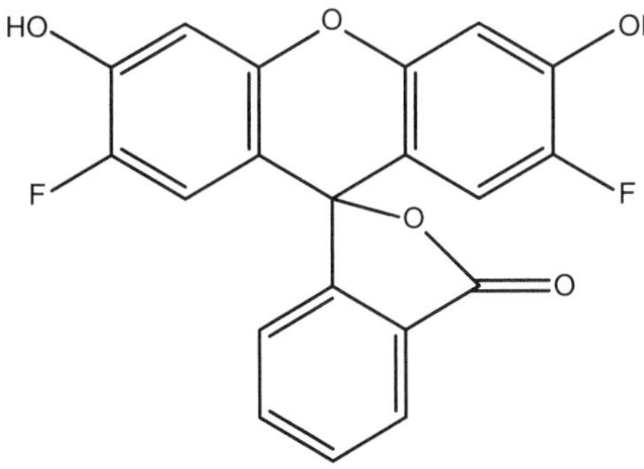

CA Index Name   Spiro[isobenzofuran-1(3*H*),9′-[9*H*]xanthen]-3-one, 2′,7′-difluoro-3′,6′-dihydroxy-

Other Names   2′,7′-Difluorofluorescein; Oregon Green 488

Merck Index Number   Not listed

Chemical/Dye Class   Xanthene

Molecular Formula $C_{20}H_{10}F_2O_5$

Molecular Weight   368.29

Physical Form   Orange powder[1–3]

Solubility   Insoluble in water; soluble in *N,N*-dimethylformamide

Boiling Point (Calcd.)   591.7 ± 50.0 °C, pressure: 760 Torr

pK$_a$ 4.8, temperature: 22 °C[1]

pK$_a$ 4.7, temperature: 22 °C[2,3]

pKa (Calcd.)   8.22 ± 0.20, most acidic, temperature: 25 °C

Absorption (λ$_{max}$)   490 nm (Buffer pH 9.0)

Emission (λ$_{max}$)   514 nm (Buffer pH 9.0)

Molar Extinction Coefficient   87,000 cm^{-1} M^{-1} (Buffer pH 9.0)

Quantum Yield   0.97 (Buffer pH 9.0)

Synthesis   Synthetic methods[1–3]

Imaging/Labeling Applications   Amyloid Aβ peptides;[4] apoplast;[5] bacteria;[6,7] cells;[8] cetuximab;[9] CNA35;[10] cortisol;[11] melatonin;[11] secretory IgA;[11] PAMAM dendrimers;[12] fimbrin;[13] fungi;[14] hair follicles;[15,16] nucleic acids;[17–23] plant cells;[24] proteins[2,3,20,25–30,] TiO$_2$ nanoparticles;[34] tumors[29]

Biological/Medical Applications   Detecting atherosclerosis;[10] detecting/monitoring amyloid Aβ production/accumulation/aggregation/disaggregation;[4] diagnosing Alzheimer's disease;[4] estimating/measuring concentration of cortisol, melatonin, secretory IgA in body fluids;[11] evaluating cetuximab-based imaging probe to target epidermal growth factor receptor (EGFR);[9] examining biodistribution of proteins;[27] investigating/measuring protein-protein and protein-nucleic acid interactions;[20] measuring pH of apoplast at the root surface;[5] visualizing hair follicles;[15,16] as a substrate for measuring β-lactamase activity,[30] phosphatases activity[31,32]

Industrial Applications   As geothermal tracers;[33] TiO$_2$ nanoparticles for light energy conversion[34]

Safety/Toxicity   Cytotoxicity;[35] mutagenicity;[35] tissue toxicity[35]

REFERENCES

1. Sun, W.; Gee, K. R.; Klaubert, D. H.; Haugland, R. P. Synthesis of fluorinated fluoresceins. *J. Org. Chem.* **1997**, *62*, 6469–6475.

2. Gee, K. R.; Poot, M.; Klaubert, D. H.; Sun, W.; Haugland, R. P.; Mao, F. Fluorinated xanthene derivatives. PCT Int. Appl. WO 9739064, 1997.

3. Gee, K. R.; Poot, M.; Klaubert, D. H.; Sun, W.; Haugland, R. P.; Mao, F. Fluorinated xanthene derivatives. U.S. Patent 6162931, 1997.

4. Glabe, C.; Garzon-Rodriguez, W. Fluorescent amyloid Aβ peptides and uses thereof. PCT Int. Appl. WO 9908695, 1999.

5. Nishiyama, H.; Ohya, T.; Tanoi, K.; Nakanishi, T. M. A simple measurement of the pH of root apoplast by the fluorescence ratio method. *Plant Root* **2008**, *2*, 3–6.

6. Holm, C.; Jespersen, L. A flow-cytometric gram-staining technique for milk-associated bacteria. *Appl. Environ. Microbiol.* **2003**, *69*, 2857–2863.

Handbook of Fluorescent Dyes and Probes, First Edition. R. W. Sabnis.
© 2015 John Wiley & Sons, Inc. Published 2015 by John Wiley & Sons, Inc.

7. Fuller, M. E.; Streger, S. H.; Rothmel, R. K.; Mailloux, B. J.; Hall, J. A.; Onstott, T. C.; Fredrickson, J. K.; Balkwill, D. L.; DeFlaun, M. F. Development of a vital fluorescent staining method for monitoring bacterial transport in subsurface environments. *Appl. Environ. Microbiol.* **2000**, *66*, 4486–4496.

8. Jones, S. A.; Shim, S.; He, J.; Zhuang, X. Fast, three-dimensional super-resolution imaging of live cells. *Nat. Methods* **2011**, *8*, 499–505.

9. Aerts, H. J. W. L.; Dubois, L.; Hackeng, T. M.; Straathof, R.; Chiu, R. K.; Lieuwes, N. G.; Jutten, B.; Weppler, S. A.; Lammering, G.; Wouters, B. G.; Lambin, P. Development and evaluation of a cetuximab-based imaging probe to target EGFR and EGFRvIII. *Radiother. Oncol.* **2007**, *83*, 326–332.

10. Megens, R. T. A.; oude Egbrink, M. G. A.; Cleutjens, J. P. M.; Kuijpers, M. J. E.; Schiffers, P. H. M.; Merkx, M.; Slaaf, D. W.; van Zandvoort, M. A. M. J. Imaging collagen in intact viable healthy and atherosclerotic arteries using fluorescently labeled CNA35 and two-photon laser scanning microscopy. *Mol. Imaging* **2007**, *6*, 247–260.

11. Cullum, M. E.; Duplessis, C. A.; Crepeau, L. J. Method for the detection of stress biomarkers including cortisol by fluorescence polarization. U.S. Pat. Appl. Publ. US 20060105397, 2006.

12. Yoo, H.; Juliano, R. L. Enhanced delivery of antisense oligonucleotides with fluorophore-conjugated PAMAM dendrimers. *Nucleic Acids Res.* **2000**, *28*, 4225–4231.

13. Kovar, D. R.; Gibbon, B. C.; McCurdy, D. W.; Staiger, C. J. Fluorescently-labeled fimbrin decorates a dynamic actin filament network in live plant cells. *Planta* **2001**, *213*, 390–395.

14. Paulitsch-Fuchs, A.; Treiber, F.; Grasser, E.; Buzina, W.; Rosker, C. New staining methods for yeast like fungi under special consideration of human pathogenic fungi. *Proc. SPIE* **2010**, *7376*, 73760W/1–73760W/7.

15. Grams, Y. Y.; Alaruikka, S.; Lashley, L.; Caussin, J.; Whitehead, L.; Bouwstra, J. A. Permeant lipophilicity and vehicle composition influence accumulation of dyes in hair follicles of human skin. *Eur. J. Pharm. Sci.* **2003**, *18*, 329–336.

16. Grams, Y. Y.; Bouwstra, J. A. Penetration and distribution of three lipophilic probes *in vitro* in human skin focusing on the hair follicle. *J. Controlled Release* **2002**, *83*, 253–262.

17. Yeh, H.; Ho, Y.; Wang, T. Quantum dot-mediated biosensing assays for specific nucleic acid detection. *Nanomedicine* **2005**, *1*, 115–121.

18. Ho, F. M.; Hall, E. A. H. A strand exchange FRET assay for DNA. *Biosens. Bioelectron.* **2004**, *20*, 1001–1010.

19. Berland, K. M. Detection of specific DNA sequences using dual-color two-photon fluorescence correlation spectroscopy. *J. Biotechnol.* **2004**, *108*, 127–136.

20. Rusinova, E.; Tretyachenko-Ladokhina, V.; Vele, O. E.; Senear, D. F.; Alexander Ross, J. B. Alexa and Oregon Green dyes as fluorescence anisotropy probes for measuring protein-protein and protein-nucleic acid interactions. *Anal. Biochem.* **2002**, *308*, 18–25.

21. Okamura, Y.; Kondo, S.; Sase, I.; Suga, T.; Mise, K.; Furusawa, I.; Kawakami, S.; Watanabe, Y. Double-labeled donor probe can enhance the signal of fluorescence resonance energy transfer (FRET) in detection of nucleic acid hybridization. *Nucleic Acids Res.* **2000**, *28*, e107/1–e107/6.

22. Henegariu, O.; Bray-Ward, P.; Ward, D. C. Custom fluorescent-nucleotide synthesis as an alternative method for nucleic acid labeling. *Nat. Biotechnol.* **2000**, *18*, 345–348.

23. Wiegant, J. C. A. G.; Van Gijlswijk, R. P. M.; Heetebrij, R. J.; Bezrookove, V.; Raap, A. K.; Tanke, H. J. ULS: a versatile method of labeling nucleic acids for FISH based on a monofunctional reaction of cisplatin derivatives with guanine moieties. *Cytogenet. Cell Genet.* **1999**, *87*, 47–52.

24. French, A. P.; Mills, S.; Swarup, R.; Bennett, M. J.; Pridmore, T. P. Colocalization of fluorescent markers in confocal microscope images of plant cells. *Nat. Protoc.* **2008**, *3*, 619–628.

25. Shen, D.; Coleman, J.; Chan, E.; Nicholson, T. P.; Dai, L.; Sheppard, P. W.; Patton, W. F. Novel cell- and tissue-based assays for detecting misfolded and aggregated protein accumulation within aggresomes and inclusion bodies. *Cell Biochem. Biophys.* **2011**, *60*, 173–185.

26. Brusnichkin, A. V.; Nedosekin, D. A.; Galanzha, E. I.; Vladimirov, Y. A.; Shevtsova, E. F.; Proskurnin, M. A.; Zharov, V. P. Ultrasensitive label-free photothermal imaging, spectral identification, and quantification of cytochrome c in mitochondria, live cells, and solutions. *J. Biophotonics* **2010**, *3*, 791–806.

27. Piepenhagen, P. A.; Vanpatten, S.; Hughes, H.; Waire, J.; Murray, J.; Andrews, L.; Edmunds, T.; O'Callaghan, M.; Thurberg, B. L. Use of direct fluorescence labeling and confocal microscopy to determine the biodistribution of two protein therapeutics, Cerezyme and Ceredase. *Microsc. Res. Tech.* **2010**, *73*, 694–703.

28. Reichel, A.; Schaible, D.; Furoukh, N. A.; Cohen, M.; Schreiber, G.; Piehler, J. Noncovalent, site-specific biotinylation of histidine-tagged proteins. *Anal. Chem.* **2007**, *79*, 8590–8600.

29. Hama, Y.; Urano, Y.; Koyama, Y.; Bernardo, M.; Choyke, P. L.; Kobayashi, H. A comparison of the emission efficiency of four common green fluorescence dyes after internalization into cancer cells. *Bioconjugate Chem.* **2006**, *17*, 1426–1431.

30. Rukavishnikov, A.; Gee, K. R.; Johnson, I.; Corry, S. Fluorogenic cephalosporin substrates for β-lactamase TEM-1. *Anal. Biochem.* **2011**, *419*, 9–16.

31. Telford, W. G.; Cox, W. G.; Singer, V. L. Detection of endogenous and antibody-conjugated alkaline phosphatase with ELF-97 phosphate in multicolor flow cytometry applications. *Cytometry* **2001**, *43*, 117–125.

32. Gee, K. R. Novel fluorogenic substrates for acid phosphatase. *Bioorg. Med. Chem. Lett.* **1999**, *9*, 1395–1396.

33. Wong, Y. L.; Rose, P. E. The testing of fluorescein derivatives as candidate geothermal tracers. *GRC Trans.* **2000**, *24*, 637–640.

34. Hasobe, T.; Hattori, S.; Kamat, P. V.; Urano, Y.; Umezawa, N.; Nagano, T.; Fukuzumi, S. Organization of supramolecular assemblies of fullerene, porphyrin and fluorescein dye derivatives on TiO_2 nanoparticles for light energy conversion. *Chem. Phys.* **2005**, *319*, 243–252.

35. Alford, R.; Simpson, H. M.; Duberman, J.; Hill, G. C.; Ogawa, M.; Regino, C.; Kobayashi, H.; Choyke, P. L. Toxicity of organic fluorophores used in molecular imaging: literature review. *Mol. Imaging* **2009**, *8*, 341–354.

OREGON GREEN 488 CADAVERINE

CAS Registry Number   886210-16-8
Chemical Structure

CA Index Name   Spiro[isobenzofuran-1(3H),9'-[9H]
xanthene]-5-carboxamide,   N-(5-aminopentyl)-2',7'-
difluoro-3',6'-dihydroxy-3-oxo-

Other Names   Oregon Green 488 cadaverine; Oregon
Green cadaverine

Merck Index Number   Not listed

Chemical/Dye Class   Xanthene

Molecular Formula $C_{26}H_{22}F_2N_2O_6$

Molecular Weight   496.46

Physical Form   Orange solid

Solubility   Insoluble in water; soluble in N,N-
dimethylformamide

Boiling Point (Calcd.) $723.7 \pm 60.0\,°C$, pressure:
760 Torr

pK$_a$ (Calcd.) 8.19 ± 0.20, most acidic, temperature:
25 °C; 10.52 ± 0.10, most basic, temperature: 25 °C

Absorption (λ_{max})   494 nm (Buffer pH 9.0)

Emission (λ_{max}) 521 nm (Buffer pH 9.0)

Molar Extinction Coefficient   75,000 $cm^{-1}\,M^{-1}$
(Buffer pH 9.0)

Synthesis   Synthetic method[1]

Imaging/Labeling Applications   Autophagosomes;[2]
dextrin;[3] N-(2-hydroxypropyl)-methacrylamide (HPMA)
copolymers;[3,4]   p(HPMA)-b-p(LLA)   copolymers;[5]
p(HPMA)-co-p(LMA) copolymers;[6] polyethylene glycol
(PEG);[3] poly(β-amino ester);[7] diamond nanoparticles;[8]
nanohydrogel particles;[9] thymosin b4;[10] vinyl polymers[11]

Biological/Medical Applications   Defining polymer
localization to late endocytic intracellular compartments;[3]
detecting/measuring amount of autophagic activity;[2] in
vitro evaluating N-(2-hydroxypropyl)-methacrylamide
(HPMA) folate conjugates;[4] exhibiting tendency to
interact with model cell membranes;[6] high-throughput
drug   screening;[2]   identifying   autophagosomes;[2]
investigating   cationic   polymers   in   promoting
surface-mediated gene delivery;[7] studying cellular
uptake in human cervix adenocarcinoma (HeLa) cells;[5]
nanodiamond as drug delivery vehicles;[8] as a platform for
proper siRNA delivery systems;[9] tracing hydrogels during
complexation with siRNA or cell uptake experiments[9]

Industrial Applications   Diamond   nanoparticles;[8]
dye-functionalized polymer[11]

Safety/Toxicity   No data available

REFERENCES

1. Haugland, R. P. *Handbook of Fluorescent Probes and
    Research Products*; Molecular Probes Inc.: Eugene,
    2002; pp 112–118.

2. Gottlieb, R.; Cole, T. E.; Perry-Garza, C. N.; Carreira,
    R. S.; Bartlett, B. J.; Finley, K. Compositions for
    labeling and identifying autophagosomes in cell
    extract, cell, tissue, organ or organism and methods
    for making and using them. PCT Int. Appl. WO
    2010045270, 2010.

3. Richardson, S. C. W.; Wallom, K.; Ferguson, E. L.;
    Deacon, S. P. E.; Davies, M. W.; Powell, A. J.;
    Piper, R. C.; Duncan, R. The use of fluorescence
    microscopy to define polymer localisation to the
    late endocytic compartments in cells that are targets
    for drug delivery. *J. Controlled Release* **2008**, *127*,
    1–11.

4. Barz, M.; Canal, F.; Koynov, K.; Zentel, R.; Vicent,
    M. J. Synthesis and *in vitro* evaluation of defined
    HPMA folate conjugates: Influence of aggregation
    on folate receptor (FR) mediated cellular uptake.
    Biomacromolecules **2010**, *11*, 2274–2282.

5. Barz, M.; Wolf, F. K.; Canal, F.; Koynov, K.;
    Vicent, M. J.; Frey, H.; Zentel, R. Synthesis,
    characterization   and   preliminary   biological
    evaluation of P(HPMA)-b-P(LLA) copolymers:
    a new type of functional biocompatible block

copolymer. *Macromol. Rapid Commun.* **2010**, *31*, 1492–1500.

6. Hemmelmann, M.; Kurzbach, D.; Koynov, K.; Hinderberger, D.; Zentel, R. Aggregation behavior of amphiphilic p(HPMA)-co-p(LMA) copolymers studied by FCS and EPR spectroscopy. *Biomacromolecules* **2012**, *13*, 4065–4072.

7. Bechler, S. L.; Lynn, D. M. Characterization of degradable polyelectrolyte multilayers fabricated using DNA and a fluorescently-labeled poly(β-amino ester): Shedding light on the role of the cationic polymer in promoting surface-mediated gene delivery. *Biomacromolecules* **2012**, *13*, 542–552.

8. Meinhardt, T.; Lang, D.; Dill, H.; Krueger, A. Pushing the functionality of diamond nanoparticles to new horizons: Orthogonally functionalized nanodiamond using click chemistry. *Adv. Funct. Mater.* **2011**, *21*, 494–500.

9. Nuhn, L.; Hirsch, M.; Krieg, B.; Koynov, K.; Fischer, K.; Schmidt, M.; Helm, M.; Zentel, R. Cationic nanohydrogel particles as potential siRNA carriers for cellular delivery. *ACS Nano* **2012**, *6*, 2198–2214.

10. Hannappel, E.; Huff, T.; Goldstein, A. L.; Crockford, D. Cell nucleus-entering compositions. U.S. Pat. Appl. Publ. US 20060100156, 2006.

11. Roth, P. J.; Haase, M.; Basche, T.; Theato, P.; Zentel, R. Synthesis of heterotelechelic α,ω dye-functionalized polymer by the RAFT process and energy transfer between the end groups. *Macromolecules* **2010**, *43*, 895–902.

OREGON GREEN 488 CARBOXYLIC ACID

CAS Registry Number   195136-52-8

Chemical Structure

CA Index Name   Spiro[isobenzofuran-1(3H),9′-[9H] xanthene]-5-carboxylic acid, 2′,7′-difluoro-3′,6′-dihydroxy-3-oxo-

Other Names   Oregon Green 488 carboxylic acid

Merck Index Number   Not listed

Chemical/Dye Class   Xanthene

Molecular Formula $C_{21}H_{10}F_2O_7$

Molecular Weight   412.30

Physical Form   Orange powder[2–4]

Solubility   Insoluble in water; soluble in N,N-dimethylformamide

Boiling Point (Calcd.)   693.9 ± 55.0 °C, pressure: 760 Torr

pK$_a$ 4.7, temperature: 22 °C

pK$_a$ (Calcd.)   3.71 ± 0.20, most acidic, temperature: 25 °C

Absorption (λ_{max})   478 nm (Buffer pH 3.0); 492 nm (Buffer pH 9.0)

Emission (λ_{max})   518 nm (Buffers pH 3.0, 9.0)

Molar Extinction Coefficient   27,000 cm^{-1} M^{-1} (Buffer pH 3.0); 85,000 cm^{-1} M^{-1} (Buffer pH 9.0)

Quantum Yield   0.92 (Buffer pH 9.0)

Synthesis   Synthetic methods[1–4]

Imaging/Labeling Applications   Actin analogs;[5] liposomes;[1–4] RNA polymerase (RNAP);[6] vasotocin (Arginine8-vasotocin (AVT))[7]

Biological/Medical Applications   Analysizing single cells;[8] detecting intracellular analytes,[9] undetectable analytes;[10] as marker for Arginine8-vasotocin (AVT) target neurons;[7] measuring phagosomal pH;[11] monitoring melting of duplex DNA;[6] temperature gradient focusing;[12,13] studying actin dynamics in plant cells[5]

Industrial Applications   Silica nanoparticles;[14] temperature gradient focusing (TGF) in a PDMS/glass hybrid microfluidic chip[15]

Safety/Toxicity   No data available

REFERENCES

1. Sabnis, R. W. *Handbook of Acid-Base Indicators*; CRC Press: Boca Raton, **2008**; p 290.

2. Sun, W.; Gee, K. R.; Klaubert, D. H.; Haugland, R. P. Synthesis of fluorinated fluoresceins. *J. Org. Chem.* **1997**, *62*, 6469–6475.

3. Gee, K. R.; Poot, M.; Klaubert, D. H.; Sun, W.; Haugland, R. P.; Mao, F. Fluorinated xanthene derivatives. PCT Int. Appl. WO 9739064, 1997.

4. Gee, K. R.; Poot, M.; Klaubert, D. H.; Sun, W.; Haugland, R. P.; Mao, F. Fluorinated xanthene derivatives. U.S. Patent 6162931, 1997.

5. Ren, H. Preparation of actin and fluorescent actin analogs from plant cells. *Zhiwu Xuebao* **1999**, *41*, 1099–1103.

6. Matlock, D. L.; Heyduk, T. A real-time fluorescence method to monitor the melting of duplex DNA during transcription initiation by RNA polymerase. *Anal. Biochem.* **1999**, *270*, 140–147.

7. Lewis, C. M.; Dolence, E. K.; Zhang, Z.; Rose, J. D. Fluorescent vasotocin conjugate for identification of the target cells for brain actions of vasotocin. *Bioconjugate Chem.* **2004**, *15*, 909–914.

8. Jiang, D.; Sims, C. E.; Allbritton, N. L. Microelectrophoresis platform for fast serial analysis of single cells. *Electrophoresis* **2010**, *31*, 2558–2565.

9. Kopelman, R.; Clark, H.; Monson, E.; Parus, S.; Philbert, M.; Thorsrud, B. Optical fiberless sensors. PCT Int. Appl. WO 9902651, 1999.

10. Santiago, J. G.; Khurana, T. Method of detecting directly undetectable analytes using directly detectable spacer molecules in form of fluorophores

for isotachophoresis. U.S. Pat. Appl. Publ. US 20080197019, 2008.

11. Vergne, I.; Constant, P.; Laneelle, G. Phagosomal pH determination by dual fluorescence flow cytometry. *Anal. Biochem.* **1998**, *255*, 127–132.

12. Munson, M. S.; Danger, G.; Shackman, J. G.; Ross, D. Temperature gradient focusing with field-amplified continuous sample injection for dual-stage analyte enrichment and separation. *Anal. Chem.* **2007**, *79*, 6201–6207.

13. Ross, D.; Locascio, L. E. Microfluidic temperature gradient focusing. *Anal. Chem.* **2002**, *74*, 2556–2564.

14. Mader, H.; Li, X.; Saleh, S.; Link, M.; Kele, P.; Wolfbeis, O. S. Fluorescent silica nanoparticles. *Ann. N. Y. Acad. Sci.* **2008**, *1130*, 218-223.

15. Matsui, T.; Franzke, J.; Manz, A.; Janasek, D. Temperature gradient focusing in a PDMS/glass hybrid microfluidic chip. *Electrophoresis* **2007**, *28*, 4606–4611.

OREGON GREEN 514 CARBOXYLIC ACID

CAS Registry Number 198139-49-0

Chemical Structure

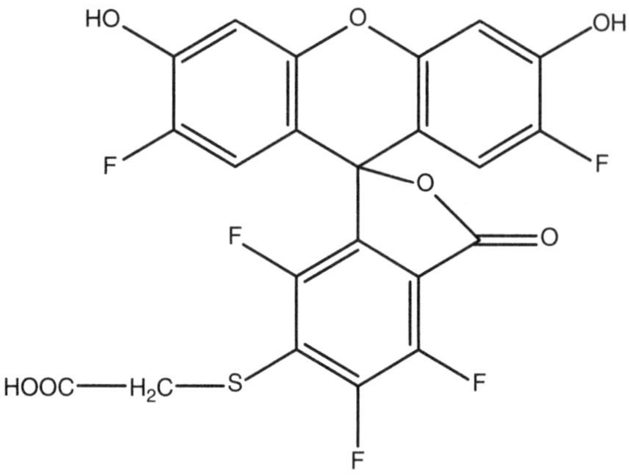

CA Index Name Acetic acid, 2-[(2′,4,5,7,7′-pentafluoro-3′,6′-dihydroxy-3-oxospiro[isobenzofuran-1(3H), 9′-[9H]xanthen]-6-yl)thio]-

Other Names Acetic acid, [(2′,4,5,7,7′-pentafluoro-3′, 6′-dihydroxy-3-oxospiro[isobenzofuran-1(3H),9′-[9H]xanthen]-6-yl)thio]-; Oregon Green 514 carboxylic acid 6-isomer

Merck Index Number Not listed

Chemical/Dye Class Xanthene

Molecular Formula $C_{22}H_9F_5O_7S$

Molecular Weight 512.36

Physical Form Orange solid

Solubility Insoluble in water; soluble in N,N-dimethylformamide

Boiling Point (Calcd.) 703.2 ± 60.0 °C, pressure: 760 Torr

pK$_a$ 4.7, temperature: 22 °C

pK$_a$ (Calcd.) 3.01 ± 0.10, most acidic, temperature: 25 °C

Absorption (λ_{max}) 489 nm (Buffer pH 3.0); 506 nm (Buffer pH 9.0)

Emission (λ_{max}) 526 nm (Buffers pH 3.0, 9.0)

Molar Extinction Coefficient 26,000 cm^{-1} M^{-1} (Buffer pH 3.0); 86,000 cm^{-1} M^{-1} (Buffer pH 9.0)

Synthesis Synthetic methods[1–3]

Imaging/Labeling Applications Beads;[4,5] cortisol;[6] melatonin;[6] secretory IgA;[6] monocytes;[7] nucleic acids;[8–10] peptides;[11–14] polynucleotides;[15] proteins[1–3,10,16,17,]

Biological/Medical Applications Analyzing protein distribution in hydrogel beads;[16] estimating/measuring concentration of cortisol, melatonin, secretory IgA in body fluids;[6] identifying beads;[4,5] investigating/measuring protein-protein and protein-nucleic acid interactions;[10] monitoring pH changes in acidic environments;[19] tracing pH in different cellular compartments;[19] treating/diagnosing disease states mediated by monocytes;[7] treating apoptosis related diseases[18]

Industrial Applications As geothermal tracers;[20] sol–gel thin films;[21–23] ultrathin liquid films[24]

Safety/Toxicity No data available

REFERENCES

1. Sabnis, R. W. *Handbook of Acid–base Indicators*; CRC Press: Boca Raton, **2008**; pp 291–292.

2. Gee, K. R.; Poot, M.; Klaubert, D. H.; Sun, W.; Haugland, R. P.; Mao, F. Fluorinated xanthene derivatives. PCT Int. Appl. WO 9739064, 1997.

3. Gee, K. R.; Poot, M.; Klaubert, D. H.; Sun, W.; Haugland, R. P.; Mao, F. Fluorinated xanthene derivatives. U.S. Patent 6162931, 1997.

4. Christensen, S. F.; Johannsen, I.; Carstensen, J. M.; Kuhlmann, L.; Meldal, M. Identification of encoded beads. U.S. Pat. Appl. Publ. US 20090210165, 2009.

5. Christensen, S. F.; Johannsen, I.; Carstensen, J. M.; Kuhlmann, L. Identification of encoded beads. PCT Int. Appl. 2005061094, 2005.

6. Cullum, M. E.; Duplessis, C. A.; Crepeau, L. J. Method for the detection of stress biomarkers including cortisol by fluorescence polarization. U.S. Pat. Appl. Publ. US 20060105397, 2006.

7. Low, P. S.; Hilgenbrink, A. R. Imaging and therapeutic method using monocytes. PCT Int. Appl. WO 2007006041, 2007.

8. Wittwer, C. T.; Gundry, C.; Abbott, R. D.; David, D. A. Genotyping by amplicon melting curve analysis. U.S. Patent 7785776, 2010.

9. MacLeod, M. C.; Aldaz, C. M.; Gaddis, S. S. Combinatorial oligonucleotide PCR: A method for rapid, global expression analysis. PCT Int. Appl. WO 2001027329, 2001.

10. Rusinova, E.; Tretyachenko-Ladokhina, V.; Vele, O. E.; Senear, D. F.; Alexander Ross, J. B. Alexa and Oregon Green dyes as fluorescence anisotropy probes for measuring protein-protein and protein-nucleic acid interactions. *Anal. Biochem.* **2002**, *308*, 18–25.

11. Nolan, G. P.; Rozinov, M. N. Fluorescent dye binding peptides. U.S. Patent 6747135, 2004.

12. Nolan, G. P.; Rozinov, M. N. Fluorescent dye binding peptides. PCT Int. Appl. WO 2000023463, 2000.

13. Delmotte, C.; Delmas, A. Synthesis and fluorescence properties of Oregon Green 514 labeled peptides. *Bioorg. Med. Chem. Lett.* **1999**, *9*, 2989–2994.

14. Rozinov, M. N.; Nolan, G. P. Evolution of peptides that modulate the spectral qualities of bound, small-molecule fluorophores. *Chem. Biol.* **1998**, *5*, 713–728.

15. Tillett, D.; Beltran, C. E.; Karunaratne, S. K.; Briedis, B. J.; Damaere, M. Z. A reagent and method for determining the size of polynucleotides. PCT Int. Appl. WO 2004053153, 2004.

16. Heinemann, M.; Wagner, T.; Doumeche, B.; Ansorge-Schumacher, M.; Buchs, J. A new approach for the spatially resolved qualitative analysis of the protein distribution in hydrogel beads based on confocal laser scanning microscopy. *Biotechnol. Lett.* **2002**, *24*, 845–850.

17. Rumbaut, R. E.; Harris, N. R.; Sial, A. J.; Huxley, V. H.; Granger, D. N. Leakage responses to L-NAME differ with the fluorescent dye used to label albumin. *Am. J. Physiol.* **1999**, *276*, H333–H339.

18. Tanuma, S.; Yoshimori, A.; Sunaga, M. Screening DNase γ inhibitors for treatment of apoptosis related diseases. Jpn. Kokai Tokkyo Koho JP 2006036711, 2006.

19. Lin, H.; Szmacinski, H.; Lakowicz, J. R. Lifetime-based pH sensors: Indicators for acidic environments. *Anal. Biochem.* **1999**, *269*, 162–167.

20. Wong, Y. L.; Rose, P. E. The testing of fluorescein derivatives as candidate geothermal tracers. *GRC Trans.* **2000**, *24*, 637–640.

21. Gilliland, J. W.; Yokoyama, K.; Yip, W. T. Solvent effect on mobility and photostability of organic dyes embedded inside silica sol–gel thin films. *Chem. Mater.* **2005**, *17*, 6702–6712.

22. Gilliland, J. W.; Yokoyama, K.; Yip, W. T. Comparative study of guest charge-charge interactions within silica sol–gel. *J. Phys. Chem. B* **2005**, *109*, 4816–4823.

23. Gilliland, J. W.; Yokoyama, K.; Yip, W. T. Effect of coulombic interactions on rotational mobility of guests in sol–gel silicate thin films. *Chem. Mater.* **2004**, *16*, 3949–3954.

24. Schuster, J.; Cichos, F.; von Borzcyskowski, C. Diffusion in ultrathin liquid films. *Eur. Polym. J.* **2004**, *40*, 993–999.

OREGON GREEN 488 CARBOXYLIC ACID DIACETATE

CAS Registry Number   195136-74-4

Chemical Structure

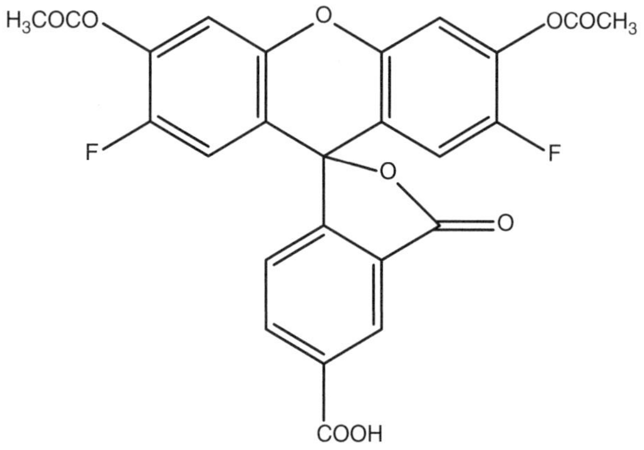

CA Index Name   Spiro[isobenzofuran-1(3H),9′-[9H] xanthene]-5-carboxylic acid, 3′,6′-bis(acetyloxy)-2′,7′-difluoro-3-oxo-

Other Names   Oregon Green 488 carboxylic acid diacetate

Merck Index Number   Not listed

Chemical/Dye Class   Xanthene

Molecular Formula $C_{25}H_{14}F_2O_9$

Molecular Weight 496.37

Physical Form   Colorless crystals;[2] Off-white solid[3,4]

Solubility   Insoluble in water; soluble in dimethyl sulfoxide, methanol

Boiling Point (Calcd.)   691.5 ± 55.0 °C, pressure: 760 Torr

pK$_a$ (Calcd.)   3.69 ± 0.20, most acidic, temperature: 25 °C

Absorption (λ_{max})   298 nm (MeOH); < 300 nm (Buffer pH 3.0)

Emission (λ_{max})   None

Molar Extinction Coefficient   8,000 cm^{-1} M^{-1} (MeOH)

Synthesis   Synthetic methods[1–4]

Imaging/Labeling Applications   Sodium ions[5]

Biological/Medical Applications   Analyzing single cells;[6–9] detecting electrical lysis of cells;[10] locating anion transport mechanism in fungi (*Pisolithus tinctorius*);[11] measuring gap junction mediated diffusion[12]

Industrial Applications   Not reported

Safety/Toxicity   No data available

REFERENCES

1. Sabnis, R. W. *Handbook of Acid–base Indicators*; CRC Press: Boca Raton, **2008**; p 293.

2. Sun, W.; Gee, K. R.; Klaubert, D. H.; Haugland, R. P. Synthesis of fluorinated fluoresceins. *J. Org. Chem.* **1997**, *62*, 6469–6475.

3. Gee, K. R.; Poot, M.; Klaubert, D. H.; Sun, W.; Haugland, R. P.; Mao, F. Fluorinated xanthene derivatives. PCT Int. Appl. WO 9739064, 1997.

4. Gee, K. R.; Poot, M.; Klaubert, D. H.; Sun, W.; Haugland, R. P.; Mao, F. Fluorinated xanthene derivatives. U.S. Patent 6162931, 1997.

5. Martin, V. V.; Rothe, A.; Gee, K. R. Fluorescent metal ion indicators based on benzoannelated crown systems: a green fluorescent indicator for intracellular sodium ions. *Bioorg. Med. Chem. Lett.* **2005**, *15*, 1851–1855.

6. Wang, Y.; Shah, P.; Phillips, C.; Sims, C. E.; Allbritton, N. L. Trapping cells on a stretchable

microwell array for single-cell analysis. *Anal. Bioanal. Chem.* **2012**, *402*, 1065–1072.

7. Hargis, A. D.; Alarie, J. P.; Ramsey, J. M. Characterization of cell lysis events on a microfluidic device for high-throughput single cell analysis. *Electrophoresis* **2011**, *32*, 3172–3179.

8. Jiang, D.; Sims, C. E.; Allbritton, N. L. Microelectrophoresis platform for fast serial analysis of single cells. *Electrophoresis* **2010**, *31*, 2558–2565.

9. Lai, H.; Quinto-Su, P. A.; Sims, C. E.; Bachman, M.; Li, G. P.; Venugopalan, V.; Allbritton, N. L. Characterization and use of laser-based lysis for cell analysis on-chip. *J. Royal Soc., Interface* **2008**, *5(Suppl. 2)*, S113–S121.

10. Han, F.; Wang, Y.; Sims, C. E.; Bachman, M.; Chang, R.; Li, G. P.; Allbritton, N. L. Fast electrical lysis of cells for capillary electrophoresis. *Anal. Chem.* **2003**, *75*, 3688–3696.

11. Cole, L.; Hyde, G. J.; Ashford, A. E. Uptake and compartmentalization of fluorescent probes by *Pisolithus tinctorius* hyphae: evidence for an anion transport mechanism at the tonoplast but not for fluid-phase endocytosis. *Protoplasma* **1997**, *199*, 18–29.

12. Bathany, C.; Beahm, D.; Felske, J. D.; Sachs, F.; Hua, S. Z. High throughput assay of diffusion through Cx43 gap junction channels with a microfluidic chip. *Anal. Chem.* **2011**, *83*, 933–939.

OREGON GREEN 488 CARBOXYLIC ACID SUCCINIMIDYL ESTER

CAS Registry Number   198139-51-4

Chemical Structure

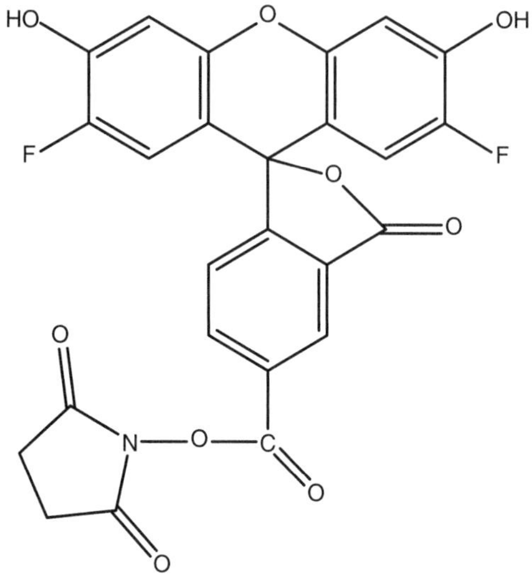

CA Index Name   Spiro[isobenzofuran-1(3H),9'-[9H] xanthene]-5-carboxylic acid, 2',7'-difluoro-3',6'-dihydroxy-3-oxo-, 2,5-dioxo-1-pyrrolidinyl ester

Other Names   2,5-Pyrrolidinedione, 1-[[(2',7'-difluoro-3',6'-dihydroxy-3-oxospiro[isobenzofuran-1(3H),9'-[9H] xanthen]-5-yl)carbonyl]oxy]-; Oregon Green 488 NHS ester; Oregon Green 488 carboxylic acid succinimidyl ester

Merck Index Number   Not listed

Chemical/Dye Class   Xanthene

Molecular Formula $C_{25}H_{13}F_2NO_9$

Molecular Weight   509.37

Physical Form   Yellow/orange solid

Solubility   Insoluble in water; soluble in N,N-dimethylformamide, dimethyl sulfoxide

Boiling Point (Calcd.)   752.9 ± 70.0 °C, pressure: 760 Torr

pK$_a$ 4.7, temperature: 22 °C

pK$_a$ (Calcd.)   8.17 ± 0.20, most acidic, temperature: 25 °C

Absorption (λ_{max})   480 nm (Buffer pH 3.0); 495 nm (Buffer pH 9.0)

Emission (λ_{max})   521 nm (Buffers pH 3.0, 9.0)

Molar Extinction Coefficient   24,000 cm^{-1} M^{-1} (Buffer pH 3.0); 76,000 cm^{-1} M^{-1} (Buffer pH 9.0)

Synthesis   Synthetic methods[1,2]

Imaging/Labeling Applications   Cells;[4] cellulose nanocrystals;[3] microorganisms;[4,5] microspheres;[1,2] nucleic acids;[6,7] proteins;[1,2,8–21] small molecules[22]

Biological/Medical Applications   Analyzing tag-fused G-protein coupled receptor (GPCR);[10] detecting cells,[4] microorganisms;[4,5] detecting/isolating/purifying proteins;[1,2,8–21] detecting/measuring nitric oxide;[20,21] detecting/monitoring protein-protein interactions in real time;[15] measuring farnesyltransferase activity;[17] visualizing formation/orientation of collagen fibers in tissue[14]

Industrial Applications   Not reported

Safety/Toxicity   No data available

REFERENCES

1. Gee, K. R.; Poot, M.; Klaubert, D. H.; Sun, W.; Haugland, R. P.; Mao, F. Fluorinated xanthene derivatives. PCT Int. Appl. WO 9739064, 1997.

2. Gee, K. R.; Poot, M.; Klaubert, D. H.; Sun, W.; Haugland, R. P.; Mao, F. Fluorinated xanthene derivatives. U.S. Patent 6162931, 1997.

3. Nielsen, L. J.; Eyley, S.; Thielemans, W.; Aylott, J. W. Dual fluorescent labelling of cellulose nanocrystals for pH sensing. *Chem. Commun.* **2010**, *46*, 8929–8931.

4. He, W.; Yang, G.; Lv, J.; Li, Y.; Guo, Z.; Han, N. Compositions and methods for quantitative detection of microorganisms and cells. PCT Int. Appl. WO 2013056501, 2013.

5. Kinbara, K.; Shimomura, Y. Microorganism bioactivity evaluation method, and kit. Jpn. Kokai Tokkyo Koho JP 2007097532, 2007.

6. Bordello, J.; Sanchez, M. I.; Vazquez, M. E.; Mascarenas, J. L.; Al-Soufi, W.; Novo, M. Single-molecule approach to DNA minor-groove association dynamics. *Angew. Chem., Int. Ed.* **2012**, *51*, 7541–7544.

7. Giller, G.; Tasara, T.; Angerer, B.; Muhlegger, K.; Amacker, M.; Winter, H. Incorporation of reporter molecule-labeled nucleotides by DNA polymerases. I. Chemical synthesis of various reporter group-labeled 2'-deoxyribonucleoside-5'-triphosphates. *Nucleic Acids Res.* **2003**, *31*, 2630–2635.

8. Tamura, T.; Kioi, Y.; Miki, T.; Tsukiji, S.; Hamachi, I. Fluorophore labeling of native FKBP12 by ligand-directed tosyl chemistry allows detection of its molecular interactions *in vitro* and in living cells. *J. Am. Chem. Soc.* **2013**, *135*, 6782–6785.

9. Ciofani, G.; Del Turco, S.; Genchi, G. G.; D'Alessandro, D.; Basta, G.; Mattoli, V. Transferrin-conjugated boron nitride nanotubes: Protein grafting, characterization, and interaction with human endothelial cells. *Int. J. Pharm.* **2012**, *436*, 444–453.

10. Nonaka, H.; Fujishima, S.; Uchinomiya, S.; Ojida, A.; Hamachi, I. Selective covalent labeling of tag-fused GPCR proteins on live cell surface with a synthetic probe for their functional analysis. *J. Am. Chem. Soc.* **2010**, *132*, 9301–9309.

11. Mingels, A. M. A.; van Dongen, J. L. J.; Merkx, M. Mapping preferred sites for fluorescent labeling by combining fluorescence and MS analysis of tryptic CNA35 protein digests. *J. Chromatogr., B* **2008**, *863*, 293–297.

12. Beechem, J.; Hagen, D.; Johnson, I. Antibody complexes and methods for immunolabeling. U.S. Pat. Appl. Publ. US 20070269902, 2007.

13. Mottram, L. F.; Maddox, E.; Schwab, M.; Beaufils, F.; Peterson, B. R. A concise synthesis of the Pennsylvania Green fluorophore and labeling of intracellular targets with O6-benzylguanine derivatives. *Org. Lett.* **2007**, *9*, 3741–3744.

14. Krahn, K. N.; Bouten, C. V. C.; Van Tuijl, S.; Van Zandvoort, M. A. M. J.; Merkx, M. Fluorescently labeled collagen binding proteins allow specific visualization of collagen in tissues and live cell culture. *Anal. Biochem.* **2006**, *350*, 177–185.

15. Lata, S.; Gavutis, M.; Tampe, R.; Piehler, J. Specific and stable fluorescence labeling of histidine-tagged proteins for dissecting multi-protein complex formation. *J. Am. Chem. Soc.* **2006**, *128*, 2365–2372.

16. Nguyen, T.; Joshi, N. S.; Francis, M. B. An affinity-based method for the purification of fluorescently-labeled biomolecules. *Bioconjugate Chem.* **2006**, *17*, 869–872.

17. Berezovski, M.; Li, W.; Poulter, C. D.; Krylov, S. N. Measuring the activity of farnesyltransferase by capillary electrophoresis with laser-induced fluorescence detection. *Electrophoresis* **2002**, *23*, 3398–3403.

18. Wee, K. E.; Lai, Z.; Auger, K. R.; Ma, J.; Horiuchi, K. Y.; Dowling, R. L.; Dougherty, C. S.; Corman, J. I.; Wynn, R.; Copeland, R. A. Steady-state kinetic analysis of human ubiquitin-activating enzyme (E1) using a fluorescently labeled ubiquitin substrate. *J. Protein Chem.* **2000**, *19*, 489–498.

19. Francis-Lang, H.; Minden, J.; Sullivan, W.; Oegema, K. Live confocal analysis with fluorescently labeled proteins. *Methods Mol. Biol.* **1999**, *122*, 223–239.

20. Barker, S. L. R.; Clark, H. A.; Swallen, S. F.; Kopelman, R.; Tsang, A. W.; Swanson, J. A. Ratiometric and fluorescence-lifetime-based biosensors incorporating cytochrome c' and the detection of extra- and intracellular macrophage nitric oxide. *Anal. Chem.* **1999**, *71*, 1767–1772.

21. Barker, S. L. R.; Kopelman, R. Development and cellular applications of fiber optic nitric oxide sensors based on a gold-adsorbed fluorophore. *Anal. Chem.* **1998**, *70*, 4902–4906.

22. Devaraj, N. K.; Hilderbrand, S.; Upadhyay, R.; Mazitschek, R.; Weissleder, R. Bioorthogonal turn-on probes for imaging small molecules inside living cells. *Angew. Chem., Int. Ed.* **2010**, *49*, 2869–2872.

OREGON GREEN 514 CARBOXYLIC ACID SUCCINIMIDYL ESTER

CAS Registry Number   198139-53-6

Chemical Structure

CA Index Name   Acetic acid, 2-[(2′,4,5,7,7′-pentafluoro-3′,6′-dihydroxy-3-oxospiro[isobenzofuran-1(3*H*),9′-[9*H*]xanthen]-6-yl)thio]-, 2,5-dioxo-1-pyrrolidinyl ester

Other Names   2,5-Pyrrolidinedione, 1-[[[(2′,4,5,7,7′-pentafluoro-3′,6′-dihydroxy-3-oxospiro[isobenzofuran-1(3*H*),9′-[9*H*]xanthen]-6-yl)thio]acetyl]oxy]-; Oregon Green 514; Oregon Green 514 carboxylic acid succinimidyl ester; Oregon Green 514 succinimidyl ester

Merck Index Number   Not listed

Chemical/Dye Class   Xanthene

Molecular Formula $C_{26}H_{12}F_5NO_9S$

Molecular Weight   609.43

Physical Form   Yellow solid

Solubility   Insoluble in water; soluble in *N,N*-dimethylformamide, dimethyl sulfoxide

Boiling Point (Calcd.)   741.4 ± 70.0 °C, pressure: 760 Torr

pK$_a$ 4.7, temperature: 22 °C

pK$_a$ (Calcd.)   8.07 ± 0.20, most acidic, temperature: 25 °C

Absorption (λ_{max})   506 nm (Buffer pH 9.0)

Emission (λ_{max})   526 nm (Buffers 9.0)

Molar Extinction Coefficient   85,000 cm^{-1} M^{-1} (Buffer pH 9.0)

Synthesis   Synthetic methods[1,2]

Imaging/Labeling Applications   Androgens;[3] bacteria;[4,5] beads;[6,7] α-bungarotoxin;[1,2] cobalamin;[8] microspheres;[1,2] microorganisms;[9] monocytes;[10] nucleic acids;[11–15] peptides;[16,17] proteins;[1,2,5,13,18–23] tumors[13]

Biological/Medical Applications   Detecting/measuring nitric oxide;[23] detecting/diagnosing human immunodeficiency virus (HIV) infection;[19] detecting/identifying nucleic acids;[11–15] diagnosing/treating diseases associated with androgens;[3] identifying beads;[6,7] *in vitro* and *in vivo* imaging of transcobalamin receptors on cancer cells;[8] investigating/measuring protein-protein and protein-nucleic acid interactions;[15] treating/diagnosing disease states mediated by monocytes;[10] treating/preventing cancers of central nervous system;[13] monolayer-functionalized microfluidics devices;[24] as temperature sensor[25,26]

Industrial Applications   Not reported

Safety/Toxicity   No data available

REFERENCES

1. Gee, K. R.; Poot, M.; Klaubert, D. H.; Sun, W.; Haugland, R. P.; Mao, F. Fluorinated xanthene derivatives. PCT Int. Appl. WO 9739064, 1997.

2. Gee, K. R.; Poot, M.; Klaubert, D. H.; Sun, W.; Haugland, R. P.; Mao, F. Fluorinated xanthene derivatives. U.S. Patent 6162931, 1997.

3. Singh, M.; Gatson, J. W. Membrane androgen receptor as a therapeutic target for the prevention/promotion of cell death. U.S. Pat. Appl. Publ. US 20070141581, 2007.

4. Hasegawa, Y.; Welch, J. L. M.; Valm, A. M.; Rieken, C. W.; Sogin, M. L.; Borisy, G. G. Imaging

marine bacteria with unique 16S rRNA V6 sequences by fluorescence *in situ* hybridization and spectral analysis. *Geomicrobiol. J.* **2010**, *27*, 251–260.

5. Bruno, J. G.; Ulvick, S. J.; Uzzell, G. L.; Tabb, J. S.; Valdes, E. R.; Batt, C. A. Novel immuno-FRET assay method for *Bacillus* spores and *Escherichia coli* O157:H7. *Biochem. Biophys. Res. Commun.* **2001**, *287*, 875–880.

6. Christensen, S. F.; Johannsen, I.; Carstensen, J. M.; Kuhlmann, L.; Meldal, M. Identification of encoded beads. U.S. Pat. Appl. Publ. US 20090210165, 2009.

7. Christensen, S. F.; Johannsen, I.; Carstensen, J. M.; Kuhlmann, L. Identification of encoded beads. PCT Int. Appl. 2005061094, 2005.

8. Smeltzer, C. C.; Cannon, M. J.; Pinson, P. R.; Munger, J. D.,Jr.; West, F. G.; Grissom, C. B. Synthesis and characterization of fluorescent cobalamin (CobalaFluor) derivatives for imaging. *Org. Lett.* **2001**, *3*, 799–801.

9. Borisy, G. Methods and compositions for identifying cells by combinatorial fluoroscence imaging. PCT Int. Appl. WO 2009102844, 2009.

10. Low, P. S.; Hilgenbrink, A. R. Imaging and therapeutic method using monocytes. PCT Int. Appl. WO 2007006041, 2007.

11. Sergeev, N. V.; Brevnov, M. G.; Furtado, M. R. Identification of nucleic acids. PCT Int. Appl. WO 2011143478, 2011.

12. Lee, M. Variant scorpion primers for nucleic acid amplification and detection. U.S. Pat. Appl. Publ. US 20090197254, 2009.

13. Tews, B.; Hahn, M.; Lichter, P.; Reifenberger, G.; Roerig, P.; Felsberg, J.; Von Deimling, A.; Hartmann, C. CITED4 as prognostic marker in oligodendroglial tumors. Eur. Pat. Appl. EP 1793003, 2007.

14. Zeglis, B. M.; Barton, J. K. A mismatch-selective bifunctional rhodium-Oregon Green conjugate: A fluorescent probe for mismatched DNA. *J. Am. Chem. Soc.* **2006**, *128*, 5654–5655.

15. Rusinova, E.; Tretyachenko-Ladokhina, V.; Vele, O. E.; Senear, D. F.; Alexander Ross, J. B. Alexa and Oregon Green dyes as fluorescence anisotropy probes for measuring protein-protein and protein-nucleic acid interactions. *Anal. Biochem.* **2002**, *308*, 18–25.

16. Cullum, M. E.; O'Connor, K. H. Method for the detection of target molecules by fluorescence polarization using peptide mimics. U.S. Pat. Appl. Publ. US 20080227222, 2008.

17. Rozinov, M. N.; Nolan, G. P. Evolution of peptides that modulate the spectral qualities of bound, small-molecule fluorophores. *Chem. Biol.* **1998**, *5*, 713–728.

18. Raphael, M. P.; Rappole, C. A.; Kurihara, L. K.; Christodoulides, J. A.; Qadri, S. N.; Byers, J. M. Iminobiotin binding induces large fluorescent enhancements in avidin and streptavidin fluorescent conjugates and exhibits diverging pH-dependent binding affinities. *J. Fluoresc.* **2011**, *21*, 647–652.

19. Goldenberg, D. M.; Chang, C. H.; Rossi, E. A.; McBride, W. J. Methods and compositions for treatment of human immunodeficiency virus infection with conjugated antibodies or antibody fragments. U.S. Pat. Appl. Publ. US 20070264265, 2007.

20. Beechem, J.; Hagen, D.; Johnson, I. Antibody complexes and methods for immunolabeling. U.S. Pat. Appl. Publ. US 20070269902, 2007.

21. Georgiou, G.; Bahra, S. S.; Mackie, A. R.; Wolfe, C. A.; O'Shea, P.; Ladha, S.; Fernandez, N.; Cherry, R. J. Measurement of the lateral diffusion of human MHC class I molecules on HeLa cells by fluorescence recovery after photobleaching using a phycoerythrin probe. *Biophys. J.* **2002**, *82*, 1828–1834.

22. Rumbaut, R. E.; Harris, N. R.; Sial, A. J.; Huxley, V. H.; Granger, D. N. Leakage responses to L-NAME differ with the fluorescent dye used to label albumin. *Am. J. Physiol.* **1999**, *276*, H333–H339.

23. Kopelman, R.; Clark, H.; Barker, S. Optical sensors for the detection of nitric oxide. PCT Int. Appl. WO 9917139, 1999.

24. Mela, P.; Onclin, S.; Goedbloed, M. H.; Levi, S.; Garcia-Parajo, M. F.; van Hulst, N. F.; Ravoo, B. J.; Reinhoudt, D. N.; van den Berg, A. Monolayer-functionalized microfluidics devices for optical sensing of acidity. *Lab Chip* **2005**, *5*, 163–170.

25. Bousseksou, A.; Salmon, L.; Molnar, G.; Cobo, S. Materials with thermochromic spin transition doped with one or more fluorescent agents for use as temperature sensor. Fr. Demande FR 2952371, 2011.

26. Bousseksou, A.; Salmon, L.; Molnar, G.; Cobo, S. Heat-sensitive spin-transition materials doped with one or more fluorescent agents for use as temperature sensor. PCT Int. Appl. WO 2011058277, 2011.

OREGON GREEN 488 MALEIMIDE

CAS Registry Number 328085-55-8

Chemical Structure

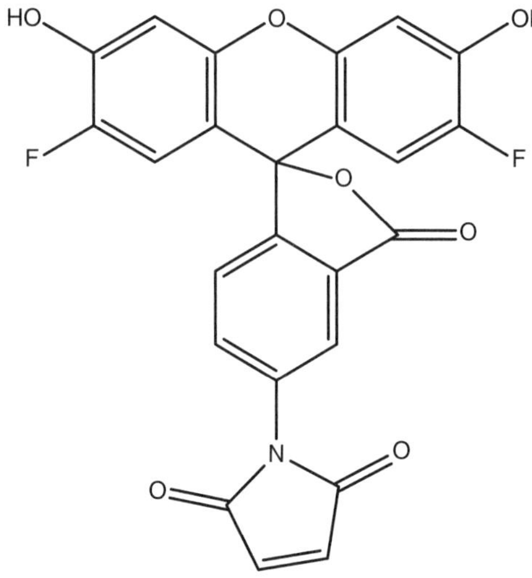

CA Index Name 1H-Pyrrole-2,5-dione, 1-(2′,7′-difluoro-3′,6′-dihydroxy-3-oxospiro[isobenzofuran-1(3H),9′-[9H]xanthen]-5-yl)-

Other Names Oregon Green 488 maleimide

Merck Index Number Not listed

Chemical/Dye Class Xanthene

Molecular Formula $C_{24}H_{11}F_2NO_7$

Molecular Weight 463.35

Physical Form Orange/yellow solid

Solubility Insoluble in water; soluble in N,N-dimethylformamide

Boiling Point (Calcd.) 740.5 ± 60.0 °C, pressure: 760 Torr

pK$_a$ 4.7, temperature: 22 °C

pK$_a$ (Calcd.) 8.17 ± 0.20, most acidic, temperature: 25 °C; −2.30 ± 0.20 most basic, temperature: 25 °C

Absorption (λ_{max}) 491 nm (Buffer pH 9.0)

Emission (λ_{max}) 515 nm (Buffer pH 9.0)

Molar Extinction Coefficient 81,000 cm^{-1} M^{-1} (Buffer pH 9.0)

Synthesis Synthetic method[1]

Imaging/Labeling Applications Actin filaments;[2] cathepsin D;[3] diamond nanoparticles;[4] metal ions (Zn(II), Cu(II), Co(II), Ni(II), Cd(II));[5] proteins;[6–21] titin molecules;[2] TonB N-terminus;[22] tumors[23]

Biological/Medical Applications Analyzing/monitoring proteins translocation;[20] detecting localized cathepsin D activity *in vivo*;[3] detecting/monitoring protein-protein interactions in real time;[15] imaging angiogenesis;[17] measuring protein expression;[18] monitoring collagen formation in live tissue cultures;[14] probing protein conformations;[10] providing affinity purification;[18] studying intranuclear protein,[11] ribonucleoprotein particle mobility,[11] protein folding at the single molecule level;[21] visualizing proteins;[18] photoluminescent sensors for detecting/quantitating metal ions in an aqueous sample;[5] viral nanoparticles (VNPs) for biomedical applications;[6,23] Western blot identification[18]

Industrial Applications Diamond nanoparticles[4]

Safety/Toxicity No data available

REFERENCES

1. Haugland, R. P. *Handbook of Fluorescent Probes and Research Products*; Molecular Probes Inc.: Eugene, **2002**; pp 84–92.
2. Kellermayer, M. S. Z.; Karsai, A.; Kengyel, A.; Nagy, A.; Bianco, P.; Huber, T.; Kulcsar, A.; Niedetzky, C.; Proksch, R.; Grama, L. Spatially and temporally synchronized atomic force and total internal reflection fluorescence microscopy for imaging and manipulating cells and biomolecules. *Biophys. J.* **2006**, *91*, 2665–2677.
3. Suchy, M.; Ta, R.; Li, A. X.; Wojciechowski, F.; Pasternak, S. H.; Bartha, R.; Hudson, R. H. E. A paramagnetic chemical exchange-based MRI probe metabolized by cathepsin D: design, synthesis and cellular uptake studies. *Org. Biomol. Chem.* **2010**, *8*, 2560–2566.
4. Jarre, G.; Liang, Y.; Betz, P.; Lang, D.; Krueger, A. Playing the surface game-Diels-Alder reactions on diamond nanoparticles. *Chem. Commun.* **2011**, *47*, 544–546.
5. Thompson, R. B.; Feliccia, V. L.; Maliwal, B. P.; Fierke, C. A. Photoluminescent sensors of chemical analytes. U.S. Patent 6197258, 2001.
6. Yildiz, I.; Tsvetkova, I.; Wen, A. M.; Shukla, S.; Masarapu, M. H.; Dragnea, B.; Steinmetz, N. F. Engineering of Brome mosaic virus for biomedical applications. *RSC Adv.* **2012**, *2*, 3670–3677.

7. Seim, K. L.; Obermeyer, A. C.; Francis, M. B. Oxidative modification of native protein residues using cerium(IV) ammonium nitrate. *J. Am. Chem. Soc.* **2011**, *133*, 16970–16976.

8. Grunwald, C.; Schulze, K.; Giannone, G.; Cognet, L.; Lounis, B.; Choquet, D.; Tampe, R. Quantum-yield-optimized fluorophores for site-specific labeling and super-resolution imaging. *J. Am. Chem. Soc.* **2011**, *133*, 8090–8093.

9. Kuiper, J. M.; Pluta, R.; Huibers, W. H. C.; Fusetti, F.; Geertsma, E. R.; Poolman, B. A method for site-specific labeling of multiple protein thiols. *Protein Sci.* **2009**, *18*, 1033–1041.

10. Strunk, J. J.; Gregor, I.; Becker, Y.; Lamken, P.; Lata, S.; Reichel, A.; Enderlein, J.; Piehler, J. Probing protein conformations by in situ non-covalent fluorescence labeling. *Bioconjugate Chem.* **2009**, *20*, 41–46.

11. Siebrasse, J. P.; Kubitscheck, U. Single molecule tracking for studying nucleocytoplasmic transport and intranuclear dynamics. *Methods Mol. Biol.* **2008**, *464*, 343–361.

12. Vogel, K.; Newman, R.; Riddle, S. Sensor proteins and assay methods. U.S. Pat. Appl. Publ. US 20080213811, 2008.

13. Miller, R. A.; Presley, A. D.; Francis, M. B. Self-assembling light-harvesting systems from synthetically modified tobacco mosaic virus coat proteins. *J. Am. Chem. Soc.* **2007**, *129*, 3104–3109.

14. Krahn, K. N.; Bouten, C. V. C.; Van Tuijl, S.; Van Zandvoort, M. A. M. J.; Merkx, M. Fluorescently labeled collagen binding proteins allow specific visualization of collagen in tissues and live cell culture. *Anal. Biochem.* **2006**, *350*, 177–185.

15. Lata, S.; Gavutis, M.; Tampe, R.; Piehler, J. Specific and stable fluorescence labeling of histidine-tagged proteins for dissecting multi-protein complex formation. *J. Am. Chem. Soc.* **2006**, *128*, 2365–2372.

16. Henderson, N. S.; So, S. S. K.; Martin, C.; Kulkarni, R.; Thanassi, D. G. Topology of the outer membrane Usher PapC determined by site-directed fluorescence labeling. *J. Biol. Chem.* **2004**, *279*, 53747–53754.

17. Dirksen, A.; Langereis, S.; de Waal, B. F. M.; van Genderen, M. H. P.; Meijer, E. W.; de Lussanet, Q. G.; Hackeng, T. M. Design and synthesis of a bimodal target-specific contrast agent for angiogenesis. *Org. Lett.* **2004**, *6*, 4857–4860.

18. La Clair, J. J.; Foley, T. L.; Schegg, T. R.; Regan, C. M.; Burkart, M. D. Manipulation of carrier proteins in antibiotic biosynthesis. *Chem. Biol.* **2004**, *11*, 195–201.

19. Qiu, H.; Edmunds, T.; Baker-Malcolm, J.; Karey, K. P.; Estes, S.; Schwarz, C.; Hughes, H.; Van Patten, S. M. Activation of human acid sphingomyelinase through modification or deletion of C-terminal cysteine. *J. Biol. Chem.* **2003**, *278*, 32744–32752.

20. de Keyzer, J. van der Does, C.; Driessen, A. J. M. Kinetic analysis of the translocation of fluorescent precursor proteins into *Eischerichia coli* membrane vesicles. *J. Biol. Chem.* **2002**, *277*, 46059–46065.

21. Zhuang, X.; Ha, T.; Kim, H. D.; Centner, T.; Labeit, S.; Chu, S. Fluorescence quenching: a tool for single-molecule protein-folding study. *Proc. Natl. Acad. Sci. U.S.A.* **2000**, *97*, 14241–14244.

22. Gresock, M. G.; Savenkova, M. I.; Larsen, R. A.; Ollis, A. A.; Postle, K. Death of the TonB shuttle hypothesis. *Front. Microbial Physiol. Metab.* **2011**, *1*, 206.

23. Wen, A. M.; Shukla, S.; Saxena, P.; Aljabali, A. A. A.; Yildiz, I.; Dey, S.; Mealy, J. E.; Yang, A. C.; Evans, D. J.; Lomonossoff, G. P. Interior engineering of a viral nanoparticle and its tumor homing properties. *Biomacromolecules* **2012**, *13*, 3990–4001.

PACIFIC BLUE (6,8-DIFLUORO-7-HYDROXYCOUMARIN-3-CARBOXYLIC ACID)

CAS Registry Number 215868-31-8

Chemical Structure

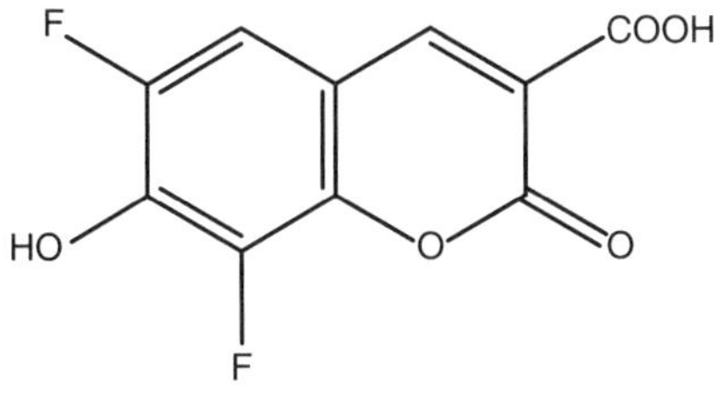

CA Index Name 2H-1-Benzopyran-3-carboxylic acid, 6,8-difluoro-7-hydroxy-2-oxo-

Other Names 3-Carboxy-6,8-difluoro-7-hydroxycoumarin; 6,8-Difluoro-7-hydroxycoumarin-3-carboxylic acid; Pacific Blue

Merck Index Number Not listed

Chemical/Dye Class Coumarin

Molecular Formula $C_{10}H_4F_2O_5$

Molecular Weight 242.13

Physical Form Green powder[2]

Solubility Soluble in ethyl acetate

Boiling Point (Calcd.) 422.0 ± 45.0 °C, pressure: 760 Torr

pK$_a$ 3.7, temperature: 22 °C

pK$_a$ (Calcd.) 1.40 ± 0.20, most acidic, temperature: 25 °C

Absorption (λ_{max}) 400 nm (Buffers pH 10.0)

Emission (λ_{max}) 447 nm (Buffer pH 10.0)

Molar Extinction Coefficient 29500 cm^{-1} M^{-1} (Buffer pH 10.0)

Quantum Yield 0.75 (Buffer pH 10.0)

Synthesis Synthetic methods[1,2]

Imaging/Labeling Applications Antibodies;[3,4] bacteria;[5] chromosomes;[6,7] histone deacetylase;[8] Hodgkin and Reed Sternberg (HRS) cells;[9] microorganisms;[10,11] nucleotides/nucleic acids;[12–19] peptides/proteins;[20–22] silver particles;[23] T lymphocyte antigens[24]

Biological/Medical Applications Analyzing JAK2 gene mutation;[25] analyzing/identifying peptides/proteins;[20–22] detecting chromosomal aberration;[6,7] diagnosing classical Hodgkin lymphoma (CHL) in lymph nodes;[9] identifying/characterizing microorganisms;[10,11] treating human immunodeficiency virus (HIV) infection,[4] ocular diseases;[26] as temperature sensors[27,28]

Industrial Applications Automotive paints[29]

Safety/Toxicity No data available

REFERENCES

1. Sun, W.; Gee, K. R.; Haugland, R. P. Synthesis of novel fluorinated coumarins: excellent UV-light excitable fluorescent dyes. *Bioorg. Med. Chem. Lett.* **1998**, *8*, 3107–3110.

2. Gee, K. R.; Haugland, R. P.; Sun, W. C. Derivatives of 6,8-difluoro-7-hydroxycoumarin. U.S. Patent 5830912, 1998.

3. Arcangeli, A.; Becchetti, A.; Pillozzi, S.; Masselli, M.; De Lorenzo, E. Method and kit for the prevention and/or the monitoring of chemoresistance of leukemia forms. PCT Int. Appl. WO 2011058509, 2011.

4. Goldenberg, D. M.; Chang, C. H.; Rossi, E. A.; McBride, W. J. Methods and compositions for treatment of human immunodeficiency virus infection with conjugated antibodies or antibody fragments. U.S. Pat. Appl. Publ. US 20070264265, 2007.

5. Tanner, M. A.; Coleman, W. J.; Everett, C. L.; Robles, S. J.; Dilworth, M. R.; Yang, M. M.; Youvan, D. C. Multispectral bacterial identification. *Proc. SPIE-Int. Soc. Opt. Eng.* **2000**, *3913*, 45–53.

6. Hauke, S. Method for detecting a chromosomal aberration. PCT Int. Appl. WO 2012150022, 2012.

7. Poulsen, T. S.; Poulsen, S. M.; Petersen, K. H. Methods for detecting chromosome aberrations. PCT Int. Appl. WO 2005111235, 2005.

8. Heidebrecht, R. W., Jr.,; Kral, A. M.; Miller, T. A. Fluorescent compounds that bind to histone deacetylase. U.S. Pat. Appl. Publ. US 20090156825, 2009.

Handbook of Fluorescent Dyes and Probes, First Edition. R. W. Sabnis.
© 2015 John Wiley & Sons, Inc. Published 2015 by John Wiley & Sons, Inc.

9. Fromm, J. R.; Thomas, A.; Wood, B. L. Flow cytometry can diagnose classical Hodgkin lymphoma in lymph nodes with high sensitivity and specificity. *Am. J. Clin. Pathol.* **2009**, *131*, 322–332.

10. Beimfohr, C.; Thelen, K.; Snaidr, J. A fluorescence *in situ* hybridization method for detection of microorganisms using FRET probes to increase signal:noise ratio. Ger. Offen. DE 102010012421, 2010.

11. Borisy, G. Methods and compositions for identifying cells by combinatorial fluoroscence imaging. PCT Int. Appl. WO 2009102844, 2009.

12. Kurose, K.; Komori, M. SLC28A1/A2/A3 gene polymorphism detection probe. Jpn. Kokai Tokkyo Koho JP 2013094116, 2013.

13. Ichikawa, K.; Hanawa, A.; Kurata, S. Nucleic acid probe for assaying nucleic acids. PCT Int. Appl. WO 2012173274, 2012.

14. Algar, W. R.; Krull, U. J. Developing mixed films of Immobilized oligonucleotides and quantum dots for the multiplexed detection of nucleic acid hybridization using a combination of fluorescence resonance energy transfer and direct excitation of fluorescence. *Langmuir* **2010**, *26*, 6041–6047.

15. Al Attar, H. A.; Monkman, A. P. FRET and competing processes between conjugated polymer and dye substituted DNA strands: A comparative study of probe selection in DNA detection. *Biomacromolecules* **2009**, *10*, 1077–1083.

16. Ichikawa, K.; Nakamura, K.; Kurata, S. Real-time PCR method using the novel type of oligonucleotide probes. Jpn. Kokai Tokkyo Koho JP 2008182974, 2008.

17. Nakamura, K.; Kanagawa, T.; Noda, T.; Tsuneda, S.; Tani, H.; Kurata, S. Novel mixtures for assaying nucleic acid, novel method for assaying nucleic acid with the use of the same and nucleic acid probe to be used therefor. PCT Int. Appl. WO 2005059548, 2005.

18. Han, X.; Tarrand, J. J.; Pham, A. S.; May, G. S. Diagnosis of mould infection. U.S. Pat. Appl. Publ. US 20050009051, 2005.

19. Ali, M. F.; Kirby, R.; Goodey, A. P.; Rodriguez, M. D.; Ellington, A. D.; Neikirk, D. P.; McDevitt, J. T. DNA hybridization and discrimination of single-nucleotide mismatches using chip-based microbead arrays. *Anal. Chem.* **2003**, *75*, 4732–4739.

20. Pashkova, A.; Moskovets, E.; Karger, B. L. Coumarin tags for improved analysis of peptides by MALDI-TOF MS and MS/MS. 1. enhancement in MALDI MS signal intensities. *Anal. Chem.* **2004**, *76*, 4550–4557.

21. Ting, A. Y.; Liu, D. S. Probe incorporation mediated by enzymes. PCT Int. Appl. WO 2013148189, 2013.

22. Maliwal, B. P.; Malicka, J.; Gryczynski, I.; Gryczynski, Z.; Lakowicz, J. R. Fluorescence properties of labeled proteins near silver colloid surfaces. *Biopolymers* **2003**, *70*, 585–594.

23. Zhang, J.; Fu, Y.; Chowdhury, M. H.; Lakowicz, J. R. Plasmon-coupled fluorescence probes: Effect of emission wavelength on fluorophore-labeled silver particles. *J. Phys. Chem. C* **2008**, *112*, 9172–9180.

24. Nakata, T.; Takayama, E.; Magari, H.; Kato, J.; Kondo, N.; Masao, I. Simultaneous detection of T lymphocyte-related antigens (CD4/CD8, CD57, TCRβ) with nuclei by fluorescence-based immunohistochemistry in paraffin-embedded human lymph node, liver cancer and stomach cancer. *Acta Cytol.* **2011**, *55*, 357–363.

25. Noda, N.; Sekiguchi, Y.; Morishita, S.; Tokita, S.; Komatsu, N.; Hasunuma, A.; Kirito, K. The JAK2 gene mutation analysis method. Jpn. Kokai Tokkyo Koho JP 2012034580, 2012.

26. Schulze, B.; Michaelis, U.; Agostini, H.; Hua, J.; Guenzi, E.; Gottfried, M.; Hansen, L. Use of a cationic colloidal preparation for the diagnosis and treatment of ocular diseases. PCT Int. Appl. WO 2008006535, 2008.

27. Bousseksou, A.; Salmon, L.; Molnar, G.; Cobo, S. Materials with thermochromic spin transition doped with one or more fluorescent agents for use as temperature sensor. Fr. Demande FR 2952371, 2011.

28. Bousseksou, A.; Salmon, L.; Molnar, G.; Cobo, S. Heat-sensitive spin-transition materials doped with one or more fluorescent agents for use as temperature sensor. PCT Int. Appl. WO 2011058277, 2011.

29. Stoecklein, W.; Fujiwara, H. The examination of UV-absorbers in 2-coat metallic and non-metallic automotive paints. *Sci. Justice* **1999**, *39*, 188–195.

PACIFIC BLUE C$_5$-MALEIMIDE

CAS Registry Number   934216-71-4
Chemical Structure

CA Index Name   2*H*-1-Benzopyran-3-carboxamide, *N*-[5-(2,5-dihydro-2,5-dioxo-1*H*-pyrrol-1-yl)pentyl]-6,8-difluoro-7-hydroxy-2-oxo-

Other Names   Pacific Blue C$_5$-Maleimide; Pacific Blue maleimide

Merck Index Number   Not listed

Chemical/Dye Class   Coumarin

Molecular Formula   C$_{19}$H$_{16}$F$_2$N$_2$O$_6$

Molecular Weight   406.34

Physical Form   Off-white solid

Solubility   Soluble in dimethyl sulfoxide

Boiling Point (Calcd.)   681.4 ± 55.0 °C, pressure: 760 Torr

pK$_a$ (Calcd.)   4.99 ± 0.20, most acidic, temperature: 25 °C; −1.54 ± 0.20, most basic, temperature: 25 °C

Absorption (λ_{max})   402 nm (Buffer pH 9.0)

Emission (λ_{max})   451 nm (Buffer pH 9.0)

Molar Extinction Coefficient   40,000 cm^{-1} M^{-1} (Buffer pH 9.0)

Quantum Yield   0.78 (Buffer pH 7.2)

Synthesis   Synthetic method[1]

Imaging/Labeling Applications   Amino acid;[2] cardiac troponin I (cTnI);[3,4] MHC tetramers;[5] pathogen;[6] proteins;[3,4] thiols[7]

Biological/Medical Applications   Analyzing/detecting of thiols;[7] detecting pathogen;[6] labeling amino acids;[2] producing MHC tetramers[5]

Industrial Applications   Not reported

Safety/Toxicity   No data available

REFERENCES

1. *The Molecular Probes Handbook: A Guide to Fluorescent Probes and Labeling Technologies*; Life Technologies Corporation: Eugene, **2010**; pp 116–121.

2. Kim, J.; Jensen, E. C.; Stockton, A. M.; Mathies, R. A. Universal microfluidic automaton for autonomous sample processing: Application to the mars organic analyzer. *Anal. Chem.* **2013**, *85*, 7682–7688.

3. Bottenus, D.; Hossan, M. R.; Ouyang, Y.; Dong, W.; Dutta, P.; Ivory, C. F. Preconcentration and detection of the phosphorylated forms of cardiac troponin I in a cascade microchip by cationic isotachophoresis. *Lab Chip* **2011**, *11*, 3793–3801.

4. Bottenus, D.; Jubery, T. Z.; Ouyang, Y.; Dong, W.; Dutta, P.; Ivory, C. F. 10000-fold concentration increase of the biomarker cardiac troponin I in a reducing union microfluidic chip using cationic isotachophoresis. *Lab Chip* **2011**, *11*, 890–898.

5. Ramachandiran, V.; Grigoriev, V.; Lan, L.; Ravkov, E.; Mertens, S. A.; Altman, J. D. A robust method for production of MHC tetramers with small molecule fluorophores. *J. Immunol. Methods* **2007**, *319*, 13–20.

6. Yung, C. W.; Ingber, D. E.; Cooper, R. M.; Vollmer, F.; Domansky, K.; Leslie, D. C.; Super, M. Rapid pathogen diagnostic device and method. PCT Int. Appl. WO 2011091037, 2011.

7. Mora, M. F.; Stockton, A. M.; Willis, P. A. Analysis of thiols by microchip capillary electrophoresis for *in situ* planetary investigations. *Electrophoresis* **2013**, *34*, 309–316.

PACIFIC BLUE SUCCINIMIDYL ESTER

CAS Registry Number   215868-33-0

Chemical Structure

CA Index Name   2H-1-Benzopyran-3-carboxylic acid, 6,8-difluoro-7-hydroxy-2-oxo-, 2,5-dioxo-1-pyrrolidinyl ester

Other Names   2,5-Pyrrolidinedione, 1-[[(6,8-difluoro-7-hydroxy-2-oxo-2H-1-benzopyran-3-yl)carbonyl]oxy]-; Pacific Blue succinimidyl ester

Merck Index Number   Not listed

Chemical/Dye Class   Coumarin

Molecular Formula $C_{14}H_7F_2NO_7$

Molecular Weight   339.20

Physical Form   Greenish powder[1] or yellow solid

Solubility   Soluble in acetonitrile, N,N-dimethylformamide, ethyl acetate

Boiling Point (Calcd.)   518.9 ± 60.0 °C, pressure: 760 Torr

pK$_a$ (Calcd.)   4.60 ± 0.20, most acidic, temperature: 25 °C

Absorption (λ_{max})   416 nm (Buffer pH 9.0)

Emission (λ_{max})   451 nm (Buffer pH 9.0)

Molar Extinction Coefficient   46,000 cm^{-1} M^{-1} (Buffer pH 9.0)

Synthesis   Synthetic method[1]

Imaging/Labeling Applications   Amines/amino acids;[2–4] cells;[5,6] mercury;[7] nucleic acids;[4] pancreatic β-cells;[8] proteins[9–12]

Biological/Medical Applications   Analyzing cells;[5,6,8] analyzing/detecting amines/amino acids;[2–4] fluorescent cell barcoding;[6] imaging/labeling proteins;[9–12] measuring/quantifying mercury concentration in a sample[7]

Industrial Applications   Not reported

Safety/Toxicity   No data available

REFERENCES

1. Gee, K. R.; Haugland, R. P.; Sun, W. C. Derivatives of 6,8-difluoro-7-hydroxycoumarin. U.S. Patent 5830912, 1998.

2. Mora, M. F.; Greer, F.; Stockton, A. M.; Bryant, S.; Willis, P. A. Toward total automation of microfluidics for extraterrestial in situ analysis. Anal. Chem. 2011, 83, 8636–8641.

3. Chiesl, T. N.; Chu, W. K.; Stockton, A. M.; Amashukeli, X.; Grunthaner, F.; Mathies, R. A. Enhanced amine and amino acid analysis using Pacific Blue and the mars organic analyzer microchip capillary electrophoresis system. Anal. Chem. 2009, 81, 2537–2544.

4. Cox, W. G.; Singer, V. L. Fluorescent DNA hybridization probe preparation using amine modification and reactive dye coupling. BioTechniques 2004, 36, 114–120,122.

5. Abrams, B.; Diwu, Z.; Guryev, O.; Aleshkov, S.; Hingorani, R.; Edinger, M.; Lee, R.; Link, J.; Dubrovsky, T. 3-Carboxy-6-chloro-7-hydroxycoumarin: A highly fluorescent, water-soluble violet-excitable dye for cell analysis. Anal. Biochem. 2009, 386, 262–269.

6. Krutzik, P. O.; Nolan, G. P. Fluorescent cell barcoding in flow cytometry allows high-throughput drug screening and signaling profiling. Nat. Methods 2006, 3, 361–368.

7. Jiao, H.; Catterall, H. Fluorescence dye tagging scheme for mercury quantification and speciation. U.S. Pat. Appl. Publ. US 20130040393, 2013.

8. Clardy, S. M.; Keliher, E. J.; Mohan, J. F.; Sebas, M.; Benoist, C.; Mathis, D.; Weissleder, R. Fluorescent exendin-4 derivatives for pancreatic β-cell analysis. Bioconjugate Chem. 2014, 25, 171–177.

9. Ting, A. Y.; Baker, D.; Richter, F.; Nivon, L.; Liu, D. S. Methods and compositions for protein labeling using lipoic acid ligases. U.S. Pat. Appl. Publ. US 20120129159, 2012.

10. Cohen, J. D.; Thompson, S.; Ting, A. Y. Structure-guided engineering of a Pacific Blue fluorophore

ligase for specific protein imaging in living cells. *Biochemistry* **2011**, *50*, 8221–8225.

11. Ting, A. Y.; Suarez, M. F.; Baruah, H.; Choi, Y. Methods and compositions for protein labeling using lipoic acid ligases. U.S. Pat. Appl. Publ. US 20090149631, 2009.

12. Ting, A. Y.; Suarez, M. F.; Baruah, H.; Choi, Y. Methods and compositions for protein labeling using lipoic acid ligases. PCT Int. Appl. WO 2009064366, 2009.

POPO 1

CAS Registry Number   169454-15-3

Chemical Structure

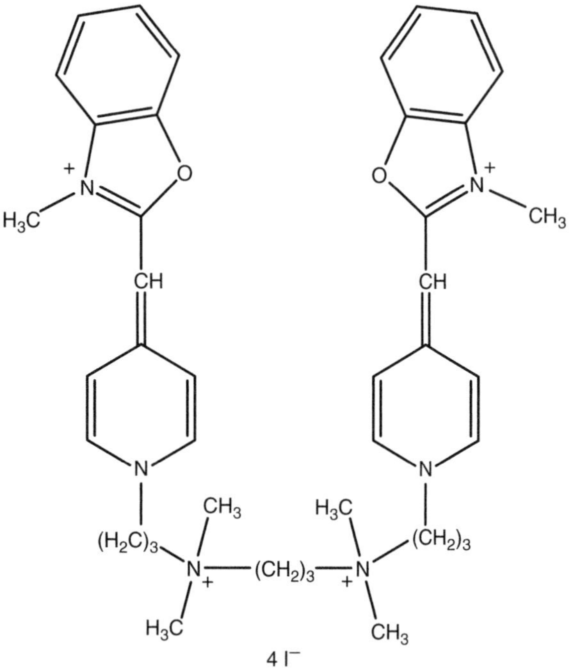

4 I⁻

CA Index Name   Benzoxazolium, 2,2′-[1,3-propanediylbis[(dimethyliminio)-3,1-propanediyl-1(4H)-pyridinyl-4-ylidenemethylidyne]]bis[3-methyl-, iodide (1 : 4)

Other Names   Benzoxazolium, 2,2′-[1,3-propanediylbis[(dimethyliminio)-3,1-propanediyl-1(4H)-pyridinyl-4-ylidenemethylidyne]]bis[3-methyl-, tetraiodide; POPO 1; POPO 1 iodide

Merck Index Number   Not listed

Chemical/Dye Class   Cyanine

Molecular Formula $C_{41}H_{54}I_4N_6O_2$

Molecular Weight   1170.53

Physical Form   Yellow-brown powder

Solubility   Soluble in dimethyl sulfoxide

Absorption (λ_{max})   434 nm (H_2O/DNA)

Emission (λ_{max})   456 nm (H_2O/DNA)

Molar Extinction Coefficient   92,400 cm⁻¹ M⁻¹ (H_2O/DNA)

Quantum Yield   0.60 (H_2O/DNA)

Synthesis   Synthetic methods[1,2]

Imaging/Labeling Applications   Nucleic acids;[1–12] cells;[2,11,12] bacteria;[12] chromosome spreads;[2] nuclei;[13,14] megakaryocytes;[15] microorganisms;[16–18] sperms;[11,19] hairs[20]

Biological/Medical Applications   Detecting nucleic acids,[1–12] cells,[2,11,12] JAK2 gene mutation;[21] classifying/counting megakaryocytes;[15] counting erythroblasts;[13,14] sperms;[11,19] detecting/measuring microorganisms;[16–18] nucleic acid hybridization;[22] nucleic acids nanochannel device;[23] nanosensor platforms for diagnostic applications;[24] as temperature sensor[25,26]

Industrial Applications   Not reported

Safety/Toxicity   No data available

REFERENCES

1. Sabnis, R. W. *Handbook of Biological Dyes and Stains*; John Wiley & Sons Inc.: Hoboken, **2010**; pp 378–379.

2. Yue, S. T.; Haugland, R. P. Dimers of unsymmetrical cyanine dyes containing pyridinium moieties. U.S. Patent 5410030, 1995.

3. Sergeev, N. V.; Brevnov, M. G.; Furtado, M. R. Identification of nucleic acids. PCT Int. Appl. WO 2011143478, 2011.

4. Miyamoto, S.; Kato, T.; Tomono, J. Nucleic acid identification method. Jpn. Kokai Tokkyo Koho JP 2010233530, 2010.

5. Exner, M.; Rogers, A. Methods for detecting nucleic acids using multiple signals. U.S. Pat. Appl. Publ. US 2007172836, 2007.

6. Erikson, G. H.; Daksis, J. I.; Kandic, I.; Picard, P. Nucleic acid multiplex formation. PCT Int. Appl. WO 2002103051, 2002.

7. Wittwer, C. T.; Dujols, V. E.; Reed, G.; Zhou, L. Amplicon melting analysis with saturation dyes. PCT Int. Appl. WO 2004038038, 2004.

8. Erikson, G. H.; Daksis, J. I. Pre-incubation method to improve signal/noise ratio of nucleic acid assays. U.S. Pat. Appl. Publ. US 2004180345, 2004.

9. Erikson, G. H. Method for modifying transcription and/or translation in an organism for therapeutic, prophylactic and/or analytic uses. U.S. Pat. Appl. Publ. US 2003181412, 2003.

10. Winter, S.; Loeber, G. DNA-Binding and fluorescence properties of the DNA *bis*-intercalating

purple oxazole dimer POPO-1. *J. Biomed. Opt.* **1997**, *2*, 125–130.

11. Anderson, A. L.; Knutson, C. R.; Mueth, D.; Plewa, J.; Tanner, E. Methods for staining cells for identification and sorting. U.S. Pat. Appl. Publ. US 2006172315, 2006.

12. Millard, P. J.; Roth, B. L.; Yue, S. T.; Haugland, R. P. Fluorescent viability assay using cyclic-substituted unsymmetrical cyanine dyes. U.S. Patent 5534416, 1996.

13. Heuven, B.; Wong, F.; Tsuji, T.; Sakata, T.; Hamaguchi, I. Method for classifying and counting erythroblasts by flow cytometry. Jpn. Kokai Tokkyo Koho JP 11326323, 1999.

14. Heuven, B.; Wong, F.; Tsuji, T.; Sakata, T.; Hamaguchi, I. Process for discriminating and counting erythroblasts. U.S. Pat. Appl. Publ. US 20020006631, 2002.

15. Minakami, T.; Mori, Y.; Tsuji, T.; Ikeuchi, Y. Megakaryocyte classification/counting method by double fluorescent staining and flow cytometry. Jpn. Kokai Tokkyo Koho JP 2006275985, 2006.

16. Eckert, R. H.; Kaplan, C.; He, J.; Yarbrough, D. K.; Anderson, M.; Sim, J. Methods and devices for the selective detection of microorganisms. U.S. Pat. Appl. Publ. US 20120003661, 2012.

17. Noda, N.; Mizutani, T. Microorganism-measuring method using multiple staining. Jpn. Kokai Tokkyo Koho JP 2006340684, 2006.

18. Vannier, E. Methods for detection of pathogens in red blood cells. PCT Int. Appl. WO 2006031544, 2006.

19. Matsumoto, T.; Okada, H.; Hamaguchi, Y. Method and reagent for counting sperm by flow cytometry. Jpn. Kokai Tokkyo Koho JP 2001242168, 2001.

20. Lagrange, A. Hair dye compositions containing a polycationic direct dye. Fr. Demande FR 2848840, 2004.

21. Park, H. G.; Choi, J. J.; Kim, H. S.; Jung, S. U. JAK2 gene mutation detection kit based on PNA mediated real-time PCR clamping. Repub. Korean Kongkae Taeho Kongbo KR 2012119571, 2012.

22. Hahn, J.; Park, N. Detection method of nucleic acid hybridization. PCT Int. Appl. WO 2003010338, 2003.

23. Tegenfeldt, J.; Reisner, W.; Flyvbjerg, H. Method for the mapping of the local AT/GC ratio along DNA. PCT Int. Appl. WO 2010042007, 2010.

24. Zhou, C.; Thompson, M. E.; Cote, R. J.; Ishikawa, F.; Curreli, M.; Chang, H. Methods of using and constructing nanosensor platforms. PCT Int. Appl. WO 2009085356, 2009.

25. Bousseksou, A.; Salmon, L.; Molnar, G.; Cobo, S. Materials with thermochromic spin transition doped with one or more fluorescent agents for use as temperature sensor. Fr. Demande FR 2952371, 2011.

26. Bousseksou, A.; Salmon, L.; Molnar, G.; Cobo, S. Heat-sensitive spin-transition materials doped with one or more fluorescent agents for use as temperature sensor. PCT Int. Appl. WO 2011058277, 2011.

POPO 3

CAS Registry Number 154757-99-0

Chemical Structure

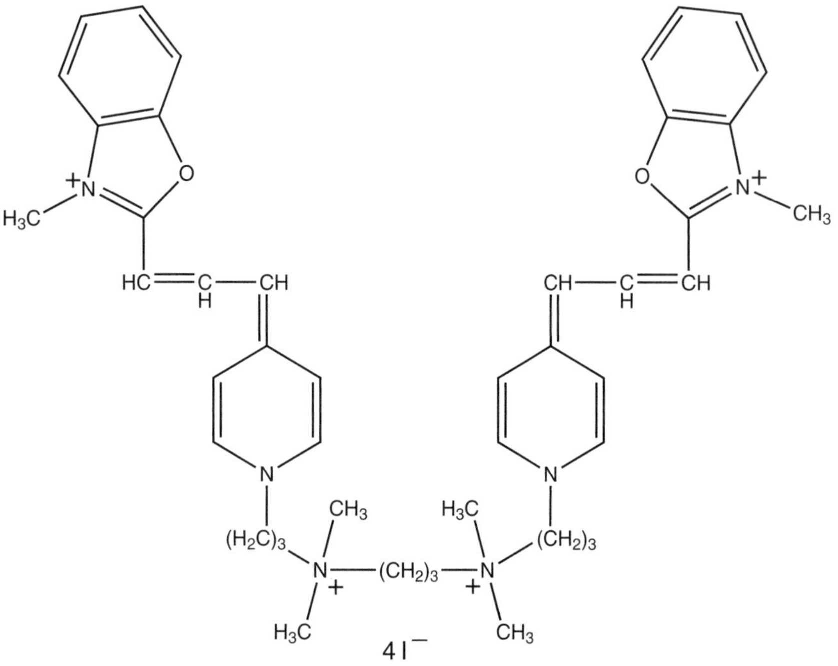

Emission (λ_{max}) 570 nm (H$_2$O/DNA)

Molar Extinction Coefficient 146,400 cm^{-1} M^{-1} (H$_2$O/DNA)

Quantum Yield 0.46 (H$_2$O/DNA)

Synthesis Synthetic methods[1,2]

CA Index Name Benzoxazolium, 2,2′-[1,3-propanediylbis[(dimethyliminio)-3,1-propanediyl-1(4*H*)-pyridinyl-4-ylidene-1-propen-1-yl-3-ylidene]]bis[3-methyl-, iodide (1 : 4)

Other Names Benzoxazolium, 2,2′-[1,3-propanediylbis[(dimethyliminio)-3,1-propanediyl-1(4*H*)-pyridinyl-4-ylidene-1-propen-1-yl-3-ylidene]]bis[3-methyl-, tetraiodide; POPO 3; POPO 3 iodide

Merck Index Number Not listed

Chemical/Dye Class Cyanine

Molecular Formula C$_{45}$H$_{58}$I$_4$N$_6$O$_2$

Molecular Weight 1222.61

Physical Form Yellow-brown powder

Solubility Soluble in *N,N*-dimethylformamide

Absorption (λ_{max}) 534 nm (H$_2$O/DNA)

Imaging/Labeling Applications Nucleic acids;[1–12] cells;[2,10–12] bacteria;[10] chromosome spreads;[2] nuclei;[13,14] megakaryocyte;[15] microorganisms;[16–18] reticulocytes;[11,19] sperms[12,20]

Biological/Medical Applications Detecting nucleic acids,[1–12] cells,[2,10–12] microorganisms,[16–18] IDH1/IDH2 mutations,[21] JAK2 gene mutation,[22] PIK3CA mutation,[23] B-type Raf Kinase (BRAF) mutation,[24] classifying/counting megakaryocytes;[15] counting erythroblasts,[13,14] reticulocytes,[11,19] sperms;[12,20] quantifying nucleic acids;[9] nucleic acid amplification;[25] nucleic acid hybridization;[26] nucleic acid nanochannel device;[27] as temperature sensor[28,29]

Industrial Applications Photonic fabric display[30]

Safety/Toxicity No data available

REFERENCES

1. Sabnis, R. W. *Handbook of Biological Dyes and Stains*; John Wiley & Sons Inc.: Hoboken, **2010**; pp 380–381.

2. Yue, S. T.; Haugland, R. P. Dimers of unsymmetrical cyanine dyes containing pyridinium moieties. U.S. Patent 5410030, 1995.

3. Sergeev, N. V.; Brevnov, M. G.; Furtado, M. R. Identification of nucleic acids. PCT Int. Appl. WO 2011143478, 2011.

4. Exner, M.; Rogers, A. Methods for detecting nucleic acids using multiple signals. U.S. Pat. Appl. Publ. US 2007172836, 2007.

5. Wittwer, C. T.; Dujols, V. E.; Reed, G.; Zhou, L. Amplicon melting analysis with saturation dyes. PCT Int. Appl. WO 2004038038, 2004.

6. Erikson, G. H.; Daksis, J. I.; Kandic, I.; Picard, P. Nucleic acid multiplex formation. PCT Int. Appl. WO 2002103051, 2002.

7. Erikson, G. H.; Daksis, J. I. Pre-incubation method to improve signal/noise ratio of nucleic acid assays. U.S. Pat. Appl. Publ. US 2004180345, 2004.

8. Erikson, G. H. Method for modifying transcription and/or translation in an organism for therapeutic, prophylactic and/or analytic uses. U.S. Pat. Appl. Publ. US 2003181412, 2003.

9. Narz, F. Quantification of RNA using internal normalization. PCT Int. Appl. WO 2010046441, 2010.

10. Millard, P. J.; Roth, B. L.; Yue, S. T.; Haugland, R. P. Fluorescent viability assay using cyclic-substituted unsymmetrical cyanine dyes. U.S. Patent 5534416, 1996.

11. Deka, C.; Gordon, K. M.; Gupta, R.; Horton, A. Methods and compositions for rapid staining of nucleic acids in whole cells. U.S. Patent 6271035, 2001.

12. Anderson, A. L.; Knutson, C. R.; Mueth, D.; Plewa, J.; Tanner, E. Methods for staining cells for identification and sorting. U.S. Pat. Appl. Publ. US 2006172315, 2006.

13. Heuven, B.; Wong, F. S.; Tsuji, T.; Sakata, T.; Hamaguchi, I. Method for classifying and counting erythroblasts by flow cytometry. Jpn. Kokai Tokkyo Koho JP 11326323, 1999.

14. Heuven, B.; Wong, F.; Tsuji, T.; Sakata, T.; Hamaguchi, I. Process for discriminating and counting erythroblasts. U.S. Pat. Appl. Publ. US 20020006631, 2002.

15. Minakami, T.; Mori, Y.; Tsuji, T.; Ikeuchi, Y. Megakaryocyte classification/counting method by double fluorescent staining and flow cytometry. Jpn. Kokai Tokkyo Koho JP 2006275985, 2006.

16. Eckert, R. H.; Kaplan, C.; He, J.; Yarbrough, D. K.; Anderson, M.; Sim, J. Methods and devices for the selective detection of microorganisms. U.S. Pat. Appl. Publ. US 20120003661, 2012.

17. Tobin, K. J.; Onstott, T. C.; DeFlaun, M. F.; Colwell, F. S.; Fredrickson, J. *In situ* imaging of microorganisms in geologic material. *J. Microbiol. Methods* **1999**, *37*, 201–213.

18. Vannier, E. Methods for detection of pathogens in red blood cells. PCT Int. Appl. WO 2006031544, 2006.

19. Yamamoto, K.; Tokita, M.; Inagami, M.; Hirako, S. Method of counting reticulocytes. U.S. Patent 5563070, 1996.

20. Matsumoto, T.; Okada, H.; Hamaguchi, Y. Method and reagent for counting sperm by flow cytometry. Jpn. Kokai Tokkyo Koho JP 2001242168, 2001.

21. Park, H. G.; Choi, J. J.; Kim, H. S.; Jung, S. U. Method and kit for detecting IDH1 and IDH2 mutations using peptide nucleic acid (PNA)-based real-time polymerase chain reaction (PCR) clamping. Repub. Korean Kongkae Taeho Kongbo KR 2012127679, 2012.

22. Park, H. G.; Choi, J. J.; Kim, H. S.; Jung, S. U. JAK2 gene mutation detection kit based on PNA mediated real-time PCR clamping. Repub. Korean Kongkae Taeho Kongbo KR 2012119571, 2012.

23. Park, H. K.; Choi, J. J.; Kim, H. S. PIK3CA mutation detection method and kit using real-time PCR clamping of PNA. PCT Int. Appl. WO 2012020965, 2012.

24. Park, H. K.; Choi, J. J.; Cho, M. H. Detection of mutation of BRAF by real-time PCR using PNA clamping probe for diagnosis of cancers. PCT Int. Appl. WO 2011093606, 2011.

25. Morrison, T. Improved selective ligation and amplification assay. PCT Int. Appl. WO 2005059178, 2005.

26. Hahn, J.; Park, N. Detection method of nucleic acid hybridization. PCT Int. Appl. WO 2003010338, 2003.

27. Tegenfeldt, J.; Reisner, W.; Flyvbjerg, H. Method for the mapping of the local AT/GC ratio along DNA. PCT Int. Appl. WO 2010042007, 2010.

28. Bousseksou, A.; Salmon, L.; Molnar, G.; Cobo, S. Materials with thermochromic spin transition doped with one or more fluorescent agents for use as temperature sensor. Fr. Demande FR 2952371, 2011.

29. Bousseksou, A.; Salmon, L.; Molnar, G.; Cobo, S. Heat-sensitive spin-transition materials doped with one or more fluorescent agents for use as temperature sensor. PCT Int. Appl. WO 2011058277, 2011.

30. Tao, X.; Cheng, X.; Yu, J.; Liu, L.; Wong, W.; Tam, W. Photonic fabric display with controlled pattern, color, luminescence intensity, scattering intensity and light self-amplification. U.S. Pat. Appl. Publ. US 20070281155, 2007.

PO-PRO 1

CAS Registry Number   157199-56-9

Chemical Structure

CA Index Name   Benzoxazolium, 3-methyl-2-[[1-[3-(trimethylammonio)-propyl]-4(1*H*)-pyridinylidene]methyl]-, iodide (1:2)

Other Names   Benzoxazolium, 3-methyl-2-[[1-[3-(trimethylammonio)-propyl]-4(1*H*)-pyridinylidene]methyl]-, diiodide; PO-PRO 1; PO-PRO 1 iodide

Merck Index Number   Not listed

Chemical/Dye Class   Cyanine

Molecular Formula $C_{20}H_{27}I_2N_3O$

Molecular Weight   579.26

Physical Form   Yellow-brown powder

Solubility   Soluble in dimethyl sulfoxide

Absorption (λ_{max})   435 nm (H_2O/DNA)

Emission (λ_{max})   455 nm (H_2O/DNA)

Molar Extinction Coefficient   50,100 $cm^{-1}\,M^{-1}$ (H_2O/DNA)

Quantum Yield   0.39 (H_2O/DNA)

Synthesis   Synthetic methods[1–3]

Imaging/Labeling Applications   Nucleic acids;[1–10] cells;[1–3,10,11] bacteria;[3] microorganisms;[12,13] proteins;[14,15] sperms[10]

Biological/Medical Applications   Detecting nucleic acids,[1–10] cells;[1–3,10,11] counting/detecting microorganisms;[12,13] nucleic acids amplification;[9] nucleic acid hybridization;[16] nucleic acid sequencing;[17] as temperature sensor[18,19]

Industrial Applications   Not reported

Safety/Toxicity   No data available

REFERENCES

1. Sabnis, R. W. *Handbook of Biological Dyes and Stains*; John Wiley & Sons Inc.: Hoboken, **2010**; pp 382–383.

2. Yue, S. T.; Johnson, I. D.; Huang, Z.; Haugland, R. P. Unsymmetrical cyanine dyes with a cationic side chain. U.S. Patent 5321130, 1994.

3. Millard, P. J.; Roth, B. L.; Yue, S. T.; Haugland, R. P. Fluorescent viability assay using cyclic-substituted unsymmetrical cyanine dyes. U.S. Patent 5534416, 1996.

4. Sergeev, N. V.; Brevnov, M. G.; Furtado, M. R. Identification of nucleic acids. PCT Int. Appl. WO 2011143478, 2011.

5. Exner, M.; Rogers, A. Methods for detecting nucleic acids using multiple signals. U.S. Pat. Appl. Publ. US 2007172836, 2007.

6. Wittwer, C. T.; Dujols, V. E.; Reed, G.; Zhou, L. Amplicon melting analysis with saturation dyes. PCT Int. Appl. WO 2004038038, 2004.

7. Erikson, G. H. Method for modifying transcription and/or translation in an organism for therapeutic, prophylactic and/or analytic uses. U.S. Pat. Appl. Publ. US 2003181412, 2003.

8. Erikson, G. H.; Daksis, J. I.; Kandic, I.; Picard, P. Nucleic acid multiplex formation. PCT Int. Appl. WO 2002103051, 2002.

9. Sutherland, J. W.; Patterson, D. R. Homogeneous method for assay of double-stranded nucleic acids using fluorescent dyes and kit useful therein. Eur. Pat. Appl. EP 684316, 1995.

10. Anderson, A. L.; Knutson, C. R.; Mueth, D.; Plewa, J.; Tanner, E. Methods for staining cells for

identification and sorting. U.S. Pat. Appl. Publ. US 2006172315, 2006.

11. Hoshi, H.; O'Brien, J.; Mills, S. L. A novel fluorescent tracer for visualizing coupled cells in neural circuits of living tissue. *J. Histochem. Cytochem.* **2006**, *54*, 1169–1176.

12. Eckert, R. H.; Kaplan, C.; He, J.; Yarbrough, D. K.; Anderson, M.; Sim, J. Methods and devices for the selective detection of microorganisms. U.S. Pat. Appl. Publ. US 20120003661, 2012.

13. Sunamura, T.; Maruyama, A.; Kurane, R. Method for detecting and counting microorganism. Jpn. Kokai Tokkyo Koho JP 2002291499, 2002.

14. Zanotti, K. J.; Silva, G. L.; Creeger, Y.; Robertson, K. L.; Waggoner, A. S.; Berget, P. B.; Armitage, B. A. Blue fluorescent dye-protein complexes based on fluorogenic cyanine dyes and single chain antibody fragments. *Org. Biomol. Chem.* **2011**, *9*, 1012–1020.

15. Ozhalici-Unal, H.; Pow, C. L.; Marks, S. A.; Jesper, L. D.; Silva, G. L.; Shank, N. I.; Jones, E. W.; Burnette, J. M., III,; Berget, P. B.; Armitage, B. A. A rainbow of fluoromodules: A Promiscuous scFv protein binds to and activates a diverse set of fluorogenic cyanine dyes. *J. Am. Chem. Soc.* **2008**, *130*, 12620–12621.

16. Hahn, J.; Park, N. Detection method of nucleic acid hybridization. PCT Int. Appl. WO 2003010338, 2003.

17. Hoser, M. J. Nucleic acid sequencing methods, kits and reagents. PCT Int. Appl. WO 2004074503, 2004.

18. Bousseksou, A.; Salmon, L.; Molnar, G.; Cobo, S. Materials with thermochromic spin transition doped with one or more fluorescent agents for use as temperature sensor. Fr. Demande FR 2952371, 2011.

19. Bousseksou, A.; Salmon, L.; Molnar, G.; Cobo, S. Heat-sensitive spin-transition materials doped with one or more fluorescent agents for use as temperature sensor. PCT Int. Appl. WO 2011058277, 2011.

PO-PRO 3

CAS Registry Number   161016-55-3

Chemical Structure

CA Index Name   Benzoxazolium, 3-methyl-2-[3-[1-[3-(trimethylammonio)-propyl]-4(1*H*)-pyridinylidene]-1-propen-1-yl]-, iodide (1 : 2)

Other Names   Benzoxazolium, 3-methyl-2-[3-[1-[3-(trimethylammonio)-propyl]-4(1*H*)-pyridinylidene]-1-propenyl]-, diiodide; PO-PRO 3; PO-PRO 3 iodide

Merck Index Number   Not listed

Chemical/Dye Class   Cyanine

Molecular Formula $C_{22}H_{29}I_2N_3O$

Molecular Weight   605.30

Physical Form   Yellow-brown powder

Solubility   Soluble in dimethyl sulfoxide

Absorption (λ_{max})   539 nm (H_2O/DNA)

Emission (λ_{max})   567 nm (H_2O/DNA)

Molar Extinction Coefficient   87,900 $cm^{-1}M^{-1}$ (H_2O/DNA)

Quantum Yield   0.57 (H_2O/DNA)

Synthesis   Synthetic methods[1–3]

Imaging/Labeling Applications   Nucleic acids;[1–11] cells;[2,3,10,11] bacteria;[3] reticulocytes;[11,12] sperms;[10] nuclei;[13] cytoplasm[13]

Biological/Medical Applications   Detecting nucleic acids,[1–11] cells;[2,3,10,11] counting reticulocytes;[11,12] nucleic acid amplification;[8] nucleic acid hybridization;[14] as temperature sensor[15,16]

Industrial Applications   Not reported

Safety/Toxicity   No data available

REFERENCES

1. Sabnis, R. W. *Handbook of Biological Dyes and Stains*; John Wiley & Sons Inc.: Hoboken, **2010**; pp 384–385.

2. Yue, S. T.; Johnson, I. D.; Huang, Z.; Haugland, R. P. Unsymmetrical cyanine dyes with a cationic side chain. U.S. Patent 5321130, 1994.

3. Millard, P. J.; Roth, B. L.; Yue, S. T.; Haugland, R. P. Fluorescent viability assay using cyclic-substituted unsymmetrical cyanine dyes. U.S. Patent 5534416, 1996.

4. Exner, M.; Rogers, A. Methods for detecting nucleic acids using multiple signals. U.S. Pat. Appl. Publ. US 2007172836, 2007.

5. Erikson, G. H.; Daksis, J. I.; Kandic, I.; Picard, P. Nucleic acid multiplex formation. PCT Int. Appl. WO 2002103051, 2002.

6. Erikson, G. H.; Daksis, J. I. Pre-incubation method to improve signal/noise ratio of nucleic acid assays. U.S. Pat. Appl. Publ. US 2004180345, 2004.

7. Erikson, G. H. Method for modifying transcription and/or translation in an organism for therapeutic, prophylactic and/or analytic uses. U.S. Pat. Appl. Publ. US 2003181412, 2003.

8. Sutherland, J. W.; Patterson, D. R. Homogeneous method for assay of double-stranded nucleic acids using fluorescent dyes and kit useful therein. Eur. Pat. Appl. EP 684316, 1995.

9. Sherwood, C. S.; Haynes, C. A.; Turner, R. F. B. Nanogram-level micro-volume DNA assay based on the monomeric cyanine dye PO-PRO-3 iodide. *BioTechniques* **1995**, *18*, 136–141.

10. Anderson, A. L.; Knutson, C. R.; Mueth, D.; Plewa, J.; Tanner, E. Methods for staining cells for identification and sorting. U.S. Pat. Appl. Publ. US 2006172315, 2006.

11. Deka, C.; Gordon, K. M.; Gupta, R.; Horton, A. Methods and compositions for rapid staining of nucleic acids in whole cells. U.S. Patent 6271035, 2001.

12. Yamamoto, K.; Tokita, M.; Inagami, M.; Hirako, S. Method of counting reticulocytes. U.S. Patent 5563070, 1996.

13. Bink, K.; Walch, A.; Feuchtinger, A.; Eisenmann, H.; Hutzler, P.; Hofler, H.; Werner, M. TO-PRO-3 is an optimal fluorescent dye for nuclear counterstaining in dual-colour FISH on paraffin sections. *Histochem. Cell Biol.* **2001**, *115*, 293–299.

14. Hahn, J.; Park, N. Detection method of nucleic acid hybridization. PCT Int. Appl. WO 2003010338, 2003.

15. Bousseksou, A.; Salmon, L.; Molnar, G.; Cobo, S. Materials with thermochromic spin transition doped with one or more fluorescent agents for use as temperature sensor. Fr. Demande FR 2952371, 2011.

16. Bousseksou, A.; Salmon, L.; Molnar, G.; Cobo, S. Heat-sensitive spin-transition materials doped with one or more fluorescent agents for use as temperature sensor. PCT Int. Appl. WO 2011058277, 2011.

PROPIDIUM IODIDE (PI)

CAS Registry Number   25535-16-4
Chemical Structure

CA Index Name   Phenanthridinium, 3,8-diamino-5-[3-(diethylmethylammonio)propyl]-6-phenyl-, iodide (1:2)

Other Names   3,8-Diamino-5-(3-diethylaminopropyl)-6-phenylphenanthridinium iodide methiodide; Phenanthridinium, 3,8-diamino-5-[3-(diethylmethylammonio)propyl]-6-phenyl-, diiodide; Ammonium, [3-(3,8-diamino-6-phenyl-5-phenanthridinio)-propyl]diethylmethyl-, diiodide; 3,8-Diamino-5-(diethylmethylaminopropyl)-6-phenyl-phenanthridinium diiodide; Propidium diiodide; Propidium iodide; PI

Merck Index Number   Not listed

Chemical/Dye Class   Heterocycle; Phenanthridine

Molecular Formula $C_{27}H_{34}I_2N_4$
Molecular Weight   668.40
Physical Form   Red powder or solid
Solubility   Soluble in water, N,N-dimethylformamide, dimethyl sulfoxide, methanol
Melting Point   220–225 °C (decompose)
Absorption (λ_{max})   535 nm (H_2O/DNA); 493 nm (H_2O)
Emission (λ_{max})   617 nm (H_2O/DNA); 636 nm (H_2O)
Molar Extinction Coefficient   5400 cm^{-1} M^{-1} (H_2O/DNA); 5900 cm^{-1} M^{-1} (H_2O)
Synthesis   Synthetic methods[1,2]
Imaging/Labeling Applications   Nucleic acids;[1–27] acetylcholinesterase;[28,29] bacteria;[30–34] cells;[35–44] chromosomes;[45,46] fungi;[47,48] liposomes;[49] leukocytes;[50,51] megakaryocytes;[52] microorganisms;[53–63] neurons;[64] nuclei;[65–68] parasites;[69–71] protozoa;[72] sperms;[73–79] stem cells;[80,81] yeasts;[82–85]
Biological/Medical Applications   Analyzing/counting cells;[35–44] detecting nucleic acids;[1–27] apoptosis assays;[86–102] cytotoxicity assays;[103–107] measuring/monitoring microorganisms;[53–63] monitoring atmospheric/indoor bioaerosols;[108,109] as temperature sensor;[110,111] treating ischemic diseases,[112] viral diseases[113]
Industrial Applications   Electroluminescent displays;[114] photoresists[115,116]
Safety/Toxicity   Cytotoxicity;[117,118] embryotoxicity;[119] mutagenicity;[120] neurotoxicity;[121,122] phototoxicity;[123] reproductive toxicity[124]

REFERENCES

1. Sabnis, R. W. *Handbook of Biological Dyes and Stains*; John Wiley & Sons Inc.: Hoboken, **2010**; pp 386–387.

2. Haugland, R. P. *Handbook of Fluorescent Probes and Research Chemicals*; Molecular Probes Inc.: Eugene, **1996**; pp 144–156.

3. Kjaerulff, S.; Glensbjerg, M. Method for analysis of cellular DNA content. PCT Int. Appl. WO 2011098085, 2011.

4. Miyamoto, S.; Kato, T.; Tomono, J. Nucleic acid identification method. Jpn. Kokai Tokkyo Koho JP 2010233530, 2010.

5. Cunningham, R. E. Fluorescent labeling of DNA. *Methods Mol. Biol.* **2010**, *588*, 341–344.

6. Swerts, K.; Van Roy, N.; Benoit, Y.; Laureys, G.; Philippe, J. DRAQ5: Improved flow cytometric DNA content analysis and minimal residual disease detection in childhood malignancies. *Clin. Chim. Acta* **2007**, *379*, 154–157.

7. Martin, R. M.; Leonhardt, H.; Cardoso, M. C. DNA labeling in living cells. *Cytometry, Part A* **2005**, *67A*, 45–52.

8. Noirot, M.; Barre, P.; Louarn, J.; Duperray, C.; Hamon, S. Consequences of stoichiometric error on nuclear DNA content evaluation in *Coffea liberica* var. dewevrei using DAPI and propidium iodide. *Ann. Bot.* **2002**, *89*, 385–389.

9. Cunningham, R. E. Fluorescent labeling of DNA. *Methods Mol. Biol.* **1999**, *115*, 271–273.

10. Rousselle, C.; Robert-Nicoud, M.; Ronot, X. Flow cytometric analysis of DNA content of living and

fixed cells: a comparative study using various fixatives. *Histochem. J.* **1998**, *30*, 773–781.

11. Suzuki, T.; Fujikura, K.; Higashiyama, T.; Takata, K. DNA staining for fluorescence and laser confocal microscopy. *J. Histochem. Cytochem.* **1997**, *45*, 49–53.

12. Haugland, R. P. Phenanthridium dye staining of nucleic acids in living cells. U.S. Patent 5437980, 1995.

13. Frankfurt, O. S. Flow cytometric analysis of double-stranded RNA content distributions. *Methods Cell Biol.* **1990**, *33*, 299–304.

14. Biggiogera, M.; Biggiogera, F. F. Ethidium bromide- and propidium iodide-PTA staining of nucleic acids at the electron microscopic level. *J. Histochem. Cytochem.* **1989**, *37*, 1161–1166.

15. Evenson, D.; Darzynkiewicz, Z.; Jost, L.; Janca, F.; Ballachey, B. Changes in accessibility of DNA to various fluorochromes during spermatogenesis. *Cytometry* **1986**, *7*, 45–53.

16. Giordano, P. A.; Mazzini, G.; Riccardi, A.; Montecucco, C. M.; Ucci, G.; Danova, M. Propidium iodide staining of cytoautoradiographic preparations for the simultaneous determination of DNA content and grain count. *Histochem. J.* **1985**, *17*, 1259–1270.

17. Davies, W.; Nordby, O. Propidium iodide and S1 nuclease: tools for studying DNA reassociation kinetics. *Anal. Biochem.* **1985**, *146*, 423–428.

18. Wallen, C. A.; Higashikubo, R.; Dethlefsen, L. A. Comparison of two flow cytometric assays for cellular RNA - acridine orange and propidium iodide. *Cytometry* **1982**, *3*, 155–160.

19. Dean, P. N.; Gray, J. W.; Dolbeare, F. A. The analysis and interpretation of DNA distributions measured by flow cytometry. *Cytometry* **1982**, *3*, 188–195.

20. Noguchi, P. D.; Johnson, J. B.; Browne, W. Measurement of DNA synthesis by flow cytometry. *Cytometry* **1981**, *1*, 390–393.

21. Tas, J.; Westerneng, G. Fundamental aspects of the interaction of propidium diiodide with nucleic acids studied in a model system of polyacrylamide films. *J. Histochem. Cytochem.* **1981**, *29*, 929–936.

22. Taylor, I. W.; Milthorpe, B. K. An evaluation of DNA fluorochromes, staining techniques, and analysis for flow cytometry. I. Unperturbed cell populations. *J. Histochem. Cytochem.* **1980**, *28*, 1224–1232.

23. Taylor, I. W. A rapid single step staining technique for DNA analysis by flow microfluorimetry. *J. Histochem. Cytochem.* **1980**, *28*, 1021–1024.

24. Hamilton, V. T.; Habbersett, M. C.; Herman, C. J. Flow microfluorometric analysis of cellular DNA: Critical comparison of mithramycin and propidium iodide. *J. Histochem. Cytochem.* **1980**, *28*, 1125–1128.

25. Mazzini, G.; Giordano, P.; Montecucco, C. M.; Riccardi, A. A rapid cytofluorometric method for quantitative DNA determination on fixed smears. *Histochem. J.* **1980**, *12*, 153–168.

26. Coulson, P. B.; Bishop, A. O.; Lenarduzzi, R. Quantitation of cellular deoxyribonucleic acid by flow microfluorometry. *J. Histochem. Cytochem.* **1977**, *25*, 1147–1153.

27. Crissman, H. A.; Oka, M. S.; Steinkamp, J. A. Rapid staining methods for analysis of deoxyribonucleic acid and protein in mammalian cells. *J. Histochem. Cytochem.* **1976**, *24*, 64–71.

28. Berman, H. A.; Becktel, W.; Taylor, P. Spectroscopic studies on acetylcholinesterase: influence of peripheral-site occupation on active-center conformation. *Biochemistry* **1981**, *20*, 4803–4810.

29. Taylor, P.; Lwebuga-Mukasa, J.; Lappi, S.; Rademacher, J. Propidium. Fluorescence probe for a peripheral anionic site on acetylcholinesterase. *Mol. Pharmacol.* **1974**, *10*, 703–708.

30. Zotta, T.; Guidone, A.; Tremonte, P.; Parente, E.; Ricciardi, A. A comparison of fluorescent stains for the assessment of viability and metabolic activity of lactic acid bacteria. *World J. Microbiol. Biotechnol.* **2012**, *28*, 919–927.

31. Bunthof, C. J.; Bloemen, K.; Breeuwer, P.; Rombouts, F. M.; Abee, T. Flow cytometric assessment of viability of lactic acid bacteria. *Appl. Environ. Microbiol.* **2001**, *67*, 2326–2335.

32. Ericsson, M.; Hanstorp, D.; Hagberg, P.; Enger, J.; Nystrom, T. Sorting out bacterial viability with optical tweezers. *J. Bacteriol.* **2000**, *182*, 5551–5555.

33. Rychlik, I.; Cardova, L.; Sevcik, M.; Barrow, P. A. Flow cytometry characterization of *Salmonella typhimurium* mutants defective in proton translocating proteins and stationary-phase growth phenotype. *J. Microbiol. Methods* **2000**, *42*, 255–263.

34. Comas, J.; Vives-Rego, J. Assessment of the effects of gramicidin, formaldehyde, and surfactants on *Escherichia coli* by flow cytometry using nucleic acid and membrane potential dyes. *Cytometry* **1997**, *29*, 58–64.

35. Shirai, M.; Azuma, Y. Method for determining existence ratio of viable cell, dead-cell, and

false viable cell. Jpn. Kokai Tokkyo Koho JP 2008054509, 2008.

36. Bestvater, F.; Spiess, E.; Stobrawa, G.; Hacker, M.; Feurer, T.; Porwol, T.; Berchner-Pfannschmidt, U.; Wotzlaw, C.; Acker, H. Two-photon fluorescence absorption and emission spectra of dyes relevant for cell imaging. *J. Microsc.* **2002**, *208*, 108–115.

37. Skyggebjerg, O.; Glensbjerg, M. A method and a system for counting cells from a plurality of species. PCT Int. Appl. WO 2002101087, 2002.

38. Foglieni, C.; Meoni, C.; Davalli, A. M. Fluorescent dyes for cell viability: an application on prefixed conditions. *Histochem. Cell Biol.* **2001**, *115*, 223–229.

39. Cunningham, R. E. Indirect immunofluorescent labeling of fixed cells. *Methods Mol. Biol.* **1999**, *115*, 265–270.

40. Cunningham, R. E. Indirect immunofluorescent labeling of viable cells. *Methods Mol. Biol.* **1999**, *115*, 261–263.

41. Ormerod, M. G. Cell cycle analysis of asynchronous populations. *Methods Mol. Biol.* **1998**, *91*, 157–165.

42. Palmeira, C. M.; Moreno, A. J.; Madeira, V. M.; Wallace, K. B. Continuous monitoring of mitochondrial membrane potential in hepatocyte cell suspensions. *J. Pharmacol. Toxicol. Methods* **1996**, *35*, 35–43.

43. Pollack, A.; Ciancio, G. Cell cycle phase-specific analysis of cell viability using Hoechst 33342 and propidium iodide after ethanol preservation. *Methods Cell Biol.* **1990**, *33*, 19–24.

44. Fried, J.; Perez, A. G.; Clarkson, B. D. Flow cytofluorometric analysis of cell cycle distributions using propidium iodide. Properties of the method and mathematical analysis of the data. *J. Cell Biol.* **1976**, *71*, 172–181.

45. Andras, S. C.; Hartman, T. P. V.; Alexander, J.; McBride, R.; Marshall, J. A.; Power, J. B.; Cocking, E. C.; Davey, M. R. Combined PI-DAPI staining (CPD) reveals NOR asymmetry and facilitates karyotyping of plant chromosomes. *Chromosome Res.* **2000**, *8*, 387–391.

46. Langlois, R. G.; Carrano, A. V.; Gray, J. W.; Van Dilla, M. A. Cytochemical studies of metaphase chromosomes by flow cytometry. *Chromosoma* **1980**, *77*, 229–251.

47. Berkes, C. A.; Chan, L. L.; Wilkinson, A.; Paradis, B. Rapid quantification of pathogenic fungi by cellometer image-based cytometry. *J. Microbiol. Methods* **2012**, *91*, 468–476.

48. Prigione, V.; Filipello Marchisio, V. Methods to maximise the staining of fungal propagules with fluorescent dyes. *J. Microbiol. Methods* **2004**, *59*, 371–379.

49. Guise, V.; Jaffray, P.; Delattre, J.; Puisieux, F.; Adolphe, M.; Couvreur, P. Comparative cell uptake of propidium iodide associated with liposomes or nanoparticles. *Cell. Mol. Biol.* **1987**, *33*, 397–405.

50. Baskic, D.; Popovic, S.; Ristic, P.; Arsenijevic, N. N. Analysis of cycloheximide-induced apoptosis in human leukocytes: Fluorescence microscopy using annexin V/propidium iodide versus acridine orange/ethidium bromide. *Cell Biol. Int.* **2006**, *30*, 924–932.

51. Sakata, T.; Mizukami, T.; Hatanaka, K. Method for classifying and counting immature leukocytes. Eur. Pat. Appl. EP 844481, 1998.

52. Minakami, T.; Mori, Y.; Tsuji, T.; Ikeuchi, Y. Megakaryocyte classification/counting method by double fluorescent staining and flow cytometry. Jpn.Kokai Tokkyo Koho JP 2006275985, 2006.

53. Nakano, K.; Kameoka, J.; Yasuike, M. Microorganism detection method. Jpn. Kokai Tokkyo Koho JP 2012231678, 2012.

54. Nakano, Y.; Kameoka, J.; Yasuike, M. Microorganism detection method. Jpn. Kokai Tokkyo Koho JP 2010130918, 2010.

55. Li, C. S.; Chia, W. C.; Chen, P. S. Fluorochrome and flow cytometry to monitor microorganisms in treated hospital wastewater. *J. Environ. Sci. Health, Part A* **2007**, *42*, 195–203.

56. Kinbara, K.; Shimomura, Y. Microorganism bioactivity evaluation method, and kit. Jpn. Kokai Tokkyo Koho JP 2007097532, 2007.

57. Noda, N.; Mizutani, T. Microorganism-measuring method using multiple staining. Jpn. Kokai Tokkyo Koho JP 2006340684, 2006.

58. Fukutome, K. Method for evaluating microorganism cell activity by flow cytometry analysis. Jpn. Kokai Tokkyo Koho JP 2006238771, 2006.

59. Chen, P.; Li, C. Real-time quantitative PCR with gene probe, fluorochrome and flow cytometry for microorganism analysis. *J. Environ. Monit.* **2005**, *7*, 257–262.

60. Inatomi, K.; Ideo, S. Microorganism-measuring apparatus/method using fluorescent-labeling and microflow cytometer. Jpn. Kokai Tokkyo Koho JP 2005245317, 2005.

61. Besson, F. I.; Hermet, J. P.; Ribault, S. Reaction medium and process for universal detection of microorganisms. Fr. Demande FR 2847589, 2004.

62. Tobin, K. J.; Onstott, T. C.; DeFlaun, M. F.; Colwell, F. S.; Fredrickson, J. *In situ* imaging of microorganisms in geologic material. *J. Microbiol. Methods* **1999**, *37*, 201–213.

63. Kaneshiro, E. S.; Wyder, M. A.; Wu, Y. P.; Cushion, M. T. Reliability of calcein acetoxy methyl ester and ethidium homodimer or propidium iodide for viability assessment of microbes. *J. Microbiol. Methods* **1993**, *17*, 1–16.

64. Wilde, G. J. C.; Sundstroem, L. E.; Iannotti, F. Propidium iodide *in vivo*: an early marker of neuronal damage in rat hippocampus. *Neurosci. Lett.* **1994**, *180*, 223–226.

65. Gayoso, R. M. Fluorescent dye mixture and procedure for staining the nucleus and cytoplasm of the cell to permit study of the morphology of cells on any type of support. Span. ES 2326061, 2009.

66. Bink, K.; Walch, A.; Feuchtinger, A.; Eisenmann, H.; Hutzler, P.; Hofler, H.; Werner, M. TO-PRO-3 is an optimal fluorescent dye for nuclear counterstaining in dual-colour FISH on paraffin sections. *Histochem. Cell Biol.* **2001**, *115*, 293–299.

67. Jones, K. H.; Kniss, D. A. Propidium iodide as a nuclear counterstain for immunofluorescence studies on cells in culture. *J. Histochem. Cytochem.* **1987**, *35*, 123–125.

68. Crompton, T.; Peitsch, M. C.; MacDonald, H. R.; Tschopp, J. Propidium iodide staining correlates with the extent of DNA degradation in isolated nuclei. *Biochem. Biophys. Res. Commun.* **1992**, *183*, 532–537.

69. Escaron, C. J.; Lees, D. M.; Tewari, R.; Smith, D. F.; Caron, E. A simple, robust and versatile method to characterize intracellular parasitism. *Mol. Biochem. Parasitol.* **2007**, *153*, 72–76.

70. Sauch, J. F.; Flanigan, D.; Galvin, M. L.; Berman, D.; Jakubowski, W. Propidium iodide as an indicator of *Giardia* cyst viability. *Appl. Environ. Microbiol.* **1991**, *57*, 3243–3247.

71. Smith, A. L.; Smith, H. V. A comparison of fluorescein diacetate and propidium iodide staining and in vitro excystation for determining *Giardia intestinalis* cyst viability. *Parasitology* **1989**, *99*, 329–331.

72. Campbell, A. T.; Robertson, L. J.; Smith, H. V. Viability of *Cryptosporidium parvum* oocysts: correlation of *in vitro* excystation with inclusion or exclusion of fluorogenic vital dyes. *Appl. Environ. Microbiol.* **1992**, *58*, 3488–3493.

73. Kamp, G.; May, T. Procedure and device for the determination of sperm fertility. Ger. Offen. DE 102010010526, 2011.

74. Shi, H.; Wang, J.; Yuan, Y.; Zhao, H. Method for measuring sperm with fluorescent staining. Faming Zhuanli Shenqing CN 101308131, 2008.

75. Siemieniuch, M.; Dubiel, A. Preservation of tomcat (*Felis catus*) semen in variable temperatures. *Anim. Reprod. Sci.* **2007**, *99*, 135–144.

76. Paniagua-Chavez, C. G.; Jenkins, J.; Segovia, M.; Tiersch, T. R. Assessment of gamete quality for the eastern oyster (*Crassostrea virginica*) by use of fluorescent dyes. *Cryobiology* **2006**, *53*, 128–138.

77. Matsumoto, T.; Okada, H.; Hamaguchi, Y. Method and reagent for counting sperm by flow cytometry. Jpn. Kokai Tokkyo Koho JP 2001242168, 2001.

78. Garner, D. L.; Johnson, L. A. Viability assessment of mammalian sperm using SYBR-14 and propidium iodide. *Biol. Reprod.* **1995**, *53*, 276–284.

79. Green, D. P. L. Comparison of Hoechst 33342 and propidium iodide as fluorescent markers for sperm fusion with hamster oocytes. *J. Reprod. Fertil.* **1992**, *96*, 581–591.

80. Prowse, A. B. J.; Wolvetang, E. J.; Gray, P. P. A rapid, cost-effective method for counting human embryonic stem cell numbers as clumps. *BioTechniques* **2009**, *47*, 599–606.

81. Nakauchi, H.; Kobayashi, T.; Yamaguchi, T.; Sato, H. The sorting method of the multipotent stem cells. Jpn. Kokai Tokkyo Koho JP 2010200676, 2010.

82. Chan, L. L. Yeast concentration and viability measurement. PCT Int. Appl. WO 2011156249, 2011.

83. Sanz, R.; Galceran, M. T.; Puignou, L. Determination of viable yeast cells by gravitational field-flow fractionation with fluorescence detection. *Biotechnol. Prog.* **2004**, *20*, 613–618.

84. Van Zandycke, S. M.; Simal, O.; Gualdoni, S.; Smart, K. A. Determination of yeast viability using fluorophores. *J. Am. Soc. Brew. Chem.* **2003**, *61*, 15–22.

85. Deere, D.; Shen, J.; Vesey, G.; Bell, P.; Bissinger, P.; Veal, D. Flow cytometry and cell sorting for yeast viability assessment and cell selection. *Yeast* **1998**, *14*, 147–160.

86. Zagariya, A. M. A novel method for detection of apoptosis. *Exp. Cell Res.* **2012**, *318*, 861–866.

87. Kabakov, A. E.; Kudryavtsev, V. A.; Gabai, V. L. Determination of cell survival or death. *Methods Mol. Biol.* **2011**, *787*, 231–244.

88. Lenardo, M. J.; McPhee, C. K.; Yu, L. Autophagic cell death. *Methods Enzymol.* **2009**, *453*, 17–31.

89. Giusti, C.; Kosta, A.; Lam, D.; Tresse, E.; Luciani, M.; Golstein, P. Analysis of autophagic and necrotic cell death in *Dictyostelium*. *Methods Enzymol.* **2008**, *446*, 1–15.

90. Darzynkiewicz, Z.; Galkowski, D.; Zhao, H. Analysis of apoptosis by cytometry using TUNEL assay. *Methods* **2008**, *44*, 250–254.

91. Buenz, E. J.; Limburg, P. J.; Howe, C. L. A high-throughput 3-parameter flow cytometry-based cell death assay. *Cytometry* **2007**, *71A*, 170–173.

92. Mattes, M. J. Apoptosis assays with lymphoma cell lines: Problems and pitfalls. *Br. J. Cancer* **2007**, *96*, 928–936.

93. Riccardi, C.; Nicoletti, I. Analysis of apoptosis by propidium iodide staining and flow cytometry. *Nat. Protoc.* **2006**, *1*, 1458–1461.

94. Miller, E. Apoptosis measurement of annexin V staining. *Methods Mol. Med.* **2004**, *88*, 191–202.

95. Steensma, D. P.; Timm, M.; Witzig, T. E. Flow cytometric methods for detection and quantification of apoptosis. *Methods Mol. Med.* **2003**, *85*, 323–332.

96. Zamai, L.; Canonico, B.; Luchetti, F.; Ferri, P.; Melloni, E.; Guidotti, L.; Cappellini, A.; Cutroneo, G.; Vitale, M.; Papa, S. Supravital exposure to propidium iodide identifies apoptosis on adherent cells. *Cytometry* **2001**, *44*, 57–64.

97. Welsh, N. Assessment of apoptosis and necrosis in isolated islets of Langerhans: Methodological considerations. *Curr. Top. Biochem. Res.* **2000**, *3*, 189–200.

98. Konkel, M. E.; Mixter, P. F. Flow cytometric detection of host cell apoptosis induced by bacterial infection. *Methods Cell Sci.* **2000**, *22*, 209–215.

99. Bossy-Wetzel, E.; Green, D. R. Detection of apoptosis by annexin V labeling. *Methods Enzymol.* **2000**, *322*, 15–18.

100. Weber, G. F.; Daley, J.; Kraeft, S.; Chen, L. B.; Cantor, H. Measurement of apoptosis in heterogeneous cell populations. *Cytometry* **1997**, *27*, 136–144.

101. Vermes, I.; Haanen, C.; Steffens-Nakken, H.; Reutelingsperger, C. A novel assay for apoptosis. Flow cytometric detection of phosphatidylserine expression on early apoptotic cells using fluorescein labeled Annexin V. *J. Immunol. Methods* **1995**, *184*, 39–51.

102. Nicoletti, I.; Migliorati, G.; Pagliacci, M. C.; Grignani, F.; Riccardi, C. A rapid and simple method for measuring thymocyte apoptosis by propidium iodide staining and flow cytometry. *J. Immunol. Methods* **1991**, *139*, 271–279.

103. Wlodkowic, D.; Faley, S.; Darzynkiewicz, Z.; Cooper, J. M. Real-time cytotoxicity assays. *Methods Mol. Biol.* **2011**, *731*, 285–291.

104. Kishi, A.; Fujita, S. Method for measuring non-specific cellular cytotoxicity. Jpn. Kokai Tokkyo Koho JP 2001174458, 2001.

105. Wrobel, K.; Claudio, E.; Segade, F.; Ramos, S.; Lazo, P. S. Measurement of cytotoxicity by propidium iodide staining of target cell DNA. Application to the quantification of murine TNF-α. *J. Immunol. Methods* **1996**, *189*, 243–249.

106. Dengler, W. A.; Schulte, J.; Berger, D. P.; Mertelsmann, R.; Fiebig, H. H. Development of a propidium iodide fluorescence assay for proliferation and cytotoxicity assays. *Anti-Cancer Drugs* **1995**, *6*, 522–532.

107. Trost, L.; Lemasters, J. J. A cytotoxicity assay for tumor necrosis factor employing a multiwell fluorescence scanner. *Anal. Biochem.* **1994**, *220*, 149–153.

108. Chi, M.; Li, C. Fluorochrome in monitoring atmospheric bioaerosols and correlations with meteorological factors and air pollutants. *Aerosol Sci. Technol.* **2007**, *41*, 672–678.

109. Li, C.; Huang, T. Fluorochrome in monitoring indoor bioaerosols. *Aerosol Sci. Technol.* **2006**, *40*, 237–241.

110. Bousseksou, A.; Salmon, L.; Molnar, G.; Cobo, S. Materials with thermochromic spin transition doped with one or more fluorescent agents for use as temperature sensor. Fr. Demande FR 2952371, 2011.

111. Bousseksou, A.; Salmon, L.; Molnar, G.; Cobo, S. Heat-sensitive spin-transition materials doped with one or more fluorescent agents for use as temperature sensor. PCT Int. Appl. WO 2011058277, 2011.

112. Wakita, H.; Igarashi, K.; Oie, K. Diagnosis and treatment of ischemic disease. PCT Int. Appl. WO 2009022756, 2009.

113. Johansen, L. M.; Owens, C. M.; Mawhinney, C.; Chappell, T. W.; Brown, A. T.; Frank, M. G.; Altmeyer, R. Compositions and methods for treatment of viral diseases. PCT Int. Appl. WO 2008033466, 2008.

114. Kinlen, P. J. Light-emitting phosphor particles and electroluminescent devices employing same. U.S. Pat. Appl. Publ. US 20040018379, 2004.

115. Garza, C. M.; Cho, S. Metrology of bilayer photoresist processes. U.S. Pat. Appl. Publ. US 20090220895, 2009.

116. Takahashi, A.; Shirakawa, H.; Adegawa, Y. Chemically-amplified negative-working resist compositions for processing with electron beam or x-ray. Jpn. Kokai Tokkyo Koho JP 2003005355, 2003.

117. Merrick, P.; Nieminen, A.L.; Harper, R. A.; Herman, B.; Lemasters, J. J. Cytotoxicity screening of surfactant-based shampoos using a multiwell fluorescence scanner: correlation with Draize eye scores. *Toxicol. in Vitro* **1996**, *10*, 101.

118. Jiang, T.; Grant, R. L.; Acosta, D. A digitized fluorescence imaging study of intracellular free calcium, mitochondrial integrity and cytotoxicity in rat renal cells exposed to ionomycin, a calcium ionophore. *Toxicology* **1993**, *85*, 41–65.

119. Kohler, M.; Kundig, A.; Reist, H. W.; Michel, C. Modification of *in vitro* mouse embryogenesis by x-rays and fluorochromes. *Radiat. Environ. Biophy.* **1994**, *33*, 341–351.

120. Fukunaga, M.; Yielding, L. W. Structure-function characterization of phenanthridinium compounds as mutagens in *Salmonella. Mutat. Res. Lett.* **1983**, *121*, 89–94.

121. de Calignon, A.; Spires-Jones, T. L.; Pitstick, R.; Carlson, G. A.; Hyman, B. T. Tangle-bearing neurons survive despite disruption of membrane integrity in a mouse model of tauopathy. *J. Neuropathol. Exp. Neurol.* **2009**, *68*, 757–761.

122. Patel, D.; Good, T. A rapid method to measure beta-amyloid induced neurotoxicity *in vitro. J. Neurosci. Methods* **2007**, *161*, 1–10.

123. Dobrucki, J. W.; Feret, D.; Noatynska, A. Scattering of exciting light by live cells in fluorescence confocal imaging: phototoxic effects and relevance for FRAP studies. *Biophys. J.* **2007**, *93*, 1778–1786.

124. Wang, X.; Shi, W.; Wu, J.; Hao, Y.; Hu, G.; Liu, H.; Han, X.; Yu, H. Reproductive toxicity of organic extracts from petrochemical plant effluents discharged to the Yangtze River, China. *J. Environ. Sci.* **2010**, *22*, 297–303.

1-PYRENEBUTANOIC ACID HYDRAZIDE (PBH)

CAS Registry Number   55486-13-0

Chemical Structure

(CH$_2$)$_3$CONHNH$_2$

CA Index Name   1-Pyrenebutanoic acid, hydrazide

Other Names   Pyrenebutyric acid hydrazide; 1-Pyrenebutyric acid hydrazide; Pyrenebutyrylhydrazine; 1-Pyrenebutyrylhydrazine; Pyrene butanoic acid hydrazide; 4-(1-Pyrene)butanoic acid hydrazide; 4-(1-Pyrene)butyric acid hydrazide; 4-(1-Pyrene)butyryl hydrazide; PBH

Merck Index Number   Not listed

Chemical/Dye Class   Pyrene

Molecular Formula   C$_{20}$H$_{18}$N$_2$O

Molecular Weight   302.37

Physical Form   Yellow needles;[2] white powder[1]

Solubility   Soluble in water, acetonitrile, N,N-dimethylformamide, dimethyl sulfoxide, methanol

Melting Point   167–171 °C

Boiling Point (Calcd.)   588.8 ± 29.0 °C, pressure: 760 Torr

pK$_a$ (Calcd.)   13.22 ± 0.35, most acidic, temperature: 25 °C; 3.23 ± 0.10, most basic, temperature: 25 °C

Absorption (λ_{max})   341 nm, 325 nm (MeOH)

Emission (λ_{max})   376 nm, 396 nm (MeOH)

Molar Extinction Coefficient   43,000 cm^{-1} M^{-1} (MeOH)

Synthesis   Synthetic methods[1,2]

Imaging/Labeling Applications   Nucleic acids;[2,3] collagen;[4,5] chitosan;[6,7] oligonucleotide;[8,9] oligosaccharides;[10–13] polynucleotide;[3] carboxylic acids[14–17]

Biological/Medical Applications   Detecting/labeling/quantifying biological carbonyl compounds (aldehydes and/or ketones);[2–5,10–13,18] analyzing/detecting/measuring carboxylic acids;[14–17] treating inflammation,[11] autoimmune diseases,[11] allergic diseases,[11] viral diseases,[11] cancer,[11] infectious diseases,[11] heart diseases[11]

Industrial Applications   Chemosensor materials;[19] lubricants[20]

Safety/Toxicity   No data available

REFERENCES

1. Li, N.; Xiang, Y.; Chen, X.; Tong, A. Salicylaldehyde hydrazones as fluorescent probes for zinc ion in aqueous solution of physiological pH. *Talanta* **2009**, *79*, 327–332.

2. Reines, S. A.; Cantor, C. R. New fluorescent hydrazide reagents for the oxidized 3′-terminus of RNA. *Nucleic Acids Res.* **1974**, *1*, 767–786.

3. Koenig, P.; Reines, S. A.; Cantor, C. R. Pyrene derivatives as fluorescent probes of conformation near the 3′ termini of polyribonucleotides. *Biopolymers* **1977**, *16*, 2231–2242.

4. Fujimori, E.; Shambaugh, N. Cross-linking and fluorescence of pyrene-labeled collagen. *Biochim. Biophys. Acta, Protein Struct. Mol. Enzymol.* **1983**, *742*, 155–161.

5. Shambaugh, N.; Fujimori, E. Fluorescent-labeled crosslinks in collagen: pyrenebutyrylhydrazine. *Biopolymers* **1982**, *21*, 79–88.

6. Xin, M.; Gao, W.; Li, M.; Li, L.; Qiu, F. Preparation, characterization and properties of a novel fluorescent chitosan derivative. *Huagong Jinzhan* **2011**, *30*, 1290–1295.

7. Xie, Q.; Xin, M.; Li, M.; Mao, Y. Preparation of pH sensitive fluorescent-labeled water-soluble chitosan. *Huagong Jinzhan* **2010**, *29*, 1943–1946.

8. Cook, P. D.; Manoharan, M.; Bruice, T. Covalently cross-linked oligonucleotides. U.S. Patent 5543507, 1996.

9. Ebata, K.; Masuko, M.; Ohtani, H.; Jibu, M. Excimer formation by hybridization using two pyrene-labeled oligonucleotide probes. *Nucleic Acids Symp. Ser.* **1995**, *34*, 187–188.

10. Zhang, Y.; Iwamoto, T.; Radke, G.; Kariya, Y.; Suzuki, K.; Conrad, A. H.; Tomich, J. M.; Conrad, G. W. On-target derivatization of keratan sulfate oligosaccharides with pyrenebutyric acid

hydrazide for MALDI-TOF/TOF-MS. *J. Mass Spectrom.* **2008**, *43*, 765–772.

11. Amano, J.; Sugawara, D. Oligosaccharide containing fucose and lactosamine or lactose or their derivatives of human milk expressing diversity. Jpn. Kokai Tokkyo Koho JP 2007297521, 2007.

12. Sugahara, D.; Amano, J.; Irimura, T. Fluorescence labeling of oligosaccharides useful in the determination of molecular interactions. *Anal. Sci.* **2003**, *19*, 167–169.

13. Lee, J. A.; Fortes, P. A. G. Labeling of the glycoprotein subunit of sodium-potassium ATPase with fluorescent probes. *Biochemistry* **1985**, *24*, 322–330.

14. Yoshida, H.; Araki, J.; Sonoda, J.; Nohta, H.; Ishida, J.; Hirose, S.; Yamaguchi, M. Screening method for organic aciduria by spectrofluorometric measurement of total dicarboxylic acids in human urine based on intramolecular excimer-forming fluorescence derivatization. *Anal. Chim. Acta* **2005**, *534*, 177–183.

15. Yoshida, H.; Horita, K.; Todoroki, K.; Nohta, H.; Yamaguchi, M. Highly selective fluorimetric determination of acidic amino acids by high-performance liquid chromatography following intramolecular excimer-forming derivatization with a pyrene-labeling reagent. *Bunseki Kagaku* **2003**, *52*, 1113–1119.

16. Nohta, H.; Sonoda, J.; Yoshida, H.; Satozono, H.; Ishida, J.; Yamaguchi, M. Liquid chromatographic determination of dicarboxylic acids based on intramolecular excimer-forming fluorescence derivatization. *J. Chromatogr., A* **2003**, *1010*, 37–44.

17. Noda, H.; Satozono, H.; Sonoda, J.; Yoshida, H.; Yamaguchi, M. Method for Polycarboxylic acid analysis. Jpn. Kokai Tokkyo Koho JP 11281578, 1999.

18. Mansano, F. V.; Kazaoka, R. M. A.; Ronsein, G. E.; Prado, F. M.; Genaro-Mattos, T. C.; Uemi, M.; Di Mascio, P.; Miyamoto, S. Highly sensitive fluorescent method for the detection of cholesterol aldehydes formed by ozone and singlet molecular oxygen. *Anal. Chem.* **2010**, *82*, 6775–6781.

19. Wu, Z.; Meng, L.; Li, C.; Lu, X.; Zhang, L.; He, Y. Amphiphilic copolymer with pendant pyrenebutyryl hydrazide group: synthesis, characterization, and recognition for carbonate anion. *J. Appl. Polym. Sci.* **2006**, *101*, 2371–2376.

20. Mishra, M. K.; Rubin, I. D. Functionalized polymers for lubricants. *Macromol. Rep.* **1994**, *A31(Suppl. 6&7)*, 995–1002.

1-PYRENEBUTANOIC ACID SUCCINIMIDYL ESTER

CAS Registry Number   114932-60-4

Chemical Structure

CA Index Name   1-Pyrenebutanoic acid, 2,5-dioxo-1-pyrrolidinyl ester

Other Names   2,5-Pyrrolidinedione, 1-[1-oxo-4-(1-pyrenyl)butoxy]-; 1-Pyrenebutanoic acid succinimidyl ester; 1-Pyrenebutyric acid N-hydroxysuccinimide ester; 4-Pyrenebutanoate N-hydroxysuccinimidyl ester; 4-(1-Pyrene)butanoic acid succimidyl ester; 4-(1-Pyrene)butyric acid N-hydroxysuccinimide ester; N-(1-Pyrene)-butyryloxysuccinimide; Succinimidyl 1-Pyrenebutanoate; 1-Succinimidyl pyrenebutyrate; PANHS; PBASE; PBSE; PNHS; PSE; PyBHS; Py-Boc; PYS

Merck Index Number   Not listed

Chemical/Dye Class   Pyrene

Molecular Formula $C_{24}H_{19}NO_4$

Molecular Weight   385.41

Physical Form   Yellow solid[1–4]

Solubility   Soluble in chloroform, N,N-dimethylformamide, dimethyl sulfoxide, methanol

Melting Point   132–136 °C; 131–133 °C[5]

Boiling Point (Calcd.)   590.9 ± 43.0 °C, pressure: 760 Torr

Absorption (λ_{max})   340 nm, 325 nm (MeOH)

Emission (λ_{max})   376 nm, 396 nm (MeOH)

Molar Extinction Coefficient   43,000 cm^{-1} M^{-1} (MeOH)

Synthesis   Synthetic methods[1–7]

Imaging/Labeling Applications   Cellulose;[8] cyclodextrin-based polyrotaxanes;[9–12] histamine/histidine;[13–17] lysozymes;[18] nucleic acids;[6,7,19–33] polyamines;[34–38] proteins;[4,39–41] poly(dimethylsiloxane) (PDMS) oligomers[58–62]

Biological/Medical Applications   Characterizing cyclodextrin-based polyrotaxanes;[9–12] detecting nucleic acids,[6,7,19–33] polyamines;[34–38] immobilizing proteins;[4,39–41] probing lysozyme in human serum;[18] as a substrate for measuring polymerases activity,[42] recombinases activity,[43] topoisomerases activity[43]

Industrial Applications   Carbon nanotubes (single-wall carbon nanotubes (SWCNT) and multi-wall carbon nanotubes (MWCNTs));[1,2,44–53] cellulose materials for bulky paper sheets;[8] graphene-based polymer nanocomposites materials/films;[54–56] photonic materials;[55] photovoltaic devices;[56] polyaniline systems;[57] poly(dimethylsiloxane) (PDMS) oligomers;[58–62] poly(propylene imine) dendrimers;[63,64] thin-films[65]

Safety/Toxicity   No data available

REFERENCES

1. Yan, Y.; Cui, J.; Zhao, S.; Zhang, J.; Liu, J.; Cheng, J. Interface molecular engineering of single-walled carbon nanotube/epoxy composites. *J. Mater. Chem.* **2012**, *22*, 1928–1936.

2. Li, F.; Zhang, B.; Li, X.; Jiang, Y.; Chen, L.; Li, Y.; Sun, L. Highly efficient oxidation of water by a molecular catalyst immobilized on carbon nanotubes. *Angew. Chem., Int. Ed.* **2011**, *50*, 12276–12279.

3. Li, D.; Song, J.; Yin, P.; Simotwo, S.; Bassler, A. J.; Aung, Y.; Roberts, J. E.; Hardcastle, K. I.; Hill, C. L.; Liu, T. Inorganic–organic hybrid vesicles with counterion- and pH-controlled fluorescent properties. *J. Am. Chem. Soc.* **2011**, *133*, 14010–14016.

4. Wu, Z. Synthesis and characterization of active ester-functionalized fluorescent polymers: new materials for protein conjugation. *J. Appl. Polym. Sci.* **2008**, *110*, 777–783.

5. Bailen, M. A.; Chinchilla, R.; Dodsworth, D. J.; Najera, C. O-Succinimidyl-1,3-dimethyl-1,3-trimethyleneuronium salts as efficient reagents in active ester synthesis. *Tetrahedron Lett.* **2002**, *43*, 1661–1664.

6. Rando, R. R.; Wang, Y. Methods and kits for RNA binding compounds. U.S. Patent 5593835, 1997.

7. Rando, R. R.; Wang, Y. Methods and kits for RNA binding compounds. PCT Int. Appl. WO 9635811, 1996.

8. Yang, Q.; Pan, X. A facile approach for fabricating fluorescent cellulose. *J. Appl. Polym. Sci.* **2010**, *117*, 3639–3644.

9. Przybylski, C.; Bonnet, V.; Jarroux, N. Further insight into the detailed characterization of a polydisperse cyclodextrin-based polyrotaxane sample by electrospray ionization mass spectrometry. *ACS Macro Lett.* **2012**, *1*, 533–536.

10. Przybylski, C.; Jarroux, N. Analysis of a polydisperse polyrotaxane based on poly(ethylene oxide) and α-cyclodextrins using nanoelectrospray and LTQ-orbitrap. *Anal. Chem.* **2011**, *83*, 8460–8467.

11. Peres, B.; Richardeau, N.; Jarroux, N.; Guegan, P.; Auvray, L. Two independent ways of preparing hypercharged hydrolyzable polyaminorotaxane. *Biomacromolecules* **2008**, *9*, 2007–2013.

12. Jarroux, N.; Guegan, P.; Cheradame, H.; Auvray, L. High conversion synthesis of pyrene end functionalized polyrotaxane based on poly(ethylene oxide) and α-cyclodextrins. *J. Phys. Chem. B* **2005**, *109*, 23816–23822.

13. Hogan, A.; Crean, C.; Barrett, U. M.; Guihen, E.; Glennon, J. D. Histamine determination in human urine using sub-2 μm C18 column with fluorescence and mass spectrometric detection. *J. Sep. Sci.* **2012**, *35*, 1087–1093.

14. Ichinose, F.; Yoshitake, T.; Yoshida, H.; Todoroki, K.; Kehr, J.; Inoue, O.; Nohta, H.; Yamaguchi, M. Determination of histamine in rat plasma and tissue extracts by intramolecular excimer-forming derivatization and LC with fluorescence detection. *Chromatographia* **2009**, *70*, 575–580.

15. Yoshida, H.; Ichinose, F.; Yoshitake, T.; Nakano, Y.; Todoroki, K.; Nohta, H.; Yamaguchi, M. Simultaneous determination of histamine and histidine by liquid chromatography following intramolecular excimer-forming fluorescence derivatization with pyrene-labeling reagent. *Anal. Sci.* **2004**, *20*, 557–559.

16. Yoshitake, T.; Ichinose, F.; Yoshida, H.; Todoroki, K.; Kehr, J.; Inoue, O.; Nohta, H.; Yamaguchi, M. A sensitive and selective determination method of histamine by HPLC with intramolecular excimer-forming derivatization and fluorescence detection. *Biomed. Chromatogr.* **2003**, *17*, 509–516.

17. Yamaguchi, M.; Nouda, H.; Yoshida, H.; Yoshitake, T.; Ichinose, F. Analytical method for detecting histamine or histidine. Jpn. Kokai Tokkyo Koho JP 2001242174, 2001.

18. Huang, J.; Zhu, Z.; Bamrungsap, S.; Zhu, G.; You, M.; He, X.; Wang, K.; Tan, W. Competition-mediated pyrene-switching aptasensor: Probing lysozyme in human serum with a monomer-excimer fluorescence switch. *Anal. Chem.* **2010**, *82*, 10158–10163.

19. Morinaga, H.; Takenaka, T.; Hashiya, F.; Kizaki, S.; Hashiya, K.; Bando, T.; Sugiyama, H. Sequence-specific electron injection into DNA from an intermolecular electron donor. *Nucleic Acids Res.* **2013**, *41*, 4724–4728.

20. Ramnani, P.; Gao, Y.; Ozsoz, M.; Mulchandani, A. Electronic detection of microRNA at attomolar level with high specificity. *Anal. Chem.* **2013**, *85*, 8061–8064.

21. Ogino, M.; Makino, Y. System for detecting nucleic acid by fluorescent substance using energy transfer. Jpn. Kokai Tokkyo Koho JP 2012147722, 2012.

22. Sasaki, S.; Onizuka, K.; Taniguchi, Y. The quick RNA modification method using functional group transition from 6-thianoguanosine of DNA hybridization probes. Jpn. Kokai Tokkyo Koho JP 2012056895, 2012.

23. Willis, B.; Arya, D. P. Triple recognition of B-DNA by a neomycin-Hoechst 33258-pyrene conjugate. *Biochemistry* **2010**, *49*, 452–469.

24. Willis, B.; Arya, D. P. Triple recognition of B-DNA. *Bioorg. Med. Chem. Lett.* **2009**, *19*, 4974–4979.

25. Benfield, A. P.; Macleod, M. C.; Liu, Y.; Wu, Q.; Wensel, T. G.; Vasquez, K. M. Targeted generation of DNA strand breaks using pyrene-conjugated triplex-forming oligonucleotides. *Biochemistry* **2008**, *47*, 6279–6288.

26. Marti, A. A.; Li, X.; Jockusch, S.; Li, Z.; Raveendra, B.; Kalachikov, S.; Russo, J. J.; Morozova, I.; Puthanveettil, S. V.; Ju, J.; Turro, N. J. Pyrene binary probes for unambiguous detection of mRNA using time-resolved fluorescence spectroscopy. *Nucleic Acids Res.* **2006**, *34*, 3161–3168.

27. Gorodetsky, A. A.; Barton, J. K. Electrochemistry using self-assembled DNA monolayers on highly oriented pyrolytic graphite. *Langmuir* **2006**, *22*, 7917–7922.

28. Zhou, Y.; Bian, G.; Wang, L.; Dong, L.; Wang, L.; Jian, K. Sensitive determination of nucleic acids using organic nanoparticle fluorescence probes. *Spectrochim. Acta, Part A* **2005**, *61A*, 1841–1845.

29. Silverman, S. K.; Deras, M. L.; Woodson, S. A.; Scaringe, S. A.; Cech, T. R. Multiple folding pathways for the P4-P6 RNA domain. *Biochemistry* **2000**, *39*, 12465–12475.

30. Silverman, S. K.; Cech, T. R. RNA tertiary folding monitored by fluorescence of covalently attached pyrene. *Biochemistry* **1999**, *38*, 14224–14237.

31. Hamasaki, K.; Rando, R. R. A high-throughput fluorescence screen to monitor the specific binding of antagonists to RNA targets. *Anal. Biochem.* **1998**, *261*, 183–190.

32. Morrison, L. E.; Halder, T. C.; Stols, L. M. Solution-phase detection of polynucleotides using interacting fluorescent labels and competitive hybridization. *Anal. Biochem.* **1989**, *183*, 231–244.

33. Morrison, L. E. Lifetime-resolved assay procedures. U.S. Patent 4822733, 1989.

34. Nishikawa, H.; Tabata, T.; Kitani, S. Simple detection method of biogenic amines in decomposed fish by intramolecular excimer fluorescence. *Food Nutr. Sci.* **2012**, *3*, 1020–1026.

35. Marks, H. S.; Anderson, C. R. Determination of putrescine and cadaverine in seafood (fin fish and shellfish) by liquid chromatography using pyrene excimer fluorescence. *J. Chromatogr., A* **2005**, *1094*, 60–69.

36. Yoshida, H.; Harada, H.; Nakano, Y.; Nohta, H.; Ishida, J.; Yamaguchi, M. Liquid chromatographic determination of polyamines in human urine based on intramolecular excimer-forming fluorescence derivatization using 4-(1-pyrene)butanoyl chloride. *Biomed. Chromatogr.* **2004**, *18*, 687–693.

37. Paproski, R. E.; Roy, K. I.; Lucy, C. A. Selective fluorometric detection of polyamines using micellar electrokinetic chromatography with laser-induced fluorescence detection. *J. Chromatogr., A* **2002**, *946*, 265–273.

38. Nohta, H.; Satozono, H.; Koiso, K.; Yoshida, H.; Ishida, J.; Yamaguchi, M. Highly selective fluorometric determination of polyamines based on intramolecular excimer-forming derivatization with a pyrene-labeling reagent. *Anal. Chem.* **2000**, *72*, 4199–4204.

39. Hu, W.; Lu, Z.; Liu, Y.; Li, C. M. *In situ* surface plasmon resonance investigation of the assembly process of multiwalled carbon nanotubes on an alkanethiol self-assembled monolayer for efficient protein immobilization and detection. *Langmuir* **2010**, *26*, 8386–8391.

40. Chen, R. J.; Zhang, Y.; Wang, D.; Dai, H. Noncovalent sidewall functionalization of single-walled carbon nanotubes for protein immobilization. *J. Am. Chem. Soc.* **2001**, *123*, 3838–3839.

41. Katz, E. Application of bifunctional reagents for immobilization of proteins on a carbon electrode surface: oriented immobilization of photosynthetic reaction centers. *J. Electroanal. Chem.* **1994**, *365*, 157–164.

42. Kolpashchikov, D. M.; Rechkunova, N. I.; Dobrikov, M. I.; Khodyreva, S. N.; Lebedeva, N. A.; Lavrik, O. I. Sensitized photomodification of mammalian DNA polymerase β. A new approach for highly selective affinity labeling of polymerases. *FEBS Lett.* **1999**, *448*, 141–144.

43. Panigrahi, G.; Zhao, B.; Krepinsky, J. J.; Sadowski, P. D. Toward a mechanism-based fluorescent assay for site-specific recombinases and topoisomerases: Assay design and syntheses of fluorescent substrates. *J. Am. Chem. Soc.* **1996**, *118*, 12004–12011.

44. Ramasamy, R. P.; Luckarift, H. R.; Ivnitski, D. M.; Atanassov, P. B.; Johnson, G. R. High electrocatalytic activity of tethered multicopper oxidase-carbon nanotube conjugates. *Chem. Commun.* **2010**, *46*, 6045–6047.

45. Liu, J.; Bibari, O.; Mailley, P.; Dijon, J.; Rouviere, E.; Sauter-Starace, F.; Caillat, P.; Vinet, F.; Marchand, G. Stable non-covalent functionalisation of multi-walled carbon nanotubes by pyrene-polyethylene glycol through π−π stacking. *New J. Chem.* **2009**, *33*, 1017–1024.

46. Zhao, L.; Shingaya, Y.; Tomimoto, H.; Huang, Q.; Nakayama, T. Functionalized carbon nanotubes for pH sensors based on SERS. *J. Mater. Chem.* **2008**, *18*, 4759–4761.

47. Zhu, J.; Brink, M.; McEuen, P. L. Single-electron force readout of nanoparticle electrometers attached to carbon nanotubes. *Nano Lett.* **2008**, *8*, 2399–2404.

48. Salzmann, C. G.; Lee, G. K. C.; Ward, M. A. H.; Chu, B. T. T.; Green, M. L. H. Highly hydrophilic and stable polypeptide/single-wall carbon nanotube conjugates. *J. Mater. Chem.* **2008**, *18*, 1977–1983.

49. Landi, B. J.; Evans, C. M.; Worman, J. J.; Castro, S. L.; Bailey, S. G.; Raffaelle, R. P. Noncovalent attachment of CdSe quantum dots to single wall carbon nanotubes. *Mater. Lett.* **2006**, *60*, 3502–3506.

50. Kim, J.; Kotagiri, N.; Kim, J.; Deaton, R. *In situ* fluorescence microscopy visualization and characterization of nanometer-scale carbon nanotubes labeled with 1-pyrenebutanoic acid,

succinimidyl ester. *Appl. Phys. Lett.* **2006**, *88*, 213110/1–213110/3.

51. Bottini, M.; Magrini, A.; Di Venere, A.; Bellucci, S.; Dawson, M. I.; Rosato, N.; Bergamaschi, A.; Mustelin, T. Synthesis and characterization of supramolecular nanostructures of carbon nanotubes and ruthenium-complex luminophores. *J. Nanosci. Nanotechnol.* **2006**, *6*, 1381–1386.

52. Okpalugo, T. I. T.; Papakonstantinou, P.; Murphy, H.; McLaughlin, J.; Brown, N. M. D. High resolution XPS characterization of chemical functionalized MWCNTs and SWCNTs. *Carbon* **2004**, *43*, 153–161.

53. Han, S.; Cagin, T.; Goddard, W. A., III,. Understanding noncovalent absorption and packing of 1-pyrene butanoic acid succinimidyl ester on single walled carbon nanotubes. *Mater. Res. Soc. Symp. Proc.* **2003**, *772*, 173–178.

54. Zhang, H.; Bao, Q.; Tang, D.; Zhao, L.; Loh, K. Large energy soliton erbium-doped fiber laser with a graphene-polymer composite mode locker. *Appl. Phys. Lett.* **2009**, *95*, 141103/1–141103/3.

55. Bao, Q.; Zhang, H.; Yang, J.; Wang, S.; Tang, D. Y.; Jose, R.; Ramakrishna, S.; Lim, C. T.; Loh, K. P. Graphene-polymer nanofiber membrane for ultrafast photonics. *Adv. Funct. Mater.* **2010**, *20*, 782–791.

56. Wang, Y.; Chen, X.; Zhong, Y.; Zhu, F.; Loh, K. P. Large area, continuous, few-layered graphene as anodes in organic photovoltaic devices. *Appl. Phys. Lett.* **2009**, *95*, 063302/1-063302/3.

57. Kang, Y. J.; Song, Y. H.; Kim, J.; Lee, C. W. Colorimetric and fluorescent changes for transition metal ions and nitroaromatic compounds based on a nano-spongy continuous structure of LB-PANI-amide-Py system. *Bull. Korean Chem. Soc.* **2011**, *32*, 2497–2500.

58. Burattini, S.; Colquhoun, H. M.; Greenland, B. W.; Hayes, W. A novel self-healing supramolecular polymer system. *Faraday Discuss.* **2009**, *143*, 251–264.

59. Gardinier, W. E.; Baker, G. A.; Baker, S. N.; Bright, F. V. Behavior of pyrene end-labeled poly(dimethylsiloxane) Polymer tails in mixtures of 1-butyl-3-methylimidazolium bis(trifluoromethyl)-sulfonylimide and toluene. *Macromolecules* **2005**, *38*, 8574–8582.

60. Gardinier, W. E.; Bright, F. V. Temperature-dependent tail-tail dynamics of pyrene-labeled poly(dimethylsiloxane) oligomers dissolved in ethyl acetate. *J. Phys. Chem. B* **2005**, *109*, 14824–14829.

61. Jones, B. A.; Torkelson, J. M. Large melting point depression of 2-3-nm length-scale nanocrystals formed by the self-assembly of an associative polymer: Telechelic, pyrene-labeled poly(dimethylsiloxane). *J. Polym. Sci., Part B: Polym. Phys.* **2004**, *42*, 3470–3475.

62. Kane, M. A.; Pandey, S.; Baker, G. A.; Perez, S. A.; Bukowski, E. J.; Hoth, D. C.; Bright, F. V. Effects of density on the intramolecular hydrogen bonding, tail-tail cyclization, and mean-free tail-to-tail distances of pyrene end-labeled poly(dimethylsiloxane) oligomers dissolved in supercritical CO_2. *Macromolecules* **2001**, *34*, 6831–6838.

63. Baker, L. A.; Sun, L.; Crooks, R. M. Cooperative effects among terminal groups of modified poly(propylene imine) dendrimers. *Polym. Mater. Sci. Eng.* **2001**, *84*, 10–11.

64. Baker, L. A.; Crooks, R. M. Photophysical properties of pyrene-functionalized poly(propylene imine) dendrimers. *Macromolecules* **2000**, *33*, 9034–9039.

65. Matsuo, Y.; Yamada, Y.; Nishikawa, M.; Fukutsuka, T.; Sugie, Y. Preparation of silylated magadiite thin-film-containing covalently attached pyrene chromophores. *J. Fluorine Chem.* **2008**, *129*, 1150–1155.

1-PYRENEDECANOIC ACID (PDA)

CAS Registry Number 64701-47-9

Chemical Structure

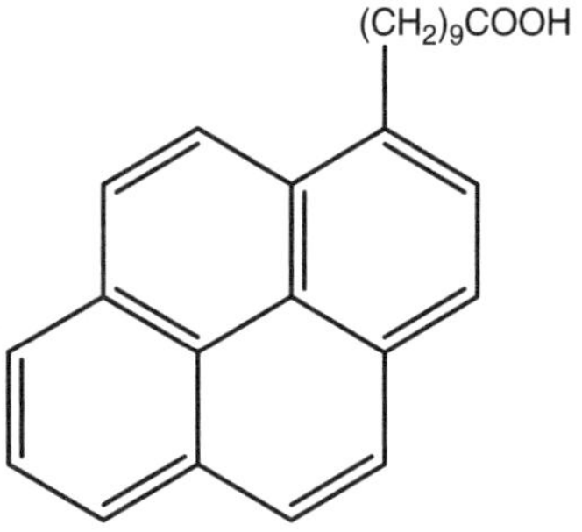

(CH$_2$)$_9$COOH

CA Index Name 1-Pyrenedecanoic acid

Other Names 10-(1-Pyrene)decanoic acid; 10-(1-Pyrenyl)decanoic acid; PDA; PyDA

Merck Index Number Not listed

Chemical/Dye Class Pyrene

Molecular Formula C$_{26}$H$_{28}$O$_2$

Molecular Weight 372.50

Physical Form Light yellow crystals[2,3]

Solubility Soluble in *N,N*-dimethylformamide, dimethyl sulfoxide, methanol

Melting Point 110 °C[3]; 104–108.5 °C[1]; 99 °C[2]

Boiling Point (Calcd.) 576.4 ± 19.0 °C, pressure: 760 Torr

pK$_a$ (Calcd.) 4.78 ± 0.10, most acidic, temperature: 25 °C

Absorption (λ$_{max}$) 341 nm, 325 nm (MeOH)

Emission (λ$_{max}$) 377 nm, 396 nm (MeOH)

Molar Extinction Coefficient 43,000 cm^{-1} M^{-1} (MeOH)

Synthesis Synthetic methods[1–3]

Imaging/Labeling Applications Liponucleotides;[1] glycosphingolipids;[4] lipids;[5–9] sphingomyelin;[10,11] sphingosines[12]

Biological/Medical Applications Analyzing lipid aggregates;[5] diagnosis of Niemann-Pick disease in cultured cells,[10,11] Wolman disease;[14] as substrates for ceramidase activity,[13] cholesteryl esterases activity,[14] lipase activity,[15–17] phospholipases A2 activity,[18,19] sphingomyelinase activity,[10] sphingosine kinases 1 and 2 activity,[12] synthetase activity[20]

Industrial Applications Electroluminescent material;[21] oxygen sensing material;[22–27] Langmuir-Blodgett films;[28–30] laser-assisted vacuum-deposited films/vacuum-deposited films[31–35]

Safety/Toxicity No data available

REFERENCES

1. Oskolkova, O. V.; Shvets, V. I.; Hermetter, A.; Paltauf, F. Synthesis and intermembrane transfer of pyrene-labelled liponucleotides: ceramide phosphothymidines. *Chem. Phys. Lipids* **1999**, *99*, 73–86.

2. Galla, H. J.; Hartmann, W. Pyrenedecanoic acid and pyrene lecithin. *Methods Enzymol.* **1981**, *72*, 471–479.

3. Galla, H. J.; Theilen, U.; Hartmann, W. Transversal mobility in bilayer membrance vesicles: Use of pyrene lecithin as optical probe. *Chem. Phys. Lipids* **1979**, *23*, 239–251.

4. Gege, C.; Oscarson, S.; Schmidt, R. R. Synthesis of fluorescence labeled sialyl Lewisx glycosphingolipids. *Tetrahedron Lett.* **2001**, *42*, 377–380.

5. Lianos, P.; Duportail, G. Time-resolved fluorescence fractal analysis in lipid aggregates. *Biophys. Chem.* **1993**, *48*, 293–299.

6. Kasurinen, J.; Somerharju, P. Metabolism of pyrenyl fatty acids in baby hamster kidney fibroblasts. Effect of the acyl chain length. *J. Biol. Chem.* **1992**, *267*, 6563–6569.

7. Pal, R.; Barenholz, Y.; Wagner, R. R. Pyrene phospholipid as a biological fluorescent probe for studying fusion of virus membrane with liposomes. *Biochemistry* **1988**, *27*, 30–36.

8. Radom, J.; Salvayre, R.; Maret, A.; Negre, A.; Douste-Blazy, L. Metabolism of 1-pyrenedecanoic acid and accumulation of neutral fluorescent lipids in cultured fibroblasts of multisystemic lipid storage myopathy. *Biochim. Biophys. Acta, Lipids Lipid Metab.* **1987**, *920*, 131–139.

9. Schenkman, S.; Araujo, P. S.; Dijkman, R.; Quina, F. H.; Chaimovich, H. Effects of temperature and lipid composition on the serum albumin-induced aggregation and fusion of small unilamellar vesicles. *Biochim. Biophys. Acta, Biomembr.* **1981**, *649*, 633–641.

10. Levade, T.; Klar, R.; Dagan, A.; Cherbu, S.; Gatt, S. Fluorescent derivatives of sphingomyelin: synthesis, use as substrates for sphingomyelinase and for diagnosis of Niemann-Pick disease in cultured cells. *NATO ASI Ser., Ser. A: Life Sci.* **1986**, *116*, 803–807.

11. Levade, T.; Gatt, S. Uptake and intracellular degradation of fluorescent sphingomyelin by fibroblasts from normal individuals and a patient with Niemann-Pick disease. *Biochim. Biophys. Acta, Lipids Lipid Metab.* **1987**, *918*, 250–259.

12. Ettmayer, P.; Billich, A.; Baumruker, T.; Mechtcheriakova, D.; Schmid, H.; Nussbaumer, P. Fluorescence-labeled sphingosines as substrates of sphingosine kinases 1 and 2. *Bioorg. Med. Chem. Lett.* **2004**, *14*, 1555–1558.

13. Nieuwenhuizen, W. F.; van Leeuwen, S.; Gotz, F.; Egmond, M. R. Synthesis of a novel fluorescent ceramide analogue and its use in the characterization of recombinant ceramidase from *Pseudomonas aeruginosa* PA01. *Chem. Phys. Lipids* **2002**, *114*, 181–191.

14. Dang, Q. Q.; Rogalle, P.; Puechmaurel, B.; Jollet, P.; Negre, A.; Vieu, C.; Salvayre, R.; Douste-Blazy, L. Microscale synthesis of fluorescent cholesteryl esters for the study of lysosomal cholesteryl esterases and diagnosis of Wolman disease. *NATO ASI Ser., Ser. A: Life Sci.* **1988**, *150*, 499–502.

15. Johnston, M.; Bhatt, S. R.; Sikka, S.; Mercier, R. W.; West, J. M.; Makriyannis, A.; Gatley, S. J.; Duclos, R. I. Assay and inhibition of diacylglycerol lipase activity. *Bioorg. Med. Chem. Lett.* **2012**, *22*, 4585–4592.

16. Duque, M.; Graupner, M.; Stuetz, H.; Wicher, I.; Zechner, R.; Paltauf, F.; Hermetter, A. New fluorogenic triacylglycerol analogs as substrates for the determination and chiral discrimination of lipase activities. *J. Lipid Res.* **1996**, *37*, 868–876.

17. Hermetter, A.; Duque, M.; Paltauf, F. Fluorescence determination of the activity of lipases using triglycerides or triglyceride analogs containing fluorophors and quenching moieties. PCT Int. Appl. WO 9532181, 1995.

18. Bayburt, Timothy.; Yu, B.; Street, I.; Ghomashchi, F.; Laliberte, F.; Perrier, H.; Wang, Z.; Homan, R.; Jain, M. K.; Gelb, M. H. Continuous, vesicle-based fluorimetric assays of 14- and 85-kDa phospholipases A2. *Anal. Biochem.* **1995**, *232*, 7–23.

19. Kusunoki, C.; Sato, S.; Kobayashi, M.; Niwa, M. Preparation of phosphatidylethanolamine derivatives as substrates for phospholipase A2 determination. Jpn. Kokai Tokkyo Koho JP 04282391, 1992.

20. Lageweg, W.; Wanders, R. J. A.; Tager, J. M. Long-chain-acyl-CoA synthetase and very-long-chain-acyl-CoA synthetase activities in peroxisomes and microsomes from rat liver. An enzymological study. *Eur. J. Biochem.* **1991**, *196*, 519–523.

21. Kawamura, F.; Ota, M.; Onuma, T.; Sakon, H.; Takahashi, T.; Yamaguchi, T.; Sasaki, M. Electroluminescent element. Jpn. Kokai Tokkyo Koho JP 05021165, 1993.

22. Serban, B.; Mihaila, M. N.; Buiu, O. Fluorescent polymers for oxygen sensing. Eur. Pat. Appl. EP 2461155, 2012.

23. Fujiwara, Y.; Amao, Y. Novel optical oxygen sensing material: 1-pyrenedecanoic acid and perfluorodecanoic acid chemisorbed onto anodic oxidized aluminium plate. *Sens. Actuators, B: Chem.* **2004**, *B99*, 130–133.

24. Fujiwara, Y.; Amao, Y. Optimizing oxygen-sensitivity of optical sensor using pyrene carboxylic acid by myristic acid co-chemisorption onto anodic oxidized aluminum plate. *Talanta* **2004**, *62*, 655–660.

25. Fujiwara, Y.; Amao, Y. An oxygen sensor based on the fluorescence quenching of pyrene chemisorbed layer onto alumina plates. *Sens. Actuators, B: Chem.* **2003**, *B89*, 187–191.

26. Fujiwara, Y.; Amao, Y. Controlling the oxygen-sensitivity of 1-pyrenedecanoic acid chemisorption layer onto anodic oxidized aluminum plate. *Bull. Chem. Soc. Jpn.* **2002**, *75*, 2697–2698.

27. Fujiwara, Y.; Amao, Y. 1-Pyrenedecanoic acid chemisorption film as a novel oxygen sensing material. *Sens. Actuators, B: Chem.* **2002**, *B85*, 175–178.

28. Itaya, A.; Masuhara, H.; Taniguchi, Y.; Imazeki, S. Fluorescence spectral change of LB films containing ω-(1-pyrenyl)alkanoic acids induced by an excimer laser. *Langmuir* **1989**, *5*, 1407–1409.

29. Fujihira, M.; Nishiyama, K.; Aoki, K. Electron transfer quenching of excited pyrene or tris(bipyridine)ruthenium(II) (2+)-ion derivative in Langmuir-Blodgett films by donor or acceptor molecules with different redox potentials. *Thin Solid Films* **1988**, *160*, 317–325.

30. Fujihira, M.; Nishiyama, K.; Yamada, H. Photoelectrochemical responses of optically transparent electrodes modified with Langmuir-Blodgett films consisting of surfactant derivatives of electron donor, acceptor and sensitizer molecules. *Thin Solid Films* **1985**, *132*, 77–82.

31. Fujita, K.; Orihashi, Y.; Itaya, A. Laser-assisted vacuum deposition process of 10-(1-pyrenyl)decanoic acid as revealed by in situ fluorescence spectroscopy. *Thin Solid Films* **1995**, *260*, 98–106.

32. Itaya, A.; Takada, S.; Masuhara, H.; Taniguchi, Y. *In-situ* observation of vacuum deposition process of 10-(1-pyrenyl)decanoic acid by fluorescence measurement. *Thin Solid Films* **1991**, *197*, 357–365.

33. Itaya, A.; Takada, S.; Masuhara, H. Laser-assisted vacuum deposition of 10-(1-pyrenyl)decanoic acid: *in situ* fluorescence observation of the process. *Chem. Mater.* **1991**, *3*, 271–275.

34. Mitsuya, M.; Kiguchi, M.; Taniguchi, Y.; Masuhara, H. Fluorescence characteristics, formation mechanism and chromophore association of ω-(1-pyrenyl)alkanoic acid films prepared by vacuum deposition. *Thin Solid Films* **1989**, *169*, 323–332.

35. Taniguchi, Y.; Mitsuya, M.; Tamai, N.; Yamazaki, I.; Masuhara, H. Fluorescence spectra of vacuum-deposited films of ω-(1-pyrenyl)alkanoic acids. *Chem. Phys. Lett.* **1986**, *132*, 516–520.

1-PYRENEDODECANOIC ACID (PDDA)

CAS Registry Number 69168-45-2

Chemical Structure

(CH$_2$)$_{11}$COOH

CA Index Name 1-Pyrenedodecanoic acid

Other Names 12-(1-Pyrene)dodecanoic acid; 12-(1-Pyrenyl)dodecanoic acid; P 12; P 12 (photosensitizer); PDDA

Merck Index Number Not listed

Chemical/Dye Class Pyrene

Molecular Formula C$_{28}$H$_{32}$O$_2$

Molecular Weight 400.55

Physical Form Pale yellow solid[2]

Solubility Soluble in acetonitrile, chloroform, N,N-dimethylformamide, dimethyl sulfoxide, methanol

Melting Point 123–125 °C[1]; 110–111 °C[2]

Boiling Point (Calcd.) 594.5 ± 19.0 °C, pressure: 760 Torr

pK$_a$ (Calcd.) 4.78 ± 0.10, most acidic, temperature: 25 °C

Absorption (λ$_{max}$) 341 nm, 325 nm (MeOH)

Emission (λ$_{max}$) 377 nm, 396 nm (MeOH)

Molar Extinction Coefficient 44,000 cm^{-1} M^{-1} (MeOH)

Synthesis Synthetic methods[1,2]

Imaging/Labeling Applications Membranes;[3,4] fatty acid;[5–11] gangliosides;[12] lipids;[13–16] sphingomyelin[17–19]

Biological/Medical Applications Lipid probe;[13–16] diagnosis of Niemann-Pick disease;[17,19] studying transport of fatty acids;[10,11] membrane probe;[3,4] as substrates for sphingomyelinase activity[19]

Industrial Applications Oxygen sensing material;[20–23] Langmuir-Blodgett films;[24–27] liquid crystal orienting film;[28] self-assembled monolayer films;[29] vacuum-deposited films[30–34]

Safety/Toxicity Cytotoxicity;[35–39] phototoxicity[35–39]

REFERENCES

1. Katusin-Razem, B. The synthesis of the fluorescence probe, 12-(1-pyrenyl)dodecanoic acid. *Croat. Chem. Acta* **1978**, *51*, 163–166.

2. Hahma, A.; Bhat, S.; Leivo, K.; Linnanto, J.; Lahtinen, M.; Rissanen, K. Pyrene derived functionalized low molecular weight organic gelators and gels. *New J. Chem.* **2008**, *32*, 1438–1448.

3. Eisinger, J.; Flores, J.; Petersen, W. P. A milling crowd model for local and long-range obstructed lateral diffusion. Mobility of excimeric probes in the membrane of intact erythrocytes. *Biophys. J.* **1986**, *49*, 987–1001.

4. Eisinger, J.; Flores, J. Fluorometry of turbid and absorbant samples and the membrane fluidity of intact erythrocytes. *Biophys. J.* **1985**, *48*, 77–84.

5. Tarshis, M.; Salman, M. Uptake of a fluorescent-labeled fatty acid by *Spiroplasma floricola* cells. *Arch. Microbiol.* **1992**, *157*, 258–263.

6. Kaneshiro, E. S.; Reuter, S. F.; Trinkle, L. Fatty acid uptake by ciliated protozoans analyzed by flow cytometry. *J. Microbiol. Methods* **1989**, *10*, 189–197.

7. Fibach, E.; Giloh, H.; Rachmilewitz, E. A.; Gatt, S. Flow cytofluorometric analysis of the uptake of the fluorescent fatty acid pyrenedodecanoic acid by human peripheral blood cells. *Cytometry* **1988**, *9*, 525–528.

8. Nahas, N.; Fibach, E.; Giloh, H.; Gatt, S. Use of the fluorescence activated cell sorter for studying uptake of fluorescent fatty acids into cultured cells. *Biochim. Biophys. Acta, Lipids Lipid Metab.* **1987**, *917*, 86–91.

9. Fibach, E.; Nahas, N.; Giloh, H.; Gatt, S. Uptake of fluorescent fatty acids by erythroleukemia cells. Effect of differentiation. *Exp. Cell Res.* **1986**, *166*, 220–228.

10. Almgren, M.; Swarup, S. Fluorescence stopped-flow study of the interaction of alkylpyridinium salts and pyrene-substituted fatty acids with lecithin vesicles. *Chem. Phys. Lipids* **1982**, *31*, 13–22.

11. Morand, O.; Fibach, E.; Dagan, A.; Gatt, S. Transport of fluorescent derivatives of fatty acids into cultured human leukemic myeloid cells and their subsequent metabolic utilization. *Biochim. Biophys. Acta, Lipids Lipid Metab.* **1982**, *711*, 539–550.

12. Ollmann, M.; Schwarzmann, G.; Sandhoff, K.; Galla, H. J. Pyrene-labeled gangliosides: micelle formation in aqueous solution, lateral diffusion, and thermotropic behavior in phosphatidylcholine bilayers. *Biochemistry* **1987**, *26*, 5943–5952.

13. Koivusalo, M.; Alvesalo, J.; Virtanen, J. A.; Somerharju, P. Partitioning of pyrene-labeled phospho- and sphingolipids between ordered and disordered bilayer domains. *Biophys. J.* **2004**, *86*, 923–935.

14. Naylor, B. L.; Picardo, M.; Homan, R.; Pownall, H. J. Effects of fluorophore structure and hydrophobicity on the uptake and metabolism of fluorescent lipid analogs. *Chem. Phys. Lipids* **1991**, *58*, 111–119.

15. Pal, R.; Barenholz, Y.; Wagner, R. R. Pyrene phospholipid as a biological fluorescent probe for studying fusion of virus membrane with liposomes. *Biochemistry* **1988**, *27*, 30–36.

16. Morand, O.; Fibach, E.; Livni, N.; Gatt, S. Induction of lipid storage in cultured leukemic myeloid cells by pyrene-dodecanoic acid. *Biochim. Biophys. Acta, Lipids Lipid Metab.* **1984**, *793*, 95–104.

17. Levade, T.; Gatt, S. Uptake and intracellular degradation of fluorescent sphingomyelin by fibroblasts from normal individuals and a patient with Niemann-Pick disease. *Biochim. Biophys. Acta, Lipids Lipid Metab.* **1987**, *918*, 250–259.

18. Ahmad, T. Y.; Sparrow, J. T.; Morrisett, J. D. Fluorine-, pyrene-, and nitroxide-labeled sphingomyelin: semi-synthesis and thermotropic properties. *J. Lipid Res.* **1985**, *26*, 1160–1165.

19. Gatt, S.; Dinur, T.; Barenholz, Y. A fluorometric determination of sphingomyelinase by use of fluorescent derivatives of sphingomyelin, and its application to diagnosis of Niemann-Pick disease. *Clin. Chem.* **1980**, *26*, 93–96.

20. Serban, B.; Mihaila, M. N.; Buiu, O. Fluorescent polymers for oxygen sensing. Eur. Pat. Appl. EP 2461155, 2012.

21. Fujiwara, Y.; Amao, Y. Optical oxygen sensing properties of 1-pyrenedodecanoic acid and perfluorocarboxylic acid chemisorbed onto anodic oxidized aluminium plate. *Sens. Lett.* **2004**, *2*, 232–237.

22. Fujiwara, Y.; Amao, Y. Optimizing oxygen-sensitivity of optical sensor using pyrene carboxylic acid by myristic acid co-chemisorption onto anodic oxidized aluminum plate. *Talanta* **2004**, *62*, 655–660.

23. Fujiwara, Y.; Amao, Y. An oxygen sensor based on the fluorescence quenching of pyrene chemisorbed layer onto alumina plates. *Sens. Actuators, B: Chem.* **2003**, *B89*, 187–191.

24. Yu, Q.; Vuorimaa, E.; Tkachenko, N. V.; Lemmetyinen, H. Interlayer energy transfer between pyrene-dodecanoic acid and NBD-dodecanoic acid in Langmuir-Blodgett films. *J. Lumin.* **1997**, *75*, 245–253.

25. Fujihira, M.; Nishiyama, K.; Kurihara, M. Luminescent Langmuir-Blodgett thin films of long-chain carboxylic acids. Jpn. Kokai Tokkyo Koho JP 03098674, 1991.

26. Itaya, A.; Masuhara, H.; Taniguchi, Y.; Imazeki, S. Fluorescence spectral change of LB films containing ω-(1-pyrenyl)alkanoic acids induced by an excimer laser. *Langmuir* **1989**, *5*, 1407–1409.

27. Murakata, T.; Miyashita, T.; Matsuda, M. Study of thermal behavior of Langmuir-Blodgett films with an emission probe. *Langmuir* **1986**, *2*, 786–788.

28. Tanaka, T.; Tani, T.; Fujisawa, K. Manufacture of liquid crystal orienting film. Jpn. Kokai Tokkyo Koho JP 04013115, 1992.

29. Chen, S. H.; Frank, C. W. Fluorescence probe studies of self-assembled monolayer films. *Langmuir* **1991**, *7*, 1719–1726.

30. Mitsuya, M.; Kiguchi, M.; Taniguchi, Y.; Masuhara, H. Fluorescence characteristics, formation mechanism and chromophore association of ω-(1-pyrenyl)alkanoic acid films prepared by vacuum deposition. *Thin Solid Films* **1989**, *169*, 323–332.

31. Itaya, A.; Kawamura, T.; Masuhara, H.; Taniguchi, Y.; Mitsuya, M. Fluorescence spectral changes of vacuum-deposited films of ω-(1-pyrenyl)alkanoic acids induced by an excimer laser: molecular aspects of laser annealing. *Chem. Phys. Lett.* **1987**, *133*, 235–238.

32. Taniguchi, Y.; Mitsuya, M.; Tamai, N.; Yamazaki, I.; Masuhara, H. Fluorescence spectra of vacuum-deposited films of ω-(1-pyrenyl)alkanoic acids. *Chem. Phys. Lett.* **1986**, *132*, 516–520.

33. Mitsuya, M.; Taniguchi, Y.; Tamai, N.; Yamazaki, I.; Masuhara, H. Vacuum-deposited films of 12-(1-pyrenyl)dodecanoic acid analyzed by fluorescence spectroscopy. *Thin Solid Films* **1985**, *129*, L45-L48.

34. Taniguchi, Y.; Mitsuya, M.; Tamai, N.; Yamazaki, I.; Masuhara, H. Time- and depth-resolved fluorescence spectra of layered organic films prepared by vacuum deposition. *J. Colloid Interface Sci.* **1985**, *104*, 596–598.

35. Spisni, E.; Cavazzoni, M.; Griffoni, C.; Calzolari, E.; Tomasi, V. Evidence that photodynamic stress kills Zellweger fibroblasts by a nonapoptotic mechanism. *Biochim. Biophys. Acta* **1998**, *1402*, 61–69.

36. Fibach, E.; Rachmilewitz, E. A.; Gatt, S. Photosensitization of human bladder carcinoma cells by pyrene-dodecanoic acid: quantitative analysis of the cytotoxicity. *Res. Exp. Med.* **1992**, *192*, 185–196.

37. Fibach, E.; Rachmilewitz, E. A.; Gatt, S. Analysis of cellular heterogeneity in the response of human leukemic cells to photosensitization induced by pyrene-containing fatty acid. *Leuk. Res.* **1989**, *13*, 1099–1104.

38. Zoeller, R. A.; Morand, O. H.; Raetz, C. R. H. A possible role for plasmalogens in protecting animal cells against photosensitized killing. *J. Biol. Chem.* **1988**, *263*, 11590–11596.

39. Fibach, E.; Morand, O.; Gatt, S. Photosensitization to ultraviolet irradiation and selective killing of cells following uptake of pyrene-containing fatty acid. *J. Cell Sci.* **1986**, *85*, 149–159.

1-PYRENEHEXADECANOIC ACID

CAS Registry Number 90936-84-8

Chemical Structure

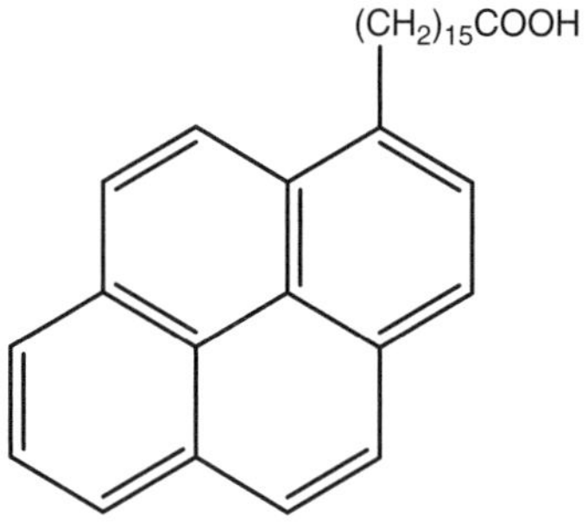

(CH$_2$)$_{15}$COOH

CA Index Name 1-Pyrenehexadecanoic acid

Other Names 16-(1-Pyrenyl)hexadecanoic acid; Pyrenehexadecanoic acid; PHA; PHD; PHDA; PyHA; PY-HDA

Merck Index Number Not listed

Chemical/Dye Class Pyrene

Molecular Formula C$_{32}$H$_{40}$O$_2$

Molecular Weight 456.66

Physical Form Solid

Solubility Soluble in *N,N*-dimethylformamide, dimethyl sulfoxide, methanol

Boiling Point (Calcd.) 631.5 ± 24.0 °C, pressure: 760 Torr

pK$_a$ (Calcd.) 4.78 ± 0.10, most acidic, temperature: 25 °C

Absorption (λ$_{max}$) 341 nm, 325 nm (MeOH)

Emission (λ$_{max}$) 377 nm, 396 nm (MeOH)

Molar Extinction Coefficient 43,000 cm^{-1} M^{-1} (MeOH)

Synthesis Synthetic method[1]

Imaging/Labeling Applications Lipids;[1–7] fatty acids;[8,9] membranes;[6,10,11] inverted micelles[12]

Biological/Medical Applications Lipid probe;[1–7] membrane probe[6,10,11]

Industrial Applications Characterizing inverted micelles;[12] electrooptical displays;[13] electronic systems for optical information processing (organic photoconductors);[14,15] Langmuir-Blodgett films;[16–21] self-assembled monolayer films;[22,23] nano-structured silica films;[24] vacuum-deposited films[25,26]

Safety/Toxicity No data available

REFERENCES

1. Haugland, R. P. *Handbook of Fluorescent Probes and Research Chemicals*; Molecular Probes Inc.: Eugene, **1996**; pp 288–297.

2. Barenholz, Y.; Cohen, T.; Korenstein, R.; Ottolenghi, M. Organization and dynamics of pyrene and pyrene lipids in intact lipid bilayers. Photo-induced charge transfer processes. *Biophys. J.* **1991**, *60*, 110–124.

3. Naylor, B. L.; Picardo, M.; Homan, R.; Pownall, H. J. Effects of fluorophore structure and hydrophobicity on the uptake and metabolism of fluorescent lipid analogs. *Chem. Phys. Lipids* **1991**, *58*, 111–119.

4. L'Heureux, G. P.; Fragata, M. Monomeric and aggregated pyrene and 16-(1-pyrenyl)hexadecanoic acid in small, unilamellar phosphatidylcholine vesicles and ethanol-buffer solutions. *J. Photochem. Photobiol., B* **1989**, *3*, 53–63.

5. L'Heureux, G. P.; Fragata, M. Micropolarities of lipid bilayers and micelles. 5. Localization of pyrene in small unilamellar phosphatidylcholine vesicles. *Biophys. Chem.* **1988**, *30*, 293–301.

6. Pal, R.; Barenholz, Y.; Wagner, R. R. Pyrene phospholipid as a biological fluorescent probe for studying fusion of virus membrane with liposomes. *Biochemistry* **1988**, *27*, 30–36.

7. L'Heureux, G. P.; Fragata, M. Micropolarities of lipid bilayers and micelles. 4. Dielectric constant determinations of unilamellar phosphatidylcholine vesicles with the probes pyrene and 16-(1-pyrenyl)hexadecanoic acid. *J. Colloid Interface Sci.* **1987**, *117*, 513–522.

8. Hoefler, G.; Paschke, E.; Hoefler, S.; Moser, A. B.; Moser, H. W. Photosensitized killing of cultured fibroblasts from patients with peroxisomal disorders due to pyrene fatty acid-mediated ultraviolet damage. *J. Clin. Invest.* **1991**, *88*, 1873–1879.

9. Fujihira, M.; Nishiyama, K.; Hamaguchi, Y.; Tatsu, Y. Fluorescence microscopic study of change in pyrene cluster size in mixed monolayers of pyrene-substituted and normal fatty acids with chain-length matching. *Chem. Lett.* **1987**, 253–256.

10. Lissi, E. A.; Caceres, T. Oxygen diffusion-concentration in erythrocyte plasma membranes studied by the fluorescence quenching of anionic and cationic pyrene derivatives. *J. Bioenerg. Biomembr.* **1989**, *21*, 375–385.

11. Eisinger, J. Fluorometry of absorbant and turbid samples and the lateral mobility in membranes of intact erythrocytes. *J. Lumin.* **1984**, *31–32*, 875–880.

12. Jao, T. C.; Kreuz, K. L. Characterization of inverted micelles of calcium alkarylsulfonates by some pyrene fluorescence probes. *ACS Symp. Ser.* **1986**, *311*, 90–99.

13. Hirai, Y.; Haruta, M.; Nishimura, Y.; Matsuda, H.; Nakagiri, T. Electrooptical displays containing organic unimolecular films. Jpn. Kokai Tokkyo Koho JP 60241273, 1985.

14. Nishimura, Y.; Hirai, Y.; Matsuda, H.; Haruta, M.; Nakagiri, T. Molecular film laminate-type electronic systems for optical information processing. Jpn. Kokai Tokkyo Koho JP 60225236, 1985.

15. Hirai, Y.; Haruta, M.; Nishimura, Y.; Matsuda, H.; Nakagiri, T. Molecular film laminate-type electronic systems for optical information processing. Jpn. Kokai Tokkyo Koho JP 60225229, 1985.

16. Sakomura, M.; Nakashima, T.; Ueda, K.; Fujihira, M. Unique fluorescence spectra of a disubstituted pyrene surfactant in Langmuir-Blodgett films. *Colloids Surf., A* **2006**, *284–285*, 528–531.

17. Yamazaki, I.; Tamai, N.; Yamazaki, T. Picosecond fluorescence spectroscopy on excimer formation and excitation energy transfer of pyrene in Langmuir-Blodgett monolayer films. *J. Phys. Chem.* **1987**, *91*, 3572–3577.

18. Ozaki, H.; Harada, Y.; Nishiyama, K.; Fujihira, M. Selective observation of molecular ends exposed outside Langmuir-Blodgett monolayer films by penning ionization electron spectroscopy. *J. Am. Chem. Soc.* **1987**, *109*, 950–951.

19. Yamazaki, I.; Tamai, N.; Yamazaki, T. Picosecond fluorescence spectroscopy on molecular association in Langmuir-Blodgett films. *Springer Ser. Chem. Phys.* **1986**, *46*, 444–446.

20. Sugi, M.; Sakai, K.; Saito, M.; Kawabata, Y.; Iizima, S. Photoelectric effects in heterojunction Langmuir-Blodgett film diodes. *Thin Solid Films* **1985**, *132*, 69–76.

21. Saito, M.; Sugi, M.; Iizima, S. Evidence for ambipolar conduction in dye-sensitized p-n junctions of Langmuir-Blodgett films. *Jpn. J. Appl. Phys., Part 1* **1985**, *24*, 379–380.

22. Uji-i, H.; Yoshidome, M.; Hobley, J.; Hatanaka, K.; Fukumura, H. Structural variations in self-assembled monolayers of 1-pyrenehexadecanoic acid and 4,4′-bipyridyl on graphite at the liquid–solid interface. *Phys. Chem. Chem. Phys.* **2003**, *5*, 4231–4235.

23. Chen, S. H.; Frank, C. W. Fluorescence probe studies of self-assembled monolayer films. *Langmuir* **1991**, *7*, 1719–1726.

24. Ghosh, K.; Rankin, S. E.; Lehmler, H.; Knutson, B. L. Processing of surfactant templated nano-structured silica films using compressed carbon dioxide as interpreted from *in situ* fluorescences. *J. Phys. Chem. B* **2012**, *116*, 11646–11655.

25. Mitsuya, M.; Kiguchi, M.; Taniguchi, Y.; Masuhara, H. Fluorescence characteristics, formation mechanism and chromophore association of ω-(1-pyrenyl)alkanoic acid films prepared by vacuum deposition. *Thin Solid Films* **1989**, *169*, 323–332.

26. Taniguchi, Y.; Mitsuya, M.; Tamai, N.; Yamazaki, I.; Masuhara, H. Fluorescence spectra of vacuum-deposited films of ω-(1-pyrenyl)alkanoic acids. *Chem. Phys. Lett.* **1986**, *132*, 516–520.

1-PYRENEHEXANOIC ACID (PHA)

CAS Registry Number   90936-85-9
Chemical Structure

(CH₂)₅COOH

CA Index Name   1-Pyrenehexanoic acid

Other Names   1-Pyrenehexanoic acid; 6-(1-Pyrenyl)-hexanoic acid; 6-(Pyren-1-yl)-hexanoic acid; 6-Pyrenylhexanoic acid; PHA

Merck Index Number   Not listed

Chemical/Dye Class   Pyrene

Molecular Formula $C_{22}H_{20}O_2$

Molecular Weight   316.39

Physical Form   Solid

Solubility   Soluble in *N,N*-dimethylformamide, dimethyl sulfoxide, methanol

Melting Point   200–201 °C[2]

Boiling Point (Calcd.)   552.2 ± 29.0 °C, pressure: 760 Torr

pK$_a$ (Calcd.)   4.77 ± 0.10, most acidic, temperature: 25 °C

Absorption (λ_{max})   341 nm, 325 nm (MeOH)

Emission (λ_{max})   377 nm, 396 nm (MeOH)

Molar Extinction Coefficient   42,000 cm^{-1} M^{-1} (MeOH)

Synthesis   Synthetic methods[1,2]

Imaging/Labeling Applications   Fibroblasts;[3,4] galactosylceramide;[14] glucose oxidase (GOx);[6] horseradish peroxidase (HRP);[6,7] laccase;[5] lipids;[3] lysozyme crystal;[8] nucleic acids;[9–11] paromomycin;[9] phosphatidylcholine;[14] proteins;[12,13] sphingomyelin;[14] sphingosines[15]

Biological/Medical Applications   Assessing role of membrane domains in lipid trafficking/sorting;[14] characterizing interactions/mass transfer of surfactants inside crystaline protein matrixes;[8] investigating transport/metabolism of cellular lipids;[3] monitoring domain formation in membranes;[14] implantable biocathodes/composite electrodes;[5–7] nucleic acid hybridization probe;[10,11] as a substrate for measuring cholesterylester hydrolase (CEH) activity,[16] sphingosine kinases 1 and 2 activity[15]

Industrial Applications   Boron-doped diamond (BDD) substrates;[17] Langmuir-Blodgett films;[18] thin films;[19] vacuum-deposited films[20,21]

Safety/Toxicity   No data available

REFERENCES

1. Templer, R. H.; Castle, S. J.; Rachael Curran, A.; Rumbles, G.; Klug, D. R. Sensing isothermal changes in the lateral pressure in model membranes using di-pyrenyl phosphatidylcholine. *Faraday Discuss.* **1999**, *111*, 41–53.

2. Reynders, P.; Kuehnle, W.; Zachariasse, K. A. Ground-state dimers in excimer-forming bichromophoric molecules: NMR and single-photon-counting data. 2. Racemic and meso dipyrenylpentanes and dipyrenylalkanes. *J. Phys. Chem.* **1990**, *94*, 4073–4082.

3. Kasurinen, J.; Somerharju, P. Metabolism of pyrenyl fatty acids in baby hamster kidney fibroblasts. Effect of the acyl chain length. *J. Biol. Chem.* **1992**, *267*, 6563–6569.

4. Hoefler, G.; Paschke, E.; Hoefler, S.; Moser, A. B.; Moser, H. W. Photosensitized killing of cultured fibroblasts from patients with peroxisomal disorders due to pyrene fatty acid-mediated ultraviolet damage. *J. Clin. Invest.* **1991**, *88*, 1873–1879.

5. Gutierrez-Sanchez, C.; Jia, W.; Beyl, Y.; Pita, M.; Schuhmann, W.; De Lacey, A. L.; Stoica, L. Enhanced direct electron transfer between laccase and hierarchical carbon microfibers/carbon nanotubes composite electrodes. Comparison of three enzyme immobilization methods. *Electrochim. Acta* **2012**, *82*, 218–223.

6. Jia, W.; Jin, C.; Xia, W.; Muhler, M.; Schuhmann, W.; Stoica, L. Glucose oxidase/horseradish peroxidase co-immobilized at a CNT-modified graphite electrode: Towards potentially implantable biocathodes. *Chem. Eur. J.* **2012**, *18*, 2783–2786.

7. Jia, W.; Schwamborn, S.; Jin, C.; Xia, W.; Muhler, M.; Schuhmann, W.; Stoica, L. Towards a high potential

biocathode based on direct bioelectrochemistry between horseradish peroxidase and hierarchically structured carbon nanotubes. *Phys. Chem. Chem. Phys.* **2010**, *12*, 10088–10092.

8. Velev, O. D.; Kaler, E. W.; Lenhoff, A. M. Surfactant diffusion into lysozyme crystal matrices Investigated by quantitative fluorescence microscopy. *J. Phys. Chem. B* **2000**, *104*, 9267–9275.

9. Tok, J. B. H.; Cho, J.; Rando, R. R. Amino glycoside hybrids as potent RNA antagonists. *Tetrahedron* **1999**, *55*, 5741–5758.

10. Masuko, M.; Ohtani, H.; Ebata, K.; Shimadzu, A. Optimization of excimer-forming two-probe nucleic acid hybridization method with pyrene as a fluorophore. *Nucleic Acids Res.* **1998**, *26*, 5409–5416.

11. Ebata, K.; Masuko, M.; Ohtani, H.; Jibu, M. Excimer formation by hybridization using two pyrene-labeled oligonucleotide probes. *Nucleic Acids Symp. Ser.* **1995**, *34*, 187–188.

12. Nishihara, M.; Perret, F.; Takeuchi, T.; Futaki, S.; Lazar, A. N.; Coleman, A. W.; Sakai, N.; Matile, S. Arginine magic with new counterions up the sleeve. *Org. Biomol. Chem.* **2005**, *3*, 1659–1669.

13. Perret, F.; Nishihara, M.; Takeuchi, T.; Futaki, S.; Lazar, A. N.; Coleman, A. W.; Sakai, N.; Matile, S. Anionic fullerenes, calixarenes, coronenes, and pyrenes as activators of oligo/polyarginines in model membranes and live cells. *J. Am. Chem. Soc.* **2005**, *127*, 1114–1115.

14. Koivusalo, M.; Alvesalo, J.; Virtanen, J. A.; Somerharju, P. Partitioning of pyrene-labeled phospho- and sphingolipids between ordered and disordered bilayer domains. *Biophys. J.* **2004**, *86*, 923–935.

15. Ettmayer, P.; Billich, A.; Baumruker, T.; Mechtcheriakova, D.; Schmid, H.; Nussbaumer, P. Fluorescence-labeled sphingosines as substrates of sphingosine kinases 1 and 2. *Bioorg. Med. Chem. Lett.* **2004**, *14*, 1555–1558.

16. Joutti, A.; Kotama, L.; Virtanen, J. A.; Kinnunen, P. K. Fluorometric assay for pancreatic cholesterylester hydrolase. *Chem. Phys. Lipids* **1985**, *36*, 335–341.

17. Mazur, M.; Krysinski, P.; Blanchard, G. J. Use of zirconium-phosphate-carbonate chemistry to immobilize polycyclic aromatic hydrocarbons on boron-doped diamond. *Langmuir* **2005**, *21*, 8802–8808.

18. Pevenage, D.; Van der Auweraer, M.; De Schryver, F. C. Determination of the photoinduced electron transfer rate constant in Langmuir-Blodgett films by time-resolved fluorescence. *Langmuir* **1999**, *15*, 4641–4647.

19. Itaya, A.; Kawamura, T.; Masuhara, H.; Taniguchi, Y. Laser-induced geometrical change of fluorescent traps in cast thin films of ω-(1-pyrenyl) alkanoic acids. *Thin Solid Films* **1990**, *185*, 307–320.

20. Mitsuya, M.; Kiguchi, M.; Taniguchi, Y.; Masuhara, H. Fluorescence characteristics, formation mechanism and chromophore association of ω-(1-pyrenyl)alkanoic acid films prepared by vacuum deposition. *Thin Solid Films* **1989**, *169*, 323–332.

21. Taniguchi, Y.; Mitsuya, M.; Tamai, N.; Yamazaki, I.; Masuhara, H. Fluorescence spectra of vacuum-deposited films of ω-(1-pyrenyl)alkanoic acids. *Chem. Phys. Lett.* **1986**, *132*, 516–520.

1-PYRENESULFONYL CHLORIDE

CAS Registry Number   61494-52-8

Chemical Structure

CA Index Name   1-Pyrenesulfonyl chloride

Other Names   3-Pyrenesulfonyl chloride; PSC; PSCL

Merck Index Number   Not listed

Chemical/Dye Class   Pyrene

Molecular Formula $C_{16}H_9ClO_2S$

Molecular Weight   300.76

Physical Form   Bright yellow crystals;[3] Orange crystals[4]

Solubility   Soluble in acetonitrile, chloroform, *N,N*-dimethylformamide, methanol, xylenes

Melting Point   172–174 °C;[5] 172–173 °C;[4] 163–166 °C (decompose)[3]

Boiling Point (Calcd.)   472.0 ± 14.0 °C, pressure: 760 Torr

Absorption (λ_{max})   350 nm (MeOH)

Emission (λ_{max})   380 nm (MeOH)

Molar Extinction Coefficient   28,000 $cm^{-1} M^{-1}$ (MeOH)

Synthesis   Synthetic methods[1–5]

Imaging/Labeling Applications   Estrogens;[6] oligosaccharides;[7,8] phospholipids;[9,10] cationic polyacrylamides (C-PAM);[11] 2-(arylsulfonyl)-methacrylates;[12] ribonucleopeptide;[13,14] sphingosines;[15,16] sugar chains[17]

Biological/Medical Applications   Quantitating estrogens;[6] as substrates for acetylcholinesterase activity,[18] arylsulfatase A activity,[19] phosphatase activity,[15] sphingosine kinases 1 and 2 activity[15,16]

Industrial Applications   Analysizing sugar chains;[17] Detecting dicarboxylic acids,[5,24,25] picric acid;[26] functionalized fluorescent polymers;[11,12] sensor for copper ions,[20,21] lead ions,[22] mercury ions;[22,23] self-assembled monolayer film;[27,28] composite materials;[30] glass fibers;[29–31] silica microfibers[32]

Safety/Toxicity   No data available

REFERENCES

1. Vollmann, H.; Becker, H.; Corell, M.; Streeck, H. Pyrene and its derivatives. *Justus Liebigs Ann. Chem.* **1937**, *531*, 1–159.

2. Shevchuk, I. N.; Paramonov, V. D. Pyreno[3,4-bc] thiopyran-3-one. U.S.S.R. SU 518497, 1976.

3. Hemgesberg, M.; Schuetz, S.; Mueller, C.; Schloerholz, M.; Latzel, H.; Sun, Y.; Ziegler, C.; Thiel, W. R. Ultra-fast photo-patterning of hydroxamic acid layers adsorbed on TiAlN: The challenge of modeling thermally induced desorption. *Appl. Surf. Sci.* **2012**, *259*, 406–415.

4. Issa, J. B.; Salameh, A. S.; Castner, E. W., Jr.,; Wishart, J. F.; Isied, S. S. Conformational analysis of the electron-transfer kinetics across oligoproline peptides using N,N-dimethyl-1,4-benzenediamine donors and pyrene-1-sulfonyl acceptors. *J. Phys. Chem. B* **2007**, *111*, 6878–6886.

5. Gao, L.; Fang, Y.; Wen, X.; Li, Y.; Hu, D. Monomolecular layers of pyrene as a sensor to dicarboxylic acids. *J. Phys. Chem. B* **2004**, *108*, 1207–1213.

6. DeSilva, K. H.; Vest, F. B.; Karnes, H. T. Pyrene sulfonyl chloride as a reagent for quantitation of estrogens in human serum using HPLC with conventional and laser-induced fluorescence detection. *Biomed. Chromatogr.* **1996**, *10*, 318–324.

7. Kurihara, T.; Min, J. Z.; Toyo'oka, T.; Fukushima, T.; Inagaki, S. Determination of fluorescence-labeled asparaginyl-oligosaccharide in glycoprotein by reversed-phase ultraperformance liquid chromatography with electrospray Ionization time-of-flight mass spectrometry. *Anal. Chem.* **2007**, *79*, 8694–8698.

8. Min, J. Z.; Kurihara, T.; Hirata, A.; Toyo'oka, T.; Inagaki, S. Identification of *N*-linked oligosaccharide labeled with 1-pyrenesulfonyl chloride by quadrupole time-of-flight tandem mass spectrometry after separation by micro- and nanoflow liquid chromatography. *Biomed. Chromatogr.* **2009**, *23*, 912–921.

9. Abidi, S. L.; Mounts, T. L. Reversed-phase retention behavior of fluorescence labeled phospholipids in ammonium acetate buffers. *J. Liq. Chromatogr.* **1994**, *17*, 105–122.

10. Abidi, S. L.; Mounts, T. L.; Rennick, K. A. Reversed-phase high-performance liquid chromatography of phospholipids with fluorescence detection. *J. Chromatogr.* **1993**, *639*, 175–184.

11. Tanaka, H.; Swerin, A.; Odberg, L.; Tanaka, M. Homogeneous fluorescent labeling of cationic polyacrylamides. *J. Appl. Polym. Sci.* **2002**, *86*, 672–675.

12. Batra, D.; Shea, K. J. Novel trifunctional building blocks for fluorescent polymers. *Org. Lett.* **2003**, *5*, 3895–3898.

13. Hasegawa, T.; Hagihara, M.; Fukuda, M.; Nakano, S.; Fujieda, N.; Morii, T. Context-dependent fluorescence detection of a phosphorylated tyrosine residue by a ribonucleopeptide. *J. Am. Chem. Soc.* **2008**, *130*, 8804–8812.

14. Fukuda, M.; Hayashi, H.; Hasegawa, T.; Morii, T. Development of a fluorescent ribonucleopeptide sensor for histamine. *Trans. Mater. Res. Soc. Jpn.* **2009**, *34*, 525–527.

15. Billich, A.; Ettmayer, P.; Mechtcheriakova, D.; Nussbaumer, P.; Wlachos, A. Fluorescent labeled sphingosines. PCT Int. Appl. WO 2005030780, 2005.

16. Ettmayer, P.; Billich, A.; Baumruker, T.; Mechtcheriakova, D.; Schmid, H.; Nussbaumer, P. Fluorescence-labeled sphingosines as substrates of sphingosine kinases 1 and 2. *Bioorg. Med. Chem. Lett.* **2004**, *14*, 1555–1558.

17. Toyooka, T.; Bin, T.; Onosawa, T.; Kumata, J.; Ishikawa, K. Novel fluorescent agents for sugar chain analysis. Jpn. Kokai Tokkyo Koho JP 2006321775, 2006.

18. Saltmarsh, J. R.; Boyd, A. E.; Rodriguez, O. P.; Radic, Z.; Taylor, P.; Thompson, C. M. Synthesis of fluorescent probes directed to the active site gorge of acetylcholinesterase. *Bioorg. Med. Chem. Lett.* **2000**, *10*, 1523–1526.

19. Marchesini, S.; Viani, P.; Cestaro, B.; Gatt, S. Synthesis of pyrene derivatives of cerebroside sulfate and their use for determining arylsulfatase A activity. *Biochim. Biophys. Acta, Lipids Lipid Metab.* **1989**, *1002*, 14–19.

20. Gao, L.; Lue, F.; Xia, H.; Ding, L.; Fang, Y. Fluorescent film sensor for copper ion based on an assembled monolayer of pyrene moieties. *Spectrochim. Acta, Part A* **2011**, *79*, 437–442.

21. Lue, F.; Gao, L.; Ding, L.; Jiang, L.; Fang, Y. Spacer layer screening effect: A novel fluorescent film sensor

for organic copper(II) salts. *Langmuir* **2006**, *22*, 841–845.

22. Neupane, L. N.; Park, J.; Park, J. H.; Lee, K. Turn-on fluorescent chemosensor based on an amino acid for Pb(II) and Hg(II) ions in aqueous solutions and role of tryptophan for sensing. *Org. Lett.* **2013**, *15*, 254–257.

23. Yang, M.; Thirupathi, P.; Lee, K. Selective and sensitive ratiometric detection of Hg(II) ions using a simple amino acid based sensor. *Org. Lett.* **2011**, *13*, 5028–5031.

24. Gao, L.; Fang, Y.; Lue, F.; Cao, M.; Ding, L. Immobilization of pyrene via diethylenetriamine on quartz plate surface for recognition of dicarboxylic acids. *Appl. Surf. Sci.* **2006**, *252*, 3884–3893.

25. Lue, F.; Fang, Y.; Gao, L.; Ding, L.; Jiang, L. Selectivity via insertion: Detection of dicarboxylic acids in water by a new film chemosensor with enhanced properties. *J. Photochem. Photobiol., A: Chem.* **2005**, *175*, 207–213.

26. Du, H.; He, G.; Liu, T.; Ding, L.; Fang, Y. Preparation of pyrene-functionalized fluorescent film with a benzene ring in spacer and sensitive detection to picric acid in aqueous phase, *J. Photochem. Photobiol., A: Chem.* **2011**, *217*, 356–362.

27. Ding, L.; Liu, Y.; Cao, Y.; Wang, L.; Xin, Y.; Fang, Y. A single fluorescent self-assembled monolayer film sensor with discriminatory power. *J. Mater. Chem.* **2012**, *22*, 11574–11582.

28. Zhang, S.; Ding, L.; Lue, F.; Liu, T.; Fang, Y. Fluorescent film sensors based on SAMs of pyrene derivatives for detecting nitroaromatics in aqueous solutions. *Spectrochim. Acta, Part A* **2012**, *97*, 31–37.

29. Turrion, S. G.; Olmos, D.; Gonzalez-Benito, J. Complementary characterization by fluorescence and AFM of polyaminosiloxane glass fibers coatings. *Polym. Test.* **2005**, *24*, 301–308.

30. Gonzalez-Benito, J.; Aznar, A. J.; Lima, J.; Bahia, F.; Macanita, A. L.; Baselga, J. Fluorescence-labeled pyrenesulfonamide response for characterizing polymeric interfaces in composite materials. *J. Fluoresc.* **2000**, *10*, 141–146.

31. Gonzalez-Benito, J.; Cabanelas, J. C.; Aznar, A.; Vigil, M. R.; Bravo, J.; Serrano, B.; Baselga, J. Photophysics of a pyrene probe grafted onto silanized glass fiber surfaces. *J. Lumin.* **1997**, *72–74*, 451–453.

32. Olmos, D.; Gonzalez-Benito, J.; Aznar, A. J.; Baselga, J. Hydrolytic damage study of the silane coupling region in coated silica microfibers: pH and coating type effects. *J. Mater. Process. Technol.* **2003**, *143–144*, 82–86.

1,3,6,8-PYRENETETRASULFONIC ACID TETRASODIUM SALT

CAS Registry Number   59572-10-0

Chemical Structure

CA Index Name   1,3,6,8-Pyrenetetrasulfonic acid, sodium salt (1:4)

Other Names   1,3,6,8-Pyrenetetrasulfonic acid, tetrasodium salt; 1,3,6,8-Pyrenetetrasulfonic acid tetra-Na salt; Tetrasodium 1,3,6,8-pyrenetetrasulfonate; Tetra-Na 1,3,6,8-pyrenetetrasulfonate; Tetrasodium pyrenetetrasulfonate; tetrasodium 1,3,6,8-pyrenetetrasulfonate; Na_4PS_4; PSA; 4- PSA; PTS; PTSA; PY; Py-4; PyTS; SPA; TPA; Trasar 23299

Merck Index Number   Not listed

Chemical/Dye Class   Pyrene

Molecular Formula $C_{16}H_6Na_4O_{12}S_4$

Molecular Weight   610.43

Physical Form   Yellow powder[1,3]

Solubility   Soluble in water

Absorption (λ_{max})   374 nm (H_2O)

Emission (λ_{max})   403 nm (H_2O)

Molar Extinction Coefficient   51,000 cm^{-1} M^{-1} (H_2O)

Synthesis   Synthetic methods[1-3]

Imaging/Labeling Applications   Cyclen 1;[4] polyelectrolytes;[5-8] metal-polyethylenimine complexes;[9] proteins/peptides[10-12]

Biological/Medical Applications   Characterizing water-in-oil-in-water (W/O/W) emulsions;[13-15] as a substrate for measuring lipase activity;[13,16] air-breathing biocathodes for zinc/oxygen batteries;[17] graphene-coated biochar for various environmental applications;[18] plants or plant parts identification labels[19]

Industrial Applications   Assessing interaction between sulfonated polystyrene and poly(ethylacrylate-co-4-vinylpyridine) ionomers;[20] carbon nanotube;[21] evaluating/studying influence of incorporation of counterions into polypyrrole;[22] examining salt group association of sulfonated polystyrene ionomers;[23] fabricating graphene-based nanoelectronic devices;[24] graphene/polymer dispersions for graphene/polymer nanocomposites,[25] organic solar cells,[25] conductive films,[25] ink-jet-printed electronic devices;[25] graphene-zeolite based microcapsules;[26] investigating electrical breakdown/permeability control of bilayer-corked capsule membrane;[27] light filters;[2] monitoring wastewater treatment;[28] poly(arylene sulfide) blends in UV weathering;[29] pulp and papermaking processes;[30] thermosensitive polymer gels;[31-33] thin films;[6,7] TiO_2-based materials;[6] tracing corrosive materials;[34] ureasil gels for water purification[35]

Safety/Toxicity   No data available

REFERENCES

1. Tietze, E.; Bayer, O. Sulfonic acids of pyrene and their derivatives. *Justus Liebigs Ann. Chem.* **1939**, *540*, 189–210.

2. Carpmael, A. Improvements in or related to light filters. GB 457708, 1936.

3. Carpmael, A. Process for the manufacture of pyrene compounds. GB 441408, 1936.

4. Winschel, C. A.; Kalidindi, A.; Zgani, I.; Magruder, J. L.; Sidorov, V. Receptor for anionic pyrene derivatives provides the basis for new biomembrane assays. *J. Am. Chem. Soc.* **2005**, *127*, 14704–14713.

5. Angelatos, A. S.; Wang, Y.; Caruso, F. Probing the conformation of polyelectrolytes in mesoporous silica spheres. *Langmuir* **2008**, *24*, 4224–4230.

6. Shi, X.; Cassagneau, T.; Caruso, F. Electrostatic interactions between polyelectrolytes and a titania precursor: Thin film and solution studies. *Langmuir* **2002**, *18*, 904–910.

7. Tedeschi, C.; Caruso, F.; Moehwald, H.; Kirstein, S. Adsorption and desorption behavior of an anionic pyrene chromophore in sequentially deposited polyelectrolyte-dye thin films. *J. Am. Chem. Soc.* **2000**, *122*, 5841–5848.

8. Caruso, F.; Donath, E.; Moehwald, H.; Georgieva, R. Fluorescence studies of the binding of anionic derivatives of pyrene and fluorescein to cationic polyelectrolytes in aqueous solution. *Macromolecules* **1998**, *31*, 7365–7377.

9. Dowling, S. D.; Mullin, J. L.; Seitz, W. R. Binding of sulfonated fluorophors by metal-polyethylenimine complexes. *Macromolecules* **1986**, *19*, 344–347.

10. Nagai, Y.; Unsworth, L. D.; Koutsopoulos, S.; Zhang, S. Slow release of molecules in self-assembling peptide nanofiber scaffold. *J. Controlled Release* **2006**, *115*, 18–25.

11. Jain, R. K.; Hamilton, A. D. Designing protein denaturants: synthetic agents induce cytochrome c unfolding at low concentrations and stoichiometries. *Angew. Chem., Int. Ed.* **2002**, *41*, 641–643.

12. Murakami, K.; Akamatsu, M.; Sano, T. A structure-activity relationship in the binding of multicharged anionic azo and pyrene dyes to serum albumin. *Bull. Chem. Soc. Jpn.* **1994**, *67*, 2647–2653.

13. Shima, M.; Tanaka, M.; Kimura, Y.; Adachi, S.; Matsuno, R. Hydrolysis of the oil phase of a W/O/W emulsion by pancreatic lipase. *J. Controlled Release* **2004**, *94*, 53–61.

14. Adachi, S.; Imaoka, H.; Hasegawa, Y.; Matsuno, R. Preparation of a water-in-oil-in-water (W/O/W) type microcapsules by a single-droplet-drying method and change in encapsulation efficiency of a hydrophilic substance during storage. *Biosci., Biotechnol., Biochem.* **2003**, *67*, 1376–1381.

15. Tokgoz, N. S.; Grossiord, J. L.; Fructus, A.; Seiller, M.; Prognon, P. Evaluation of two fluorescent probes for the characterization of W/O/W emulsions. *Int. J. Pharm.* **1996**, *141*, 27–37.

16. Koller, E.; Wolfbeis, O. S. Preparation of pyrenesulfonic acid derivatives for photometric determination of enzyme activity. Austrian AT 385755, 1988.

17. Zloczewska, A.; Jonsson-Niedziolka, M. Efficient air-breathing biocathodes for zinc/oxygen batteries. *J. Power Sources* **2013**, *228*, 104–111.

18. Zhang, M.; Gao, B.; Yao, Y.; Xue, Y.; Inyang, M. Synthesis, characterization, and environmental implications of graphene-coated biochar. *Sci. Total Environ.* **2012**, *435–436*, 567–572.

19. Van Der Krieken, W. M.; Kok, C. J. Identification labels in plants or plant parts. PCT Int. Appl. WO 2004065945, 2004.

20. Bakeev, K. N.; MacKnight, W. J. Fluorescent molecular probe technique for assessing the interaction between sulfonated polystyrene and poly (ethylacrylate-co-4-vinylpyridine) ionomers in tetrahydrofuran. *Macromolecules* **1991**, *24*, 4575–4577.

21. Artyukhin, A. B.; Bakajin, O.; Stroeve, P.; Noy, A. Layer-by-layer electrostatic self-assembly of polyelectrolyte nanoshells on individual carbon nanotube templates. *Langmuir* **2004**, *20*, 1442–1448.

22. Cheung, K. M.; Bloor, D.; Stevens, G. C. The influence of unusual counterions on the electrochemistry and physical properties of polypyrrole. *J. Mater. Sci.* **1990**, *25*, 3814–3837.

23. Bakeev, K. N.; MacKnight, W. J. Fluorescence as a probe for examining salt group association of sulfonated polystyrene ionomers in tetrahydrofuran. *Macromolecules* **1991**, *24*, 4578–4582.

24. Dong, X.; Su, C.; Zhang, W.; Zhao, J.; Ling, Q.; Huang, W.; Chen, P.; Li, L. Ultra-large single-layer graphene obtained from solution chemical reduction and its electrical properties. *Phys. Chem. Chem. Phys.* **2010**, *12*, 2164–2169.

25. Parviz, D.; Das, S.; Ahmed, H. S. T.; Irin, F.; Bhattacharia, S.; Green, M. J. Dispersions of non-covalently functionalized graphene with minimal stabilizer. *ACS Nano* **2012**, *6*, 8857–8867.

26. Zhang, M.; Gao, B.; Pu, K.; Yao, Y.; Inyang, M. Graphene-mediated self-assembly of zeolite-based microcapsules. *Chem. Eng. J.* **2013**, *223*, 556–562.

27. Okahata, Y.; Hachiya, S.; Ariga, K.; Seki, T. Functional capsule membranes. Part 22. The electrical breakdown and permeability control of a bilayer-corked capsule membrane in an external electric field. *J. Am. Chem. Soc.* **1986**, *108*, 2863–2869.

28. Sivakumar, A.; Shah, J.; Rao, N. M.; Budd, S. S. Fluorescent tracer in a water treatment process. U.S. Patent 5413719, 1995.

29. Koehler, B.; Heywang, G.; Zirngiebl, E. Poly(arylene sulfide) blends with decreased radical formation in UV weathering. Ger. Offen. DE 3827644, 1990.

30. Meade, R. J.; Grier, J. C.; Strand, M. A.; Begala, A. J. Use of fluorescence in pulp or papermaking process control. PCT Int. Appl. WO 9951817, 1999.

31. Ito, K.; Chuang, J.; Alvarez-Lorenzo, C.; Watanabe, T.; Ando, N.; Grosberg, A. Y. Multiple-contact adsorption of target molecules by heteropolymer gels *Macromol. Symp.* **2004**, *207*, 1–16.

32. Watanabe, T.; Ito, K.; Alvarez-Lorenzo, C.; Grosberg, A. Y.; Tanaka, T. Salt effects on multiple-point adsorption of target molecules by hetero-polymer gel. *J. Chem. Phys.* **2001**, *115*, 1596–1600.

33. Alvarez-Lorenzo, C.; Hiratani, H.; Tanaka, K.; Stancil, K.; Grosberg, A. Y.; Tanaka, T. Simultaneous multiple-point adsorption of aluminum ions and charged molecules by a polyampholyte thermosensitive gel: Controlling frustrations in a heteropolymer gel. *Langmuir* **2001**, *17*, 3616–3622.

34. Hoots, J. E.; Davis, B. H. Method of tracing corrosive materials. U.S. Pat. Appl. Publ. US 20080160626, 2008.

35. Bekiari, V.; Lianos, P. Ureasil gels as a highly efficient adsorbent for water purification. *Chem. Mater.* **2006**, *18*, 4142–4146.

N-(1-PYRENYL)IODOACETAMIDE

CAS Registry Number   76936-87-3

Chemical Structure

CA Index Name   Acetamide, 2-iodo-*N*-1-pyrenyl-

Other Names   *N*-(1-Pyrenyl)iodoacetamide; NPIA; PIA; PIAA

Merck Index Number   Not listed

Chemical/Dye Class   Pyrene

Molecular Formula $C_{18}H_{12}INO$

Molecular Weight   385.20

Physical Form   Yellow–green solid

Solubility   Soluble in *N,N*-dimethylformamide, dimethyl sulfoxide, methanol

Boiling Point (Calcd.) $587.1 \pm 33.0\,°C$, pressure: 760 Torr

pK$_a$ (Calcd.) 12.90 ± 0.30, most acidic, temperature: 25 °C; -0.01 ± 0.70, most basic, temperature: 25 °C

Absorption (λ_{max})   339 nm (MeOH)

Emission (λ_{max})   384 nm (MeOH)

Molar Extinction Coefficient $26{,}000\,cm^{-1}\,M^{-1}$ (MeOH)

Synthesis   Synthetic method[1]

Imaging/Labeling Applications   Actin;[2–10] RNA polymerase;[11] gelsolin;[12] histone H4;[13] α-lipoic acid;[14] myosin;[15] nucleic acid;[16] oxytocin;[17] proteins;[18–22] spectrin;[23] tropomyosins;[24–26] thiols[1–26]

Biological/Medical Applications   Analyzing α-lipoic acid;[14] characterizing/monitoring spectrin structural changes;[23] determing oxytocin in human urine;[17] monitoring association-dissociation equilibria of protein subunits transitions;[20] nucleic acid hybridization probe;[16] studying local/global tropomyosin unfolding[24]

Industrial Applications   Not reported

Safety/Toxicity   No data available

REFERENCES

1. Haugland, R. P. *Handbook of Fluorescent Probes and Research Chemicals*; Molecular Probes Inc.: Eugene, **1996**; pp 55–58.

2. Wazawa, T.; Sagawa, T.; Ogawa, T.; Morimoto, N.; Kodama, T.; Suzuki, M. Hyper-mobility of water around actin filaments revealed using pulse-field gradient spin-echo 1H NMR and fluorescence spectroscopy. *Biochem. Biophys. Res. Commun.* **2011**, *404*, 985–990.

3. Ooi, A.; Yano, F.; Okagaki, T. Thermal stability of carp G-actin monitored by loss of polymerization activity using an extrinsic fluorescent probe. *Fish. Sci.* **2008**, *74*, 193–199.

4. Ikkai, T.; Arii, T.; Shimada, K. Excimer fluorescence as a tool for monitoring protein domain dynamics applied to actin conformation changes based on circulatory polarized fluorescence spectroscopy. *J. Fluoresc.* **2006**, *16*, 367–374.

5. Sasaki, Y.; Tsunomori, F.; Yamashita, T.; Horie, K.; Ushiki, H.; Ishikawa, R.; Kohama, K. Local environmental change from the G- to F-form of

the actin molecule detected on anisotropy decay measurement. *J. Biochem.* **1994**, *116*, 236–238.

6. Criddle, A. H.; Geeves, M. A.; Jeffries, T. The use of actin labeled with N-(1-pyrenyl)iodoacetamide to study the interaction of actin with myosin subfragments and troponin/tropomyosin. *Biochem. J.* **1985**, *232*, 343–349.

7. Grazi, E. Polymerization of N-(1-pyrenyl) iodoacetamide-labeled actin: the fluorescence signal is not directly proportional to the incorporation of the monomer into the polymer. *Biochem. Biophys. Res. Commun.* **1985**, *128*, 1058–1063.

8. Cooper, J. A.; Walker, S. B.; Pollard, T. D. Pyrene actin: documentation of the validity of a sensitive assay for actin polymerization. *J. Muscle Res. Cell Motil.* **1983**, *4*, 253–262.

9. Lin, T. I.; Dowben, R. M. Fluorescence spectroscopic studies of pyrene-actin adducts. *Biophys. Chem.* **1982**, *15*, 289–298.

10. Kouyama, T.; Mihashi, K. Fluorimetry study of N-(1-pyrenyl)iodoacetamide-labeled F-actin. Local structural change of actin protomer both on

polymerization and on binding of heavy meromyosin. *Eur. J. Biochem.* **1981**, *114*, 33–38.

11. Johnson, R. S.; Bowers, M.; Eaton, Q. Preparation and characterization of N-(1-pyrenyl) iodoacetamide-labeled *Escherichia coli* RNA polymerase. *Biochemistry* **1991**, *30*, 189–198.

12. Silva, B. E. R.; Koepf, E. K.; Burtnick, L. D. Monomer and excimer fluorescence of horse plasma gelsolin labeled with N-(1-pyrenyl)iodoacetamide. *Biochem. Cell Biol.* **1992**, *70*, 573–578.

13. Chung, D. G.; Lewis, P. N. Intermolecular histone H4 interactions in core nucleosomes. *Biochemistry* **1986**, *25*, 2048–2054.

14. Inoue, T.; Sudo, M.; Yoshida, H.; Todoroki, K.; Nohta, H.; Yamaguchi, M. Liquid chromatographic determination of polythiols based on pre-column excimer fluorescence derivatization and its application to α-lipoic acid analysis. *J. Chromatogr. A* **2009**, *1216*, 7564–7569.

15. Ikkai, T.; Mihashi, K. The excimer fluorescence of N-(1-pyrenyl)iodoacetamide labeled to myosin and its subfragment 1. *FEBS Lett.* **1986**, *207*, 177–180.

16. Masuko, M.; Ohtani, H.; Ebata, K.; Shimadzu, A. Optimization of excimer-forming two-probe nucleic acid hybridization method with pyrene as a fluorophore. *Nucleic Acids Res.* **1998**, *26*, 5409–5416.

17. Inoue, T.; Umemura, H.; Yoshida, H.; Todoroki, K.; Nohta, H.; Yamaguchi, M. Sensitive determination of oxytocin-related compounds by liquid chromatography with pre-column excimer fluorescence derivatization. *Chromatography* **2009**, *30*(*Suppl. 1*), 17–18.

18. De Lorimier, R. M.; Smith, J. J.; Dwyer, M. A.; Looger, L. L.; Sali, K. M.; Paavola, C. D.; Rizk, S. S.; Sadigov, S.; Conrad, D. W.; Loew, L. Construction of a fluorescent biosensor family. *Protein Sci.* **2002**, *11*, 2655–2675.

19. Lehrer, S. S. Intramolecular pyrene excimer fluorescence: a probe of proximity and protein conformational change. *Methods Enzymol.* **1997**, *278*, 286–295.

20. Ikkai, T.; Kondo, H. Excimer fluorescence to monitor transitions of the association-dissociation equilibria induced by dilution of proteins composed of subunits. *J. Biochem. Biophys. Methods* **1996**, *33*, 55–58.

21. Clark, I. D.; MacManus, J. P.; Banville, D.; Szabo, A. G. A study of sensitized lanthanide luminescence in an engineered calcium-binding protein. *Anal. Biochem.* **1993**, *210*, 1–6.

22. Mani, R. S.; Kay, C. M. A fluorimetry study of N-(1-pyrenyl)iodoacetamide-labeled bovine brain S-100a protein. *FEBS Lett.* **1985**, *181*, 275–280.

23. Yamaguchi, T.; Furukawa, Y.; Terada, S. Excimer fluorescence of N-(1-pyrenyl)iodoacetamide-labeled spectrin. *Bull. Chem. Soc. Jpn.* **1999**, *72*, 2509–2513.

24. Ishii, Y. The local and global unfolding of coiled-coil tropomyosin. *Eur. J. Biochem.* **1994**, *221*, 705–712.

25. Burtnick, L. D.; Sanders, C.; Smillie, L. B. Fluorescence from pyrene-labeled native and reconstituted chicken gizzard tropomyosins. *Arch. Biochem. Biophys.* **1988**, *266*, 622–627.

26. Lin, T. I. Excimer fluorescence of pyrene-tropomyosin adducts. *Biophys. Chem.* **1982**, *15*, 277–288.

N-(1-PYRENYL)MALEIMIDE

CAS Registry Number   42189-56-0

Chemical Structure

CA Index Name   1*H*-Pyrrole-2,5-dione, 1-(1-pyrenyl)-

Other Names   *N*-(1-Pyrenyl)maleimide;   MPy; MalPy; Mal-pyrene*;* NPM; *N*-(1-Pyrene)maleimide*; N*-(3-pyrene)maleimide; *N*-(3-pyrenyl)maleimide; PM; PMI; pyrene-TM

Merck Index Number   Not listed

Chemical/Dye Class   Pyrene

Molecular Formula $C_{20}H_{11}NO_2$

Molecular Weight   297.31

Physical Form   Yellow needles;[2] Gold needles[1]

Solubility   Soluble in chloroform, *N,N*-dimethylformamide, dimethyl sulfoxide, ethanol, methanol

Melting Point   239–240 °C;[2] 235–237 °C; 223–225 °C[1]

Boiling Point (Calcd.)   526.8 ± 19.0 °C, pressure: 760 Torr

pK$_a$ (Calcd.)   −1.21 ± 0.20, most basic, temperature: 25 °C

Absorption (λ_{max})   338 nm, 325 nm (MeOH)

Emission (λ_{max})   375 nm, 396 (MeOH)

Molar Extinction Coefficient   40,000 cm^{-1} M^{-1} (MeOH)

Synthesis   Synthetic methods[1–4]

Imaging/Labeling Applications   Nicotinic acetylcholine receptor (AChR);[5–7] actin;[8–11] antibodies;[12,13] cells;[14,37] chloroplast coupling factor;[15,16] cellobiohydrolase I (CBH I);[17] dehydrogenases;[18–23] luciferase;[24] 4-aminobutyrate aminotransferase;[25] glutathione S-transferase;[26] histone acetyltransferases (HATs);[27] phosphotransferase;[28] immunoglobulin;[29,30] myosin;[31] nucleic acids;[3,35,41] profilin;[32] proteins;[1,3,33–42] ribosomes;[43,44] radio-fluorogenic gel;[67] sarcoplasmic reticulum ATPase;[45–51] α-synuclein;[52] tropomyosins;[53–56] tubulin;[57,58] thiols[1–60,65]

Biological/Medical Applications   Analyzing molecular aspects of muscle contraction;[31] detecting nucleic acids,[3,35,41] proteins;[1,3,33–42] determining/evaluating cell activity;[14,37] diagnosing/treating Parkinson's disease;[52] monitoring/studying ribosomal structure and conformation;[43,44] studying nicotinic acetylcholine receptor (AChR) structure;[5] sulfhydryl reagent[1–60,65]

Industrial Applications   Dendrimers;[61,62] dye functionalized single-wall carbon nanotube (SWNT) conjugated polymer devices;[63,64] photovoltaic cells;[63,64] polymer-coated gold electrodes;[65] fluorescent polymers;[66] radio-fluorogenic gel;[67] in sequence-controlled polymerizations;[68] rubber[69]

Safety/Toxicity   Cytotoxicity;[70] peroxidative damage[71]

REFERENCES

1. Weltman, J. K.; Szaro, R. P.; Frackelton, A. R., Jr.,; Dowben, R. M.; Bunting, J. R.; Cathou, R. E. N-(3-pyrene)maleimide, a long, lifetime fluorescent sulfhydryl reagent. *J. Biol. Chem.* **1973**, *248*, 3173–3177.

2. Reddy, P. Y.; Kondo, S.; Fujita, S.; Toru, T. Efficient synthesis of fluorophore-linked maleimide derivatives. *Synthesis* **1998**, 999–1002.

3. Fujita, S.; Reddi, P. J.; Toru, T. Preparation of polycyclic amic acids and maleimides. Jpn. Kokai Tokkyo Koho JP 10330337, 1998.

4. Fujita, S.; Reddi, P. J.; Toru, T. Efficient method for the synthesis of *N*-cyclic maleamic acids and *N*-cyclic maleimides. U.S. Patent 5965746, 1999.

5. Kim, J.; McNamee, M. G. Topological disposition of Cys 222 in the α-subunit of nicotinic acetylcholine receptor analyzed by fluorescence-quenching and electron paramagnetic resonance measurements. *Biochemistry* **1998**, *37*, 4680–4686.

6. Narayanaswami, V.; Kim, J.; McNamee, M. G. Protein-lipid interactions and *Torpedo californica* nicotinic acetylcholine receptor function. 1. Spatial disposition of cysteine residues in the γ subunit analyzed by fluorescence-quenching and

energy-transfer measurements. *Biochemistry* **1993**, *32*, 12413–12419.

7. Clarke, J. H.; Martinez-Carrion, M. Labeling of functionally sensitive sulfhydryl-containing domains of acetylcholine receptor from *Torpedo californica* membranes. *J. Biol. Chem.* **1986**, *261*, 10063–10072.

8. Pengelly, K.; Loncar, A.; Perieteanu, A. A.; Dawson, J. F. Cysteine engineering of actin self-assembly interfaces. *Biochem. Cell Biol.* **2009**, *87*, 663–675.

9. Chen, W.; Wen, K.; Sens, A. E.; Rubenstein, P. A. Differential interaction of cardiac, skeletal muscle, and yeast tropomyosins with fluorescent (pyrene235) yeast actin. *Biophys. J.* **2006**, *90*, 1308–1318.

10. Lin, T. I.; Dowben, R. M. Fluorescence spectroscopic studies of pyrene-actin adducts. *Biophys. Chem.* **1982**, *15*, 289–298.

11. Kawasaki, Y.; Mihashi, K.; Tanaka, H.; Ohnuma, H. Fluorescence study of N-(3-pyrene)maleimide conjugated to rabbit skeletal F-actin and plasmodium actin polymers. *Biochim. Biophys. Acta, Protein Struct.* **1976**, *446*, 166–178.

12. Lacy, E. R.; Baker, M.; Brigham-Burke, M. Free sulfhydryl measurement as an indicator of antibody stability. *Anal. Biochem.* **2008**, *382*, 66–68.

13. Konishi, S.; Imai, T.; Wakabayashi, G.; Kishioka, H. Novel method for manufacture of fluorescent-labeled antibodies for immunoassays. Jpn. Kokai Tokkyo Koho JP 01032170, 1989.

14. Oonishi, T.; Takahashi, T. Maleimide derivative for determining cell activity. Jpn. Kokai Tokkyo Koho JP 06319594, 1994.

15. Snyder, B.; Hammes, G. G. Structural organization of chloroplast coupling factor. *Biochemistry* **1985**, *24*, 2324–2331.

16. Holowka, D. A.; Hammes, G. G. Chemical modification and fluorescence studies of chloroplast coupling factor. *Biochemistry* **1977**, *16*, 5538–5545.

17. Woodward, J.; Tate, J.; Herrmann, P. C.; Evans, B. R. Comparison of Ellman's reagent with N-(1-pyrenyl)maleimide for the determination of free sulfhydryl groups in reduced cellobiohydrolase I from *Trichoderma reesei*. *J. Biochem. Biophys. Methods* **1993**, *26*, 121–129.

18. Santra, M. K.; Dasgupta, D.; Panda, D. Pyrene excimer fluorescence of yeast alcohol dehydrogenase: a sensitive probe to investigate ligand binding and unfolding pathway of the enzyme. *Photochem. Photobiol.* **2006**, *82*, 480–486.

19. Dallocchio, F.; Matteuzzi, M.; Bellini, T. Evidence for the proximity of a cysteine and a lysine residue in the active site of 6-phosphogluconate dehydrogenase from *Candida utilis*. *Ital. J. Biochem.* **1983**, *32*, 124–130.

20. Waskiewicz, D. E.; Hammes, G. G. Fluorescence polarization study of the α-ketoglutarate dehydrogenase complex from *Escherichia coli*. *Biochemistry* **1982**, *21*, 6489–6496.

21. Angelides, K. J.; Hammes, G. G. Fluorescence studies of the pyruvate dehydrogenase multienzyme complex from *Escherichia coli*. *Biochemistry* **1979**, *18*, 1223–1229.

22. Scouten, W. H.; De Graaf-Hess, A. C.; De Kok, A.; Grande, H. J.; Visser, A. J. W. G.; Veeger, C. Fluorescence energy-transfer studies on the pyruvate dehydrogenase complex isolated from *Azotobacter vinelandii*. *Eur. J. Biochem.* **1978**, *84*, 17–25.

23. Shepherd, G. B.; Hammes, G. G. Fluorescence energy transfer measurements in the pyruvate dehydrogenase multienzyme complex from *Escherichia coli* with chemically modified lipoic acid. *Biochemistry* **1977**, *16*, 5234–5240.

24. Tu, S.; Wu, C.; Hastings, J. W. Structural studies on bacterial luciferase using energy transfer and emission anisotropy. *Biochemistry* **1978**, *17*, 987–993.

25. Choi, S. Y.; Churchich, J. E. 4-Aminobutyrate aminotransferase reaction of sulfhydryl residues connected with catalytic activity. *J. Biol. Chem.* **1985**, *260*, 993–997.

26. Piemonte, F.; Caccuri, A. M.; Morgenstern, R.; Rosato, N.; Federici, G. Aggregation of pyrene-labeled microsomal glutathione S-transferase. Effect of concentration. *Eur. J. Biochem.* **1993**, *217*, 661–663.

27. Gao, T.; Yang, C.; Zheng, Y. G. Comparative studies of thiol-sensitive fluorogenic probes for HAT assays. *Anal. Bioanal. Chem.* **2013**, *405*, 1361–1371.

28. Han, M. K.; Roseman, S.; Brand, L. Sugar transport by the bacterial phosphotransferase system. Characterization of the sulfhydryl groups and site-specific labeling of enzyme I. *J. Biol. Chem.* **1990**, *265*, 1985–1995.

29. Liburdy, R. P. Rabbit immunoglobulin-N-(3-pyrene)-maleimide conjugate for fluorescent immunoassay. U.S. Patent 4207075, 1980.

30. Liburdy, R. P. Antibody induced fluorescence enhancement of an N-(3-pyrene)maleimide conjugate of rabbit anti-human immunoglobulin G: quantitation of human IgG. *J. Immunol. Methods* **1979**, *28*, 233–242.

31. Liburdy, R. P.; Weltman, J. K. Preparation and characterization of fluorescent

N-(3-pyrene)maleimide adducts of myosin. *J. Mechanochem. Cell Motil.* **1976**, *3*, 229–234.

32. Janmey, P. A. Polyproline affinity method for purification of platelet profilin and modification with pyrene-maleimide. *Methods Enzymol.* **1991**, *196*, 92–99.

33. Ruan, Q.; Chen, Y.; Kong, X.; Hua, Y. Comparative studies on sulfhydryl determination of soy protein using two aromatic disulfide reagents and two fluorescent reagents. *J. Agric. Food Chem.* **2013**, *61*, 2661–2668.

34. Jain, N.; Bhattacharya, M.; Mukhopadhyay, S. Chain collapse of an amyloidogenic intrinsically disordered protein. *Biophys. J.* **2011**, *101*, 1720–1729.

35. Fujita, S.; Kagiyama, N.; Kondo, Y.; Reddi, P. J.; Toru, T. Detection method for nucleic acid and protein. Jpn. Kokai Tokkyo Koho JP 10332695, 1998.

36. Karim, A. S.; Johansson, C. S.; Weltman, J. K. Maleimide-mediated protein conjugates of a nucleoside triphosphate gamma-S and an internucleotide phosphorothioate diester. *Nucleic Acids Res.* **1995**, *23*, 2037–2040.

37. Benci, S.; Bottiroli, G.; Schianchi, G.; Vaccari, S.; Vaghi, P. Luminescence and anisotropy decays of N-3-pyrene maleimide labeling IgG proteins and cells. *J. Fluoresc.* **1993**, *3*, 223–227.

38. Karim, A. S.; Weltman, J. K. Formation of protein conjugates of phosphorothioate nucleoside diphosphate beta-S. *Nucleic Acids Res.* **1993**, *21*, 5281–5282.

39. Kinsland, L. N.; Wiechelman, K. J. Use of β-cyclodextrin fluorophore complexes to improve the efficiency of fluorescent label incorporation into proteins. *J. Biochem. Biophys. Methods* **1984**, *9*, 81–83.

40. Lux, B.; Gerard, D. Reappraisal of the binding processes of N-(3-pyrene)maleimide as a fluorescent probe of proteins. *J. Biol. Chem.* **1981**, *256*, 1767–1771.

41. Brown, R. D.; Matthews, K. S. Chemical modification of lactose repressor protein using N-substituted maleimides. *J. Biol. Chem.* **1979**, *254*, 5128–5134.

42. Wu, C.; Yarbrough, L. R.; Wu, F. Y. H. N-(1-Pyrene)maleimide: a fluorescent crosslinking reagent. *Biochemistry* **1976**, *15*, 2863–2868.

43. Spitnik-Elson, P.; Schechter, N.; Abramovitz, R.; Elson, D. Conformational changes of 30S ribosomes measured by intrinsic and extrinsic fluorescence. *Biochemistry* **1976**, *15*, 5246–5253.

44. Lee, T.; Heintz, R. L. Reaction of N-(3-pyrene)maleimide with thiol groups of reticulocyte ribosomes. *Eur. J. Biochem.* **1976**, *66*, 105–114.

45. Suzuki, T.; Kawakita, M. Sites of labeling with N-(3-pyrene)maleimide on Ca2+−transporting ATPase of the sarcoplasmic reticulum. *J. Biochem.* **1995**, *117*, 881–887.

46. Suzuki, T.; Kawakita, M. Uncoupling of ATP splitting from calcium-transport in calcium-transporting ATPase of the sarcoplasmic reticulum as a result of modification by N-(3-pyrene)maleimide: Activation of a channel with a specificity for alkaline earth metal ions. *J. Biochem.* **1993**, *114*, 203–209.

47. Papp, S.; Kracke, G.; Joshi, N.; Martonosi, A. The reaction of N-(1-pyrene)maleimide with sarcoplasmic reticulum. *Biophys. J.* **1986**, *49*, 411–424.

48. Papp, S.; Rutzke, M.; Martonosi, A. The effect of chelating agents on the elemental composition of sarcoplasmic reticulum: the reactivity of SH groups with N-(1-pyrene)maleimide. *Arch. Biochem. Biophys.* **1985**, *243*, 254–263.

49. Gupte, S. S.; Lane, L. K. Reaction of sodium-potassium ATPase with fluorescent maleimide derivatives. Probes for studying ATP site(s) function. *J. Biol. Chem.* **1983**, *258*, 5005–5012.

50. Prokop'eva, V. D.; Lopina, O. D.; Boldyrev, A. A. Effect of temperature on the fluorescence of labels and probes at various localization in sarcoplasmic reticulum preparations. *Biofizika* **1983**, *28*, 40–44.

51. Luedi, H.; Hasselbach, W. Excimer formation of ATPase from sarcoplasmic reticulum labeled with N-(3-pyrene)maleinimide. *Eur. J. Biochem.* **1983**, *130*, 5–8.

52. Thirunavukkuarasu, S.; Jares-Erijman, E. A.; Jovin, T. M. Multiparametric fluorescence detection of early stages in the amyloid protein aggregation of pyrene-labeled α-synuclein. *J. Mol. Biol.* **2008**, *378*, 1064–1073.

53. Burtnick, L. D.; Sanders, C.; Smillie, L. B. Fluorescence from pyrene-labeled native and reconstituted chicken gizzard tropomyosins. *Arch. Biochem. Biophys.* **1988**, *266*, 622–627.

54. Ishii, Y.; Lehrer, S. S. Effects of the state of the succinimido-ring on the fluorescence and structural properties of pyrene maleimide-labeled αα-tropomyosin. *Biophys. J.* **1986**, *50*, 75–80.

55. Ishii, Y.; Lehrer, S. S. Fluorescence studies of the conformation of pyrene-labeled tropomyosin: effects

of F-actin and myosin subfragment. *Biochemistry* **1985**, *24*, 6631–6638.

56. Betcher-Lange, S. L.; Lehrer, S. S. Pyrene excimer fluorescence in rabbit skeletal αα-tropomyosin labeled with N-(1-pyrene)maleimide. A probe of sulfhydryl proximity and local chain separation. *J. Biol. Chem.* **1978**, *253*, 3757–3760.

57. Basusarkar, P.; Chandra, S.; Bhattacharyya, B. The colchicine-binding and pyrene-excimer-formation activities of tubulin involve a common cysteine residue in the β subunit. *Eur. J. Biochem.* **1997**, *244*, 378–383.

58. Panda, D.; Bhattacharyya, B. Excimer fluorescence of pyrene-maleimide-labeled tubulin. *Eur. J. Biochem.* **1992**, *204*, 783–787.

59. Ridnour, L. A.; Winters, R. A.; Ercal, N.; Spitz, D. R. Measurement of glutathione, glutathione disulfide, and other thiols in mammalian cell and tissue homogenates using high-performance liquid chromatography separation of N-(1-pyrenyl)maleimide derivatives. *Methods Enzymol.* **1999**, *299*, 258–267.

60. Parmentier, C.; Leroy, P.; Wellman, M.; Nicolas, A. Determination of cellular thiols and glutathione-related enzyme activities: versatility of high-performance liquid chromatography-spectrofluorimetric detection. *J. Chromatogr., B: Biomed. Sci. Appl.* **1998**, *719*, 37–46.

61. Wang, B.; Zhang, X.; Yang, L.; Jia, X.; Ji, Y.; Li, W.; Wei, Y. Poly(amidoamine) dendrimers bearing electron-donating chromophores: Fluorescence and electrochemical properties. *Polym. Bull.* **2006**, *56*, 63–74.

62. Wang, B.; Zhang, X.; Jia, X.; Li, Z.; Ji, Y.; Yang, L.; Wei, Y. Fluorescence and aggregation behavior of poly(amidoamine) dendrimers peripherally modified with aromatic chromophores: the effect of dendritic architectures. *J. Am. Chem. Soc.* **2004**, *126*, 15180–15194.

63. Kymakis, E.; Bhattacharyya, S.; Amaratunga, G. A. J. Photovoltaic cells based on dye functionalized single-wall carbon nanotubes. *Polym. Prepr.* **2005**, *46*, 213–214.

64. Bhattacharyya, S.; Kymakis, E.; Amaratunga, G. A. J. Photovoltaic properties of dye functionalized single-wall carbon nanotube/conjugated polymer devices. *Chem. Mater.* **2004**, *16*, 4819–4823.

65. Luo, N.; Hatchett, D. W.; Rogers, K. R. Recognition of pyrene using molecularly imprinted electrochemically deposited poly(2-mercaptobenzimidazole) or poly(resorcinol) on gold electrodes. *Electroanalysis* **2007**, *19*, 2117–2124.

66. Scales, C. W.; Convertine, A. J.; McCormick, C. L. Room-temperature polymerization of N-isopropylacrylamide via RAFT and subsequent conjugation of fluorescent labels. *Polym. Prepr.* **2005**, *46*, 393–394.

67. Warman, J. M.; de Haas, M. P.; Luthjens, L. H.; Hom, M. L. High-energy radiation monitoring based on radio-fluorogenic co-polymerization III: Fluorescent images of the cross-section and depth-dose profile of a 3 MV electronbeam. *Radiat. Phys. Chem.* **2013**, *84*, 129–135.

68. Srichan, S.; Chan-Seng, D.; Lutz, J. Influence of strong electron-donor monomers in sequence-controlled polymerizations. *ACS Macro Lett.* **2012**, *1*, 589–592.

69. Ohshima, N. Rubber composition. U.S. Pat. Appl. Publ. US 20070191533, 2007.

70. Huang, P.; Yeh, Y.; Pao, C.; Chen, C.; Wang, T. V. N-(1-Pyrenyl) maleimide inhibits telomerase activity in a cell free system and induces apoptosis in Jurkat cells. *Mol. Biol. Rep.* **2012**, *39*, 8899–8905.

71. Ohyashiki, T.; Sakata, N.; Kamata, K.; Matsui, K. A study on peroxidative damage of the porcine intestinal brush-border membranes using a fluorogenic thiol reagent, N-(1-pyrene)maleimide. *Biochim. Biophys. Acta* **1991**, *1067*, 159–165.

QSY 7 CARBOXYLIC ACID SUCCINIMIDYL ESTER

CAS Registry Number   304014-12-8
Chemical Structure

CA Index Name   Xanthylium, 9-[2-[[4-[[(2,5-dioxo-1-pyrrolidinyl)oxy]-carbonyl]-1-piperidinyl]sulfonyl]phenyl]-3,6-bis(methylphenylamino)-, chloride (1:1)

Other Names   Xanthylium, 9-[2-[[4-[[(2,5-dioxo-1-pyrrolidinyl)oxy]-carbonyl]-1-piperidinyl]sulfonyl]phenyl]-3,6-bis(methylphenylamino)-, chloride; QSY 7; QSY 7 carboxylic acid succinimidyl ester; QSY 7 succinimidyl ester; QSY 7NHS; QSY 7SE

Merck Index Number   Not listed

Chemical/Dye Class   Xanthene

Molecular Formula $C_{43}H_{39}ClN_4O_7S$

Molecular Weight   791.32

Physical Form   Purple solid

Solubility   Soluble in dimethyl sulfoxide, methanol

Absorption (λ_{max})   560 nm (MeOH)

Emission (λ_{max})   None

Molar Extinction Coefficient   90,000 cm^{-1} M^{-1} (MeOH)

Quantum Yield   0.0009 (MeOH)

Synthesis   Synthetic methods[1–4]

Imaging/Labeling Applications   Nucleic acids;[1–10] cells;[11] oligonucleotides;[1–4,20] avidin;[12,13] trastuzumab;[12] bacteria;[14] spores;[14] cathepsin;[15–17] G-protein-coupled receptors (GPCRs);[18] microorganisms;[19] proteins;[11,21,22] polynucleotides;[23] tumors[12,15–17]

Biological/Medical Applications   Fluorescence quencher;[1–4,11–34] detecting nucleic acids,[1–10] microorganisms,[19] membrane protein internalization,[21,22] analytes;[33] imaging/detecting/identifying cancer;[12,15–17] diagnosing active wound infection;[27] assays for measuring/detecting/imaging nuclease activity,[3,4] ligase activity,[3,4] topoisomerase activity,[3,4] phospholipase activity,[3,4] clostridial toxin activity,[25,26] human neutrophil elastase activity,[27] protease activity,[15–17,25,26,30,31] peptidylarginine deiminase (PAD) activity,[28] phosphoinositide kinase activity,[29] phosphatase activity,[29] gamma-secretase activity;[32] nucleic acid sequencing[34]

Industrial Applications   Not reported

Safety/Toxicity   No data available

Handbook of Fluorescent Dyes and Probes, First Edition. R. W. Sabnis.
© 2015 John Wiley & Sons, Inc. Published 2015 by John Wiley & Sons, Inc.

REFERENCES

1. Sabnis, R. W. *Handbook of Biological Dyes and Stains*; John Wiley & Sons Inc.: Hoboken, **2010**; pp 393–395.

2. Singer, V. L.; Haugland, R. P. Quenching oligonucleotides. U.S. Patent 6323337, 2001.

3. Haugland, R. P.; Singer, V. L.; Yue, S. T. Xanthene dyes and their application as luminescence quenching compounds. PCT Int. Appl. WO 2000064988, 2000.

4. Haugland, R. P.; Singer, V. L.; Yue, S. T. Xanthene dyes and their application as luminescence quenching compounds. U.S. Patent 6399392, 2002.

5. Le Reste, L.; Hohlbein, J.; Gryte, K.; Kapanidis, A. N. Characterization of dark quencher chromophores as nonfluorescent acceptors for single-molecule FRET. *Biophys. J.* **2012**, *102*, 2658–2668.

6. Sergeev, N. V.; Brevnov, M. G.; Furtado, M. R. Identification of nucleic acids. PCT Int. Appl. WO 2011143478, 2011.

7. Tan, W.; Tang, Z. Novel nucleic acid-based molecular probes. PCT Int. Appl. WO 2010011884, 2010.

8. Tao, S.; Cheng, J.; Max, X.; Zhou, Y. Lab-on-chip system for analyzing nucleic acid. PCT Int. Appl. WO 2004079002,2004.

9. Nicklas, J. A.; Buel, E. Development of an Alu-based, QSY 7-labeled primer PCR method for quantitation of human DNA in forensic samples. *J. Forensic Sci.* **2003**, *48*, 282–291.

10. Schmitt-John, T.; Palmisano, R.; Plessow, R.; Brockhinke, A.; Weidner, J. Detection of a nucleic acid by hybridization with pairs of probes labeled with dyes that interact by FRET. Ger. Offen. DE 10133308, 2003.

11. Hintersteiner, M.; Weidemann, T.; Kimmerlin, T.; Filiz, N.; Buehler, C.; Auer, M. Covalent fluorescence labeling of his-tagged proteins on the surface of living cells. *ChemBioChem* **2008**, *9*, 1391–1395.

12. Ogawa, M.; Kosaka, N.; Longmire, M. R.; Urano, Y.; Choyke, P. L.; Kobayashi, H. Fluorophore-quencher based activatable targeted optical probes for detecting *in vivo* cancer metastases. *Mol. Pharm.* **2009**, *6*, 386–395.

13. Adamczyk, M.; Moore, J. A.; Shreder, K. Quenching of biotinylated aequorin bioluminescence by dye-labeled avidin conjugates: Application to homogeneous bioluminescence resonance energy transfer assays. *Org. Lett.* **2001**, *3*, 1797–1800.

14. Bruno, J. G.; Ulvick, S. J.; Uzzell, G. L.; Tabb, J. S.; Valdes, E. R.; Batt, C. A. Novel immuno-FRET assay method for bacillus spores and *Escherichia coli* O157:H7. *Biochem. Biophys. Res. Commun.* **2001**, *287*, 875–880.

15. Lee, W. D.; Bawendi, M. G.; Ferrer, J. Methods and systems for spatially identifying abnormal cells. PCT Int. Appl. WO 2012075075, 2012.

16. Lee, W. D.; Bawendi, M. G.; Ferrer, J. Methods and systems for spatially identifying abnormal cells. U.S. Pat. Appl. Publ. US 20110104071, 2011.

17. Bogyo, M. S.; Blum, G.; Von Degenfeld, G. Imaging of protease activity in live cells using activity based probes. U.S. Pat. Appl. Publ. US 20070036725, 2007.

18. McMurchie, E. J.; Leifert, W. R. Cell free G-protein coupled receptor and ligand assay. PCT Int. Appl. WO 2005121755, 2005.

19. Beimfohr, C.; Thelen, K.; Snaidr, J. A fluorescence in situ hybridization method for detection of micro-organisms using FRET probes to increase signal:noise ratio. Ger. Offen. DE 102010012421, 2010.

20. Moreira, B. G.; You, Y.; Behlke, M. A.; Owczarzy, R. Effects of fluorescent dyes, quenchers, and dangling ends on DNA duplex stability. *Biochem. Biophys. Res. Commun.* **2005**, *327*, 473–484.

21. Zwier, J.; Poole, R.; Ansanay, H.; Fink, M.; Trinquet, E. Method for detecting membrane protein internalization. PCT Int. Appl. WO 2010012962, 2010.

22. Zwier, J.; Poole, R.; Ansanay, H.; Fink, M.; Trinquet, E. Method for detecting membrane protein internalization. Fr. Demande FR 2934684, 2010.

23. Mirkin, C. A.; Thaxton, C. S.; Giljohann, D. A.; Cutler, J. I. Crosslinked polynucleotide structure. PCT Int. Appl. WO 2011113054, 2011.

24. Vortmeyer, A. O.; Li, J. Biomarkers for assessment of the molecular quality in biospecimens. PCT Int. Appl. WO 2012170669, 2012.

25. Williams, D. J.; Gilmore, M.; Steward, L.; Verhagen, M.; Aoki, K. R. Fluorescence polarization assays for determining clostridial toxin activity. U.S. Pat. Appl. Publ. US 20060063222, 2006.

26. Steward, L. E.; Fernandez-Salas, E.; Aoki, K. R. FRET protease assays for clostridial toxins. U.S. Pat. Appl. Publ. US 20030143651, 2003.

27. Wardell, M. R. Methods for using human neutrophil elastase as an indicator of active wound infection. PCT Int. Appl. WO 2010022281, 2010.

28. Werneburg, B. G.; Brown, M. P.; Freeman, D. M.; Yingling, J. D. High throughput assay for modulators of peptidylarginine deiminase activity. PCT Int. Appl. WO 2007076302, 2007.

29. Drees, B. E.; Neilsen, P. O.; Branch, A. M.; Weipert, A.; Hudson, H. A.; Feng, L.; Prestwich, G. Assays for detection of phosphoinositide kinase and phosphatase activity. U.S. Pat. Appl. Publ. US 2005009124, 2005.

30. Blum, G.; Mullins, S. R.; Keren, K.; Fonovic, M.; Jedeszko, C.; Rice, M. J.; Sloane, B. F.; Bogyo, M. Dynamic imaging of protease activity with fluorescently quenched activity-based probes. *Nat. Chem. Biol.* **2005**, *1*, 203–209.

31. Kumaraswamy, S.; Bergstedt, T.; Shi, X.; Rininsland, F.; Kushon, S.; Xia, W.; Ley, K.; Achyuthan, K.; McBranch, D.; Whitten, D. Fluorescent-conjugated polymer superquenching facilitates highly sensitive detection of proteases. *Proc. Natl. Acad. Sci. U.S.A.* **2004**, *101*, 7511–7515.

32. Roberts, S. B.; Hendrick, J. P.; Vinitsky, A.; Lewis, M.; Smith, D. W.; Pak, R. Isolation of functionally active gamma-secretase protein complex and methods for detection of activity and inhibitors thereof. PCT Int. Appl. WO 2001075435, 2001.

33. Volland, H.; Creminon, C.; Neuburger, L. M.; Grassi, J. Apparatus and process for the continuous detection of an analyte using a trifunctional detection reagent. Fr. Demande FR 2847984, 2004.

34. Kapanidis, A. Polymerase-based single molecule sequencing. PCT Int. Appl. WO 2009056831, 2009.

QSY 9 CARBOXYLIC ACID SUCCINIMIDYL ESTER

CAS Registry Number   700834-40-8

Chemical Structure

CA Index Name   Xanthylium, 9-[2-[[4-[[(2,5-dioxo-1-pyrrolidinyl)oxy]-carbonyl]-1-piperidinyl]sulfonyl]phenyl]-3,6-bis[methyl(4-sulfophenyl)amino]-, chloride (1:1)

Other Names   Xanthylium, 9-[2-[[4-[[(2,5-dioxo-1-pyrrolidinyl)oxy]-carbonyl]-1-piperidinyl]sulfonyl]phenyl]-3,6-bis[methyl(4-sulfophenyl)amino]-, chloride; QSY 9; QSY 9 carboxylic acid succinimidyl ester

Merck Index Number   Not listed

Chemical/Dye Class   Xanthene

Molecular Formula $C_{43}H_{39}ClN_4O_{13}S_3$

Molecular Weight   951.43

Physical Form   Purple solid

Solubility   Soluble in water, dimethyl sulfoxide, methanol

Absorption (λ_{max})   562 nm (MeOH)

Emission (λ_{max})   None

Molar Extinction Coefficient   88,000 cm^{-1} M^{-1} (MeOH)

Quantum Yield   0.002 (MeOH)

Synthesis   Synthetic methods[1–4]

Imaging/Labeling Applications   Nucleic acids;[1–6] oligonucleotides;[1–4] antibody;[7] G-protein-coupled receptors (GPCRs);[8] microorganisms;[9] peptides;[10] proteins;[11–13] polynucleotides[14]

Biological/Medical Applications   Fluorescence quencher;[1–4,7,8,13–20] detecting nucleic acids,[1–6] microorganisms,[9] membrane protein internalization;[11,12] G-protein-coupled receptors (GPCRs) assay;[8] ligand assays;[7,8,13] bioaffinity assay;[18] peptide substrates;[10] measuring/detecting analytes;[18,19] nucleic acid sequencing;[20] assays for measuring/detecting nuclease activity,[3,4] ligase activity,[3,4] topoisomerase activity,[3,4] phospholipase activity,[3,4] caspase activity,[16] phosphoinositide kinase activity,[17] phosphatase activity[17]

Industrial Applications   Not reported

Safety/Toxicity   No data available

REFERENCES

1. Sabnis, R. W. *Handbook of Biological Dyes and Stains*; John Wiley & Sons Inc.: Hoboken, 2010; pp 396–397.

2. Singer, V. L.; Haugland, R. P. Quenching oligonucleotides. U.S. Patent 6323337, 2001.

3. Haugland, R. P.; Singer, V. L.; Yue, S. T. Xanthene dyes and their application as luminescence quenching compounds. PCT Int. Appl. WO 2000064988, 2000.

4. Haugland, R. P.; Singer, V. L.; Yue, S. T. Xanthene dyes and their application as luminescence

quenching compounds. U.S. Patent 6399392, 2002.

5. Sergeev, N. V.; Brevnov, M. G.; Furtado, M. R. Identification of nucleic acids. PCT Int. Appl. WO 2011143478, 2011.

6. Chiuman, W.; Li, Y. Efficient signaling platforms built from a small catalytic DNA and doubly labeled fluorogenic substrates. *Nucleic Acids Res.* **2007**, *35*, 401–405.

7. Beechem, J.; Hagen, D.; Johnson, I. Antibody complexes and methods for immunolabeling. U.S. Pat. Appl. Publ. US 2007269902, 2007.

8. McMurchie, E. J.; Leifert, W. R. Cell free G-protein coupled receptor and ligand assay. PCT Int. Appl. WO 2005121755, 2005.

9. Beimfohr, C.; Thelen, K.; Snaidr, J. A fluorescence in situ hybridization method for detection of microorganisms using FRET probes to increase signal:noise ratio. Ger. Offen. DE 102010012421, 2010.

10. Hills, R.; Mazzarella, R.; Fok, K.; Liu, M.; Nemirovskiy, O.; Leone, J.; Zack, M. D.; Arner, E. C.; Viswanathan, M.; Abujoub, A. Identification of an ADAMTS-4 cleavage motif using phage display leads to the development of fluorogenic peptide substrates and reveals matrilin-3 as a novel substrate. *J. Biol. Chem.* **2007**, *282*, 11101–11109.

11. Zwier, J.; Poole, R.; Ansanay, H.; Fink, M.; Trinquet, E. Method for detecting membrane protein internalization. PCT Int. Appl. WO 2010012962, 2010.

12. Zwier, J.; Poole, R.; Ansanay, H.; Fink, M.; Trinquet, E. Method for detecting membrane protein internalization. Fr. Demande FR 2934684, 2010.

13. Beechem, J.; Gee, K.; Hagen, D.; Johnson, I.; Kang, H. C.; Pastula, C. Competitive immunoassay. PCT Int. Appl. WO 2005050206, 2005.

14. Mirkin, C. A.; Thaxton, C. S.; Giljohann, D. A.; Cutler, J. I. Crosslinked polynucleotide structure. PCT Int. Appl. WO 2011113054, 2011.

15. Vortmeyer, A. O.; Li, J. Biomarkers for assessment of the molecular quality in biospecimens. PCT Int. Appl. WO 2012170669, 2012.

16. Kindermann, M.; Miniejew, C.; Wendt, K. Caspase imaging probes. PCT Int. Appl. WO 2009019115, 2009.

17. Drees, B. E.; Neilsen, P. O.; Branch, A. M.; Weipert, A.; Hudson, H. A.; Feng, L.; Prestwich, G. Assays for detection of phosphoinositide kinase and phosphatase activity. U.S. Pat. Appl. Publ. US 2005009124, 2005.

18. Kokko, T.; Kokko, L.; Soukka, T.; Loevgren, T. Homogeneous non-competitive bioaffinity assay based on fluorescence resonance energy transfer. *Anal. Chim. Acta* **2007**, *585*, 120–125.

19. Volland, H.; Creminon, C.; Neuburger, L. M.; Grassi, J. Apparatus and process for the continuous detection of an analyte using a trifunctional detection reagent. Fr. Demande FR 2847984, 2004.

20. Kapanidis, A. Polymerase-based single molecule sequencing. PCT Int. Appl. WO 2009056831, 2009.

QSY 21 CARBOXYLIC ACID SUCCINIMIDYL ESTER

CAS Registry Number   304014-13-9

Chemical Structure

CA Index Name   Xanthylium, 3,6-bis(2,3-dihydro-1*H*-indol-1-yl)-9-[2-[[4-[[(2,5-dioxo-1-pyrrolidinyl)oxy]carbonyl]-1-piperidinyl]-sulfonyl]phenyl]-,   chloride (1:1)

Other Names   Xanthylium, 3,6-bis(2,3-dihydro-1*H*-indol-1-yl)-9-[2-[[4-[[(2,5-dioxo-1-pyrrolidinyl)oxy]carbonyl]-1-piperidinyl]-sulfonyl]phenyl]-,   chloride; QSY 21; QSY 21 carboxylic acid succinimidyl ester; QSY 21NHS

Merck Index Number   Not listed

Chemical/Dye Class   Xanthene

Molecular Formula $C_{45}H_{39}ClN_4O_7S$

Molecular Weight   815.34

Physical Form   Purple solid

Solubility   Soluble in dimethyl sulfoxide, methanol

Absorption (λ_{max})   661 nm (MeOH)

Emission (λ_{max})   None

Molar Extinction Coefficient $90,000\,cm^{-1}\,M^{-1}$ (MeOH)

Quantum Yield   0.0002 (MeOH)

Synthesis   Synthetic methods[1-4]

Imaging/Labeling Applications   Nucleic acids;[1-8] cells;[9] oligonucleotides;[1-4] bacteria;[10,11] cathepsin;[12-15] protein-coupled receptors (GPCRs);[19] microorganisms;[20] oligopeptides;[21] proteins;[22,23] polynucleotides[24]

Biological/Medical Applications   Fluorescence quencher;[1-4,9-19,21,24-32] detecting nucleic acids,[1-8] microorganisms,[20] membrane protein internalization,[22,23] analytes;[30] detecting/imaging cells,[9] bacteria,[10,11] apoptosis;[32] imaging/detecting/identifying cancer;[12-18,27] diagnosing active wound infection;[28] assays for measuring/detecting/imaging nuclease activity,[3,4] ligase activity,[3,4] topoisomerase activity,[3,4] phospholipase activity,[3,4] β-lactamase activity,[9-11] caspase activity,[26] lysophospholipase D (LysoPLD) activity,[27] human neutrophil elastase activity,[28] saccharide activity;[29] nucleic acid sequencing;[31] as a quantum dot quencher;[33] glucose monitoring[34,35]

Industrial Applications   Not reported

Safety/Toxicity   No data available

REFERENCES

1. Sabnis, R. W. *Handbook of Biological Dyes and Stains*; John Wiley & Sons Inc.: Hoboken, **2010**; pp 398–399.

2. Singer, V. L.; Haugland, R. P. Quenching oligonucleotides. U.S. Patent 6323337, 2001.

3. Haugland, R. P.; Singer, V. L.; Yue, S. T. Xanthene dyes and their application as luminescence quenching compounds. PCT Int. Appl. WO 2000064988, 2000.

4. Haugland, R. P.; Singer, V. L.; Yue, S. T. Xanthene dyes and their application as luminescence

quenching compounds. U.S. Patent 6399392, 2002.

5. Tan, W.; Tang, Z. Novel nucleic acid-based molecular probes. PCT Int. Appl. WO 2010011884, 2010.

6. Kabelac, M.; Zimandl, F.; Fessl, T.; Chval, Z.; Lankas, F. A comparative study of the binding of QSY 21 and Rhodamine 6G fluorescence probes to DNA: structure and dynamics. *Phys. Chem. Chem. Phys.* **2010**, *12*, 9677–9684.

7. Schwartz, D. E.; Gong, P.; Shepard, K. L. Time-resolved Foerster-resonance-energy-transfer DNA assay on an active CMOS microarray. *Biosens. Bioelectron.* **2008**, *24*, 383–390.

8. Chiuman, W.; Li, Y. Efficient signaling platforms built from a small catalytic DNA and doubly labeled fluorogenic substrates. *Nucleic Acids Res.* **2007**, *35*, 401–405.

9. Xing, B.; Khanamiryan, A.; Rao, J. Cell-permeable near-infrared fluorogenic substrates for imaging β-lactamase activity. *J. Am. Chem. Soc.* **2005**, *127*, 4158–4159.

10. Kong, Y.; Yao, H.; Ren, H.; Subbian, S.; Cirillo, S. L. G.; Sacchettini, J. C.; Rao, J.; Cirillo, E. D. Imaging tuberculosis with endogenous β-lactamase reporter enzyme fluorescence in live mice. *Proc. Natl. Acad. Sci. U.S.A.* **2010**, *107*, 12239–12244

11. Cirillo, J. D.; Rao, J. Use of bacterial β-lactamase for in vitro diagnostics and in vivo imaging, diagnostics and therapeutics. PCT Int. Appl. WO 2010016911, 2010.

12. Bogyo, M. S.; Edgington, L. E.; Verdoes, M. Non-peptidic quenched fluorescent imaging probes. PCT Int. Appl. WO 2012118715, 2012.

13. Lee, W. D.; Bawendi, M. G.; Ferrer, J. Methods and systems for spatially identifying abnormal cells. PCT Int. Appl. WO 2012075075, 2012.

14. Lee, W. D.; Bawendi, M. G.; Ferrer, J. Methods and systems for spatially identifying abnormal cells. U.S. Pat. Appl. Publ. US 20110104071, 2011.

15. Blum, G.; von Degenfeld, G.; Merchant, M. J.; Blau, H. M.; Bogyo, M. Noninvasive optical imaging of cysteine protease activity using fluorescently quenched activity-based probes. *Nat. Chem. Biol.* **2007**, *3*, 668–677.

16. Wang, X.; Qian, X.; Beitler, J. J.; Chen, Z. G.; Khuri, F. R.; Lewis, M. M.; Shin, H. J. C.; Nie, S.; Shin, D. M. Detection of circulating tumor cells in human peripheral blood using surface-enhanced Raman scattering nanoparticles. *Cancer Res.* **2011**, *71*, 1526–1532.

17. Ogawa, M.; Kosaka, N.; Choyke, P. L.; Kobayashi, H. Tumor-specific detection of an optically targeted antibody combined with a quencher-conjugated neutravidin "quencher-chaser": A dual "quench and chase" strategy to improve target to nontarget ratios for molecular imaging of cancer. *Bioconjugate Chem.* **2009**, *20*, 147–154.

18. Razkin, J.; Josserand, V.; Boturyn, D.; Jin, Z.; Dumy, P.; Favrot, M.; Coll, J.; Texier, I. Activatable fluorescent probes for tumour-targeting imaging in live mice. *ChemMedChem* **2006**, *1*, 1069–1072.

19. McMurchie, E. J.; Leifert, W. R. Cell free G-protein coupled receptor and ligand assay. PCT Int. Appl. WO 2005121755, 2005.

20. Beimfohr, C.; Thelen, K.; Snaidr, J. A fluorescence in situ hybridization method for detection of microorganisms using FRET probes to increase signal:noise ratio. Ger. Offen. DE 102010012421, 2010.

21. Sadler, J. E.; Muia, J.; Weiqiang, G. Fluorogenic substrates for ADAM-TS13. U.S. Pat. Appl. Publ. US 20130023004, 2013.

22. Zwier, J.; Poole, R.; Ansanay, H.; Fink, M.; Trinquet, E. Method for detecting membrane protein internalization. PCT Int. Appl. WO 2010012962, 2010.

23. Zwier, J.; Poole, R.; Ansanay, H.; Fink, M.; Trinquet, E. Method for detecting membrane protein internalization. Fr. Demande FR 2934684, 2010.

24. Mirkin, C. A.; Thaxton, C. S.; Giljohann, D. A.; Cutler, J. I. Crosslinked polynucleotide structure. PCT Int. Appl. WO 2011113054, 2011.

25. Vortmeyer, A. O.; Li, J. Biomarkers for assessment of the molecular quality in biospecimens. PCT Int. Appl. WO 2012170669, 2012.

26. Kindermann, M.; Miniejew, C.; Wendt, K. Caspase imaging probes. PCT Int. Appl. WO 2009019115, 2009.

27. Ferguson, C.; Prestwich, G.; Madan, D. Fluorogenic assay for lysophospholipase D activity using fluorogenic lysophospholipid derivatives as substrates, and diagnostic and screening applications. U.S. Pat. Appl. Publ. US 20100260682, 2010.

28. Wardell, M. R. Methods for using human neutrophil elastase as an indicator of active wound infection. PCT Int. Appl. WO 2010022281, 2010.

29. Texier-Nogues, I.; Robert, V.; Coll, J.; Imberty, A. Saccharide fluorescent substrates, preparation method and uses thereof. PCT Int. Appl. WO 2007010145, 2007.

30. Volland, H.; Creminon, C.; Neuburger, L. M.; Grassi, J. Apparatus and process for the continuous detection of an analyte using a trifunctional detection reagent. Fr. Demande FR 2847984, 2004.

31. Kapanidis, A. Polymerase-based single molecule sequencing. PCT Int. Appl. WO 2009056831, 2009.

32. Bullok, K.; Piwnica-Worms, D. Synthesis and characterization of a small, membrane-permeant, caspase-activatable far-red fluorescent peptide for imaging apoptosis. *J. Med. Chem.* **2005**, *48*, 5404–5407.

33. Jablonski, A. E.; Kawakami, T.; Ting, A. Y.; Payne, C. K. Pyrenebutyrate leads to cellular binding, not intracellular delivery, of polyarginine quantum dots. *J. Phys. Chem. Lett.* **2010**, *1*, 1312–1315.

34. Chaudhary, A.; Harma, H.; Hanninen, P.; McShane, M. J.; Srivastava, R. Glucose response of near-infrared alginate-based microsphere sensors under dynamic reversible conditions. *Diabetes Technol. Ther.* **2011**, *13*, 827–835.

35. Ballerstadt, R.; Gowda, A.; McNichols, R. Fluorescence resonance energy transfer-based near-infrared fluorescence sensor for glucose monitoring. *Diabetes Technol. Ther.* **2004**, *6*, 191–200.

REDOXSENSOR RED CC-1

CAS Registry Number 296277-09-3
Chemical Structure

CA Index Name 9H-Xanthene-3,6-diamine, N^3,N^3,N^6,N^6-tetramethyl-9-(2,3,4,5,6-pentafluorophenyl)-

Other Names 9H-Xanthene-3,6-diamine, N,N,N',N'-tetramethyl-9-(pentafluorophenyl)-; R 14060; RedoxSensor; RedoxSensor CC 1; RedoxSensor Red; RedoxSensor Red CC 1

Merck Index Number Not listed

Chemical/Dye Class Xanthene

Molecular Formula $C_{23}H_{19}F_5N_2O$

Molecular Weight 434.40

Physical Form Colorless solid

Solubility Soluble in dimethyl sulfoxide, methanol

Boiling Point (Calcd.) 445.1 ± 45.0 °C, pressure: 760 Torr

pK$_a$ (Calcd.) 4.83 ± 0.40, most basic, temperature: 25 °C

Absorption (λ_{max}) 239 nm (MeOH)

Emission (λ_{max}) None

Molar Extinction Coefficient 52,000 cm^{-1} M^{-1} (MeOH)

Synthesis Synthetic methods[1,2]

Imaging/Labeling Applications Mitochondria;[1–4] lysosomes;[1,2,5] cells;[5] semiconductor nanocrystal (quantum dot)[5]

Biological/Medical Applications Detecting prostate cancer[4] generating/detecting reactive oxygen species (ROS)[1,2,6–9]

Industrial Applications Assaying reactive oxidants in smoke[9]

Safety/Toxicity No data available

REFERENCES

1. Sabnis, R. W. *Handbook of Biological Dyes and Stains*; John Wiley & Sons Inc.: Hoboken, **2010**; p 405.

2. Chen, C. S.; Gee, K. R. Redox-dependent trafficking of 2,3,4,5,6-pentafluorodihydrotetramethylrosamine, a novel fluorogenic indicator of cellular oxidative activity. *Free Radical Biol. Med.* **2000**, *28*, 1266–1278.

3. Hattori, F.; Fukuda, K. Method of selecting myocardial cells by using intracellular mitochondria as indication. PCT Int. Appl. WO 2006022377, 2006.

4. Dickman, D. Methods of detecting prostate cancer. PCT Int. Appl. WO 2006054296, 2006.

5. Nadeau, J. L.; Cohen, N. Use of quantum dots for biological labels and sensors. PCT Int. Appl. WO 2006037226, 2006.

6. Puthiyaveetil, S.; Ibrahim, I. M.; Allen, J. F. Oxidation-reduction signalling components in regulatory pathways of state transitions and photosystem stoichiometry adjustment in chloroplasts. *Plant, Cell Environ.* **2012**, *35*, 347–359.

7. McLaughlin, K. J.; Strain-Damerell, C. M.; Xie, K.; Brekasis, D.; Soares, A. S.; Paget, M. S. B.; Kielkopf, C. L. Structural basis for NADH/NAD+ redox sensing by a Rex family repressor. *Mol. Cell* **2010**, *38*, 563–575.

8. Zhang, Q.; Wang, S.; Nottke, A. C.; Rocheleau, J. V.; Piston, D. W.; Goodman, R. H. Redox sensor CtBP mediates hypoxia-induced tumor cell migration. *Proc. Natl. Acad. Sci. U.S.A.* **2006**, *103*, 9029–9033.

9. Huang, D.; Ou, B. Method for assaying reactive oxidants in smoke. U.S. Pat. Appl. Publ. US 2004126891, 2004.

Handbook of Fluorescent Dyes and Probes, First Edition. R. W. Sabnis.
© 2015 John Wiley & Sons, Inc. Published 2015 by John Wiley & Sons, Inc.

RESAZURIN SODIUM SALT (ALAMAR BLUE)

CAS Registry Number   62758-13-8

Chemical Structure

CA Index Name   3*H*-Phenoxazin-3-one, 7-hydroxy-, 10-oxide, sodium salt (1:1)

Other Names   Alamar Blue; 3*H*-Phenoxazin-3-one, 7-hydroxy-, 10-oxide, sodium salt; 7-hydroxy-3*H*-Phenoxazin-3-one 10-oxide, sodium salt; Resazurin sodium salt; Sodium resazurin

Merck Index Number   Not listed

Chemical/Dye Class   Heterocycle; Phenoxazine

Molecular Formula $C_{12}H_6NNaO_4$

Molecular Weight   251.17

Physical Form   Dark green to black powder or purple solid

Solubility   Soluble in water, methanol

Absorption (λ_{max})   604 nm (MeOH)

Emission (λ_{max})   None

Molar Extinction Coefficient   60,000 cm^{-1} M^{-1} (MeOH)

Synthesis   Synthetic methods[1–3]

Imaging/Labeling Applications   D-Arabinitol;[4] bacteria;[5–9] bile acids;[10–14] cells;[32–35] hairs;[15] hypoxic cells;[16] microorganisms;[17–19] sperms;[20–29] yeasts[30,31]

Biological/Medical Applications   Analyzing/evaluating sperm parameters;[20–29] detecting bacteria,[5–9] D-arabinitol in vaginal fluid,[4] yeasts;[30,31] determining malt quality for use in beer production;[49] diagnosing *Candida* vaginitis;[4] evaluating/quantitating/measuring glucose, 2-deoxy-D-glucose (2DG), glucose 6-phosphate, 2-deoxy-D-glucose 6- phosphate (DG6P);[45,46] locating hidden microorganism contaminated surfaces in industrial water systems;[17] measuring mitochondrial metabolic activity;[47,48] as cell viability assay reagent;[32–35] as cytotoxicity assay reagent;[36–44] as a substrate for measuring dehydrogenases activity,[50,51] β-glucosidase activity,[52] glutathione transferase activity,[53] glycosidases activity,[54] hydrolases activity;[55,56] testing/characterizing microorganisms[18,19]

Industrial Applications   Electrochemichromic solutions/devices (mirrors, glazings, partitions, filters, displays, lenses);[57,58] oxygen-barrier packaging materials;[59] oxidation-reduction (redox) indicator[1–3,5–9,17–19,30,31,49,57,58]

Safety/Toxicity   No data available

Certification/Approval   Certified by Biological Stain Commission (BSC)

REFERENCES

1. Sabnis, R. W. *Handbook of Biological Dyes and Stains*; John Wiley & Sons Inc.: Hoboken, **2010**; pp 406–407.

2. Green, F. J. *The Sigma-Aldrich Handbook of Stains, Dyes and Indicators*; Aldrich Chemical Company Inc.: Milwaukee, **1991**; pp 622–623.

3. Welcher, F. J. *Organic Analytical Reagents;* D. Van Nostrand Company: New York, **1948**; Vol. *4*, pp 548–549.

4. Anderson-Mauser, L. Diagnosis of *Candida vaginitis* by detecting D-arabinitol in vaginal fluid. Eur. Pat. Appl. EP 556725, 1993.

5. Nishinaga, E.; Suzuki, N.; Hama, T.; Fukuta, I. Method for determining color change in oxidation-reduction indicator. PCT Int. Appl. WO 2013046995, 2013.

6. Ukaji, F. Dental plaque-dyeing solution containing glycanase and dental caries-related bacteria-recognizing dyes. Jpn. Kokai Tokkyo Koho JP 2010202583, 2010.

7. Yano, Y.; Kiyotaki, K.; Nakagawa, M.; Kano, K.; Sasaki, K. Method and kit for detecting coliform bacteria. Jpn. Kokai Tokkyo Koho JP 2004187588, 2004.

8. Toyama, K.; Fukuwatari, Y.; Yano, Y.; Kiyotaki, K.; Nakagawa, M.; Kano, K.; Sasaki, K. Method and kit for detecting Coliform bacteria. Jpn. Kokai Tokkyo Koho JP 2002360296, 2002.

9. Reinheimer, J. A.; Demkow, M. R. Comparison of rapid tests for assessing UHT milk sterility. *J. Dairy Res.* **1990**, *57*, 239–243.

10. Nakasuga, A. Enzymic-chromatographic-spectrometric determination of bile acids. Jpn. Kokai Tokkyo Koho JP 61260896, 1986.

11. Beher, W. T.; Stradnieks, S.; Lin, G. J.; Sanfield, J. Rapid analysis of human fecal bile acids. *Steroids* **1981**, *38*, 281–295.

12. Barnes, S.; Spenney, J. G. Improved enzymatic assays for bile acids using resazurin and NADH oxidoreductase from *Clostridium kluyveri*. *Clin. Chim. Acta* **1980**, *102*, 241–245.

13. Ohsuga, T.; Mashige, F.; Imai, K. Microquantification of bile acids in body fluids. Jpn. Kokai Tokkyo Koho JP 52084796, 1977.

14. Mashige, F.; Imai, K.; Osuga, T. A simple and sensitive assay of total serum bile acids. *Clin. Chim. Acta* **1976**, *70*, 79–86.

15. Kobayashi, S. Hair preparations containing direct dyes. Jpn. Kokai Tokkyo Koho JP 2006265158, 2006.

16. Hodgkiss, R. J.; Begg, A. C.; Middleton, R. W.; Parrick, J.; Stratford, M. R. L.; Wardman, P.; Wilson, G. D. Fluorescent markers for hypoxic cells: A study of novel heterocyclic compounds that undergo bioreductive binding. *Biochem. Pharmacol.* **1991**, *41*, 533–541.

17. Cooper, A. J.; Enzien, M. V.; Hatch, S. R.; Ho, B. P.; Wu, M. M. Method for locating hidden microorganism contaminated surfaces in industrial water systems. U.S. Pat. Appl. Publ. US 20040029211, 2004.

18. Bochner, B. R.; Naleway, J. J. Gel matrix with redox purple for testing and characterizing microorganisms. U.S. Patent 5882882, 1999.

19. Bochner, B. R.; Naleway, J. J. Gel matrix with redox purple. PCT Int. Appl. WO 9826270, 1998.

20. Foote, R. H. Resazurin reduction and other tests of semen quality and fertility of bulls. *Asian J. Androl.* **1999**, *1*, 109–114.

21. Reddy, K. V.; Bordekar, A. D. Spectrophotometric analysis of resazurin reduction test and semen quality in men. *Indian J. Exp. Biol.* **1999**, *37*, 782–786.

22. Wang, S.; Holyoak, G. R.; Panter, K. E.; Liu, Y.; Evans, R. C.; Bunch, T. D. Resazurin reduction assay for ram sperm metabolic activity measured by spectrophotometry. *Proc. Soc. Exp. Biol. Med.* **1998**, *217*, 197–202.

23. Zalata, A. A.; Lammertijn, N.; Christophe, A.; Comhaire, F. H. The correlates and alleged biochemical background of the resazurin reduction test in semen. *Int. J. Androl.* **1998**, *21*, 289–294.

24. Reddy, K. V.; Meherji, P. K.; Gokral, J. S.; Shahani, S. K. Resazurin reduction test to evaluate sperm quality. *Indian J. Exp. Biol.* **1997**, *35*, 369–373.

25. Comhaire, F.; Vermeulen, L. Human semen analysis. *Hum. Reprod. Update* **1995**, *1*, 343–362.

26. Dart, M. G.; Mesta, J.; Crenshaw, C.; Ericsson, S. A. Modified resazurin reduction test for determining the fertility potential of bovine spermatozoa. *Arch. Androl.* **1994**, *33*, 71–75.

27. Mahmoud, A. M.; Comhaire, F. H.; Vermeulen, L.; Andreou, E. Comparison of the resazurin test, adenosite triphosphate in semen, and various sperm parameters. *Hum. Reprod.* **1994**, *9*, 1688–1693.

28. Fuse, H.; Okumura, M.; Kazama, T.; Katayama, T. Comparison of resazurin test results with various sperm parameters. *Andrologia* **1993**, *25*, 153–157.

29. Glass, R. H.; Ericsson, S. A.; Ericsson, R. J.; Drouin, M. T.; Marcoux, L. J.; Sullivan, H. The resazurin reduction test provides an assessment of sperm activity. *Fertil. Steril.* **1991**, *56*, 743–746.

30. Tiballi, R. N.; He, X.; Zarins, L. T.; Revankar, S. G.; Kaurfman, C. A. Use of a colorimetric system for yeast susceptibility testing. *J. Clin. Microbiol.* **1995**, *33*, 915–917.

31. Visser, W.; Scheffers, W. A.; Batenburg-van der Vegte, W. H.; van Dijken, J. P. Oxygen requirements of yeasts. *Appl. Environ. Microbiol.* **1990**, *56*, 3785–3792.

32. Gloeckner, H.; Jonuleit, T.; Lemke, H. D. Monitoring of cell viability and cell growth in a hollow-fiber bioreactor by use of the dye Alamar Blue. *J. Immunol. Methods* **2001**, *252*, 131–138.

33. Nakayama, G. R.; Caton, M. C.; Nova, M. P.; Parandoosh, Z. Assessment of the Alamar Blue assay for cellular growth and viability *in vitro*. *J. Immunol. Methods* **1997**, *204*, 205–208.

34. Larson, E. M.; Doughman, D. J.; Gregerson, D. S.; Obritsch, W. F. A new, simple, nonradioactive, nontoxic *in vitro* assay to monitor corneal endothelain cell viability. *Invest. Ophthalmol. Vis. Sci.* **1997**, *38*, 1929–1933.

35. Nikolaychik, V. V.; Samet, M. M.; Lelkes, P. I. A new method for continual quantitation of viable cells on endothelialized polyurethanes. *J. Biomater. Sci. Polym. Ed.* **1996**, *7*, 881–891.

36. Riss, T. L.; Moravec, R. A. Use of multiple assay endpoints to investigate the effects of incubation time, dose of toxin, and plating density in cell-based cytotoxicity assays. *Assay Drug Dev. Technol.* **2004**, *2*, 51–62.

37. Khromykh, L. M.; Anfalova, T. V.; Kazanskii, D. B. Colorimetric *in vitro* evaluation of T-cell cytotoxicity. *Bull. Exp. Biol. Med.* **2003**, *136*, 314–317.

38. McMillian, M. K.; Li, L.; Parker, J. B.; Patel, L.; Zhong, Z.; Gunnett, J. W.; Powers, W. J.; Johnson, M. D. An improved resazurin-based cytotoxicity assay for hepatic cells. *Cell Biol. Toxicol.* **2002**, *18*, 157–173.

39. O'Brien, J.; Wilson, I.; Orton, T.; Pognan, F. Investigation of the Alamar Blue (resazurin) fluorescent dye for the assessment of mammalian cell cytotoxicity. *Eur. J. Biochem.* **2000**, *267*, 5421–5426.

40. Lee, J. K.; Kim, D. B.; Kim, J. I.; Kim, P. Y. *In vitro* cytotoxicity tests on cultured human skin fibroblasts to predict skin irritation potential of surfactants. *Toxicol. In Vitro* **2000**, *14*, 345–349.

41. Mikus, J.; Steverding, D. A simple colorimetric method to screen drug cytotoxicity against *Leishmania* using the dye Alamar Blue. *Parasitol. Int.* **2000**, *48*, 265–269.

42. Rao, S.; Shirata, K.; Furukawa, K. S.; Ushida, T.; Tateishi, T.; Kanazawa, M.; Katsube, S.; Janna, S. Evaluation of cytotoxicity of UHMWPE wear debris. *Biomed. Mater. Eng.* **1999**, *9*, 209–217.

43. Nociari, M. M.; Shalev, A.; Benias, P.; Rucco, C. A novel one-step, highly sensitive fluorometric assay to evaluate cell-mediated cytotoxicity. *J. Immunol. Methods* **1998**, *213*, 157–167.

44. Gazzano-Santoro, H.; Ralph, P.; Ryskamp, T. C.; Chen, A. B.; Mukku, V. R. A non-radioactive complement-dependent cytotoxicity assay for anti-CD20 monoclonalantibody. *J. Immunol. Methods* **1997**, *202*, 163–171.

45. Yamamoto, N.; Kawasaki, K.; Kawabata, K.; Ashida, H. An enzymatic fluorimetric assay to quantitate 2-deoxyglucose and 2-deoxyglucose-6-phosphate for *in vitro* and *in vivo* use. *Anal. Biochem.* **2010**, *404*, 238–240.

46. Yamamoto, N.; Kawasaki, K.; Sato, T.; Hirose, Y.; Muroyama, K. A nonradioisotope, enzymatic microplate assay for *in vivo* evaluation of 2-deoxyglucose uptake in muscle tissue. *Anal. Biochem.* **2008**, *375*, 397–399.

47. Zhang, H. X.; Du, G. H.; Zhang, J. T. Assay of mitochondrial functions by resazurin in vitro. *Acta Pharmacol. Sin.* **2004**, *25*, 385–389.

48. Springer, J. E.; Azbill, R. D.; Carlson, S. L. A rapid and sensitive assay for measuring mitochondrial metabolic activity in isolated neural tissue. *Brain Res. Brain Res. Protoc.* **1998**, *2*, 259–263.

49. Nishikawa, N.; Kamata, K. Determination of the quality of malt for use in beer production. Jpn. Kokai Tokkyo Koho JP 62019099, 1987.

50. Strotmann, U. J.; Butz, B.; Bias, W. R. The dehydrogenase assay with resazurin: practical performance as a monitoring system and pH-dependent toxicity of phenolic compounds. *Ecotoxicol. Environ. Saf.* **1993**, *25*, 79–89.

51. De Jong, D. W.; Woodlief, W. G. Fluorimetric assay of tobacco leaf dehydrogenases with resazurin. *Biochim. Biophys. Acta* **1977**, *484*, 249–259.

52. Tokutake, S.; Kasai, K.; Tomikura, T.; Yamaji, N.; Kato, M. Glycosides having chromophores as substrates for sensitive enzyme analysis. II. Synthesis of phenolindophenyl-β-D-glucopyranosides having an electron-withdrawing substituent as substrates for β-glucosidase. *Chem. Pharm. Bull.* **1990**, *38*, 3466–3470.

53. Suvorov, A. A.; Stulovskij, A. V.; Vilyatser, A. Y.; Voznyj, I. V.; Rozengart, E. V.; Khovanskikh, A. E. A chromogenic substrate for glutathione transferase assay. U.S.S.R. SU 1759874, 1992.

54. Klein, C.; Batz, H. G.; Sernetz, M.; Hofmann, J. Glycosides of resorufin derivatives, useful in determining the activity of glycosidases. Ger. Offen. DE 3411574, 1985.

55. Guder, H. J.; Von der Eltz, H.; Eltz, H. V. New dihydroresorufin derivatives for use as hydrolase substrates. Ger. Offen. DE 3644401, 1988.

56. Wallenfels, K.; Fathy, A. M. Gycosyloxy and peptidyloxy heterocycles as substrates for hydrolases and their use. Ger. Offen. DE 3412939, 1985.

57. Varaprasad, D. V.; Looman, S. D.; Zhao, M.; Habibi, H. R.; Lynam, N. R. Electrochemichromic solutions, processes for preparing and using the same, and devices manufactured with the same. U.S. Patent 5500760, 1996.

58. Varaprasad, D. V.; Habibi, H. R.; Looman, S. D.; Lynam, N. R.; Zhao, M. Electrochemichromic solutions, processes for preparing and using the same, and devices manufactured with the same. Eur. Pat. Appl. EP 531143, 1993.

59. Kamiyama, M. Oxygen-barrier packaging materials. Jpn. Kokai Tokkyo Koho JP 09124076, 1997.

RHODAMINE 6G (RH 6G)

CAS Registry Number   989-38-8

Chemical Structure

CA Index Name   Xanthylium, 9-[2-(ethoxycarbonyl) phenyl]-3,6-bis(ethylamino)-2,7-dimethyl-, chloride (1:1)

Other Names   Benzoic acid, *o*-[6-(ethylamino)-3-(ethylimino)-2,7-dimethyl-3*H*-xanthen-9-yl]-, ethyl ester, monohydrochloride; Rhodamine 6GCP; Xanthylium, 9-[2-(ethoxycarbonyl)phenyl]-3,6-bis(ethylamino)-2,7-dimethyl-, chloride; Aizen Rhodamine 6GCP; Basic Red 1; Basonyl Red 482; Basonyl Red 483; C.I. 45160; C.I. Basic Red 1; Calcozine Red 6G; Calcozine Rhodamine 6GX; Eljon Pink Toner; Exciton 590; Fanal Pink B; Fanal Pink GFK; Fanal Red 25532; Flexo Red 482; Heliostable Brilliant Pink B extra; Mitsui Rhodamine 6GCP; NSC 36345; Nyco Liquid Red GF; R 634; R 6G; Rh 6G; Rhodamin 6G; Rhodamine 4GD; Rhodamine 4GH; Rhodamine 5GDN; Rhodamine 5GDN Extra; Rhodamine 5GL; Rhodamine 6G; Rhodamine 6G Extra; Rhodamine 6G Extra Base; Rhodamine 6G chloride; Rhodamine 6GB; Rhodamine 6GBN; Rhodamine 6GD; Rhodamine 6GDN; Rhodamine 6GDN Extra; Rhodamine 6GEx ethyl ester; Rhodamine 6GH; Rhodamine 6GO; Rhodamine 6GX; Rhodamine 6JH; Rhodamine 6JH-SA; Rhodamine 6JH-SA Extra 1150; Rhodamine 6Zh-DN; Rhodamine F 5G; Rhodamine F 5GL; Rhodamine GDN; Rhodamine GDN Extra; Rhodamine Y 20–7425; Rhodamine Zh; Silosuper Pink B; Vali Fast Red 1308

Merck Index Number   Not listed

Chemical/Dye Class   Xanthene

Molecular Formula $C_{28}H_{31}ClN_2O_3$

Molecular Weight   479.01

Physical Form   Red crystals, powder or solid

Solubility   Soluble in water, ethanol, methanol

Absorption (λ_{max})   528 nm (MeOH)

Emission (λ_{max})   551 nm (MeOH)

Molar Extinction Coefficient   105,000 $cm^{-1} M^{-1}$ (MeOH)

Quantum Yield   0.95 (1.0 N H_2SO_4)[5]

Synthesis   Synthetic methods[1–4]

Imaging/Labeling Applications   Mitochondria;[1,6–10] cells;[11,12] hairs;[13,14] metal ions: copper ions;[15–17] gold ions;[18] iron ions;[19,20] mercury ions;[21–25] palladium ions;[26] rhodium ions;[27] vanadium ions;[28] nitrite ions;[29] microorganisms;[30] nucleic acids;[31–34] proteins[35,36]

Biological/Medical Applications   Analyzing cells;[11,12] analyzing/labeling nucleic acids;[31–34] characterizing multidrug resistance;[37] as a substrate for measuring phospholipase activity,[38] protease activity;[39] measuring membrane potential;[40] treating amyloidosis,[41] cancer[41,42]

Industrial Applications   Clays;[43–46] colored bubbles;[47,48] color filters;[49–58] detergents;[59,60] drawing materials;[61] dye lasers;[62–73] electroluminescent (EL) displays;[52,53,55,58,74–77] glass materials;[78,79] inks;[80–94] laser devices/laser materials;[95–97] light-emitting diode devices/materials;[54,56,98–100] liquid crystals;[101] liquid crystal displays;[50,102] luminescent materials;[103] metal oxide particles;[104] petroleum products;[105] photographic materials;[106] photoresists;[107] photovoltaic devices;[108] polymer films;[109] printing materials;[110] recording materials;[111,112] silica gel;[113] solar cells;[114,115] sol–gel materials;[116] sputtered gold films;[117] textiles;[118] thin films;[119,120] toners;[121–123] tracers for hydrology;[124] waveguides[125–127]

Safety/Toxicity   Carcinogenicity;[128–132] cardiotoxicity;[133] DNA damage;[134] ecotoxicity;[135] genotoxicity;[136] mutagenicity;[130,134,137–139] phototocicity;[140,141] reproductive toxicity[142]

REFERENCES

1. Sabnis, R. W. *Handbook of Biological Dyes and Stains*; John Wiley & Sons Inc.: Hoboken, **2010**; pp 415–417.

2. Sekima, H. Method for manufacture of granular rhodamine dyes. Jpn. Kokai Tokkyo Koho JP 2000273342, 2000.

3. Xiao, G.; Zhu, D. Bulk-scale preparation of ethyl 2-[3,6-bis(N-ethylamino)-2,7-dimethyl-3-hydroxanth-9-yl]benzoate hydrochloride. Faming Zhuanli Shenqing Gongkai Shuomingshu CN 1083058, 1994.

4. Aburada, K.; Akagi, M. Process for the preparation of rhodamines. Eur. Pat. Appl. EP 0468821, 1992.

5. Kubin, R. F.; Fletcher, A. N. Fluorescence quantum yields of some rhodamine dyes. *J. Luminesc.* **1983**, *27*, 455–462.

6. Mottram, L. F.; Forbes, S.; Ackley, B. D.; Peterson, B. R. Hydrophobic analogues of rhodamine B and rhodamine 101: potent fluorescent probes of mitochondria in living *C. elegans. Beilstein J. Org. Chem.* **2012**, *8*, 2156–2165.

7. Hu, S.; Zhao, H.; Yin, X. J.; Ma, J. K. H. Role of mitochondria in silica-induced apoptosis of alveolar macrophages: Inhibition of apoptosis by rhodamine 6G and N-acetyl-L-cysteine. *J. Toxicol. Environ. Health, Part A* **2007**, *70*, 1403–1415.

8. Bunting, J. R. Influx and efflux kinetics of cationic dye binding to respiring mitochondria. *Biophys. Chem.* **1992**, *42*, 163–175.

9. Rashid, F.; Horobin, R. W. Interaction of molecular probes with living cells and tissues. Part 2. A structure-activity analysis of mitochondrial staining by cationic probes, and a discussion of the synergistic nature of image-based and biochemical approaches. *Histochemistry* **1990**, *94*, 303–308.

10. Bunting, J. R.; Phan, T. V.; Kamali, E.; Dowben, R. M. Fluorescent cationic probes of mitochondria. Metrics and mechanism of interaction. *Biophys. J.* **1989**, *56*, 979–993.

11. Byassee, T. A.; Chan, W. C. W.; Nie, S. Probing single molecules in single living cells. *Anal. Chem.* **2000**, *72*, 5606–5611.

12. Nakamoto, H.; Fujiwara, C. Reagent and method for analyzing cells in urine. Eur. Pat. Appl. EP 0513762, 1992.

13. Hercouet, L. Hair dye composition comprising a derivative of diamino-N,N-dihydropyrazolone, a coupler, and a heterocyclic direct dye. Fr. Demande FR 2886132, 2006.

14. Moeller, H.; Meinigke, B. Hair dye compositions containing xanthenes. Ger. Offen. DE 19926377, 2000.

15. Xu, W.; Mu, L.; Miao, R.; Zhang, T.; Shi, W. Fluorescence sensor for Cu(II) based on R6G derivatives modified silicon nanowires. *J. Luminesc.* **2011**, *131*, 2616–2620.

16. Huang, L.; Hou, F.; Xi, P.; Bai, D.; Xu, M.; Li, Z.; Xie, G.; Shi, Y.; Liu, H.; Zeng, Z. A rhodamine-based "turn-on" fluorescent chemodosimeter for Cu^{2+} and its application in living cell imaging. *J. Inorg. Biochem.* **2011**, *105*, 800–805.

17. Xiang, Y.; Li, Z.; Chen, X.; Tong, A. Highly sensitive and selective optical chemosensor for determination of Cu^{2+} in aqueous solution. *Talanta* **2008**, *74*, 1148–1153.

18. Yang, Y.; Lee, S.; Tae, J. A gold(III) ion-selective fluorescent probe and its application to bioimagings. *Org. Lett.* **2009**, *11*, 5610–5613.

19. Zhang, L.; Wang, J.; Fan, J.; Guo, K.; Peng, X. A highly selective, fluorescent chemosensor for bioimaging of Fe^{3+}. *Bioorg. Med. Chem. Lett.* **2011**, *21*, 5413–5416.

20. Mao, J.; He, Q.; Liu, W. An rhodamine-based fluorescence probe for iron(III) ion determination in aqueous solution. *Talanta* **2010**, *80*, 2093–2098.

21. Ganbold, E.; Park, J.; Ock, K.; Joo, S. Gold nanoparticle-based detection of Hg(II) in an aqueous solution: fluorescence quenching and surface-enhanced Raman scattering study. *Bull. Korean Chem. Soc.* **2011**, *32*, 519–523.

22. Zhou, P.; Meng, Q.; He, G.; Wu, H.; Duan, C.; Quan, X. Highly sensitive fluorescence probe based on functional SBA-15 for selective detection of Hg^{2+} in aqueous media. *J. Environ. Monit.* **2009**, *11*, 648–653.

23. Chen, X.; Nam, S.; Jou, M. J.; Kim, Y.; Kim, S.; Park, S.; Yoon, J. Hg^{2+} Selective fluorescent and colorimetric sensor: Its crystal structure and application to bioimaging. *Org. Lett.* **2008**, *10*, 5235–5238.

24. Chen, J.; Zheng, A.; Chen, A.; Gao, Y.; He, C.; Kai, X.; Wu, G.; Chen, Y. A functionalized gold nanoparticles and Rhodamine 6G based fluorescent sensor for high sensitive and selective detection of mercury(II) in environmental water samples. *Anal. Chim. Acta* **2007**, *599*, 134–142.

25. Yang, Y.; Yook, K.; Tae, J. A rhodamine-based fluorescent and colorimetric chemodosimeter for the rapid detection of Hg^{2+} Ions in aqueous media. *J. Am. Chem. Soc.* **2005**, *127*, 16760–16761.

26. Goswami, S.; Sen, D.; Das, N. K.; Fun, H.; Quah, C. K. A new rhodamine based colorimetric 'off-on' fluorescence sensor selective for Pd^{2+} along with the first bound X-ray crystal structure. *Chem. Commun.* **2011**, *47*, 9101–9103.

27. Jaya, S.; Rao, T. P.; Ramakrishna, T. V. Spectrophotometric determination of rhodium(III) in thermocouple wires using thiocyanate and Rhodamine 6G. *Analyst* **1983**, *108*, 1151–1155.

28. Jie, N.; Zhang, Q.; Wei, Y.; Jiang, Q. Study on the catalytic determination of vanadium (V) using the rhodamine 6G-periodate redox reaction and its analytical application. *Mikrochim. Acta* **2002**, *140*, 103–107.

29. Jie, N.; Si, Z.; Yang, J.; Miao, Z.; Huang, X.; Zhang, Q.; Song, Z. Fluorometric determination of traces of nitrite with rhodamine 6G. *Microchem. J.* **1997**, *55*, 351–356.

30. Nashimoto, K.; Ishida, K.; Ikeda, Y.; Hanaoka, Y. Microorganism staining agent, and its use. Jpn. Tokkyo Koho JP 4370118, 2009.

31. Khan, H. A. The effect of DNA labeling with the fluorescent dyes R110 and R6G on genotype analysis using capillary electrophoresis. *Cell. Mol. Biol. Lett.* **2005**, *10*, 247–253.

32. Xiao, M.; Kwok, P. DNA analysis by fluorescence quenching detection. *Genome Res.* **2003**, *13*, 932–939.

33. Itakura, M. DNA fragment labeling. Jpn. Kokai Tokkyo Koho JP 08140700, 1996.

34. Shmurun, R. I. Staining of nucleic acids. U.S.S.R. SU 219114, 1968.

35. Das, D. K.; Mondal, T.; Mandal, A. K.; Bhattacharyya, K. Binding of organic dyes with human serum albumin: A single-molecule study. *Chem. Asian J.* **2011**, *6*, 3097–3103.

36. Nizomov, N.; Ismailov, Z. F.; Kurtaliev, E. N.; Nizamov, Sh. N.; Khodzhaev, G.; Patsenker, L. D. Luminescent spectral properties of rhodamine derivatives while binding to serum albumin. *J. Appl. Spectrosc.* **2006**, *73*, 432–436.

37. Kessel, D.; Beck, W. T.; Kukuruga, D.; Schulz, V. Characterization of multidrug resistance by fluorescent dyes. *Cancer Res.* **1991**, *51*, 4665–4670.

38. Graham, R. J. Fluorescent phospholipase assays and compositions. PCT Int. Appl. WO 2005005977, 2005.

39. Liu, G. L.; Rosa-Bauza, Y. T.; Salisbury, C. M.; Craik, C.; Ellman, J. A.; Chen, F. F.; Lee, L. P. Peptide-nanoparticle hybrid SERS probes for optical detection of protease activity. *J. Nanosci. Nanotechnol.* **2007**, *7*, 2323–2330.

40. Mandalà, M.; Serck-Hanssen, G.; Martino, G.; Helle, K. B. The fluorescent cationic dye rhodamine 6G as a probe for membrane potential in bovine aortic endothelial cells. *Anal Biochem.* **1999**, *274*, 1–6.

41. Kutushov, M. V. Use of Rhodamine 6G for the treatment of malignant neoplasms and amyloidosis. Russ. RU 2354369, 2009.

42. Haghighat, S.; Castro, D. J.; Lufkin, R. B.; Fetterman, H. R.; Castro, D. J.; Soudant, J.; Ward, P. H.; Saxton, R. E. Laser dyes for experimental phototherapy of human cancer: comparison of three rhodamines. *Laryngoscope* **1992**, *102*, 81–87.

43. Gitis, V.; Dlugy, C.; Ziskind, G.; Sladkevich, S.; Lev, O. Fluorescent clays - Similar transfer with sensitive detection. *Chem. Eng. J.* **2011**, *174*, 482–488.

44. Tani, S.; Yamaki, H.; Sumiyoshi, A.; Suzuki, Y.; Hasegawa, S.; Yamazaki, S.; Kawamata, J. Enhanced photodegradation of organic dyes adsorbed on a clay. *J. Nanosci. Nanotechnol.* **2009**, *9*, 658–661.

45. Lotsch, B. V.; Ozin, G. A. Photonic clays: A new family of functional 1D photonic crystals. *ACS Nano* **2008**, *2*, 2065–2074.

46. Martinez Martinez, V.; Lopez Arbeloa, F.; Banuelos Prieto, J.; Lopez Arbeloa, I. Orientation of adsorbed dyes in the interlayer space of clays. 1. Anisotropy of rhodamine 6G in laponite films by vis-absorption with polarized light. *Chem. Mater.* **2005**, *17*, 4134–4141.

47. Sabnis, R. W.; Kehoe, T. D. Composition and method for producing colored bubbles. U.S. Pat. Appl. Publ. US 20120244777, 2012.

48. Sabnis, R. W.; Kehoe, T. D. Composition and method for producing colored bubbles. U.S. Pat. Appl. Publ. US 20060004110, 2006.

49. Ando, M.; Yamamoto, Y.; Sakamoto, S. Color compositions with high contrast for color filters. Repub. Korean Kongkae Taeho Kongbo KR 2013039697, 2013.

50. Shin, Y. C.; Kwon, Y. S.; Kim, B. G. Colored photosensitive resin composition, colored pattern, color filter and liquid crystal display device having the same. Repub. Korean Kongkae Taeho Kongbo KR 2013048169, 2013.

51. Fujita, T. Dye compounds and coloring composition for color filter. Repub. Korean Kongkae Taeho Kongbo KR 2012001629, 2012.

52. Hama, T. Color conversion filter substrates with stabilized fluorescent dyes and organic electroluminescent displays using them. Jpn. Kokai Tokkyo Koho JP 2007207578, 2007.

53. Fukuhara, S.; Asano, M. Color filter plate for organic electroluminescent device. Jpn. Kokai Tokkyo Koho JP 2007250437, 2007.

54. Hama, T. Methods for manufacture color conversion filters, their substrates, and multicolor light-emitting devices. Jpn. Kokai Tokkyo Koho JP 2007213993, 2007.

55. Kawamura, Y. Color changing filters and organic EL displays therewith having high outcoupling efficiency. Jpn. Kokai Tokkyo Koho JP 2005123088, 2005.

56. Matsuzaki, S.; Fukuda, M.; Eida, T. Light-emitting devices and manufacture of color-changing films and color filters. Jpn. Kokai Tokkyo Koho JP 2004006133, 2004.

57. Suzuki, N. Photosensitive magenta color composition, color filter, and its manufacture. Jpn. Kokai Tokkyo Koho JP 2003344998, 2003.

58. Kawamura, Y.; Kawaguchi, T.; Shiraishi, Y. Color filter substrate employing color changing method (CCM) material and organic EL color display using the same. Jpn. Kokai Tokkyo Koho JP 2002216962, 2002.

59. Lant, N. J. Detergent compositions. U.S. Pat. Appl. Publ. US 20100093591, 2010.

60. Kabuto, S.; Tamura, N.; Nagayasu, K.; Nagata, S. Laundry detergent compositions containing coated particles containing pigments and/or fluorescent agents. Jpn. Kokai Tokkyo Koho JP 2005179530, 2005.

61. Seki, K. Solid drawing materials for overwriting. Jpn. Kokai Tokkyo Koho JP 2009286860, 2009.

62. Chen, F.; Gindre, D.; Nunzi, J. M. First order distributed feedback dye laser effect in reflection pumping geometry for nonlinear optical measurements. Proc. SPIE-Int. Soc. Opt. Eng. 2007, 6653, 665304/1–665304/5.

63. Matsuyama, N.; Kanamori, Y.; Ye, J.; Hane, K. Micromachined surface emitting dye laser with a self-suspended guided mode resonant grating. J. Opt. A: Pure Appl. Opt. 2007, 9, 940–944.

64. Jiang, X.; Song, Q.; Xu, L.; Fu, J.; Tong, L. Microfiber knot dye laser based on the evanescent-wave-coupled gain. Appl. Phys. Lett. 2007, 90, 233501/1-233501/3.

65. Shopova, S. I.; Zhou, H.; Fan, X.; Zhang, P. Optofluidic ring resonator based dye laser. Appl. Phys. Lett. 2007, 90, 221101/1–221101/3.

66. Enmanji, K.; Kato, K. Dye lasers. Jpn. Kokai Tokkyo Koho JP 2000183435, 2000.

67. Iijima, T.; Fukazawa, T.; Tanaka, Y. Dye laser. Jpn. Kokai Tokkyo Koho JP 04045590, 1992.

68. Mitachi, N.; Nishi, T.; Hiratsuka, H. Solid-state dye laser. Jpn. Kokai Tokkyo Koho JP 03029384, 1991.

69. Spears, K. G.; Zhu, X.; Yang, X.; Wang, L. Picosecond infrared generation from neodymium-doped YAG and a visible, short cavity dye laser. Opt. Commun. 1988, 66, 167–171.

70. Seki, K.; Nakanishi, H. Dye lasers. Jpn. Kokai Tokkyo Koho JP 63237493, 1988.

71. Singh, S. Short pulse generation from a flashlamp-pumped Rhodamine 6G ring dye laser using the colliding pulse mode-locking technique. Appl. Opt. 1987, 26, 66–69.

72. Wirth, M. J.; Sanders, M. J.; Koskelo, A. C. Generation of picosecond pulses from a cavity-dumped synchronously pumped dye laser. Appl. Phys. Lett. 1981, 38, 295–296.

73. Ippen, E. P.; Shank, C. V.; Dienes, A. Rapid photobleaching of organic laser dyes in continuously operated devices. IEEE J. Quantum Electron. 1971, 7, 178–179.

74. Ogino, S. Organic EL (electroluminescent) display having color conversion layer and its manufacture. Jpn. Kokai Tokkyo Koho JP 2008047493, 2008.

75. Sato, H. Manufacture of organic electroluminescent (EL) displays with suppressed leak and dark spot formation. Jpn. Kokai Tokkyo Koho JP 2005149811, 2005.

76. Matsukaze, N.; Terao, Y. Manufacture of organic EL devices. Jpn. Kokai Tokkyo Koho JP 2001093664, 2001.

77. Oonishi, T.; Noguchi, M.; Doi, H. Organic electroluminescent. Jpn. Kokai Tokkyo Koho JP 05029078, 1993.

78. Ovechko, V.; Schur, O.; Mygashko, V. Optical properties of the porous glass composite material. Opt. Appl. 2008, 38, 75–82.

79. Obara, K.; Kawamura, H.; Tomita, Y.; Kawamura, T.; Misumi, A. Coloring of glass articles. Jpn. Kokai Tokkyo Koho JP 63030346, 1988.

80. Asada, K. Oil-based inks and writing instruments therewith. Jpn. Kokai Tokkyo Koho JP 2013142110, 2013.

81. Matsuo, E.; Onuki, Y. Oil-based ink compositions for ball-point pens and ball-point pens using same. Jpn. Kokai Tokkyo Koho JP 2013095845, 2013.

82. Kurosawa, Y. Inks for ball-point pens with good writability. Jpn. Kokai Tokkyo Koho JP 2013028669, 2013.

83. Nasukawa, R. Oil-based ink compositions preventing ink blots for ball-point pens. Jpn. Kokai Tokkyo Koho JP 2012031229, 2012.

84. Nasukawa, R. Oil-based ink compositions with bleeding prevention and smooth writing properties for ball-point pens. Jpn. Kokai Tokkyo Koho JP 2011213863, 2011.

85. Yadoiwa, T. Ink compositions for writing with good storage stability. Jpn. Kokai Tokkyo Koho JP 2001172541, 2001.

86. Nishimoto, T. Oil-based red ink. Jpn. Kokai Tokkyo Koho JP 2000212495, 2000.

87. Nakamura, H. Ink compositions for oily marking pens. Jpn. Kokai Tokkyo Koho JP 08157763, 1996.

88. Maeda, M.; Oka, R.; Nakai, M.; Shiromae, S.; Fujiwara, Y. Oil-based ink compositions for felt-tip pens. Jpn. Kokai Tokkyo Koho JP 02150472, 1990.

89. Sasage, D. Erasable marking inks for writing boards. Jpn. Kokai Tokkyo Koho JP 01289881, 1989.

90. Kobayashi, Y.; Saito, S. Marking inks with pen-point drying resistance. Jpn. Kokai Tokkyo Koho JP 01087675, 1989.

91. Reichelt, H. Ink for marking plastic films in wound electric capacitors. Ger. (East) DD 259534, 1988.

92. Kosaka, T. Oil-based ink compositions with suppressed pen tip drying. Jpn. Kokai Tokkyo Koho JP 63178177, 1988.

93. Obara, K.; Kawamura, H.; Tomita, Y.; Kawamura, T.; Misumi, A. Coloring of glass articles. Jpn. Kokai Tokkyo Koho JP 63030346, 1988.

94. Zychlinski, B. V. Dyes for ball-point-pen inks. FR 1397267, 1965.

95. Nenchev, M.; Deneva, M. Laser device for generating fixed frequency radiation along a reference atomic absorption line. Bulg. Pat. Appl. BG 109300, 2007.

96. Serova, V. N.; Vasilev, A. A.; Mukmeneva, N. A.; Cherkasova, O. A.; Dubinskij, M. A.; Naumov, A. K.; Akhmetzyanova, L. K. Laser material. U.S.S.R. SU 1820809, 1996.

97. Volkin, H. C. Direct solar pumped laser. U.S. Patent 4281294, 1981.

98. Hanada, Y.; Maeda, T.; Sakagami, M. Wavelength converter material for light-emitting device. Jpn. Kokai Tokkyo Koho JP 2003163376, 2003.

99. Maeda, T.; Hanada, Y. Semiconductor LED devices. Jpn. Kokai Tokkyo Koho JP 2003046136, 2003.

100. Mizoshita, N.; Goto, Y.; Horii, M.; Tani, T.; Inagaki, S. Light-emitting materials containing organic silicon compound polymers having pyrene rings. Jpn. Kokai Tokkyo Koho JP 2009249504, 2009.

101. Ilchishin, I. P.; Maslov, P. Yu.; Tikhonov, E. A.; Lipnitsky, S. O.; Stepanov, A. A. Lasing in dye-doped nematic liquid crystals at a dynamic distributed feedback for two-scheme excitation. *Mol. Cryst. Liq. Cryst.* **2007**, *467*, 235–245.

102. Hino, K.; Tawaraya, S. Liquid crystal display having visual characteristic adjusting layer. Jpn. Kokai Tokkyo Koho JP 2013160943, 2013.

103. Rohwer, L. S.; Martin, J. E. Measuring the absolute quantum efficiency of luminescent materials. *J. Luminesc.* **2005**, *115*, 77–90.

104. Ocana, M.; Levy, D.; Serna, C. J. Preparation and optical properties of spherical metal oxide particles containing fluorescent dyes. *J. Non-Cryst. Solids* **1992**, *147–148*, 621–626.

105. Riedel, G.; Vamvakaris, C. Marking of petroleum products by basic dyes. Ger. Offen. DE 4001662, 1991.

106. Riester, O. Xanthylium sensitizers for photographic material. Ger. Offen. DE 1946263, 1971.

107. Garza, C. M.; Cho, S. Metrology of bilayer photoresist processes. U.S. Pat. Appl. Publ. US 20090220895, 2009.

108. Jung, B.; Lee, M.; Kim, D.; Lee, C.; Kim, J. Photovoltaic device. U.S. Pat. Appl. Publ. US 20070144579, 2007.

109. Rocha, L.; Dumarcher, V.; Denis, C.; Raimond, P.; Fiorini, C.; Nunzi, J. Laser emission in periodically modulated polymer films. *J. Appl. Phys.* **2001**, *89*, 3067–3069.

110. Sumi, T.; Inamura, N.; Kida, A. Dyeing with cationic dyes and transfer printing materials. Jpn. Kokai Tokkyo Koho JP 55001345, 1980.

111. Shimada, K. Thermal-transfer recording material. Jpn. Kokai Tokkyo Koho JP 01159290, 1989.

112. Ikoma, K.; Miura, K.; Kawade, I.; Oguchi, Y.; Myagawa, M. Optical recording material contg. light-emitting dye, and recording method. Jpn. Kokai Tokkyo Koho JP 63062791, 1988.

113. Takahashi, S.; Nakai, M. Coloration of silica gel. Jpn. Kokai Tokkyo Koho JP 05032411, 1993.

114. El Zayat, M. Y.; Saed, A. O.; El-Dessouki, M. S. Dye sensitization of antimony-doped CdS photoelectrochemical solar cell. *Sol. Energy Mater. Sol. Cells* **2002**, *71*, 27–39.

115. Skyllas Kazacos, M.; McHenry, E. J.; Heller, A.; Miller, B. Fluorescent window for liquid junction solar cells. *Sol. Energy Mater.* **1980**, *2*, 333–342.

116. Canva, M.; Roger, G.; Cassagne, F.; Levy, Y.; Brun, A.; Chaput, F.; Boilot, J.; Rapaport, A.; Heerdt, C.; Bass, M. Dye-doped sol–gel materials for two-photon absorption induced fluorescence. *Opt. Mater.* **2002**, *18*, 391–396.

117. Maya, L.; Vallet, C. E.; Lee, Y. H. Sputtered gold films for surface-enhanced Raman scattering. *J. Vac. Sci. Technol., A* **1997**, *15*, 238–242.

118. Bendak, A. Cationic dyeing of base-treated polyester fibers. *Am. Dyestuff Rep.* **1989**, *78*, 39–43, 45.

119. Zhao, Y.; Xie, Y.; Hui, Y. Y.; Tang, L.; Jie, W.; Jiang, Y.; Xu, L.; Lau, S. P.; Chai, Y. Highly impermeable and transparent graphene as an ultra-thin protection barrier for Ag thin films. *J. Mater. Chem. C* **2013**, *1*, 4956–4961.

120. Palomino-Merino, R.; Torres-Kauffman, J.; Lozada-Morales, R.; Portillo-Moreno, O.; Garcia-Rocha, M.; Zelaya-Angel, O. Photoluminescence of rhodamine 6G-doped amorphous TiO_2 thin films grown by sol–gel. *Vacuum* **2007**, *81*, 1480–1483.

121. Uchida, M.; Yasuda, S. Color electrophotographic toners. Jpn. Kokai Tokkyo Koho JP 62015555, 1987.

122. Imai, E.; Tomari, S. Color electrophotographic toners. Jpn. Kokai Tokkyo Koho JP 54005733, 1979.

123. Tomari, S.; Imai, E. Electrophotographic magenta toners. Jpn. Kokai Tokkyo Koho JP 53144741, 1978.

124. Viriot, M. L.; Andre, J. C. Fluorescent dyes: a search for new tracers for hydrology. *Analusis* **1989**, *17*, 97–111.

125. Kuzyk, A.; Pettersson, M.; Toppari, J. J.; Hakala, T. K.; Tikkanen, H.; Kunttu, H.; Torma, P. Molecular coupling of light with plasmonic waveguides. *Opt. Express* **2007**, *15*, 9908–9917.

126. Shiota, T. Polymer optical waveguide and its production method. Jpn. Kokai Tokkyo Koho JP 2004212775, 2004.

127. Yoshimura, A.; Imamura, S.; Izawa, T. Functional polymer optical waveguide. Jpn. Kokai Tokkyo Koho JP 04012333, 1992.

128. Johnson, F. M. Carcinogenic chemical-response "fingerprint" for male F344 rats exposed to a series of 195 chemicals: Implications for predicting carcinogens with transgenic models. *Environ. Mol. Mutagen.* **1999**, *34*, 234–245.

129. Albert, R. E. Allergic contact sensitizing chemicals as environmental carcinogens. *Environ. Health Perspect.* **1997**, *105*, 940–948.

130. Ashby, J.; Tennant, R. W. Definitive relationships among chemical structure, carcinogenicity and mutagenicity for 301 chemicals tested by the U.S. NTP. *Mutat. Res., Rev. Genet. Toxicol.* **1991**, *257*, 229–306.

131. Tennant, R. W.; Ashby, J. Classification according to chemical structure, mutagenicity to *Salmonella* and level of carcinogenicity of a further 39 chemicals tested for carcinogenicity by the U.S. National Toxicology Program. *Mutat. Res., Rev. Genet. Toxicol.* **1991**, *257*, 209–227.

132. Rao, G. N.; Piegorsch, W. W.; Crawford, D. D.; Edmondson, J.; Haseman, J. K. Influence of viral infections on body weight, survival, and tumor prevalence of B6C3F1 (C57BL/6N × C3H/HeN) mice in carcinogenicity studies. *Fundam. Appl. Toxicol.* **1989**, *13*, 156–164.

133. Lampidis, T. J.; Salet, C.; Moreno, G.; Chen, L. B. Effects of the mitochondrial probe rhodamine 123 and related analogs on the function and viability of pulsating myocardial cells in culture. *Agents Actions* **1984**, *14*, 751–757.

134. Nestmann, E. R.; Douglas, G. R.; Matula, T. I.; Grant, C. E.; Kowbel, D. J. Mutagenic activity of rhodamine dyes and their impurities as detected by mutation induction in *Salmonella* and DNA damage in Chinese hamster ovary cells. *Cancer Res.* **1979**, *39*, 4412–4417.

135. Benoit-Guyod, J. L.; Rochat, J.; Alary, J.; Andre, C.; Taillandier, G. Correlations between physicochemical properties and ecotoxicity of fluorescent xanthenic water tracers. *Toxicol. Eur. Res.* **1979**, *2*, 241–246.

136. Kirkland, D.; Aardema, M.; Henderson, L.; Mueller, L. Evaluation of the ability of a battery of three *in vitro* genotoxicity tests to discriminate rodent carcinogens and non-carcinogens. I. Sensitivity, specificity and relative predictivity. *Mutat. Res., Genetic Toxicol. Environ. Mutagen.* **2005**, *584*, 1–256.

137. Serafimova, R.; Todorov, M.; Pavlov, T.; Kotov, S.; Jacob, E.; Aptula, A.; Mekenyan, O. Identification of the structural requirements for mutagenicity, by incorporating molecular flexibility and metabolic activation of chemicals. II. General Ames mutagenicity model. *Chem. Research Toxicol.* **2007**, *20*, 662–676.

138. Zeiger, E.; Anderson, B.; Haworth, S.; Lawlor, T.; Mortelmans, K.; Speck, W. *Salmonella* mutagenicity tests: III. Results from the testing of 255 chemicals. *Environ. Mutagen.* **1987**, *9*, 1–109.

139. Wuebbles, B. J. Y.; Felton, J. S. Evaluation of laser dye mutagenicity using the Ames/Salmonella

microsome test. *Environ. Mutagen.* **1985**, *7*, 511–522.

140. Harris, A. G.; Sinitsina, I.; Messmer, K. Intravital fluorescence microscopy and phototocicity: effects on leukocytes. *Eur. J. Med. Res.* **2002**, *7*, 117–124.

141. Saetzler, R. K.; Jallo, J.; Lehr, H. A.; Philips, C.M.; Vasthare, U.; Arfors, K. E.; Tuma, R. F. Intravital fluorescence microscopy: impact of light-induced phototoxicity on adhesion of fluorescently labeled leukocytes. *J. Histochem. Cytochem.* **1997**, *45*, 505–573.

142. Zhang, W. W.; Hood, R. D.; Smith-Sommerville, H. E. Effects of rhodamine 6G on the mitochondrial ultrastructure of mouse spermatocytes. *Toxicol. Lett.* **1990**, *51*, 35–40.

RHODAMINE 110 (RH 110)

CAS Registry Number   13558-31-1

Chemical Structure

CA Index Name   Xanthylium, 3,6-diamino-9-(2-carboxyphenyl)-, chloride (1:1)

Other Names   Benzoic acid, 2-(6-amino-3-imino-3*H*-xanthen-9-yl)-, monohydrochloride; Benzoic acid, *o*-(6-amino-3-imino-3*H*-xanthen-9-yl)-, monohydrochloride; Xanthylium, 3,6-diamino-9-(2-carboxyphenyl)-, chloride; R 110; RH 110; Rh 110; rh 110; Rho 110; Rhod 110; Rhodamine 110; Rhodamine 560; Rhodamine N

Merck Index Number   Not listed

Chemical/Dye Class   Xanthene

Molecular Formula $C_{20}H_{15}ClN_2O_3$

Molecular Weight   366.80

Physical Form   Green or red or reddish-green solid

Solubility   Soluble in water, *N,N*-dimethylformamide, dimethyl sulfoxide, ethanol, methanol

Melting Point   >380 °C (decompose)[2]

Absorption (λ_{max})   499 nm (MeOH); 496 nm (H_2O)

Emission (λ_{max})   521 nm (MeOH); 520 nm (H_2O)

Molar Extinction Coefficient   92,000 cm^{-1} M^{-1} (MeOH); 83,000 cm^{-1} M^{-1} (H_2O)

Quantum Yield   0.92 (1.0 N H_2SO_4)[3]

Synthesis   Synthetic methods[1,2]

Imaging/Labeling Applications   Actin filaments;[4] amino acids;[5] amyloid Aβ peptides;[6] bacteria;[7] cells;[8] glycopeptide;[9] hairs;[10,11] metal ions (copper ions,[12] iron ions,[12] chromium ions,[13] lead ions,[13] mercury(II) ions,[14] nitrite ions[15]); microorganisms;[16] monosaccharides;[17] nucleic acids;[18–25] proteins;[26,27] protoplasts;[28] sperms;[29] starch granules;[30] superoxide anions;[31] thiols[32]

Biological/Medical Applications   Detecting risk of Alzheimer's disease and stroke;[33] evaluating/testing sperm quality;[29] identifying bacteria;[7] as a substrate for measuring aromatase activity,[34] azoreductase activity,[35] phospholipase activity,[36] proteases activity (caspase activity, cathepsin C activity, elastase activity proteinase activity);[37–49] implantable drug-delivery devices[50]

Industrial Applications   Color filters;[51,52] liquid crystal displays;[51,52] dye lasers;[53–63] electroluminescent displays;[64–67] inks;[68,69] light-emitting diode (LED);[70–72] papermaking process;[73] recording materials;[74,75] solar cells;[76] silica thin films;[77] sol–gel titania films;[78] waveguides[79]

Safety/Toxicity   Mutagenicity[80]

REFERENCES

1. Cruickshank, K. H.; Bittner, M. L. Green fluorescent labeled nucleotides for use in probes. PCT Int. Appl. WO 9406812, 1994.

2. Ioffe, I. S.; Otten, V. F. Rhodamine dyes and related compounds. I. Progenitor of rhodamines, its preparation and properties. *Zh. Obshch. Khim.* **1961**, *31*, 1511–1516.

3. Kubin, R. F.; Fletcher, A. N. Fluorescence quantum yields of some rhodamine dyes. *J. Luminesc.* **1983**, *27*, 455–462.

4. Kakimoto, T.; Shibaoka, H. A new method for preservation of actin filaments in higher plant cells. *Plant Cell Physiol.* **1987**, *28*, 1581–1585.

5. Fujita, S.; Reddi, P. J.; Toru, T. Fluorescent labeling compounds for amino acid-containing compounds. Jpn. Kokai Tokkyo Koho JP 10330299, 1988.

6. Glabe, C.; Garzon-Rodriguez, W. Fluorescent amyloid Aβ peptides and uses thereof. PCT Int. Appl. WO 9908695, 1999.

7. Oka, A. Dyeability discriminant method for identifying gram negative bacteria. Jpn. Kokai Tokkyo Koho JP 2004208526, 2004.

8. Nowinski, R. C. Method of enhancing direct immunofluorescence staining of cells. U.S. Patent 5068178, 1991.

9. Sugawara, S.; Osumi, K. Labeled glycopeptide derivatives, their manufacture, and kits for detection

of influenza viruses. Jpn. Kokai Tokkyo Koho JP 2011231293, 2011.

10. Guerin, F. Use of a chromoionophore and/or a fluoroionophore compositions for dyeing of human keratinic fibers. Fr. Demande FR 2841468, 2004.

11. Moeller, H.; Meinigke, B. Hair dye compositions containing xanthenes. Ger. Offen. DE 19926377, 2000.

12. Franz, K. J.; Hyman, L. M. Fluorescent prochelators for cellular iron detection. U.S. Pat. Appl. Publ. US 20090253161, 2009.

13. Mo, Z.; Fan, Y.; Wen, Z.; Xiang, X. Colorimetric determination method for simultaneously determining hexavalent chromium and lead ion. Faming Zhuanli Shenqing CN 102288600, 2011.

14. Lee, S. H.; Parthasarathy, A.; Schanze, K. S. A sensitive and selective mercury(II) sensor based on amplified fluorescence quenching in a conjugated polyelectrolyte/spiro-cyclic rhodamine system. *Macromol. Rapid Commun.* **2013**, *34*, 791–795.

15. Zhang, X.; Wang, H.; Fu, N.; Zhang, H. A fluorescence quenching method for the determination of nitrite with Rhodamine 110. *Spectrochim. Acta, Part A: Mol. Biomol. Spectrosc.* **2003**, *59A*, 1667–1672.

16. Nashimoto, K.; Ishida, K.; Ikeda, Y.; Hanaoka, Y. Microorganism staining agent, and its use. Jpn. Tokkyo Koho JP 4370118, 2009.

17. Ijiri, S.; Todoroki, K.; Yoshida, H.; Yoshitake, T.; Nohta, H.; Yamaguchi, M. Sensitive determination of rhodamine 110-labeled monosaccharides in glycoprotein by capillary electrophoresis with laser-induced fluorescence detection. *J. Chromatogr., A* **2010**, *1217*, 3161–3166.

18. Abe, H.; Wang, J.; Furukawa, K.; Oki, K.; Uda, M.; Tsuneda, S.; Ito, Y. A reduction-triggered fluorescence probe for sensing nucleic acids. *Bioconjugate Chem.* **2008**, *19*, 1219–1226.

19. Kimoto, Y. Method of detecting nucleic acid using single fluorescent labeled oligonucleotide primer having non-complementary bases at the 5 ' or 3 ' end. Jpn. Kokai Tokkyo Koho JP 2007075052, 2007.

20. Khan, H. A. The effect of DNA labeling with the fluorescent dyes R110 and R6G on genotype analysis using capillary electrophoresis. *Cell. Mol. Biol. Lett.* **2005**, *10*, 247–253.

21. Brakmann, S.; Lobermann, S. High-density labeling of DNA: preparation and characterization of the target material for single-molecule sequencing. *Angew. Chem., Int. Ed.* **2001**, *40*, 1427–1429.

22. Fujita, S.; Kagiyama, N.; Momiyama, M.; Kondo, Y.; Nishiyauchi, M. Fluorometric detection of substances like nucleic acids immobilized on solid carriers. Jpn. Kokai Tokkyo Koho JP 09152433, 1997.

23. Ansorge, W.; Voss, H.; Stegemann, J.; Wiemann, S. Multiplex DNA sequencing using dye-labeled primers. Ger. Offen. DE 19515552, 1996.

24. Itakura, M. DNA fragment labeling. Jpn. Kokai Tokkyo Koho JP 08140700, 1996.

25. Reeves, R. H.; Bennison, B. W.; LaRock, P. A. Species-specific DNA probes for *Vibrio vulnificus* methods and kits. U.S. Patent 5426025, 1995.

26. Watkins, R. W.; Lavis, L. D.; Kung, V. M.; Los, G. V.; Raines, R. T. Fluorogenic affinity label for the facile, rapid imaging of proteins in live cells. *Org. Biomol. Chem.* **2009**, *7*, 3969–3975.

27. Koller, E. Fluorescent protein labels and labeling protocols. 2. Rhodamine derivatives. *Appl. Fluoresc. Technol.* **1991**, *3*, 20–22.

28. Walko, R. M.; Furtula, V.; Nothnagel, E. A. Analysis of labeling of plant protoplast surface by fluorophore-conjugated lectins. *Protoplasma* **1987**, *141*, 33–46.

29. Schatten, G. P.; Zoran, S. S.; Simerly, C. R.; Navara, C. S. Assay for sperm quality. U.S. Patent 6103481, 2000.

30. Van de Velde, F.; Van Riel, J.; Tromp, R. H. Visualisation of starch granule morphologies using confocal scanning laser microscopy (CSLM). *J. Sci. Food Agric.* **2002**, *82*, 1528–1536.

31. Teranishi, K. Chemiluminescent substance for detecting superoxide anion. Jpn. Kokai Tokkyo Koho JP 2007099967, 2007.

32. Shibata, A.; Furukawa, K.; Abe, H.; Tsuneda, S.; Ito, Y. Rhodamine-based fluorogenic probe for imaging biological thiol. *Bioorg. Med. Chem. Lett.* **2008**, *18*, 2246–2249.

33. Solvason, N. W.; Kell, S. H. Diagnostic tests and reagents for detecting risk of Alzheimer's disease and stroke. PCT Int. Appl. WO 9516787, 1995.

34. Whateley, J. G.; Ismail, R. A.; Laughton, P. G. Method for measuring aromatase activity. PCT Int. Appl. WO 2005012901, 2005.

35. Chevalier, A.; Mercier, C.; Saurel, L.; Orenga, S.; Renard, P.; Romieu, A. The first latent green fluorophores for the detection of azoreductase activity in bacterial cultures. *Chem. Commun.* **2013**, *49*, 8815–8817.

36. Graham, R. J. Fluorescent phospholipase assays and compositions. PCT Int. Appl. WO 2005005977, 2005.

37. Li, J.; Petrassi, H. M.; Tumanut, C.; Masick, B. T.; Trussell, C.; Harris, J. L. Substrate optimization for monitoring cathepsin C activity in live cells. *Bioorg. Med. Chem.* **2009**, *17*, 1064–1070.

38. Steinfeld, R.; Fuhrmann, J. C.; Gaertner, J. Detection of tripeptidyl peptidase I activity in living cells by fluorogenic substrates. *J. Histochem. Cytochem.* **2006**, *54*, 991–996.

39. Shiosaka, S.; Tamura, H. Protease activity assay method by using polymeric membrane. Jpn. Kokai Tokkyo Koho JP 2005253436, 2005.

40. Wang, Z.; Liao, J.; Diwu, Z. N-DEVD-N'-morpholinecarbonyl-rhodamine 110: novel caspase-3 fluorogenic substrates for cell-based apoptosis assay. *Bioorg. Med. Chem. Lett.* **2005**, *15*, 2335–2338.

41. Zhang, H.; Kasibhatla, S.; Guastella, J.; Tseng, B.; Drewe, J.; Cai, S. X. N-Ac-DEVD-N'-(Polyfluorobenzoyl)-R110: Novel cell-permeable fluorogenic caspase substrates for the detection of caspase activity and apoptosis. *Bioconjugate Chem.* **2003**, *14*, 458–463.

42. Cai, S. X.; Zhang, H. Z.; Guastella, J.; Drewe, J.; Yang, W.; Weber, E. Design and synthesis of Rhodamine 110 derivative and caspase-3 substrate for enzyme and cell-based fluorescent assay. *Bioorg. Med. Chem. Lett.* **2000**, *11*, 39–42.

43. Guzikowski, A. P.; Naleway, J. J.; Shipp, C. T.; Schutte, R. C. Synthesis of a macrocyclic rhodamine 110 enzyme substrate as an intracellular probe for caspase 3 activity. *Tetrahedron Lett.* **2000**, *41*, 4733–4735.

44. Neubert, K.; Ansorge, S.; Faust, J.; Buehling, F.; Lorey, S. Method for the determination of protease activity on cell surfaces using fluorescence-labeled peptide substrates. Ger. Offen. DE 19843873, 2000.

45. Liu, J.; Bhalgat, M.; Zhang, C.; Diwu, Z.; Hoyland, B.; Klaubert, D. H. Fluorescent molecular probes V: a sensitive caspase-3 substrate for fluorometric assays. *Bioorg. Med. Chem. Lett.* **1999**, *9*, 3231–3236.

46. Diouri, M.; Geoghegan, K. F.; Weber, J. M. Functional characterization of the adenovirus proteinase using fluorogenic substrates. *Protein Pept. Lett.* **1995**, *2*, 363–370.

47. Klingel, S.; Rothe, G.; Kellermann, W.; Valet, G. Flow cytometric determination of cysteine and serine proteinase activities in living cells with rhodamine 110 substrates. *Methods Cell Biol.* **1994**, *41*, 449–459.

48. Johnson, A. F.; Struthers, M. D.; Pierson, K. B.; Mangel, W. F.; Smith, L. M. Nonisotopic DNA detection system employing elastase and a fluorogenic rhodamine substrate. *Anal. Chem.* **1993**, *65*, 2352–2359.

49. Mangel, W. F.; Leytus, S.; Melhado, L. L. Rhodamine derivatives as chromophoric substrates for proteinases. U.S. Patent 4557862, 1985.

50. Meng, E.; Humayun, M.; Lo, R.; Li, P.; Saati, S. Implantable drug-delivery devices, and apparatus and methods for refilling the devices. U.S. Pat. Appl. Publ. US 20090192493, 2009.

51. Shin, Y. C.; Kwon, Y. S.; Kim, B. G. Colored photosensitive resin composition, colored pattern, color filter and liquid crystal display device having the same. Repub. Korean Kongkae Taeho Kongbo KR 2013048169, 2013.

52. Zhao, M.; Lu, J.; Xue, J.; Sun, W.; Qi, Y.; Xu, C. Color photoresist, color optical filter for trans flective liquid crystal display. Faming Zhuanli Shenqing CN 102654731, 2012.

53. Kornev, A. F.; Pokrovskii, V. P.; Soms, L. N.; Stupnikov, V. K. Full-color laser projectors and blue dye laser sources for laser projectors. Russ. RU 2254649, 2005.

54. Enmanji, K.; Kato, K. Dye lasers. Jpn. Kokai Tokkyo Koho JP 2000183435, 2000.

55. Azim, S. A.; Ghazy, R.; Shaheen, M.; El-Mekawey, F. Investigations of energy transfer from some diolefinic laser dyes to Rhodamine 110. *J. Photochem. Photobiol., A: Chem.* **2000**, *133*, 185–188.

56. Avramopoulos, H.; French, P. M. W.; New, G. H. C.; Opalinska, M. M.; Taylor, J. R.; Williams, J. A. R. Temporal and spectral behavior of passively mode locked dye lasers. *Opt. Commun.* **1990**, *76*, 229–234.

57. French, P. M. W.; Taylor, J. R. Passively mode-locked continuous-wave Rhodamine 110 dye laser. *Opt. Lett.* **1986**, *11*, 297–299.

58. Broyer, M.; Chevaleyre, J.; Delacretaz, G.; Woeste, L. CVL-pumped dye laser for spectroscopic application. *Appl. Phys. B: Photophys. Laser Chem.* **1984**, *B35*, 31–36.

59. Fletcher, A. N.; Hollins, R. A.; Kubin, R. F.; Henry, R. A.; Atienza Moore, T. M.; Pietrak, M. E. Luminescent coolants for solid-state lasers. *Appl. Phys. B: Photophys. Laser Chem.* **1983**, *B30*, 195–202.

60. Lucatorto, T. B.; McIlrath, T. J.; Mayo, S.; Furumoto, H. W. High-stability coaxial flashlamp-pumped dye laser. *Appl. Opt.* **1980**, *19*, 3178–3180.

61. Jain, R. K.; Dienes, A. Polychromatic molecular nitrogen laser-pumped dye lasers. *Spectrosc. Lett.* **1974**, *7*, 491–501.

62. Yarborough, J. M. A cw [continuous wave] dye laser emission spanning the visible spectrum. *Appl. Phys. Lett.* **1974**, *24*, 629–630.

63. Marowsky, G. Tunable flashlamp-pumped dye ring laser of extremely narrow bandwidth. *IEEE J. Quantum Electron.* **1973**, *9*, 245–246.

64. Yamada, K.; Suzuki, T. Color films having curable (meth)acrylate ester copolymer adhesive layers for organic electroluminescent devices. Jpn. Kokai Tokkyo Koho JP 2011023323, 2011.

65. Hasegawa, K. Dispersed electroluminescent devices suppressing UV degradation of fluorescent pigments and manufacture thereof. Jpn. Kokai Tokkyo Koho JP 2005228670, 2005.

66. Oh, H. Y.; Lee, S. K.; Park, C. G.; Seo, J. D.; Kim, M. S.Organic electroluminescent display having red light-emitting layer. Jpn. Kokai Tokkyo Koho JP 2003142269, 2003.

67. Sato, H. Dispersion-type electroluminescent devices. Jpn. Kokai Tokkyo Koho JP 06349580, 1994.

68. Hyodo, M.; Aoyama, Y. Biomolecule ink printed matter and its manufacture. Jpn. Kokai Tokkyo Koho JP 2013202966, 2013.

69. Ren, E.; Feng, L.; Zhao, Z. Colorless ultraviolet fluorescent composition for printing inks. Faming Zhuanli Shenqing Gongkai Shuomingshu CN 1097452, 1995.

70. Kraeuter, G. Methods for producing a light-emitting diode. Ger. Offen. DE 102009022682, 2010.

71. Kraeuter, G. Methods for producing a light-emitting diode. PCT Int. Appl. WO 2010136252, 2010.

72. Butterworth, M. M.; Helbing, R. P. Fluorescent dye added to epoxy of light emitting diode lens. U.S. Patent 5847507, 1998.

73. Thomas, J. L.; Clot, O.; Banks, R. H. Fluorometric method for monitoring surface additives in a papermaking process. U.S. Pat. Appl. Publ. US 20090126889, 2009.

74. Sakota, K.; Maeda, Y.; Iwamoto, M. Optical recording material using photochemical hole burning. Jpn. Kokai Tokkyo Koho JP 05139048, 1993.

75. Aihara, H. Heat-sensitive recording materials. Jpn. Kokai Tokkyo Koho JP 01133782, 1989.

76. Nattestad, A.; Ferguson, M.; Kerr, R.; Cheng, Y.; Bach, U. Dye-sensitized nickel(II)oxide photocathodes for tandem solar cell applications. *Nanotechnology* **2008**, *19*, 295304/1–295304/9.

77. Synak, A.; Bojarski, P.; Grobelna, B.; Kulak, L.; Lewkowicz, A. Determination of local dye concentration in hybrid porous silica thin films. *J. Phys. Chem. C* **2013**, *117*, 11385–11392.

78. Nishikiori, H.; Setiawan, R. A.; Kawamoto, S.; Takagi, S.; Teshima, K.; Fujii, T. Dimerization of xanthene dyes in sol–gel titania films. *Catal. Sci. Technol.* **2013**, *3*, 2786–2792.

79. Sriram, S.; Jackson, Howard E.; Boyd, J. T. Distribution-feedback dye laser integrated with a channel waveguide formed on silicon. *Appl. Phys. Lett.* **1980**, *36*, 721–723.

80. Wuebbles, B. J. Y.; Felton, J. S. Evaluation of laser dye mutagenicity using the Ames/*Salmonella* microsome test. *Environ. Mutagen.* **1985**, *7*, 511–522.

RHODAMINE 123 (RH 123)

CAS Registry Number   62669-70-9

Chemical Structure

CA Index Name   Xanthylium, 3,6-diamino-9-[2-(methoxycarbonyl)phenyl]-, chloride (1:1)

Other Names   Benzoic acid, 2-(6-amino-3-imino-3H-xanthen-9-yl)-, methyl ester, monohydrochloride; Xanthylium, 3,6-diamino-9-[2-(methoxycarbonyl)phenyl]-, chloride; R 22420; R 302; R 123; RH 123; Rh 123; rh 123; Rho 123; Rhod 123; Rhodamine 123

Merck Index Number   8306

Chemical/Dye Class   Xanthene

Molecular Formula $C_{21}H_{17}ClN_2O_3$

Molecular Weight   380.83

Physical Form   Red powder or solid Brown solid[3]

Solubility   Soluble in water, N,N-dimethylformamide, dimethyl sulfoxide, ethanol, ether, methanol

Absorption (λ_{max})   507 nm (MeOH); 500 nm (H_2O)

Emission (λ_{max})   529 nm (MeOH); 530 nm (EtOH)

Molar Extinction Coefficient   101,000 cm^{-1} M^{-1} (MeOH); 75,000 cm^{-1} M^{-1} (H_2O)

Quantum Yield   0.90 (1.0 N H_2SO_4)[99]

Synthesis   Synthetic methods[1–3]

Imaging/Labeling Applications   Mitochondria;[4–16] bacteria;[17–19] cells;[20–22] mercury ions;[23,24] microorganisms;[25] P-glycoprotein;[26–40] retina;[41] sperm;[42–45] stem cells;[46–52] tumors[54,55]

Biological/Medical Applications   Characterizing/identifying stem cells;[46–52] detecting Hg(II) ions;[23,24] evaluating sperm quality/motility;[42–45] measuring/monitoring membrane potential;[4–16] as a substrate for measuring P-glycoprotein activity;[26–40] anticancer drug;[55–57] apoptosis assay;[58,59] cytotoxicity assay;[60] implantable medical device;[61–63] for photodynamic therapy;[64] treating disc degenerative diseases,[51] erectile dysfunction,[52] epilepsy,[53] prostate cancer;[65–67] as temperature sensor[68,69]

Industrial Applications   Colored bubbles;[70,71,76] color filters;[72] liquid crystal displays;[72] light-emitting diodes;[73] luminescent materials;[74] nanophotonic composites;[75] paints;[76,77] semiconductor electrodes;[78] solar cells[79]

Safety/Toxicity   Acute toxicity;[80] aluminum toxicity;[81] cardiotoxicity;[82] cellular toxicity;[83] cytotoxicity;[84–89] metabolic toxicity;[90] mitochondrial toxicity;[84,91] mutagenicity;[92] nephrotoxicity;[93] neurotoxicity;[94–96] ocular toxicity;[97] phototoxicity;[3] reproductive toxicity[98]

REFERENCES

1. Sabnis, R. W. *Handbook of Biological Dyes and Stains*; John Wiley & Sons Inc.: Hoboken, **2010**; pp 418–420.

2. Ross, J. A.; Ross, B. P.; Rubinsztein-Dunlop, H.; McGeary, R. P. Facile synthesis of rhodamine esters using acetyl chloride in alcohol solution. *Synth. Commun.* **2006**, *36*, 1745–1750.

3. Pal, P.; Zeng, H.; Durocher, G.; Girard, D.; Li, T.; Gupta, A. K.; Giasson, R.; Blanchard, L.; Gaboury, L.; Balassy, A.; Turmel, C.; Laperriere, A.; Villeneuve, L. Phototoxicity of some bromine-substituted rhodamine dyes: synthesis, photophysical properties and application as photosensitizers. *Photochem. Photobiol.* **1996**, *63*, 161–168.

4. Ward, M. W. Quantitative analysis of membrane potentials. *Methods Mol. Biol.* **2010**, *591*, 335–351.

5. Kahlert, S.; Zuendorf, G.; Reiser, G. Detection of de- and hyperpolarization of mitochondria of cultures astrocytes and neurons by the cationic fluorescent dye rhodamine 123. *J. Neurosci. Methods* **2008**, *171*, 87–92.

6. Swayne, T. C.; Gay, A. C.; Pon, L. A. Visualization of mitochondria in budding yeast. *Methods Cell Biol.* **2007**, *80*, 591–626.

7. Rueck, A.; Huelshoff, C.; Kinzler, I.; Becker, W.; Steiner, R. SLIM: a new method for molecular imaging. *Microsc. Res. Tech.* **2007**, *70*, 485–492.

8. Lemasters, J. J.; Ramsheh, V. K. Imaging of mitochondrial polarization of depolarization with cationic fluorophores. *Methods Cell Biol.* **2007**, *80*, 283–295.

9. Hattori, F.; Fukuda, K. Method for selecting myocardial cells using intracellular mitochondria as indication. PCT Int. Appl. WO 2006022377, 2006.

10. Baracca, A.; Sgarbi, G.; Solaini, G.; Lenaz, G. Rhodamine 123 as a probe of mitochondrial membrane potential: evaluation of proton flux through F0 during ATP synthesis. *Biochim. Biophys. Acta, Bioenerg.* **2003**, *1606*, 137–146.

11. Feeney, C. J.; Pennefather, P. S.; Gyulkhandanyan, A. V. A cuvette-based fluorometric analysis of mitochondrial membrane potential measured in cultured astrocyte monolayers. *J. Neurosci. Methods* **2003**, *125*, 13–25.

12. Ludovico, P.; Sansonetty, F.; Corte-Real, M. Assessment of mitochondrial membrane potential in yeast cell populations by flow cytometry. *Microbiology* **2001**, *147*, 3335–3343.

13. Mathur, A.; Hong, Y.; Kemp, B. K.; Barrientos, A. A.; Erusalimsky, J. D. Evaluation of fluorescent dyes for the detection of mitochondrial membrane potential changes in cultured cardiomyocytes. *Cardiovasc. Res.* **2000**, *46*, 126–138.

14. Schneckenburger, H.; Gschwend, M. H.; Strauss, W. S. L.; Sailer, R.; Steiner, R. Time-gated microscopic energy transfer measurements for probing mitochondrial metabolism. *J. Fluoresc.* **1997**, *7*, 3–10.

15. Sureda, F. X.; Escubedo, E.; Gabriel, C.; Comas, J.; Camarasa, J.; Camins, A. Mitochondrial membrane potential measurement in rat cerebellar neurons by flow cytometry. *Cytometry* **1997**, *28*, 74–80.

16. Chen, L. B. Fluorescent labeling of mitochondria. *Methods Cell Biol.* **1989**, *29*, 103–123.

17. Rychlik, I.; Cardova, L.; Sevcik, M.; Barrow, P. A. Flow cytometry characterization of *Salmonella typhimurium* mutants defective in proton translocating proteins and stationary-phase growth phenotype. *J. Microbiol. Methods* **2000**, *42*, 255–263.

18. Comas, J.; Vives-Rego, J. Assessment of the effects of gramicidin, formaldehyde, and surfactants on *Escherichia coli* by flow cytometry using nucleic acid and membrane potential dyes. *Cytometry* **1997**, *29*, 58–64.

19. Nexmann J., C.; Rasmussen, J.; Jakobsen, M. Viability staining and flow cytometric detection of *Listeria monocytogenes. J. Microbiol. Methods* **1997**, *28*, 35–43.

20. Hatz, S.; Lambert, J. D. C.; Ogilby, P. R. Measuring the lifetime of singlet oxygen in a single cell: Addressing the issue of cell viability. *Photochem. Photobiol. Sci.* **2007**, *6*, 1106–1116.

21. Bestvater, F.; Spiess, E.; Stobrawa, G.; Hacker, M.; Feurer, T.; Porwol, T.; Berchner-Pfannschmidt, U.; Wotzlaw, C.; Acker, H. Two-photon fluorescence absorption and emission spectra of dyes relevant for cell imaging. *J. Microsc.* **2002**, *208*, 108–115.

22. Wong, C. K. C.; Chan, D. K. O. Isolation of viable cell types from the gill epithelium of Japanese eel *Anguilla japonica. Am. J. Physiol.* **1999**, *276*, R363-R372.

23. Ju, S. U.; Lee, S. Y.; Cho, G. C. Method for detecting Hg(II) concentration in water solution using colloidal gold nanoparticles through detecting fluorescence and surface enhanced Raman scattering intensity change. Repub. Korean Kongkae Taeho Kongbo KR 2011047921, 2011.

24. Ganbold, E.; Park, J.; Ock, K.; Joo, S. Gold nanoparticle-based detection of Hg(II) in an aqueous solution: fluorescence quenching and surface-enhanced Raman scattering study. *Bull. Korean Chem. Soc.* **2011**, *32*, 519–523.

25. Wikstrom, P.; Johansson, T.; Lundstedt, S.; Hagglund, L.; Forsman, M. Phenotypic biomonitoring using multivariate flow cytometric analysis of multi-stained microorganisms. *FEMS Microbiol. Ecol.* **2001**, *34*, 187–196.

26. Al-Jayyoussi, G.; Price, D. F.; Francombe, D.; Taylor, G.; Smith, M. W.; Morris, C.; Edwards, C. D.; Eddershaw, P.; Gumbleton, M. Selectivity in the impact of P-glycoprotein upon pulmonary absorption of airway-dosed substrates: A study in *ex vivo* lung models using chemical inhibition and genetic knockout. *J. Pharm. Sci.* **2013**, *102*, 3382–3394.

27. Lee, S. D.; Osei-Twum, J.; Wasan, K. M. Dose-dependent targeted suppression of P-glycoprotein expression and function in Caco-2 cells. *Mol. Pharm.* **2013**, *10*, 2323–2330.

28. Oh, S.; Han, H.; Kang, K.; Lee, Y.; Lee, M. Menadione serves as a substrate for P-glycoprotein: Implication in chemosensitizing activity. *Arch. Pharm. Res.* **2013**, *36*, 509–516.

29. Kirthivasan, B.; Singh, D.; Bommana, M. M.; Raut, S. L.; Squillante, E.; Sadoqi, M. Active brain targeting of a fluorescent P-gp substrate using polymeric magnetic nanocarrier system. *Nanotechnology* **2012**, *23*, 255102/1–255102/9.

30. He, L.; Zhao, C.; Yan, M.; Zhang, L.; Xia, Y. Inhibition of P-glycoprotein function by procyanidine on blood–brain barrier. *Phytother. Res.* **2009**, *23*, 933–937.

31. Lu, Y.; Pang, T.; Wang, J.; Xiong, D.; Ma, L.; Li, B.; Li, Q.; Wakabayashi, S. Down-regulation of P-glycoprotein expression by sustained intracellular acidification in K562/Dox cells. *Biochem. Biophys. Res. Commun.* **2008**, *377*, 441–446.

32. Senthilkumari, S.; Velpandian, T.; Biswas, N. R.; Saxena, R.; Ghose, S. Evaluation of the modulation of P-glycoprotein (P-gp) on the intraocular disposition of its substrate in rabbits. *Curr. Eye Res.* **2008**, *33*, 333–343.

33. Foeger, F.; Hoyer, H.; Kafedjiiski, K.; Thaurer, M.; Bernkop-Schnuerch, A. *In vivo* comparison of various polymeric and low molecular mass inhibitors of intestinal P-glycoprotein. *Biomaterials* **2006**, *27*, 5855–5860.

34. Chaoui, D.; Faussat, A. M.; Majdak, P.; Tang, R.; Perrot, J. Y.; Pasco, S.; Klein, C.; Marie, J. P.; Legrand, O. JC-1, a sensitive probe for a simultaneous detection of P-glycoprotein activity and apoptosis in leukemic cells. *Cytometry* **2006**, *70B*, 189–196.

35. Constable, P. A.; Lawrenson, J. G.; Dolman, D. E. M.; Arden, G. B.; Abbott, N. J. P-Glycoprotein expression in human retinal pigment epithelium cell lines. *Exp. Eye Res.* **2006**, *83*, 24–30.

36. Zastre, J.; Jackson, J.; Bajwa, M.; Liggins, R.; Iqbal, F.; Burt, H. Enhanced cellular accumulation of a P-glycoprotein substrate, rhodamine-123, by Caco-2 cells using low molecular weight methoxypolyethylene glycol-block-polycaprolactone diblock copolymers. *Eur. J. Pharm. Biopharm.* **2002**, *54*, 299–309.

37. van der Sandt, I. C. J.; Blom-Roosemalen, M. C. M.; de Boer, A. G.; Breimer, D. D. Specificity of doxorubicin versus rhodamine-123 in assessing P-glycoprotein functionality in the LLC-PK1, LLC-PK1:MDR1 and Caco-2 cell lines. *Eur. J. Pharm. Sci.* **2000**, *11*, 207–214.

38. Broxterman, H. J. Measurement of P-glycoprotein function. *Methods Mol. Med.* **1999**, *28*, 53–61.

39. Bosch, I.; Crankshaw, C. L.; Piwnica-Worms, D.; Croop, J. M. Characterization of functional assays of multidrug resistance P-glycoprotein transport activity. *Leukemia* **1997**, *11*, 1131–1137.

40. Petriz, J.; Garcia-Lopez, J. Flow cytometric analysis of P-glycoprotein function using rhodamine 123. *Leukemia* **1997**, *11*, 1124–1130.

41. Calzia, D.; Bianchini, P.; Ravera, S.; Bachi, A.; Candiano, G.; Diaspro, A.; Panfoli, I. Imaging of living mammalian retina *ex vivo* by confocal laser scanning microscopy. *Anal. Methods* **2010**, *2*, 1816–1818.

42. Fraser, L.; Dziekonska, A.; Strzezek, R.; Strzezek, J. Dialysis of boar semen prior to freezing-thawing: Its effects on post-thaw sperm characteristics. *Theriogenology* **2007**, *67*, 994–1003.

43. Paniagua-Chavez, C. G.; Jenkins, J.; Segovia, M.; Tiersch, T. R. Assessment of gamete quality for the eastern oyster (*Crassostrea virginica*) by use of fluorescent dyes. *Cryobiology* **2006**, *53*, 128–138.

44. Aziz, D. M.; Ahlswede, L.; Enbergs, H. Application of MTT reduction assay to evaluate equine sperm viability. *Theriogenology* **2005**, *64*, 1350–1356.

45. Garner, D. L.; Thomas, C. A.; Joerg, H. W.; DeJarnette, J. M.; Marshall, C. E. Fluorometric assessments of mitochondrial function and viability in cryopreserved bovine spermatozoa. *Biol. Reprod.* **1997**, *57*, 1401–1406.

46. Notta, F.; Doulatov, S.; Laurenti, E.; Poeppl, A.; Jurisica, I.; Dick, J. E. Isolation of single human hematopoietic stem cells capable of long-term multilineage engraftment. *Science* **2011**, *333*, 218–221.

47. Donnenberg, V. S.; Meyer, E. M.; Donnenberg, A. D. Measurement of multiple drug resistance transporter activity in putative cancer stem/progenitor cells. *Methods Mol. Biol.* **2009**, *568*, 261–279.

48. McKenzie, J. L.; Takenaka, K.; Gan, O. I.; Doedens, M.; Dick, J. E. Low rhodamine 123 retention identifies long-term human hematopoietic stem cells within the Lin-CD34+CD38- population. *Blood* **2007**, *109*, 543–545.

49. Uchida, N.; Dykstra, B.; Lyons, K.; Leung, F.; Kristiansen, M.; Eaves, C. ABC transporter activities of murine hematopoietic stem cells vary according to their developmental and activation status. *Blood* **2004**, *103*, 4487–4495.

50. Bertoncello, I.; Williams, B. Hematopoietic stem cell characterization by Hoechst 33342 and rhodamine 123 staining. *Methods Mol. Biol.* **2004**, *263*, 181–200.

51. Ichim, T. E. Treatment of disc degenerative disease and compositions for same. PCT Int. Appl. WO 2007136673, 2007.

52. Ichim, T. E. Treatment of erectile dysfunction by stem cell therapy. PCT Int. Appl. WO 2007149548, 2007.

53. Nedergaard, M.; Tian, G. F. Treatment and prevention of epilepsy. PCT Int. Appl. WO 2006062683, 2006.

54. Minakami, T.; Tsuji, T.; Oguni, S.; Hamaguchi, Y.; Tsuruta, K.; Kamihiro, T. Method for detecting leukocyte tumor cells. Jpn. Kokai Tokkyo Koho JP 2002207036, 2002.

55. Trapp, S.; Horobin, R. W. A predictive model for the selective accumulation of chemicals in tumor cells. *Eur. Biophys. J.* **2005**, *34*, 959–966.

56. Park, S. M.; Han, S. B.; Hong, D. H.; Lee, C. W.; Park, S. H.; Jeon, Y. J.; Kim, H. M. Effect of extracellular cations on the chemotherapeutic efficacy of anticancer drugs. *Arch. Pharm. Res.* **2000**, *23*, 59–65.

57. Bodden, W. L.; Palayoor, S. T.; Hait, W. N. Selective antimitochondrial agents inhibit calmodulin. *Biochem. Biophys. Res. Commun.* **1986**, *135*, 574–582.

58. Ferlini, C.; Scambia, G. Assay for apoptosis using the mitochondrial probes, Rhodamine123 and 10-N-nonyl acridine orange. *Nat. Protoc.* **2007**, *2*, 3111–3114.

59. Gorczyca, W.; Melamed, M. R.; Darzynkiewicz, Z. Analysis of apoptosis by flow cytometry. *Methods Mol. Biol.* **1998**, *91*, 217–238.

60. Muraoka, S.; Ochi, C. A fluorometric method for measuring cytotoxicity. Jpn. Kokai Tokkyo Koho JP 2000014399, 2000.

61. Fischer, F. J.; Miller, J. W.; Andrews, M. O. Implantable medical device with anti-neoplastic drug. U.S. Pat. Appl. Publ. US 20060030826, 2006.

62. Ragheb, A. O.; Bates, B. L.; Fearnot, N. E.; Kozma, T. G.; Voorhees, W. D., III,; Gershlick, A. H. Coated implantable medical device. PCT Int. Appl. WO 9836784, 1998.

63. Bates, B. L.; Osborne, T. A.; Roberts, J. W.; Fearnot, N. E.; Kozma, T. G.; Ragheb, A. O.; Voorhees, W. D. Silver implantable medical device. PCT Int. Appl. WO 9817331, 1998.

64. Castro, D. J.; Saxton, R. E.; Rodgerson, D. O.; Fu, Y. S.; Bhuta, S. M.; Fetterman, H. R.; Castro, D. J.; Tartell, P. B.; Ward, P. H. Rhodamine-123 as a new laser dye: *in vivo* study of dye effects on murine metabolism, histology and ultrastructure. *Laryngoscope* **1989**, *99*, 1057–1062.

65. Arcadi, J. A. Composition and method for treating carcinoma. U.S. Patent 7008961, 2006.

66. Jones, L. W.; Narayan, K. S.; Shapiro, C. E.; Sweatman, T. W. Rhodamine-123: Therapy for hormone refractory prostate cancer, A phase I clinical trial. *J. Chemother.* **2005**, *17*, 435–440.

67. Arcadi, J. A. The effect of rhodamine-123 on 3 prostate tumors from the rat. *J. Urol.* **1998**, *160*, 2402–2406.

68. Bousseksou, A.; Salmon, L.; Molnar, G.; Cobo, S. Materials with thermochromic spin transition doped with one or more fluorescent agents for use as temperature sensor. Fr. Demande FR 2952371, 2011.

69. Bousseksou, A.; Salmon, L.; Molnar, G.; Cobo, S. Heat-sensitive spin-transition materials doped with one or more fluorescent agents for use as temperature sensor. PCT Int. Appl. WO 2011058277, 2011.

70. Sabnis, R. W.; Kehoe, T. D. Composition and method for producing colored bubbles. U.S. Pat. Appl. Publ. US 20120244777, 2012.

71. Sabnis, R. W.; Kehoe, T. D. Composition and method for producing colored bubbles. U.S. Pat. Appl. Publ. US 20060004110, 2006.

72. Shin, Y. C.; Kwon, Y. S.; Kim, B. G. Colored photosensitive resin composition, colored pattern, color filter and liquid crystal display device having the same. Korean Kongkae Taeho Kongbo KR 2013048169, 2013.

73. Kim, J.; Kim, K.; Yoo, S. Il; Sohn, B. Dispersion of micelle-encapsulated fluorophores in a polymer matrix for control of color of light emitted by light-emitting diodes. *Thin Solid Films* **2011**, *519*, 8161–8165.

74. Reisfeld, R.; Levchenko, V.; Saraidarov, T. Interaction of luminescent dyes with noble metal nanoparticles in organic–inorganic glasses for future luminescent materials. *Polym. Adv. Technol.* **2011**, *22*, 60–64.

75. Levitsky, I. A.; Liang, J.; Xu, J. M. Highly ordered arrays of organic–inorganic nanophotonic composites. *Appl. Phys. Lett.* **2002**, *81*, 1696–1698.

76. Sabnis, R. W.; Kehoe, T. D.; Balchunis, R. J. Novelty compositions with color changing indicator. PCT Int. Appl. WO 2006105191, 2006.

77. Fujishima, A.; Tada, K. Photocatalytic paint containing pigment fading or discoloring by photocatalytic activity and method for coating using same. Jpn. Kokai Tokkyo Koho JP 2001321676, 2001.

78. Sayama, K.; Arakawa, H.; Sugihara, H. Organic pigment sensitized oxide semiconductor electrodes and photoelectrochemical cells using the electrodes. Jpn. Kokai Tokkyo Koho JP 10092477, 1998.

79. Nanba, N.; Kadota, A.; Tanabe, J. Dye-sensitized solar cells having low internal resistance. Jpn. Kokai Tokkyo Koho JP 2001076777, 2001.

80. Hartig, S.; Fries, S.; Balcarcel, R. R. Reduced mitochondrial membrane potential and metabolism correspond to acute chloroform toxicity of *in vitro* hepatocytes. *J. Appl. Toxicol.* **2005**, *25*, 310–317.

81. Yamamoto, Y.; Kobayashi, Y.; Devi, S. R.; Rikiishi, S.; Matsumoto, H. Aluminum toxicity is associated with mitochondrial dysfunction and the production of reactive oxygen species in plant cells. *Plant Physiol.* **2002**, *128*, 63–72.

82. Lampidis, T. J.; Salet, C.; Moreno, G.; Chen, L. B. Effects of the mitochondrial probe rhodamine 123 and related analogs on the function and viability of pulsating myocardial cells in culture. *Agents Actions* **1984**, *14*, 751–757.

83. Gupta, R. S.; Dudani, A. K. Species-specific differences in the toxicity of rhodamine 123 toward cultured mammalian cells. *J. Cell. Physiol.* **1987**, *130*, 321–327.

84. Pourahmad, J.; Rabiei, M.; Jokar, F.; O'brien, P. J. A comparison of hepatocyte cytotoxic mechanisms for chromate and arsenite. *Toxicology* **2005**, *206*, 449–460.

85. Palmeira, C. M.; Moreno, A. J.; Madeira, V. M.; Wallace, K. B. Continuous monitoring of mitochondrial membrane potential in hepatocyte cell suspensions. *J. Pharmacol. Toxicol. Methods* **1996**, *35*, 35–43.

86. VanDerWal, J.; Lagerberg, J. W. M.; Dubbelman, T. M. A. R.; VanStevennick, J. Interaction of photodynamically induced cell killing and dark cytotoxicity of rhodamine 123. *Photochem. Photobiol.* **1995**, *62*, 757–763.

87. Shinomiya, N.; Shinomiya, M.; Wakiyama, H.; Katsura, Y.; Rokutanda, M. Enhancement of CDDP cytotoxicity by caffeine is characterized by apoptotic cell death. *Exp. Cell Res.* **1994**, *210*, 236–242.

88. Jiang, T.; Grant, R. L.; Acosta, D. A digitized fluorescence imaging study of intracellular free calcium, mitochondrial integrity and cytotoxicity in rat renal cells exposed to ionomycin, a calcium ionophore. *Toxicology* **1993**, *85*, 41–65.

89. Rahn, C. A.; Bombick, D. W.; Doolittle, D. J. Assessment of mitochondrial membrane potential as an indicator of cytotoxicity. *Fundam. Appl. Toxicol.* **1991**, *16*, 435–448.

90. Downing, T. W.; Garner, D. L.; Ericsson, S. A.; Redelman, D. Metabolic toxicity of fluorescent stains on thawed cryopreserved bovine sperm cells. *J. Histochem. Cytochem.* **1991**, *39*, 485–489.

91. Zhang, H.; Chen, Q.; Xiang, M.; Ma, C.; Huang, Q.; Yang, S. *In silico* prediction of mitochondrial toxicity by using GA-CG-SVM approach. *Toxicol. in Vitro* **2009**, *23*, 134–140.

92. Ferguson, L. R.; Baguley, B. C. Verapamil as a co-mutagen in the *Salmonella*/mammalian microsome mutagenicity test. *Mutat. Res. Lett.* **1988**, *209*, 57–62.

93. Zhang, J. G.; Lindup, W. E. Cisplatin nephrotoxicity: decreases in mitochondrial protein sulphydryl concentration and calcium uptake by mitochondria from rat renal cortical slices. *Biochem. Pharmacol.* **1994**, *47*, 1127–1135.

94. de Arriba, S. G.; Krugel, U.; Regenthal, R.; Vissiennon, Z.; Verdaguer, E.; Lewerenz, A.; Garcia-Jorda, E.; Pallas, M.; Camins, A.; Munch, G.; Nieber, K.; Allgaier, C. Carbonyl stress and NMDA receptor activation contribute to methylglyoxal neurotoxicity. *Free Radical Biol. Med.* **2006**, *40*, 779–790.

95. Uchida, K.; Yamada, M.; Hayashi, T.; Mine, Y.; Kawase, T. Possible harmful effects on central nervous system cells in the use of physiological saline as an irrigant during neurosurgical procedures. *Surg. Neurol.* **2004**, *62*, 96–105;

96. Chute, S. K.; Flint, O. P.; Durham, S. K. Analysis of the steady-state dynamics of organelle motion in cultured neurites: putative indicator of neurotoxic effect. *Clin. Exp. Pharmacol. Physiol.* **1995**, *22*, 360–361.

97. Bantseev, V.; McCanna, D.; Banh, A.; Wong, W. W.; Moran, K. L.; Dixon, D. G.; Trevithick, J. R.; Sivak, J. G. Mechanisms of ocular toxicity using the *in vitro* bovine lens and sodium dodecyl sulfate as a chemical model. *Toxicol. Sci.* **2003**, *73*, 98–107.

98. O'Connell, M.; McClure, N.; Lewis, S. E. M. The effects of cryopreservation on sperm morphology, motility and mitochondrial function. *Hum. Reprod.* **2002**, *17*, 704–709.

99. Kubin, R. F.; Fletcher, A. N. Fluorescence quantum yields of some rhodamine dyes. *J. Luminesc.* **1983**, *27*, 455–462.

5,5′,6,6′-TETRACHLORO-1,1′,3,3′-TETRAETHYLBENZIMIDAZOLO-CARBOCYANINE IODIDE (JC-1)

CAS Registry Number 3520-43-2
Chemical Structure

CA Index Name 1*H*-Benzimidazolium, 5,6-dichloro-2-[3-(5,6-dichloro-1,3-diethyl-1,3-dihydro-2*H*-benzimidazol-2-ylidene)-1-propen-1-yl]-1,3-diethyl-, iodide (1:1)

Other Names 1*H*-Benzimidazolium, 5,6-dichloro-2-[3-(5,6-dichloro-1,3-diethyl-1,3-dihydro-2*H*-benzimidazol-2-ylidene)-1-propenyl]-1,3-diethyl-, iodide; 5,6-Dichloro-2-[3-(5,6-dichloro-1,3-diethyl-2-benzimidazol-inylidene)propenyl]-1,3-diethylbenzimidazolium iodide; Benzimidazolium, 5,6-dichloro-2-[3-(5,6-dichloro-1,3-diethyl-2-benzimidazolinylidene)propenyl]-1,3-diethyl-, iodide; Benzimidazolocarbocyanine iodide, 5,5′,6,6′-tetrachloro-1,1′,3,3′-tetraethyl-; Imidacarbocyanine iodide, 1,1′,3,3′-tetraethyl-5,5′,6,6′-tetrachloro-; 1,1′,3,3′-Tetraethyl-5,5′,6,6′-tetrachlorobenzimida-zolocarbocyanine iodide; 1,1′,3,3′-Tetraethyl-5,5′,6,6′-tetrachloroimidacarbocyanine iodide; 5,5′,6,6′-Tetrachloro-1,1′,3,3′-tetraethylbenzimidazolocarbocyanine iodide; Bis(5,6-dichloro-1,3-di-ethyl-2-benzimidazole)trimethine-cyanine iodide; CBIC$_2$; CBIC$_2$(3); JC 1; NK 1420

Merck Index Number Not listed
Chemical/Dye Class Cyanine
Molecular Formula C$_{25}$H$_{27}$Cl$_4$IN$_4$
Molecular Weight 652.23
Physical Form Reddish lustrous crystals[7,9]

Solubility Soluble in *N,N*-dimethylformamide, dimethyl sulfoxide, methanol
Melting Point 275–278 °C[7,9]
Absorption (λ_{max}) 514 nm (MeOH)
Emission (λ_{max}) 529 nm (MeOH)
Molar Extinction Coefficient 195,000 cm^{-1} M^{-1} (MeOH)
Synthesis Synthetic methods[1–10]
Imaging/Labeling Applications Amyloid plaque;[11] astrocytes;[12,30] bacteria;[13] fungi;[13–15] hairs;[16] insect cells;[17] mitochondria;[1,17–43,46,47,52–54,56–58,60–63,65,78,79] neurons;[45–47] P-glycoprotein (P-gp);[66–68] proteins;[48] retina tissue;[49] sperms;[50–58] α-synuclein;[59] yeast cells[48]

Biological/Medical Applications Analyzing mitochondrial morphology and function;[19,20,36,54,56,58,65] assessing sperm integrity;[56] detecting prostate cancer;[19] diagnosing/treating Alzheimer's disease;[60] evaluating sperm motility and chemotaxis;[51] measuring *in vivo* hematotoxicity;[26] measuring/monitoring membrane potential;[1,13,17,21–44,46,52,53,56–58,61–64,78] apoptosis assay;[26,61–63] cytotoxicity assay;[23,64,65] P-glycoprotein (P-gp) activity acute myeloid leukemia (AML) assay;[66–68] multidrug resistance assay;[69,70] as temperature sensor[71,72]

Industrial Applications Langmuir-Blodgett films;[73] lasing systems;[74] nonlinear optical materials;[75] photographic materials[1,2,7–10,76–79]

Safety/Toxicity Hepatotoxicity;[80] mitochondrial toxicity[81]

REFERENCES

1. Sabnis, R. W. *Handbook of Biological Dyes and Stains*; John Wiley & Sons Inc.: Hoboken, **2010**; pp 252–255.

2. Peng, Z. H.; Geise, H. J.; Zhou, X. F.; Peng, B. X.; Carleer, R.; Dommisse, R. The structure of benzimidazole cyanine dyes, their spectroscopy, and

Handbook of Fluorescent Dyes and Probes, First Edition. R. W. Sabnis.
© 2015 John Wiley & Sons, Inc. Published 2015 by John Wiley & Sons, Inc.

their performance in photographic emulsions. *Liebigs Ann./Rec.* **1997**, 27–33.

3. Gandino, M.; Baldassarri, A. Benzimidazacarbocyanines. Fr. FR 1525450, 1968.

4. Yagupol'skii, L. M.; Troitskaya, V. I.; Levkoev, I. I.; Lifshits, E. B.; Yufa, P. A.; Barvyn, N. S. Cyanine dyes containing fluorine. XIV. Some ditri- and tetra-substituted benzimidazolocyanines. *Zh. Obshch. Khim.* **1967**, *37*, 191–198.

5. Kodak, Soc. Anon. Photographic emulsions. BE 659415, 1965.

6. Gevaert Photo-Producten N.V. Benzimidazole cyanines. BE 510948, 1952.

7. Van Lare, E. J. Symmetrical carbocyanine dyes for optically sensitizing gelatin silver halide emulsions. GB 754546, 1956.

8. Allen, C. F. H.; Kennard, K. C. Stabilized photographic silver halide emulsions. U.S. Patent 2776211, 1957.

9. Van Lare, E. J. Symmetrical carbocyanin dyes for optically sensitizing gelatin silver halide emulsions. U.S. Patent 2739149, 1956.

10. Carroll, B. H.; Jones, J. E. Supersensitization of photographic emulsions. U.S. Patent 2688545, 1954.

11. Bertoncini, C. W.; Celej, M. S. Small molecule fluorescent probes for the detection of amyloid self-assembly *in vitro* and *in vivo*. *Curr. Protein Pept. Sci.* **2011**, *12*, 206–220.

12. Chen, X.; Schluesener, H. J. Mode of dye loading affects staining outcomes of fluorescent dyes in astrocytes exposed to multiwalled carbon nanotubes. *Carbon* **2010**, *48*, 730–743.

13. Little, R. G., II; Abrahamson, S.; Wong, P. Identification of novel antimicrobial agents using membrane potential indicator dyes. PCT Int. Appl. WO 2000018951, 2000.

14. Pina-Vaz, C.; Rodrigues, A. G. Evaluation of antifungal susceptibility using flow cytometry. *Methods Mol. Biol.* **2010**, *638*, 281–289.

15. Pina-Vaz, C.; Sansonetty, F.; Rodrigues, A. G.; Costa-Oliveira, S.; Tavares, C.; Martinez-De-Oliveira, J. Cytometric approach for a rapid evaluation of susceptibility of *Candida* strains to antifungals. *Clin. Microbiol. Infect.* **2001**, *7*, 609–618.

16. Ohashi, Y.; Miyabe, H.; Matsunaga, K. Hair dye composition. Eur. Pat. Appl. EP 1166753, 2002.

17. Kong, M.; Xu, M.; He, Y.; Zhang, Y. Expression of *Helicobacter pylori* ggt gene in baculovirus expression system and activity analysis of its products. *Pol. J. Microbiol.* **2011**, *60*, 203–207.

18. Brickley, M. R.; Lawrie, E.; Weise, V.; Hawes, C.; Cobb, A. H. Use of a potentiometric vital dye to determine the effect of the herbicide bromoxynil octanoate on mitochondrial bioenenergetics in *Chlamydomonas reinhardtii*. *Pest Manag. Sci.* **2012**, *68*, 580–586.

19. Dickman, D. Methods of detecting prostate cancer. PCT Int. Appl. WO 2006054296, 2006.

20. Poot, M.; Zhang, Y. Z.; Kraemer, J. A.; Wells, K. S.; Jones, L. J.; Hanzel, D. K.; Lugade, A. G.; Singer, V. L.; Haugland, R. P. Analysis of mitochondrial morphology and function with novel fixable fluorescent stains. *J. Histochem. Cytochem.* **1996**, *44*, 1363–1372.

21. Lin, H.; Liu, S.; Lai, H.; Lai, I. Isolated mitochondria infusion mitigates ischemia-reperfusion injury of the liver in rats. *Shock* **2013**, *39*, 304–310.

22. Baban, B.; Liu, J. Y.; Mozaffari, M. S. Pressure overload regulates expression of cytokines, γH2AX, and growth arrest- and DNA-damage inducible protein 153 via glycogen synthase kinase-3β in ischemic-reperfused hearts. *Hypertension* **2013**, *61*, 95–104.

23. Sathishkumar, K.; Gao, X.; Raghavamenon, A. C.; Murthy, S. N.; Kadowitz, P. J.; Uppu, R. M. Determination of glutathione, mitochondrial transmembrane potential, and cytotoxicity in H9c2 cardiomyoblasts exposed to reactive oxygen and nitrogen species. *Methods Mol. Biol.* **2010**, *610*, 51–61.

24. Widlansky, M. E.; Wang, J.; Shenouda, S. M.; Hagen, T. M.; Smith, A. R.; Kizhakekuttu, T. J.; Kluge, M. A.; Weihrauch, D.; Gutterman, D. D.; Vita, J. A. Altered mitochondrial membrane potential, mass, and morphology in the mononuclear cells of humans with type 2 diabetes. *Transl. Res.* **2010**, *156*, 15–25.

25. Pietilae, M.; Lehtonen, S.; Naerhi, M.; Hassinen, I. E.; Leskelae, H.; Aranko, K.; Nordstroem, K.; Vepsaelaeinen, A.; Lehenkari, P. Mitochondrial function determines the viability and osteogenic potency of human mesenchymal stem *Cells. Tissue Eng.*, *Part C: Methods* **2010**, *16*, 435–445.

26. Dertinger, S. D.; Bemis, J. C.; Bryce, S. M. Method for measuring *in vivo* hematotoxicity with an emphasis on radiation exposure assessment. U.S. Pat. Appl. Publ. US 20080311586, 2008.

27. Dressler, C.; Beuthan, J.; Mueller, G.; Zabarylo, U.; Minet, O. Fluorescence imaging of heat-stress induced mitochondrial long-term depolarization in breast cancer cells. *J. Fluoresc.* **2006**, *16*, 689–695.

28. Szilagyi, G.; Simon, L.; Koska, P.; Telek, G.; Nagy, Z. Visualization of mitochondrial membrane potential and reactive oxygen species via double staining. *Neurosci. Lett.* **2006**, *399*, 206–209.

29. Lecoeur, H.; Langonne, A.; Baux, L.; Rebouillat, D.; Rustin, P.; Prevost, M. C.; Brenner, C.; Edelman, L.; Jacotot, E. Real-time flow cytometry analysis of permeability transition in isolated mitochondria. *Exp. Cell Res.* **2004**, *294*, 106–117.

30. Feeney, C. J.; Pennefather, P. S.; Gyulkhandanyan, A. V. A cuvette-based fluorometric analysis of mitochondrial membrane potential measured in cultured astrocyte monolayers. *J. Neurosci. Methods* **2003**, *125*, 13–25.

31. Bernas, T.; Dobrucki, J. Mitochondrial and nonmitochondrial reduction of MTT: Interaction of MTT with TMRE, JC-1, and NAO mitochondrial fluorescent probes. *Cytometry* **2002**, *47*, 236–242.

32. Mathur, A.; Hong, Y.; Kemp, B. K.; Barrientos, A. A.; Erusalimsky, J. D. Evaluation of fluorescent dyes for the detection of mitochondrial membrane potential changes in cultured cardiomyocytes. *Cardiovas. Res.* **2000**, *46*, 126–138.

33. Zamzami, N.; Metivier, D.; Kroemer, G. Quantitation of mitochondrial transmembrane potential in cells and in isolated mitochondria. *Methods Enzymol.* **2000**, *322*, 208–213.

34. Dykens, J. A.; Velicelebi, G.; Ghosh, S. S. Compositions and methods for assaying subcellular conditions and processes using energy transfer for drug screening. PCT Int. Appl. WO 2000079274, 2000.

35. Nuydens, R.; Novalbos, J.; Dispersyn, G.; Weber, C.; Borgers, M.; Geerts, H. A rapid method for the evaluation of compounds with mitochondria-protective properties. *J. Neurosci. Methods* **1999**, *92*, 153–159.

36. Sick, T. J.; Perez-Pinzon, M. A. Optical methods for probing mitochondrial function in brain slices. *Methods* **1999**, *18*, 104–108.

37. Salvioli, S.; Maseroli, R.; Pazienza, T. L.; Bobyleva, V.; Cossarizza, A. Use of flow cytometry as a tool to study mitochondrial membrane potential in isolated, living hepatocytes. *Biochemistry (Moscow)* **1998**, *63*, 235–238.

38. Salvioli, S.; Ardizzoni, A.; Franceschi, C.; Cossarizza, A. JC-1, but not DiOC6(3) or rhodamine 123, is a reliable fluorescent probe to assess ΔΨ changes in intact cells: implications for studies on mitochondrial functionality during apoptosis. *FEBS Lett.* **1997**, *411*, 77–82.

39. Cossarizza, A.; Ceccarelli, D.; Masini, A. Functional heterogeneity of an isolated mitochondrial population revealed by cytofluorometric analysis at the single organelle level. *Exp. Cell Res.* **1996**, *222*, 84–94.

40. Reers, M.; Smiley, S. T.; Mottola-Hartshorn, C.; Chen, A.; Lin, M.; Chen, L. B. Mitochondrial membrane potential monitored by JC-1 dye. *Methods Enzymol.* **1995**, *260*, 406–417.

41. Cossarizza, A.; Cooper, E. L.; Quaglino, D.; Salvioli, S.; Kalachnikova, G.; Franceschi, C. Mitochondrial mass and membrane potential in coelomocytes from the earthworm *Eisenia foetida*: Studies with fluorescent probes in single intact cells. *Biochem. Biophys. Res. Commun.* **1995**, *214*, 503–510.

42. Cossarizza, A.; Baccaranai-Contri, M.; Kalashnikova, G.; Franceschi, C. A new method for the cytofluorimetric analysis of mitochondrial membrane potential using the J-aggregate-forming lipophilic cation 5,5′,6,6′-tetrachloro-1,1′,3,3′-tetraethylbenzimidazolcarbocyanine iodide (JC-1). *Biochem. Biophys. Res. Commun.* **1993**, *197*, 40–45.

43. Smiley, S. T.; Reers, M.; Mottola-Hartshorn, C.; Lin, M.; Chen, A.; Smith, T. W.; Steele, G. D., Jr.; Chen, L. B. Intracellular heterogeneity in mitochondrial membrane potentials revealed by a J-aggregate-forming lipophilic cation JC-1. *Proc. Natl. Acad. Sci. USA* **1991**, *88*, 3671–3675.

44. Steele, G. D., Jr.; Chen, L. B.; Smiley, S. T. Fluorometric method of measuring cell membrane potential using J-aggregate-forming cyanine dyes. PCT Int. Appl. WO 9015317, 1990.

45. Sung, D. K.; Chang, Y. S.; Kang, S.; Song, H. Y.; Park, W. S.; Lee, B. H. Comparative evaluation of hypoxic-ischemic brain injury by flow cytometric analysis of mitochondrial membrane potential with JC-1 in neonatal rats. *J. Neurosci. Methods* **2010**, *193*, 232–238.

46. Safiulina, D.; Kaasik, A.; Seppet, E.; Peet, N.; Zharkovsky, A.; Seppet, E. Method for *in situ* detection of the mitochondrial function in neurons. *J. Neurosci. Methods* **2004**, *137*, 87–95.

47. Dedov, V. N.; Cox, G. C.; Roufogalis, B. D. Visualization of mitochondria in living neurons with single- and two-photon fluorescence laser microscopy. *Micron* **2001**, *32*, 653–660.

48. Kim, E. J.; Barker, L.; Burnet, M.; Guse, J.; Luyten, K.; Tsotsou, G. Method for identifying transport proteins. PCT Int. Appl. WO 2003038092, 2003.

49. Watanabe, K.; Shindo, T.; Nomoto, T.; Miyazaki, T.; Tanaka, T.; Nishimura, A.; Shimada, Y.; Nishimura, K. Retina tissue staining composition, retina tissue staining method. Jpn. Kokai Tokkyo Koho JP 2010148447, 2010.

50. Becchetti, E.; Calafiore, R.; Calvitti, M.; De Toni, L.; Ferlin, A.; Foresta, C.; Garolla, A.; Luca, G.; Mancuso, F.; Menegazzo, M. Prolongation and improvement of viability and functionality of spermatozoa *in vitro* in co-culture with Sertoli cells. Ital. IT 1396725, 2012.

51. Xie, L.; Ma, R.; Han, C.; Su, K.; Zhang, Q.; Qiu, T.; Wang, L.; Huang, G.; Qiao, J.; Wang, J. Integration of sperm motility and chemotaxis screening with a microchannel-based device. *Clin. Chem.* **2010**, *56*, 1270–1278.

52. Guthrie, H. D.; Welch, G. R. Determination of high mitochondrial membrane potential in spermatozoa loaded with the mitochondrial probe 5,5′-6,6′-tetrachloro-1,1′,3,3′-tetraethylbenzimidazo-lyl-carbocyanine iodide (JC-1) by using fluorescence-activated flow cytometry. *Methods Mol. Biol.* **2008**, *477*, 89–97.

53. Martinez-Pastor, F.; Johannisson, A.; Gil, J.; Kaabi, M.; Anel, L.; Paz, P.; Rodriguez-Martinez, H. Use of chromatin stability assay, mitochondrial stain JC-1, and fluorometric assessment of plasma membrane to evaluate frozen-thawed ram semen. *Anim. Reprod. Sci.* **2004**, *84*, 121–133.

54. Gravance, C. G.; Garner, D. L.; Miller, M. G.; Berger, T. Flow cytometric assessment of changes in rat sperm mitochondrial function after treatment with pentachlorophenol. *Toxicol. in Vitro* **2003**, *17*, 253–257.

55. De Pauw, I. M. C.; Van Soom, A.; Laevens, H.; Verberckmoes, S.; De Kruif, A. Sperm binding to epithelial oviduct explants in bulls with different nonreturn rates investigated with a new *in vitro* model. *Biol. Reprod.* **2002**, *67*, 1073–1079.

56. Gravance, C. G.; Garner, D. L.; Miller, M. G.; Berger, T. Fluorescent probes and flow cytometry to assess rat sperm integrity and mitochondrial function. *Reprod. Toxicol.* **2001**, *15*, 5–10.

57. Troiano, L.; Granata, A. R. M.; Cossarizza, A.; Kalashnikova, G.; Bianchi, R.; Pini, G.; Tropea, F.; Carani, C.; Franceschi, C. Mitochondrial membrane potential and DNA stainability in human sperm cells: A flow cytometry analysis with implications for male infertility. *Exp. Cell Res.* **1998**, *241*, 384–393.

58. Garner, D. L.; Thomas, C. A.; Joerg, H. W.; DeJarnette, J. M.; Marshall, C. E. Fluorometric assessments of mitochondrial function and viability in cryopreserved bovine spermatozoa. *Biol. Reprod.* **1997**, *57*, 1401–1406.

59. Lee, J.; Lee, I.; Choe, Y.; Kang, Sungsoo; K., Hui Y.; Gai, W.; Hahn, J.; Paik, S. R. Real-time analysis of amyloid fibril formation of α-synuclein using a fibrillation-state-specific fluorescent probe of JC-1. *Biochem. J.* **2009**, *418*, 311–323.

60. Herrnstadt, C.; Parker, W. D. Method of targeting conjugate molecules to mitochondria. U.S. Patent 6171859, 2001.

61. Troiano, L.; Ferraresi, R.; Lugli, E.; Nemes, E.; Roat, E.; Nasi, M.; Pinti, M.; Cossarizza, A. Multiparametric analysis of cells with different mitochondrial membrane potential during apoptosis by polychromatic flow cytometry. *Nat. Protoc.* **2007**, *2*, 2719–2727.

62. Morrison, M. L.; Williamson, K.; Arthur, K.; Price, G. J.; Hamilton, P. W.; Maxwell, P. Phenotypic changes in mitochondrial membrane potential ($\Delta\psi$m) during valinomycin-induced depolarisation and apoptosis. *Cell. Oncol.* **2005**, *27*, 231–236.

63. Lecoeur, H.; Chauvier, D.; Langonne, A.; Rebouillat, D.; Brugg, B.; Mariani, J.; Edelman, L.; Jacotot, E. Dynamic analysis of apoptosis in primary cortical neurons by fixed- and real-time cytofluorometry. *Apoptosis* **2004**, *9*, 157–169.

64. Murakami, T. Cytotoxicity test method by measuring membrane electric potential. Jpn. Kokai Tokkyo Koho JP 2000300290, 2000.

65. Lilius, H.; Haestbacka, T.; Isomaa, B. A combination of fluorescent probes for evaluation of cytotoxicity and toxic mechanisms in isolated rainbow trout hepatocytes. *Toxicology in Vitro* **1996**, *10*, 341–348.

66. Benderra, Z.; Faussat, A. M.; Sayada, L.; Perrot, J.; Tang, R.; Chaoui, D.; Morjani, H.; Marzac, C.; Marie, J.; Legrand, O. MRP3, BCRP, and P-Glycoprotein activities are prognostic factors in adult acute myeloid leukemia. *Clin. Cancer Res.* **2005**, *11*, 7764–7772.

67. Swerts, K.; De Moerloose, B.; Dhooge, C.; Noens, L.; Laureys, G.; Benoit, Y.; Philippe, J. Comparison of two functional flow cytometric assays to assess P-gp activity in acute leukemia. *Leuk. Lymphoma* **2004**, *45*, 2221–2228.

68. Legrand, O.; Perrot, J. Y.; Simonin, G.; Baudard, M.; Marie, J. P. JC-1: a very sensitive fluorescent probe to test Pgp activity in adult acute myeloid leukemia. *Blood* **2001**, *97*, 502–508.

69. Loken, M. R. Functional multi-drug resistance assay. U.S. Pat. Appl. Publ. US 20060263834, 2006.

70. Kuhnel, J. M.; Perrot, J. Y.; Faussat, A. M.; Marie, J. P.; Schwaller, M. A. Functional assay of multidrug

resistant cells using JC-1, a carbocyanine fluorescent probe. *Leukemia* **1997**, *11*, 1147–1155.

71. Bousseksou, A.; Salmon, L.; Molnar, G.; Cobo, S. Materials with thermochromic spin transition doped with one or more fluorescent agents for use as temperature sensor. Fr. Demande FR 2952371, 2011.

72. Bousseksou, A.; Salmon, L.; Molnar, G.; Cobo, S. Heat-sensitive spin-transition materials doped with one or more fluorescent agents for use as temperature sensor. PCT Int. Appl. WO 2011058277, 2011.

73. Miyano, K.; Ishikawa, H.; Tomioka, A. Homogeneous width of excitons in Langmuir-Blodgett films. *Mater. Sci. Eng., B* **1997**, *B48*, 122–125.

74. Ozcelik, S.; Akins, D. L. Extremely low excitation threshold, superradiant, molecular aggregate lasing system. *Appl. Phys. Lett.* **1997**, *71*, 3057–3059.

75. Kato, T. Composition used as nonlinear optical material. Jpn. Kokai Tokkyo Koho JP 2005128152, 2005.

76. Fenton, D. E.; Buitano, L. A.; Link, S. G.; Johnston, S. G.; Moorehouse, D. B. Photographic film element containing an emulsion with dual peak green responsivity. U.S. Pat. Appl. Publ. US 20050130083, 2005.

77. Ikeda, T. Silver halide emulsion containing cyanine dye for improved storage stability and high sensitivity and heat-developable photographic material. Jpn. Kokai Tokkyo Koho JP 2004094015, 2004.

78. Miura, T.; Tanaka, A. Direct-positive silver halide photographic material containing cyanine dye and desensitizer with quaternary nitrogen atom. Jpn. Kokai Tokkyo Koho JP 63256947, 1988.

79. Miura, T.; Tanaka, A. Direct-positive silver halide photographic material containing cyanine dye and onium-type desensitizer. Jpn. Kokai Tokkyo Koho JP 63256946, 1988.

80. Oshima, R.; Nakano, H.; Katayama, M.; Sakurai, J.; Wu, W.; Koizumi, S.; Asano, T.; Watanabe, T.; Asakura, T.; Ohta, T.; Otsubo, T. Modification of the hepatic mitochondrial proteome in response to ischemic preconditioning following ischemia-reperfusion injury of the rat liver. *Eur. Surg. Res.* **2008**, *40*, 247–255.

81. Zhang, H.; Chen, Q.; Xiang, M.; Ma, C.; Huang, Q.; Yang, S. *In silico* prediction of mitochondrial toxicity by using GA-CG-SVM approach. *Toxicol. in Vitro* **2009**, *23*, 134–140.

TETRAMETHYLRHODAMINE ETHYL ESTER PERCHLORATE (TMRE)

CAS Registry Number   115532-52-0
Chemical Structure

CA Index Name   Xanthylium, 3,6-bis(dimethylamino)-9-[2-(ethoxycarbonyl)phenyl]-, perchlorate (1:1)

Other Names   Xanthylium, 3,6-bis(dimethylamino)-9-[2-(ethoxycarbonyl)phenyl]-, perchlorate; T 669; Tetramethylrhodamine ethyl ester perchlorate; TMRE

Merck Index Number   Not listed
Chemical/Dye Class   Xanthene
Molecular Formula $C_{26}H_{27}ClN_2O_7$
Molecular Weight   514.95
Physical Form   Dark green crystals[2]
Solubility   Soluble in *N,N*-dimethylformamide, dimethyl sulfoxide, ethanol, methanol
Melting Point   264–266 °C (decompose)[2]
Absorption (λ_{max})   549 nm (MeOH)
Emission (λ_{max})   574 nm (MeOH)
Molar Extinction Coefficient   109,000 cm^{-1} M^{-1} (MeOH)
Synthesis   Synthetic methods[1,2]
Imaging/Labeling Applications   Mitochondria;[1–14,16,17] cells[1–4,15,20]
Biological/Medical Applications   Monitoring/measuring membrane potential;[1–8] detecting/measuring analytes or metabolites;[15] cell viability assays;[1–4,15,20] apoptosis assays;[16–19] multidrug resistance assays[20]
Industrial Applications   Semi-conducting polymer nanoparticles[21]
Safety/Toxicity   Carcinogenicity[22]

REFERENCES

1. Sabnis, R. W. *Handbook of Biological Dyes and Stains*; John Wiley & Sons Inc.: Hoboken, **2010**; pp 467–468.

2. Ehrenberg, B.; Montana, V.; Wei, M. D.; Wuskell, J. P.; Loew, L. M. Membrane potential can be determined in individual cells from the Nernstian distribution of cationic dyes. *Biophys. J.* **1988**, *53*, 785–794.

3. Hattori, F.; Fukuda, K. Method of selecting myocardial cells by using intracellular mitochondria as indication. PCT Int. Appl. WO 2006022377, 2006.

4. Farkas, D. L.; Wei, M. D.; Febbroriello, P.; Carson, J. H.; Loew, L. M. Simultaneous imaging of cell and mitochondrial membrane potentials. *Biophys. J.* **1989**, *56*, 1053–1069.

5. Yu, Q.; Wang, H. Method for detecting reactive oxygen species (ros) and mitochondrial membrane potential of heart in isolated heart perfusion model. Faming Zhuanli Shenqing CN 102608089, 2012.

6. Prochazkova, J.; Kubala, L.; Kotasova, H.; Gudernova, I.; Sramkova, Z.; Pekarova, M.; Sarkadi, B.; Pachernik, J. ABC transporters affect the detection of intracellular oxidants by fluorescent probes. *Free Radical Res.* **2011**, *45*, 779–787.

7. Chalmers, S.; McCarron, J. G. The mitochondrial membrane potential and Ca2+ oscillations in smooth muscle. *J. Cell Sci.* **2008**, *121*, 75–85.

8. Nicholls, D. G.; Ward, M. W. Mitochondrial membrane potential and neuronal glutamate excitotoxicity: Mortality and millivolts. *Trends Neurosci.* **2000**, *23*, 166–174.

9. Bernas, T.; Dobrucki, J. Mitochondrial and nonmitochondrial reduction of MTT: Interaction of MTT with TMRE, JC-1, and NAO mitochondrial fluorescent probes. *Cytometry* **2002**, *47*, 236–242.

10. Yin, Z.; Aschner, J. L.; dos Santos, A. P.; Aschner, M. Mitochondrial-dependent manganese neurotoxicity in rat primary astrocyte cultures. *Brain Res.* **2008**, *1203*, 1–11.

11. Mironov, S. L. ADP regulates movements of mitochondria in neurons. *Biophys. J.* **2007**, *92*, 2944–2952.

12. Spees, J. L.; Olson, S. D.; Whitney, M. J.; Prockop, D. J. Mitochondrial transfer between cells can rescue aerobic respiration. *Proc. Natl. Acad. Sci. U.S.A.* **2006**, *103*, 1283–1288.

13. Collins, T. J.; Bootman, M. D. Mitochondria are morphologically heterogeneous within cells. *J. Exp. Biol.* **2003**, *206*, 1993–2000.

14. Collins, T. J.; Berridge, M. J.; Lipp, P.; Bootman, M. D. Mitochondria are morphologically and functionally heterogeneous within cells. *EMBO J.* **2002**, *21*, 1616–1627.

15. Workman, J. J.; Lambert, C. R.; Coleman, R. L. Non-invasive measurement of analytes. PCT Int. Appl. WO 2004044557, 2004.

16. Yu, W. R.; Liu, T.; Fehlings, T. K.; Fehlings, M. G. Involvement of mitochondrial signaling pathways in the mechanism of Fas-mediated apoptosis after spinal cord injury. *Eur. J. Neurosci.* **2009**, *29*, 114–131.

17. Galluzzi, L.; Zamzami, N.; de La Motte Rouge, T.; Lemaire, C.; Brenner, C.; Kroemer, G. Methods for the assessment of mitochondrial membrane permeabilization in apoptosis. *Apoptosis* **2007**, *12*, 803–813.

18. Goldstein, J. C.; Muñoz-Pinedo, C.; Ricci, J. E.; Adams, S. R.; Kelekar, A.; Schuler, M.; Tsien, R. Y.; Green, D. R. Cytochrome c is released in a single step during apoptosis. *Cell Death Differ.* **2005**, *12*, 453–462.

19. MacKenzie, F.; Duriez, P.; Wong, F.; Noseda, M.; Karsan, A. Notch4 inhibits endothelial apoptosis via RBP-Jkappa-dependent and -independent pathways. *J. Biol. Chem.* **2004**, *279*, 11657–11663.

20. Eytan, G.; Assaraf, Y. Assay for multidrug resistance. PCT Int. Appl. WO 9807034, 1998.

21. Grigalevicius, S.; Forster, M.; Ellinger, S.; Landfester, K.; Scherf, U. Excitation energy transfer from semi-conducting polymer nanoparticles to surface-bound fluorescent dyes. *Macromol. Rapid Commun.* **2006**, *27*, 200–202.

22. Young, A.; Lou, D.; McCormick, F. Oncogenic and wild-type Ras play divergent roles in the regulation of mitogen-activated protein kinase signaling. *Cancer Discov.* **2013**, *3*, 112–123.

TETRAMETHYLRHODAMINE METHYL ESTER PERCHLORATE (TMRM)

CAS Registry Number   115532-50-8

Chemical Structure

CA Index Name   Xanthylium, 3,6-bis(dimethylamino)-9-[2-(methoxycarbonyl)phenyl]-, perchlorate (1:1)

Other Names   Xanthylium, 3,6-bis(dimethylamino)-9-[2-(methoxycarbonyl)phenyl]-, perchlorate; T 668; TMRM; Tetramethylrhodamine methyl ester perchlorate

Merck Index Number   Not listed

Chemical/Dye Class   Xanthene

Molecular Formula $C_{25}H_{25}ClN_2O_7$

Molecular Weight   500.93

Physical Form   Dark green crystals[2]

Solubility   Soluble in *N,N*-dimethylformamide, dimethyl sulfoxide, ethanol, methanol

Melting Point   274–276 °C (decompose)[2]

Absorption (λ_{max})   549 nm (MeOH)

Emission (λ_{max})   573 nm (MeOH)

Molar Extinction Coefficient   115,000 cm^{-1} M^{-1} (MeOH)

Synthesis   Synthetic methods[1,2]

Imaging/Labeling Applications   Mitochondria;[1–7,10,11] cells[1–6,8,9,12]

Biological/Medical Applications   Monitoring/measuring membrane potential;[1–7] detecting/measuring analytes or metabolites;[9] cell viability assays;[1–6,8,9,12] apoptosis assays;[10,11] multidrug resistance assays[12]

Industrial Applications   Not reported

Safety/Toxicity   No data available

REFERENCES

1. Sabnis, R. W. *Handbook of Biological Dyes and Stains*; John Wiley & Sons Inc.: Hoboken, **2010**; p 469.

2. Ehrenberg, B.; Montana, V.; Wei, M. D.; Wuskell, J. P.; Loew, L. M. Membrane potential can be determined in individual cells from the Nernstian distribution of cationic dyes. *Biophys. J.* **1988**, *53*, 785–794.

3. Hattori, F.; Fukuda, K. Method of selecting myocardial cells by using intracellular mitochondria as indication. PCT Int. Appl. WO 2006022377, 2006.

4. Farkas, D. L.; Wei, M. D.; Febbroriello, P.; Carson, J. H.; Loew, L. M. Simultaneous imaging of cell and mitochondrial membrane potentials. *Biophys. J.* **1989**, *56*, 1053–1069.

5. Distelmaier, F.; Koopman, W. J.; Testa, E. R.; de Jong, A. S.; Swarts, H. G.; Mayatepek, E.; Smeitink, J. A.; Willems, P. H. Life cell quantification of mitochondrial membrane potential at the single organelle level. *Cytometry* **2008**, *73A*, 129–138.

6. Zhang, H.; Huang, H. M.; Carson, R. C.; Mahmood, J.; Thomas, H. M.; Gibson, G. E. Assessment of membrane potentials of mitochondrial populations in living cells. *Anal. Biochem.* **2001**, *298*, 170–180.

7. Diaz, G.; Liu, S.; Isola, R.; Diana, A.; Falchi, A. M. Mitochondrial localization of reactive oxygen species by dihydrofluorescein probes. *Histochem. Cell Biol.* **2003**, *120*, 319–325.

8. Wu, Z.; Zhang, L.; Song, Z.; Lin, Y.; Luo, M.; Huang, Q.; Liao, Y.; Jin, L.; Li, C.; Casalida, W. S.; Chen, J. Method for detecting cellular potassium electrode properties based on Nernst potential fluorescent dye. Faming Zhuanli Shenqing CN 102393451, 2012.

9. Workman, J. J.; Lambert, C. R.; Coleman, R. L. Non-invasive measurement of analytes. PCT Int. Appl. WO 2004044557, 2004.

10. Galluzzi, L.; Zamzami, N.; de La Motte Rouge, T.; Lemaire, C.; Brenner, C.; Kroemer, G. Methods for the assessment of mitochondrial membrane permabilization in apoptosis. *Apoptosis* **2007**, *12*, 803–813.

Tetramethylrhodamine methyl ester perchlorate (TMRM)

11. Gottlieb, R. A.; Granville, D. J. Analyzing mitochondrial changes during apoptosis. *Methods* **2002**, *26*, 341–347.

12. Eytan, G.; Assaraf, Y. Assay for multidrug resistance. PCT Int. Appl. WO 9807034, 1998.

TO-PRO 1

CAS Registry Number   157199-59-2

Chemical Structure

CA Index Name   Quinolinium, 4-[(3-methyl-2(3*H*)-benzothiazolylidene)-methyl]-1-[3-(trimethylammonio) propyl]-, iodide (1:2)

Other Names   Quinolinium, 4-[(3-methyl-2(3*H*)-benzothiazolylidene)-methyl]-1-[3-(trimethylammonio) propyl]-, diiodide; TO-PRO 1; TO-PRO 1 iodide

Merck Index Number   Not listed

Chemical/Dye Class   Cyanine

Molecular Formula $C_{24}H_{29}I_2N_3S$

Molecular Weight   645.38

Physical Form   Red-brown powder

Solubility   Soluble in dimethyl sulfoxide

Absorption (λ_{max})   515 nm (H_2O/DNA)

Emission (λ_{max})   531 nm (H_2O/DNA)

Molar Extinction Coefficient   62,800 $cm^{-1} M^{-1}$ (H_2O/DNA)

Quantum Yield   0.25 (H_2O/DNA)

Synthesis   Synthetic methods[1–4]

Imaging/Labeling Applications   Nucleic acids;[1–17] cells;[1–3,17] nuclei;[18–20] bacteria;[3,4,21,22] cytoplasm;[20] eggs of zooplankton species;[23] leukocytes;[24] megakaryocyte;[25] microorganisms;[26,27] reticulocytes;[28] sperms[17,29]

Biological/Medical Applications   Detecting nucleic acids,[1–17] cells,[1–3,17] classifying/counting leukocytes,[24] megakaryocytes;[25] counting erythroblasts,[18,19] sperms;[17,29] detecting/measuring microorganisms;[26,27] detecting location of cancer cells/tissues;[30] nucleic acids amplification;[16] nucleic acid sequencing;[31] as temperature sensor[32,33]

Industrial Applications   Not reported

Safety/Toxicity   No data available

REFERENCES

1. Sabnis, R. W. *Handbook of Biological Dyes and Stains*; John Wiley & Sons Inc.: Hoboken, **2010**; pp 475–476.

2. Yue, S. T.; Johnson, I. D.; Huang, Z.; Haugland, R. P. Unsymmetrical cyanine dyes with a cationic side chain. U.S. Patent 5321130, 1994.

3. Millard, P. J.; Roth, B. L.; Yue, S. T.; Haugland, R. P. Fluorescent viability assay using cyclic-substituted unsymmetrical cyanine dyes. U.S. Patent 5534416, 1996.

4. Roth, B. L.; Millard, P. J.; Yue, S. T.; Wells, K. S.; Haugland, R. P. Fluorescent assay for bacterial gram reaction. U.S. Patent 5545535, 1996.

5. Sergeev, N. V.; Brevnov, M. G.; Furtado, M. R. Identification of nucleic acids. PCT Int. Appl. WO 2011143478, 2011.

6. Exner, M.; Rogers, A. Methods for detecting nucleic acids using multiple signals. U.S. Pat. Appl. Publ. US 2007172836, 2007.

7. Wittwer, C. T.; Dujols, V. E.; Reed, G.; Zhou, L. Amplicon melting analysis with saturation dyes. PCT Int. Appl. WO 2004038038, 2004.

8. Erikson, G. H.; Daksis, J. I.; Kandic, I.; Picard, P. Nucleic acid multiplex formation. PCT Int. Appl. WO 2002103051, 2002.

9. Erikson, G. H.; Daksis, J. I. Pre-incubation method to improve signal/noise ratio of nucleic acid assays. U.S. Pat. Appl. Publ. US 2004180345, 2004.

10. Erikson, G. H. Method for modifying transcription and/or translation in an organism for therapeutic, prophylactic and/or analytic uses. U.S. Pat. Appl. Publ. US 2003181412, 2003.

11. Liu, M.; Chen, F. Nucleic acid separation and detection by electrophoresis with a counter-migrating high-affinity intercalating dye. U.S. Pat. Appl. Publ. US 2003198964, 2003.

12. Tomita, N.; Mori, Y. Method for efficiently detecting double-stranded nucleic acid. PCT Int. Appl. WO 2002103053, 2002.

13. Kricka, L. J. Stains, labels and detection strategies for nucleic acids assays. *Ann. Clin. Biochem.* **2002**, *39*, 114–129.

14. Beisker, W.; Weller-Mewe, E. M.; Nusse, M. Fluorescence enhancement of DNA-bound TO-PRO-3 by incorporation of bromodeoxyuridine to monitor cell cycle kinetics. *Cytometry* **1999**, *37*, 221–229.

15. Wagner, B.; Mathis, H.; Schmidt, K.; Kalusche, G.; McCaskill, J. Single molecule detection in microstructures. *Nucleosides Nucleotides* **1997**, *16*, 635–642.

16. Sutherland, J. W.; Patterson, D. R. Homogeneous method for assay of double-stranded nucleic acids using fluorescent dyes and kit useful therein. Eur. Pat. Appl. EP 684316, 1995.

17. Anderson, A. L.; Knutson, C. R.; Mueth, D.; Plewa, J.; Tanner, E. Methods for staining cells for identification and sorting. U.S. Pat. Appl. Publ. US 2006172315, 2006.

18. Heuven, B.; Wong, F. S.; Tsuji, T.; Sakata, T.; Hamaguchi, I. Method for classifying and counting erythroblasts by flow cytometry. Jpn. Kokai Tokkyo Koho JP 11326323, 1999.

19. Heuven, B.; Wong, F.; Tsuji, T.; Sakata, T.; Hamaguchi, I. Process for discriminating and counting erythroblasts. U.S. Pat. Appl. Publ. US 20020006631, 2002.

20. Bink, K.; Walch, A.; Feuchtinger, A.; Eisenmann, H.; Hutzler, P.; Hofler, H.; Werner, M. TO-PRO-3 is an optimal fluorescent dye for nuclear counterstaining in dual-colour FISH on paraffin sections. *Histochem. Cell Biol.* **2001**, *115*, 293–299.

21. Mortimer, F. C.; Mason, D. J.; Gant, V. A. Flow cytometric monitoring of antibiotic-induced injury in *Escherichia coli* using cell-impermeant fluorescent probes. *Antimicrob. Agents Chemother.* **2000**, *44*, 676–681.

22. Li, W. K. W.; Jellett, J. F.; Dickie, P. M. DNA distributions in planktonic bacteria stained with TOTO or TO-PRO. *Limnol. Oceanogr.* **1995**, *40*, 1485–1495.

23. Gorokhova, E. A single-step staining method to evaluate egg viability in zooplankton. *Limnol. Oceanogr.: Methods* **2010**, *8*, 414–423.

24. Sakata, T.; Mizukami, T.; Hatanaka, K. Method for classifying and counting immature leukocytes. Eur. Pat. Appl. EP 844481, 1998.

25. Minakami, T.; Mori, Y.; Tsuji, T.; Ikeuchi, Y. Megakaryocyte classification/counting method by double fluorescent staining and flow cytometry. Jpn. Kokai Tokkyo Koho JP 2006275985, 2006.

26. Eckert, R. H.; Kaplan, C.; He, J.; Yarbrough, D. K.; Anderson, M.; Sim, J. Methods and devices for the selective detection of microorganisms. U.S. Pat. Appl. Publ. US 20120003661, 2012.

27. Noda, N.; Mizutani, T. Microorganism-measuring method using multiple staining. Jpn. Kokai Tokkyo Koho JP 2006340684, 2006.

28. Veriac, S. Colour staining reagent for blood cells. Eur. Pat. Appl. EP 856735, 1998.

29. Matsumoto, T.; Okada, H.; Hamaguchi, Y. Method and reagent for counting sperm by flow cytometry. Jpn. Kokai Tokkyo Koho JP 2001242168, 2001.

30. Moudgil, B. M.; Varshney, M.; Grobmyer, S. R. Targeted cellular selectivity of surface active molecules. PCT Int. Appl. WO 2009100011, 2009.

31. Hoser, M. J. Nucleic acid sequencing methods, kits and reagents. PCT Int. Appl. WO 2004074503, 2004.

32. Bousseksou, A.; Salmon, L.; Molnar, G.; Cobo, S. Materials with thermochromic spin transition doped with one or more fluorescent agents for use as temperature sensor. Fr. Demande FR 2952371, 2011.

33. Bousseksou, A.; Salmon, L.; Molnar, G.; Cobo, S. Heat-sensitive spin-transition materials doped with one or more fluorescent agents for use as temperature sensor. PCT Int. Appl. WO 2011058277, 2011.

TO-PRO 3

CAS Registry Number   157199-63-8

Chemical Structure

CA Index Name   Quinolinium, 4-[3-(3-methyl-2(3H)-benzothiazolylidene)-1-propen-1-yl]-1-[3-(trimethylammonio)propyl]-, iodide (1:2)

Other Names   Quinolinium, 4-[3-(3-methyl-2(3H)-benzothiazolylidene)-1-propenyl]-1-[3-(trimethylammonio)propyl]-, diiodide; TO-PRO 3; TO-PRO 3 iodide

Merck Index Number   Not listed

Chemical/Dye Class   Cyanine

Molecular Formula $C_{26}H_{31}I_2N_3S$

Molecular Weight   671.42

Physical Form   Red-brown powder

Solubility   Soluble in dimethyl sulfoxide

Absorption (λ_{max})   642 nm (H_2O/DNA)

Emission (λ_{max})   661 nm (H_2O/DNA)

Molar Extinction Coefficient   102,000 $cm^{-1}\,M^{-1}$ (H_2O/DNA)

Quantum Yield   0.11 (H_2O/DNA)

Synthesis   Synthetic methods[1-4]

Imaging/Labeling Applications   Nucleic acids;[1-19] cells;[1-3,18-20] nuclei;[21-23] bacteria;[3,24-26] leukocytes;[27] liposomes;[28,29] microorganisms;[30-32] reticulocytes;[18,33] sperms[19]

Biological/Medical Applications   Detecting nucleic acids,[1-19] cells,[1-3,18-20] microorganisms;[30-32] IDH1/IDH2 mutations,[34] JAK2 gene mutation,[35] PIK3CA mutation,[36] B-type Raf Kinase (BRAF) mutation;[37] classifying/counting leukocytes;[27] counting reticulocytes;[18,33] detecting/counting bacteria;[3,24-26] cytotoxicity assay;[38-40] drug delivery;[28,29] as temperature sensor[41,42]

Industrial Applications   Not reported

Safety/Toxicity   No data available

REFERENCES

1. Sabnis, R. W. *Handbook of Biological Dyes and Stains*; John Wiley & Sons Inc.: Hoboken, **2010**; pp 477–478.

2. Yue, S. T.; Johnson, I. D.; Huang, Z.; Haugland, R. P. Unsymmetrical cyanine dyes with a cationic side chain. U.S. Patent 5321130, 1994.

3. Millard, P. J.; Roth, B. L.; Yue, S. T.; Haugland, R. P. Fluorescent viability assay using cyclic-substituted unsymmetrical cyanine dyes. U.S. Patent 5534416, 1996.

4. Soper, S. A.; Davidson, Y. Y.; Flanagan, J. H.; Legendre, B. L.; Owens, C.; Williams, D. C.; Hammer, R. P. Micro-DNA sequence analysis using capillary electrophoresis and near-IR fluorescence detection. *Proc. SPIE-Int. Soc. Opt. Eng.* **1996**, *2680*, 235–246.

5. Kroneis, T.; Geigl, J. B.; El-Heliebi, A.; Auer, M.; Ulz, P.; Schwarzbraun, T.; Dohr, G.; Sedlmayr, P. Combined molecular genetic and cytogenetic analysis from single cells after isothermal whole-genome amplification. *Clin. Chem.* **2011**, *57*, 1032–1041.

6. Ploeger, L. S.; Dullens, H. F. J.; Huisman, A.; van Diest, P. J. Fluorescent stains for quantification of DNA by confocal laser scanning microscopy in 3-D. *Biotech. Histochem.* **2008**, *83*, 63–69.

7. Exner, M.; Rogers, A. Methods for detecting nucleic acids using multiple signals. U.S. Pat. Appl. Publ. US 2007172836, 2007.

8. Martin, R. M.; Leonhardt, H.; Cardoso, M. C. DNA labeling in living cells. *Cytometry, Part A* **2005**, *67A*, 45–52.

9. Ploeger, L. S.; Huisman, A.; van der Gugten, J.; van der Giezen, D. M.; Belien, J. A. M.; Abbaker, A. Y.; Dullens, H. F. J.; Grizzle, W.; Poulin, N. M.; Meijer, G. A. Implementation of accurate and fast DNA cytometry by confocal microscopy in 3D. *Cell. Oncol.* **2005**, *27*, 225–230.

10. Sovenyhazy, K. M.; Bordelon, J. A.; Petty, J. T. Spectroscopic studies of the multiple binding modes of a trimethine-bridged cyanine dye with DNA. *Nucleic Acids Res.* **2003**, *31*, 2561–2569.

11. Erikson, G. H.; Daksis, J. I.; Kandic, I.; Picard, P. Nucleic acid multiplex formation. PCT Int. Appl. WO 2002103051, 2002.

12. Erikson, G. H.; Daksis, J. I. Pre-incubation method to improve signal/noise ratio of nucleic acid assays. U.S. Pat. Appl. Publ. US 2004180345, 2004.

13. Erikson, G. H. Method for modifying transcription and/or translation in an organism for therapeutic, prophylactic and/or analytic uses. U.S. Pat. Appl. Publ. US 2003181412, 2003.

14. Liu, M.; Chen, F. Nucleic acid separation and detection by electrophoresis with a counter-migrating high-affinity intercalating dye. U.S. Pat. Appl. Publ. US 2003198964, 2003.

15. Vanickova, M.; Labuda, J.; Buckova, M.; Surugiu, I.; Mecklenburg, M.; Danielsson, B. Investigation of catechin and acridine derivatives using voltammetric and fluorimetric DNA-based sensors. *Collect. Czech. Chem. Commun.* **2000**, *65*, 1055–1066.

16. Suzuki, T.; Fujikura, K.; Higashiyama, T.; Takata, K. DNA staining for fluorescence and laser confocal microscopy. *J. Histochem. Cytochem.* **1997**, *45*, 49–53.

17. Sutherland, J. W.; Patterson, D. R. Homogeneous method for assay of double-stranded nucleic acids using fluorescent dyes and kit useful therein. Eur. Pat. Appl. EP 684316, 1995.

18. Deka, C.; Gordon, K. M.; Gupta, R.; Horton, A. Methods and compositions for rapid staining of nucleic acids in whole cells. U.S. Patent 6271035, 2001.

19. Anderson, A. L.; Knutson, C. R.; Mueth, D.; Plewa, J.; Tanner, E. Methods for staining cells for identification and sorting. U.S. Pat. Appl. Publ. US 2006172315, 2006.

20. Gjelsnes, O.; Ronning, O. Method and device for counting cells in urine. PCT Int. Appl. WO 2001016595, 2001.

21. Gauer, C.; Mann, W.; Alunni-Fabbroni, M. Method for carrying out an enzymic reactions. Ger. DE 102006056694, 2010.

22. Gauer, C.; Mann, W.; Alunni-Fabbroni, M. Method for carrying out an enzymic reaction. PCT Int. Appl. WO 2008064730, 2008.

23. Bink, K.; Walch, A.; Feuchtinger, A.; Eisenmann, H.; Hutzler, P.; Hofler, H.; Werner, M. TO-PRO-3 is an optimal fluorescent dye for nuclear counterstaining in dual-colour FISH on paraffin sections. *Histochem. Cell Biol.* **2001**, *115*, 293–299.

24. Sakai, Y.; Kawashima, Y.; Inoue, J.; Ikeuchi, Y. Method of staining, detecting and counting bacteria, and a diluent for bacterial stain. Eur. Pat. Appl. EP 1203825, 2002.

25. Inoue, J.; Ikeuchi, Y.; Kawashima, Y. Method for staining and detecting bacteria. Jpn. Kokai Tokkyo Koho JP 2001258590, 2001.

26. Shapiro, H. M. Multiparameter flow cytometry of bacteria: Implications for diagnostics and therapeutics. *Cytometry* **2001**, *43*, 223–226.

27. Sakata, T.; Mizukami, T.; Hatanaka, K. Method for classifying and counting immature leukocytes. Eur. Pat. Appl. EP 844481, 1998.

28. Leung, S. J.; Romanowski, M. Molecular catch and release: Controlled delivery using optical trapping with light-responsive liposomes. *Adv. Mater.* **2012**, *24*, 6380–6383.

29. Yudina, A.; de Smet, M.; Lepetit-Coiffe, M.; Langereis, S.; Van Ruijssevelt, L.; Smirnov, P.; Bouchaud, V.; Voisin, P.; Gruell, H.; Moonen, C. T. W. Ultrasound-mediated intracellular drug delivery using microbubbles and temperature-sensitive liposomes. *J. Controlled Release* **2011**, *155*, 442–448.

30. Eckert, R. H.; Kaplan, C.; He, J.; Yarbrough, D. K.; Anderson, M.; Sim, J. Methods and devices for the selective detection of microorganisms. U.S. Pat. Appl. Publ. US 20120003661, 2012.

31. Henneberger, R.; Birch, D.; Bergquist, P.; Walter, M.; Anitori, R. P. The fluorescent dyes TO-PRO-3 and TOTO-3 iodide allow detection of microbial cells in soil samples without interference from background fluorescence. *BioTechniques* **2011**, *51*, 190–192.

32. Vannier, E. Methods for detection of pathogens in red blood cells. PCT Int. Appl. WO 2006031544, 2006.

33. Yamamoto, K.; Tokita, M.; Inagami, M.; Hirako, S. Method of counting reticulocytes. U.S. Patent 5563070, 1996.

34. Park, H. G.; Choi, J. J.; Kim, H. S.; Jung, S. U. Method and kit for detecting IDH1 and IDH2 mutations using peptide nucleic acid (PNA)-based real-time polymerase chain reaction (PCR) clamping.

Repub. Korean Kongkae Taeho Kongbo KR 2012127679, 2012.

35. Park, H. G.; Choi, J. J.; Kim, H. S.; Jung, S. U. JAK2 gene mutation detection kit based on PNA mediated real-time PCR clamping. Repub. Korean Kongkae Taeho Kongbo KR 2012119571, 2012.

36. Park, H. K.; Choi, J. J.; Kim, H. S. PIK3CA mutation detection method and kit using real-time PCR clamping of PNA. PCT Int. Appl. WO 2012020965, 2012.

37. Park, H. K.; Choi, J. J.; Cho, M. H. Detection of mutation of BRAF by real-time PCR using PNA clamping probe for diagnosis of cancers. PCT Int. Appl. WO 2011093606, 2011.

38. Gerritsen, A. F.; Bosch, M.; de Weers, M.; van de Winkel, J. G. J.; Parren, P. W. H. I. High throughput screening for antibody induced complement-dependent cytotoxicity in early antibody discovery using homogeneous macroconfocal fluorescence imaging. *J. Immunol. Methods* **2010**, *352*, 140–146.

39. Wilkinson, R. W.; Lee-MacAry, A. E.; Davies, D.; Snary, D.; Ross, E. L. Antibody-dependent cell-mediated cytotoxicity: a flow cytometry-based assay using fluorophores. *J. Immunol. Methods* **2001**, *258*, 183–191.

40. Lee-MacAry, A. E.; Ross, E. L.; Davies, D.; Laylor, R.; Honeychurch, J.; Glennie, M. J.; Snary, D.; Wilkinson, R. W. Development of a novel flow cytometric cell-mediated cytotoxicity assay using the fluorophores PKH-26 and TO-PRO-3 iodide. *J. Immunol. Methods* **2001**, *252*, 83–92.

41. Bousseksou, A.; Salmon, L.; Molnar, G.; Cobo, S. Materials with thermochromic spin transition doped with one or more fluorescent agents for use as temperature sensor. Fr. Demande FR 2952371, 2011.

42. Bousseksou, A.; Salmon, L.; Molnar, G.; Cobo, S. Heat-sensitive spin-transition materials doped with one or more fluorescent agents for use as temperature sensor. PCT Int. Appl. WO 2011058277, 2011.

TO-PRO 5

CAS Registry Number   177027-61-1

Chemical Structure

CA Index Name   Quinolinium, 4-[5-(3-methyl-2(3*H*)-benzothiazolylidene)-1,3-pentadien-1-yl]-1-[3-(trimethylammonio)propyl]-, iodide (1:2)

Other Names   Quinolinium, 4-[5-(3-methyl-2(3*H*)-benzothiazolylidene)-1,3-pentadienyl]-1-[3-(trimethylammonio)propyl]-, diiodide; 3-methyl-2-[5-[1-[3-(trimethylammonio)propyl]-4(1*H*)-quinolinylidene]-1,3-pentadienyl]benzothiazolium diiodide; TO-PRO 5; TO-PRO 5 iodide

Merck Index Number   Not listed

Chemical/Dye Class   Cyanine

Molecular Formula $C_{28}H_{33}I_2N_3S$

Molecular Weight   697.46

Physical Form   Red-brown powder

Solubility   Soluble in dimethyl sulfoxide

Absorption (λ_{max})   748 nm (H_2O/DNA)

Emission (λ_{max})   768 nm (H_2O/DNA)

Molar Extinction Coefficient   108,500 $cm^{-1} M^{-1}$ (H_2O/DNA)

Synthesis   Synthetic methods[1–4]

Imaging/Labeling Applications   Nucleic acids;[1–11] cells;[1–3,11] bacteria;[3] proteins;[12] sperms[11]

Biological/Medical Applications   Detecting nucleic acids,[1–11] cells[1–3,11]

Industrial Applications   Not reported

Safety/Toxicity   No data available

REFERENCES

1. Sabnis, R. W. *Handbook of Biological Dyes and Stains*; John Wiley & Sons Inc.: Hoboken, **2010**; p 479.

2. Yue, S. T.; Johnson, I. D.; Huang, Z.; Haugland, R. P. Unsymmetrical cyanine dyes with a cationic side chain. U.S. Patent 5321130, 1994.

3. Millard, P. J.; Roth, B. L.; Yue, S. T.; Haugland, R. P. Fluorescent viability assay using cyclic-substituted unsymmetrical cyanine dyes. U.S. Patent 5534416, 1996.

4. Soper, S. A.; Davidson, Y. Y.; Flanagan, J. H.; Legendre, B. L.; Owens, C.; Williams, D. C.; Hammer, R. P. Micro-DNA sequence analysis using capillary electrophoresis and near-IR fluorescence detection. *Proc. SPIE-Int. Soc. Opt. Eng.* **1996**, *2680*, 235–246.

5. Exner, M.; Rogers, A. Methods for detecting nucleic acids using multiple signals. U.S. Pat. Appl. Publ. US 2007172836, 2007.

6. Liu, M.; Chen, F. Nucleic acid separation and detection by electrophoresis with a counter-migrating high-affinity intercalating dye. U.S. Pat. Appl. Publ. US 2003198964, 2003.

7. Erikson, G. H.; Daksis, J. I.; Kandic, I.; Picard, P. Nucleic acid multiplex formation. PCT Int. Appl. WO 2002103051, 2002.

8. Erikson, G. H.; Daksis, J. I. Pre-incubation method to improve signal/noise ratio of nucleic acid assays. U.S. Pat. Appl. Publ. US 2004180345, 2004.

9. Erikson, G. H. Method for modifying transcription and/or translation in an organism for therapeutic, prophylactic and/or analytic uses. U.S. Pat. Appl. Publ. US 2003181412, 2003.

10. Beisker, W.; Weller-Mewe, E. M.; Nusse, M. Fluorescence enhancement of DNA-bound TO-PRO-3 by incorporation of bromodeoxyuridine to monitor cell cycle kinetics. *Cytometry* **1999**, *37*, 221–229.

11. Anderson, A. L.; Knutson, C. R.; Mueth, D.; Plewa, J.; Tanner, E. Methods for staining cells for identification and sorting. U.S. Pat. Appl. Publ. US 2006172315, 2006.

12. Ozhalici-Unal, H.; Pow, C. L.; Marks, S. A.; Jesper, L. D.; Silva, G. L.; Shank, N. I.; Jones, E. W.; Burnette, J. M., III; Berget, P. B.; Armitage, B. A. A rainbow of fluoromodules: A promiscuous scFv protein binds to and activates a diverse set of fluorogenic cyanine dyes. *J. Am. Chem. Soc.* **2008**, *130*, 12620–12621.

TOTO 1

CAS Registry Number 143413-84-7

Chemical Structure

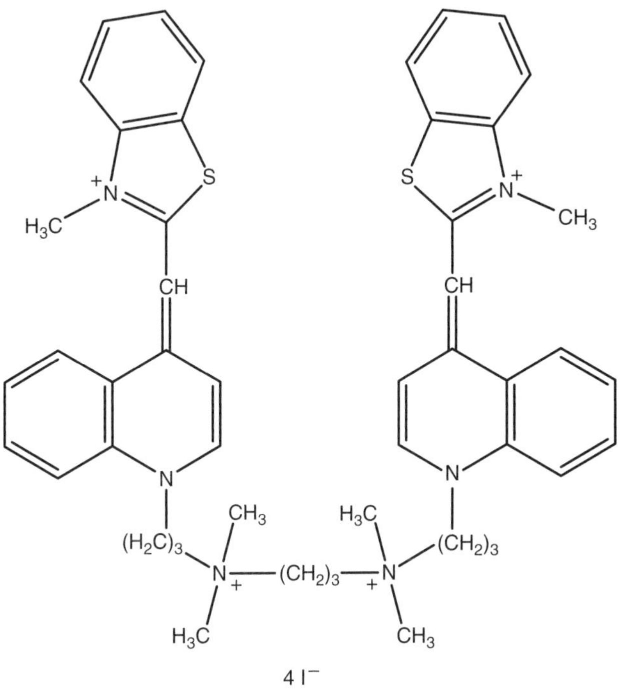

4 I⁻

CA Index Name Quinolinium, 1,1′-[1,3-propanediylbis-[(dimethyliminio)-3,1-propanediyl]]bis[4-[(3-methyl-2 (3*H*)-benzothiazolylidene)methyl]-, iodide (1:4)

Other Names Quinolinium, 1,1′-[1,3-propanediylbis [(dimethyliminio)-3,1-propanediyl]]bis[4-[(3-methyl-2 (3*H*)-benzothiazolylidene)methyl]-, tetraiodide; TOTO 1; TOTO 1 iodide

Merck Index Number 9460

Chemical/Dye Class Cyanine

Molecular Formula $C_{49}H_{58}I_4N_6S_2$

Molecular Weight 1302.78

Physical Form Red powder[3,4]

Solubility Soluble in dimethyl sulfoxide

Melting Point 281 °C[2]

Absorption (λ_{max}) 514 nm (H₂O/DNA)

Emission (λ_{max}) 533 nm (H₂O/DNA)

Molar Extinction Coefficient 117,000 cm⁻¹ M⁻¹ (H₂O/DNA)

Quantum Yield 0.34 (H₂O/DNA)

Synthesis Synthetic methods[1–7]

Imaging/Labeling Applications Nucleic acids;[1–23] cells;[4,6,21–24] nuclei;[25,26] bacteria;[3,21,27–29] chromosoms;[30] chromosome spreads;[4] leukocytes;[31] micronuclei;[32] megakaryocyte;[33] microorganisms;[34,35] reticulocytes;[22] sperms[23,36]

Biological/Medical Applications Detecting nucleic acids,[1–23] cells,[4,6,21–24] characterizing chromosoms;[30] classifying/counting leukocytes,[31] megakaryocytes;[33] counting erythroblasts,[25,26] sperms;[23,36] detecting/measuring microorganisms;[34,35] detecting location of cancer cells/tissues;[37] nucleic acids amplification;[38] nanosensor platforms for diagnostic applications;[39] detecting/quantifying nuclease activity;[40,41] as temperature sensor[42,43]

Industrial Applications Marker systems for product authentication[44]

Safety/Toxicity No data available

REFERENCES

1. Sabnis, R. W. *Handbook of Biological Dyes and Stains*; John Wiley & Sons Inc.: Hoboken, **2010**; pp 480–482.

2. Kabatc, J. Multicationic monomethine dyes as sensitizers in two- and three-component photoinitiating systems for multiacrylate monomers. *J. Photochem. Photobiol., A: Chem.* **2010**, *214*, 74–85.

3. Roth, B. L.; Millard, P. J.; Yue, S. T.; Wells, K. S.; Haugland, R. P. Fluorescent assay for bacterial gram reaction. U.S. Patent 5545535, 1996.

4. Yue, S. T.; Haugland, R. P. Dimers of unsymmetrical cyanine dyes containing pyridinium moieties. U.S. Patent 5410030, 1995.

5. Glazer, A. N.; Benson, S. C. DNA complexes with dyes designed for energy transfer as fluorescent markers. PCT Int. Appl. WO 9417397, 1994.

6. Yue, S. T.; Johnson, I. D.; Haugland, R. P. Dimers of unsymmetrical cyanine dyes. PCT Int. Appl. WO 9306482, 1993.

7. Rye, H. S.; Yue, S.; Wemmer, D. E.; Quesada, M. A.; Haugland, R. P.; Mathies, R. A.; Glazer, A. N. Stable fluorescent complexes of double-stranded DNA with bis-intercalating asymmetric cyanine dyes: Properties and applications. *Nucleic Acids Res.* **1992**, *20*, 2803–2812.

8. McLenachan, S.; Sarsero, J. P.; Ioannou, P. A. Flow-cytometric analysis of mouse embryonic stem cell lipofection using small and large DNA constructs. *Genomics* **2007**, *89*, 708–720.

9. Exner, M.; Rogers, A. Methods for detecting nucleic acids using multiple signals. U.S. Pat. Appl. Publ. US 2007172836, 2007.

10. Erikson, G. H.; Daksis, J. I.; Kandic, I.; Picard, P. Nucleic acid multiplex formation. PCT Int. Appl. WO 2002103051, 2002.

11. Erikson, G. H.; Daksis, J. I. Pre-incubation method to improve signal/noise ratio of nucleic acid assays. U.S. Pat. Appl. Publ. US 2004180345, 2004.

12. Erikson, G. H. Method for modifying transcription and/or translation in an organism for therapeutic, prophylactic and/or analytic uses. U.S. Pat. Appl. Publ. US 2003181412, 2003.

13. Kim, K.; Min, J.; Lee, I.; Kim, A. Method for highly sensitive nucleic acid detection using nanopore and non-specific nucleic acid-binding agent. U.S. Pat. Appl. Publ. US 2006292605, 2006.

14. Petersen, M.; Hamed, A. A.; Pedersen, E. B.; Jacobsen, J. P. *Bis*-intercalation of homodimeric thiazole orange dye derivatives in DNA. *Bioconjugate Chem.* **1999**, *10*, 66–74.

15. Sailer, B. L.; Nastasi, A. J.; Valdez, J. G.; Steinkamp, J. A.; Crissman, H. A. Differential effects of deuterium oxide on the fluorescence lifetimes and intensities of dyes with different modes of binding to DNA. *J. Histochem. Cytochem.* **1997**, *45*, 165–175.

16. Hansen, L. F.; Jensen, L. K.; Jacobsen, J. P. *Bis*-intercalation of a homodimeric thiazole orange dye in DNA in symmetrical pyrimidine-pyrimidine-purine-purine oligonucleotides. *Nucleic Acids Res.* **1996**, *24*, 859–867.

17. Rye, H. S.; Glazer, A. N. Interaction of dimeric intercalating dyes with single-stranded DNA. *Nucleic Acids Res.* **1995**, *23*, 1215–1222.

18. Tekola, P.; Baak, J. P. A.; Belien, J. A. M.; Brugghe, J. Highly sensitive, specific, and stable new fluorescent DNA stains for confocal laser microscopy and image processing of normal paraffin sections. *Cytometry* **1994**, *17*, 191–195.

19. Hirons, G. T.; Fawcett, J. J.; Crissman, H. A. TOTO and YOYO: new very bright fluorochromes for DNA content analyses by flow cytometry. *Cytometry* **1994**, *15*, 129–140.

20. Rye, H. S.; Dabora, J. M.; Quesada, M. A.; Mathies, R. A.; Glazer, A. N. Fluorometric assay using dimeric dyes for double- and single-stranded DNA and RNA with picogram sensitivity. *Anal. Biochem.* **1993**, *208*, 144–150.

21. Millard, P. J.; Roth, B. L.; Yue, S. T.; Haugland, R. P. Fluorescent viability assay using cyclic-substituted unsymmetrical cyanine dyes. U.S. Patent 5534416, 1996.

22. Deka, C.; Gordon, K. M.; Gupta, R.; Horton, A. Methods and compositions for rapid staining of nucleic acids in whole cells. U.S. Patent 6271035, 2001.

23. Anderson, A. L.; Knutson, C. R.; Mueth, D.; Plewa, J.; Tanner, E. Methods for staining cells for identification and sorting. U.S. Pat. Appl. Publ. US 2006172315, 2006.

24. Skyggebjerg, O.; Glensbjerg, M. A method and a system for counting cells from a plurality of species. PCT Int. Appl. WO 2002101087, 2002.

25. Heuven, B.; Wong, F.; Tsuji, T.; Sakata, T.; Hamaguchi, I. Method for classifying and counting erythroblasts by flow cytometry. Jpn. Kokai Tokkyo Koho JP 11326323, 1999.

26. Heuven, B.; Wong, F.; Tsuji, T.; Sakata, T.; Hamaguchi, I. Process for discriminating and counting erythroblasts. U.S. Pat. Appl. Publ. US 20020006631, 2002.

27. Luppens, S. B. I.; Barbaras, B.; Breeuwer, P.; Rombouts, F. M.; Abee, T. Selection of fluorescent probes for flow cytometric viability assessment of *Listeria monocytogenes* exposed to membrane-active and oxidizing disinfectants. *J. Food Prot.* **2003**, *66*, 1393–1401.

28. Bunthof, C. J.; Abee, T. Development of a flow cytometric method to analyze subpopulations of bacteria in probiotic products and dairy starters. *Appl. Environ. Microbiol.* **2002**, *68*, 2934–2942.

29. Bunthof, C. J.; Bloemen, K.; Breeuwer, P.; Rombouts, F. M.; Abee, T. Flow cytometric assessment of viability of lactic acid bacteria. *Appl. Environ. Microbiol.* **2001**, *67*, 2326–2335.

30. Crissman, H. A.; Hirons, G. T. Chromosome characterization using single fluorescent dye. U.S. Patent 5418169, 1995.

31. Sakata, T.; Mizukami, T.; Hatanaka, K. Method for classifying and counting immature leukocytes. Eur. Pat. Appl. EP 844481, 1998.

32. Dertinger, S. D.; Cairns, S. E.; Avlasevich, S. L.; Torous, D. K. Method for enumerating mammalian cell micronuclei with an emphasis on differentially staining micronuclei and the chromatin of dead and dying cells. PCT Int. Appl. WO 2006007479, 2006.

33. Minakami, T.; Mori, Y.; Tsuji, T.; Ikeuchi, Y. Megakaryocyte classification/counting method by double fluorescent staining and flow cytometry. Jpn. Kokai Tokkyo Koho JP 2006275985, 2006.

34. Eckert, R. H.; Kaplan, C.; He, J.; Yarbrough, D. K.; Anderson, M.; Sim, J. Methods and devices for the selective detection of microorganisms. U.S. Pat. Appl. Publ. US 20120003661, 2012.

35. Noda, N.; Mizutani, T. Microorganism-measuring method using multiple staining. Jpn. Kokai Tokkyo Koho JP 2006340684, 2006.

36. Matsumoto, T.; Okada, H.; Hamaguchi, Y. Method and reagent for counting sperm by flow cytometry. Jpn. Kokai Tokkyo Koho JP 2001242168, 2001.

37. Moudgil, B. M.; Varshney, M.; Grobmyer, S. R. Targeted cellular selectivity of surface active molecules. PCT Int. Appl. WO 2009100011, 2009.

38. Morrison, T. Improved selective ligation and amplification assay. PCT Int. Appl. WO 2005059178, **2005**.

39. Zhou, C.; Thompson, M. E.; Cote, R. J.; Ishikawa, F.; Curreli, M.; Chang, H. Methods of using and constructing nanosensor platforms. PCT Int. Appl. WO 2009085356, 2009.

40. Mitsuhashi, M.; Ogura, M. Process for determining nuclease activity. U.S. Patent 5554502, 1996.

41. Mitsuhashi, M.; Ogura, M. Process for determining nuclease activity. PCT Int. Appl. WO 9508623, 1995.

42. Bousseksou, A.; Salmon, L.; Molnar, G.; Cobo, S. Materials with thermochromic spin transition doped with one or more fluorescent agents for use as temperature sensor. Fr. Demande FR 2952371, 2011.

43. Bousseksou, A.; Salmon, L.; Molnar, G.; Cobo, S. Heat-sensitive spin-transition materials doped with one or more fluorescent agents for use as temperature sensor. PCT Int. Appl. WO 2011058277, 2011.

44. Van Asbrouck, J.; Draaijer, A. Multi-level markers. PCT Int. Appl. WO 2009139631, 2009.

TOTO 3

CAS Registry Number   166196-17-4

Chemical Structure

CA Index Name   Quinolinium, 1,1'-[1,3-propanediylbis
[(dimethyliminio)-3,1-propanediyl]]bis[4-[3-(3-methyl-2
(3H)-benzothiazolylidene)-1-propen-1-yl]-, iodide (1:4)

Other Names   Quinolinium, 1,1'-[1,3-propanediylbis
[(dimethyliminio)-3,1-propanediyl]]bis[4-[3-(3-methyl-2
(3H)-benzothiazolylidene)-1-propenyl]-, tetraiodide;
TOTO 3; TOTO 3 iodide

Merck Index Number   Not listed

Chemical/Dye Class   Cyanine

Molecular Formula $C_{53}H_{62}I_4N_6S_2$

Molecular Weight   1354.85

Physical Form   Red powder

Solubility   Soluble in dimethyl sulfoxide

Absorption (λ_{max})   642 nm (H_2O/DNA)

Emission (λ_{max})   660 nm (H_2O/DNA)

Molar Extinction Coefficient   154,100 cm^{-1} M^{-1}
(H_2O/DNA)

Quantum Yield   0.06 (H_2O/DNA)

Synthesis   Synthetic methods[1,2]

Imaging/Labeling Applications   Nucleic
acids;[1-14] cells;[2,12-14] nuclei;[15-19] bacteria;[12,20,21]
chromosome spreads;[2] cytoplasm;[19] leukocytes;[22]
megakaryocyte;[23] microorganisms;[24-26] porous wall
hollow glass microspheres (PW-HGMs);[27] proteins;[28]
reticulocytes;[13,29] sperms;[14,30] hairs[31]

Biological/Medical Applications   Detecting nucleic
acids,[1-14] cells,[2,12-14] microorganisms,[24-26] IDH1/IDH2
mutations,[32] JAK2 gene mutation,[33] PIK3CA
mutation,[34] B-type Raf Kinase (BRAF) mutation;[35]
classifying/counting leukocytes,[22] megakaryocytes;[23]
counting erythroblasts,[17,18] reticulocytes;[13,29] sperms;[14,30]
detecting/counting bacteria;[12,20,21] detecting/quantifying
nuclease activity;[36,37] quantifying proteins;[28] monitoring
ultrasound-mediated local drug delivery *in vivo*;[38] as
temperature sensor[39,40]

Industrial Applications   Semiconductor devices[41]

Safety/Toxicity   No data available

REFERENCES

1. Sabnis, R. W. *Handbook of Biological Dyes and Stains*; John Wiley & Sons Inc.: Hoboken, **2010**; pp 483–484.

2. Yue, S. T.; Haugland, R. P. Dimers of unsymmetrical cyanine dyes containing pyridinium moieties. U.S. Patent 5410030, 1995.

3. Xiang, D.; Zhang, C.; Chen, L.; Ji, X.; He, Z. Tricolour fluorescence detection of sequence-specific DNA with a new molecular beacon and a nucleic acid dye TOTO-3. *Analyst* **2012**, *137*, 5898–5905.

4. Sergeev, N. V.; Brevnov, M. G.; Furtado, M. R. Identification of nucleic acids. PCT Int. Appl. WO 2011143478, 2011.

5. Exner, M.; Rogers, A. Methods for detecting nucleic acids using multiple signals. U.S. Pat. Appl. Publ. US 2007172836, 2007.

6. Erikson, G. H.; Daksis, J. I.; Kandic, I.; Picard, P. Nucleic acid multiplex formation. PCT Int. Appl. WO 2002103051, 2002.

7. Martin, R. M.; Leonhardt, H.; Cardoso, M. C. DNA labeling in living cells. *Cytometry, Part A* **2005**, *67A*, 45–52.

8. Wittwer, C. T.; Dujols, V. E.; Reed, G.; Zhou, L. Amplicon melting analysis with saturation dyes. PCT Int. Appl. WO 2004038038, 2004.

9. Erikson, G. H.; Daksis, J. I. Pre-incubation method to improve signal/noise ratio of nucleic acid assays. U.S. Pat. Appl. Publ. US 2004180345, 2004.

10. Erikson, G. H. Method for modifying transcription and/or translation in an organism for therapeutic, prophylactic and/or analytic uses. U.S. Pat. Appl. Publ. US 2003181412, 2003.

11. Suzuki, T.; Fujikura, K.; Higashiyama, T.; Takata, K. DNA staining for fluorescence and laser confocal microscopy. *J. Histochem. Cytochem.* **1997**, *45*, 49–53.

12. Millard, P. J.; Roth, B. L.; Yue, S. T.; Haugland, R. P. Fluorescent viability assay using cyclic-substituted unsymmetrical cyanine dyes. U.S. Patent 5534416, 1996.

13. Deka, C.; Gordon, K. M.; Gupta, R.; Horton, A. Methods and compositions for rapid staining of nucleic acids in whole cells. U.S. Patent 6271035, 2001.

14. Anderson, A. L.; Knutson, C. R.; Mueth, D.; Plewa, J.; Tanner, E. Methods for staining cells for identification and sorting. U.S. Pat. Appl. Publ. US 2006172315, 2006.

15. Gauer, C.; Mann, W.; Alunni-Fabbroni, M. Method for carrying out an enzymic reactions. Ger. DE 102006056694, 2010.

16. Gauer, C.; Mann, W.; Alunni-Fabbroni, M. Method for carrying out an enzymic reaction. PCT Int. Appl. WO 2008064730, 2008.

17. Heuven, B.; Wong, F. S.; Tsuji, T.; Sakata, T.; Hamaguchi, I. Method for classifying and counting erythroblasts by flow cytometry. Jpn. Kokai Tokkyo Koho JP 11326323, 1999.

18. Heuven, B.; Wong, F.; Tsuji, T.; Sakata, T.; Hamaguchi, I. Process for discriminating and counting erythroblasts. U.S. Pat. Appl. Publ. US 20020006631, 2002.

19. Bink, K.; Walch, A.; Feuchtinger, A.; Eisenmann, H.; Hutzler, P.; Hofler, H.; Werner, M. TO-PRO-3 is an optimal fluorescent dye for nuclear counterstaining in dual-colour FISH on paraffin sections. *Histochem. Cell Biol.* **2001**, *115*, 293–299.

20. Sakai, Y.; Kawashima, Y.; Inoue, J.; Ikeuchi, Y. Method of staining, detecting and counting bacteria, and a diluent for bacterial stain. Eur. Pat. Appl. EP 1203825, 2002.

21. Inoue, J.; Ikeuchi, Y.; Kawashima, Y. Method for staining and detecting bacteria. Jpn. Kokai Tokkyo Koho JP 2001258590, 2001.

22. Sakata, T.; Mizukami, T.; Hatanaka, K. Method for classifying and counting immature leukocytes. Eur. Pat. Appl. EP 844481, 1998.

23. Minakami, T.; Mori, Y.; Tsuji, T.; Ikeuchi, Y. Megakaryocyte classification/counting method by double fluorescent staining and flow cytometry. Jpn. Kokai Tokkyo Koho JP 2006275985, 2006.

24. Eckert, R. H.; Kaplan, C.; He, J.; Yarbrough, D. K.; Anderson, M.; Sim, J. Methods and devices for the selective detection of microorganisms. U.S. Pat. Appl. Publ. US 20120003661, 2012.

25. Henneberger, R.; Birch, D.; Bergquist, P.; Walter, M.; Anitori, R. P. The fluorescent dyes TO-PRO-3 and TOTO-3 iodide allow detection of microbial cells in soil samples without interference from background fluorescence. *BioTechniques* **2011**, *51*, 190–192.

26. Vannier, E. Methods for detection of pathogens in red blood cells. PCT Int. Appl. WO 2006031544, 2006.

27. Cai, W. Method for parallel amplification of nucleic acids using PCR or isothermal amplification by encapsulation in microdroplets. U.S. Pat. Appl. Publ. US 20130005591, 2013.

28. Kai, K.; Kitajima, Y.; Miyazaki, K.; Tokunaga, O. Protein quantitation method by fluorescent multiple staining. Jpn. Kokai Tokkyo Koho JP 2008268167, 2008.

29. Yamamoto, K.; Tokita, M.; Inagami, M.; Hirako, S. Method of counting reticulocytes. U.S. Patent 5563070, 1996.

30. Matsumoto, T.; Okada, H.; Hamaguchi, Y. Method and reagent for counting sperm by flow cytometry. Jpn. Kokai Tokkyo Koho JP 2001242168, 2001.

31. Lagrange, A. Hair dye compositions containing a polycationic direct dye. Fr. Demande FR 2848840, 2004.

32. Park, H. G.; Choi, J. J.; Kim, H. S.; Jung, S. U. Method and kit for detecting IDH1 and IDH2 mutations using peptide nucleic acid (PNA)-based real-time polymerase chain reaction (PCR) clamping. Repub. Korean Kongkae Taeho Kongbo KR 2012127679, 2012.

33. Park, H. G.; Choi, J. J.; Kim, H. S.; Jung, S. U. JAK2 gene mutation detection kit based on PNA mediated real-time PCR clamping. Repub. Korean Kongkae Taeho Kongbo KR 2012119571, 2012.

34. Park, H. K.; Choi, J. J.; Kim, H. S. PIK3CA mutation detection method and kit using real-time PCR clamping of PNA. PCT Int. Appl. WO 2012020965, 2012.

35. Park, H. K.; Choi, J. J.; Cho, M. H. Detection of mutation of BRAF by real-time PCR using PNA clamping probe for diagnosis of cancers. PCT Int. Appl. WO 2011093606, 2011.

36. Mitsuhashi, M.; Ogura, M. Process for determining nuclease activity. U.S. Patent 5554502, 1996.

37. Mitsuhashi, M.; Ogura, M. Process for determining nuclease activity. PCT Int. Appl. WO 9508623, 1995.

38. Deckers, R.; Yudina, A.; Cardoit, L. C.; Moonen, C. T. W. A fluorescent chromophore TOTO-3 as a smart probe' for the assessment of ultrasound-mediated local drug delivery *in vivo. Contrast Media Mol. Imaging* **2011**, *6*, 267–274.

39. Bousseksou, A.; Salmon, L.; Molnar, G.; Cobo, S. Materials with thermochromic spin transition doped with one or more fluorescent agents for use as temperature sensor. Fr. Demande FR 2952371, 2011.

40. Bousseksou, A.; Salmon, L.; Molnar, G.; Cobo, S. Heat-sensitive spin-transition materials doped with one or more fluorescent agents for use as temperature sensor. PCT Int. Appl. WO 2011058277, 2011.

41. Porta, P. A.; Summers, H. D. Vertical-cavity semiconductor devices for generation and detection of fluorescence emission on a single chip. *Appl. Phys. Lett.* **2004**, *85*, 1889–1891.

TRUE BLUE (TRUE BLUE CHLORIDE)

CAS Registry Number 71431-30-6

Chemical Structure

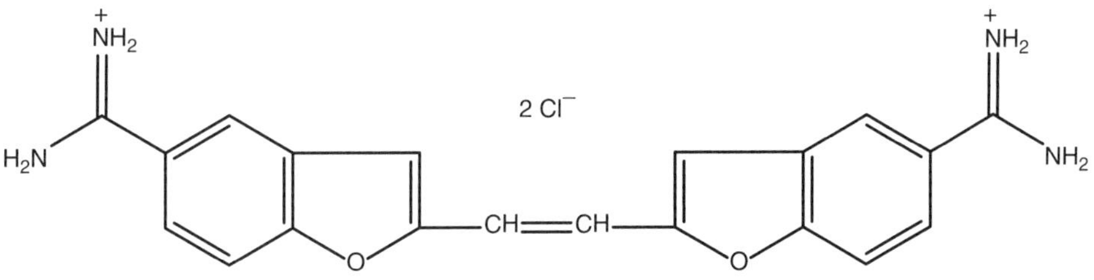

CA Index Name 5-Benzofurancarboximidamide, 2,2'-(1E)-1,2-ethenediylbis-, hydrochloride (1:2)

Other Names 5-Benzofurancarboximidamide, 2,2'-(1,2-ethenediyl)bis-, dihydrochloride, (E)-; 5-Benzofurancarboximidamide, 2,2'-(1E)-1,2-ethenediylbis-, dihydrochloride; NCI 240899; True Blue; True Blue chloride

Merck Index Number Not listed

Chemical/Dye Class Heterocycle; Benzofuran

Molecular Formula $C_{20}H_{18}Cl_2N_4O_2$

Molecular Weight 417.29

Physical Form Yellow solid

Solubility Soluble in water, N,N-dimethylformamide, dimethyl sulfoxide, methanol

Absorption (λ_{max}) 375 nm (H_2O)

Emission (λ_{max}) 403 nm (H_2O)

Molar Extinction Coefficient 68,000 cm^{-1} M^{-1} (H_2O)

Synthesis Synthetic method[1]

Imaging/Labeling Applications Analytes;[2] antibodies;[3] antigens;[3] axons;[4,5] cytoplasm;[5] endotoxins (lipopolysaccharides) (LPS);[6,7] keratin;[8] class III β-tubulin;[8] motoneural projections;[9] neurons[4,5,10–12,14–28]

Biological/Medical Applications Retrograde neuronal tracer;[4,5,10–12,14–28] detecting analytes,[2] axonal transport;[4,5] detecting/removing/inactivating/detoxifying endotoxins (lipopolysaccharides) (LPS);[6,7] evaluating a test element for antigens or antibodies;[3] assays for gene induction;[13] identifying genes related to malfunctions of central nervous system;[14] reducing the risk of iatrogenic nerve injury during a surgical procedure of the head or neck;[10] studying changes in motoneural projections;[9] preventing/treating septic shoc;[7] as temperature sensor[29,30]

Industrial Applications Not reported

Safety/Toxicity No data available

REFERENCES

1. Dann, O.; Char, H.; Fleischmann, P.; Fricke, H. Synthesis of antileukemic 2-[2-(indol-2-yl)vinyl]-1-benzofuran and -benzo[b]thiophenes. *Liebigs Ann. Chem.* **1986**, 438–455.

2. Halushka, P. V.; Watson, D. K.; Moussa, O.; Dickie, R. G. Devices and methods for concentration and analysis of fluids. PCT Int. Appl. WO 2012078308, 2012.

3. Mitoh, A.; Hiraide, T. Method for evaluating a test element for antigens or antibodies. Ger. Offen. DE 19648304, 1997.

4. Weidner, C.; Miceli, D.; Reperant, J. Orthograde axonal and transcellular transport of different fluorescent tracers in the primary visual system of the rat. *Brain Res.* **1983**, *272*, 129–136.

5. Aschoff, A.; Fritz, N.; Illert, M. Axonal transport of fluorescent compounds in the brain and spinal cord of cat and rat. *Axoplasmic Transp. Physiol. Pathol.* **1982**, 177–187.

6. Guo, J.; Wood, S. J.; David, S. A.; Lushington, G. H. Molecular modeling analysis of the interaction of novel bis-cationic ligands with the lipid A moiety of lipopolysaccharide. *Bioorg. Med. Chem. Lett.* **2006**, *16*, 714–717.

7. Evans, D.; Jindal, S. Methods and compositions for binding endotoxins. PCT Int. Appl. WO 9641185, 1996.

8. Vinores, S. A.; Vinores, M. A.; Chiu, C.; Woerner, T. M.; Campochiaro, P. A. Double-labeling for keratin and class III β-tubulin within cultured retinal pigment

epithelial cells: Comparison of chromogens to yield maximum resolution of two structural proteins within the same cell. *J. Histotechnol.* **1997**, *20*, 19–25.

9. Katada, A.; Vos, J. D.; Swelstad, B. B.; Zealear, D. L. A sequential double labeling technique for studying changes in motoneural projections to muscle following nerve injury and reinnervation. *J. Neurosci. Methods* **2006**, *155*, 20–27.

10. Mangat, G.; Brzozowski, L. Intra-operative head and neck nerve mapping. PCT Int. Appl. WO 2007016790, 2007.

11. Kuchiiwa, S.; Kuchiiwa, T. Agent for targeting drug to cerebral neuron. PCT Int. Appl. WO 2007086587, 2007.

12. Nylen, A.; Larsson, B.; Skagerberg, G. A sequential fluorescence method for neurotransmitter-specific retrograde tracing in the central nervous system of the rat: utilizing True Blue and immunohistochemistry in combination with computer-assisted photography. *Brain Res. Protoc.* **2005**, *15*, 30–37.

13. Byrne, C.; Hardman, M. J. Whole-mount assays for gene induction and barrier formation in the developing epidermis. *Methods Mol. Biol.* **2004**, *289*, 127–136.

14. Larsen, L. K.; Vrang, N.; Larsen, P. J. Methods for identifying genes related to malfunctions of central nervous system. PCT Int. Appl. WO 2004009842, 2004.

15. Zhang, L.; McClellan, A. D. Fluorescent tracers as potential candidates for double labeling of descending brain neurons in larval lamprey. *J. Neurosci. Methods* **1998**, *85*, 51–62.

16. Garrett, W. T.; McBride, R. L.; Williams, J. K., Jr.; Feringa, E. R. Fluoro-Gold's toxicity makes it inferior to True Blue for long-term studies of dorsal root ganglion neurons and motoneurons. *Neurosci. Lett.* **1991**, *128*, 137–139.

17. Bentivoglio, M.; Su, H. S. Photoconversion of fluorescent retrograde tracers. *Neurosci. Lett.* **1990**, *113*, 127–133.

18. Tanabe, T.; Ueda, S.; Sano, Y. Combined method of fluorescence tracer technique and PAP immunohistochemistry for discrimination of the transplanted cells. *Histochemistry* **1989**, *91*, 191–194.

19. Sripanidkulchai, K.; Wyss, J. M. Two rapid methods of counterstaining fluorescent dye tracer containing sections without reducing the fluorescence. *Brain Res.* **1986**, *397*, 117–129.

20. Skagerberg, G.; Bjoerklund, A.; Lindvall, O. Further studies on the use of the fluorescent retrograde tracer True Blue in combination with monoamine histochemistry. *J. Neurosci. Methods* **1985**, *14*, 25–40.

21. Skirboll, L.; Hoekfelt, T.; Norell, G.; Phillipson, O.; Kuypers, H. G. J. M.; Bentivoglio, M.; Catsman-Berrevoets, C. E.; Visser, T. J.; Steinbusch, H. A method for specific transmitter identification of retrogradely labeled neurons: immunofluorescence combined with fluorescence tracing. *Brain Res. Rev.* **1984**, *8*, 99–127.

22. Cavada, C.; Huisman, A. M.; Kuypers, H. G. J. M. Retrograde double labeling of neurons: the combined use of horseradish peroxidase and Diamidino Yellow Dihydrochloride (DY·2HCl) compared with True Blue and DY·2HCl in rat descending brain stem pathways. *Brain Res.* **1984**, *308*, 123–136.

23. Skirboll, L.; Hoekfelt, T. Transmitter specific mapping of neuronal pathways by immunohistochemistry combined with fluorescent dyes. *IBRO Handb. Ser.* **1983**, *3*, 465–476.

24. Van der Krans, A.; Hoogland, P. V. Labeling of neurons following intravenous injections of fluorescent tracers in mice. *J. Neurosci. Methods* **1983**, *9*, 95–103.

25. Keizer, K.; Kuypers, H. G. J. M.; Huisman, A. M.; Dann, O. Diamidino yellow dihydrochloride (DY·2HCl); a new fluorescent retrograde neuronal tracer, which migrates only very slowly out of the cell. *Exp. Brain Res.* **1983**, *51*, 179–191.

26. Albanese, A.; Bentivoglio, M. Retrograde fluorescent neuronal tracing combined with acetylcholinesterase histochemistry. *J. Neurosci. Methods* **1982**, *6*, 121–127.

27. Sawchenko, P. E.; Swanson, L. W. A method for tracing biochemically defined pathways in the central nervous system using combined fluorescence retrograde transport and immunohistochemical techniques. *Brain Res.* **1981**, *210*, 31–51.

28. Bentivoglio, M.; Kuypers, H. G. J. M.; Catsman-Berrevoets, C. E.; Dann, O. Fluorescent retrograde neuronal labeling in rat by means of substances binding specifically to adenine-thymine rich DNA. *Neurosci. Lett.* **1979**, *12*, 235–240.

29. Bousseksou, A.; Salmon, L.; Molnar, G.; Cobo, S. Heat-sensitive spin-transition materials doped with one or more fluorescent agents for use as temperature sensor. PCT Int. Appl. WO 2011058277, 2011.

30. Bousseksou, A.; Salmon, L.; Molnar, G.; Cobo, S. Materials with thermochromic spin transition doped with one or more fluorescent agents for use as temperature sensor. Fr. Demande FR 2952371, 2011.

VITA BLUE

CAS Registry Number   122079-36-1

Chemical Structure

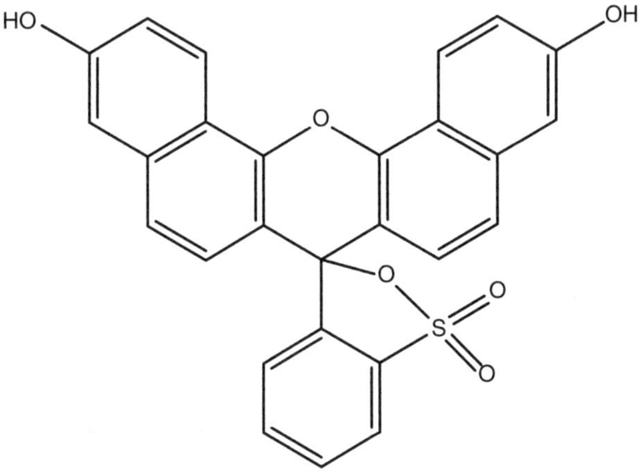

CA Index Name   Spiro[3H-2,1-benzoxathiole-3,7'-[7H]dibenzo[c,h]xan-thene]-3',11'-diol, 1,1-dioxide

Other Names   Vita Blue

Merck Index Number   Not listed

Chemical/Dye Class   Xanthene

Molecular Formula $C_{27}H_{16}O_6S$

Molecular Weight   468.48

Physical Form   Purple powder[1,2]

Solubility   Soluble in dimethyl sulfoxide

Boiling Point (Calcd.) $756.2 \pm 60.0\,°C$, pressure: 760 Torr

pKₐ 7.56 ± 0.03, temperature: 22 °C

pKₐ (Calcd.) 8.73 ± 0.20, most acidic, temperature: 25 °C

Absorption (λ_{max})   609 nm (Buffer pH 9.5)

Emission (λ_{max})   665 nm (Buffer pH 9.5)

Molar Extinction Coefficient $36,000\,cm^{-1}\,M^{-1}$ (Buffer pH 9.5)

Quantum Yield   0.15 (Buffer pH 9.5)

Synthesis   Synthetic methods[1,2]

Imaging/Labeling Applications   Cells;[1,2] stem cells[3–6]

Biological/Medical Applications   Detecting/identifying cells;[1,2] discriminating between live and dead cells;[1,2] enhancing muscle stem cell activity;[3] increasing stem cell proliferation;[3–6] preventing/repairing cell damage due to aging, injury or disease (stroke, heart attack or coronary artery disease);[4–6] as a substrate for measuring alkaline phosphatases activity,[7] esterase activity,[1,2,7] peptidases/proteases activity,[7] kinases activity,[7] sulfatases activity;[7] treating stroke[6]

Industrial Applications   Not reported

Safety/Toxicity   No data available

REFERENCES

1. Lee, L. G.; Berry, G. M.; Chen, C. H. Vita blue: a new 633-nm excitable fluorescent dye for cell analysis. *Cytometry* **1989**, *10*, 151–164.

2. Lee, L. Xanthene dyes that emit to the red of fluorescein. U.S. Patent 5066580, 1991.

3. Fernyhough, M. E.; Bucci, L. R.; Feliciano, J.; Dodson, M. V. The effect of nutritional supplements on muscle-derived stem cells *in vitro*. *Int. J. Stem Cells* **2010**, *3*, 63–67.

4. Sanberg, C. D.; Bickford, P. C.; Sanberg, P. R.; Tan, J.; Shytle, R. D.; Anderson, J. Compounds for stimulating stem cell proliferation including an ethanol extract of Aphanizomenon flos-aquae omega product (AFA-omega (EtOH)). U.S. Pat. Appl. Publ. US 20080089905, 2008.

5. Sanberg, C. D.; Bickford, P.; Sanberg, P.; Tan, J.; Shytle, R. D. Compounds for stimulating stem cell proliferation including Spirulina. U.S. Pat. Appl. Publ. US 20080085330, 2008.

6. Sanberg, C. D.; Sanberg, P.; Bickford, P.; Shytle, R. D.; Tan, J. Combined effects of nutrients on proliferation of stem cells. U.S. Pat. Appl. Publ. US 20060275512, 2006.

7. Palmer, D. A.; French, M. T. A composition for use in fluorescence assay systems. PCT Int. Appl. WO 9934219, 1999.

Handbook of Fluorescent Dyes and Probes, First Edition. R. W. Sabnis.
© 2015 John Wiley & Sons, Inc. Published 2015 by John Wiley & Sons, Inc.

YO-PRO 1

CAS Registry Number   152068-09-2

Chemical Structure

CA Index Name   Quinolinium, 4-[(3-methyl-2(3*H*)-benzoxazolylidene)-methyl]-1-[3-(trimethylammonio)propyl]-, iodide (1:2)

Other Names   Quinolinium, 4-[(3-methyl-2(3*H*)-benz-oxazolylidene) methyl]-1-[3-(trimethylammonio)propyl]-, diiodide; Oxazole yellow; YO-PRO 1; YO-PRO 1 iodide

Merck Index Number   Not listed

Chemical/Dye Class   Cyanine

Molecular Formula $C_{24}H_{29}I_2N_3O$

Molecular Weight   629.32

Physical Form   Orange-red powder

Solubility   Soluble in dimethyl sulfoxide

Absorption (λ_{max})   491 nm (H_2O/DNA)

Emission (λ_{max})   509 nm (H_2O/DNA)

Molar Extinction Coefficient   52,000 $cm^{-1}M^{-1}$ (H_2O/DNA)

Quantum Yield   0.44 (H_2O/DNA)

Synthesis   Synthetic methods[1–4]

Imaging/Labeling Applications   Nucleic acids;[1–21] cells;[3,4,20–22] nuclei;[23–26] bacteria;[2,3,27] islets;[28] marine prokaryotes;[29] megakaryocyte;[30] micronuclei;[31] microorganisms;[32–36] reticulocytes;[20,37] sperms;[21,38–41] viruses[42,43]

Biological/Medical Applications   Detecting nucleic acids,[1–21] cells,[3,4,20–22] bacteria,[2,3,27] microorganisms,[32–36] IDH1/IDH2 mutations,[44] PIK3CA mutation,[45] B-type Raf Kinase (BRAF) mutation;[46] analyzing marine prokaryotes;[29] classifying/counting megakaryocytes;[30] counting erythroblasts,[25,26] reticulocytes,[20,37] sperms,[21,38–41] viruses;[42,43] nucleic acids amplification;[19] nucleic acid hybridization;[47] nucleic acid sequencing;[48] apoptosis assay;[49–52] monitoring atmospheric/indoor bioaerosols;[53–55] as temperature sensor[56,57]

Industrial Applications   Not reported

Safety/Toxicity   Cytotoxicity;[58] neurotoxicity;[59] phototoxicity;[60] vasotoxicity[61]

REFERENCES

1. Sabnis, R. W. *Handbook of Biological Dyes and Stains*; John Wiley & Sons Inc.: Hoboken, **2010**; pp 495–496.

2. Roth, B. L.; Millard, P. J.; Yue, S. T.; Wells, K. S.; Haugland, R. P. Fluorescent assay for bacterial gram reaction. U.S. Patent 5545535, 1996.

3. Millard, P. J.; Roth, B. L.; Yue, S. T.; Haugland, R. P. Fluorescent viability assay using cyclic-substituted unsymmetrical cyanine dyes. U.S. Patent 5534416, 1996.

4. Yue, S. T.; Johnson, I. D.; Huang, Z.; Haugland, R. P. Unsymmetrical cyanine dyes with a cationic side chain. U.S. Patent 5321130, 1994.

5. Sergeev, N. V.; Brevnov, M. G.; Furtado, M. R. Identification of nucleic acids. PCT Int. Appl. WO 2011143478, 2011.

6. Miyamoto, S.; Kato, T.; Tomono, J. Nucleic acid identification method. Jpn. Kokai Tokkyo Koho JP 2010233530, 2010.

Handbook of Fluorescent Dyes and Probes, First Edition. R. W. Sabnis.
© 2015 John Wiley & Sons, Inc. Published 2015 by John Wiley & Sons, Inc.

7. Une, K.; Saito, T.; Hayashi, T. The detection method and detection reagent of norovirus RNA. Jpn. Kokai Tokkyo Koho JP 2010004793, 2010.

8. Exner, M.; Rogers, A. Methods for detecting nucleic acids using multiple signals. U.S. Pat. Appl. Publ. US 2007172836, 2007.

9. Erikson, G. H.; Daksis, J. I.; Kandic, I.; Picard, P. Nucleic acid multiplex formation. PCT Int. Appl. WO 2002103051, 2002.

10. Guillo, C.; Ferrance, J. P.; Landers, J. P. Use of a capillary electrophoresis instrument with laser-induced fluorescence detection for DNA quantitation. Comparison of YO-PRO-1 and PicoGreen assays. *J. Chromatogr., A* **2006**, *1113*, 239–243.

11. Al-Gubory, K. H. Fibered confocal fluorescence microscopy for imaging apoptotic DNA fragmentation at the single-cell level *in vivo*. *Exp. Cell Res.* **2005**, *310*, 474–481.

12. Suzuki, T.; Fujikura, K.; Higashiyama, T.; Takata, K. DNA staining for fluorescence and laser confocal microscopy. *J. Histochem. Cytochem.* **1997**, *45*, 49–53.

13. Wittwer, C. T.; Dujols, V. E.; Reed, G.; Zhou, L. Amplicon melting analysis with saturation dyes. PCT Int. Appl. WO 2004038038, 2004.

14. Erikson, G. H.; Daksis, J. I. Pre-incubation method to improve signal/noise ratio of nucleic acid assays. U.S. Pat. Appl. Publ. US 2004180345, 2004.

15. Erikson, G. H. Method for modifying transcription and/or translation in an organism for therapeutic, prophylactic and/or analytic uses. U.S. Pat. Appl. Publ. US 2003181412, 2003.

16. Garcia-Canas, V.; Gonzalez, R.; Cifuentes, A. Ultrasensitive detection of genetically modified maize DNA by capillary gel electrophoresis with laser-induced fluorescence using different fluorescent intercalating dyes. *J. Agric. Food Chem.* **2002**, *50*, 4497–4502.

17. Tomita, N.; Mori, Y. Method for efficiently detecting double-stranded nucleic acid. PCT Int. Appl. WO 2002103053, 2002.

18. Sodmergen, N. J.; Li, L.; He, J.; Guo, F. Application of YO-PRO-1 as an epifluorescent dye for *in situ* detection of small amount DNA in plant cells. *J. Plant Res.* **1999**, *112*, 117–122.

19. Sutherland, J. W.; Patterson, D. R. Homogeneous method for assay of double-stranded nucleic acids using fluorescent dyes and kit useful therein. Eur. Pat. Appl. EP 684316, 1995.

20. Deka, C.; Gordon, K. M.; Gupta, R.; Horton, A. Methods and compositions for rapid staining of nucleic acids in whole cells. U.S. Patent 6271035, 2001.

21. Anderson, A. L.; Knutson, C. R.; Mueth, D.; Plewa, J.; Tanner, E. Methods for staining cells for identification and sorting. U.S. Pat. Appl. Publ. US 2006172315, 2006.

22. Skyggebjerg, O.; Glensbjerg, M. A method and a system for counting cells from a plurality of species. PCT Int. Appl. WO 2002101087, 2002.

23. Gauer, C.; Mann, W.; Alunni-Fabbroni, M. Method for carrying out an enzymic reactions. Ger. DE 102006056694, 2010.

24. Gauer, C.; Mann, W.; Alunni-Fabbroni, M. Method for carrying out an enzymic reaction. PCT Int. Appl. WO 2008064730, 2008.

25. Heuven, B.; Wong, F. S.; Tsuji, T.; Sakata, T.; Hamaguchi, I. Method for classifying and counting erythroblasts by flow cytometry. Jpn. Kokai Tokkyo Koho JP 11326323, 1999.

26. Heuven, B.; Wong, F.; Tsuji, T.; Sakata, T.; Hamaguchi, I. Process for discriminating and counting erythroblasts. U.S. Pat. Appl. Publ. US 20020006631, 2002.

27. Stopa, P. J.; Mastromanolis, S. A. The use of blue-excitable nucleic-acid dyes for the detection of bacteria in well water using a simple field fluorometer and a flow cytometer. *J. Microbiol. Methods* **2001**, *45*, 143–153.

28. Boffa, D. J.; Waka, J.; Thomas, D.; Suh, S.; Curran, K.; Sharma, V. K.; Besada, M.; Muthukumar, T.; Yang, H.; Suthanthiran, M.; Manova, K. Measurement of apoptosis of intact human islets by confocal optical sectioning and stereologic analysis of YO-PRO-1-stained islets. *Transplantation* **2005**, *79*, 842–845.

29. Marie, D.; Vaulot, D.; Partensky, F. Application of the novel nucleic acid dyes YOYO-1, YO-PRO-1, and PicoGreen for flow cytometric analysis of marine prokaryotes. *Appl. Environ. Microbiol.* **1996**, *62*, 1649–1655.

30. Minakami, T.; Mori, Y.; Tsuji, T.; Ikeuchi, Y. Megakaryocyte classification/counting method by double fluorescent staining and flow cytometry. Jpn. Kokai Tokkyo Koho JP 2006275985, 2006.

31. Dertinger, S. D.; Cairns, S. E.; Avlasevich, S. L.; Torous, D. K. Method for enumerating mammalian cell micronuclei with an emphasis on differentially

staining micronuclei and the chromatin of dead and dying cells. PCT Int. Appl. WO 2006007479, 2006.

32. Eckert, R. H.; Kaplan, C.; He, J.; Yarbrough, D. K.; Anderson, M.; Sim, J. Methods and devices for the selective detection of microorganisms. U.S. Pat. Appl. Publ. US 20120003661, 2012.

33. Li, C. S.; Chia, W. C.; Chen, P. S. Fluorochrome and flow cytometry to monitor microorganisms in treated hospital wastewater. *J. Environ. Sci. Health, Part A* **2007**, *42*, 195–203.

34. Besson, F. I.; Hermet, J. P.; Ribault, S. Reaction medium and process for universal detection of microorganisms. Fr. Demande FR 2847589, 2004.

35. Sunamura, T.; Maruyama, A.; Kurane, R. Method for detecting and counting microorganism. Jpn. Kokai Tokkyo Koho JP 2002291499, 2002.

36. Schut, F.; Tan, P. S. T. Rapid detection and identification of microorganisms. PCT Int. Appl. WO 9910533, 1999.

37. Yamamoto, K.; Tokita, M.; Inagami, M.; Hirako, S. Method of counting reticulocytes. U.S. Patent 5563070, 1996.

38. Kumaresan, A.; Kadirvel, G.; Bujarbaruah, K. M.; Bardoloi, R. K.; Das, A.; Kumar, S.; Naskar, S. Preservation of boar semen at 18° induces lipid peroxidation and apoptosis like changes in spermatozoa. *Anim. Reprod. Sci.* **2009**, *110*, 162–171.

39. Cohen, B. A. Assay for semen optimization. PCT Int. Appl. WO 2009146043, 2009.

40. Hallap, T.; Nagy, S.; Jaakma, U.; Johannisson, A.; Rodriguez-Martinez, H. Usefulness of a triple fluorochrome combination Merocyanine 540/Yo-Pro 1/Hoechst 33342 in assessing membrane stability of viable frozen-thawed spermatozoa from Estonian Holstein AI bulls. *Theriogenology* **2006**, *65*, 1122–1136.

41. Matsumoto, T.; Okada, H.; Hamaguchi, Y. Method and reagent for counting sperm by flow cytometry. Jpn. Kokai Tokkyo Koho JP 2001242168, 2001.

42. Bettarel, Y.; Sime-Ngando, T.; Amblard, C.; Laveran, H. A comparison of methods for counting viruses in aquatic systems. *Appl. Environ. Microbiol.* **2000**, *66*, 2283–2289.

43. Hennes, K. P.; Suttle, C. A. Direct counts of viruses in natural waters and laboratory cultures by epifluorescence microscopy. *Limnol. Oceanogr.* **1995**, *40*, 1050–1055.

44. Park, H. G.; Choi, J. J.; Kim, H. S.; Jung, S. U. Method and kit for detecting IDH1 and IDH2 mutations using peptide nucleic acid (PNA)-based

real-time polymerase chain reaction (PCR) clamping. Repub. Korean Kongkae Taeho Kongbo KR 2012127679, 2012.

45. Park, H. K.; Choi, J. J.; Kim, H. S. PIK3CA mutation detection method and kit using real-time PCR clamping of PNA. PCT Int. Appl. WO 2012020965, 2012.

46. Park, H. K.; Choi, J. J.; Cho, M. H. Detection of mutation of BRAF by real-time PCR using PNA clamping probe for diagnosis of cancers. PCT Int. Appl. WO 2011093606, 2011.

47. Hahn, J.; Park, N. Detection method of nucleic acid hybridization. PCT Int. Appl. WO 2003010338, 2003.

48. Hoser, M. J. Nucleic acid sequencing methods, kits and reagents. PCT Int. Appl. WO 2004074503, 2004.

49. Wlodkowic, D.; Skommer, J.; Hillier, C.; Darzynkiewicz, Z. Multiparameter detection of apoptosis using red-excitable SYTO probes. *Cytometry, Part A* **2008**, *73A*, 563–569.

50. Plantin-Carrenard, E.; Bringuier, A.; Derappe, C.; Pichon, J.; Guillot, R.; Bernard, M.; Foglietti, M. J.; Feldmann, G.; Aubery, M.; Braut-Boucher, F. A Fluorescence microplate assay using YOPRO-1 to measure apoptosis: Application to HL60 cells subjected to oxidative stress. *Cell Biology Toxicol.* **2003**, *19*, 121–133.

51. Wronski, R.; Golob, N.; Grygar, E.; Windisch, M. Two-color, fluorescence-based microplate assay for apoptosis detection. *BioTechniques* **2002**, *32*, 666–668.

52. Idziorek, T.; Estaquier, J.; De Bels, F.; Ameisen, J. YOPRO-1 permits cytofluorometric analysis of programmed cell death (apoptosis) without interfering with cell viability. *J. Immunol. Methods* **1995**, *185*, 249–258.

53. Chi, M.; Li, C. Fluorochrome in monitoring atmospheric bioaerosols and correlations with meteorological factors and air pollutants. *Aerosol Sci. Technol.* **2007**, *41*, 672–678.

54. Li, C.; Huang, T. Fluorochrome in monitoring indoor bioaerosols. *Aerosol Sci. Technol.* **2006**, *40*, 237–241.

55. Chen, P.; Li, C. Bioaerosol characterization by flow cytometry with fluorochrome. *J. Environ. Monit.* **2005**, *7*, 950–959.

56. Bousseksou, A.; Salmon, L.; Molnar, G.; Cobo, S. Materials with thermochromic spin transition doped with one or more fluorescent agents for use as temperature sensor. Fr. Demande FR 2952371, 2011.

57. Bousseksou, A.; Salmon, L.; Molnar, G.; Cobo, S. Heat-sensitive spin-transition materials doped with

one or more fluorescent agents for use as temperature sensor. PCT Int. Appl. WO 2011058277, 2011.

58. Reilly, T. P.; MacArthur, R. D.; Farrough, M. J.; Crane, L. R.; Woster, P. M.; Svensson, C. K. Is hydroxylamine-induced cytotoxicity a valid marker for hypersensitivity reactions to sulfamethoxazole in human immunodeficiency virus-infected individuals? *J. Pharmacol. Exp. Ther.* **1999**, *291*, 1356–1364.

59. Shimazawa, M.; Yamashima, T.; Agarwal, N.; Hara, H. Neuroprotective effects of minocycline against *in vitro* and *in vivo* retinal ganglion cell damage. *Brain Res.* **2005**, *1053*, 185–194.

60. Tycon, M. A.; Dial, C. F.; Faison, K.; Melvin, W.; Fecko, C. J. Quantification of dye-mediated photodamage during single-molecule DNA imaging. *Anal. Biochem.* **2012**, *426*, 13–21.

61. Liao, S. D.; Puro, D. G. NAD+−induced vasotoxicity in the pericyte-containing microvasculature of the rat retina: effect of diabetes. *Invest. Ophthalmol. Vis. Sci.* **2006**, *47*, 5032–5038.

YO-PRO 3

CAS Registry Number 157199-62-7
Chemical Structure

CA Index Name Quinolinium, 4-[3-(3-methyl-2(3*H*)-benzoxazolylidene)-1-propen-1-yl]-1-[3-(trimethylammonio)propyl]-, iodide (1:2)

Other Names Quinolinium, 4-[3-(3-methyl-2(3*H*)-benzoxazolylidene)-1-propenyl]-1-[3-(trimethylammonio)propyl]-, diiodide; YO-PRO 3; YO-PRO 3 iodide

Merck Index Number Not listed

Chemical/Dye Class Cyanine

Molecular Formula $C_{26}H_{31}I_2N_3O$

Molecular Weight 655.36

Physical Form Orange-red powder

Solubility Soluble in dimethyl sulfoxide

Absorption (λ_{max}) 612 nm (H_2O/DNA)

Emission (λ_{max}) 631 nm (H_2O/DNA)

Molar Extinction Coefficient 100,100 $cm^{-1}\,M^{-1}$ (H_2O/DNA)

Quantum Yield 0.16 (H_2O/DNA)

Synthesis Synthetic methods[1–3]

Imaging/Labeling Applications Nucleic acids;[1–11] cells;[1–3,11] nuclei;[12] bacteria;[3] cytoplasm;[12] microorganisms;[13] reticulocytes;[14] sperms[11]

Biological/Medical Applications Detecting nucleic acids;[1–11] cells;[1–3,11] microorganisms;[13] counting reticulocytes;[14] nucleic acids amplification;[10] nucleic acid hybridization[15]

Industrial Applications Not reported

Safety/Toxicity No data available

REFERENCES

1. Sabnis, R. W. *Handbook of Biological Dyes and Stains*; John Wiley & Sons Inc.: Hoboken, **2010**; p 497.

2. Yue, S. T.; Johnson, I. D.; Huang, Z.; Haugland, R. P. Unsymmetrical cyanine dyes with a cationic side chain. U.S. Patent 5321130, 1994.

3. Millard, P. J.; Roth, B. L.; Yue, S. T.; Haugland, R. P. Fluorescent viability assay using cyclic-substituted unsymmetrical cyanine dyes. U.S. Patent 5534416, 1996.

4. Exner, M.; Rogers, A. Methods for detecting nucleic acids using multiple signals. U.S. Pat. Appl. Publ. US 2007172836, 2007.

5. Miller, B. L.; Krauss, T. D.; Du, H.; Crnkovich, N.; Strohsahl, C. M. Hybridization-based biosensor containing hairpin probes and use thereof. PCT Int. Appl. WO 2004061127, 2004.

6. Erikson, G. H.; Daksis, J. I.; Kandic, I.; Picard, P. Nucleic acid multiplex formation. PCT Int. Appl. WO 2002103051, 2002.

7. Erikson, G. H.; Daksis, J. I. Pre-incubation method to improve signal/noise ratio of nucleic acid assays. U.S. Pat. Appl. Publ. US 2004180345, 2004.

8. Erikson, G. H. Method for modifying transcription and/or translation in an organism for therapeutic, prophylactic and/or analytic uses. U.S. Pat. Appl. Publ. US 2003181412, 2003.

9. Beisker, W.; Weller-Mewe, E. M.; Nusse, M. Fluorescence enhancement of DNA-bound TO-PRO-3 by incorporation of bromodeoxyuridine to monitor cell cycle kinetics. *Cytometry* **1999**, *37*, 221–229.

10. Sutherland, J. W.; Patterson, D. R. Homogeneous method for assay of double-stranded nucleic acids

using fluorescent dyes and kit useful therein. Eur. Pat. Appl. EP 684316, 1995.

11. Anderson, A. L.; Knutson, C. R.; Mueth, D.; Plewa, J.; Tanner, E. Methods for staining cells for identification and sorting. U.S. Pat. Appl. Publ. US 2006172315, 2006.

12. Bink, K.; Walch, A.; Feuchtinger, A.; Eisenmann, H.; Hutzler, P.; Hofler, H.; Werner, M. TO-PRO-3 is an optimal fluorescent dye for nuclear counterstaining in dual-colour FISH on paraffin sections. *Histochem. Cell Biol.* **2001**, *115*, 293–299.

13. Eckert, R. H.; Kaplan, C.; He, J.; Yarbrough, D. K.; Anderson, M.; Sim, J. Methods and devices for the selective detection of microorganisms. U.S. Pat. Appl. Publ. US 20120003661, 2012.

14. Yamamoto, K.; Tokita, M.; Inagami, M.; Hirako, S. Method of counting reticulocytes. U.S. Patent 5563070, 1996.

15. Hahn, J.; Park, N. Detection method of nucleic acid hybridization. PCT Int. Appl. WO 2003010338, 2003.

YOYO 1

CAS Registry Number   143413-85-8

Chemical Structure

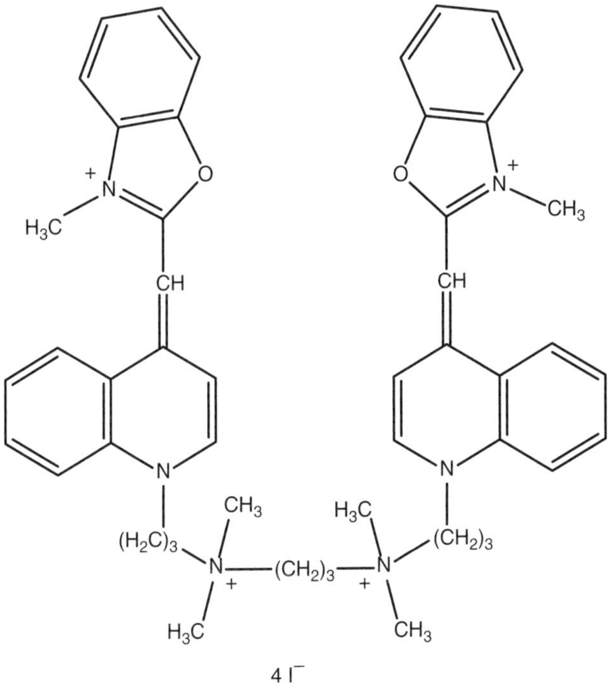

4 I⁻

CA Index Name   Quinolinium, 1,1′-[1,3-propanediylbis-[(dimethyliminio)-3,1-propanediyl]]bis[4-[(3-methyl-2(3*H*)-benzoxazolylidene)methyl]-, iodide (1:4)

Other Names   Quinolinium, 1,1′-[1,3-propanediylbis[(dimethyliminio)-3,1-propanediyl]]bis[4-[(3-methyl-2(3*H*)-benzoxazolylidene)methyl]-, tetraiodide; YOYO 1; YOYO 1 iodide

Merck Index Number   Not listed

Chemical/Dye Class   Cyanine

Molecular Formula $C_{49}H_{58}I_4N_6O_2$

Molecular Weight   1270.65

Physical Form   Orange-red powder

Solubility   Soluble in dimethyl sulfoxide

Absorption (λ_{max})   491 nm (H_2O/DNA)

Emission (λ_{max})   509 nm (H_2O/DNA)

Molar Extinction Coefficient   98,900 $cm^{-1}\,M^{-1}$ (H_2O/DNA)

Quantum Yield   0.52 (H_2O/DNA)

Synthesis   Synthetic methods[1-5]

Imaging/Labeling Applications   Nucleic acids;[1-31] cells;[1-3,29-33] nuclei;[34-38] atherosclerotic plaque;[39] bacteria;[4,29,40] chromosomes;[41] chromosome spreads;[3] cytoplasm;[38] erythrocytes;[42-44] marine prokaryotes;[45] megakaryocyte;[46] micronuclei;[47] microorganisms;[48-50] reticulocytes;[30,51] sperms[31,52,53]

Biological/Medical Applications   Detecting nucleic acids,[1-31] cells,[1-3,29-33] bacteria;[4,29,40] analyzing marine prokaryotes;[45] characterizing atherosclerotic plaque,[39] chromosomes;[41] classifying/counting megakaryocytes;[46] counting erythroblasts,[36,37] reticulocytes,[30,51] sperms;[31,52,53] detecting/measuring microorganisms;[48-50] detecting/quantifying nuclease activity;[54-56] evaluating freshness of a fish product;[57] nucleic acid amplification;[58] nucleic acid hybridization;[59] nucleic acid mapping;[60,61] nucleic acid sequencing;[62] nanosensor platforms for diagnostic applications;[63] as temperature sensor[64,65]

Industrial Applications   Not reported

Safety/Toxicity   Double-strand breaks in DNA/chromatin;[66] phototoxicity[67]

REFERENCES

1. Sabnis, R. W. *Handbook of Biological Dyes and Stains*; John Wiley & Sons Inc.: Hoboken, **2010**; pp 498–499.

2. Yue, S. T.; Johnson, I. D.; Haugland, R. P. Dimers of unsymmetrical cyanine dyes. PCT Int. Appl. WO 9306482, 1993.

3. Yue, S. T.; Haugland, R. P. Dimers of unsymmetrical cyanine dyes containing pyridinium moieties. U.S. Patent 5410030, 1995.

4. Roth, B. L.; Millard, P. J.; Yue, S. T.; Wells, K. S.; Haugland, R. P. Fluorescent assay for bacterial gram reaction. U.S. Patent 5545535, 1996.

5. Rye, H. S.; Yue, S.; Wemmer, D. E.; Quesada, M. A.; Haugland, R. P.; Mathies, R. A.; Glazer, A. N. Stable fluorescent complexes of double-stranded DNA with bis-intercalating asymmetric cyanine dyes: Properties and applications. *Nucleic Acids Res.* **1992**, *20*, 2803–2812.

6. Oh, D.; Lee, S.; Kang, S. H. Native and fluorescent dye-dependent single-DNA molecule microchip dynamics as measured by differential interference contrast microscopy. *Chem. Commun.* **2011**, *47*, 9137–9139.

7. Reuter, M.; Dryden, D. T. F. The kinetics of YOYO-1 intercalation into single molecules of double-stranded DNA. *Biochem. Biophys. Res. Commun.* **2010**, *403*, 225–229.

8. Li, Q.; Liu, X.; He, Y. Single molecule fluorescence imaging of DNA at agarose surfaces by total internal reflection fluorescence microscopy. *J. Biomed. Nanotechnol.* **2009**, *5*, 565–572.

9. Flors, C.; Ravarani, C. N. J.; Dryden, D. T. F. Super-resolution imaging of DNA labelled with intercalating dyes. *ChemPhysChem* **2009**, *10*, 2201–2204.

10. Perry, H. A. D.; Saleh, A. F. A.; Aojula, H.; Pluen, A. YOYO as a dye to track penetration of LK15 DNA complexes in spheroids: use and limits. *J. Fluores.* **2008**, *18*, 155–161.

11. Exner, M.; Rogers, A. Methods for detecting nucleic acids using multiple signals. U.S. Pat. Appl. Publ. US 2007172836, 2007.

12. Erikson, G. H.; Daksis, J. I.; Kandic, I.; Picard, P. Nucleic acid multiplex formation. PCT Int. Appl. WO 2002103051, 2002.

13. Erikson, G. H.; Daksis, J. I. Pre-incubation method to improve signal/noise ratio of nucleic acid assays. U.S. Pat. Appl. Publ. US 2004180345, 2004.

14. Erikson, G. H. Method for modifying transcription and/or translation in an organism for therapeutic, prophylactic and/or analytic uses. U.S. Pat. Appl. Publ. US 2003181412, 2003.

15. Kim, K.; Min, J.; Lee, I.; Kim, A. Method for highly sensitive nucleic acid detection using nanopore and non-specific nucleic acid-binding agent. U.S. Pat. Appl. Publ. US 2006292605, 2006.

16. Ploeger, L. S.; Huisman, A.; van der Gugten, J.; van der Giezen, D. M.; Belien, J. A. M.; Abbaker, A. Y.; Dullens, H. F. J.; Grizzle, W.; Poulin, N. M.; Meijer, G. A. Implementation of accurate and fast DNA cytometry by confocal microscopy in 3D. *Cell. Oncol.* **2005**, *27*, 225–230.

17. Shimizu, M.; Sasaki, S.; Tsuruoka, M. DNA length evaluation using cyanine dye and fluorescence correlation spectroscopy. *Biomacromolecules* **2005**, *6*, 2703–2707.

18. Yoshikawa, Y.; Suzuki, M.; Yamada, N.; Yoshikawa, K. Double-strand break of giant DNA: protection by glucosyl-hesperidin as evidenced through direct observation on individual DNA molecules. *FEBS Lett.* **2004**, *566*, 39–42.

19. Zheng, J.; Yeung, E. S. Counting single DNA molecules in a capillary with radial focusing. *Aust. J. Chem.* **2003**, *56*, 149–153.

20. Kricka, L. J. Stains, labels and detection strategies for nucleic acids assays. *Ann. Clin. Biochem.* **2002**, *39*, 114–129.

21. Wong, M.; Kong, S.; Dragowska, W. H.; Bally, M. B. Oxazole yellow homodimer YOYO-1-labeled DNA: a fluorescent complex that can be used to assess structural changes in DNA following formation and cellular delivery of cationic lipid DNA complexes. *Biochim. Biophys. Acta, Gen. Sub.* **2001**, *1527*, 61–72.

22. Okamoto, H.; Suzuki, T.; Yamamoto, N. Method for detecting/quantitating target nucleic acid by dry fluorometry. Jpn. Kokai Tokkyo Koho JP 2001033439, 2001.

23. Fasco, M. J. Analysis of amplified DNA molecules by capillary electrophoresis and laser induced fluorescence. *Methods Mol. Med.* **1999**, *26*, 131–146.

24. Davis, W. P.; Janssen, I. M. W.; Mossman, B. T.; Taatjes, D. J. Simultaneous triple fluorescence detection of mRNA localization, nuclear DNA, and apoptosis in cultured cells using confocal scanning laser microscopy. *Histochem. Cell Biol.* **1997**, *108*, 307–311.

25. Gurrieri, S.; Wells, K. S.; Johnson, I. D.; Bustamante, C. Direct visualization of individual DNA molecules by fluorescence microscopy: characterization of the factors affecting signal/background and optimization of imaging conditions using YOYO. *Anal. Biochem.* **1997**, *249*, 44–53.

26. Tekola, P.; Baak, J. P. A.; Belien, J. A. M.; Brugghe, J. Highly sensitive, specific, and stable new fluorescent DNA stains for confocal laser microscopy and image processing of normal paraffin sections. *Cytometry* **1994**, *17*, 191–195.

27. Hirons, G. T.; Fawcett, J. J.; Crissman, H. A. TOTO and YOYO: new very bright fluorochromes for DNA content analyses by flow cytometry. *Cytometry* **1994**, *15*, 129–140.

28. Rye, H. S.; Dabora, J. M.; Quesada, M. A.; Mathies, R. A.; Glazer, A. N. Fluorometric assay using dimeric dyes for double- and single-stranded DNA and RNA with picogram sensitivity. *Anal. Biochem.* **1993**, *208*, 144–150.

29. Millard, P. J.; Roth, B. L.; Yue, S. T.; Haugland, R. P. Fluorescent viability assay using cyclic-substituted unsymmetrical cyanine dyes. U.S. Patent 5534416, 1996.

30. Deka, C.; Gordon, K. M.; Gupta, R.; Horton, A. Methods and compositions for rapid staining of nucleic acids in whole cells. U.S. Patent 6271035, 2001.

31. Anderson, A. L.; Knutson, C. R.; Mueth, D.; Plewa, J.; Tanner, E. Methods for staining cells for identification and sorting. U.S. Pat. Appl. Publ. US 2006172315, 2006.

32. Raska, I.; Koberna, K.; Stanek, D.; Malinsky, J.; Ctrnacta, V.; Cermanova, S. Method of introducing molecules and ions into living cells. Czech Rep. CZ 295681, 2005.

33. Becker, B.; Clapper, J.; Harkins, K. R.; Olson, J. A. *In situ* screening assay for cell viability using a dimeric cyanine nucleic acid stain. *Anal. Biochem.* **1994**, *221*, 78–84.

34. Gauer, C.; Mann, W.; Alunni-Fabbroni, M. Method for carrying out an enzymic reactions. Ger. DE 102006056694, 2010.

35. Gauer, C.; Mann, W.; Alunni-Fabbroni, M. Method for carrying out an enzymic reaction. PCT Int. Appl. WO 2008064730, 2008.

36. Heuven, B.; Wong, F.; Tsuji, T.; Sakata, T.; Hamaguchi, I. Method for classifying and counting erythroblasts by flow cytometry. Jpn. Kokai Tokkyo Koho JP 11326323, 1999.

37. Heuven, B.; Wong, F.; Tsuji, T.; Sakata, T.; Hamaguchi, I. Process for discriminating and counting erythroblasts. U.S. Pat. Appl. Publ. US 20020006631, 2002.

38. Bink, K.; Walch, A.; Feuchtinger, A.; Eisenmann, H.; Hutzler, P.; Hofler, H.; Werner, M. TO-PRO-3 is an optimal fluorescent dye for nuclear counterstaining in dual-colour FISH on paraffin sections. *Histochem. Cell Biol.* **2001**, *115*, 293–299.

39. Taatjes, D. J.; Wadsworth, M. P.; Schneider, D. J.; Sobel, B. E. Improved quantitative characterization of atherosclerotic plaque composition with immunohistochemistry, confocal fluorescence microscopy, and computer-assisted image analysis. *Histochem. Cell Biol.* **2000**, *113*, 161–173.

40. Stopa, P. J.; Mastromanolis, S. A. The use of blue-excitable nucleic-acid dyes for the detection of bacteria in well water using a simple field fluorometer and a flow cytometer. *J. Microbiol. Methods* **2001**, *45*, 143–153.

41. Crissman, H. A.; Hirons, G. T. Chromosome characterization using single fluorescent dye. U.S. Patent 5418169, 1995.

42. Campo, J. J.; Aponte, J. J.; Nhabomba, A. J.; Sacarlal, J.; Angulo-Barturen, I.; Jimenez-Diaz, M. B.; Alonso, P. L.; Dobano, C. Feasibility of flow cytometry for measurements of *Plasmodium falciparum* parasite burden in studies in areas of malaria endemicity by use of bidimensional assessment of YOYO-1

and autofluorescence. *J. Clin. Microbiol.* **2011**, *49*, 968–974.

43. Schuck, D. C.; Ribeiro, R. Y.; Nery, A. A.; Ulrich, H.; Garcia, C. R. S. Flow cytometry as a tool for analyzing changes in *Plasmodium falciparum* cell cycle following treatment with indol compounds. *Cytometry, Part A* **2011**, *79A*, 959–964.

44. Barkan, D.; Ginsburg, H.; Golenser, J. Optimisation of flow cytometric measurement of parasitaemia in *Plasmodium*-infected mice. *Int. J. Parasitol.* **2000**, *30*, 649–653.

45. Marie, D.; Vaulot, D.; Partensky, F. Application of the novel nucleic acid dyes YOYO-1, YO-PRO-1, and PicoGreen for flow cytometric analysis of marine prokaryotes. *Appl. Environ. Microbiol.* **1996**, *62*, 1649–1655.

46. Minakami, T.; Mori, Y.; Tsuji, T.; Ikeuchi, Y. Megakaryocyte classification/counting method by double fluorescent staining and flow cytometry. Jpn. Kokai Tokkyo Koho JP 2006275985, 2006.

47. Dertinger, S. D.; Cairns, S. E.; Avlasevich, S. L.; Torous, D. K. Method for enumerating mammalian cell micronuclei with an emphasis on differentially staining micronuclei and the chromatin of dead and dying cells. PCT Int. Appl. WO 2006007479, 2006.

48. Eckert, R. H.; Kaplan, C.; He, J.; Yarbrough, D. K.; Anderson, M.; Sim, J. Methods and devices for the selective detection of microorganisms. U.S. Pat. Appl. Publ. US 20120003661, 2012.

49. Noda, N.; Mizutani, T. Microorganism-measuring method using multiple staining. Jpn. Kokai Tokkyo Koho JP 2006340684, 2006.

50. Vannier, E. Methods for detection of pathogens in red blood cells. PCT Int. Appl. WO 2006031544, 2006.

51. Yamamoto, K.; Tokita, M.; Inagami, M.; Hirako, S. Method of counting reticulocytes. U.S. Patent 5563070, 1996.

52. Duty, S. M.; Singh, N. P.; Ryan, L.; Chen, Z.; Lewis, C.; Huang, T.; Hauser, R. Reliability of the comet assay in cryopreserved human sperm. *Hum. Reprod.* **2002**, *17*, 1274–1280.

53. Matsumoto, T.; Okada, H.; Hamaguchi, Y. Method and reagent for counting sperm by flow cytometry. Jpn. Kokai Tokkyo Koho JP 2001242168, 2001.

54. Mitsuhashi, M.; Ogura, M. Process for determining nuclease activity. U.S. Patent 5554502, 1996.

55. Mitsuhashi, M.; Ogura, M. Process for determining nuclease activity. PCT Int. Appl. WO 9508623, 1995.

56. Ogura, M.; Mitsuhashi, M. Fluorometric method for the measurement of nuclease activity on plastic plates. *BioTechniques* **1995**, *18*, 231–233.

57. Liberman, B. A method of evaluating freshness of a fish product. PCT Int. Appl. WO 2005089254, 2005.

58. Morrison, T. Improved selective ligation and amplification assay. PCT Int. Appl. WO 2005059178, 2005.

59. Hahn, J.; Park, N. Detection method of nucleic acid hybridization. PCT Int. Appl. WO 2003010338, 2003.

60. Nyberg, L. K.; Persson, F.; Berg, J.; Bergstroem, J.; Fransson, E.; Olsson, L.; Persson, M.; Stalnacke, A.; Wigenius, J.; Tegenfeldt, J. O. A single-step competitive binding assay for mapping of single DNA molecules. *Biochem. Biophys. Res. Commun.* **2012**, *417*, 404–408.

61. Xiao, M.; Phong, A.; Ha, C.; Chan, T.; Cai, D.; Leung, L.; Wan, E.; Kistler, A. L.; DeRisi, J. L.; Selvin, P. R. Rapid DNA mapping by fluorescent single molecule detection. *Nucleic Acids Res.* **2007**, *35*, e16/1–e16/12.

62. Williams, J. G. K.; Anderson, J. P. Field-switch sequencing. PCT Int. Appl. WO 2005111240, 2005.

63. Zhou, C.; Thompson, M. E.; Cote, R. J.; Ishikawa, F.; Curreli, M.; Chang, H. Methods of using and constructing nanosensor platforms. PCT Int. Appl. WO 2009085356, 2009.

64. Bousseksou, A.; Salmon, L.; Molnar, G.; Cobo, S. Materials with thermochromic spin transition doped with one or more fluorescent agents for use as temperature sensor. Fr. Demande FR 2952371, 2011.

65. Bousseksou, A.; Salmon, L.; Molnar, G.; Cobo, S. Heat-sensitive spin-transition materials doped with one or more fluorescent agents for use as temperature sensor. PCT Int. Appl. WO 2011058277, 2011.

66. Yoshikawa, Y.; Hizume, K.; Oda, Y.; Takeyasu, K.; Araki, S.; Yoshikawa, K. Protective effect of vitamin C against double-strand breaks in reconstituted chromatin visualized by single-molecule observation. *Biophys. J.* **2006**, *90*, 993–999.

67. Tycon, M. A.; Dial, C. F.; Faison, K.; Melvin, W.; Fecko, C. J. Quantification of dye-mediated photodamage during single-molecule DNA imaging. *Anal. Biochem.* **2012**, *426*, 13–21.

YOYO 3

CAS Registry Number   156312-20-8

Chemical Structure

Absorption (λ_{max})   612 nm (H_2O/DNA)

Emission (λ_{max})   631 nm (H_2O/DNA)

Molar Extinction Coefficient   167,000 $cm^{-1} M^{-1}$ (H_2O/DNA)

CA Index Name   Quinolinium, 1,1′-[1,3-propanediylbis-[(dimethyliminio)-3,1-propanediyl]]bis[4-[3-(3-methyl-2(3H)-benzoxazolylidene)-1-propen-1-yl]-, iodide (1:4)

Other Names   Quinolinium, 1,1′-[1,3-propanediylbis-[(dimethyliminio)-3,1-propanediyl]]bis[4-[3-(3-methyl-2(3H)-benzoxazolylidene)-1-propenyl]-,   tetraiodide; YOYO 3; YOYO 3 iodide

Merck Index Number   Not listed

Chemical/Dye Class   Cyanine

Molecular Formula $C_{53}H_{62}I_4N_6O_2$

Molecular Weight   1322.73

Physical Form   Orange-red powder

Solubility   Soluble in dimethyl sulfoxide

Quantum Yield   0.15 (H_2O/DNA)

Synthesis   Synthetic methods[1,2]

Imaging/Labeling Applications   Nucleic acids;[1–11] cells;[1–3,11] bacteria;[3] chromosome spreads;[2] nuclei;[12–14] cytoplasm;[14] microorganisms;[15,16] porous wall hollow glass microspheres;[17] reticulocytes;[18] sperms[11,19]

Biological/Medical Applications   Detecting nucleic acids,[1–11] cells,[1–3,11] microorganisms;[15,16] counting erythroblasts,[12,13] reticulocytes,[18] sperms;[11,19] nucleic acid hybridization;[20] nucleic acid nanochannel device;[21] nucleic acid sequencing[22,23]

Industrial Applications   Analyzing polymers[17]

Safety/Toxicity   No data available

REFERENCES

1. Sabnis, R. W. *Handbook of Biological Dyes and Stains*; John Wiley & Sons Inc.: Hoboken, **2010**; pp 500–501.

2. Yue, S. T.; Haugland, R. P. Dimers of unsymmetrical cyanine dyes containing pyridinium moieties. U.S. Patent 5410030, 1995.

3. Millard, P. J.; Roth, B. L.; Yue, S. T.; Haugland, R. P. Fluorescent viability assay using cyclic-substituted unsymmetrical cyanine dyes. U.S. Patent 5534416, 1996.

4. Sergeev, N. V.; Brevnov, M. G.; Furtado, M. R. Identification of nucleic acids. PCT Int. Appl. WO 2011143478, 2011.

5. Exner, M.; Rogers, A. Methods for detecting nucleic acids using multiple signals. U.S. Pat. Appl. Publ. US 2007172836, 2007.

6. Wittwer, C. T.; Dujols, V. E.; Reed, G.; Zhou, L. Amplicon melting analysis with saturation dyes. PCT Int. Appl. WO 2004038038, 2004.

7. Erikson, G. H.; Daksis, J. I.; Kandic, I.; Picard, P. Nucleic acid multiplex formation. PCT Int. Appl. WO 2002103051, 2002.

8. Erikson, G. H.; Daksis, J. I. Pre-incubation method to improve signal/noise ratio of nucleic acid assays. U.S. Pat. Appl. Publ. US 2004180345, 2004.

9. Erikson, G. H. Method for modifying transcription and/or translation in an organism for therapeutic, prophylactic and/or analytic uses. U.S. Pat. Appl. Publ. US 2003181412, 2003.

10. Beisker, W.; Weller-Mewe, E. M.; Nusse, M. Fluorescence enhancement of DNA-bound TO-PRO-3 by incorporation of bromodeoxyuridine to monitor cell cycle kinetics. *Cytometry* **1999**, *37*, 221–229.

11. Anderson, A. L.; Knutson, C. R.; Mueth, D.; Plewa, J.; Tanner, E. Methods for staining cells for identification and sorting. U.S. Pat. Appl. Publ. US 2006172315, 2006.

12. Heuven, B.; Wong, F. S.; Tsuji, T.; Sakata, T.; Hamaguchi, I. Method for classifying and counting erythroblasts by flow cytometry. Jpn. Kokai Tokkyo Koho JP 11326323, 1999.

13. Heuven, B.; Wong, F.; Tsuji, T.; Sakata, T.; Hamaguchi, I. Process for discriminating and counting erythroblasts. U.S. Pat. Appl. Publ. US 20020006631, 2002.

14. Bink, K.; Walch, A.; Feuchtinger, A.; Eisenmann, H.; Hutzler, P.; Hofler, H.; Werner, M. TO-PRO-3 is an optimal fluorescent dye for nuclear counterstaining in dual-colour FISH on paraffin sections. *Histochem. Cell Biol.* **2001**, *115*, 293–299.

15. Eckert, R. H.; Kaplan, C.; He, J.; Yarbrough, D. K.; Anderson, M.; Sim, J. Methods and devices for the selective detection of microorganisms. U.S. Pat. Appl. Publ. US 20120003661, 2012.

16. Vannier, E. Methods for detection of pathogens in red blood cells. PCT Int. Appl. WO 2006031544, 2006.

17. Cai, W. Method for parallel amplification of nucleic acids using PCR or isothermal amplification by encapsulation in microdroplets. U.S. Pat. Appl. Publ. US 20130005591, 2013.

18. Yamamoto, K.; Tokita, M.; Inagami, M.; Hirako, S. Method of counting reticulocytes. U.S. Patent 5563070, 1996.

19. Matsumoto, T.; Okada, H.; Hamaguchi, Y. Method and reagent for counting sperm by flow cytometry. Jpn. Kokai Tokkyo Koho JP 2001242168, 2001.

20. Hahn, J.; Park, N. Detection method of nucleic acid hybridization. PCT Int. Appl. WO 2003010338, 2003.

21. Tegenfeldt, J.; Reisner, W.; Flyvbjerg, H. Method for the mapping of the local AT/GC ratio along DNA. PCT Int. Appl. WO 2010042007, 2010.

22. Williams, J. G. K.; Anderson, J. P. Field-switch sequencing. U.S. Patent 7462452, 2008.

23. Williams, J. G. K.; Anderson, J. P. Field-switch sequencing. PCT Int. Appl. WO 2005111240, 2005.

APPENDIX A
INDEX OF CAS REGISTRY NUMBERS

Handbook of Fluorescent Dyes and Probes, First Edition. R. W. Sabnis.
© 2015 John Wiley & Sons, Inc. Published 2015 by John Wiley & Sons, Inc.

CAS Registry Number	Dye	Page Number
69168-45-2	1-Pyrenedodecanoic acid (PDDA)	343
71431-30-6	True Blue (True Blue chloride)	411
74681-68-8	Nuclear Yellow (Hoechst S 769121)	295
74802-04-3	7-Dimethylamino-4-methylcoumarin-3-isothiocyanate (DACITC)	180
75168-11-5	Acridine Orange 10-nonyl bromide (Nonyl-Acridine Orange (NAO))	12
76823-03-5	5-Carboxyfluorescein	110
76863-28-0	Fluorescein-5-thiosemicarbazide	224
76936-87-3	N-(1-Pyrenyl)iodoacetamide	355
79955-27-4	5-Carboxyfluorescein diacetate	119
80883-54-1	7-Dimethylaminocoumarin-4-acetic acid (DMACA)	176
82412-15-5	7-Anilinocoumarin-4-acetic acid (ACAA)	65
82962-86-5	8-Methoxypyrene-1,3,6-trisulfonic acid trisodium salt (MPTS)	287
85985-44-0	9-Anthroylnitrile	66
86701-10-2	NBD C_6-ceramide	289
88404-25-5	4-Bromomethyl-6,7-dimethoxycoumarin	91
88404-26-6	6,7-Dimethoxycoumarin-4-acetic acid	175
90936-84-8	1-Pyrenehexadecanoic acid	346
90936-85-9	1-Pyrenehexanoic acid (PHA)	348
92557-80-7	5-Carboxyfluorescein succinimidyl ester	125
92557-81-8	6-Carboxyfluorescein succinimidyl ester	129
96686-59-8	7-Dimethylaminocoumarin-4-acetic acid succinimidyl ester (DMACA SE)	178
97632-67-2	7-Methoxycoumarin-3-carbonyl azide	280
100343-98-4	7-Diethylaminocoumarin-3-carboxylic acid hydrazide (DCCH)	153
101821-61-8	LysoSensor Blue DND 167	269
105802-46-8	4-[4-(Diethylamino)styryl]-N-methylpyridinium iodide (4-Di-2-ASP)	157
109244-58-8	Dihydrorhodamine 123 (DHR 123)	169
114932-60-4	1-Pyrenebutanoic acid succinimidyl ester	336
115532-50-8	Tetramethylrhodamine methyl ester perchlorate (TMRM)	396
115532-52-0	Tetramethylrhodamine ethyl ester perchlorate (TMRE)	394
117620-77-6	3-Cyano-7-ethoxycoumarin (7-Ethoxycoumarin-3-carbonitrile)	139
122079-36-1	Vita Blue	413
133867-52-4	BODIPY FL C_5-succinimidyl ester	83
133867-53-5	BODIPY FL C_5-ceramide (C_5-DMB-ceramide)	74
134471-24-2	7-Hydroxycoumarin-3-carboxylic acid succinimidyl ester	250
137182-38-8	Cascade Blue hydrazide trisodium salt	138
138039-58-4	Cascade Blue acetyl azide trisodium salt	136
139346-57-9	7-Diethylaminocoumarin-3-carboxylic acid succinimidyl ester (DEAC SE)	155
142975-81-3	5-Carboxy-2′,7′-dichlorofluorescein	102
143413-84-7	TOTO 1	405
143413-85-8	YOYO 1	421
144316-86-9	6-Carboxy-2′,7′-dichlorofluorescein	104
144489-09-8	5-Carboxy-2′,7′-dichlorofluorescein diacetate	106
144489-10-1	6-Carboxy-2′,7′-dichlorofluorescein diacetate	108
145103-60-2	5-Carboxynaphthofluorescein	132
145103-61-3	6-Carboxynaphthofluorescein	134
148942-72-7	5-(Bromomethyl)fluorescein	93
150321-92-9	7-Methoxycoumarin-3-carboxylic acid succinimidyl ester	285
152068-09-2	YO-PRO 1	415
152584-38-8	LysoSensor Green DND 189	273
154757-99-0	POPO 3	322

CAS Registry Number	Dye	Page Number
156312-20-8	YOYO 3	425
157199-56-9	PO-PRO 1	324
157199-57-0	BO-PRO 1	87
157199-59-2	TO-PRO 1	398
157199-62-7	YO-PRO 3	419
157199-63-8	TO-PRO 3	400
157673-16-0	7-Diethylaminocoumarin-3-carbonyl azide	148
161016-55-3	PO-PRO 3	326
166196-17-4	TOTO 3	408
166821-89-2	LysoSensor Blue DND 192	271
168482-84-6	Calcein Blue AM	101
169454-13-1	BOBO 1	69
169454-15-3	POPO 1	320
169454-17-5	BOBO 3	71
173357-16-9	BO-PRO 3	89
177027-61-1	TO-PRO 5	403
180389-01-9	Ethidium Homodimer-2 (EthD-2)	203
191853-29-9	BODIPY FL C_5-sphingomyelin (C_5-DMB-sphingomyelin)	81
195136-52-8	Oregon Green 488 carboxylic acid	302
195136-58-4	Oregon Green 488 (2′,7′-Difluorofluorescein)	297
195136-74-4	Oregon Green 488 carboxylic acid diacetate	306
196504-57-1	8-Aminopyrene-1,3,6-trisulfonic acid trisodium salt (APTS)	63
198139-49-0	Oregon Green 514 carboxylic acid	304
198139-51-4	Oregon Green 488 carboxylic acid succinimidyl ester	308
198139-53-6	Oregon Green 514 carboxylic acid succinimidyl ester	310
200554-19-4	Alexa Fluor 350 carboxylic acid succinimidyl ester (AMCA-S)	15
211566-66-4	Hexidium iodide	229
214147-22-5	NBD methylhydrazine (MNBDH)	293
215868-23-8	Marina Blue (6,8-Difluoro-7-hydroxy-4-methylcoumarin) (DiFMU)	277
215868-31-8	Pacific Blue (6,8-Difluoro-7-hydroxycoumarin-3-carboxylic acid)	315
215868-33-0	Pacific Blue succinimidyl ester	318
217176-83-5	Dihydrorhodamine 6G (DHR 6G)	173
222164-96-7	Alexa Fluor 488 carboxylic acid succinimidyl ester	20
231632-16-9	LysoSensor Green DND 153	272
231632-18-1	LysoSensor Yellow/Blue DND 160 (PDMPO)	275
251955-11-0	BODIPY FL C_5-lactosylceramide	79
254109-20-1	4-Amino-5-methylamino-2′,7′-difluorofluorescein (DAF-FM)	50
254109-22-3	4-Amino-5-methylamino-2′,7′-difluorofluorescein diacetate (DAF-FM DA)	52
295348-87-7	Alexa Fluor 594 carboxylic acid succinimidyl ester	30
296277-09-3	RedoxSensor Red CC-1	369
304014-12-8	QSY 7 carboxylic acid succinimidyl ester	361
304014-13-9	QSY 21 carboxylic acid succinimidyl ester	366
305801-86-9	JO-PRO 1	265
305801-87-0	JOJO 1	263
305802-06-6	LOLO 1	267
328085-55-8	Oregon Green 488 maleimide	312
422309-89-5	Alexa Fluor 660 carboxylic acid succinimidyl ester	35
467233-94-9	Alexa Fluor 430 carboxylic acid succinimidyl ester	17
477876-64-5	Alexa Fluor 532 carboxylic acid succinimidyl ester	25
522592-13-8	3,3′-Dimethyl-α-naphthoxacarbocyanine iodide (JC-9)	186

CAS Registry Number	Dye	Page Number
571186-05-5	BODIPY TR ceramide	85
697795-06-5	Alexa Fluor 750 carboxylic acid succinimidyl ester	42
700834-40-8	QSY 9 carboxylic acid succinimidyl ester	364
875756-97-1	Hoechst 33342	238
886046-94-2	Alexa Fluor 555 carboxylic acid succinimidyl ester	28
886210-16-8	Oregon Green 488 cadaverine	300
908143-55-5	BODIPY FL C_5-ganglioside GM1	77
918946-23-3	Alexa Fluor 514 carboxylic acid succinimidyl ester	23
934216-71-4	Pacific Blue C_5-maleimide	317
948558-33-6	Alexa Fluor 680 carboxylic acid succinimidyl ester	38
950891-33-5	Alexa Fluor 790 carboxylic acid succinimidyl ester	45
1132773-86-4	Alexa Fluor 633 carboxylic acid succinimidyl ester	33
1246956-22-8	Alexa Fluor 700 carboxylic acid succinimidyl ester	40

APPENDIX B
INDEX OF ACRIDINES

Handbook of Fluorescent Dyes and Probes, First Edition. R. W. Sabnis.
© 2015 John Wiley & Sons, Inc. Published 2015 by John Wiley & Sons, Inc.

APPENDIX C
INDEX OF ANTHRACENES

Handbook of Fluorescent Dyes and Probes, First Edition. R. W. Sabnis.
© 2015 John Wiley & Sons, Inc. Published 2015 by John Wiley & Sons, Inc.

APPENDIX D
INDEX OF BORON CO-ORDINATION COMPOUNDS/DYES

Handbook of Fluorescent Dyes and Probes, First Edition. R. W. Sabnis.
© 2015 John Wiley & Sons, Inc. Published 2015 by John Wiley & Sons, Inc.

APPENDIX E
INDEX OF COUMARINS

Handbook of Fluorescent Dyes and Probes, First Edition. R. W. Sabnis.
© 2015 John Wiley & Sons, Inc. Published 2015 by John Wiley & Sons, Inc.

APPENDIX F
INDEX OF CYANINES/STYRYLS

Handbook of Fluorescent Dyes and Probes, First Edition. R. W. Sabnis.
© 2015 John Wiley & Sons, Inc. Published 2015 by John Wiley & Sons, Inc.

APPENDIX G
INDEX OF HETEROCYCLES

Handbook of Fluorescent Dyes and Probes, First Edition. R. W. Sabnis.
© 2015 John Wiley & Sons, Inc. Published 2015 by John Wiley & Sons, Inc.

APPENDIX H
INDEX OF PYRENES

Handbook of Fluorescent Dyes and Probes, First Edition. R. W. Sabnis.
© 2015 John Wiley & Sons, Inc. Published 2015 by John Wiley & Sons, Inc.

APPENDIX I
INDEX OF XANTHENES

Handbook of Fluorescent Dyes and Probes, First Edition. R. W. Sabnis.
© 2015 John Wiley & Sons, Inc. Published 2015 by John Wiley & Sons, Inc.